Audio Signal Processing

Special Issue Editor
Vesa Välimäki

MDPI

Special Issue Editor
Vesa Välimäki
Department of Signal Processing and Acoustics
School of Electrical Engineering
Aalto University
Finland

Editorial Office
MDPI AG
St. Alban-Anlage 66
Basel, Switzerland

This edition is a reprint of the Special Issue published online in the open access journal *Applied Sciences* (ISSN 2076-3417) in 2016 (available at: http://www.mdpi.com/journal/applsci/special_issues/audio_signal_processing).

For citation purposes, cite each article independently as indicated on the article page online and as indicated below:

Author 1; Author 2; Author 3 etc. Article title. *Journal Name*. **Year**. Article number/page range.

ISBN 978-3-03842-350-8 (Pbk)
ISBN 978-3-03842-351-5 (PDF)

Table of Contents

About the Guest Editor

Vesa Välimäki is a full professor at Aalto University in Espoo, Finland. He is a Fellow of the IEEE, a Fellow of the Audio Engineering Society (AES), and a Life Member of the Acoustical Society of Finland. In 2008–2009 he was a Visiting Scholar at Stanford University. He was the General Chair of the 2008 International Conference on Digital Audio Effects DAFX-08. Currently, he is a Senior Area Editor of the IEEE/ACM Transactions on Audio, Speech and Language Processing. He will be the General Chair of the SMC-17.

Preface to "Audio Signal Processing"

This Special Issue gathers 20 fine contributions on audio signal processing, a topic which belongs to the Applied Acoustics section in this journal, *Applied Sciences*. These articles include revised and extended versions of three papers which won Best Paper Awards at the 2015 International Conference on Digital Audio Effects (DAFX-15) and also new versions of three papers which were presented at the 2015 International Computer Music Conference (ICMC-15).

Submissions were received from many parts of the world, such as Asia, Europe, and North America, and from some of the largest research units in this field, such as CCRMA (Stanford University, CA, USA), IRCAM (Paris, France), the International Audio Laboratories Erlangen (Erlangen, Germany), and the Center for Digital Music (Queen Mary University of London, UK). All manuscripts went through a rigorous peer review. Many fundamental topics in audio signal processing are dealt with in this collection, including active noise control, audio effects processing, automatic mixing, audio content analysis, equalizers, machine listening, music information retrieval, physical modeling of musical instruments, sound reproduction using headphones and loudspeakers, sound synthesis, spectral analysis, and virtual analog modeling.

The first three papers in this collection are review articles, which focus on time-scale modification (by Driedger and Müller), equalization (by Välimäki and Reiss), and feature extraction (by Alías et al.) applied to audio signals. The paper by Pieren et al. discusses auralization of traffic noise produced by a single car. Chun and Kim explain how to improve stereo sound reproduction using frequency-dependent amplitude panning. Stasis et al. present a second paper on audio equalization, concentrating on user-friendly control using high-level descriptors. An earlier version of this work received a Best Paper Award at the DAFX-15 conference.

The contribution by Gutierrez-Parera and Lopez investigates the effect of headphone quality on the perceived spatial sound. Antoñanzas et al. study active noise control in a room using several microphones and loudspeakers. Kim et al. propose methods to actively modify the sonic environment using acoustic feedback. Caetano et al. explore the synthesis of musical instrument sounds using a novel adaptive quasi-harmonic sinusoidal model. Gebhardt et al. won the First Prize at the DAFX-15 conference with their paper on harmonic music mixing using psychoacoustic principles. Desvages and Bilbao also received a Best Paper Award at DAFX-15 with their paper on detailed physical modeling of the violin using a finite difference scheme.

In the second paper on physical modeling, Medine considers two intrinsic problems in acoustic and audio system models: nonlinearities and delay-free loops. Rao et al. propose a new technique for automatic chord recognition from recorded music. Mesaros et al. tackle the evaluation of sound event detection systems. Werner and Abel develop sound effects processing methods, which are inspired by the Hammond organ. Thoret et al. suggest a control strategy for friction sound synthesis capable of producing noises similar to a creaky door or a singing glass, for example. Xu et al. write about a machine listening technique, which can be used for acoustic monitoring in a coal mine. Falaize and Hélie discuss the port-Hamiltonian approach to physical modeling of analog audio circuits, such as the diode clipper and the wah-wah pedal. In the last paper, Werner and Germain discover a way to improve the estimation of the frequency and amplitude of a sinusoid with the help of power scaling of the spectral data.

I am grateful to all contributors who made this Special Issue a success. The organizers of DAFX-15 and ICMC-15 conferences, especially Prof. Sigurd Saue (General Chair, DAFX-15) and Prof. Richard Dudas (Paper Chair, ICMC-15), deserve special thanks for their collaboration, which made it possible to get some excellent works presented in those conferences published in this Special Issue. I also want to thank all reviewers; although they were busy with many tasks, they tirelessly helped to improve these manuscripts. Finally, thanks go to Jennifer Li and the rest of the *Applied Sciences* editorial team for their effective and friendly collaboration. I hope that this collection will serve as an inspiration for future research in audio signal processing.

Vesa Välimäki
Guest Editor

applied
sciences

Review

A Review of Time-Scale Modification of Music Signals †

Jonathan Driedger *,‡ and Meinard Müller *,‡

International Audio Laboratories Erlangen, 91058 Erlangen, Germany
* Correspondence: jonathan.driedger@audiolabs-erlangen.de (J.D.);
   meinard.mueller@audiolabs-erlangen.de(M.M.); Tel.: +49-913-185-20519 (J.D.); +49-913-185-20504 (M.M.);
   Fax: +49-913-185-20524 (J.D. & M.M.)
† This paper is an extended version of our paper published in [44].
‡ These authors contributed equally to this work.

Academic Editor: Vesa Valimaki
Received: 22 December 2015; Accepted: 25 January 2016; Published: 18 February 2016

Abstract: Time-scale modification (TSM) is the task of speeding up or slowing down an audio signal's playback speed without changing its pitch. In digital music production, TSM has become an indispensable tool, which is nowadays integrated in a wide range of music production software. Music signals are diverse—they comprise harmonic, percussive, and transient components, among others. Because of this wide range of acoustic and musical characteristics, there is no single TSM method that can cope with all kinds of audio signals equally well. Our main objective is to foster a better understanding of the capabilities and limitations of TSM procedures. To this end, we review fundamental TSM methods, discuss typical challenges, and indicate potential solutions that combine different strategies. In particular, we discuss a fusion approach that involves recent techniques for harmonic-percussive separation along with time-domain and frequency-domain TSM procedures.

Keywords: digital signal processing; overlap-add; WSOLA; phase vocoder; harmonic-percussive separation; transient preservation; pitch-shifting; music synchronization

1. Introduction

Time-scale modification (TSM) procedures are digital signal processing methods for stretching or compressing the duration of a given audio signal. Ideally, the time-scale modified signal should sound as if the original signal's content was performed at a different tempo while preserving properties like pitch and timbre. TSM procedures are applied in a wide range of scenarios. For example, they simplify the process of creating music remixes. Music producers or DJs apply TSM to adjust the durations of music recordings, enabling synchronous playback [1,2]. Nowadays TSM is built into music production software as well as hardware devices. A second application scenario is adjusting an audio stream's duration to that of a given video clip. For example, when generating a slow motion video, it is often desirable to also slow down the tempo of the associated audio stream. Here, TSM can be used to synchronize the audio material with the video's visual content [3].

A main challenge for TSM procedures is that music signals are complex sound mixtures, consisting of a wide range of different sounds. As an example, imagine a music recording consisting of a violin playing together with castanets. When modifying this music signal with a TSM procedure, both the harmonic sound of the violin as well as the percussive sound of the castanets should be preserved in the output signal. To keep the violin's sound intact, it is essential to maintain its pitch as well as its timbre. On the other hand, the clicking sound of the castanets does not have a pitch—it is much more important to maintain the crisp sound of the single clicks, as well as their exact relative time positions, in order to preserve the original rhythm. Retaining these contrasting characteristics usually requires conceptually different TSM approaches. For example, classical TSM procedures based on waveform similarity overlap-add (WSOLA) [4] or on the phase vocoder (PV-TSM) [5–7] are capable of preserving the perceptual quality of harmonic signals to a high degree, but introduce noticeable artifacts when modifying percussive signals. However, it is possible to substantially reduce artifacts by combining different TSM approaches. For example, in [8], a given audio signal is first separated into a harmonic and a percussive component. Afterwards, each component is processed with a different TSM procedure that preserves its respective characteristics. The final output signal is then obtained by superimposing the two intermediate output signals.

   Our goals in this article are two-fold. First, we aim to foster an understanding of fundamental challenges and algorithmic approaches in the field of TSM by reviewing well-known TSM methods and discussing their respective advantages and drawbacks in detail. Second, having identified the core issues of these classical procedures, we show—through an example—how to improve on them by combining different algorithmic ideas. We begin the article by introducing a fundamental TSM strategy as used in many TSM procedures (Section 2) and discussing a simple TSM approach based on overlap-add (Section 3). Afterwards, we review two conceptually different TSM methods: the time-domain WSOLA (Section 4) as well as the frequency-domain PV-TSM (Section 5). We then review the state-of-the-art TSM procedure from [8] that improves on the quality of both WSOLA as well as PV-TSM by incorporating harmonic-percussive separation (Section 6). Finally, we point out different application scenarios for TSM (such as music synchronization and pitch-shifting), as well as various freely available TSM implementations (Section 7).

2. Fundamentals of Time-Scale Modification (TSM)

   As mentioned above, a key requirement for time-scale modification procedures is that they change the time-scale of a given audio signal without altering its pitch content. To achieve this goal, many TSM procedures follow a common fundamental strategy which is sketched in Figure 1. The core idea is to decompose the input signal into short *frames*. Having a fixed length, usually in the range of 50 to 100 milliseconds of audio material, each frame captures the local pitch content of the signal. The frames are then relocated on the time axis to achieve the actual time-scale modification, while, at the same time, preserving the signal's pitch.

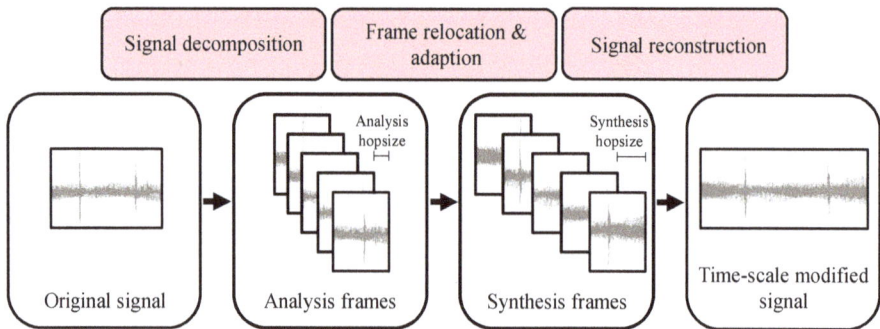

Figure 1. Generalized processing pipeline of Time-scale modification (TSM) procedures.

   More precisely, this process can be described as follows. The input of a TSM procedure is a discrete-time audio signal $x : \mathbb{Z} \to \mathbb{R}$, equidistantly sampled at a sampling rate of F_s. Note that although audio signals typically have a finite length of $L \in \mathbb{N}$ samples $x(r)$ for $r \in [0 : L-1] := \{0, 1, \ldots, L-1\}$, for the sake of simplicity, we model them to have an infinite support by defining $x(r) = 0$ for $r \in \mathbb{Z} \setminus [0 : L-1]$. The first step of the TSM procedure is to split x into short *analysis frames* x_m, $m \in \mathbb{Z}$, each of them having a length of N samples (in the literature, the analysis frames are sometimes also referred to as *grains*, see [9]). The analysis frames are spaced by an *analysis hopsize* H_a:

$$x_m(r) = \begin{cases} x(r + mH_a), & \text{if } r \in [-N/2 : N/2 - 1], \\ 0, & \text{otherwise.} \end{cases} \tag{1}$$

In a second step, these frames are relocated on the time axis with regard to a specified *synthesis hopsize* H_s. This relocation accounts for the actual modification of the input signal's time-scale by a *stretching factor* $\alpha = H_s/H_a$. Since it is often desirable to have a specific overlap of the relocated frames, the synthesis hopsize H_s is often fixed (common choices are $H_s = N/2$ or $H_s = N/4$) while the analysis hopsize is given by $H_a = H_s/\alpha$. However, simply superimposing the overlapping relocated frames would lead to undesired artifacts such as phase discontinuities at the frame boundaries and amplitude fluctuations. Therefore, prior to signal reconstruction, the analysis frames are suitably adapted to form *synthesis frames* y_m. In the final step, the synthesis frames are superimposed in order to reconstruct the actual time-scale modified output signal $y : \mathbb{Z} \to \mathbb{R}$ of the TSM procedure:

$$y(r) = \sum_{m \in \mathbb{Z}} y_m(r - mH_s) . \tag{2}$$

Although this fundamental strategy seems straightforward at a first glance, there are many pitfalls and design choices that may strongly influence the perceptual quality of the time-scale modified output signal. The most obvious question is how to adapt the analysis frames x_m in order to form the synthesis frames y_m. There are many ways to approach this task, leading to conceptually different TSM procedures. In the following, we discuss several strategies.

3. TSM Based on Overlap-Add (OLA)

3.1. The Procedure

In the general scheme described in the previous section, a straightforward approach would be to simply define the synthesis frames y_m to be equal to the unmodified analysis frames x_m. This strategy, however, immediately leads to two problems which are visualized in Figure 2. First, when reconstructing the output signal by using Equation (2), the resulting waveform typically shows discontinuities—perceivable as clicking sounds—at the unmodified frames' boundaries. Second, the synthesis hopsize H_s is usually chosen such that the synthesis frames are overlapping. When superimposing the unmodified frames—each of them having the same amplitude as the input signal—this typically leads to an undesired increase of the output signal's amplitude.

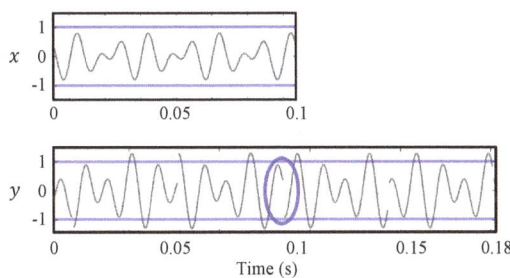

Figure 2. Typical artifacts that occur when choosing the synthesis frames y_m to be equal to the analysis frames x_m. The input signal x is stretched by a factor of $\alpha = 1.8$. The output signal y shows discontinuities (blue oval) and amplitude fluctuations (indicated by blue lines).

A basic TSM procedure should both enforce a smooth transition between frames as well as compensate for unwanted amplitude fluctuations. The idea of the *overlap-add* (OLA) TSM procedure is to apply a window function w to the analysis frames, prior to the reconstruction of the output signal y. The task of the window function is to remove the abrupt waveform discontinuities at the the analysis frames' boundaries. A typical choice for w is a *Hann window* function

$$w(r) = \begin{cases} 0.5 \left(1 - \cos\left(\frac{2\pi(r+N/2)}{N-1}\right)\right), & \text{if } r \in [-N/2 : N/2 - 1], \\ 0, & \text{otherwise.} \end{cases} \tag{3}$$

The Hann window has the nice property that

$$\sum_{n\in\mathbb{Z}} w\left(r - n\frac{N}{2}\right) = 1, \tag{4}$$

for all $r \in \mathbb{Z}$. The principle of the iterative OLA procedure is visualized in Figure 3. For the frame index $m \in \mathbb{Z}$, we first use Equation (1) to compute the mth analysis frame x_m (Figure 3a). Then, we derive the synthesis frame y_m by

$$y_m(r) = \frac{w(r)\, x_m(r)}{\sum_{n\in\mathbb{Z}} w(r - nH_s)}. \tag{5}$$

The nominator of Equation (5) constitutes the actual windowing of the analysis frame by multiplying it pointwise with the given window function. The denominator normalizes the frame by the sum of the overlapping window functions, which prevents amplitude fluctuations in the output signal. Note that, when choosing w to be a Hann window and $H_s = N/2$, the denominator always reduces to one by Equation (4). This is the case in Figure 3b where the synthesis frame's amplitude is not scaled before being added to the output signal y. Proceeding to the next analysis frame x_{m+1}, (Figure 3c), this frame is again windowed, overlapped with the preceding synthesis frame, and added to the output signal (Figure 3d). Note that Figure 3 visualizes the case where the original signal is compressed ($H_a > H_s$). Stretching the signal ($H_a < H_s$) works in exactly the same fashion. In this case, the analysis frames overlap to a larger degree than the synthesis frames.

OLA is an example of a *time-domain* TSM procedure where the modifications to the analysis frames are applied purely in the time-domain. In general, time-domain TSM procedures are not only efficient but also preserve the timbre of the input signal to a high degree. On the downside, output signals produced by OLA often suffer from other artifacts, as we explain next.

3.2. Artifacts

The OLA procedure is in general not capable of preserving local periodic structures that are present in the input signal. This is visualized in Figure 4 where a periodic input signal x is stretched by a factor of $\alpha = 1.8$ using OLA. When relocating the analysis frames, the periodic structures of x may not align any longer in the superimposed synthesis frames. In the resulting output signal y, the periodic patterns are distorted. These distortions are also known as *phase jump artifacts*. Since local periodicities in the waveforms of audio signals correspond to harmonic sounds, OLA is not suited to modify signals that contain harmonic components. When applied to harmonic signals, the output signals of OLA have a characteristic *warbling* sound, which is a kind of periodic frequency modulation [6]. Since most music signals contain at least some harmonic sources (as for example singing voice, piano, violins, or guitars), OLA is usually not suited to modify music.

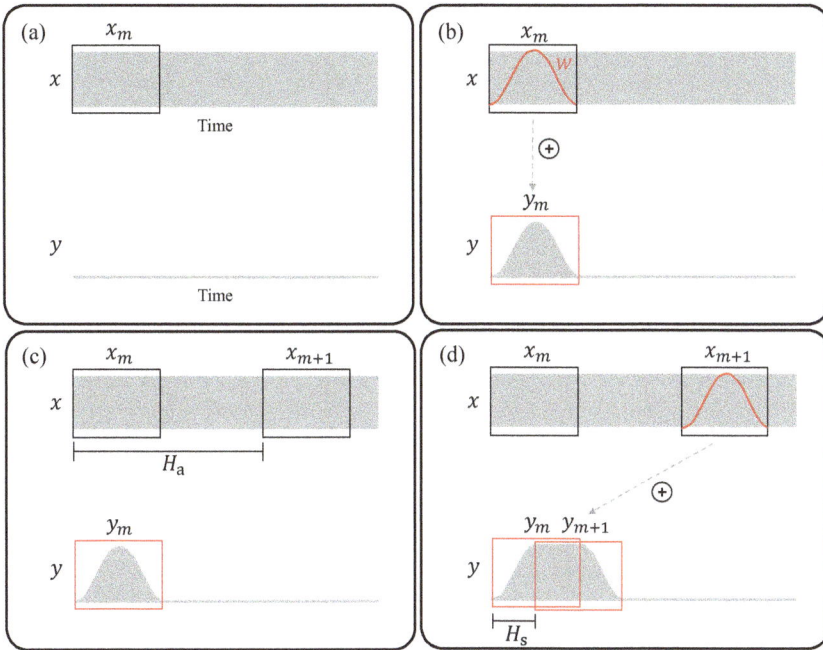

Figure 3. The principle of TSM based on overlap-add (OLA). (**a**) Input audio signal x with analysis frame x_m. The output signal y is constructed iteratively; (**b**) Application of Hann window function w to the analysis frame x_m resulting in the synthesis frame y_m; (**c**) The next analysis frame x_{m+1} having a specified distance of H_a samples from x_m; (**d**) Overlap-add using the specified synthesis hopsize H_s.

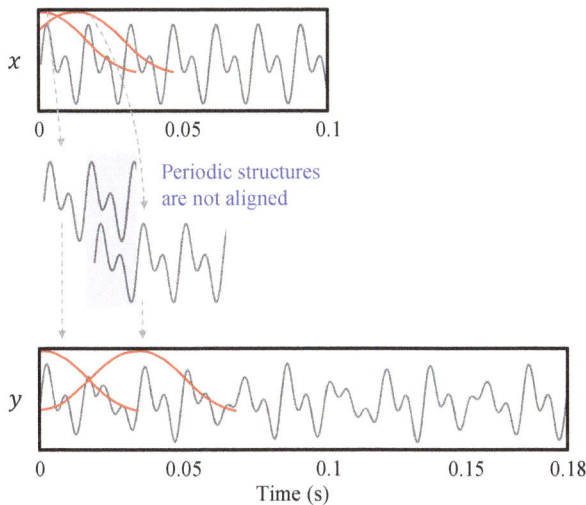

Figure 4. Illustration of a typical artifact for an audio signal modified with OLA. In this example, the input signal x is stretched by a factor of $\alpha = 1.8$. The OLA procedure is visualized for two frames indicated by window functions. The periodic structure of x is not preserved in the output signal y.

3.3. Tricks of the Trade

While OLA is unsuited for modifying audio signals with harmonic content, it delivers high quality results for purely percussive signals, as is the case with drum or castanet recordings [8]. This is because audio signals with percussive content seldom have local periodic structures. The phase jump artifacts introduced by OLA are therefore not noticeable in the output signals. In this scenario, it is important to choose a very small frame length N (corresponding to roughly 10 milliseconds of audio material) in order to reduce the effect of *transient doubling*, an artifact that we discuss at length in Section 4.2.

4. TSM Based on Waveform Similarity Overlap-Add (WSOLA)

4.1. The Procedure

One of OLA's problems is that it lacks signal sensitivity: the windowed analysis frames are copied from fixed positions in the input signal to fixed positions in the output signal. In other words, the input signal has no influence on the procedure. One time-domain strategy to reduce phase jump artifacts as caused by OLA is to introduce some flexibility in the TSM process, achieved by allowing some tolerance in the analysis or synthesis frames' positions. The main idea is to adjust successive synthesis frames in such a way that periodic structures in the frames' waveforms are aligned in the overlapping regions. Periodic patterns in the input signal are therefore maintained in the output. In the literature, several variants of this idea have been described, such as *synchronized OLA* (SOLA) [10], *time-domain pitch-synchronized OLA* (TD-PSOLA) [11], or autocorrelation-based approaches [12].

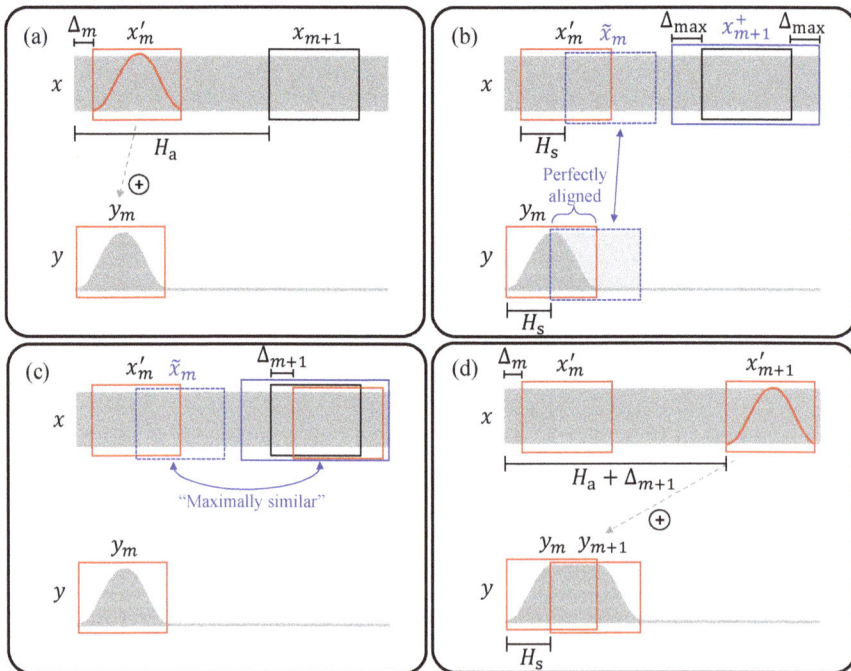

Figure 5. The principle of Waveform Similarity Overlap-Add (WSOLA). (**a**) Input audio signal x with the adjusted analysis frame x'_m. The frame was already windowed and copied to the output signal y; (**b,c**) Retrieval of a frame from the extended frame region x^+_{m+1} (solid blue box) that is as similar as possible to the natural progression $\tilde{x}_m$ (dashed blue box) of the adjusted analysis frame x'_m; (**d**) The adjusted analysis frame x'_{m+1} is windowed and copied to the output signal y.

Another well-known procedure is *waveform similarity-based OLA* (WSOLA) [4]. This method's core idea is to allow for slight shifts (of up to $\pm\Delta_{\max} \in \mathbb{Z}$ samples) of the analysis frames' positions. Figure 5 visualizes the principle of WSOLA. Similar to OLA, WSOLA proceeds in an iterative fashion. Assume that in the *m*th iteration the position of the analysis frame x_m was shifted by $\Delta_m \in [-\Delta_{\max} : \Delta_{\max}]$ samples. We call this frame the *adjusted analysis frame* x'_m (Figure 5a):

$$x'_m(r) = \begin{cases} x(r + mH_a + \Delta_m), & \text{if } r \in [-N/2 : N/2 - 1], \\ 0, & \text{otherwise.} \end{cases} \tag{6}$$

The adjusted analysis frame x'_m is windowed and copied to the output signal y as in OLA (Figure 5a). Now, we need to adjust the position of the next analysis frame x_{m+1}. This task can be interpreted as a constrained optimization problem. Our goal is to find the optimal shift index $\Delta_{m+1} \in [-\Delta_{\max} : \Delta_{\max}]$ such that periodic structures of the adjusted analysis frame x'_{m+1} are optimally aligned with structures of the previously copied synthesis frame y_m in the overlapping region when superimposing both frames at the synthesis hopsize H_s. The *natural progression* $\tilde{x}_m$ of the adjusted analysis frame x'_m (dashed blue box in Figure 5b) would be an optimal choice for the adjusted analysis frame x'_{m+1} in an unconstrained scenario:

$$\tilde{x}_m(r) = \begin{cases} x(r + mH_a + \Delta_m + H_s), & \text{if } r \in [-N/2 : N/2 - 1], \\ 0, & \text{otherwise.} \end{cases} \tag{7}$$

This is the case, since the adjusted analysis frame x'_m and the synthesis frame y_m are essentially the same (up to windowing). As a result, the structures of the natural progression $\tilde{x}_m$ are perfectly aligned with the structures of the synthesis frame y_m when superimposing the two frames at the synthesis hopsize H_s (Figure 5b). However, due to the constraint $\Delta_{m+1} \in [-\Delta_{\max} : \Delta_{\max}]$, the adjusted frame x'_{m+1} must be located inside of the *extended frame region* x^+_{m+1} (solid blue box in Figure 5b):

$$x^+_{m+1}(r) = \begin{cases} x(r + (m+1)H_a), & \text{if } r \in [-N/2 - \Delta_{\max} : N/2 - 1 + \Delta_{\max}], \\ 0, & \text{otherwise.} \end{cases} \tag{8}$$

Therefore, the idea is to retrieve the adjusted frame x'_{m+1} as the frame in x^+_{m+1} whose waveform is most similar to $\tilde{x}_m$. To this end, we must define a measure that quantifies the similarity of two frames. One possible choice for this metric is the *cross-correlation*

$$c(q, p, \Delta) = \sum_{r \in \mathbb{Z}} q(r)\, p(r + \Delta) \tag{9}$$

of a signal q and a signal p shifted by $\Delta \in \mathbb{Z}$ samples. We can then compute the optimal shift index Δ_{m+1} that maximizes the cross-correlation of $\tilde{x}_m$ and x^+_{m+1} by

$$\Delta_{m+1} = \operatorname*{argmax}_{\Delta \in [-\Delta_{\max} : \Delta_{\max}]} c(\tilde{x}_m, x^+_{m+1}, \Delta) . \tag{10}$$

The shift index Δ_{m+1} defines the position of the adjusted analysis frame x'_{m+1} inside the extended frame region x^+_{m+1} (Figure 5c). Finally, we compute the synthesis frame y_{m+1}, similarly to OLA, by

$$y_{m+1}(r) = \frac{w(r)\, x(r + (m+1)H_a + \Delta_{m+1})}{\sum_{n \in \mathbb{Z}} w(r - nH_s)} \tag{11}$$

and use Equation (2) to reconstruct the output signal y (Figure 5d). In practice, we start the iterative optimization process at the frame index $m = 0$ and assume $\Delta_0 = 0$.

4.2. Artifacts

WSOLA can improve on some of the drawbacks of OLA as illustrated by Figure 6 (WSOLA) and Figure 4 (OLA). However, it still causes certain artifacts that are characteristic for time-domain TSM procedures.

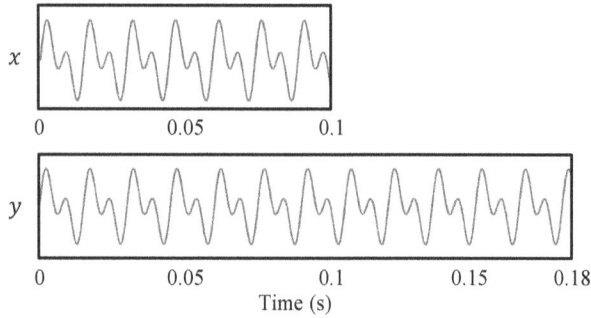

Figure 6. Preservation of periodic structures by WSOLA. The input signal x is the same as in Figure 4, this time modified with WSOLA (time-stretch with $\alpha = 1.8$).

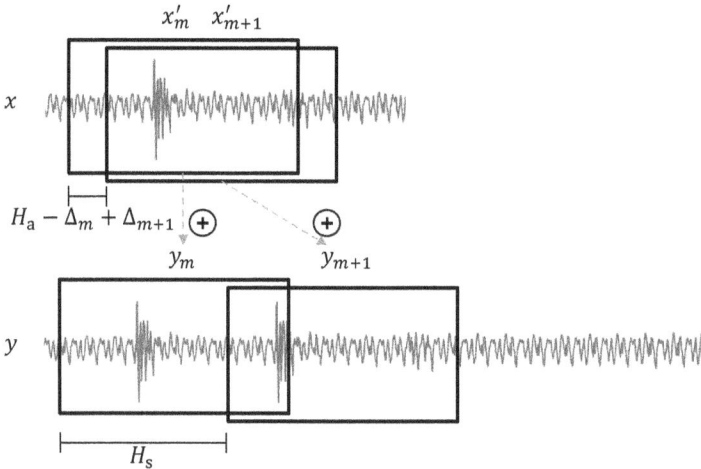

Figure 7. Transient doubling artifact as it typically occurs in signals modified with WSOLA.

A prominent problem with WSOLA-like methods is known as *transient doubling* or *stuttering*. This artifact is visualized in Figure 7, where we can see a single transient in the input signal x that is contained in the overlapping region of two successive adjusted analysis frames x'_m and x'_{m+1}. When the frames are relocated and copied to the output signal, the transient is duplicated and can be heard two times in quick succession. A related artifact is *transient skipping* where transients get lost in the modification process since they are not contained in any of the analysis frames. While transient doubling commonly occurs when stretching a signal ($\alpha > 1$), transient skipping usually happens when the signal is compressed ($\alpha < 1$). These artifacts are particularly noticeable when modifying signals with percussive components, such as recordings of instruments with strong onsets (e.g., drums, piano).

Furthermore, WSOLA has particular problems when modifying polyphonic input signals such as recordings of orchestral music. For these input signals, the output often still contains considerable warbling. The reason for this is that WSOLA can, by design, only preserve the most prominent periodic pattern in the input signal's waveform. Therefore, when modifying recordings with multiple harmonic sound sources, only the sound of the most dominant source is preserved in the output, whereas the remaining sources can still cause phase jump artifacts. While WSOLA is well-suited to modify monophonic input signals, this is often not the case with more complex audio.

4.3. Tricks of the Trade

In order to assure that WSOLA can adapt to the most dominant periodic pattern in the waveform of the input signal, one frame must be able to capture at least a full period of the pattern. In addition, the tolerance parameter Δ_{max} must be large enough to allow for an appropriate adjustment. Therefore, it should be set to at least half a period's length. Assuming that the lowest frequency that can be heard by humans is roughly 20 Hz, a common choice is a frame length N corresponding to 50 ms and a tolerance parameter of 25 ms.

One possibility to reduce transient doubling and skipping artifacts in WSOLA is to apply a *transient preservation* scheme. In [13], the idea is to first identify the temporal positions of transients in the input signal by using a transient detector. Then, in the WSOLA process, the analysis hopsize is temporarily fixed to be equal to the synthesis hopsize whenever an analysis frame is located in a neighborhood of an identified transient. This neighborhood, including the transient, is therefore copied without modification to the output signal, preventing WSOLA from doubling or skipping it. The deviation in the global stretching factor is compensated dynamically in the regions between transients.

5. TSM Based on the Phase Vocoder (PV-TSM)

5.1. Overview

As we have seen, WSOLA is capable of maintaining the most prominent periodicity in the input signal. The next step to improve the quality of time-scale modified signals is to preserve the periodicities of all signal components. This is the main idea of *frequency-domain* TSM procedures, which interpret each analysis frame as a weighted sum of sinusoidal components with known frequency and phase. Based on these parameters, each of these components is manipulated individually to avoid phase jump artifacts across all frequencies in the reconstructed signal.

A fundamental tool for the frequency analysis of the input signal is the *short-time Fourier transform* [14] (Section 5.2). However, depending on the chosen discretization parameters, the resulting frequency estimates may be inaccurate. To this end, the *phase vocoder* (Although the term "phase vocoder" refers to a specific technique for the estimation of instantaneous frequencies, it is also frequently used in the literature as the name of the TSM procedure itself.) technique [5,7] is used to improve on the the short-time Fourier transform's coarse frequency estimates by deriving the sinusoidal components' *instantaneous frequencies* (Section 5.3). In TSM procedures based on the phase vocoder (PV-TSM), these improved estimates are used to update the phases of an input signal's sinusoidal components in a process known as *phase propagation* (Section 5.4).

5.2. The Short-Time Fourier Transform

The most important tool of PV-TSM is the short-time Fourier transform (STFT), which applies the Fourier transform to every analysis frame of a given input signal. The STFT X of a signal x is given by

$$X(m,k) = \sum_{r=-N/2}^{N/2-1} x_m(r)\, w(r)\, \exp(-2\pi i k r/N)\,, \qquad (12)$$

where $m \in \mathbb{Z}$ is the frame index, $k \in [0 : N-1]$ is the frequency index, N is the frame length, x_m is the mth analysis frame of x as defined in Equation (1), and w is a window function. Given the input signal's sampling rate F_s, the frame index m of $X(m,k)$ is associated to the physical time

$$T_{\text{coef}}(m) = \frac{m\, H_a}{F_s} \qquad (13)$$

given in seconds, and the frequency index k corresponds to the physical frequency

$$F_{\text{coef}}(k) = \frac{k\, F_s}{N} \qquad (14)$$

given in Hertz. The complex number $X(m,k)$, also called a *time-frequency bin*, denotes the kth Fourier coefficient for the mth analysis frame. It can be represented by a magnitude $|X(m,k)| \in \mathbb{R}^+$ and a phase $\varphi(m,k) \in [0,1)$ as

$$X(m,k) = |X(m,k)|\, \exp(2\pi i\, \varphi(m,k))\,. \qquad (15)$$

The magnitude of an STFT X is also called a *spectrogram* which is denoted by Y:

$$Y = |X|\,. \qquad (16)$$

In the context of PV-TSM, it is necessary to reconstruct the output signal y from a modified STFT X^{Mod}. Note that a modified STFT is in general not invertible [15]. In other words, there might be no signal y that has X^{Mod} as its STFT. However, there exist methods that aim to reconstruct a signal y from X^{Mod} whose STFT is close to X^{Mod} with respect to some distance measure. Following the procedure described in [16], we first compute time-domain frames x_m^{Mod} by using the inverse Fourier transform.

$$x_m^{\text{Mod}}(r) = \frac{1}{N} \sum_{k=0}^{N-1} X^{\text{Mod}}(m,k)\, \exp(2\pi i k r/N)\,. \qquad (17)$$

From these frames, we then derive synthesis frames

$$y_m(r) = \frac{w(r)\, x_m^{\text{Mod}}(r)}{\sum_{n \in \mathbb{Z}} w(r - n H_s)^2} \qquad (18)$$

and reconstruct the output signal y by Equation (2). It can be shown that, when computing the synthesis frames by Equation (18), the STFT of y (when choosing $H_a = H_s$) minimizes a squared error distance measure defined in [16].

5.3. The Phase Vocoder

Each time-frequency bin $X(m,k)$ of an STFT can be interpreted as a sinusoidal component with amplitude $|X(m,k)|$ and phase $\varphi(m,k)$ that contributes to the mth analysis frame of the input signal x. However, the Fourier transform yields coefficients only for a discrete set of frequencies that are sampled linearly on the frequency axis, see Equation (14). The STFT's frequency resolution therefore does not suffice to assign a precise frequency value to this sinusoidal component. The phase vocoder is a technique that refines the STFT's coarse frequency estimate by exploiting the given phase information.

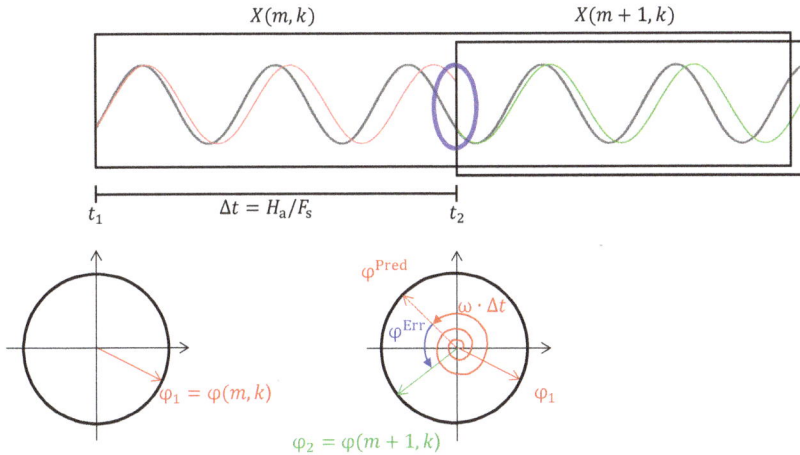

Figure 8. Principle of the phase vocoder.

In order to understand the phase vocoder, let us have a look at the scenario shown in Figure 8. Assume we are given two phase estimates $\varphi_1 = \varphi(m, k)$ and $\varphi_2 = \varphi(m + 1, k)$ at the time instances $t_1 = T_{\text{coef}}(m)$ and $t_2 = T_{\text{coef}}(m + 1)$ of a sinusoidal component for which we only have a coarse frequency estimate $\omega = F_{\text{coef}}(k)$. Our goal is to estimate the sinusoid's "real" instantaneous frequency IF(ω). Figure 8 shows this sinusoidal component (gray) as well as two sinusoids that have a frequency of ω (red and green). In addition, we also see phase representations at the time instances t_1 and t_2. The red sinusoid has a phase of φ_1 at t_1 and the green sinusoid a phase of φ_2 at t_2. One can see that the frequency ω of the red and green sinusoids is slightly lower than the frequency of the gray sinusoid. Intuitively, while the phases of the gray and the red sinusoid match at t_1, they diverge over time, and we can observe a considerable discrepancy after $\Delta t = t_2 - t_1$ seconds (blue oval). Since we know the red sinusoid's frequency, we can compute its *unwrapped phase advance*, i.e., the number of oscillations that occur over the course of Δt seconds:

$$\omega \, \Delta t \, . \tag{19}$$

Knowing that its phase at t_1 is φ_1, we can predict its phase after Δt seconds:

$$\varphi^{\text{Pred}} = \varphi_1 + \omega \, \Delta t \, . \tag{20}$$

At t_2 we again have a precise phase estimate φ_2 for the gray sinusoid. We therefore can compute the *phase error* φ^{Err} (also called the *heterodyned phase increment* [17]) between the phase actually measured at t_2 and the predicted phase when assuming a frequency of ω:

$$\varphi^{\text{Err}} = \Psi(\varphi_2 - \varphi^{\text{Pred}}) \, , \tag{21}$$

where Ψ is the *principal argument function* that maps a given phase into the range $[-0.5, 0.5]$. Note that by mapping φ^{Err} into this range, we assume that the number of oscillations of the gray and red sinusoids differ by at most half a period. In the context of instantaneous frequency estimation, this means that the coarse frequency estimate ω needs to be close to the actual frequency of the sinusoid, and that the interval Δt should be small. The unwrapped phase advance of the gray sinusoid can then be computed by the sum of the unwrapped phase advance of the red sinusoid with frequency ω (red spiral arrow) and the phase error (blue curved arrow):

$$\omega \, \Delta t + \varphi^{\mathrm{Err}} \, . \tag{22}$$

This gives us the number of oscillations of the gray sinusoid over the course of Δt seconds. From this we can derive the instantaneous frequency of the gray sinusoid by

$$\mathrm{IF}(\omega) = \frac{\omega \, \Delta t + \varphi^{\mathrm{Err}}}{\Delta t} = \omega + \frac{\varphi^{\mathrm{Err}}}{\Delta t} \, . \tag{23}$$

The frequency $\varphi^{\mathrm{Err}} / \Delta t$ can be interpreted as the small offset of the gray sinusoid's actual frequency from the rough frequency estimate ω.

We can use this approach to refine the coarse frequency resolution of the STFT by computing instantaneous frequency estimates $F_{\mathrm{coef}}^{\mathrm{IF}}(m, k)$ for all time-frequency bins $X(m, k)$:

$$F_{\mathrm{coef}}^{\mathrm{IF}}(m, k) = \mathrm{IF}(\omega) = \omega + \frac{\Psi \left(\varphi_2 - (\varphi_1 + \omega \, \Delta t) \right)}{\Delta t} \tag{24}$$

with $\omega = F_{\mathrm{coef}}(k)$, $\Delta t = H_a / F_s$ (the analysis hopsize given in seconds), $\varphi_1 = \varphi(m, k)$, and $\varphi_2 = \varphi(m + 1, k)$. For further details, we refer to ([15], Chapter 8).

5.4. PV-TSM

The principle of PV-TSM is visualized in Figure 9. Given an input audio signal x, the first step of PV-TSM is to compute the STFT X. Figure 9a depicts the two successive frequency spectra of the *mth* and $(m+1)th$ analysis frames, denoted by $X(m)$ and $X(m+1)$, respectively (here, a frame's Fourier spectrum is visualized as a set of sinusoidals, representing the sinusoidal components that are captured in the time-frequency bins). Our goal is to compute a modified STFT X^{Mod} with adjusted phases φ^{Mod} from which we can reconstruct a time-scale modified signal without phase jump artifacts:

$$X^{\mathrm{Mod}}(m, k) = |X(m, k)| \, \exp(2\pi i \, \varphi^{\mathrm{Mod}}(m, k)) \, . \tag{25}$$

We compute the adjusted phases φ^{Mod} in an iterative process that is known as phase propagation. Assume that the phases of the *mth* frame have already been modified (see the red sinusoid's phase in Figure 9b being different from its phase in Figure 9a). As indicated by Figure 9b, overlapping the *mth* and $(m+1)th$ frame at the synthesis hopsize H_s may lead to phase jumps. Knowing the instantaneous frequencies $F_{\mathrm{coef}}^{\mathrm{IF}}$ derived by the phase vocoder, we can predict the phases of the sinusoidal components in the *mth* frame after a time interval corresponding to H_s samples. To this end, we set $\varphi_1 = \varphi^{\mathrm{Mod}}(m, k)$, $\omega = F_{\mathrm{coef}}^{\mathrm{IF}}(m, k)$, and $\Delta t = H_s / F_s$ (the synthesis hopsize given in seconds) in Equation (20). This allows us to replace the phases of the $(m+1)th$ frame with the predicted phase:

$$\varphi^{\mathrm{Mod}}(m + 1, k) = \varphi^{\mathrm{Mod}}(m, k) + F_{\mathrm{coef}}^{\mathrm{IF}}(m, k) \frac{H_s}{F_s} \tag{26}$$

for $k \in [0 : N - 1]$. Assuming that the estimates of the instantaneous frequencies $F_{\text{coef}}^{\text{IF}}$ are correct, there are no phase jumps any more when overlapping the modified spectra at the synthesis hopsize H_s (Figure 9c). In practice, we start the iterative phase propagation with the frame index $m = 0$ and set

$$\varphi^{\text{Mod}}(0,k) = \varphi(0,k) , \tag{27}$$

for all $k \in [0 : N - 1]$. Finally, the output signal y can be computed using the signal reconstruction procedure described in Section 5.2 (Figure 9d).

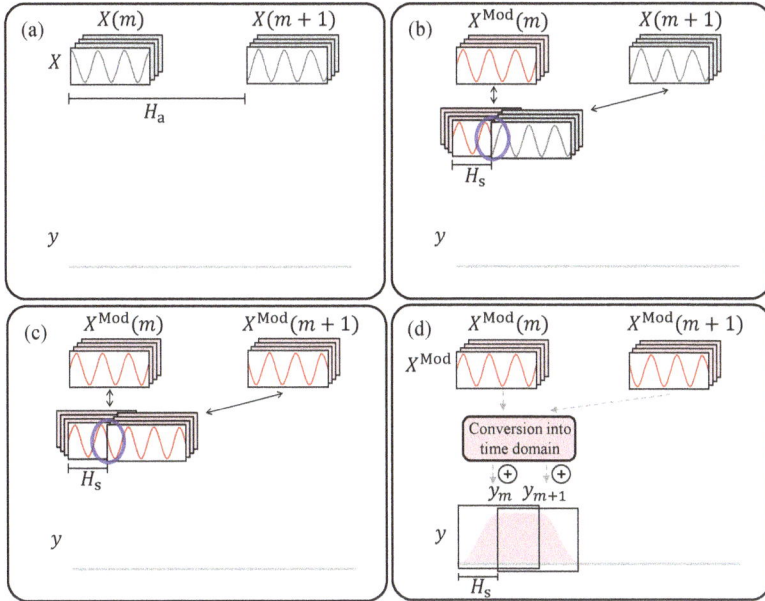

Figure 9. The principle of TSM Based on the Phase Vocoder (PV-TSM). (**a**) STFT X (time-frequency bins are visualized by sinusoidals); (**b**) Using the original phase of $X(m + 1, k)$ leads to phase jumps when overlapping the frames at the synthesis hopsize H_s (blue oval); (**c**) Update of the sinusoids' phases via phase propagation. Phase jumps are reduced (blue oval); (**d**) Signal reconstruction from the modified STFT X^{Mod}.

5.5. Artifacts

By design, PV-TSM can achieve phase continuity for all sinusoidal components contributing to the output signal. This property, also known as *horizontal phase coherence*, ensures that there are no artifacts related to phase jumps. However, the *vertical phase coherence*, i.e., the phase relationships of sinusoidal components within one frame, is usually destroyed in the phase propagation process. The loss of vertical phase coherence affects the time localization of events such as transients. As a result, transients are often *smeared* in signals modified with PV-TSM (Figure 10). Furthermore, output signals of PV-TSM commonly have a very distinct sound coloration known as *phasiness* [18], an artifact also caused by the loss of vertical phase coherence. Signals suffering from phasiness are described as being reverberant, having "less bite," or a "loss of presence" [6].

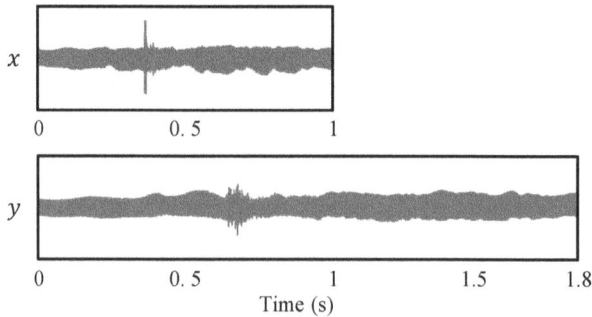

Figure 10. Illustration of a smeared transient as a result of applying PV-TSM. The input signal x has been stretched by a factor of $\alpha = 1.8$ to yield the output signal y.

5.6. Tricks of the Trade

The phase vocoder technique highly benefits from frequency estimates that are already close to the instantaneous frequencies as well as from a small analysis hopsize (resulting in a small Δt). In PV-TSM, the frame length N is therefore commonly set to a relatively large value. In practice, N typically corresponds to roughly 100 ms in order to achieve a high frequency resolution of the STFT.

To reduce the loss of vertical phase coherence in the phase vocoder, Laroche and Dolson proposed a modification to the standard PV-TSM in [6]. Their core observation is that a sinusoidal component may affect multiple adjacent time-frequency bins of a single analysis frame. Therefore, the phases of these bins should not be updated independently, but in a joint fashion. A peak in a frame's magnitude spectrum is assumed to be representative of a particular sinusoidal component. The frequency bins with lower magnitude values surrounding the peak are assumed to be affected by the same component as well. In the phase propagation process, only the time-frequency bins with a peak magnitude are updated in the usual fashion described in Equation (26). The phases of the remaining frequency bins are *locked* to the phase of the sinusoidal component corresponding to the closest peak. This procedure allows to locally preserve the signal's vertical phase coherence. This technique, also known as *identity phase locking*, leads to reduced phasiness artifacts and less smearing of transients.

Another approach to reduce phasieness artifacts is the *phase vocoder with synchronized overlap-add* (PVSOLA) and similar formulations [19–21]. The core observation of these methods is that the original vertical phase coherence is usually lost gradually over the course of phase-updated synthesis frames. Therefore, these procedures "reset" the vertical phase coherence repeatedly after a fixed number of frames. The resetting is done by copying an analysis frame unmodified to the output signal and adjusting it in a WSOLA-like fashion. After having reset the vertical phase coherence, the next synthesis frames are again computed by the phase propagation strategy.

To reduce the effect of transient smearing, other approaches include transient preservation schemes, similar to those for WSOLA. In [22], transients are identified in the input signal, cut out from the waveform, and temporarily stored. The remaining signal, after filling the gaps using linear prediction techniques, is modified using PV-TSM. Finally, the stored transients are relocated according to the stretching factor and reinserted into the modified signal.

6. TSM Based on Harmonic-Percussive Separation

6.1. The Procedure

As we have seen in the previous sections, percussive sound components and transients in the input signal cause issues with WSOLA and PV-TSM. On the other hand, OLA is capable of modifying percussive and transient sounds rather well, while causing phase jump artifacts in harmonic signals.

The idea of a recent TSM procedure proposed in [8] is to combine PV-TSM and OLA in order to modify both harmonic and percussive sound components with high quality. The procedure's principle is visualized in Figure 11. The shown input signal consists of a tone played on a violin (harmonic) superimposed with a single click of castanets halfway through the signal (percussive). The first step is to decompose the input signal into two sound components: a *harmonic component* and a *percussive component*. This is done by using a *harmonic-percussive separation* (HPS) technique. A simple and effective method for this task was proposed by Fitzgerald in [23], which we review in Section 6.2. As we can see in Figure 11, the harmonic component captures the violin's sound while the percussive component contains the castanets' click. A crucial observation is that the harmonic component usually does not contain percussive residues and vice versa. After having decomposed the input signal, the idea of [8] is to apply PV-TSM with identity phase locking (see Section 5.6) to the harmonic component and OLA with a very short frame length (see Section 3.3) to the percussive component. By treating the two components independently with the two different TSM procedures, their respective characteristics can be preserved to a high degree. Finally, the superposition of the two modified signal components forms the output of the procedure. A conceptually similar strategy for TSM has been proposed by Verma and Meng in [24] where they use a *sines+transients+noise* signal model [25–28] to parameterize a given signal's sinusoidal, transient, and noise-like sound components. The estimated parameters are then modified in order to synthesize a time-scaled output signal. However, their procedure relies on an explicit transient detection in order to estimate appropriate parameters for the transient sound components.

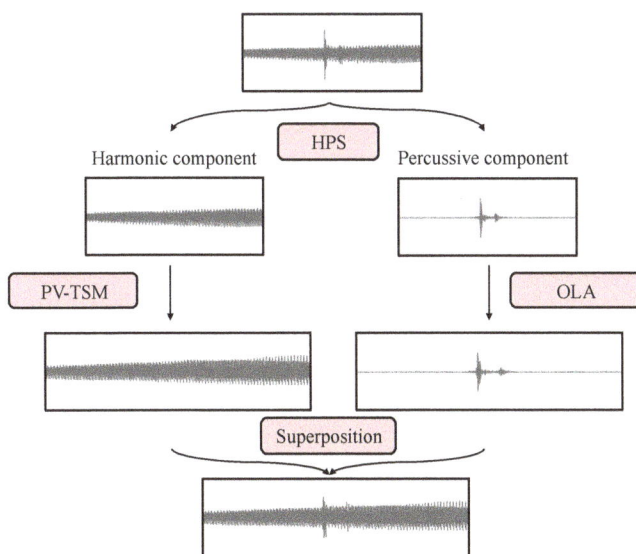

Figure 11. Principle of TSM based on harmonic-percussive separation.

In contrast to other transient preserving TSM procedures such as [13,22,24,29], the approach based on HPS has the advantage that an explicit detection and preservation of transients is not necessary. Figure 12 shows the same input signal x as in Figure 10. Here, the signal is stretched by a factor of $\alpha = 1.8$ using TSM based on HPS. Unlike Figure 10, the transient from the input signal x is also clearly visible in the output signal y. When looking closely, one can see that the transient is actually doubled by OLA (Section 4.2). However, due to the short frame length used in the procedure, the transient repeats so quickly that the doubling is not noticeable when listening to the output signal.

The TSM procedure based on HPS was therefore capable of preserving the transient without an explicit transient detection. Since both the explicit detection and preservation of transients are non-trivial and error-prone tasks, the achieved implicit transient preservation yields a more robust TSM procedure. The TSM approach based on HPS is therefore a good example of how the strengths of different TSM procedures can be combined in a beneficial way.

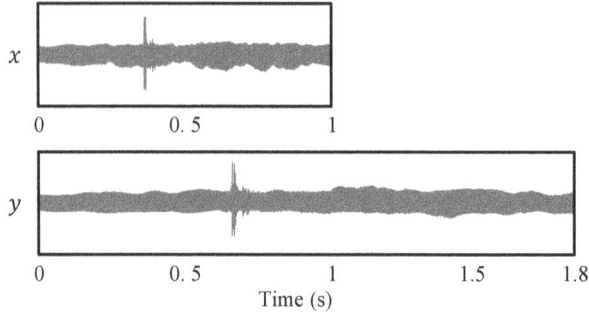

Figure 12. Transient preservation in TSM based on harmonic-percussive separation (HPS). The input signal x is the same as in Figure 10 and is stretched by a factor of $\alpha = 1.8$ using TSM based on HPS.

6.2. Harmonic-Percussive Separation

The goal of harmonic-percussive separation (HPS) is to decompose a given audio signal x into a signal x_h consisting of all harmonic sound components and a signal x_p consisting of all percussive sound components. While there exist numerous approaches for this task [30–35], a particularly simple and effective method was proposed by Fitzgerald in [23], which we review in the following. To illustrate the procedure, let us have a look at Figure 13. The input is an audio signal as shown in Figure 13a. Here, we again revert to our example of a tone played on the violin, superimposed with a single click of castanets. The first step is to compute the STFT X of the given audio signal x as defined in Equation (15). A critical observation is that, in the spectrogram $Y = |X|$, harmonic sounds form structures in the time direction, while percussive sounds yield structures in the frequency direction. In the spectrogram shown in Figure 13b we can see horizontal lines, reflecting the harmonic sound of the violin, as well as a single vertical line in the middle of the spectrogram, stemming from a click of the castanets. By applying a median filter to Y—once horizontally and once vertically—we get a horizontally enhanced spectrogram $\tilde{Y}_h$ and a vertically enhanced spectrogram $\tilde{Y}_p$:

$$\tilde{Y}_h(m,k) \;=\; \text{median}\left(Y(m-\ell_h,k),\dots,Y(m+\ell_h,k)\right) \tag{28}$$

$$\tilde{Y}_p(m,k) \;=\; \text{median}\left(Y(m,k-\ell_p),\dots,Y(m,k+\ell_p)\right) \tag{29}$$

for $\ell_h, \ell_p \in \mathbb{N}$, where $2\ell_h + 1$ and $2\ell_p + 1$ are the lengths of the median filters, respectively (Figure 13c). We assume a time-frequency bin of the original STFT $X(m,k)$ to be part of the harmonic component if $\tilde{Y}_h(m,k) > \tilde{Y}_p(m,k)$ and of the percussive component if $\tilde{Y}_p(m,k) \geq \tilde{Y}_h(m,k)$. Using this principle, we can define binary masks $\mathcal{M}_h$ and $\mathcal{M}_p$ for the harmonic and the percussive components (Figure 13d):

$$\mathcal{M}_h(m,k) = \begin{cases} 1, & \text{if } \tilde{Y}_h(m,k) > \tilde{Y}_p(m,k), \\ 0, & \text{otherwise,} \end{cases} \tag{30}$$

$$\mathcal{M}_p(m,k) = \begin{cases} 1, & \text{if } \tilde{Y}_p(m,k) \geq \tilde{Y}_h(m,k), \\ 0, & \text{otherwise.} \end{cases} \tag{31}$$

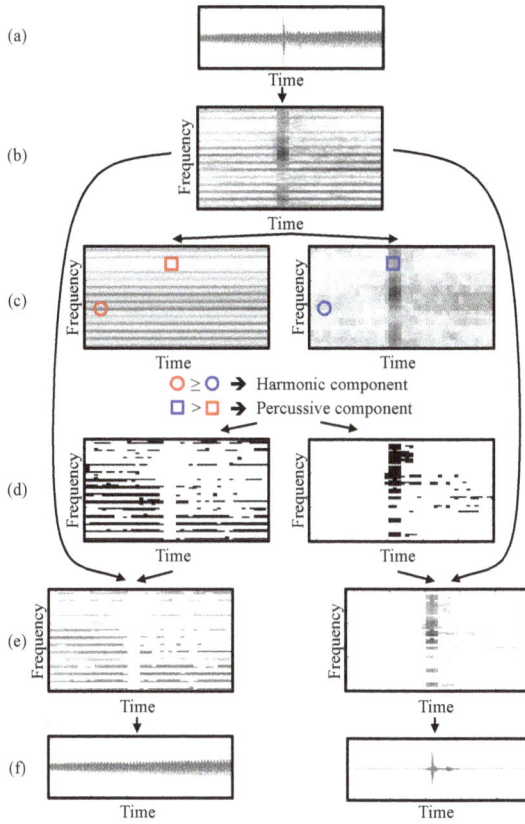

Figure 13. The HPS procedure presented in [23]. (**a**) Input audio signal x; (**b**) STFT X of x; (**c**) Horizontally and vertically enhanced spectrograms $\tilde{Y}_h$ and $\tilde{Y}_p$; (**d**) Binary masks $\mathcal{M}_h$ and $\mathcal{M}_p$; (**e**) Spectrograms of the harmonic and the percussive component X_h and X_p; (**f**) Harmonic and percussive components x_h and x_p.

Applying these masks to the original STFT X yields modified STFTs corresponding to the harmonic and percussive components (Figure 13e):

$$X_h(m,k) = X(m,k)\,\mathcal{M}_h(m,k)\,, \tag{32}$$

$$X_p(m,k) = X(m,k)\,\mathcal{M}_p(m,k)\,. \tag{33}$$

These modified STFTs can then be transformed back to the time-domain by using the reconstruction strategy discussed in Section 5.2. This yields the desired signals x_h and x_p, respectively. As we can see in Figure 13f, the harmonic component x_h contains the tone played by the violin, while the percussive component x_p shows the waveform of the castanets' click.

7. Applications

7.1. Music Synchronization—Non-Linear TSM

In scenarios like *automated soundtrack generation* [36] or *automated DJing* [1,2], it is often necessary to synchronize two or more music recordings by aligning musically related beat positions in time.

However, music recordings do not necessarily have a constant tempo. In this case, stretching or compressing the recordings by a constant factor α is insufficient to align their beat positions. Instead, the recordings' time-scales need to be modified in a non-linear fashion. The goal of non-linear TSM is to modify an audio signal according to a strictly monotonously increasing *time-stretch function* $\tau : \mathbb{R} \to \mathbb{R}$, which defines a mapping between points in time (given in seconds) of the input and output signals. Figure 14 shows an example of such a function. The first part, shown in red, has a slope greater than one. As a consequence, the red region in the input signal is mapped to a larger region in the output, resulting in a stretch. The slope of the function's second part, shown in blue, is smaller than one, leading to a compression of this region in the output. One possibility to realize this kind of non-linear modifications is to define the positions of the analysis frames according to the time-stretch function τ, instead of a constant analysis hopsize H_a. The process of deriving the analysis frames' positions is presented in Figure 14. Given τ, we first fix a synthesis hopsize H_s as we did for linear TSM (see Section 2). From this, we derive the *synthesis instances* $s_m \in \mathbb{R}$ (given in seconds), which are the positions of the synthesis frames in the output signal:

$$s_m = \frac{m \, H_s}{F_s} \, . \tag{34}$$

By inverting τ, we compute the *analysis instances* $a_m \in \mathbb{R}$ (given in seconds):

$$a_m = \tau^{-1}(s_m) \, . \tag{35}$$

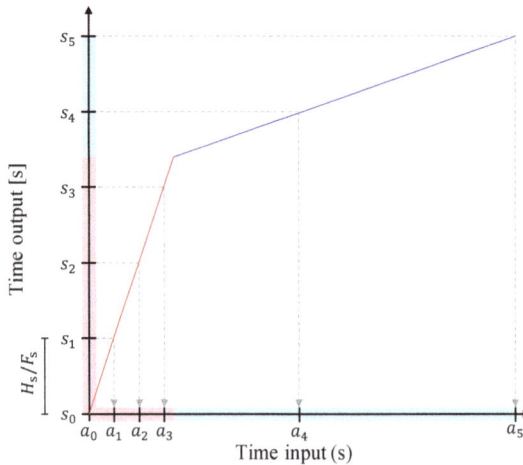

Figure 14. Example of a non-linear time-stretch function τ.

When using analysis frames indicated by the analysis instances for TSM with a procedure of our choice, the resulting output signal is modified according to the time-stretch function τ. To this end, all of the previously discussed TSM procedures can also be used for non-linear TSM.

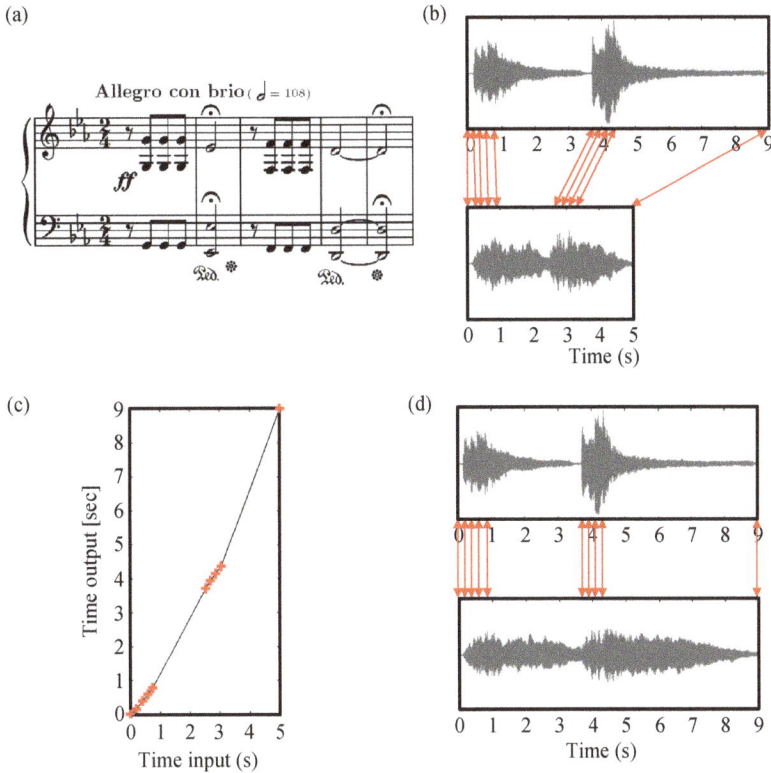

Figure 15. (**a**) Score of the first five measures of Beethoven's Symphony No. 5; (**b**) Waveforms of two performances. Corresponding onset positions are indicated by red arrows; (**c**) Set of anchor points (indicated in red) and inferred time-stretch function τ; (**d**) Onset-synchronized waveforms of the two performances.

A very convenient way of defining a time-stretch function is by introducing a set of *anchor points*. An anchor point is a pair of time positions where the first entry specifies a time position in the input signal and the second entry is a time position in the output signal. The actual time-stretch function τ is then obtained by a linear interpolation between the anchor points. Figure 15 shows a real-world example of a non-linear modification. In Figure 15b, we see the waveforms of two recorded performances of the first five measures of Beethoven's Symphony No. 5. The corresponding time positions of the note onsets are indicated by red arrows. Obviously, the first recording is longer than the second. However, the performances' tempi do not differ by a constant factor. While the eighth notes in the first and third measure are played at almost the same tempo in both performances, the durations of the half notes (with fermata) in measures two and five are significantly longer in the first recording. The mapping between the note onsets of the two performances is therefore non-linear. In Figure 15c, we define eight anchor points that map the onset positions of the second performance to the onset positions of the first performance (plus two additional anchor points aligning the recordings' start times and end times, respectively). Based on these anchor points and the derived time-stretch function τ, we then apply the TSM procedure of our choice to the second performance in order to obtain a version that is onset-synchronized with the first performance (Figure 15d).

7.2. Pitch-Shifting

Pitch-shifting is the task of changing an audio recording's pitch without altering its length—it can be seen as the dual problem to TSM. While there are specialized pitch-shifting procedures [37,38], it is also possible to approach the problem by combining TSM with resampling. The core observation is that resampling a given signal and playing it back at the original sampling rate changes the length and the pitch of the signal at the same time (The same effect can be simulated with vinyl records by changing the rotation speed of the record player). To pitch-shift a given signal, it is therefore first resampled and afterwards time-scale modified in order to compensate for the change in length. More precisely, an audio signal, sampled at a rate of $F_s^{(in)}$, is first resampled to have a new sampling rate of $F_s^{(out)}$. When playing back the signal at its original sampling rate $F_s^{(in)}$, this operation changes the signal's length by a factor of $F_s^{(out)}/F_s^{(in)}$ and scales its frequencies by the term's inverse. For example, musically speaking, a factor of $F_s^{(out)}/F_s^{(in)} = \frac{1}{2}$ increases the pitch content by one octave. To compensate for the change in length, the signal needs to be stretched by a factor of $\alpha = F_s^{(in)}/F_s^{(out)}$, using a TSM procedure.

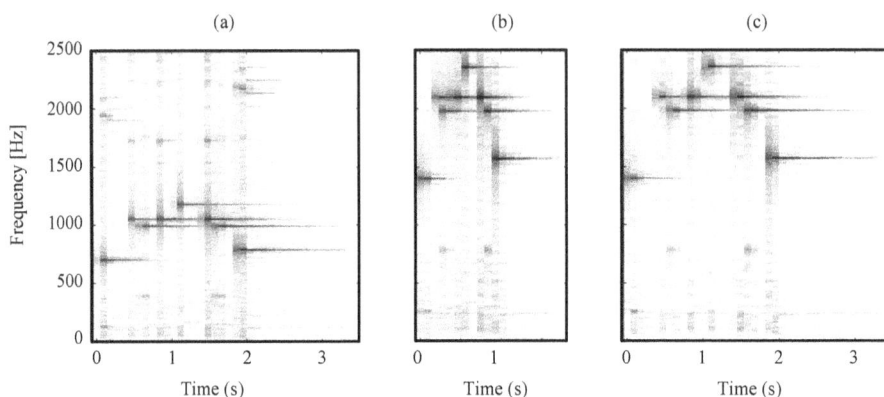

Figure 16. Pitch-shifting *via* resampling and TSM. (**a**) Spectrogram of an input audio signal; (**b**) Spectrogram of the resampled signal; (**c**) Spectrogram after TSM application.

To demonstrate this, we show an example in Figure 16, where the goal is to apply a pitch-shift of one octave to the input audio signal. The original signal has a sampling rate of $F_s^{(in)} = 44,100$ Hz (Figure 16a). To achieve the desired pitch-shift, the signal is resampled to $F_s^{(out)} = 22,050$ Hz (Figure 16b). One can see that the resampling changed the pitch of the signal as well as its length when interpreting the signal as still being sampled at $F_s^{(in)} = 44,100$ Hz. While the change in pitch is desired, the change in length needs to be compensated for. Thus, we stretch the signal by a factor of $\alpha = 44,100$ Hz$/22,050$ Hz $= 2$ to its original length (Figure 16c).

The quality of the pitch-shifting result crucially depends on various factors. First, artifacts that are produced by the applied TSM procedure are also audible in the pitch-shifted signal. However, even when using a high-quality TSM procedure, the pitch-shifted signals generated by the method described above often sound unnatural. For example, when pitch-shifting singing voice upwards by several semitones, the modified voice has an artificial sound, sometimes referred to as the *chipmunk* effect [39]. Here, the problem is that the *spectral envelope*, which is the rough shape of a frequency spectrum, is of central importance for the timbre of a sound. In Figure 17a, we see a frame's frequency spectrum from a singing voice recording. Due to the harmonic nature of the singing voice, the spectrum shows a comb-like pattern where the energy is distributed at multiples of a certain frequency—in this example roughly 250 Hz. The peaks in the pattern are called the *harmonics* of the singing voice. The magnitudes of the harmonics follow a certain shape which is specified by the spectral envelope (marked in red). Peaks in the spectral envelope are known as *formants*. In the example from Figure 17a, we see four formants at around 200 Hz, 2200 Hz, 3100 Hz, and 5900 Hz. The frequency positions of these formants are closely related to the singing voice's timbre. In Figure 17b, we see the spectrum of the same frame after being pitch-shifted by four semitones with the previously described pitch-shifting procedure. Due to the resampling, the harmonics' positions are scaled. The spectral envelope is therefore scaled as well, relocating the positions of the formants. This leads to a (usually undesired) change in timbre of the singing voice.

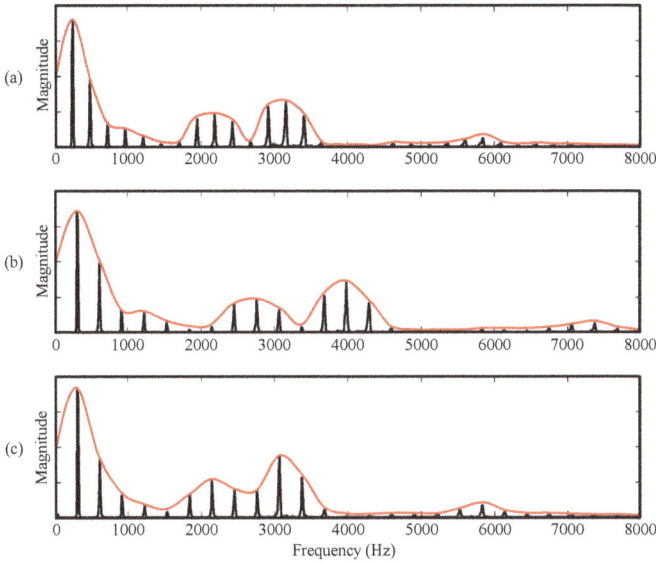

Figure 17. Frequency spectra for a fixed frame m of different versions of a singing voice recording. The spectral envelopes are marked in red. (**a**) Original spectrum $X(m)$ with spectral envelope Γ_m; (**b**) Pitch-shifted spectrum $X^{\text{Shift}}(m)$ with spectral envelope Γ_m^{Shift}; (**c**) Pitch-shifted spectrum $X^{\text{Mod}}(m)$ with adapted spectral envelope.

One strategy to compensate for this change is outlined in [17]. Let X and X^{Shift} be the STFTs of a given input signal x and its pitch-shifted version x^{Shift}, respectively. Fixing a frame index m, let $X(m)$ and $X^{\text{Shift}}(m)$ be the frequency spectra of the mth frames in X and X^{Shift}. Furthermore, let $\Gamma_m : [0 : N-1] \to \mathbb{R}$ and $\Gamma_m^{\text{Shift}} : [0 : N-1] \to \mathbb{R}$ denote the spectral envelopes of $X(m)$ and $X^{\text{Shift}}(m)$, respectively. Our goal is to compute modified spectra $X^{\text{Mod}}(m)$ that have the frequency content of $X^{\text{Shift}}(m)$ but the original spectral envelopes Γ_m. To this end, we normalize the magnitudes of $X^{\text{Shift}}(m)$ with respect to its spectral envelope Γ_m^{Shift} and then scale them by the original envelope Γ_m:

$$X^{\text{Mod}}(m,k) = X^{\text{Shift}}(m,k) \, \frac{\Gamma_m(k)}{\Gamma_m^{\text{Shift}}(k)} \, . \qquad (36)$$

In Figure 17c we see the frame's spectrum from Figure 17b after the envelope adaption, leading to a signal that sounds more natural.

There exist several approaches to estimate spectral envelopes, many of them either based on *linear predictive coding* (LPC) [40] or on the *cepstrum* (the inverse Fourier transform of the logarithmic magnitude spectrum) [41,42]. Generally, the task of spectral envelope estimation is highly complex and error-prone. In envelope or formant-preserving pitch-shifting methods, it is often necessary to manually specify parameters (e.g., the pitch range of the sources in the recording) to make the envelope estimation more robust.

7.3. Websources

Various free implementations of TSM procedures are available in different programming languages. MATLAB implementations of OLA, WSOLA, PV-TSM, and TSM based on HPS, as well as additional test and demo material can be found in the TSM Toolbox [43,44]. A further MATLAB implementation of PV-TSM can be found at [45]. PV-TSM implemented in Python is included in LibROSA [46]. Furthermore, there are open source C/C++ audio processing libraries that offer TSM functionalities. For example, the Rubber Band Library [47] includes a transient preserving PV-TSM and the SoundTouch Audio Processing Library [48] offers a WSOLA-like TSM procedure. Finally, there also exist proprietary commercial implementations such as the élastique algorithm by zplane [49].

8. Conclusions

In this article, we reviewed time-scale modification of music signals. We presented fundamental principles and discussed various well-known time and frequency-domain TSM procedures. Additionally, we pointed out more involved procedures proposed in the literature, which are—to varying extents—based on the fundamental approaches we reviewed. In particular, we discussed a recent approach that involves harmonic-percussive separation and combines the advantages of two fundamental TSM methods in order to attenuate artifacts and improve the quality of time-scale modified signals. Finally, we introduced some applications of TSM, including music synchronization and pitch-shifting, and gave pointers to freely available TSM implementations.

A major goal of this contribution was to present fundamental concepts in the field of TSM. Beyond discussing technical details, our main motivation was to give illustrative explanations in a tutorial-like style. Furthermore, we aimed to foster a deep understanding of the strengths and weaknesses of the various TSM approaches by discussing typical artifacts and the importance of parameter choices. We hope that this work constitutes a good starting point for becoming familiar with the field of TSM and developing novel TSM techniques.

Acknowledgments: This work has been supported by the German Research Foundation (DFG MU 2686/6-1). The International Audio Laboratories Erlangen are a joint institution of the Friedrich-Alexander-Universität Erlangen-Nürnberg (FAU) and Fraunhofer Institut für Integrierte Schaltungen. We also would like to thank Patricio López-Serrano for his careful proofreading and comments on the article.

Author Contributions: The authors contributed equally to this work. The choice of focus, didactical preparation of the material, design of the figures, as well as writing the paper were all done in close collaboration.

Conflicts of Interest: The authors declare no conflict of interest.

References

1. Cliff, D. *Hang the DJ: Automatic Sequencing and Seamless Mixing of Dance-Music Tracks*; Technical Report; HP Laboratories Bristol: Bristol, Great Britain, 2000.
2. Ishizaki, H.; Hoashi, K.; Takishima, Y. Full-automatic DJ mixing system with optimal tempo adjustment based on measurement function of user discomfort. In Proceedings of the International Society for Music Information Retrieval Conference (ISMIR), Kobe, Japan, 26–30 October 2009; pp. 135–140.
3. Moinet, A.; Dutoit, T.; Latour, P. Audio time-scaling for slow motion sports videos. In Proceedings of the International Conference on Digital Audio Effects (DAFx), Maynooth, Ireland, 2–5 September 2013.
4. Verhelst, W.; Roelands, M. An overlap-add technique based on waveform similarity (WSOLA) for high quality time-scale modification of speech. In Proceedings of the IEEE International Conference on Acoustics, Speech, and Signal Processing (ICASSP), Minneapolis, MN, USA, 27–30 April 1993.
5. Flanagan, J.L.; Golden, R.M. Phase vocoder. *Bell Syst. Tech. J.* **1966**, *45*, 1493–1509.

6.   Laroche, J.; Dolson, M. Improved phase vocoder time-scale modification of audio. *IEEE Trans. Speech Audio Process.* **1999**, *7*, 323–332.

7.   Portnoff, M.R. Implementation of the digital phase vocoder using the fast Fourier transform. *IEEE Trans. Acoust. Speech Signal Process.* **1976**, *24*, 243–248.

8.   Driedger, J.; Müller, M.; Ewert, S. Improving time-scale modification of music signals using harmonic-percussive separation. *IEEE Signal Process. Lett.* **2014**, *21*, 105–109.

9.   Zölzer, U. *DAFX: Digital Audio Effects*; John Wiley & Sons, Inc.: New York, NY, USA, 2002.

10. Roucos, S.; Wilgus, A.M. High quality time-scale modification for speech. In Proceedings of the IEEE International Conference on Acoustics, Speech, and Signal Processing (ICASSP), Tampa, Florida, USA, 26–29 April 1985; Volume 10, pp. 493–496.

11. Moulines, E.; Charpentier, F. Pitch-synchronous waveform processing techniques for text-to-speech synthesis using diphones. *Speech Commun.* **1990**, *9*, 453–467.

12. Laroche, J. Autocorrelation method for high-quality time/pitch-scaling. In Proceedings of the IEEE Workshop Applications of Signal Processing to Audio and Acoustics (WASPAA), Mohonk, NY, USA, 17–20 October 1993; pp. 131–134.

13. Grofit, S.; Lavner, Y. Time-scale modification of audio signals using enhanced WSOLA with management of transients. *IEEE Trans. Audio Speech Lang. Process.* **2008**, *16*, 106–115.

14. Gabor, D. Theory of communication. *J. Inst. Electr. Eng. IEE* **1946**, *93*, 429–457.

15. Müller, M. *Fundamentals of Music Processing*; Springer International Publishing: Cham, Switzerland, 2015.

16. Griffin, D.W.; Lim, J.S. Signal estimation from modified short-time Fourier transform. *IEEE Trans. Acoust. Speech Signal Process.* **1984**, *32*, 236–243.

17. Dolson, M. The phase vocoder: A tutorial. *Comput. Music J.* **1986**, *10*, 14–27.

18. Laroche, J.; Dolson, M. Phase-vocoder: About this phasiness business. In Proceedings of the IEEE ASSP Workshop on Applications of Signal Processing to Audio and Acoustics, New Paltz, NY, USA, 19–22 October 1997.

19. Dorran, D.; Lawlor, R.; Coyle, E. A hybrid time-frequency domain approach to audio time-scale modification. *J. Audio Eng. Soc.* **2006**, *54*, 21–31.

20. Kraft, S.; Holters, M.; von dem Knesebeck, A.; Zölzer, U. Improved PVSOLA time stretching and pitch shifting for polyhonic audio. In Proceedings of the International Conference on Digital Audio Effects (DAFx), York, UK, 17–21 September 2012.

21. Moinet, A.; Dutoit, T. PVSOLA: A phase vocoder with synchronized overlapp-add. In Proceedings of the International Conference on Digital Audio Effects (DAFx), Paris, France, 19–23 September 2011; pp. 269–275.

22. Nagel, F.; Walther, A. A novel transient handling scheme for time stretching algorithms. In Proceedings of the AES Convention, New York, NY, USA, 2009; pp. 185–192.

23. Fitzgerald, D. Harmonic/percussive separation using median filtering. In Proceedings of the International Conference on Digital Audio Effects (DAFx), Graz, Austria, 6–10 September 2010; pp. 246–253.

24. Verma, T.S.; Meng, T.H. Time scale modification using a sines+transients+noise signal model. In Proceedings of the Digital Audio Effects Workshop (DAFx98), Barcelona, Spain, 19–21 November 1998.

25. Levine, S. N.; Smith, J.O., III. A sines+transients+noise audio representation for data compression and time/pitch scale modications. In Proceedings of the AES Convention, Amsterdam, The Netherlands, 16–19 May 1998.

26. Verma, T.S.; Meng, T.H. An analysis/synthesis tool for transient signals that allows a flexible sines+transients+noise model for audio. In Proceedings of the IEEE International Conference on Acoustics, Speech, and Signal Processing (ICASSP), Seattle, WA, USA, 12–15 May 1998; pp. 3573–3576.

27. Verma, T.S.; Meng, T.H. Extending spectral modeling synthesis with transient modeling synthesis. *Comput. Music J.* **2000**, *24*, 47–59.

28. Serra, X.; Smith, J.O. Spectral modeling synthesis: A sound analysis/synthesis system based on a deterministic plus stochastic decomposition. *Comput. Music J.* **1990**, *14*, 12–24.

29. Duxbury, C.; Davies, M.; Sandler, M. Improved time-scaling of musical audio using phase locking at transients. In Proceedings of Audio Engineering Society Convention, Munich, Germany, 10–13 May 2002.

30. Cañadas-Quesada, F.J.; Vera-Candeas, P.; Ruiz-Reyes, N.; Carabias-Orti, J.J.; Molero, P.C. Percussive/harmonic sound separation by non-negative matrix factorization with smoothness/sparseness constraints. *EURASIP J. Audio Speech Music Process.* **2014**, doi:10.1186/s13636-014-0026-5.

31. Duxbury, C.; Davies, M.; Sandler, M. Separation of transient information in audio using multiresolution analysis techniques. In Proceedings of the International Conference on Digital Audio Effects (DAFx), Limerick, Ireland, 17–22 September 2001.

32. Gkiokas, A.; Papavassiliou, V.; Katsouros, V.; Carayannis, G. Deploying nonlinear image filters to spectrograms for harmonic/percussive separation. In Proceedings of the International Conference on Digital Audio Effects (DAFx), York, UK, 17–21 September 2012.

33. Ono, N.; Miyamoto, K.; LeRoux, J.; Kameoka, H.; Sagayama, S. Separation of a monaural audio signal into harmonic/percussive components by complementary diffusion on spectrogram. In Proceedings of the European Signal Processing Conference, Lausanne, Switzerland, 25–29 August 2008; pp. 240–244.

34. Park, J.; Lee, K. Harmonic-percussive source separation using harmonicity and sparsity constraints. In Proceedings of the International Conference on Music Information Retrieval (ISMIR), Málaga, Spain, 26–30 October 2015; pp. 148–154.

35. Tachibana, H.; Ono, N.; Kameoka, H.; Sagayama, S. Harmonic/percussive sound separation based on anisotropic smoothness of spectrograms. *IEEE/ACM Trans. Audio Speech Lang. Process.* **2014**, *22*, 2059–2073.

36. Müller, M.; Driedger, J. Data-driven sound track generation. In *Multimodal Music Processing*; Müller, M., Goto, M., Schedl, M., Eds.; Schloss Dagstuhl–Leibniz-Zentrum für Informatik: Dagstuhl, Germany, 2012; Volume 3, pp. 175–194.

37. Haghparast, A.; Penttinen, H.; Välimäki, V. Real-time pitch-shifting of musical signals by a time-varying factor using normalized filtered correlation time-scale modification. In Proceedings of the International Conference on Digital Audio Effects (DAFx), Bordeaux, France, 10–15 September 2007; pp. 7–14.

38. Schörkhuber, C.; Klapuri, A.; Sontacchi, A. Audio pitch shifting using the constant-q transform. *J. Audio Eng. Soc.* **2013**, *61*, 562–572.

39. Alvin and the Chipmunks—Recording Technique. Available online: https://en.wikipedia.org/wiki/Alvin_and_the_Chipmunks#Recording_technique (accessed on 3 December 2015).

40. Markel, J.D.; Gray, A.H. *Linear Prediction of Speech*; Springer Verlag: Secaucus, NJ, USA, 1976.

41. Röbel, A.; Rodet, X. Efficient spectral envelope estimation an its application to pitch shifting and envelope preservation. In Proceedings of the International Conference on Digital Audio Effects (DAFx), Madrid, Spain, 20–22 September 2005.

42. Röbel, A.; Rodet, X. Real time signal transposition with envelope preservation in the phase vocoder. In Proceedings of the International Computer Music Conference (ICMC), Barcelona, Spain, 5–9 September 2005; pp. 672–675.

43. Driedger, J.; Müller, M. TSM Toolbox. Available online: http://www.audiolabs-erlangen.de/resources/matlab/pvoc/ (accessed on 5 February 2016).

44. Driedger, J.; Müller, M. TSM Toolbox: MATLAB implementations of time-scale modification algorithms. In Proceedings of the International Conference on Digital Audio Effects (DAFx), Erlangen, Germany, 1–5 September 2014; pp. 249–256.

45. Ellis, D.P.W. A Phase Vocoder in Matlab. Available online: http://www.ee.columbia.edu/dpwe/resources/matlab/pvoc/ (accessed on 3 December 2015).

46. McFee, B. Librosa—Time Stretching. Available online: https://bmcfee.github.io/librosa/generated/librosa.effects.time_stretch.html (accessed on 3 December 2015).

47. Breakfast Quay. Rubber Band Library. Aailable online: http://breakfastquay.com/rubberband/ (accessed on 26 November2015).

48. Parviainen, O. Soundtouch Audio Processing Library. Available online: http://www.surina.net/ soundtouch/ (accessed on 27 November 2015).

49. Zplane Development. Élastique Time Stretching & Pitch Shifting SDKs. Available online: http://www.zplane.de/index.php?page=description-elastique (accessed on 5 February 2016).

applied
sciences

MDPI

Review

All About Audio Equalization: Solutions and Frontiers

Vesa Välimäki [1],* and Joshua D. Reiss [2]

[1]   Department of Signal Processing and Acoustics, Aalto University, Espoo 02150, Finland
[2]   Centre for Digital Music, Queen Mary University London, London E1 4NS, UK; joshua.reiss@qmul.ac.uk
*   Correspondence: vesa.valimaki@aalto.fi; Tel.: +358-50-569-1176

Academic Editor: Gino Iannace
Received: 15 March 2016; Accepted: 19 April 2016; Published: 6 May 2016

Abstract: Audio equalization is a vast and active research area. The extent of research means that one often cannot identify the preferred technique for a particular problem. This review paper bridges those gaps, systemically providing a deep understanding of the problems and approaches in audio equalization, their relative merits and applications. Digital signal processing techniques for modifying the spectral balance in audio signals and applications of these techniques are reviewed, ranging from classic equalizers to emerging designs based on new advances in signal processing and machine learning. Emphasis is placed on putting the range of approaches within a common mathematical and conceptual framework. The application areas discussed herein are diverse, and include well-defined, solvable problems of filter design subject to constraints, as well as newly emerging challenges that touch on problems in semantics, perception and human computer interaction. Case studies are given in order to illustrate key concepts and how they are applied in practice. We also recommend preferred signal processing approaches for important audio equalization problems. Finally, we discuss current challenges and the uncharted frontiers in this field. The source code for methods discussed in this paper is made available at https://code.soundsoftware.ac.uk/projects/allaboutaudioeq.

Keywords: acoustic signal processing; audio systems; digital filters; digital signal processing; equalizers; infinite impulse response filters; music technology

---

1. Introduction

The term *equalizer* (EQ) has its origins in early telephone engineering, when high frequency losses over long distances had to be corrected so that the spectrum of the sound at the receiver matched the sound spectrum that was initially transmitted. Ideally, the system's net frequency response has an equal response to all frequencies, and thus the term 'equalization.' Since then, the term has been used for any procedure that involves altering or adjusting the magnitude frequency response.

This paper reviews developments and applications in audio equalization. A main emphasis is on design methods for digital equalizing filters. Although these filters were originally constructed using analog electronics, we want to show how to apply them using Digital Signal Processing (DSP).

To our knowledge this is the first review article on red digital audio equalization which gathers all historical developments, methods, and applications, as well as state-of-the-art approaches. However, in 1988 Bohn wrote an overview of this topic, covering the history and the analog era [1]. Additionally, some audio signal processing books have a section devoted to digital equalizers. For example, Orfanidis has a wide section on Infinite Impulse Response (IIR) filters, which includes parametric equalizers and shelving filters [2]. Zölzer's textbook has a large and excellent chapter on this topic, covering shelving, peak, and notch filters [3]. Reiss and McPherson's recent book reviews digital tone control and peak equalizing filter designs, among others, and provides a history of the invention of the parametric equalizer [4].

This review article is organized as follows: Section 2 presents the history of audio equalization, Section 3 focuses on parametric equalizers and shelving filters, Section 4 is devoted to graphic equalizers, Section 5 discusses other audio equalizer designs, Section 6 looks into applications in sound reproduction, and Section 7 addresses applications in audio content creation, such as in music production and mixing. Finally, Section 8 concludes this review. The source code for the methods discussed in this paper and for many of the figures are available online at https://code. soundsoftware.ac.uk/projects/allaboutaudioeq, as described in Appendix.

2. History of Audio Equalization

The concept of filtering audio frequencies was understood at least as far back as the 1870s, and was used in designs for harmonic telegraphs. Telegraph operator keys activated vibrating electromechanical reeds, each one assigned a specific frequency. Filtering at the receiving operator was achieved by tuning a similar reed to the same frequency, so that multiple independent connections could be established over a single telegraph line [5].

Telephone lines were equalized in repeaters using wave filters [6,7] to cancel resonances caused by impedance mismatches or the attenuation of high frequencies in long cables. The response of microphones was flattened by appropriately selecting the resistance and other component parameters [8]. Early equalizers were fixed and integrated into the circuits of audio receivers, and later into phonograph playback systems. The advent of motion picture sound saw the emergence of variable equalization. Notably, John Volkman's external equalizer design featured a set of selectable frequencies with boosts and cuts, and is sometimes considered to be the first operator-variable equalizer. *Tone controls*—boost and cut applied to bass and treble—were first introduced for use with gramophones in 1949 [9]. In 1952 Baxandall [10] devised tone controls using potentiometers as opposed to switches, thus allowing full user control.

In the 1950s, equalization circuits became a standard part of LP disc production and playback systems. The Recording Industry Association of America (RIAA) proposed a pre-emphasis equalization curve for recording and its complementary de-emphasis equalization curve for playback, see e.g., [11]. The purpose of the RIAA pre-emphasis is to reduce the groove width on the disc by suppressing low frequencies while at the same time improving the signal-to-noise ratio of treble by boosting higher frequencies. Similarly, the National Association of Broadcasters (NAB) suggested a playback equalization curve for C cassettes. It boosted frequencies below 1 kHz, using a first order filter having a pole just below the audio range and a zero above 1 kHz, to obtain a mostly flat frequency response during playback [12].

In what is now considered to be a founding work in the field of sound system equalization, Rudmose [13] applied emerging equalization methods to the Dallas Love Field Airport. Advances in the theory of acoustic feedback led to the development of an equalization system with very narrow notch filters that could be manually tuned to the frequency at which feedback occurs [14,15]. Expanding and extending this work, Conner established the theory and methodology of sound system equalization [16].

Throughout the 1950s and 1960s, equalizers grew in popularity, finding applications in sound post-production and speech enhancement. The Langevin Model EQ-251A, a program equalizer with slide controls, was a precursor to the graphic equalizer [1]. One 15-position slider controlled the gain of a bass shelving filter, and the other adjusted the boost or cut of a peak/notch filter. Each filter had switchable frequencies. Cinema Engineering introduced the first graphic equalizer [1]. It could adjust six bands with a boost or cut range of ±8 dB in 1 dB increments. However, with graphic equalizers, engineers were still limited to the constraints imposed by the number and location of frequency bands. Around 1967, Walker introduced the API 550A equalizer in which the gain appropriately changes the bandwidth of the peak or notch. As was typical in early equalizers, this device had a fixed selection of frequencies, and variable boost or cut controls at those frequencies. In 1971, Daniel Flickinger invented an important tunable equalizer [17]. His circuit, known as "sweepable EQ", allowed arbitrary selection of frequency and gain in three overlapping bands.

In 1966, Burgess Macneal and George Massenburg, who was still a teenager, began work on a new recording console. They conceptualized an idea for a sweep-tunable EQ that would avoid inductors and switches. Soon after, Bob Meushaw, a friend of Massenburg, built a three-band, frequency adjustable, fixed-Q equalizer. When asked who invented the parametric equalizer, Massenburg stated "Four people could possibly lay claim to the modern concept: Bob Meushaw, Burgess Macneal, Daniel Flickinger, and myself... Our (Bob's, Burgess' and my) sweep-tunable EQ was borne, more or less, out of an idea that Burgess and I had around 1966 or 1967 for an EQ... three controls adjusting, independently, the parameters for each of three bands for a recording console... I wrote and delivered the AES paper on Parametrics at the Los Angeles show in 1972 [18]... It's the first mention of 'Parametric' associated with sweep-tunable EQ" [4].

Digital audio equalizers emerged in the late 1970s and early 1980s [19–22], leading to commercial deployment in Yamaha's 1987 DEQ7 Digital Equalizer, the first DSP-based parametric equalizer. A year later, Roland rolled out its variable digital equalizer, the E-660. Around the same time, important digital equalizer designs appeared: White published design formulas for digital biquadratic equalizing filters [23], and Regalia and Mitra proposed a useful allpass-filter based structure for digital shelving and parametric equalizing filters [24]. Both of these groundbreaking works are based on discretization of analog prototypes. Since then, there has been a proliferation of advances and improvements, including the octave and 1/3-octave graphic equalizer, constant-Q (bandwidth) graphic equalizers, and many other advances discussed throughout this paper.

3. Parametric Equalizers

The parametric equalizer is the most powerful and flexible of the equalizer types. Midrange bands in a parametric equalizer have three adjustments: gain, center frequency, and quality factor Q (or bandwidth). A parametric equalizer allows the operator to add a peak or a notch at an arbitrary location in the audio spectrum. At other frequencies, far away from the peak or notch, the parametric equalizer does not modify the spectral content, as its magnitude response there is unity (0 dB). Adding a peak can be useful to help an instrument be heard in a complex mix, or to deliberately add coloration to an instrument's sound by boosting or reducing a particular frequency range. Notches can be used to attenuate unwanted sounds, including removing power line hum (50 Hz or 60 Hz and sometimes their harmonics) [25] and reducing feedback [26]. To remove artifacts without affecting the rest of the sound, a narrow bandwidth would be used.

A single section of a parametric equalizer is created from a second order peaking/notch filter, or in certain cases, a first or second order shelving filter for the lowest and highest frequency bands. When multiple sections are used, they are always connected *in cascade* so that the effects of each subfilter are cumulative on a decibel scale. In this section, we derive the transfer functions for the first and second order shelving filter and for the second order peaking/notch filter.

3.1. First Order Shelving Filters

In shelving mode, gain G is applied to all frequencies below or above a *crossover frequency f_c*. These are called the low shelving and the high shelving filter, respectively, and are sometimes used in the lowest-frequency and highest-frequency bands. These shelving filters are similar to those found in the basic tone control, but where the tone control usually uses first order shelving filters, the parametric equalizer uses either first or second order filters. First order shelving filters are used to create smooth transitions (about 6 dB/octave) between affected and unaffected frequency regions. The crossover frequency is sometimes called the corner, cut-off, or transition frequency, but we prefer to use the term crossover herein.

We first derive the transfer function $H_{LS}(z)$ for a low-frequency shelving filter. We would like its gain to be G at low frequencies (such as at the zero frequency, or $z = 1$) and 1 at the highest frequencies (such as at the Nyquist limit, or $z = -1$), that is:

$$H(1) = G, \tag{1}$$

$$H(-1) = 1. \tag{2}$$

There are several possible choices for the gain at the crossover frequency, such as 3 dB lower (in the boost case) or higher (in the cut case) than the filter gain, e.g., [3,27], but it leads to a confusing situation when the filter gain or loss is smaller than 3 dB [28]. Alternative choices are the arithmetic or geometric mean of the extreme gain factors. However, if the geometric mean is used, then a boost and a cut by the same decibel amount do not cancel. It is also problematic since the arithmetic mean loses its relevance when a decibel scale is used, and it means that the shelving filter will not match the design of peaking and notch filters in parametric equalizers, which often use the geometric mean (see Section 3.3).

We choose to use the midpoint gain $\sqrt{G}$, which corresponds to the geometric mean of the extreme gains (1 and G), and is the arithmetic mean of the extreme gains on a decibel scale. This is one way to retain the overall magnitude response shape when gain G is varied. This requirement may be written as

$$|H(e^{j\omega_c})| = \sqrt{G}, \tag{3}$$

where $\omega_c = 2\pi f_c / f_s$ is the crossover frequency in radians ($0 \leq \omega_c \leq \pi$), f_s is the sampling frequency, and j is the imaginary unit.

We start by defining a prototype low-frequency first order shelving filter, which has its crossover frequency at the midpoint of the frequency range, $\pi/2$, and has the transfer function of the form

$$H_P(z) = g\frac{z - q}{z - p}, \tag{4}$$

where p and q are the pole and the zero, respectively, and g is the scaling factor. By evaluating this transfer function at $\omega_c = \pi/2$ and using the requirement (3), it is seen that a solution is obtained by setting $q = -p$, or by placing the pole and the zero symmetrically w.r.t. the origin on the z plane. The scaling factor must then be $g = \sqrt{G}$:

$$H_P(z) = \sqrt{G}\frac{z + p}{z - p}. \tag{5}$$

The pole location in terms of gain G can be solved by combining Equations (5) and (1):

$$p = \frac{G - \sqrt{G}}{G + \sqrt{G}}, \tag{6}$$

which also satisfies Equation (2). Substituting this result to Equation (5) completes the prototype first order low shelf design:

$$H_P(z) = \sqrt{G}\frac{G + \sqrt{G} + (G - \sqrt{G})z^{-1}}{G + \sqrt{G} - (G - \sqrt{G})z^{-1}}. \tag{7}$$

To shift the crossover frequency to ω_c, which we restrict to be $0 \leq \omega_c \leq \pi$, we apply the lowpass-to-lowpass transformation [4,29]

$$z \rightarrow \frac{z - \beta}{1 - \beta z}, \tag{8}$$

where

$$\beta = \frac{1 - \tan(\omega_c/2)}{1 + \tan(\omega_c/2)}. \tag{9}$$

Rearranging the terms leads to the transfer function of the first order low-frequency shelving filter

$$H_{LS}(z) = \frac{G\tan(\omega_c/2) + \sqrt{G} + [G\tan(\omega_c/2) - \sqrt{G}]z^{-1}}{\tan(\omega_c/2) + \sqrt{G} + [\tan(\omega_c/2) - \sqrt{G}]z^{-1}}. \tag{10}$$

Due to the choice of gain at crossover frequency, (3), this produces symmetric magnitude frequency responses, above and below the unit gain level, for gains G and $1/G$ (*i.e.*, equal boost or cut in dB), as wanted. Figure 1 illustrates this for three pairs of low shelf filters. In this and all other examples, the sample rate is f_s = 44,100 Hz and the responses are shown up to the Nyquist limit, f_N = 22,050 Hz. When $G = 1$, the pole and the zero of the transfer function (10) coincide, the magnitude response becomes unity, and the filter's impulse response becomes a non-delayed unit impulse.

This way it was possible to derive transfer function coefficients, which are the same for the boost ($G > 1$) and cut cases ($0 < G < 1$), whereas in the past several shelving filter designs have required separate formulas for the two cases, e.g., [3,30–33]. The transfer function (10) is the same as that given in [4] (Equation (4.1)). Jot recently proposed a modified Regalia-Mitra shelving filter design, which also leads to exactly the same transfer function [34].

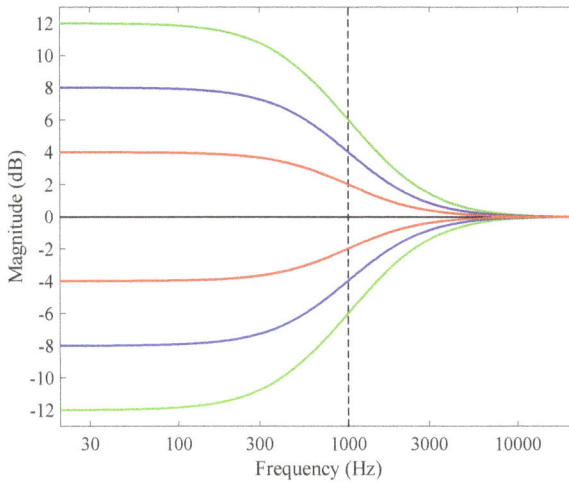

Figure 1. Magnitude responses of low-frequency shelving filters with complementary gains: −12 dB and 12 dB (green), −8 dB and 8 dB (blue), −4 dB and 4 dB (red). The vertical dashed line indicates the crossover frequency (1 kHz) where midpoint gain is reached.

The complementary high-frequency shelving filter can be obtained by replacing the gain G in Equation (10) by $1/G$, which yields the inverse filter, and by multiplying the transfer function (numerator) by G, which shifts the magnitude response vertically so that the gain at zero frequency becomes unity (0 dB), and gain G is achieved at high frequencies as desired. Additionally, we multiply both the numerator and the denominator by $\sqrt{G}$ to cancel divisions by G:

$$H_{HS}(z) = \frac{\sqrt{G}\tan(\omega_c/2) + G + [\sqrt{G}\tan(\omega_c/2) - G]z^{-1}}{\sqrt{G}\tan(\omega_c/2) + 1 + [\sqrt{G}\tan(\omega_c/2) - 1]z^{-1}}. \tag{11}$$

Comparison with Equation (10) shows that these transfer functions are related as $H_{HS}(z) = G/H_{LS}(z)$, which means that the pole and the zero of the filter can be interchanged to convert a low shelf to a high shelf or vice versa. Furthermore, a scaling by G must be applied. This implies that a cascade of a low and high shelf with the same crossover frequency ω_c and gain G will produce a constant gain G across all frequencies. Figure 2 shows this for a pair of low- and high-frequency shelving filters.

An alternative way to convert a low shelf filter to a high one is to replace z with $-z$ in the transfer function (*i.e.*, change the sign of every second coefficient), which turns the response over in frequency so that the zero and Nyquist frequencies are interchanged [28]. Additionally, this low-to-high mapping requires replacing ω_c with $\pi - \omega_c$ to restore the crossover frequency.

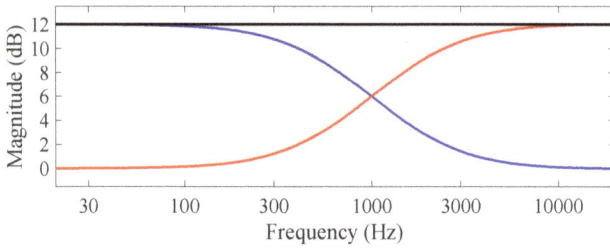

Figure 2. Magnitude responses of a low (blue) and high (red) shelf filter with a 12-dB gain and 1-kHz crossover frequency, and the cascade of the two filters (black).

3.2. Second Order Shelving Filters

Here, we show a technique to design a second order low shelving filter, which is often used in parametric equalizer design [19,28,31,32,35]. The design method differs slightly from the approach for first order design since the lowpass to lowpass transformation is performed in the s domain, to keep the maths slightly simpler.

The transfer function of a second order analog shelving filter with crossover frequency f_c is given by [3],

$$H_{LS,1}(s) = \frac{s^2 + \sqrt{2G}\omega_0 s + G\omega_0^2}{s^2 + \sqrt{2}\omega_0 s + \omega_0^2},\tag{12}$$

where $s = j\omega, \omega_0 = 2\pi f_c$, and G is the gain at DC ($\omega = 0$). Note that if the $\sqrt{2}$ in the numerator and denominator of Equation (12) is replaced with a larger number, then resonant low shelving filters are produced, which have a bump in the magnitude response before the roll-off. The square magnitude of this filter is

$$|H_{LS,1}(s)|^2 = \frac{G^2\omega_0^4 + \omega^4}{\omega_0^4 + \omega^4},\tag{13}$$

so that the square magnitude at the crossover frequency is equal to the arithmetic mean of the square magnitudes at $\omega = 0$ and $\omega = \infty$. However, as discussed in Section 3.1 generally this is not what is wanted. To correct this, we need to find the frequency where the low shelf's square magnitude is the geometric mean of 1 and G^2, which is G. So we solve

$$G = |H_{LS,1}(s)|^2 = \frac{G^2\omega_0^4 + \omega^4}{\omega_0^4 + \omega^4} \rightarrow \omega = G^{1/4}\omega_0.\tag{14}$$

So make the replacement $\omega \rightarrow G^{1/4}\omega$ in the low shelf filter, to create a new low shelf $H_{LS,2}(s)$. That way, $|H_{LS,2}(\omega_0)|^2 = |H_{LS,1}(G^{1/4}\omega_0)|^2 = G$. This results in

$$H_{LS,2}(s) = \frac{\left(G^{1/4}s\right)^2 + \sqrt{2G}\omega_0(G^{1/4}s) + G\omega_0^2}{\left(G^{1/4}s\right)^2 + \sqrt{2}\omega_0(G^{1/4}s) + \omega_0^2} = \frac{s^2 + \sqrt{2}\omega_0 G^{1/4}s + G^{1/2}\omega_0^2}{s^2 + \sqrt{2}\omega_0 G^{-1/4}s + G^{-1/2}\omega_0^2}. \tag{15}$$

To convert to the digital domain, we use the bilinear transform. We make the following substitutions

$$s = \frac{z-1}{z+1}, \omega_0 \to \Omega = \tan(\omega_c/2), \tag{16}$$

So

$$H_{LS}(z) = \frac{\left(\frac{z-1}{z+1}\right)^2 + \sqrt{2}\Omega G^{1/4}\left(\frac{z-1}{z+1}\right) + G^{1/2}\Omega^2}{\left(\frac{z-1}{z+1}\right)^2 + \sqrt{2}\Omega G^{-1/4}\left(\frac{z-1}{z+1}\right) + G^{-1/2}\Omega^2} \tag{17}$$

$$= G^{1/2}\frac{G^{1/2}\Omega^2 + \sqrt{2}\Omega G^{1/4} + 1 + 2(G^{1/2}\Omega^2 - 1)z^{-1} + (G^{1/2}\Omega^2 - \sqrt{2}\Omega G^{1/4} + 1)z^{-2}}{G^{1/2} + \sqrt{2}\Omega G^{1/4} + \Omega^2 + 2(\Omega^2 - G^{1/2})z^{-1} + (G^{1/2} - \sqrt{2}\Omega G^{1/4} + \Omega^2)z^{-2}}, \tag{18}$$

which satisfies all the conditions for the desired low shelving filter. The second order high shelving filter can be derived in the same way, and as with first order shelving filter designs, is equal to the gain G times the reciprocal of the second order low shelving filter:

$$H_{HS}(z) = G^{1/2}\frac{G^{1/2} + \sqrt{2}\Omega G^{1/4} + \Omega^2 - 2[G^{1/2} - \Omega^2]z^{-1} + (G^{1/2} - \sqrt{2}\Omega G^{1/4} + \Omega^2)z^{-2}}{G^{1/2}\Omega^2 + \sqrt{2}\Omega G^{1/4} + 1 + 2[G^{1/2}\Omega^2 - 1]z^{-1} + (G^{1/2}\Omega^2 - \sqrt{2}\Omega G^{1/4} + 1)z^{-2}} \tag{19}$$

$$= G/H_{LS}(z). \tag{20}$$

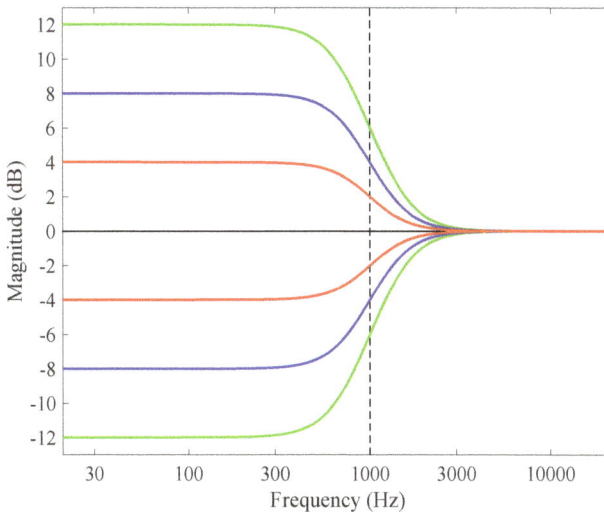

Figure 3. Magnitude responses of second order low shelving filters with complementary gains: ±12 dB (green), ±8 dB (blue), and ±4 dB (red). The vertical dashed line indicates the crossover frequency (1 kHz). Cf. Figure 1.

Figure 3 shows example magnitude responses of the second order low shelving filter. It is instructive to consider whether the second order shelf filter is better than cascading two first order filters. Figure 4 gives an example where we compare the magnitude responses of a single and two cascaded first order shelving filters against that of a second order one. It becomes evident that the second order filter has a steeper transition than the other filters. In fact, cascading two first order filters has a very similar magnitude response as one of them, since the two cascaded filters have a gain of $\sqrt{G}$.

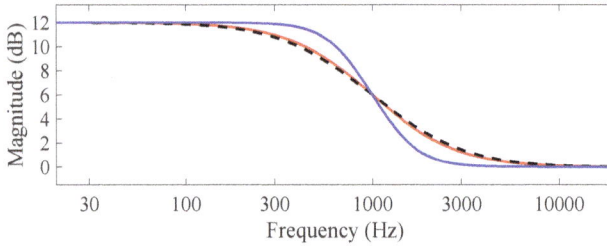

Figure 4. Magnitude responses of a single low shelf filter (red), cascaded two first order low shelf filters (black dotted line), and a second order shelf filter (blue) with the same gain (12 dB) and crossover frequency (1 kHz).

3.3. Second Order Peaking and Notch Filters

In a typical parametric equalizer all sections but the low and high shelving filters are comprised of second order peaking or notch filters, which are IIR filters having two poles and two zeros. The aim is to obtain a filter with the magnitude frequency response, which has a bump of gain G ($G > 1$ for a peaking filter and $0 < G < 1$ for a notch filter) and width specified by Q at the center frequency ω_c, but which has a unity gain (*i.e.*, 0 dB) at frequencies far away from the bump, especially at the zero frequency and at the Nyquist limit. (An alternative approach was taken by Orfanidis who presented a peak/notch filter design in which the specified gain at the Nyquist limit $f_s/2$ was not unity but the same as that of a corresponding analog equalizing filter [36]. The magnitude responses of his filters are also less asymmetric than the ones designed here, when the center frequency is high, close to the Nyquist limit. See Section 5.2.)

These specifications lead to the following four constraints [37,38] for the filter transfer function $H(z)$:

$$H(1) = 1, \tag{21}$$

$$H(-1) = 1, \tag{22}$$

$$|H(e^{j\omega_c})| = G, \tag{23}$$

$$\frac{d|H(e^{j\omega})|}{d\omega}\Big|_{\omega=\omega_c} = 0. \tag{24}$$

The bandwidth of the resonance in a parametric equalizer can be defined in one of several ways [2,37]. As with the gain at the crossover frequency of the shelving filter, we choose to define the gain at the upper and lower crossover frequencies ω_u and ω_l to be equal to $\sqrt{G}$. This leads to the fifth constraint:

$$|H(e^{j\omega_u})| = |H(e^{j\omega_l})| = \sqrt{G}. \tag{25}$$

To shift the crossover frequency to ω_c, we set the crossover frequency of our low shelving filter to the bandwidth B of the peaking notch filter,

$$H_{LS}(z) = \frac{G \tan(B/2) + \sqrt{G} + [G \tan(B/2) - \sqrt{G}]z^{-1}}{\tan(B/2) + \sqrt{G} + [\tan(B/2) - \sqrt{G}]z^{-1}}, \tag{26}$$

and then apply the well-known lowpass-to-bandpass transformation [4,29]

$$z^{-1} \rightarrow -z^{-1}\frac{z^{-1} - \alpha}{1 - \alpha z^{-1}}, \tag{27}$$

where

$$\alpha = \cos\omega_c, \tag{28}$$

which yields the transfer function of the second order peaking or notch filter:

$$H_{PN}(z) = \frac{\sqrt{G} + G \tan(B/2) - [2\sqrt{G}\cos(\omega_c)]z^{-1} + [\sqrt{(G)} - G \tan(B/2)]z^{-2}}{\sqrt{G} + \tan(B/2) - [2\sqrt{G}\cos(\omega_c)]z^{-1} + [\sqrt{G} - \tan(B/2)]z^{-2}}. \tag{29}$$

The same solution was originally provided by Bristow-Johnson [37] and also recently by Jot [34].

Instead of bandwidth B, peak filters are often parameterized using quality factor Q. These two terms are related by

$$Q = \frac{\omega_c}{B}. \tag{30}$$

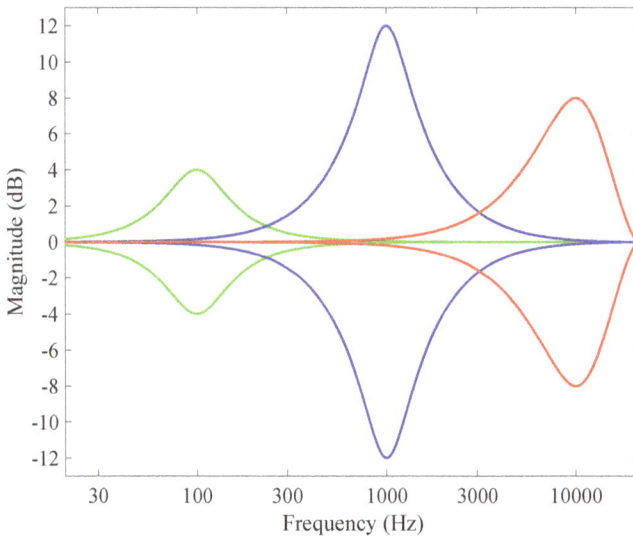

Figure 5. Magnitude responses of peak and notch filters at center frequencies 100 Hz (green), 1000 Hz (blue), and 10,000 Hz (red), when $Q = 1$. The gains of the peak and notch filters are complementary, *i.e.*, G and $1/G$, respectively.

Figure 5 presents examples of peak and notch filters with fixed Q but varying the other parameters, showing that they are symmetric on the dB scale for G and $1/G$. It is also seen in Figure 5 that when the center frequency is high, such as $f_c = 10\,\text{kHz}$, the magnitude response itself becomes asymmetric so that the upper *skirt* (*i.e.,* the magnitude response of a peaking filter on either side of the center frequency) is steeper than the lower one. This feature is caused by the requirement that the digital equalizing filters have a unity gain at the Nyquist limit.

3.4. High Order Filter Designs

In most audio production applications, low order filters are used. The smooth transitions are often advantageous in that the effect of equalization is subtle and not perceived as an artefact. However, they lack the ability for fine control. To address this, higher order IIR filters can be employed. They provide steep transitions at crossover frequencies for shelving filters, or at the upper and lower crossover frequencies for peaking and notch filters.

Orfanidis suggested techniques for the design of high order minimum phase filters based on a variety of classic filters: Butterworth, Chebyshev I, Chebyshev II, and elliptic polynomials [39]. Holters and Zölzer derived high order Butterworth shelving filters [40]. Fernandez-Vazquez *et al.* focused on the design of such filters based on a parallel connection of two stable allpass filters [41]. Särkkä and Huovilainen modeled analog parametric equalizer responses accurately using high order IIR filters [42]. However, all these high order designs begin with analog prototypes. But it is possible to design high order filters entirely in the digital domain, as described in [4]. We summarize the digital approach next.

Consider a prototype first order, lowpass filter with crossover frequency $\omega_c = \pi/2$;

$$H(z) = \frac{1 + z^{-1}}{2z^{-1}}. \tag{31}$$

Replace the pole at 0 by N poles given by

$$p_k = j\tan\left(\frac{(2k - N - 1)\pi}{4N}\right) \tag{32}$$

and scale the transfer function to again have unity gain at $z = 1$. So our Nth order prototype filter becomes

$$H_P(z) = \prod_{k=1}^{N} \frac{1}{2\cos\gamma_k} \frac{1 + z^{-1}}{-j\tan\gamma_k + z^{-1}}, \quad \text{where } \gamma_k = \frac{\pi[(2k - 1)/N - 1]}{4}, \text{with } k = 1, 2, ..., N. \tag{33}$$

This filter has the same crossover frequency and same behaviour at DC and Nyquist as the first order filter. But now the poles result in a sharper transition at $\pi/2$. Then, simple transformations may be applied to each first order section in order to transform this prototype filter into any of the standard designs. The steps to follow in these transformations are depicted in Figure 6.

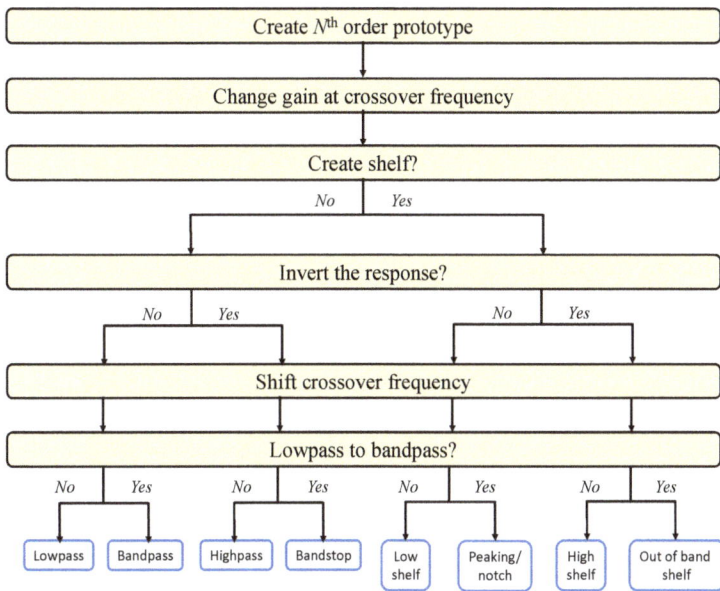

Figure 6. Flow diagram of design technique for high order IIR filters. Starting with a simple Nth order prototype filter, a series of transformations are made on each first order section to transform this into any of the main high order filter types.

3.5. FIR Shelving Filter Designs

All types of equalizers discussed herein were originally designed as analog filters, and their digital implementation as IIR filters follows the analog designs as closely as possible. However, FIR (Finite Impulse Response) shelving filters have also been proposed. Lim and Lian [43,44] and Yang [45] applied clever methods to design linear phase FIR shelving filters using long delay lines and a few mathematical operations per output sample. Vieira and Ferreira used a multirate approach to implement the bass shelving filter at one fifth of the audio sample rate while the treble shelving filter was running at the full rate in a linear phase tone control system [46].

Nevertheless, the FIR shelving filter designs have not become popular. One reason is that processing delay is a concern whenever any equalizer is used in live performance. While the maximum allowable total delay in an audio system may be a matter of discussion, it is safe to require each individual device to have the lowest possible processing delay, so as to allow cascading of several devices. This suggests using minimum phase IIR filters instead of linear phase FIR filters, since FIR filters typically exhibit higher latency for the same performance. Another motivation to use minimum phase equalizers is that they emulate more closely the behavior of analog filters.

4. Graphic Equalizers

The graphic equalizer is a tool for independently adjusting the gain of multiple frequency regions in an audio signal. Common graphic equalizer designs can provide up to about 30 controls for manipulating the frequency response of each audio channel. Structurally, a graphic EQ is a set of filters, each with a fixed center frequency and bandwidth. The only user control is the *command gain*, or the amount of boost or cut, in each frequency band, which is often controlled with vertical sliders. The gain of each frequency band can usually be adjusted within the range ±12 dB, corresponding to approximately $0.25 < G < 4$ for each filter [19,47–49]. The term *graphic* refers to the fact that the position of the sliders' knobs can be understood as a graph of the equalizer's magnitude response versus frequency, which makes the graphic equalizer intuitive to use despite the number of controls. For this reason it is a very popular device for sound enhancement, although it is more restricted than a parametric equalizer. Digital music players, such as those available in mobile phones, usually have several preset settings for the graphic equalizer for different music styles, such as pop, rock, and jazz, see e.g., [50].

A graphic equalizer can be implemented using either a cascade of equalizing filters [21,47,51–53] or a parallel bank of bandpass filters [47,54–56]. Additionally, hybrid structures have been suggested in which two or more sets of parallel filters are cascaded [57,58]. In a cascade implementation, each equalizing filter determines the magnitude response of the system mainly around its center frequency according to its own command gain, and, ideally, the gain of each equalizing filter should be unity at all other center frequencies to allow independent control of gain on each band. In a parallel implementation, each bandpass filter produces at its center frequency a gain determined by its command gain; the magnitude response of the bandpass filters should be close to zero at all other frequencies to avoid interaction between the filters. Nonetheless, both types of graphic equalizers suffer from interaction between neighboring filters [51,56,59,60]. In practice, a change in one command gain affects the magnitude response on a fairly wide frequency band, and this complicates the design of accurate graphic equalizers. This section first discusses the choice of frequency bands and then continues to the design of cascade and parallel graphic equalizers.

4.1. Bands in Graphic Equalizers

The basic unit of the graphic equalizer is the band. A graphic equalizer typically has more bands than a parametric equalizer. The bands are usually distributed logarithmically in frequency, to match human perception. Let us denote the normalized lower and upper crossover frequencies of the mth band with $\omega_{l,m}$ and $\omega_{u,m}$ respectively. Bandwidth is the difference between the upper and lower crossover frequencies, $B_m = \omega_{u,m} - \omega_{l,m}$. The frequency bands are adjacent [52], so the upper crossover of band m will be the lower crossover of band $m + 1$, $\omega_{u,m} = \omega_{l,m+1}$. That is, input audio frequencies below this crossover will be primarily affected by the gain control for band m, whereas input frequencies above it will be primarily affected by the gain control for band $m + 1$.

The logarithmic distribution of the frequency bands can be specified using a fixed ratio R between each band, so $\omega_{l,m+1} = R\omega_{l,m}$, $\omega_{u,m+1} = R\omega_{u,m}$, or $B_{m+1} = RB_m$. We also consider the geometric mean of the two crossover frequencies,

$$\omega_{M,m} = \sqrt{\omega_{l,m}\omega_{u,m}}, \tag{34}$$

where it can be seen that the same relationship holds for these values, $\omega_{M,m+1} = R\omega_{M,m}$. Additionally, the bandwidth can be related to the geometric mean $\omega_{M,m}$, because according to Equation (34), $\omega_{M,m} = \sqrt{R}\omega_{l,m} = \omega_{l,m}/\sqrt{R}$:

$$B_m = \omega_{u,m} - \omega_{l,m} = \left(\sqrt{R} - \frac{1}{\sqrt{R}} \right) \omega_{M,m}. \tag{35}$$

Two common designs are octave and 1/3-octave graphic equalizers. An octave is a musical interval defined by a doubling in frequency, so octave graphic equalizers will have the ratio $R = 2$ between each band. In a 1/3-octave design, each octave contains three bands, which implies $R^3 = 2$ or $R \approx 1.26$. So starting at 1000 Hz, an octave spacing would have geometric mean frequencies at 2000 Hz, 4000 Hz, 8000 Hz etc. and a 1/3-octave spacing would have filters centered at 1260 Hz, 1587 Hz, 2000 Hz etc.

The number of bands is determined by their spacing and the requirement to cover the entire audible spectrum. Octave graphic equalizers usually have 10 bands, ranging from about 31 Hz at the lowest to 16 kHz at the highest. Third-octave designs usually have 30 bands ranging from 25 Hz to 20 kHz. These frequencies, shown in Tables 1 and 2, are standardized by the ISO (International

Standards Organization) [61]. Note that some frequencies have been rounded coarsely in the standard in order to have simpler numbers.

Table 1. Preferred octave frequency bands according to the ISO standard [61].

Lower frequency f_l (Hz)	Geometric mean frequency f_c (Hz)	Upper frequency f_u (Hz)
22	31.5	44
44	63	88
88	125	177
177	250	355
355	500	710
710	1,000	1,420
1,420	2,000	2,840
2,840	4,000	5,680
5,680	8,000	11,360
11,360	16,000	22,720

Table 2. ISO standard for one-third octave frequency bands [61]. The center frequencies are highlighted with bold font.

f_l (Hz)	f_c (Hz)	f_u (Hz)	f_l (Hz)	f_c (Hz)	f_u (Hz)
22.4	**25**	28.2	708	**800**	891
28.2	**31.5**	35.5	891	**1,000**	1,122
35.5	**40**	44.7	1,122	**1,250**	1,413
44.7	**50**	56.2	1,413	**1,600**	1,778
56.2	**63**	70.8	1,778	**2,000**	2,239
70.8	**80**	89.1	2,239	**2,500**	2,818
89.1	**100**	112	2,818	**3,150**	3,548
112	**125**	141	3,548	**4,000**	4,467
141	**160**	178	4,467	**5,000**	5,623
178	**200**	224	5,623	**6,300**	7,079
224	**250**	282	7,079	**8,000**	8,913
282	**315**	355	8,913	**10,000**	11,220
355	**400**	447	11,220	**12,500**	14,130
447	**500**	562	14,130	**16,000**	17,780
562	**630**	708	17,780	**20,000**	22,390

The geometric mean of the crossover frequencies of a filter, $\omega_{M,m}$, is not exactly the true center frequency where the filter reaches its maximum or minimum value, $\omega_{c,m}$. We can find a relationship between the upper and lower crossover frequencies and the center frequency of a bandpass, bandstop, peaking or notch filter,

$$\tan^2\left(\frac{\omega_{c,m}}{2}\right) = \tan\left(\frac{\omega_{u,m}}{2}\right)\tan\left(\frac{\omega_{l,m}}{2}\right). \tag{36}$$

However, the geometric mean is usually quite close to the center frequency. Thus, the bandwidth scales roughly proportionally with the center frequency and higher bands will have a larger bandwidth than lower ones. Since $Q = \omega_c/B$, this is another way of saying that the Q factor is nearly constant for each band in a graphic equalizer (such constant Q graphic equalizers were first described in [62]). From Equation (35), we can estimate Q as

$$Q = \frac{\omega_{c,m}}{B_m} \approx \frac{\omega_{M,m}}{B_m} = \frac{\sqrt{R}}{R-1}. \tag{37}$$

So for an octave equalizer, $Q = \omega_{c,m}/B_m \approx \sqrt{2} = 1.41$, since $R = 2$. For a third-octave equalizer, $Q \approx \sqrt[6]{2}/(\sqrt[3]{2}-1) = 4.32$, since $R = \sqrt[3]{2}$.

4.2. Cascade Graphic Equalizers

One way to construct a graphic equalizer is by cascading several peaking/notch filters, as shown in Figure 7, where M is the number of filters. Usually the lowest and highest band filters, $H_1(z)$ and $H_M(z)$, are a low and a high shelf filter, respectively [63]. It is possible to include an optional overall gain term G_0, which may be set to the average or median command gain value [63]. This helps in the cases when many command gains are set to the same value.

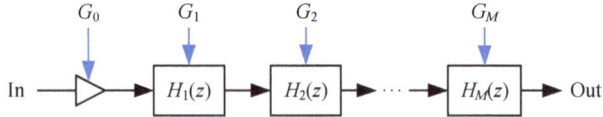

Figure 7. Cascade implementation of a graphic equalizer. Equalizing filters $H_m(z)$ are controlled with command gain parameters G_m, as indicated with blue arrows.

The transfer function of the cascade graphic equalizer can be written as

$$H_{\text{CGEQ}}(z) = G_0 \prod_{m=1}^{M} H_m(z), \tag{38}$$

where $H_m(z)$ is the equalizing filter for the mth band ($m = 1, 2, ..., M$). From this form it can be deduced that the total transfer function $H_{\text{CGEQ}}(z)$ of the cascade structure has all the same poles and zeros as the individual filters $H_m(z)$. This implies that when all band filters are minimum phase, the whole graphic equalizer will also have the minimum-phase property. Ideally, each equalizing filter $H_m(z)$ has magnitude of the desired gain G_m inside the band and gain $\sqrt{G_m}$ at band edges, so that at the crossover frequency, when $G_m = G_{m+1} = G$, we have $|H_m(e^{j\omega_{l,m}})H_{m+1}(e^{j\omega_{u,m+1}})| = G$. However, in practice other band filters also affect the gain at that frequency, causing interaction, so the total gain may differ from this value.

Figure 8 shows an example magnitude response of a 10-band cascade graphic equalizer. The lowest and highest filters are second order shelving filters designed using Equations (18) and (19), respectively. The rest of the filters are peak/notch filters whose center frequencies have been selected according to Table 1 and bandwidths according to Equation (37) with $R = 2$. The bandwidths of band filters $H_2(z)$ to $H_9(z)$ are 44.5 Hz, 88.4 Hz, 176.8 Hz, 353.6 Hz, 707.1 Hz, 1414 Hz, 2828 Hz, and 5657 Hz. The crossover frequency of the low shelf has been chosen to be 46 Hz (instead of 44 Hz, which is the nominal upper band edge frequency) and that of the high shelf is 11,360 Hz, the nominal lower band edge of the highest band (see Table 1).

In Figure 8, the filter gains are set equal to the command gains. Here we neglect the overall gain factor, which is equivalent to fixing it to $G_0 = 1$. It is seen that using command gains directly as filter gains leads to severe gain buildup, as the overall magnitude response exceeds the command points. The error is about 5 dB at several center frequencies. This is caused by the skirts of the neighboring band filter responses, which is visible in Figure 8. For example, at 63 Hz (the second command point from the left in Figure 8), the low shelf filter and the 125-Hz peak filter contribute a 2.4 dB and 2.1 dB gain increase, respectively. Additionally, a few other peak filters centered at higher frequencies contribute a few tenths of a dB to that frequency. Together they lead to a 5.0-dB overshoot at 63 Hz. To reduce the overshoot, it seems tempting to change the filter bandwidths to be narrower. In fact, this helps in reducing the gain buildup but increases the ripple by making the valleys between the command frequencies deeper.

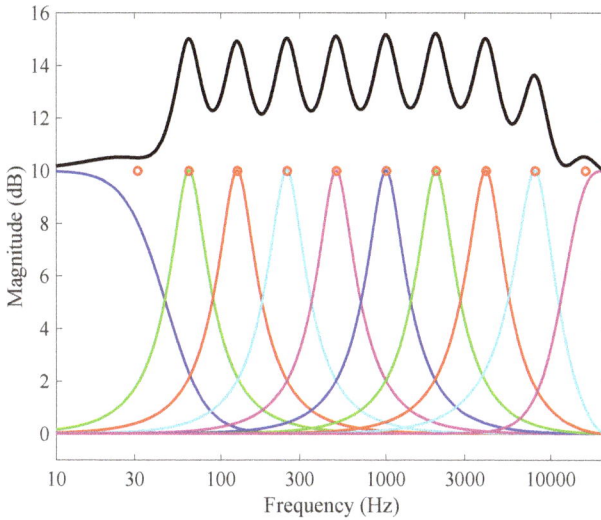

Figure 8. Magnitude responses of an octave graphic equalizer implemented using the cascade structure (thick line) and its two shelving and eight band filters with $Q = 1.41$ (colored curves). The red circles indicate the command gains, which are now all at 10 dB.

The accuracy of a graphic equalizer can be improved by allowing filter gains to be different from command gains, and by then optimizing the filter gains for each command gain configuration. First Abel and Berners [51] and recently Oliver and Jot [63] have proposed to optimize the filter gains of the cascade graphic equalizer by solving a system of linear equations. This is made possible by the fact that the magnitude responses of the peak/notch filters at different gain settings (but with fixed center frequency and Q) are self-similar on the dB scale [51]. (However, Abel and Berners [51] used parametric sections characterized by crossover frequencies at which the filter gain is the square root of its extreme value. Doing so slightly better maintains the self similarity property, as Q varies a little with filter center frequency.)

Thus, the magnitude responses of the peak/notch filters can be used as basis functions to approximate the magnitude response of the graphic equalizer at the center frequencies. This can be written in the form $\hat{\mathbf{h}} = \mathbf{Bg}$, where $\hat{\mathbf{h}}$ is an M-by-1 vector of estimated dB magnitude response values at command frequencies, $\mathbf{B}$ is the M-by-M interaction matrix representing how much the response of each band filter leaks to other center frequencies in dB, and $\mathbf{g}$ is an M-by-1 vector of command gains in dB,

$$\mathbf{g} = \left[\begin{array}{cccc} 20\log(G_1) & 20\log(G_2) & \cdots & 20\log(G_M) \end{array} \right]^T, \tag{39}$$

where T denotes transposition. The $\mathbf{B}$ matrix models the leakage of the filters to other center frequencies when all filter gains are 1 dB [51].

We have estimated the following ten-by-ten $\mathbf{B}$ matrix by setting all filter gains of the octave graphic equalizer to 10 dB (an arbitrarily chosen fairly large gain), estimating the dB magnitude response of the filters at the 10 center frequencies, and by dividing the gain values by 10:

$$\mathbf{B} = \begin{bmatrix} 0.80 & 0.23 & 0.02 & 0 & 0 & 0 & 0 & 0 & 0 & 0 \\ 0.19 & 1 & 0.21 & 0.04 & 0.01 & 0 & 0 & 0 & 0 & 0 \\ 0.04 & 0.21 & 1 & 0.20 & 0.04 & 0.01 & 0 & 0 & 0 & 0 \\ 0.01 & 0.04 & 0.20 & 1 & 0.20 & 0.04 & 0.01 & 0 & 0 & 0 \\ 0 & 0.01 & 0.04 & 0.20 & 1 & 0.20 & 0.04 & 0.01 & 0 & 0 \\ 0 & 0 & 0.01 & 0.04 & 0.20 & 1 & 0.20 & 0.04 & 0.01 & 0 \\ 0 & 0 & 0 & 0.01 & 0.04 & 0.20 & 1 & 0.20 & 0.03 & 0 \\ 0 & 0 & 0 & 0 & 0.01 & 0.04 & 0.21 & 1 & 0.18 & 0.01 \\ 0 & 0 & 0 & 0 & 0 & 0.01 & 0.06 & 0.25 & 1 & 0.10 \\ 0 & 0 & 0 & 0 & 0 & 0 & 0 & 0.01 & 0.14 & 0.94 \end{bmatrix}. \tag{40}$$

Using the inverse matrix of **B**, it is now possible to determine the filter gains $\mathbf{g}_{opt}$ which approximate in the least squares sense the magnitude response specified by command gain values **g** [51]:

$$\mathbf{g}_{opt} = \mathbf{B}^{-1}\mathbf{g}. \tag{41}$$

The inverse matrix $\mathbf{B}^{-1}$ can be computed off-line and then stored. It is applied every time a command gain is changed. This means that all filter gains will be modified even if one command gain is changed, and so all band filters must also be redesigned using Equation (29).

Figure 9 shows an example usage of the above method. The gain of all peak filters is about 6 dB although all command gains are at 10 dB, because the modeling technique accounts for the gain buildup caused by interaction. The overall magnitude response of the graphic equalizer is, however, quite close to the command points. The maximum error, which occurs at the 125-Hz point, is only 0.5 dB—a remarkable improvement over the example of Figure 8. Nevertheless, the overall fit is not highly accurate, because the response still oscillates between the command points. The peak-to-peak extent of this ripple (about 2 dB in this case) is proportional to filter gains. Oliver and Jot [63] mitigated this issue by setting the global gain G_0 to the median of all command gain values and offsetting these values accordingly.

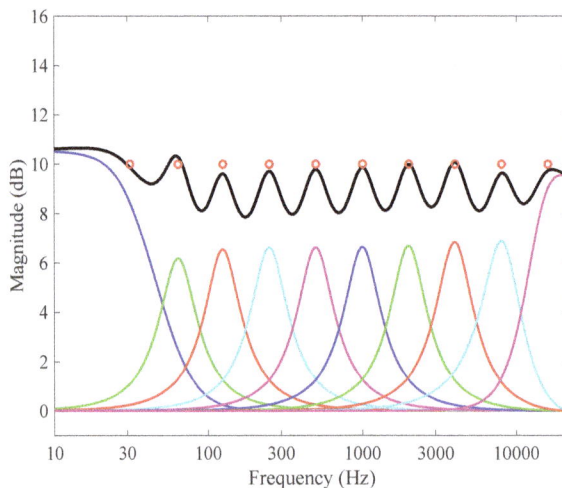

Figure 9. Magnitude response of the cascade graphic equalizer when its filter gains have been solved accounting for the interference between band filters. Note that peak gains of band filters are not the same as command gains (red circles). Cf. Figure 8.

To reduce the ripple in a cascade graphic equalizer's magnitude response, Azizi [59] has proposed to use extra filter sections, which he calls 'opposite' filters, between the band filters in a cascade graphic equalizer. When the gain and Q value of the opposite filters are set appropriately, the ripple seen in the overall response can be reduced further. Lee *et al.* refined this method by allowing the Q value of the band filters to change with gain [64]. McGrath *et al.* developed an accurate graphic equalizing method by constructing each band filter using several cascaded second order filters so that its magnitude response approximates a cosine pulse, spanning one band only [65]. Such high order filters have reduced interference, while the overall computational cost of the graphic equalizer is increased. Chen *et al.* have proposed an optimization method, which needs good initial values and definition of the desired gain at command points as well as at additional intermediate frequency points and which then iteratively adjust the coefficients of the cascaded biquad filters to reduce the mean squared error [66].

Holters and Zölzer showed how cascaded high order equalizing filters can be effectively used in graphic and parametric equalizer design [52]. Each high order band filter has a steep roll-off at its crossover frequencies, so the interference between bands gets reduced in comparison to an equalizer based on second order band filters. Even a graphic equalizer based on fourth order band filters is considerably more accurate than the one based on second order filters, as the examples shown in [56] demonstrate. Rämö and Välimäki presented a filter optimization algorithm for the high order graphic equalizer, which minimized errors in the transition bands by iteratively optimizing the orders of adjacent band filters [53].

4.3. Parallel Graphic Equalizers

Instead of using a cascade structure, a graphic equalizer often uses a collection of bandpass filters arranged in parallel, as shown in Figure 10 [47,54–56]. The audio input is split and sent to the input of every bandpass filter. It is common to include a direct path, which passes the scaled input signal to the output, as seen in Figure 10. Each filter allows only a narrow frequency band to pass through. Ideally, the center frequencies and bandwidths are configured so that if all the outputs were added together, the original signal would be reconstructed. The controls on the graphic equalizer are then implemented by changing the gain of each bandpass filter output before the signals are summed together. In theory, the parallel and cascade implementations are mathematically identical but their design must be treated differently. Parallel connection of bandpass filters avoids the accumulating phase errors and, potentially, quantization noise, found in the cascade. The parallel filter structure is also well suited to parallel computing [67].

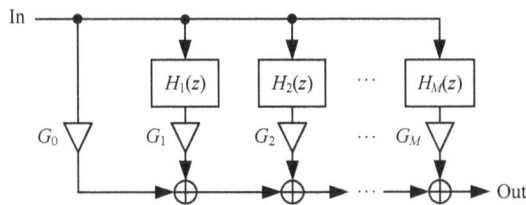

Figure 10. Parallel implementation of a graphic equalizer in which all bandpass filters receive the same input signal. The output is obtained as a sum of individual filter outputs weighted by command gains.

In the parallel structure, the direct path with gain G_0 allows easy implementation of flat command gain configurations, just like in the cascade structure. For example, the constant 10-dB gain at all command bands, which was shown in Figures 8 and 9, can be easily implemented using the parallel graphic equalizer by setting $G_0 = 3.2$ and by zeroing all other filter gains. However, the parallel structure is equally prone to interaction between band filters as the cascade structure, when command gains are set to a non-flat configuration. The transfer function of the parallel graphic equalizer is obtained as the sum of all band filters and the direct path:

$$H_{\text{PGEQ}}(z) = G_0 + \sum_{m=1}^{M} G_m H_m(z). \tag{42}$$

This implies that also the phase, not only the magnitude response, of each bandpass filter affects the total frequency response of the parallel graphic equalizer. Therefore, for good accuracy it is necessary to optimize the gain and phase of all filters for each command gain setting. This is inevitably more complicated than design of a cascade structure.

Rämö *et al.* recently developed a parallel graphic equalizer, which can accurately follow the command gain settings [56]. This method is based on Bank's fixed-pole parallel filters [68,69]. In this structure, each IIR filter transfer function $H_m(z)$ has a second order denominator but a first order numerator:

$$H_m(z) = \frac{b_{m,0} + b_{m,1}z^{-1}}{1 - 2|p_m| \cos(\theta_m)z^{-1} + |p_m|^2 z^{-2}}, \text{ for } m = 1, 2, ..., M, \tag{43}$$

where p_m and θ_m are the pole radius and pole angle of the mth bandpass filter. The poles are set in advance at pre-designed frequencies determined by the frequency resolution of the graphic equalizer. When an octave graphic equalizer is designed, the poles are set at the 10 standard frequencies from 31.5 Hz to 16 kHz, which are listed in Table 1. To obtain high accuracy, additional poles are assigned between each two standard center frequencies (at their geometric mean frequency) and below the lowest standard frequency (for example at 20 Hz), so that there will be altogether 20 predesigned poles. The pole radii are chosen so that the magnitude responses associated with neighboring poles meet at their −3-dB points:

$$p_m = e^{\frac{\Delta\theta_m}{2}} e^{\pm j\theta_m}, \tag{44}$$

where $\Delta\theta_m$ are the differences between the neighboring pole frequencies [56].

This graphic equalizer design uses least squares optimization to adjust the direct path gain G_0 and the numerator coefficients $b_{m,0}$ and $b_{m,1}$ for each filter. This method requires a target frequency response $H_t(\omega_k)$ to be constructed by interpolating between the points defined by the command gains. Hermite interpolation has been shown to be suitable for this, as it does not introduce overshoot [56]. A computationally efficient method combining linear interpolation and constant segments reduces the computational complexity of this design without much reduction in the accuracy [70]. For good accuracy, the target response needs to be evaluated at 10 times as many points as there are pole frequencies [56].

In addition to the target magnitude response, the target phase response must be generated. A natural choice for a graphic equalizer is a minimum phase response. This requires resampling the target response to the linear frequency scale and computing the Hilbert transform of the log magnitude response using FFT and IFFT operations. In the end, the data is sampled again at logarithmic frequency points.

A modeling matrix $\mathbf{M}$ is constructed by sampling the complex frequency responses of the denominators and their delayed versions. A non-negative weighting function $W(\omega_k) = 1/|H_t(\omega_k)|^2$ is needed to ensure that attenuation is implemented as accurately as amplification [56]. Finally, the optimal parameters $\mathbf{p}_{\text{opt}}$ are obtained as

$$\mathbf{p}_{\text{opt}} = (\mathbf{M}^H \mathbf{W} \mathbf{M})^{-1} \mathbf{M}^H \mathbf{W} \mathbf{h}_t, \tag{45}$$

where $\mathbf{M}$ is the modeling matrix, $\mathbf{M}^H$ is its conjugate transpose, $\mathbf{W}$ is the non-negative diagonal weighting matrix, and $\mathbf{h}_t$ is the column vector containing the target response. As a result, every time a command gain is changed, both the target response and the weighting matrix also change and matrix inversion needs to be executed [56].

Figure 11 shows a design example in which the command gains are set to ±10 dB. For the octave equalizer, the target response was interpolated at 100 points on the logarithmic frequency axis from 20 Hz to 22,050 Hz. Now the match to the command points is excellent. The largest deviation at a command point is 0.64 dB, which occurs at the lowest point 31.5 Hz.

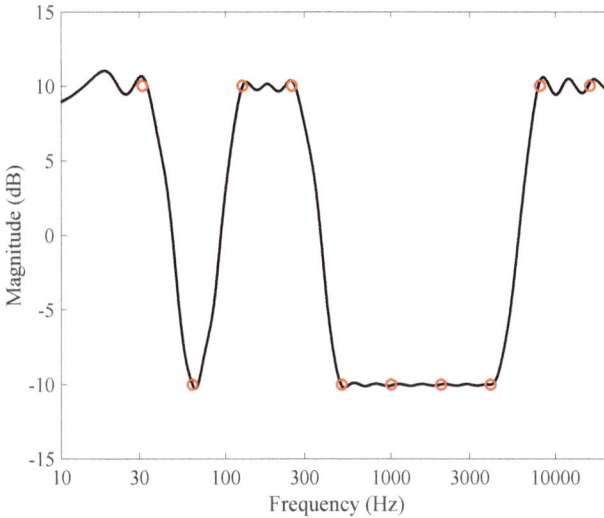

Figure 11. Magnitude response of an optimized parallel graphic equalizer, when command gains (red circles) at octave bands are set to either 10 dB or −10 dB.

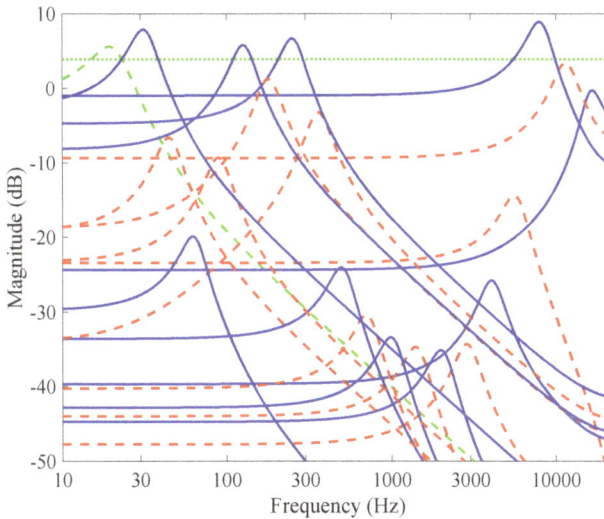

Figure 12. Parallel subfilter responses forming the magnitude response of Figure 11: filters at command bands (solid blue lines), slave filters between bands (red dashed lines), extra filter at 20 Hz (green dashed line), and the direct path gain (green dotted line).

Figure 12 illustrates the subfilters used to obtain this result. It is interesting to see that even the gains of the main filters having a pole at command frequencies (blue solid lines) do not coincide with the command gains, but are smaller. The intermediate (dashed red lines) filters usually have a gain midway between the neighboring main filters. Additionally, the responses related to the extra pole at 20 Hz and the direct path gain (3.9 dB) are shown. Since only magnitude responses of the subfilters are shown, Figure 12 fails to illustrate the effects of phase.

The computational load of the above parallel graphic equalizer is modest: its operation count per sample is only 23% larger than that of a basic graphic equalizer consisting of cascaded biquad filter sections having second order transfer function numerator and denominator [56]. However, the design phase of the parallel structure is quite complicated, requiring many more operations than designing a cascade graphic equalizer. Though this may be an important factor in interactive real-time use, the computational cost is not very critical when designing equalizer preset settings off-line.

Alternatively, it is possible to design a parallel graphic equalizer based on higher order IIR filters, although this will increase the computational cost in comparison to designs based on second order band filters. Virgulti *et al.* [71] have proposed an equalizer based on a multirate filterbank with critical downsampling of the bands. In their system, the band filters are cosine modulated versions of a high order IIR prototype filter, yielding very good frequency resolution and extremely low ripple. Interestingly, their IIR band filters have an approximately linear phase response in their passband [71].

4.4. FIR Graphic Equalizers

FIR-type graphic equalizers have been developed since the early years of digital equalizers [72–75]. An advantage of FIR filters here is the possibility to realize a linear phase response, causing no phase distortion and thus largely retaining the waveform of the signal [73,76]. However, many audio engineers doubt whether linear phase filters are a good choice for audio signal processing, as they may lead to pre-echos or ringing of transients [76–78]. Furthermore, just like in the case of tone controls, linear phase filters produce more delay than minimum phase filters, which may be noticeable in live sound reproduction. Nevertheless, it is also possible to convert an FIR filter to be minimum phase, which reduces the delay [79].

For FIR graphic equalizers, there are two basic implementation strategies to choose from: a single high order FIR filter and a parallel structure. These choices are different from those used in the IIR case, mainly because the cascade structure is pointless in the case of standard FIR filters: cascading several low order FIR filters leads to a less powerful system than a single high order filter with the same number of coefficients.

In principle, a single high order FIR filter could approximate the frequency response specification dictated by the command gains of a graphic equalizer [75,77]. However, as the command points are usually located logarithmically in frequency, it will be necessary to use interpolation to obtain a smooth target response [70,75,79,80]. Still, a fundamental problem remains: the order of the FIR filter must be large, at least several thousand, to approximate well the desired magnitude response at the lowest band [75,77,79]. That is clearly more costly than using a biquad IIR filter per band. To reduce the computational cost of long FIR equalizing filters at low frequencies, Waters *et al.* have suggested a multirate system in which downsampled filters running at a low rate improve the approximation performed by a fullband filter running at the audio sample rate [75]. Oliver describes in his patent how a graphic equalizer can be implemented using a frequency-warped FIR filter [81].

4.4.1. Graphic Equalization Using Fast Convolution

The *Fast Fourier Transform* (FFT) algorithm provides a way to reduce the computational cost of a high order FIR filter in computers using sequential processing. This is called *fast convolution* and was first introduced by Stockham [82]. In fast convolution, FIR filtering is implemented as a complex multiplication of the discrete Fourier transforms of the signal and the filter, and then using the inverse FFT algorithm to obtain the output signal in the time domain. The FFT of the FIR filter can be computed in advance and stored, but the input signal must be transformed in frames (buffers) using the FFT. For long filters, the complex multiplication of spectra and the inverse FFT require much less operations than a direct convolution. Schöpp and Hetzel have proposed to create a graphic equalizer using this approach, allowing the user to control the gain at 512 linearly spaced frequency points when the FFT-length is 1024 [83]. Fernandes *et al.* implemented a 20-band equalizer [84], and Ries and Frieling implemented a 31-band equalizer [85] using the fast convolution method, using 512-point and 32768-point FFTs, respectively. A disadvantage of this method is the processing latency, which is generally twice the impulse response length.

Kulp [86] proposed to divide the impulse response of the equalizer into several segments, which are processed using the FFT technique. This reduces the latency considerably, but can still be quite efficient, as only one inverse FFT is needed, when all segments have the same length. It is also possible to process the first segment of the impulse response using the direct convolution to avoid the delay completely [87]. This approach has become popular in artificial reverberation, where very long FIR filters are used [88–91].

4.4.2. Parallel and Multirate FIR Graphic Equalizers

A parallel FIR filter structure leads to a filterbank in which bandpass filters handle each individual frequency band [72,73,92]. This implies that all bands are first separated, and this requires in principle a high order FIR filter for each band. McGrath proposed a straightforward method in which he approximated the responses of the IIR peak/notch filters of a parallel graphic equalizer using FIR filters [92]: the impulse responses of IIR band filters were simply sampled and truncated to the length of 1024 samples, when the sample rate was 48 kHz. Since all band filters have the same input signal, as seen in Figure 10, the FIR equalizer could be implemented as a single 1024-tap filter, which was obtained as the weighted sum (command gains were used as weights) of the 30 truncated band filter responses [92].

As a general method reducing the computational burden of FIR graphic equalizers, multirate filterbank techniques can be used [74,93]: the sample rate of the system is reduced in stages, so that the highest band filter uses the largest sample rate while the lowest band filter uses the smallest sample rate. After applying all filters, the sample rate of each branch is elevated back to the original rate, and the outputs of all branches are added to obtain the output signal. Cecchi *et al.* have shown how to design a cosine modulated multirate FIR filterbank for implementing an octave graphic equalizer [94]. Väänänen and Hiipakka have discussed how to implement graphic equalizers using a quadrature mirror filterbank [77]. An alternative way is to allow some overlap between band filters, such as in the cubic B-spline FIR graphic equalizer proposed by Kraght [80].

5. Other Equalization Filter Designs

Sections 3 and 4 described the two most common forms of user-controlled equalizers, parametric and graphic equalizers. However, there are many other approaches to equalization, especially if the goal is less focused on interactive user control. For instance, equalizers may be designed to match an arbitrary magnitude and phase response, or closely match analog designs, or for the filters to be IIR yet still maintain linear phase. In this section, we consider a few alternative equalization filter design approaches.

5.1. Matched EQ and Optimal Design Techniques

A common problem that arises in many equalization applications is targeted or matched EQ. The goal is to find a transfer function that corresponds to a given frequency response. Most FIR filter design methods (impulse response truncation, windowing design, and optimal filter design) aim to do exactly this. However, IIR designs are often preferred and there exist several important methods to design an IIR filter for an arbitrary frequency response. These methods are considered optimal design techniques since they apply numerical optimization to the design problem. They may be classified into those where one seeks to match a complex response, and those where only the magnitude response needs to be matched.

The general problem is as follows. We need to find a filter $H(z) = B(z)/A(z)$ that is a close match to a desired frequency response $D(\omega)$. However, finding the optimal values for the IIR filter coefficients is not easy since the frequency response is a nonlinear function of the coefficients of $H(z)$. We generally want to minimize the solution error;

$$E_S(\omega) = W_S(\omega) \left[\frac{B(z)}{A(z)} - D(\omega) \right]. \tag{46}$$

But in practice the equation error is often used since it gives rise to linear equations,

$$E_E(\omega) = W_E(\omega)[B(z) - D(\omega)A(z)], \tag{47}$$

and either the complex values $D(\omega)$ or the magnitudes $|D(\omega)|$ are known. Given this, the optimization problem is usually formulated as minimize

$$\int_{-1}^{1} |E_*(\omega)|^k d\omega, \tag{48}$$

where k is most often chosen as 2 for a least squares approach, and ∞ is sometimes used for a minimax approach. The weighting functions $W_*(\omega)$ control the relative importance of errors at different frequencies. Since $D(\omega)$ is known at more values than the coefficients of $H(\omega)$, this gives rise to an overdetermined set of equations, for which linear least squares can be used. Once the solution is found, we can then enforce stability by replacing any pole p in $A(z)$ having a magnitude larger than 1 with $1/p^*$.

If we seek only to match a magnitude response, and the phase response is not specified, then the equation error becomes,

$$E_E(\omega) = W_E(\omega) \left(|B(z)| - |D(\omega)||A(z)| \right), \tag{49}$$

and similarly for the solution error. Iterative approaches can be used to find an appropriate target phase corresponding to the optimal magnitude specification, as in [95].

There are three main choices that need to be made when using this technique; the order of the filter (or alternatively, a tolerance in the error function), the weighting function to use, and the amount and locations of frequency sampling. The choice of frequency sampling is related to the appropriate choice of weighting function. Since perception is related to frequency on a log scale, we may require much more accuracy at low frequencies. Yet if our target spectrum comes from the Fourier transform of finite data, we may have uniform sampling of the spectrum. This suggests using more samples at low frequencies, and with a higher weighting. Furthermore, if computational time is important, then we may choose to use fewer samples. On the other hand, if the desired frequency response is derived from the Fourier transform of a short set of time domain samples, then we may choose to use all available samples of the frequency response, in order to maximize our information regarding the desired response.

This technique is known as the *equation error* method [96], and is based around early work on complex curve fitting [97]. A similar least squares approach was described in [98]. Improved performance may also be achieved using the output-error algorithm, which performs a damped Gauss-Newton iterative search [99], where initial conditions are chosen from the output of the equation error method. Another related approach is the Yule-Walker algorithm [100], which uses separate algorithms, applied in a recursive manner, to estimate the numerator and denominator coefficients. Further iterative approaches include the Steiglitz-McBride method, and its frequency domain implementations [101]. Vargas and Burrus provide an excellent overview of the field [102].

Frequency warping can be used to stretch the frequency axis at low frequencies and compress it at high frequencies, which is useful in audio [103,104]. The frequency warping method can be applied to the target response before the filter design. This is done in the time domain by interpreting the target impulse response as an FIR filter, and by replacing each unit delay in its delay line by a first order allpass filter. The allpass filter coefficient λ determines the amount of warping. Finally, the designed filter can be dewarped by applying warping with the complementary coefficient $-\lambda$. Warping has been combined successively with linear prediction, leading to a perceptually meaningful estimation of the spectral envelope, which is more accurate at low than at high frequencies [104–106]. A related early filter design method proposed by Kautz [107] first sets the poles of an IIR filter and then optimizes their weights. The Kautz method has been shown to suit well to audio filter design, since the distribution of poles can be used to improve the approximation at low frequencies in the same way as frequency warping [108].

Recently, Bank developed an IIR filter design method in which the poles of the IIR filters are first chosen and the weights are then optimized using the least squares method. This method leads to a similar accuracy as the warped and Kautz filters and is easier to apply. The IIR filter designed this way can be implemented directly using the parallel structure. In Section 4.3 we showed how this method can be applied to parallel graphic equalizer design [56]. The parallel IIR filter design with fixed poles can be recommended for audio filter design.

5.2. Digital Equalizer Design Matching Analog Prototypes

Digital filters are frequently derived from analog prototypes, either because analog filter design methods are easily used and understood, or because one is attempting to model analog systems. For example, digital versions of RIAA and NAB equalizers have been proposed [109,110]. The bilinear transform is often used since it offers the desirable properties of preserving order and stability of the analog prototype while mapping the entire continuous-time response onto the unit circle in the z-plane. But it also introduces severe warping as the crossover frequency approaches the Nyquist limit.

The optimization approaches described in Section 5.1 could be used to model an analog equalizer. But they require considerable additional processing and thus are generally performed off-line or at least with latency, or require coefficient tables of sufficient resolution and a suitable interpolation scheme. One could also model and approximate the analog equalizer using differential equations [42], but this is again computationally expensive. Thus these methods are inadequate for controllable and responsive digital equalizer design that matches the analog prototype. Another approach is to use oversampling such that the response over audible frequencies has reduced warping. But this may introduce latency and requires additional anti-aliasing filters.

In this section, we describe a method, first given in [111], for digital equalizer design which anticipates the warping effects and compensates before applying the transform. This give a near optimal match to the analog magnitude response and closed-form expressions for the filter coefficients. Furthermore, the crossover frequency can be specified to be greater than the Nyquist limit.

To illustrate the design process, we show how a lowpass filter matching the analog frequency response may be designed. The first order analog lowpass transfer function with crossover frequency f_c is defined as

$$H_{\mathrm{LP}}(s) = \frac{1}{s/\omega_c + 1}, \tag{50}$$

where $\omega_c = 2\pi f_c$ and $s = j\omega$. The square magnitude response $|H_{\mathrm{LP}}(\omega)|^2 = 1/[1 + (\omega/\omega_c)^2]$ is zero at infinity. But with a standard bilinear transform and prewarping, the digital filter's zero gain occurs at Nyquist frequency [112] (pp. 535–537). Instead, we start with a first order analog prototype featuring a gain at infinity equal to the desired Nyquist gain. This is achieved using a standard first order analog shelving filter,

$$H_{LS}(s) = \frac{g_1 s + \Omega_s}{s + \Omega_s}. \tag{51}$$

We want to set the asymptotic high frequency gain g_1 to be the original magnitude response $|H_{LP}|$ at the Nyquist frequency $f_s/2$. The required high frequency gain is then given by

$$g_1 = |H_{LP}(s = 2\pi f_s/2)| = 2/\sqrt{4 + f_s^2/f_c^2}. \tag{52}$$

Ordinarily, we would also want to match the gain at the crossover frequency. But when the crossover frequency is close to the Nyquist limit, we observe a steep slope in the digital magnitude response due to the response curve being forced through two close points. So instead, we match the point where the magnitude reaches half Nyquist gain on a Decibel scale. This leads to a matching gain of $g_m = \sqrt{g_1}$ and the associated frequency at which $|H_{LP}(\omega_m)| = \sqrt{g_1}$ is $\omega_m = \omega_c\sqrt{1/g_1 - 1}$. The matching frequency specification needs to be prewarped with

$$\Omega_m = \tan[\omega_m/(2f_s)] \tag{53}$$

to ensure it will be placed correctly after the bilinear transform with $z = e^{j\omega f_s}$. The shelving frequency that provides this match is found by solving $|H_{LS}(\Omega_m)| = g_m$, giving

$$\Omega_s = \Omega_m\sqrt{g_1}. \tag{54}$$

Applying the bilinear transform leads to the desired lowpass transfer function,

$$H(z) = H_{LS}\left(\frac{1 - z^{-1}}{1 + z^{-1}}\right) = \frac{(\Omega_s + g_1) + (\Omega_s - g_1)z^{-1}}{(\Omega_s + 1) + (\Omega_s - 1)z^{-1}}. \tag{55}$$

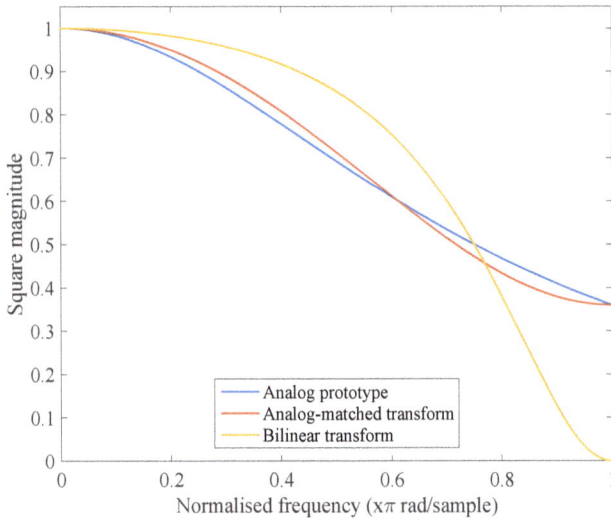

Figure 13. The square magnitude responses of first order lowpass filters with crossover frequency 0.75π for an analog prototype, an analog-matched digital filter, and a digital filter derived using the bilinear transform.

Figure 13 depicts the square magnitude response for an analog prototype lowpass filter, the digital filter derived from this using the bilinear transform, and the analog-matched digital filter derived using the method herein. Frequency and square magnitude are given on linear scales in order to more easily see the warping with the bilinear transform and the small differences between the analog prototype and the matched-analog digital filter.

Similar approaches for matched-analog digital designs are described in [36] for peaking and notch filters, and in [31] for shelving filters. Or newly proposed analog-to-digital transforms [113] may be used to also give a close match to an analog magnitude response [114]. However, all the standard filters used in digital equalizer design may be matched to analog prototypes by applying the usual transformations in Figure 6 to Equation (55).

5.3. Linear Phase IIR Filters

It is well-known that IIR filters generally are not linear phase. However, several approaches exist for linear phase IIR filter design. Powell and Chau's real-time linear phase IIR filters [115] used a modification of a well-known time reversal technique. Willson and Orchard described a variation on this approach that yields higher performance (more stopband loss, less passband ripple and/or narrower transition band) [116]. Kurosu *et al.* described performance issues in Powell and Chau's original design; a sinusoidal variation in the group delay, and a harmonic distortion with sinusoidal input [117]. To alleviate these issues, they reduce the filter's overall processing delay by using shorter sections with truncated impulse response [117]. Azizi proposed an efficient arbitrary sample rate converter using zero phase IIR filtering, which later led to a patented signal interpolator [118]. Another approach to linear phase IIR filtering was proposed in [119]. To perform time reversal methods in real time, the signal is divided into blocks of samples. The impulse response of the time-reversed IIR filter must be truncated, but only after it has decayed sufficiently. The cumulative energy of the impulse response can be used for determining an appropriate impulse response length [120].

A basic linear phase IIR method is shown in Figure 14. The input signal passes through a filter $H(z)$, then the order of the samples is reversed, and then this procedure is repeated on the output, i.e, $H(z)$ applied, then time reversal. In the time reversal operation, the input signal $x[n]$ becomes $x[-n]$. The corresponding z-domain transformation can be written as

$$X(z) \rightarrow X(z^{-1}). \tag{56}$$

The intermediate transfer functions in Figure 14 can be written as $Y_1(z) = H(z)X(z)$, $Y_2(z) = H(z^{-1})X(z^{-1})$, and $Y_3(z) = H(z)H(z^{-1})X(z^{-1})$. Thus, the transfer function for the linear phase IIR filter given in Figure 14 is

$$Y(z) = H(z^{-1})H(z)X(z) = |H(z)|^2X(z), \tag{57}$$

which corresponds to zero phase filtering by the magnitude response of $H(z)$ twice.

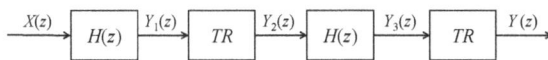

Figure 14. Implementation of a real-time, causal linear phase IIR filter using time reversal (TR).

5.4. Dynamic Equalization

Equalization and dynamics processing are essential signal processing operations in audio engineering. The equalizer is the conventional tool to manipulate the spectral characteristics of the audio signal to achieve frequency balance. In contrast, dynamics processors such as compressors, limiters, gates, expanders and duckers, control the level variation and dynamic envelope of the signal. Among them, the dynamic range compressor [121] is one of the most important tools in mixing, and its use defines much of the sound of contemporary mixes.

Equalization and dynamics processing are often considered to dominate exclusive domains, with equalization controlling amplitude in the spectral domain, and dynamics processing controlling amplitude in the time domain, especially in regard to the input level. Yet those domains are not fully independent. Previous research [122,123] has shown that it is good practice to set dynamic range compressor parameters based on the frequency content in the signal, and many problems in audio production can be addressed by using combinations of filtering and dynamics processing. Thus variants often address specific functionality such as de-essing or hum removal, and as such have limited configurability beyond their applications.

The multiband compressor operates differently and independently on different frequency bands of a signal, offering more precise adjustment of dynamics than a single band compressor. The processing on each frequency band is controlled by its own compression parameters, and output signals of each frequency band are combined as a final step. Unwanted gain changes or artefacts (such as pumping and breathing) are avoided when applying compression on one frequency band. The crossover frequencies are often adjustable.

The dynamic equalizer [4] provides the ballistic control of a dynamic range compressor like threshold, attack and release, to the conventional equalizer allowing time-varying adjustment of equalization curve. In other words, the equalization stage is able to respond dynamically to the input signal level. Many of these dynamic equalizer implementations are used for noise reduction in audio restoration [124], hearing-loss correction [125], and compliance with broadcasting regulations. Other dynamic equalizers employ automatic gain adjustment of a fixed FIR or IIR filter. The modulation can be gated, as in de-hum and de-ess processors [126]. Still other dynamic equalizers allow the filter to be configurable in the band it operates on. The dynamics that most of these systems offer to the engineer are constrained to the point that not all of the details are controllable. Yet the dynamic equalizer is the closest design currently available to the concept of a general frequency and dynamics tool.

Wise proposed a General Dynamic Parametric Equalizer that, in principle, could cover multiband compressors and dynamic equalizers [127]. Conventional multiband compressors compress frequency bands differently through band-pass filters. The general processor utilizes this concept, but replaces these filters with parametric equalizer filters [127]. It offers larger control over the dynamics of specific frequencies or frequency bands in audio signals. It can adjust the frequency, gain, and bandwidth of each filter, with controls common to dynamic range compression controls. The attack and release times determine how fast the dynamic equalization acts towards the defined amount of boost or cut. The characteristic of the processing on each frequency band is controlled by four parameters: gain, threshold, attack, and release. Assuming all parameters are adjustable, this could be configured to a conventional equalizer, dynamic range compressor or multiband compressor. The control characteristics of a multiband compressor and a dynamic equalizer are shown in Figure 15.

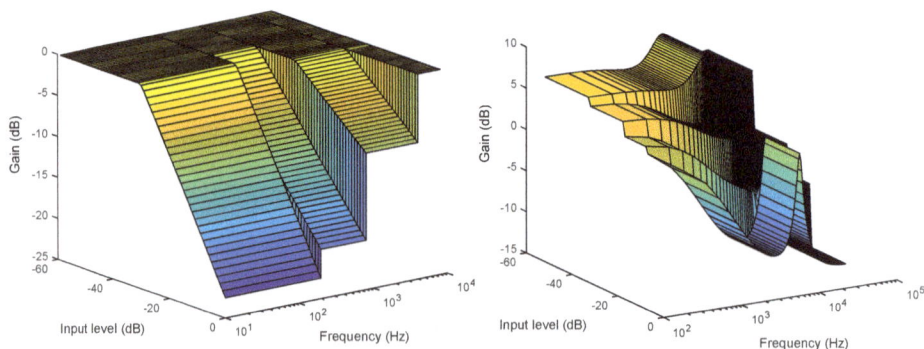

Figure 15. Input-output characteristics of a 4-band compressor (**left**) and a 4-band dynamic equalizer (**right**), showing dependence on both input signal level and frequency.

6. Sound Reproduction Applications

In this section, we discuss applications in which equalization is used for enhancing audio reproduction, including tone controls, loudspeaker and headphone equalization, room-loudspeaker equalization, loudness equalization, and noise-based equalization.

6.1. Tone Control

Most stereo systems feature tone controls, which provide a quick way to adjust the sound to suit the listener's taste and compensate for the frequency response of the room. Tone controls are the simplest and possibly most common equalization system. A basic version consists of two knobs, typically labeled "bass" and "treble". These knobs are used to control the gain of low and high frequencies, respectively, through the use of shelving filters (see Section 3). In each case, the maximum gain of the shelf, G, is adjustable. Typical tone controls have an adjustment range of ±12 dB. The crossover frequency is usually fixed. The values for bass and treble vary by manufacturer, but typical crossover frequencies for the bass control might range from 100 Hz to 300 Hz, and the treble control from 3 kHz to 10 kHz. In two-knob tone controls, the midrange frequencies (between bass and treble) are usually left unchanged.

On some units, in addition to control of the bass and treble, there may be a 'midrange' or 'mid' control. This control is usually implemented as a peaking or notch filter. The knob on the midrange control affects the gain G at center frequency, which generally takes the same range of values as the bass and treble controls, e.g., ±12 dB. The center frequency is generally fixed to be midway between the bass and treble controls, and the bandwidth is chosen so that the midrange control mainly affects the frequencies which are left unadjusted by the bass and treble controls.

Loudness compensation, or loudness control, at low listening levels is another application for equalizing filters in home audio systems [128,129]. The loudness control aims to automatically compensate for the nonlinear behavior of the human hearing system particularly at low frequencies: the equal loudness curves of hearing are compressed at low frequencies with respect to middle and high frequencies. Without the loudness control, as the sound level is turned down, the low-frequency content in music becomes inaudible while middle and higher frequencies are still present [128–132]. Additionally, even when the low-frequency sounds may be heard at soft listening levels, they appear to be more quiet than higher frequencies. A loudness control switch may add one or two shelving or peak filters to the signal path to amplify low frequencies to cancel this nonlinear effect. Alternatively, it is possible to implement a continuous loudness control [128,131,132]. Then the gain and crossover frequency of low-frequency shelving filters are adjusted together with the volume control to modify the rate of level change of low frequencies with volume control. A similar nonlinear compression effect takes place also at very high frequencies, but to a smaller extent. However, it is possible to use high frequency shelving filters for loudness compensation at high frequencies, if desired.

Musical instrument amplifiers use different type of tone control units than other sound systems. They can be specific to the instrument, helping to produce a personal sound quality. For example in a tone stack of a guitar amplifier studied by Yeh and Smith, there are three controls (low, mid, high), but they all control parameters of a third order filter so that their effects overlap [133]. Some electronic keyboard instruments have a built-in tone control circuit, which may have similarly overlapping bands, such as the Clavinet, which has four independent tone equalizing circuits (soft, medium, treble, brilliant) connected in parallel [134].

6.2. Loudspeaker Equalization

One common use of digital equalizers in audio is the flattening of the magnitude response of loudspeakers [78,135–139]. In addition to hi-fi and studio loudspeakers, there are several special categories, such as flat panel speakers [140–144], PA (public address) systems [145–147], and micro speakers of mobile devices [148], which benefit from equalization.

The response of the speaker is first measured, usually with the sine-sweep method, in an anechoic room or another special space in which the room effect can be deduced from the impulse response. Unless the loudspeaker under study is a monitor speaker, which is usually aimed directly at the listener, it may be of interest to average the response from a few angles in the frontal sector [149]. The measured magnitude response is usually smoothed using 1/3-octave averaging, which approximately imitates the resolution of human hearing [150]. This implies that very narrow peaks or valleys may not need to be corrected. It is also known that a local peak in the speaker's magnitude response is more noticeable than a notch [151], and this may be accounted for in the processing of the response.

Loudspeaker response equalization always requires a *target response* against which the measured and smoothed speaker response is compared [108,138]. The target response generally cannot be constant across all frequencies. Loudspeakers are inherently highpass systems, having a low-frequency crossover point below which they cannot produce sound effectively. An attempt to boost the response beyond this point leads to increased distortion and may even destroy the speaker, if its cone is forced to exceed the maximum deviation. One easy way to obtain a target curve at low frequencies is to use the magnitude response of a Butterworth highpass filter, which is smooth and does not overshoot. A reasonable crossover frequency, in the range 30 to 100 Hz, should be selected, which is near the natural low-frequency point of the speaker. A second order Butterworth highpass filter having an approximately 12-dB per octave slope is suitable for producing the target response for regular dynamic loudspeakers, while a fourth order highpass filter with a 24-dB/oct. slope should be used for vented (bass reflex) speakers. The ratio of the measured and target responses, or the difference between the dB magnitude responses, yields the necessary correction as a function of frequency. Matched filter design methods discussed in Section 5.1 and the high-precision graphic equalizer of Section 4.3 are good candidate methods for designing the equalizer.

Usually loudspeakers divide the audio band for two or three speaker elements, because a single element cannot reproduce the whole audio band with good quality [78]. The input signal is delivered to each element via a crossover network, which implements a selective filter (lowpass, bandpass, or highpass) for each element. The most common crossover filter design is the Linkwitz-Riley network [78,152]. Each element can be equalized separately after the crossover filtering [78,136,149]. One possibility is to combine the equalizers and the crossover filters, as suggested by Ramos and López [153].

Various equalization filters and design methods have been proposed for loudspeaker equalization, including automatic adjustment of parametric equalizers [154,155]. Karjalainen *et al.* first suggested the use of warped FIR and IIR filters for loudspeaker equalization to ease the approximation at low frequencies [138,156]. Warped filters then became a popular method for equalizing loudspeakers [138,157–159]. Ramos *et al.* have proposed to use a cascade connection of a regular and a warped FIR filter [160]. The balanced model truncation method is applied in [161].

In addition to magnitude-only equalization, also phase equalization of loudspeakers is sometimes considered, especially at low frequencies [138,162]. This can be combined with the magnitude response equalization, which leads to complex frequency response equalization, or it can be realized separately using a delay equalizer, which is usually based on allpass filter design. Herzog and Hilsamer have proposed to simulate the group delay at low frequencies using an allpass filter and then to implement the phase equalization by filtering the input signal in short blocks backwards in time [162], similar to the linear phase IIR filter design described in Section 5.3. This solution produces an inevitable processing delay. Alternatively, it is possible to design a high order allpass filter, for example using the method proposed by Abel and Smith [163].

6.3. Room Equalization

One common use of graphic equalization is to "tune" a room, adjusting the equalizer to roughly compensate for room resonances. The goal is to achieve a desired magnitude response, flattening out extremes, reducing coloration in the sound and achieving greater sonic consistency among listening spaces. Graphic equalization is more commonly found in live performance and recording studios than in most home stereo systems. However, graphic equalizers are occasionally found in consumer stereo systems and even in digital music player software, where they can be used as a more flexible form of tone control for refining the sound according to one's taste. In addition to studios, performance venues, and living rooms, equalization of a listening space is highly relevant in car cabins [49,164–168].

A traditional way to equalize a loudspeaker in a room is to consider loudspeaker equalization and room equalization separately [169]. First the anechoic response of the loudspeaker is equalized using an equalizer, and then the equalized speaker is measured in a room, and a second equalizer is designed. In more recent approaches, a measured anechoic speaker response is no longer required. It is now known that acoustic impulse responses of typical listening spaces are short at high and middle frequencies, but can be very long at low frequencies. For this reason, one idea is to use a gated (truncated) version of the loudspeaker impulse response in a room to model the first equalizer [108,170]. After that another equalizer may be designed to correct the response at low frequencies. Bank and Ramos have proposed techniques for automatically selecting the poles of parallel IIR filters, which has the same structure as the parallel graphic equalizer in Section 4.3, to equalize the loudspeaker-room response [171].

There are two radically different scenarios in room equalization: the single-point and multiple-point cases [164]. In the single-point case, the loudspeaker-room response is equalized for a 'sweet spot,' which may be the main listening position or a microphone location. However, this approach usually decreases the sound quality at other locations in the room, and thus is rarely a good solution [164,172,173]. In multi-point room equalization, the response is measured at two or more listening positions, and the equalizer is designed to improve the response at all of them. Naturally, this may not be achieved fully at any location, so the equalizer design becomes an optimization problem [164]. One useful idea is to suppress the room modes appearing at all measured locations, which are called common acoustical poles [174,175]. Other approaches to multi-point room equalization include spatial averaging [176] and effort variation [177].

Related problems are the shortening of the room impulse response and reduction of the room reverberation, which is called dereverberation [173,178]. It is known that when the reproduced sound in a room becomes boomy, it may not be enough to suppress the level of sound at that frequency with an equalizer, since that mode will still remain ringing, although less loudly. A special method for shortening low-frequency room modes is called *modal equalization* [178,179]. It involves estimating, canceling, and replacing single ringy room modes, which appear at frequencies below about 200 Hz. The cancellation filter is a digital notch filter while a resonator is used for replacing it with a more quickly decaying response [179]. Karjalainen *et al.* have suggested an iterative method to find the longest decaying modes at low frequencies using high order pole-zero modeling on a downsampled band, which is called frequency-zooming ARMA modeling [180]. Fazenda *et al.* have studied the audibility of low-frequency modes, which is important to know when designing a modal equalizer [181].

In acoustic impulse response shortening, the aim is to modify the decay rate of room reverberation, but not to cancel it fully [182,183]. This can improve the sound quality and intelligibility. Impulse response reshaping can be combined with crosstalk cancellation in spatial sound reproduction systems in which it is necessary to control the sounds reaching the ears of a listener [184]. Dereverberation is particularly important for speech, because room reverberation reduces speech intelligibility in telephone and teleconferencing systems and complicates automatic speech recognition [185,186].

6.4. Headphone Equalization

Equalization of headphones is largely based on the same general principle as loudspeaker equalization with a few differences [187,188]: measure the impulse response of the system, compute the corresponding smoothed magnitude response, compare it against a target response, design an equalizer to minimize the difference between the measured and target responses, and implement the equalizing filter, which should process the input signal of the system. A sine sweep is commonly used for measuring the impulse response of headphones. The main differences between loudspeaker and headphone equalization are in the measurement of the impulse response and in the shape of the target response.

In the case of headphones, there are only two impulse responses to be measured per headphone, those from the left and right earpiece. However, variations in these responses can easily appear since the fit of the headphones can have a major effect on their frequency response. If the headphones are not tightly fitted, the low-frequency response can suffer considerable attenuation. This is especially prominent in in-ear headphones, in which the earpiece is similar to an earplug [189]. The headphone response is easiest to measure using a dummy head. Miniature microphones inside the ear canal must be mounted when headphones are measured while they are worn by a user. Bolaños and Pulkki have proposed a method to measure the headphone response of the user at the ear canal entrance and map it to the eardrum [190]. This method requires measuring both the acoustic velocity and the pressure.

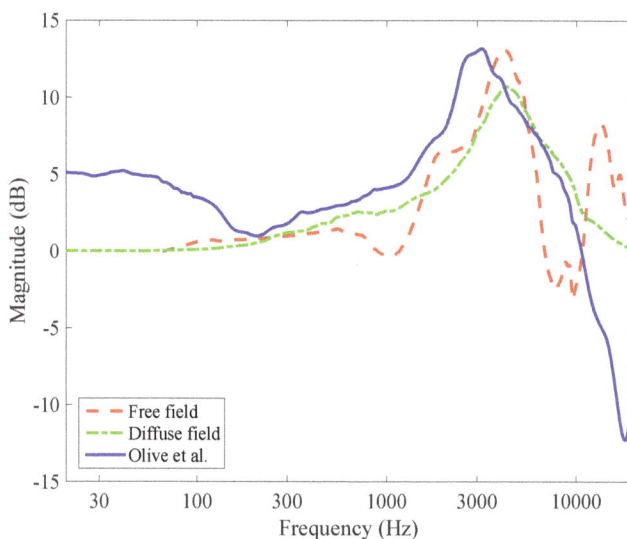

Figure 16. Various target responses for headphones: free field, diffuse field, and preferred response simulating in-room loudspeaker response proposed by Olive *et al.* [191].

There are different options for the headphone target response, which are usually not flat, as shown in Figure 16. The free field response corresponds to listening to a sound source with a flat magnitude response located in front of the listener in an anechoic room. The free field response is seen to contain resonances, the highest of which appears at approximately 3 kHz and compensates for the open ear canal resonance, which is suppressed when the headphones cover the entrance to the ear. Additionally, the free-field response is attenuated around 8 kHz to cancel a closed ear canal resonance, which appears when wearing headphones. The diffuse-field response, also shown in Figure 16, can be measured with a dummy head in an echo chamber. It corresponds to listening to an omnidirectional loudspeaker located far away from the listener, so that in the echo chamber the sound appears to come from all directions, *i.e.*, the sound field is diffuse.

Olive *et al.* [191] have conducted extensive studies trying to find out the preferred target response for headphone listening. The blue curve in Figure 16 is their new target response, which corresponds to listening to a high-quality speaker in a room. They suggest that this curve should be used as the prototype curve for good headphones for music listening [191]. It may be noticed that this target curve has a large boost at approximately 3 kHz and also a shelf-like boost of about 5 dB at frequencies below 100 Hz. High frequencies are attenuated more than in free or diffuse-field responses.

The magnitude response of headphones often contains notches at high frequencies, which also change their location when the headphones move in the head of the listener. For this reason, it is typical not to attempt to equalize headphones at frequencies higher than about 10 kHz. To avoid the spectral notches from causing large gain peaks or ringing in the equalization filter response, regularization techniques can be used [192,193].

In addition to correcting the headphone response to be closer to a chosen target response, it is possible to design an equalizer to imitate the response of another pair of headphones. This idea was first proposed by Briolle and Voinier, who simulated various headphones by convolving the input signal with the headphone impulse response [194]. Olive *et al.* [195] have compared headphone responses in a similar fashion while Rämö and Välimäki [192] also included an equalizer modeling the acoustic isolation response of the headphones, so that the ambient noise could be included in the simulation.

Another headphone application based on equalization filters is the modeling of head-related transfer functions (HRTF), which leads way to three-dimensional sound reproduction [196,197]. Various digital filter design methods, such as balanced model truncation [198] and cascaded peak/notch filters [51,199], have been used for modeling HRTF responses [200]. Simplification of HRTF filters has been used for demonstrating which spectral details are required for hearing directions correctly [201]. Hiekkanen *et al.* have developed a method to compare loudspeakers using headphones with equalization filters, which are based on HRTF responses [202].

In *augmented reality audio*, a headset having microphones near the user's ears is used for combining reproduced and ambient sounds [203–206]. The ambient sounds captured by the microphones must be equalized to enhance their naturalness [203,205,207]. Also the isolation of the headphones must be accounted for, since the leaked and reproduced ambient sound both reach the user's ear [208,209]. The equalization of the microphone signal requires shelving and peak/notch filters [188,208]. A related application is a real-time audio equalizer, which processes the external microphone sounds to be played instantly through the headphones, for example to enhance the perceived sound in a concert [210].

6.5. Equalization to Combat Ambient Noise

When music listening takes place in a noisy environment, such as in a speeding car or bus, the ambient noise can considerably disturb the hearing experience [211]. A traditional solution to this in car radios and audio systems has been the automatic gain control, which adjusts the volume according to car speed or background noise level inside the car cabin [212–214]. Additionally, dynamic range compression [121] may be used to reduce loudness changes, since the softest passages of music easily become inaudible in noise [212,215]. However, since the ambient noise hardly ever has a flat spectrum but usually contains more energy at low frequencies, like pink noise, simply increasing the gain is not the best way to combat this: the gain should usually be increased more at low frequencies than at high frequencies.

It is well understood that background noise disturbs the listening experience through the auditory masking phenomenon. It causes the hearing threshold to increase mostly at those frequencies where the noise signal has energy. Additionally, the masking phenomenon spreads to other frequencies, so that higher frequencies than the noise component are more affected and lower frequencies less. Rämö *et al.* have convincingly simulated this phenomenon for headphone listening in noise [216]: a graphic equalizer having a resolution of the auditory Bark bands controlled by an auditory model imitates the increase of the hearing threshold. This simulation of noise-based masking allows one to listen to music samples without both the noise and those components that get masked by the noise.

Miller and Barish have described a Dynamic Sound Optimization system for cars, which analyzes the ambient noise from signals captured by microphones [217]. Figure 17(a) shows a single-channel version of this system. Adaptive filtering is first used to separate the background noise from the microphone signal by using the music signal as the reference. This is the same principle as in adaptive echo canceling (AEC), which is used for suppressing the near-end signal in telephone and teleconferencing systems [218]. Based on the analysis of the extracted noise signal, a low-frequency shelving filter, a variable equalizer, and a dynamic range compressor are controlled to effectively cancel the auditory masking experienced by the passengers. When more loudspeakers are used, their signals can be canceled from the microphone signal one after the other [217]. Tzur and Goldin [219], Kim and Cho [50], and Christoph [214] have proposed a noise-based equalizer using an FFT filterbank. The music or speech signal to be played in a noisy environment, such as in a car, is processed with the filterbank so that the gain of each Bark-band depends on estimated masking caused by the ambient noise signal.

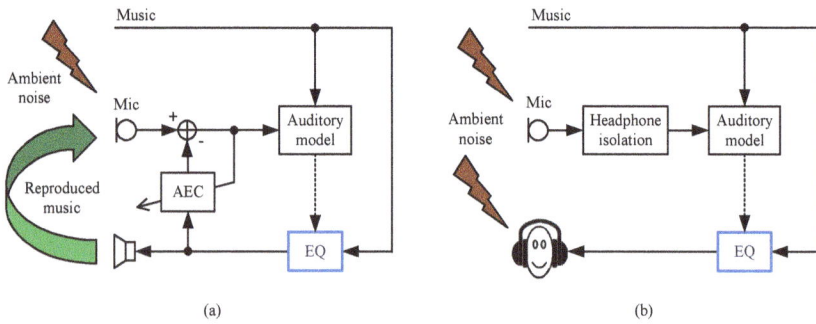

Figure 17. Block diagrams of noise-based equalization for (**a**) loudspeaker [217] and (**b**) headphone reproduction [206,220]. The dashed line indicates that the auditory model controls the parameters of the equalizer.

Headphones have become highly popular in recent years, because they are now the common way of listening to music and other sounds from mobile phones and tablet computers. The mobile use of headphones mostly happens in noisy environments, such as in cities with heavy traffic and on buses, trains, and airplanes [215]. Additionally, headphones are used in noisy work places for personal audio and for sound shading, which refers to pleasant masking noise to hide for example conversations. In all these use cases, the background noise disturbs the listening experience, and so it becomes important to select headphones with good noise attenuation, possibly combined with active noise control [192]. Still, some ambient noise may leak to the user's ear, masking the music signal at some frequencies.

Rämö *et al.* [220] and Välimäki *et al.* [206] have described a noise-based equalization system for headphone listening, which is sketched in Figure 17(b). It captures the ambient noise using an external microphone, such as the one in the cord of a hands-free headset, typically used for the user's own voice. This ambient noise signal is processed with a filter, which models the estimated isolation of noise through the headphone to the ear. By using the processed ambient noise signal as the masker and the music signal played through the headphones as the desired signal, it is possible with an auditory model to estimate the masking as a function of frequency. This information is used for controlling an equalizer, which aims to cancel the masking phenomenon. A graphic equalizer with Bark-band division was used in which the filter bandwidth is constant 100 Hz for the lowest five band and approximately one-third octave for the other bands [220].

7. Audio Content Creation

Equalization and its use in the music production process has become well-established, with standardized designs for graphic and parametric equalizers, and well-established operations to achieve desired tasks [221]. However, in recent years there has been an emergence of adaptive, intelligent or autonomous audio equalizers, which provide novel approaches to their design and use. This section gives an overview of these new directions in equalization for content creation.

7.1. Adaptive and Intuitive Interfaces

Here, we provide an overview of the state of the art concerning interfaces for equalization, with an emphasis on perceptual adaptive and intuitive controls. Various approaches for learning a listener's preferences for an equalization curve with a small number of frequency bands have been applied to research in the setting of hearing aids [222,223] and cochlear implants [224], and the modified simplex procedure [225,226] is now an established approach for selecting hearing aid frequency responses. However, many recent innovations have emerged in the field of music production.

In [227], Dewey and Wakefield evaluated various novel equalizer interface designs that incorporated spectral information in addition to or as a replacement for equalization curves and parametric controls. Their subjective evaluation suggested that such designs may be more efficient and preferred over traditional interfaces for many equalization tasks. Loviscach presented an interface for a five-band parametric equalizer where the user simply freehand draws the desired magnitude response and an evolutionary optimization strategy (chosen for real-time interaction) finds the closest match [228]. Informal testing suggested that this interface reduced the set-up time for a parametric equalizer compared to more traditional interfaces. Building on this, Heise *et al.* proposed a procedure to achieve equalization and other effects using a black-box genetic optimization strategy [229]: Users are confronted with a series of comparisons of two differently processed sound examples. Parameter settings are optimized by learning from the users' choices. Though these interfaces are novel and easy to use by the nonexpert, they make no use of semantics or descriptors.

Considerable research has aimed at the development of technologies that let musicians or sound engineers perform equalization using perceptually relevant or intuitive terms, e.g., brightness, warmth, or presence. Reed presented an assistive sound equalization expert system [230]. Inductive learning based on nearest neighbor pattern recognition was used to acquire expert skills. These were then applied to adjust the timbral qualities of sound in a context-dependent fashion. They emphasized that the system must be context-dependent, that is, the equalization depends on the input signal system and hence operates as an adaptive audio effect. In [231], a self-organizing map was trained to represent common equalizer settings in a two-dimensional (2-D) space organized by similarity. The space was hand-labelled with descriptors that the researchers considered to be intuitive. However, informal subjective evaluation suggested that users would like to choose their own descriptors.

The work of Bryan Pardo and his collaborators has focused on new, intelligent and adaptive interfaces for equalization tasks. They address the challenge that complex interfaces for equalizers can prevent novices from achieving their desired modifications. Sabin and Pardo described and evaluated an algorithm to rapidly learn a listener's desired equalization curve [232–235]. Listeners were asked to indicate how well an equalized sound could be described by a perceptual term. After rating, weightings for each frequency band were found by correlating the gain at each frequency band with listener responses, thus providing a mapping from the descriptors to audio processing parameters. Listeners reported that the resultant sounds captured their intended meanings of descriptors, and machine ratings generated by computing the similarity of a given curve to the weighting function were highly correlated to listener responses. This allows automated construction of a simple and intuitive audio equalizer interface. In [236], active and transfer learning techniques were applied to exploit knowledge from prior concepts taught to the system from prior users, greatly enhancing the performance of the equalization learning algorithm.

The early work on intelligent equalization based on intuitive descriptors was hampered by a limited set of descriptors with a limited set of training data to map those descriptors to equalizer settings. Cartwright and Pardo addressed this with SocialEQ, a web-based crowdsourcing application aimed at learning the vocabulary of audio equalization descriptors [237]. To date, 633 participants have participated in a total of 1102 training sessions (one session per learned word), of which

731 sessions were deemed reliable in the sense that users were self-consistent in their answers (Personal communication with B. Pardo, 2015). This resulted in 324 distinct terms, and data on these terms is made available for download.

Building on the mappings from descriptors to equalization curves, Sabin and Pardo [233] described a simple equalizer where the entire set of curves were represented in a 2-D space (similar to [231]), thus assigning spatial locations to each descriptor. Equalization is performed by the user dragging a single dot around the interface, which simultaneously manipulates 40 bands of a graphic equalizer. This approach was extended to multitrack equalization in [238], which provided an interface that, by varying simple graphic equalizers applied to each track in a multitrack, allowed the user to intuitively explore a diverse set of mixes.

The concepts of perceptual control, learned from crowdsourcing, intuitive interface design and mapping of a high dimensional parameter space to a lower dimensional representation were all employed in [239,240]. This approach scaled equalizer parameters to spectral features of the input signal, then mapped the equalizer's 13 controls to a 2-D space. The system was trained with a large set of parameter space data representing warmth and brightness, measured across a range of musical instrument samples, allowing users to perform equalization using a perceptually and semantically relevant, simple interface.

7.2. Autonomous and Intelligent Systems for Equalization

As a recent example application of targeted equalization (see Section 5.1), Ma *et al.* [241] described an intelligent equalization tool that, in real-time, equalized an incoming audio stream towards a target frequency spectrum. The target spectrum was derived from analysis of 50 years of commercially successful recordings [242]. The target equalization curve is thus determined by the difference in spectrum between an incoming signal and this target spectrum. A hysteresis gate is first applied on the incoming signal, to ensure that only active content (*i.e.*, not silence or low level noise) is used to estimate the input signal spectrum. Since the input signal to be equalized is continually changing, the desired magnitude response of the target filter is also changing (though the target output spectrum remains the same). Thus, smoothing was applied from frame to frame on the desired magnitude response $D(\omega)$ and on the applied filter $H(\omega)$. Targeting was achieved using the Yule-Walker method, and testing showed that Yule-Walker offered superior performance to the method proposed by Lee [98]. Figure 18 depicts a block diagram of this system.

Perez Gonzalez and Reiss described a method for automatically equalizing a multitrack mixture [243]. The method aimed to achieve equal average perceptual loudness on all frequencies amongst all tracks within a multitrack audio stream. Accumulative spectral decomposition techniques were used together with cross-adaptive audio effects to apply graphic equalizer settings to each track. Analysis demonstrated that this automatic equalization method was able to achieve a well-balanced and equalized final mix.

An alternative approach to autonomous multitrack equalization was provided by Hafezi and Reiss [244]. They created a multitrack intelligent equalizer that used a measure of auditory masking and rules based on best practices from the literature to apply, in real-time, different multiband equalization curves to each track. The method is intended as a component of an automatic mixing system that applies equalization as it might be applied manually as part of the mixing process. Results of objective and subjective evaluation were mixed and showed room for improvement, but they indicated that masking was reduced and the resultant mixes were preferred over amateur, manual mixes.

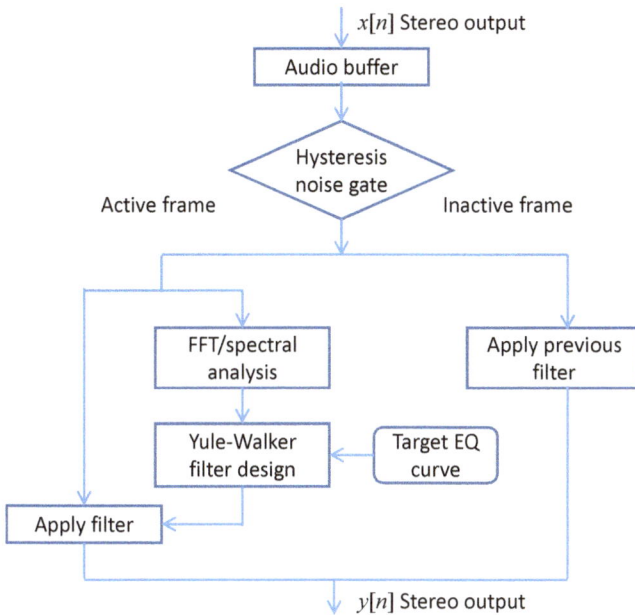

Figure 18. Block diagram of the real-time targeted equalization [241].

Finally, Barchiesi and Reiss showed how to reverse engineer the equalization settings applied to each track in a multitrack mix [245]. That is, given a multitrack recording and the mixed output track, the settings of all time-varying equalizers applied to each track can be derived. Assuming that mixing involves the processing of each input track by a linear time-invariant (LTI) system (slowly time varying processing may be valid if a frame-by-frame approach is applied), then a least squares approach may be used to find the impulse response applied to each track. This impulse response represents all the LTI effects that might be applied during the mixing process, including gains, delays, stereo panners, and equalization filters. Equalization is distinguished from delays, gains and panning by assuming that equalization introduces the minimum possible delay, that the equalization does not affect the signal level, and that level differences between channels are due to stereo panning. If the multitrack recording was mixed using FIR filters, then the estimation order can be increased until an almost exact solution is obtained. If IIR filters were used, then either one estimates the IIR filter best approximating the finite impulse response, or nonlinear or iterative least squares approaches are used to derive the filter coefficients directly. Reported results showed the effectiveness of their approach with 4, 6 and 8 track multitracks, when both IIR and (up to 128th order) FIR filters were applied, and when there was no knowledge of the characteristics of the (third party, commercial) filters that were applied.

8. Conclusions

In this review paper, we described audio equalization in terms of its historical perspective, the established designs and approaches, and the frontier research directions and emerging challenges. Though low order parametric and graphic equalizers are well-established, variations and extensions of these designs, such as high order parametric equalizers or convolution-based graphic equalizers, offer almost limitless ability to shape the spectral content of a signal. These design choices allow the user or designer to favour precise control, low computational cost and/or minimal latency. Moving away from graphic and parametric equalizers, other approaches allow for the design of equalization filters that combine the most appealing features of IIR and FIR filters, or provide the means to match an arbitrary magnitude (or magnitude and phase) response. A summary of approaches to equalization and their relevant features and applications is given in Table 3.

Table 3. A summary of approaches to equalization and their relevant features and applications. Note that Yule-Walker and the Equation Error method can be implemented in real-time if performed on a frame-by-frame basis.

Method		Section in text	Compu- tation	Latency	Real- time	Precision of control	Ease of use	Applications/notes
Parametric EQ		3.1-3.3	Low	Low	Yes	Low	Easy	Standard control, music production
High order parametric EQ		3.4	Mid	Low	Yes	High	Easy	Arbitrary spectral shaping
FIR parametric EQ		3.5	Mid	High	Yes	Low	Easy	Linear phase design
Linear phase IIR EQ		5.3	Mid	High	Yes	Low	Easy	Linear phase design
Graphic EQ		4.1-4.3	Low	Low	Yes	Mid	Easy	Standard control, room EQ
FIR graphic EQ		4.4	High	High	Yes	Low	Easy	Linear phase design
	FFT-based	4.4.1	Mid	High	Yes	High	Mid	Fast FIR graphic EQ
	Parallel/multirate	4.4.2	High	High	Yes	High	Mid	Nonuniform frequency resolution
High order graphic EQ		4.2	High	Low	Yes	High	Easy	Reduced band interference
Matched EQ	Yule-Walker	5.1	High	Low	No*	High	Easy	Arbitrary frequency response fitting
	Equation error	5.1	High	Low	No*	High	Easy	
	Warped filters	5.1	Mid	Low	Yes	High	Mid	Gain accuracy at low frequencies
Analog-matched EQ		5.2	Low	Low	Yes	Low	Easy	Analog emulation
Dynamic EQ		5.4	Low	Low	Yes	Low	Difficult	Music production
Noise-based EQ		6.5	Mid	Low	Yes	Mid	Mid	Listening in a noisy environment
Bank's parallel filters		4.3, 5.1	Mid	Low	Yes	High	Mid	General method

The applications of these designs are vast, with equalization being relevant to any situation where it may be of benefit to shape a spectrum. Of note, loudspeaker, room and headphone equalization are common, but equalizers have uses throughout the music recording, production and playback chain. A particularly relevant application in today's busy world is the noise-based equalization, which can enhance the audibility of music in high ambient noise. Finally, Section 7 considered new and emerging approaches to equalization. It focused on more relevant and intuitive interfaces, often based on machine learning, and on autonomous and intelligent audio equalization systems that attempt to automate aspects of their use. This area is still in its infancy, and might see great advances with improved knowledge of the semantics and perceptual aspects of equalization.

Acknowledgments: We would like to thank our colleagues Riitta Väänänen, Jonathan S. Abel, Fabián Esqueda, Ville Pulkki, German Ramos, Jussi Rämö, Julius O. Smith, Miikka Tikander, and Sampo Vesa for helpful comments and discussions on audio equalizers. Special thanks go to Juho Liski for his help on producing the headphone target response figure.

Author Contributions: The authors contributed equally to this work.

Conflicts of Interest: The authors declare no conflict of interest.

Appendix

The source code for the methods discussed in this paper and for many of the figures are available online at https://code.soundsoftware.ac.uk/projects/allaboutaudioeq. It includes;

- Matlab code demonstrating a parametric equalizer comprised of a first order low shelving filter, second order peaking/notch filter and first order high shelving filter, see Section 3, especially Sections 3.1 and 3.3;
- Matlab code for second order low and high shelving filter, see Section 3.2.
- Matlab code for generating a wide variety of high order filter designs of arbitrary order and for differing definitions of the gain at crossover frequencies, see Section 3.4;

- Matlab code for cascade and parallel graphic equalizer design, see Section 4;
- Matlab code for a lowpass filter matching the analog frequency response, see Section 5.2;
- Matlab code for illustrating dynamic equalization and multiband dynamic range compression, see Section 5.4;
- A Powerpoint file containing original versions of many of the figures; and
- Any additions or errata since initial publication.

References

1. Bohn, D.A. Operator adjustable equalizers: An overview. In Proceedings of the Audio Engineering Society 6th International Conference on Sound Reinforcement, Nashville, Tennessee, 5–8 May 1988; pp. 369–381.
2. Orfanidis, S.J. IIR digital filter design (Chapter 11). In *Introduction to Signal Processing*; Prentice Hall: Englewood Cliffs, NJ, USA, 1996; pp. 573–643.
3. Zölzer, U. Equalizers (Chapter 5). In *Digital Audio Signal Processing*, 2nd ed.; Wiley: Chichester, UK, 2008; pp. 115–190.
4. Reiss, J.D.; McPherson, A. Filter effects (Chapter 4). In *Audio Effects: Theory, Implementation and Application*; CRC Press: Boca Raton, FL, USA, 2015; pp. 89–124.
5. Lundheim, L. On Shannon and Shannon's formula. *Telektronikk* **2002**, *98*, 20–29.
6. Campbell, G.A. Physical theory of the electric wave-filter. *Bell Syst. Tech. J.* **1922**, *1*, 1–32.
7. Zobel, O.J. Theory and design of uniform and composite electric wave-filters. *Bell Syst. Tech. J.* **1923**, *2*, 1–46.
8. Bauer, B. A century of microphones. *Proc. IRE* **1962**, *50*, 719–729.
9. Williamson, D.T.N. Design of tone controls and auxiliary gramophone circuits. *Wireless World* **1949**, *55*, 20–29.
10. Baxandall, P.J. Negative-feedback tone control. *Wireless World* **1952**, *58*, 402–405.
11. Stanton, W.O. Magnetic pickups and proper playback equalization. *J. Audio. Eng. Soc.* **1955**, *3*, 202–205.
12. Hoff, P. Tape (NAB) equalization. In *Consumer Electronics for Engineers*; Cambridge University Press: Cambridge, UK, 2008; pp. 131–135.
13. Rudmose, W. Equalization of sound systems. *Noise Control* **1958**, *24*, 82–85.
14. Boner, C.P.; Boner, C.R. Minimizing feedback in sound systems and room-ring modes with passive networks. *J. Acoust. Soc. Am.* **1965**, *37*, 131–135.
15. Boner, C.P.; Boner, C.R. Behavior of sound system response immediately below feedback. *J. Audio. Eng. Soc.* **1966**, *14*, 200–203.
16. Conner, W.K. Theoretical and practical considerations in the equalization of sound systems. *J. Audio Eng. Soc.* **1967**, *15*, 194–198.
17. Flickinger, D. Amplifier System Utilizing Regenerative and Degenerative Feedback to Shape the Frequency Response. U.S. Patent #3,752,928, 14 August 1973.
18. Massenburg, G. Parametric equalization. In Proceedings of the 42nd Convention of the Audio Engineering Society, Los Angeles, CA, USA, 2–5 May 1972.
19. McNally, G.W. Microprocessor mixing and processing of digital audio signals. *J. Audio Eng. Soc.* **1979**, *27*, 793–803.
20. Guarino, C.R. Audio equalization using digital signal processing. In Proceedings of the 63rd Convention of the Audio Engineering Society, Los Angeles, CA, USA, 15–18 May 1979.
21. Hirata, Y. Digitalization of conventional analog filters for recording use. *J. Audio Eng. Soc.* **1981**, *29*, 333–337.
22. Berkovitz, R. Digital equalization of audio signals. In Proceedings of the Audio Engineering Society 1st International Conference on Digital Audio, Rye, NY, USA, 3–6 June 1982.
23. White, S.A. Design of a digital biquadratic peaking or notch filter for digital audio equalization. *J. Audio Eng. Soc.* **1986**, *34*, 479–483.
24. Regalia, P.; Mitra, S. Tunable digital frequency response equalization filters. *IEEE Trans. Acoust. Speech Signal Process.* **1987**, *35*, 118–120.
25. Brandt, M.; Bitzer, J. Hum removal filters: Overview and analysis. In Proceedings of the 132nd Convention of the Audio Engineering Society, Budapest, Hungary, 26–29 April 2012.
26. Van Waterschoot, T.; Moonen, M. Fifty years of acoustic feedback control: State of the art and future challenges. *Proc. IEEE* **2011**, *99*, 288–327.

27. Harris, f.; Brooking, E. A versatile parametric filter using an imbedded all-pass sub-filter to independently adjust bandwidth, center frequency, and boost or cut. In Proceedings of the Asilomar Conference on Signals, Systems and Computers, Pacific Grove, CA, USA, 26–28 October 1992; pp. 269–273.

28. Moorer, J.A. The manifold joys of conformal mapping: Applications to digital filtering in the studio. *J. Audio Eng. Soc.* **1983**, *31*, 826–841.

29. Constantinides, A. Spectral transformations for digital filters. *Proc. IEE* **1970**, *117*, 1585–1590.

30. Fontana, F.; Karjalainen, M. A digital bandpass/bandstop complementary equalization filter with independent tuning characteristics. *IEEE Signal Process. Lett.* **2003**, *10*, 119–122.

31. Berners, D.P.; Abel, J.S. Discrete-time shelf filter design for analog modeling. In Proceedings of the 115th Convention of the Audio Engineering Society, New York, NY, USA, 10–13 October 2003.

32. Keiler, F.; Zölzer, U. Parametric second- and fourth-order shelving filters for audio applications. In Proceedings of the IEEE Workshop on Multimedia Signal Processing, Siena, Italy, 29 September–1 October 2004; pp. 231–234.

33. Christensen, K.B. A generalization of the biquadratic parametric equalizer. In Proceedings of the 115th Convention of the Audio Engineering Society, New York, NY, USA, 10–13 October 2003.

34. Jot, J.M. Proportional parametric equalizers—Application to digital reverberation and environmental audio processing. In Proceedings of the 139th Convention of the Audio Engineering Society, New York, NY, USA, 29 September–1 October 2015.

35. Clark, R.J.; Ifeachor, E.C.; Rogers, G.M.; Van Eetvelt, P.W. Techniques for generating digital equalizer coefficients. *J. Audio Eng. Soc.* **2000**, *48*, 281–298.

36. Orfanidis, S.J. Digital parametric equalizer design with prescribed Nyquist-frequency gain. *J. Audio Eng. Soc.* **1997**, *45*, 444–455.

37. Bristow-Johnson, R. The equivalence of various methods of computing biquad coefficients for audio parametric equalizers. In Proceedings of the 97th Convention of the Audio Engineering Society, San Francisco, CA, USA, 10–13 November 1994.

38. Reiss, J.D. Design of audio parametric equalizer filters directly in the digital domain. *IEEE Trans. Audio Speech Lang. Process.* **2011**, *19*, 1843–1848.

39. Orfanidis, S.J. High-order digital parametric equalizer design. *J. Audio Eng. Soc.* **2005**, *53*, 1026–1046.

40. Holters, M.; Zölzer, U. Parametric higher-order shelving filters. In Proceedings of the European Signal Process. Conference (EUSIPCO), Florence, Italy, 4–8 September 2006; pp. 1–4.

41. Fernandez-Vazquez, A.; Rosas-Romero, R.; Rodriguez-Asomoza, J. A new method for designing flat shelving and peaking filters based on allpass filters. In Proceedings of the International Conference on Electronics, Communications and Computers (CONIELECOMP-07), Cholula, Puebla, Mexico, 26–28 February 2007.

42. Särkkä, S.; Huovilainen, A. Accurate discretization of analog audio filters with application to parametric equalizer design. *IEEE Trans. Audio Speech Lang. Process.* **2011**, *19*, 2486–2493.

43. Lim, Y.C. Linear-phase digital audio tone control. *J. Audio Eng. Soc.* **1987**, *35*, 38–40.

44. Lian, Y.; Lim, Y.C. Linear-phase digital audio tone control using multiplication-free FIR filter. *J. Audio Eng. Soc.* **1993**, *41*, 791–794.

45. Yang, R.H. Linear-phase digital audio tone control using dual RRS structures. *Electron. Lett.* **1989**, *25*, 360–362.

46. Vieira, J.M.N.; Ferreira, A.J.S. Digital tone control. In Proceedings of the 98th Convention of the Audio Engineering Society, Paris, France, 25–28 February 1995.

47. Greiner, R.A.; Schoessow, M. Design aspects of graphic equalizers. *J. Audio Eng. Soc.* **1983**, *31*, 394–407.

48. Takahashi, S.; Kameda, H.; Tanaka, Y.; Miyazaki, H.; Chikashige, T.; Furukawa, M. Graphic equalizer with microprocessor. *J. Audio Eng. Soc.* **1983**, *31*, 25–28.

49. Kontro, J.; Koski, A.; Sjöberg, J.; Väänänen, M. Digital car audio system. *IEEE Trans. Consum. Electron.* **1993**, *35*, 514–521.

50. Kim, H.G.; Cho, J.M. Car audio equalizer system using music classification and loudness compensation. In Proceedings of the International Conference on ICT Convergence (ICTC), Seoul, Korea, 28–30 September 2011; pp. 553–558.

51. Abel, J.S.; Berners, D.P. Filter design using second-order peaking and shelving sections. In Proceedings of the International Computer Music Conference, Coral Gables, FL, USA, 1–6 November 2004.

52. Holters, M.; Zölzer, U. Graphic equalizer design using higher-order recursive filters. In Proceedings of the International Conference on Digital Audio Effects (DAFX), Montreal, QC, Canada, 18–20 September 2006; pp. 37–40.

53. Rämö, J.; Välimäki, V. Optimizing a high-order graphic equalizer for audio processing. *IEEE Signal Process. Lett.* **2014**, *21*, 301–305.

54. Motorola Inc. Digital stereo 10-band graphic equalizer using the DSP56001. *Appl. Note* **1988**, *APR2/D*.

55. Tassart, S. Graphical equalization using interpolated filter banks. *J. Audio Eng. Soc.* **2013**, *61*, 263–279.

56. Rämö, J.; Välimäki, V.; Bank, B. High-precision parallel graphic equalizer. *IEEE/ACM Trans. Audio Speech Lang. Process.* **2014**, *22*, 1894–1904.

57. Adams, R.W. An automatic equalizer/analyzer. In Proceedings of the 67th Convention of the Audio Engineering Society, New York, NY, USA, 31 October–3 November 1980.

58. Erne, M.; Heidelberger, C. Design of a DSP-based 27 band digital equalizer. In Proceedings of the 90th Convention of the Audio Engineering Society, Paris, France, 19–22 February 1991.

59. Azizi, S.A. A new concept of interference compensation for parametric and graphic equalizer banks. In Proceedings of the 111th Convention of the Audio Engineering Society, New York, NY, USA, 30 November–3 December 2001.

60. Miller, R. Equalization methods with true response using discrete filters. In Proceedings of the 116th Convention of the Audio Engineering Society, Berlin, Germany, 8–11 May 2004.

61. ISO. ISO 266, Acoustics—Preferred frequencies for measurements. **1975**.

62. Bohn, D.A. Constant-Q graphic equalizers. *J. Audio Eng. Soc.* **1986**, *34*, 611–626.

63. Oliver, R.J.; Jot, J.M. Efficient multi-band digital audio graphic equalizer with accurate frequency response control. In Proceedings of the 139th Convention of the Audio Engineering Society, New York, NY, USA, 29 September–1 October 2015.

64. Lee, Y.; Kim, R.; Cho, G.; Choi, S.J. An adjusted-Q digital graphic equalizer employing opposite filters. In *Lect. Notes Comput. Sci.*; Springer-Verlag: London, UK, 2005; Volume 3768, pp. 981–992.

65. McGrath, D.; Baird, J.; Jackson, B. Raised cosine equalization utilizing log scale filter synthesis. In Proceedings of the 117th Convention of the Audio Engineering Society, San Francisco, CA, USA, 28–31 October 2004.

66. Chen, Z.; Lin, Y.; Geng, G.; Yin, J. Optimal design of digital audio parametric equalizer. *J. Inform. Comput. Sci.* **2014**, *11*, 57–66.

67. Belloch, J.A.; Bank, B.; Savioja, L.; Gonzalez, A.; Välimäki, V. Multi-channel IIR filtering of audio signals using a GPU. In Proceedings of the IEEE International Conference Acoustics, Speech and Signal Processing (ICASSP), Florence, Italy, 4–9 May 2014; pp. 6692–6696.

68. Bank, B. Perceptually motivated audio equalization using fixed-pole parallel second-order filters. *IEEE Signal Process. Lett.* **2008**, *15*, 477–480.

69. Bank, B. Audio equalization with fixed-pole parallel filters: An efficient alternative to complex smoothing. *J. Audio Eng. Soc.* **2013**, *61*, 39–49.

70. Belloch, J.A.; Välimäki, V. Efficient target response interpolation for a graphic equalizer. In Proceedings of the IEEE International Conference Acoustics, Speech and Signal Processing (ICASSP), Shanghai, China, 20–25 March 2016; pp. 564–568.

71. Virgulti, M.; Cecchi, S.; Piazza, F. IIR filter approximation of an innovative digital audio equalizer. In Proceedings of the International Symposium on Image and Signal Processing and Analysis (ISPA), Trieste, Italy, 4–6 September 2013; pp. 410–415.

72. Jensen, J.A. A new principle for an all-digital preamplifier and equalizer. *J. Audio Eng. Soc.* **1987**, *35*, 994–1003.

73. Henriquez, J.A.; Riemer, T.E.; Trahan, Jr., R.E. A phase-linear audio equalizer: Design and implementation. *J. Audio Eng. Soc.* **1990**, *38*, 653–666.

74. Principe, J.; Gugel, K.; Adkins, A.; Eatemadi, S. Multi-rate sampling digital audio equalizer. In Proceedings of the IEEE Southeastcon-91, Williamsburg, VA, USA, 7–10 April 1991; pp. 499–502.

75. Waters, M.; Sandler, M.; Davies, A.C. Low-order FIR filters for audio equalization. In Proceedings of the 91st Convention of the Audio Engineering Society, New York, NY, USA, 4–8 October 1991.

76. Slump, C.H.; van Asma, C.G.M.; Barels, J.K.P.; Brunink, W.J.A.; Drenth, F.B.; Pol, J.V.; Schouten, D.; Samsom, M.M.; Herrmann, O.E. Design and implementation of a linear-phase equalizer in digital audio signal processing. In Proceedings of the Workshop on VLSI Signal Processing, Napa Valley, CA, USA, 28–30 October 1992; pp. 297–306.

77. Väänänen, R.; Hiipakka, J. Efficient audio equalization using multirate processing. *J. Audio Eng. Soc.* **2008**, *56*, 255–266.

78. Mäkivirta, A.V. Loudspeaker design and performance evaluation. In *Handbook of Signal Processing in Acoustics*; Springer: New York, NY, USA, 2008; pp. 649–667.

79. McGrath, D.S. A new approach to digital audio equalization. In Proceedings of the 97th Convention of the Audio Engineering Society, San Francisco, CA, USA, 10–13 November 1994.

80. Kraght, P.H. A linear-phase digital equalizer with cubic-spline frequency response. *J. Audio Eng. Soc.* **1992**, *40*, 403–414.

81. Oliver, R.J. Frequency-Warped Audio Equalizer. U.S. Patent #7,764,802 B2, 27 July 2010.

82. Stockham, T.G., Jr. High-speed convolution and correlation. In Proceedings of the Spring Joint Computer Conference, New York, NY, USA, 26–28 April 1966; pp. 229–233.

83. Schöpp, H.; Hetze, H. A linear-phase 512-band graphic equalizer using the fast-Fourier transform. In Proceedings of the 96th Convention of the Audio Engineering Society, Amsterdam, The Netherlands, 26 February–1 March 1994.

84. Fernandes, G.F.P.; Martins, L.G.P.M.; Sousa, M.F.M.; Pinto, F.S.; Ferreira, A.J.S. Implementation of a new method to digital audio equalization. In Proceedings of the 106th Convention of the Audio Engineering Society, Munich, Germany, 8–11 May 1999.

85. Ries, S.; Frieling, G. PC-based equalizer with variable gain and delay in 31 frequency bands. In Proceedings of the 108th Convention of the Audio Engineering Society, Paris, France, 19–22 February 2000.

86. Kulp, B.D. Digital equalization using Fourier transform techniques. In Proceedings of the 85th Convention of the Audio Engineering Society, Los Angeles, CA, USA, 3–6 November 1988.

87. Gardner, W.G. Efficient convolution without input-output delay. *J. Audio Eng. Soc.* **1995**, *43*, 127–136.

88. Garcia, G. Optimal filter partition for efficient convolution with short input/output delay. In Proceedings of the 113th Convention of the Audio Engineering Society, Los Angeles, CA, USA, 5–8 October 2002.

89. Välimäki, V.; Parker, J.D.; Savioja, L.; Smith, J.O.; Abel, J.S. Fifty years of artificial reverberation. *IEEE Trans. Audio Speech Lang. Process.* **2012**, *20*, 1421–1448.

90. Wefers, F. Partitioned Convolution Algorithms for Real-Time Auralization. Ph.D. Thesis, RWTH Aachen University, Institute of Technical Acoustics, Aachen, Germany, 2014.

91. Primavera, A.; Cecchi, S.; Romoli, L.; Peretti, P.; Piazza, F. A low latency implementation of a non-uniform partitioned convolution algorithm for room acoustic simulation. *Signal, Image Video Process.* **2014**, *8*, 985–994.

92. McGrath, D.S. An efficient 30-band graphic equalizer implementation for a low-cost DSP processor. In Proceedings of the 95th Convention of the Audio Engineering Society, New York, NY, USA, 7–10 October 1993.

93. Vieira, J.M.N. Digital five-band equalizer with linear phase. In Proceedings of the 100th Convention of the Audio Engineering Society, Copenhagen, Denmark, 11–14 May 1996.

94. Cecchi, S.; Palestini, L.; Moretti, E.; Piazza, F. A new approach to digital audio equalization. In Proceedings of the IEEE Workshop on Applications of Signal Processing to Audio and Acoustics (WASPAA), New Paltz, NY, USA, 21–24 October 2007; pp. 62–65.

95. Soewito, A.W. Least Square Digital Filter Design in the Frequency Domain. Ph.D. Thesis, Rice University, Houston, TX, USA, 1991.

96. Smith, J.O. Recursive digital filter design (Appendix I). In *Introduction to Digital Filters with Audio Applications, Second Printing*; W3K Publishing: Palo Alto, CA, USA, 2008.

97. Levi, E.C. Complex-curve fitting. *IRE Trans. Automatic Control* **1959**, *4*, 37–44.

98. Lee, R. Simple arbitrary IIRs. In Proceedings of the 125th Convention of the Audio Engineering Society, San Francisco, CA, USA, 2–5 October 2008.

99. Dennis, J.E.; Schnabel, R.B. *Numerical Methods for Unconstrained Optimization and Nonlinear Equations*; Prentice-Hall: Englewood Cliffs, NJ, USA, 1983.

100. Friedlander, B.; Porat, B. The modified Yule-Walker method of ARMA spectral estimation. *IEEE Trans. Aerospace Electron. Syst.* **1984**, *20*, 158–173.

101. Jackson, L.B. Frequency-domain Steiglitz-McBride method for least-squares IIR filter design, ARMA modeling, and periodogram smoothing. *IEEE Signal Process. Lett.* **2008**, *15*, 49–52.

102. Vargas, R.; Burrus, C. The direct design of recursive or IIR digital filters. In Proceedings of the Third International Symposium on Communications, Control and Signal Processing, St. Julian's, Malta, 12–14 March 2008; pp. 188–192.

103. Oppenheim, A.V.; Johnson, D.H. Discrete representation of signals. *Proc. IEEE* **1972**, *60*, 681–691.

104. Härmä, A.; Karjalainen, M.; Savioja, L.; Välimäki, V.; Laine, U.K.; Huopaniemi, J. Frequency-warped signal processing for audio applications. *J. Audio Eng. Soc.* **2000**, *48*, 1011–1031.

105. Strube, H.W. Linear prediction on a warped frequency scale. *J. Acoust. Soc. Am.* **1980**, *68*, 1071–1076.

106. Asavathiratham, C.; Beckmann, P.E.; Oppenheim, A.V. Frequency warping in the design and implementation of fixed-point audio equalizers. In Proceedings of the IEEE Workshop on Applications of Signal Processing to Audio and Acoustics (WASPAA), New Paltz, NY, USA, 17–20 October 1999; pp. 55–58.

107. Kautz, W. Transient synthesis in the time domain. *Trans. IRE Prof. Group Circuit Theory* **1954**, *1*, 29–39.

108. Karjalainen, M.; Paatero, T. Equalization of loudspeaker and room responses using Kautz filters: Direct least squares design. *EURASIP J. Adv. Signal Process.* **2007**.

109. Backman, J. Digital realisation of phono (RIAA) equalisers. *IEEE Trans. Consum. Electron.* **1991**, *37*, 659–662.

110. Xia, J. A digital implementation of tape equalizers. *IEEE Trans. Consum. Electron.* **1994**, *40*, 114–118.

111. Massberg, M. Digital low-pass filter design with analog-matched magnitude response. In Proceedings of the 131st Convention of the Audio Engineering Society, New York, NY, USA, 20–23 October 2011.

112. Oppenheim, A.V.; Schafer, R.W. *Discrete-Time Signal Processing*, 3rd ed.; Pearson: Upper Saddle River, NJ, USA, 2010.

113. Al-Alaoui, M.A. Novel approach to analog-to-digital transforms. *IEEE Trans. Circ. Syst. I: Regular Papers* **2007**, *54*, 338–350.

114. Al-Alaoui, M.A. Improving the magnitude responses of digital filters for loudspeaker equalization. *J. Audio Eng. Soc.* **2010**, *58*, 1064–1082.

115. Powell, S.R.; Chau, P.M. A technique for realizing linear phase IIR filters. *IEEE Trans. Signal Process.* **1991**, *39*, 2425–2435.

116. Willson, A.N.; Orchard, H.J. An improvement to the Powell and Chau linear phase IIR filters. *IEEE Trans. Signal Process.* **1994**, *42*, 2842–2848.

117. Kurosu, A.; Miyase, S.; Tomiyama, S.; Takebe, T. A technique to truncate IIR filter impulse response and its application to real-time implementation of linear-phase IIR filters. *IEEE Trans. Signal Process.* **2003**, *51*, 1284–1292.

118. Azizi, S.A. Efficient arbitrary sample rate conversion using zero phase IIR filters. In Proceedings of the 116th Convention of the Audio Engineering Society, Berlin, Germany, 8–11 May 2004.

119. Mouffak, A.; Belbachir, M.F. Noncausal forward/backward two-pass IIR digital filters in real time. *Turkish J. Electr. Eng. Comp. Sci.* **2012**, *20*, 769–789.

120. Laakso, T.I.; Välimäki, V. Energy-based effective length of the impulse response of a recursive filter. *IEEE Trans. Instr. Meas.* **1999**, *48*, 7–17.

121. Giannoulis, D.; Massberg, M.; Reiss, J.D. Digital dynamic range compressor design—A tutorial and analysis. *J. Audio Eng. Soc.* **2012**, *60*, 399–408.

122. Pestana, P.D.; Reiss, J.D. Intelligent audio production strategies informed by best practices. In Proceedings of the Audio Engineering Society 53rd International Conference on Semantic Audio, London, UK, 27–29 January 2014.

123. Ma, Z.; De Man, B.; Pestana, P.D.L.; Black, D.A.A.; Reiss, J.D. Intelligent multitrack dynamic range compression. *J. Audio Eng. Soc.* **2015**, *63*, 412–426.

124. Godsill, S.; Rayner, P.; Cappé, O. Digital audio restoration. In *Applications of Digital Signal Processing to Audio and Acoustics*; Kahrs, M., Brandenburg, K., Eds.; Kluwer: New York, NY, USA, 2002; pp. 133–194.

125. Lindemann, E. The continuous frequency dynamic range compressor. In Proceedings of the IEEE Workshop on Applications of Signal Processing to Audio and Acoustics (WASPAA), New Paltz, NY, USA, 21–24 October 1997.

126. Zölzer, U. (Ed.) *DAFX: Digital Audio Effects*, 2nd ed.; Wiley: Chichester, UK, 2011.

127. Wise, D.K. Concept, design, and implementation of a general dynamic parametric equalizer. *J. Audio Eng. Soc.* **2009**, *57*, 16–28.

128. Newcomb, A.L., Jr.; Young, R.N. Practical loudness: An active circuit design approach. *J. Audio Eng. Soc.* **1976**, *24*, 32–35.

129. Holman, T.; Kampmann, F. Loudness compensation: Use and abuse. *J. Audio Eng. Soc.* **1978**, *26*, 526–536.

130. Grimes, J.; Doran, S. Equal loudness contour circuit using an operational amplifier. *IEEE Trans. Audio Electroacoust.* **1970**, *18*, 313–315.

131. Seefeldt, A. Loudness domain signal processing. In Proceedings of the 123rd Convention of the Audio Engineering Society, New York, NY, USA, 5–8 October 2007.

132. Hawker, O.; Wang, Y. A method of equal loudness compensation for uncalibrated listening systems. In Proceedings of the 139th Convention of the Audio Engineering Society, New York, NY, USA, 29 September–1 October 2015.

133. Yeh, D.T.; Smith, J.O. Discretization of the '59 Fender Bassman tone stack. In Proceedings of the International Conference on Digital Audio Effects (DAFX), Montreal, QC, Canada, 18–20 September 2006; pp. 1–5.

134. Gabrielli, L.; Välimäki, V.; Penttinen, H.; Squartini, S.; Bilbao, S. A digital waveguide based approach for Clavinet modeling and synthesis. *EURASIP J. Appl. Signal Process.* **2013**, *2013*, 1–14.

135. Clarkson, P.M.; Mourjopoulos, J.; Hammond, J.K. Spectral, phase, and transient equalization for audio systems. *J. Audio Eng. Soc.* **1985**, *33*, 127–132.

136. Wilson, R.; Adams, G.; Scott, J. Application of digital filters to loudspeaker crossover networks. *J. Audio Eng. Soc.* **1989**, *37*, 455–464.

137. Greenfield, R.; Hawksford, M.J. Efficient filter design for loudspeaker equalization. *J. Audio Eng. Soc.* **1991**, *39*, 739–751.

138. Karjalainen, M.; Piirilä, E.; Järvinen, A.; Huopaniemi, J. Comparison of loudspeaker equalization methods based on DSP techniques. *J. Audio Eng. Soc.* **1999**, *47*, 15–31.

139. MacDonald, J.A.; Tran, P.K. Loudspeaker equalization for auditory research. *Behav. Res. Methods* **2007**, *39*, 133–136.

140. Berndtsson, G. Acoustical properties of wooden loudspeakers used in an artificial reverberation system. *Appl. Acoust.* **1995**, *44*, 7–23.

141. Pueo, B.; López, J.J.; Ramos, G.; Escolano, J. Efficient equalization of multi-exciter distributed mode loudspeakers. *Appl. Acoust.* **2009**, *70*, 737–746.

142. Höchens, L.; de Vries, D. Comparison of measurement methods for the equalization of loudspeaker panels based on bending wave radiation. In Proceedings of the 130th Convention of the Audio Engineering Society, London, UK, 13–16 May 2011.

143. Lähdeoja, O.; Haapaniemi, A.; Välimäki, V. Sonic scenography—Equalized structure-borne sound for aurally active set design. In Proceedings of the International Computer Music Conference Joint with the Sound Music Computing Conference, Athens, Greece, 14–20 September 2014.

144. Ho, J.H.; Berkhoff, A.P. Flat acoustic sources with frequency response correction based on feedback and feed-forward distributed control. *J. Acoust. Soc. Am.* **2015**, *137*, 2080–2088.

145. Kitson, A.B. Equalisation of sound systems by ear and instruments: Similarities and differences. In Proceedings of the 5th Australian Regional Convention of the Audio Engineering Society, Sydney, Australia, 26–28 April 1995.

146. Terrell, M.; Sandler, M. Optimizing the controls of homogeneous loudspeaker arrays. In Proceedings of the 129th Convention of the Audio Engineering Society, San Francisco, CA, USA, 4–7 November 2010.

147. Vidal Wagner, F.; Välimäki, V. Automatic calibration and equalization of a line array system. In Proceedings of the International Conference on Digital Audio Effects (DAFX), Trondheim, Norway, 30 November–3 December 2015; pp. 123–130.

148. Cecchi, S.; Virgulti, M.; Primavera, A.; Piazza, F.; Bettarelli, F.; Li, J. Investigation on audio algorithms architecture for stereo portable devices. *J. Audio Eng. Soc.* **2016**, *64*, 175–188.

149. Wilson, R. Equalization of loudspeaker drive units considering both on- and off-axis responses. *J. Audio Eng. Soc.* **1991**, *39*, 127–139.

150. Hatziantoniou, P.D.; Mourjopoulos, J.N. Generalized fractional-octave smoothing of audio and acoustic responses. *J. Audio Eng. Soc.* **2000**, *48*, 259–280.

151. Toole, F.E. Loudspeaker measurements and their relationship to listener preferences: Part 1. *J. Audio Eng. Soc.* **1986**, *34*, 227–235.

152. Linkwitz, S.H. Active crossover networks for noncoincident drivers. *J. Audio Eng. Soc.* **1976**, *24*, 2–8.

153. Ramos, G.; López, J.J. Filter design method for loudspeaker equalization based on IIR parametric filters. *J. Audio Eng. Soc.* **2006**, *54*, 1162–1178.

154. Ramos, G.; Tomas, P. Improvements on automatic parametric equalization and cross-over alignment of audio systems. In Proceedings of the 126th Convention of the Audio Engineering Society, Munich, Germany, 7–10 May 2009.

155. Behrends, H.; von dem Knesebeck, A.; Bradinal, W.; Neumann, P.; Zölzer, U. Automatic equalization using parametric IIR filters. *J. Audio Eng. Soc.* **2011**, *59*, 102–109.

156. Karjalainen, M.; Piirilä, E.; Järvinen, A. Loudspeaker response equalization using warped digital filters. In Proceedings of the Nordic Signal Processing Conference (NORSIG), Espoo, Finland, 24–27 September 1996; pp. 367–370.

157. Wang, P.; Ser, W.; Zhang, M. A dual-band equalizer for loudspeakers. *J. Audio Eng. Soc.* **2000**, *48*, 917–921.

158. Wang, P.; Ser, W.; Zhang, M. Bark scale equalizer design using warped filter. In Proceedings of the IEEE International Conference Acoustics, Speech and Signal Processing (ICASSP), Salt Lake City, UT, USA, 7–11 May 2001; Volume 5, pp. 3317–3320.

159. Tyril, M.; Pedersen, J.A.; Rubak, P. Digital filters for low-frequency equalization. *J. Audio Eng. Soc.* **2001**, *49*, 36–43.

160. Ramos, G.; López, J.J.; Pueo, B. Cascaded warped-FIR and FIR filter structure for loudspeaker equalization with low computational cost requirements. *Digital Signal Process.* **2009**, *19*, 393–409.

161. Li, X.; Cai, Z.; Zheng, C.; Li, X. Equalization of loudspeaker response using balanced model truncation. *J. Acoust. Soc. Am.* **2015**, *137*, EL241–EL247.

162. Herzog, S.; Hilsamer, M. Low frequency group delay equalization of vented boxes using digital correction filters. In Proceedings of the International Conference on Digital Audio Effects (DAFX), Erlangen, Germany, 1–5 September 2014.

163. Abel, J.; Smith, J.O. Robust design of very high-order allpass dispersion filters. In Proceedings of the International Conference on Digital Audio Effects (DAFX), Montreal, QC, Canada, 18–20 September 2006.

164. Elliott, S.J.; Nelson, P.A. Multiple-point equalization in a room using adaptive digital filters. *J. Audio Eng. Soc.* **1989**, *37*, 899–907.

165. Bellini, A.; Cibelli, G.; Armelloni, E.; Ugolotti, E.; Farina, A. Car cockpit equalization by warping filters. *IEEE Trans. Consum. Electron.* **2001**, *47*, 108–116.

166. Kim, L.H.; Lim, J.S.; Choi, C.; Sung, K.M. Equalization of low frequency response in automobile. *IEEE Trans. Consum. Electron.* **2003**, *49*, 243–252.

167. Cecchi, S.; Palestini, L.; Peretti, P.; Piazza, F.; Bettarelli, F.; Toppi, R. Automotive audio equalization. In Proceedings of the Audio Engineering Society 36th International Conference on Automotive Audio, Dearborn, MI, USA, 2–4 June 2009.

168. Bahne, A.; Ahlén, A. Optimizing the similarity of loudspeaker-room responses in multiple listening positions. *IEEE/ACM Trans. Audio Speech Lang. Process.* **2016**, *24*, 340–353.

169. Craven, P.G.; Gerzon, M.A. Practical adaptive room and loudspeaker equalizer for hi-fi use. In Proceedings of the 92nd Convention of the Audio Engineering Society, Vienna, Austria, 24–27 March 1992.

170. Bank, B. Combined quasi-anechoic and in-room equalization of loudspeaker responses. In Proceedings of the 134th Convention of the Audio Engineering Society, Rome, Italy, 4–7 May 2013.

171. Bank, B.; Ramos, G. Improved pole positioning for parallel filters based on spectral smoothing and multiband warping. *IEEE Signal Process. Lett.* **2011**, *18*, 299–302.

172. Mourjopoulos, J. On the variation and invertibility of room impulse response functions. *J. Sound Vibr.* **1985**, *102*, 217–228.

173. Mourjopoulos, J.N. Digital equalization of room acoustics. *J. Audio Eng. Soc.* **1994**, *42*, 884–900.

174. Haneda, Y.; Makino, S.; Kaneda, Y. Multiple-point equalization of room transfer functions by using common acoustical poles. *IEEE Trans. Speech Audio Process.* **1997**, *5*, 325–333.

175. Fontana, F.; Gibin, L.; Rocchesso, D.; Ballan, O. Common pole equalization of small rooms using a two-step real-time digital equalizer. In Proceedings of the IEEE Workshop on Applications of Signal Processing to Audio and Acoustics (WASPAA), New Paltz, NY, USA, 17–20 October 1999; pp. 195–198.

176. Bharitkar, S.; Hilmes, P.; Kyriakakis, C. Robustness of spatial average equalization: A statistical reverberation model approach. *J. Acoust. Soc. Am.* **2004**, *116*, 3491–3497.

177. Stefanakis, N.; Sarris, J.; Jacobsen, F. Regularization in global sound equalization based on effort variation. *J. Acoust. Soc. Am.* **2009**, *126*, 666–675.

178. Karjalainen, M.; Paatero, T.; Mourjopoulos, J.N.; Hatziantoniou, P.D. About room response equalization and dereverberation. In Proceedings of the IEEE Workshop on Applications of Signal Processing to Audio and Acoustics (WASPAA), New Paltz, NY, USA, 16–19 October 2005; pp. 183–186.
179. Mäkivirta, A.; Antsalo, P.; Karjalainen, M.; Välimäki, V. Modal equalization of loudspeaker-room responses at low frequencies. *J. Audio Eng. Soc.* **2003**, *51*, 324–343.
180. Karjalainen, M.; Esquef, P.A.A.; Antsalo, P.; Mäkivirta, A.; Välimäki, V. Frequency-zooming ARMA modeling of resonant and reverberant systems. *J. Audio Eng. Soc.* **2002**, *50*, 1012–1029.
181. Fazenda, B.M.; Stephenson, M.; Goldberg, A. Perceptual thresholds for the effects of room modes as a function of modal decay. *J. Acoust. Soc. Am.* **2015**, *137*, 1088–1098.
182. Mertins, A.; Mei, T.; Kallinger, M. Room impulse response shortening/reshaping with infinity- and p-norm optimization. *IEEE Trans. Audio Speech Lang. Process.* **2010**, *18*, 249–259.
183. Krishnan, L.; Teal, P.D.; Betlehem, T. A robust sparse approach to acoustic impulse response shaping. In Proceedings of the IEEE International Conference on Acoustics, Speech and Signal Processing (ICASSP), Brisbane, Australia, 19–24 April 2015; pp. 738–742.
184. Jungmann, J.O.; Mazur, R.; Kallinger, M.; Mei, T.; Mertins, A. Combined acoustic MIMO channel crosstalk cancellation and room impulse response reshaping. *IEEE Trans. Audio Speech Lang. Process.* **2012**, *20*, 1829–1842.
185. Kodrasi, I.; Goetze, S.; Doclo, S. Regularization for partial multichannel equalization for speech dereverberation. *IEEE Trans. Audio Speech Lang. Process.* **2013**, *21*, 1879–1890.
186. Lim, F.; Zhang, W.; Habets, E.A.P.; Naylor, P.A. Robust multichannel dereverberation using relaxed multichannel least squares. *IEEE/ACM Trans. Audio Speech Lang. Process.* **2014**, *22*, 1379–1390.
187. Liem, H.M.; Gan, W.S. Headphone equalization using DSP approaches. In Proceedings of the 107th Convention of the Audio Engineering Society, New York, NY, USA, 24–27 September 1999.
188. Rämö, J. Equalization Techniques for Headphone Listening. Ph.D. Thesis, Aalto University, Espoo, Finland, 2014.
189. Hiipakka, M.; Tikander, M.; Karjalainen, M. Modeling the external ear acoustics for insert headphone usage. *J. Audio Eng. Soc.* **2010**, *58*, 269–281.
190. Gómez Bolaños, J.; Pulkki, V. Estimation of pressure at the eardrum in magnitude and phase for headphone equalization using pressure-velocity measurements at the ear canal entrance. In Proceedings of the IEEE Workshop on Applications of Signal Processing to Audio and Acoustics (WASPAA), New Paltz, NY, USA, 18–21 October 2015.
191. Olive, S.; Welti, T.; McMullin, E. Listener preferences for in-room loudspeaker and headphone target responses. In Proceedings of the 135th Convention of the Audio Engineering Society, New York, NY, USA, 17–20 October 2013.
192. Rämö, J.; Välimäki, V. Signal processing framework for virtual headphone listening tests in a noisy environment. In Proceedings of the 132nd Convention of the Audio Engineering Society, Budapest, Hungary, 26–29 April 2012.
193. Gómez Bolaños, J.; Mäkivirta, A.; Pulkki, V. Automatic regularization parameter for headphone transfer function inversion. *J. Audio Eng. Soc.* **2016**, unpublished work.
194. Briolle, F.; Voinier, T. Transfer function and subjective quality of headphones: Part 2, subjective quality evaluations. In Proceedings of the Audio Engineering Society 11th International Conference on Test and Measurement, Portland, OR, USA, 29–31 May 1992.
195. Olive, S.E.; Welti, T.; McMullin, E. A virtual headphone listening test methodology. In Proceedings of the Audio Engineering Society 51st International Conference on Loudspeakers and Headphones, Helsinki, Finland, 21–24 August 2013.
196. Blauert, J. *Spatial Hearing: The Psychophysics of Human Sound Localization*; MIT Press: Cambridge, MA, USA, 1997.
197. Sunder, K.; He, J.; Tan, E.L.; Gan, W.S. Natural sound rendering for headphones: Integration of signal processing techniques. *IEEE Signal Process. Mag.* **2015**, *32*, 100–113.
198. Mackenzie, J.; Huopaniemi, J.; Välimäki, V.; Kale, I. Low-order modeling of head-related transfer functions using balanced model truncation. *IEEE Signal Process. Lett.* **1997**, *4*, 39–41.
199. Ramos, G.; Cobos, M. Parametric head-related transfer function modeling and interpolation for cost-efficient binaural sound applications. *J. Acoust. Soc. Am.* **2013**, *134*, 1735–1738.

200. Huopaniemi, J.; Zacharov, N.; Karjalainen, M. Objective and subjective evaluation of head-related transfer function filter design. *J. Audio Eng. Soc.* **1999**, *47*, 218–239.
201. Kulkarni, A.; Colburn, H.S. Role of spectral detail in sound-source localization. *Nature* **1998**, *396*, 747–749.
202. Hiekkanen, T.; Mäkivirta, A.; Karjalainen, M. Virtualized listening tests for loudspeakers. *J. Audio Eng. Soc.* **2009**, *57*, 237–251.
203. Härmä, A.; Jakka, J.; Tikander, M.; Karjalainen, M.; Lokki, T.; Hiipakka, J.; Lorho, G. Augmented reality audio for mobile and wearable appliances. *J. Audio Eng. Soc.* **2004**, *52*, 618–639.
204. Lindeman, R.W.; Noma, H.; De Barros, P.G. Hear-through and mic-through augmented reality: Using bone conduction to display spatialized audio. In Proceedings of the IEEE and ACM International Symposium on Mixed and Augmented Reality, Nara, Japan, 13–16 November 2007; pp. 173–176.
205. Tikander, M.; Karjalainen, M.; Riikonen, V. An augmented reality audio headset. In Proceedings of the International Conference on Digital Audio Effects (DAFX), Espoo, Finland, 1–4 September 2008.
206. Välimäki, V.; Franck, A.; Rämö, J.; Gamper, H.; Savioja, L. Assisted listening using a headset: Enhancing audio perception in real, augmented, and virtual environments. *IEEE Signal Process. Mag.* **2015**, *32*, 92–99.
207. Hoffmann, P.F.; Christensen, F.; Hammershøi, D. Insert earphone calibration for hear-through options. In Proceedings of the Audio Engineering Society 51st International Conference on Loudspeakers and Headphones, Helsinki, Finland, 21–24 August 2013.
208. Riikonen, V.; Tikander, M.; Karjalainen, M. An augmented reality audio mixer and equalizer. In Proceedings of the 124th Convention of the Audio Engineering Society, Amsterdam, The Netherlands, 17–20 May 2008.
209. Rämö, J.; Välimäki, V. Digital augmented reality audio headset. *J. Electr. Comput. Eng.* **2012**.
210. Rämö, J.; Välimäki, V.; Tikander, M. Live sound equalization and attenuation with a headset. In Proceedings of the Audio Engineering Society 51st International Conference on Loudspeakers and Headphones, Helsinki, Finland, 21–24 August 2013.
211. Goldin, A.A.; Budkin, A.; Kib, S. Automatic volume and equalization control in mobile devices. In Proceedings of the 121st Convention of the Audio Engineering Society, San Francisco, CA, USA, 5–8 October 2006.
212. Kitzen, W.J.W.; Kemna, J.W.; Druyvesteyn, W.F.; Knibbeler, C.L.C.M.; van de Voort, A.T.A.M. Noise-dependent sound reproduction in a car: Application of a digital audio signal processor. *J. Audio Eng. Soc.* **1988**, *36*, 18–26.
213. Al-Jarrah, O.; Shaout, A. Automotive volume control using fuzzy logic. *J. Intell. Fuzzy Syst.* **2007**, *18*, 329–343.
214. Christoph, M. Noise dependent equalization control. In Proceedings of the Audio Engineering Society 48th International Conference on Automotive Audio, Munich, Germany, 21–23 September 2012.
215. Sack, M.C.; Buchinger, S.; Robitza, W.; Hummelbrunner, P.; Nezveda, M.; Hlavacs, H. Loudness and auditory masking compensation for mobile TV. In Proceedings of the IEEE International Symposium on Broadband Multimedia Systems and Broadcasting (BMSB), Shanghai, China, 24–26 March 2010; pp. 1–6.
216. Rämö, J.; Välimäki, V.; Alanko, M.; Tikander, M. Perceptual frequency response simulator for music in noisy environments. In Proceedings of the Audio Engineering Society 45th Conference on Applications of Time-Frequency Processing in Audio, Helsinki, Finland, 1–4 March 2012; pp. 269–278.
217. Miller, T.E.; Barish, J. Optimizing sound for listening in the presence of road noise. In Proceedings of the 95th Convention of the Audio Engineering Society, New York, NY, USA, 7–10 October 1993.
218. Sondhi, M.M.; Berkley, D.A. Silencing echoes on the telephone network. *Proc. IEEE* **1980**, *68*, 948–963.
219. Tzur (Zibulski), M.; Goldin, A. Sound equalization in a noisy environment. In Proceedings of the 110th Convention of the Audio Engineering Society, Amsterdam, The Netherlands, 12–15 May 2001.
220. Rämö, J.; Välimäki, V.; Tikander, M. Perceptual headphone equalization for mitigation of ambient noise. In Proceedings of the IEEE International Conference on Acoustics, Speech and Signal Processing (ICASSP), Vancouver, BC, Canada, 26–31 May 2013; pp. 724–728.
221. Hodgson, J. A field guide to equalisation and dynamics processing on rock and electronica records. *Pop. Music* **2010**, *29*, 283–297.
222. Neuman, A.C.; Levitt, H.; Mills, R.; Schwander, T. An evaluation of three adaptive hearing aid selection strategies. *J. Acoust. Soc. Am.* **1987**, *82*, 1967–1976.
223. Durant, E.A.; Wakefield, G.H.; Van Tasell, D.J.; Rickert, M.E. Efficient perceptual tuning of hearing aids with genetic algorithms. *IEEE Trans. Speech Audio Process.* **2004**, *12*, 144–155.

224. Wakefield, G.H.; van den Honert, C.; Parkinson, W.; Lineaweaver, S. Genetic algorithms for adaptive psychophysical procedures: Recipient-directed design of speech-processor MAPs. *Ear Hear.* **2005**, *26*, 57S–72S.
225. Kuk, F.K.; Pape, N.M. The reliability of a modified simplex procedure in hearing aid frequency response selection. *J. Speech Hear. Res.* **1992**, *35*, 418–429.
226. Stelmachowicz, P.G.; Lewis, D.E.; Carney, E. Preferred hearing-aid frequency responses in simulated listening environments. *J. Speech Hear. Res.* **1994**, *37*, 712–719.
227. Dewey, C.; Wakefield, J. Novel designs for the parametric peaking EQ user interface. In Proceedings of the 134th Convention of the Audio Engineering Society, Rome, Italy, 4–7 May 2013.
228. Loviscach, J. Graphical control of a parametric equalizer. In Proceedings of the 124th Convention of the Audio Engineering Society, Amsterdam, the Netherlands, 17–20 May 2008.
229. Heise, S.; Hlatky, M.; Loviscach, J. A computer-aided audio effect setup procedure for untrained users. In Proceedings of the 128th Convention of the Audio Engineering Society, London, UK, 22–25 May 2010.
230. Reed, D. A perceptual assistant to do sound equalization. In Proceedings of the 5th International Conference on Intelligent User Interfaces, New Orleans, LA, USA, 9–12 January 2000; pp. 212–218.
231. Mecklenburg, S.; Loviscach, J. SubjEQt: Controlling an equalizer through subjective terms. In Proceedings of the Human Factors in Computing Systems (CHI-06), Montreal, QC, Canada, 22–27 April 2006.
232. Sabin, A.; Pardo, B. Rapid learning of subjective preference in equalization. In Proceedings of the 125th Convention of the Audio Engineering Society, San Francisco, CA, USA, 2–5 October 2008.
233. Sabin, A.; Pardo, B. 2DEQ: An intuitive audio equalizer. In Proceedings of the ACM Conference on Creativity and Cognition, Berkeley, CA, USA, 27–30 October 2009; pp. 435–436.
234. Sabin, A.; Pardo, B. A method for rapid personalization of audio equalization parameters. In Proceedings of the ACM Multimedia Conference, Beijing, China, 19–24 October 2009.
235. Sabin, A.T.; Rafii, Z.; Pardo, B. Weighted-function-based rapid mapping of descriptors to audio processing parameters. *J. Audio Eng. Soc.* **2011**, *59*, 419–430.
236. Pardo, B.; Little, D.; Gergle, D. Towards speeding audio EQ interface building with transfer learning. In Proceedings of the International Conference on New Interfaces for Musical Expression (NIME), Ann Arbor, MI, USA, 21–23 May 2012.
237. Cartwright, M.; Pardo, B. Social-EQ: Crowdsourcing an equalization descriptor map. In Proceedings of the International Conference Music Information Retrieval (ISMIR), Curitiba, Brazil, 4–8 November 2013.
238. Cartwright, M.; Pardo, B.; Reiss, J.D. Mixploration: Rethinking the audio mixer interface. In Proceedings of the International Conference on Intelligent User Interfaces, Haifa, Israel, 24–27 February 2014.
239. Stasis, S.; Stables, R.; Hockman, J. A model for adaptive reduced-dimensionality equalisation. In Proceedings of the International Conference on Digital Audio Effects (DAFX), Trondheim, Norway, 30 November–3 December 2015; pp. 315–320.
240. Stasis, S.; Stables, R.; Hockman, J. Semantically controlled adaptive equalization in reduced dimensionality parameter space. *Appl. Sci.* **2016**, *6*, 116.
241. Ma, Z.; Reiss, J.D.; Black, D. Implementation of an intelligent equalization tool using Yule-Walker for music mixing and mastering. In Proceedings of the 134th Convention of the Audio Engineering Society, Rome, Italy, 4–7 May 2013.
242. Pestana, P.D.; Ma, Z.; Reiss, J.D.; Barbosa, A.; Black, D. Spectral characteristics of popular commercial recordings 1950-2010. In Proceedings of the 135th Convention of the Audio Engineering Society, New York, NY, USA, 17–20 October 2013.
243. Perez Gonzalez, E.; Reiss, J.D. Automatic equalization of multi-channel audio using cross-adaptive methods. In Proceedings of the 127th Convention of the Audio Engineering Society, New York, NY, USA, 9–12 October 2009.
244. Hafezi, S.; Reiss, J.D. Autonomous multitrack equalization based on masking reduction. *J. Audio Eng. Soc.* **2015**, *63*, 312–323.
245. Barchiesi, D.; Reiss, J.D. Reverse engineering the mix. *J. Audio. Eng. Soc.* **2010**, *58*, 563–576.

applied
sciences

MDPI

Review

A Review of Physical and Perceptual Feature Extraction Techniques for Speech, Music and Environmental Sounds

Francesc Alías *, Joan Claudi Socoró and Xavier Sevillano

GTM - Grup de recerca en Tecnologies Mèdia, La Salle-Universitat Ramon Llull, Quatre Camins, 30, 08022 Barcelona, Spain; jclaudi@salleurl.edu (J.C.S.); xavis@salleurl.edu (X.S.)
*   Correspondence: falias@salleurl.edu; Tel.: +34-93-290-24-40

Academic Editor: Vesa Välimäki
Received: 15 March 2016; Accepted: 28 April 2016; Published: 12 May 2016

Abstract: Endowing machines with sensing capabilities similar to those of humans is a prevalent quest in engineering and computer science. In the pursuit of making computers sense their surroundings, a huge effort has been conducted to allow machines and computers to acquire, process, analyze and understand their environment in a human-like way. Focusing on the sense of hearing, the ability of computers to sense their acoustic environment as humans do goes by the name of machine hearing. To achieve this ambitious aim, the representation of the audio signal is of paramount importance. In this paper, we present an up-to-date review of the most relevant audio feature extraction techniques developed to analyze the most usual audio signals: speech, music and environmental sounds. Besides revisiting classic approaches for completeness, we include the latest advances in the field based on new domains of analysis together with novel bio-inspired proposals. These approaches are described following a taxonomy that organizes them according to their physical or perceptual basis, being subsequently divided depending on the domain of computation (time, frequency, wavelet, image-based, cepstral, or other domains). The description of the approaches is accompanied with recent examples of their application to machine hearing related problems.

Keywords: audio feature extraction; machine hearing; audio analysis; music; speech; environmental sound

PACS: 43.60.Lq; 43.60.-c; 43.50.Rq; 43.64.-q

---

1. Introduction

Endowing machines with sensing capabilities similar to those of humans (such as vision, hearing, touch, smell and taste) is a long pursued goal in several engineering and computer science disciplines. Ideally, we would like machines and computers to be aware of their immediate surroundings as human beings are. This way, they would be able to produce the most appropriate response for a given operational environment, taking one step forward towards full and natural human–machine interaction (e.g., making fully autonomous robots aware of their environment), improve the accessibility of people with special needs (e.g., through the design of hearing aids with environment recognition capabilities), or even as a means for substituting humans beings in different tasks (e.g., autonomous driving, in potentially hazardous situations, *etc.*).

One of the main avenues of human perception is hearing. Therefore, in the quest for making computers sense their environment in a human-like way, sensing the acoustic environment in broad sense is a key task. However, the acoustic surroundings of a particular point in space can be extremely complex to decode for machines, be it due to the presence of simultaneous sound sources of highly diverse nature (from a natural or artificial origin), or due to many other causes such as the presence of high background noise, or the existence of a long distance to the sound source, to name a few.

This challenging problem goes by the name of *machine hearing*, as defined by Lyon [1]. Machine hearing is the ability of computers to hear as humans do, e.g., by distinguishing speech from music and background noises, pulling the two former out for special treatment due to their origin. Moreover, it includes the ability to analyze environmental sounds to discern the direction of arrival of sound events (e.g., a car pass-by), besides detecting which of them are usual or unusual in that specific context (e.g., a gun shot in the street), together with the recognition of acoustic objects such as actions, events, places, instruments or speakers. Therefore, an ideal hearing machine will face a wide variety of hearable sounds, and should be able to deal successfully with all of them. To further illustrate the complexity of the scope of the problem, Figure 1 presents a general sound classification scheme, which was firstly proposed by Gerhard [2] and more recently used in the works by Temko [3] and Dennis [4].

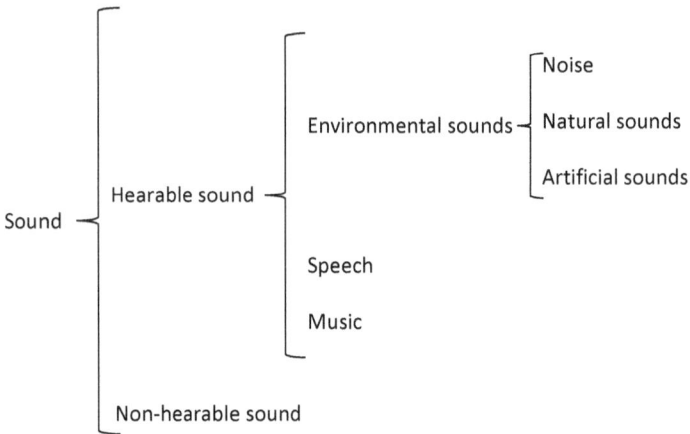

Figure 1. General sound classification scheme (adapted from [4]).

As the reader may have deduced, machine hearing is an extremely complex and daunting task given the wide diversity of possible audio inputs and application scenarios. For this reason, it is typically subdivided into smaller subproblems, and most research efforts are focused on solving simpler, more specific tasks. Such simplification can be achieved from different perspectives. One of these perspectives has to do with the nature of the audio signal of interest. Indeed, devising a generic machine hearing system capable of dealing successfully with different types of sounds regardless of their nature is a truly challenging endeavor. In contrast, it becomes easier to develop systems capable of accomplishing a specific task but limited to signals of a particular nature, as the system design can be adapted and optimized to take into account the signal characteristics.

For instance, we can focus on speech signals, that is, the sounds produced through the human vocal tract that entail some linguistic content. Speech has a set of very distinctive traits that make it different from other types of sounds, ranging from its characteristic spectral distribution to its phonetic structure. In this case, the literature contains plenty of works dealing with speech-sensing related topics such as speech detection (Bach *et al.* [5]), speaker recognition and identification (Kinnunen and Li [6]), and speech recognition (Pieraccini [7]), to name a few.

As in the case of speech, music also is a structured sound that has a set of specific and distinguishing traits (such as repeated stationary pattern structures as melody and rhythm) that make it rather unique, being generated by humans with some aesthetic intent. Following an analogous pathway to that of speech, music is another type of sound that has received attention from researchers in the development of machine hearing systems, including those targeting specific tasks such as artist and song identification (Wang [8]), genre classification (Wang *et al.* [9]), instrument recognition

(Benetos *et al.* [10], Liu and Wan [11]), mood classification (Lu *et al.* [12]) or music annotation and recommendation (Fu [13]).

Thus, speech and music, which up to now have been by far the most extensively studied types of sound sources in the context of machine hearing, present several particular rather unique characteristics. In contrast, other kind of sound sources coming from our environment (e.g., traffic noise, sounds from animals in the nature, *etc.*) do not exhibit such particularities, or at least not in such in a clear way. Nevertheless, these non-speech nor music related sounds (hereafter denoted as environmental sounds) should be also detectable and recognizable by hearing machines as individual events (Chu *et al.* [14]) or as acoustic scenes (Valero and Alías [15]) (the latter can also be found in the literature denoted as soundscapes, as in the work by Schafer [16]).

Regardless of its specific goal, any machine hearing system requires performing an in-depth analysis of the incoming audio signal, aiming at making the most of its particular characteristics. This analysis starts with the extraction of appropriate parameters of the audio signal that inform about its most significant traits, a process that usually goes by the name of *audio feature extraction*.

Logically, extracting the right features from an audio signal is a key issue to guarantee the success of machine hearing applications. Indeed, the extracted features should provide a compact yet descriptive vision of the parametrized signal, highlighting those signal characteristics that are most useful to accomplish the task at hand, be it detection, identification, classification, indexing, retrieval or recognition. And of course, depending on the nature of the signal (*i.e.*, speech, music or environmental sound) and the targeted application, it will be more interesting that these extracted features reflect the characteristics of the signal from a physical or perceptual point of view.

This paper presents an up-to-date state-of-the-art review of the main audio feature extraction techniques applied to machine hearing. We build on the complete review about features for audio retrieval by Mitrović *et al.* [17], and we have included the classic approaches in that work for the sake of completeness. In addition, we present the latest advances on audio feature extraction techniques together with new examples of their application to the analysis of speech, music and environmental sounds. It is worth noting that most of the recently developed audio feature techniques introduced in the last decade have entailed the definition of new approaches of analysis beyond the classic domains (*i.e.*, temporal, frequency-based and cepstral), such as the ones developed on the wavelet domain, besides introducing image-based and multilinear or non-linear representations, together with a significant increase of bio-inspired proposals.

This paper is organized as follows. Section 2 describes the main constituting blocks of any machine hearing system, focusing the attention on the audio feature extraction process. Moreover, given the importance of relating the nature of the signal with the type of extracted features, we detail the primary characteristics of the three most frequent types of signals involved in machine hearing applications: speech, music and environmental sounds. Next, Section 3 describes the followed taxonomy to describe both classic and recently defined audio feature extraction techniques. Next, the description of the rationale and main principles of approaches that are based on the physical characteristics of the audio signal are described in Section 4, while those that try to somehow include perception in the parameterization process are explained in Section 5. Finally, Section 6 discusses the conclusions of this review.

2. Machine Hearing

As mentioned earlier, the problem of endowing machines with the ability of sensing their acoustic environment is typically addressed by facing specific subproblems such as the detection, identification, classification, indexing, retrieval or recognition of particular types of sound events, scenes or compositions. Among them, speech, music and environmental sounds constitute the vast majority of acoustic stimuli we can ultimately find in a given context of a machine hearing system.

In this section, we first present a brief description of the primary goals and characteristics of the constituting blocks of the generic architecture of machine hearing systems. Then, the main characteristics of the audio sources those systems process, that is, speech, music and environmental sounds, are detailed.

2.1. Architecture of Machine Hearing Systems

Regardless of the specific kind of problem addressed, the structure of the underlying system can be described by means of a generic and common architecture design that is depicted in Figure 2.

Figure 2. Generic architecture of a typical machine hearing system.

In a first stage, the continuous audio stream captured by a microphone is segmented into shorter signal chunks by means of a windowing process. This is achieved by sliding a window function over the theoretically infinite stream of samples of the input signal, and ends up by converting it into a continuous sequence of finite blocks of samples. Thanks to windowing, the system will be capable of operating on sample chunks of finite length. Moreover, depending on the length of the window function, the typically non-stationary audio signal can be assumed to be *quasi-stationary* within each frame, thus facilitating subsequent signal analysis.

The choice of the type and length of the window function, as well as the overlap between consecutive signal frames, is intimately related to the machine hearing application at hand. It seems logical that, for instance, the length of the window function should be proportional to the minimum length of the acoustic events of interest. Therefore, window lengths between 10 and 50 milliseconds are typically employed to process speech or to detect transient noise events [13], while windows of several seconds are used in computational auditory scene analysis (CASA) applications (as in the works by Peltonen *et al.* [18], Chu *et al.* [14], Valero and Alías [15], or Geiger *et al.* [19]). Further discussion about the windowing process and its effect on the windowed signal lies beyond the scope of this work. The interested reader is referred to classic digital signal processing texts (e.g., see the book by Oppenheim and Schafer [20]).

Once the incoming audio stream has been segmented into finite length chunks, audio features are extracted from each one of them. The goal of feature extraction is to obtain a compact representation of the most salient acoustic characteristics of the signal, converting a N samples long frame into K scalar coefficients (with $K << N$), thus attaining a data compaction that allows increasing the efficiency of subsequent processes [13]. To that effect, these features may consider the physical or perceptual impact of signal contents computed in the time, frequency, *etc.* domains.

In this sense, modeling the time evolution of audio signals has been found to be of paramount importance when it comes to perform some types of machine hearing tasks, such as the recognition of environmental sounds (as described by Gygi [21]) or the identification of rhythmic patterns in music (Foote and Uchihashi [22]) for example. To keep this time information, the features extracted from several subsequent signal frames can be merged into a single feature vector. It should be noted that, due to this feature merging process, the feature vectors acquire a very high dimensionality that may represent a hurdle to the subsequent audio analysis process, with the so-called *curse of dimensionality* problem, as described by Bellman [23]. In order to compact the feature vectors, feature extraction techniques are sometimes followed by a data dimensionality reduction process. To this end, several approaches may be considered: from representing vectors in terms of some of their statistics (as done in the works by Rabaoui *et al.* [24] or by Hurst [25]) to more complex approaches like analyzing the principal components of the feature vector (Eronen *et al.* [26]), thus projecting the data onto a transformed space.

And finally, an audio analysis task must be conducted upon the feature vectors obtained in the previous step. Of course, *audio analysis* is a generic label that tries to encompass any audio processing necessary to tackle the specific machine hearing application at hand. For instance, in case that recognizing a specific type of sound was the goal of our hearing machine, this audio analysis block would consist of a supervised machine learning algorithm that should first build representative acoustic models upon multiple samples from each sound class that we want the system to recognize, to subsequently classify any incoming unknown sound signal into one of the predefined classes based on the information acquired during the algorithm's training phase.

Of course, each machine hearing application will require that the audio analysis block is designed according to the application-specific needs and requirements. Although, providing the reader with a comprehensive view of specific machine hearing problems exceeds the goals of this work, the interested reader will find diverse examples of the machine hearing applications throughout the paper. Examples include speaker identification (like in Yuo *et al.* [27]), music genre classification (Tzanetakis and Cook [28]), environmental sound recognition (e.g., the works by Ando [29] and Valero and Alías [30]), audio indexing and retrieval (Richard *et al.* [31]), or CASA (as in the works by Peltonen *et al.* [18], Chu *et al.* [14], Valero and Alías [15]).

2.2. Key Differences among Speech, Music and Environmental Sounds

In what concerns the audio input that the machine hearing system is asked to process, speech, music and environmental sounds present specific characteristics. The key differences can be directly observed both in the time and the frequency domains, as well as in the structure and the semantics of the signal. These differences can be then parametrized following a physical or perceptual approach depending on the targeted application.

Firstly, music and speech signals present a certain periodicity that can be observed when analyzing these signals in the time domain (see Figure 3). Although with some exceptions (e.g., some natural sounds such as bird chirps or cricket sounds), the periodicity in environmental sounds may not be so evident.

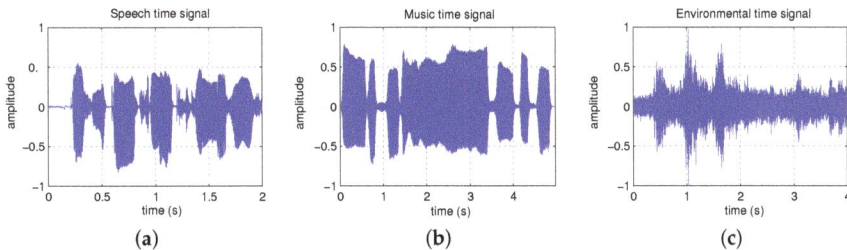

Figure 3. Time envelope of a: (**a**) speech signal; (**b**) music signal (trumpet); (**c**) environmental sound signal (traffic street).

Secondly, when analyzed in the frequency domain, it can be generally determined that the complexity of the spectrum of environmental sounds (e.g., the sound of a passing car) is notably larger than that of speech or music signals, as depicted in Figure 4. Moreover, it can be observed that speech and music signals usually present harmonic structures in their spectra, a trait that is not that common in environmental sounds, as mentioned before.

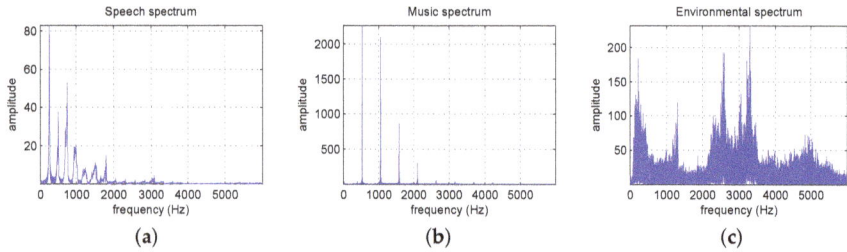

Figure 4. Normalized power spectral density of the: (**a**) speech signal; (**b**) music signal (trumpet); (**c**) environmental sound signal (traffic street) of representative regions extracted from Figure 3.

Thirdly, notice that both speech and music sounds are composed of a limited dictionary of sound units: phonemes and notes, respectively. On the contrary, the range of environmental sounds is theoretically infinite, since any occurring sound in the environment may be included in this category (*i.e.*, originated from noise, artificial or natural sound sources, see Figure 1).

Furthermore, there exists a key difference between these types of signals. In speech and music, phonemes and musical notes are combined so as to obtain meaningful sequences that are actually transmitting a particular semantic or aesthetic message. As opposed, the sequences on environmental sounds do not follow any rule or predefined grammar, although they may convey some kind of meaning (e.g., bird chirps or cricket sounds). Unlike speech and music, also other important information is unknown, such as the duration of the sound events or the proportion between harmonic and non-harmonic spectral structure.

Finally, Table 1 presents a summary of the specific characteristics of speech, music and environmental sounds in terms of several factors. Given the noticeable differences between the nature of these sounds, the research community has proposed diverse feature extraction techniques adapted to the particularities of these sounds. However, some works also make use of well-established approaches to build analogous systems in related research fields, e.g., by borrowing features showing good performances for speech and/or music sounds analysis to parametrize environmental sounds.

Table 1. List of features that characterize speech, music and environmental sounds (adapted from [21]).

Features	Speech	Music	Environmental Sounds
Type of audio units	Phones	Notes	Any other type of audio event
Source	Human speech production system	Instruments or Human speech production system	Any other source producing an audio event
Temporal and spectral characteristics	Short durations (40–200 ms), constrained and steady timming and variability, largely harmonic content around 500 Hz to 2 KHz with some noise-like sounds.	From short to long durations (40–1200ms), with a mix of steady and transient sounds organized in periodic structures, largely harmonic content in the full 20 Hz to 20 KHz audio band, with some inharmonic components.	From short to very long durations (40–3000ms), with wide range of steady and transient type of sounds, ans also a wide range of harmonic to inharmonic balances.

The following sections put the focus on the central topic of this work, presenting an in-depth review of audio feature extraction techniques divided according to their physical or perceptual basis, together with some specific applications of machine hearing focused on the analysis of speech, music or environmental sounds.

3. Audio Features Taxonomy and Review of Extraction Techniques

There exists a myriad of approaches to extract significant features from the audio input of a machine hearing system. On the one hand, we can find those approaches that are only devoted to extract physical features of the audio input. These extraction techniques differ on the domain of operation, ranging from the classic time, frequency or cepstral domains to the derivation of features based on other recent representations. Specifically, speech, music and environmental sounds typically present rich time-varying characteristics with very diverse contents (as shown in Figure 3), which can be parameterized in that domain, e.g., by computing from the analyzed input frame the sign-change rate, the fundamental periodicity, the signal power or amplitude, *etc.* Moreover, the dynamic variations of those audio signals can present relevant information in a transformed domain, e.g., through a Fourier transform (see Figure 4), in the cepstral or Wavelet domains, or from eigenspaces or even through non-linear representations, from which specific features related to e.g., spectrum, harmonicity, line prediction or phase-space can be extracted.

On the other hand, we can find those techniques that try to explicitly integrate perception in the parameterization process or derive it through the computation of signal features capable of extracting perceptually relevant aspects from the input audio, as described by Richard *et al.* [31]. The former typically include in the parameterization process simplified audition models of the hearing system (e.g., by considering from Bark, Mel or Gammatone filter-banks to more complex models based on electroencephalograms). This bio-inspired approach has to take into account the target species of the machine hearing system, being adapted to the cochlear response of that species, e.g., human beings or animals (see the work by Clemins *et al.* [32,33]). The latter approach to embed perception during the feature extraction process is based on the computation of low-level features that somehow explain a high-level sensation of sound similarity, which has been validated perceptually (Richard *et al.* [31]), such the ones related to temporal or frequency-based domains (e.g., loudness, pitch, rhythm, *etc.*), or the ones derived from the computation of the autocorrelation function and the auditory image model for example.

In this work, we organize the review of the most relevant and recent audio feature extracting techniques found in the literature following the hierarchical taxonomy depicted in Figure 5. This taxonomy builds on the one introduced in the review by Mitrović *et al.* [17]. We fist classify the techniques by differentiating physically-based approaches from those with a perceptual basis, and subsequently dividing them according to the domain of parameterization: time, frequency, wavelet, image-based, cepstral, or other domains.

It is important to highlight that the main goal of this paper is to provide the reader with a broad view of the existing approaches to audio feature extraction. The detailed mathematical analysis and critical comparison between features lies beyond the scope and objectives of our work. The reader interested in a mathematical description of audio features is referred to the works by Peeters [34] and by Sharan and Moir [35]. Additionally, comparisons between several types of features can be found in other works. Some of these works are focused on comparing the performance of several features in the context of different machine hearing applications, such as sound recognition [36] or music retrieval [37]. Finally, the work by Hengel and Krijnders [38] presents a comparison of characteristics of audio features, such as their robustness to noise and spectro-temporal detail.

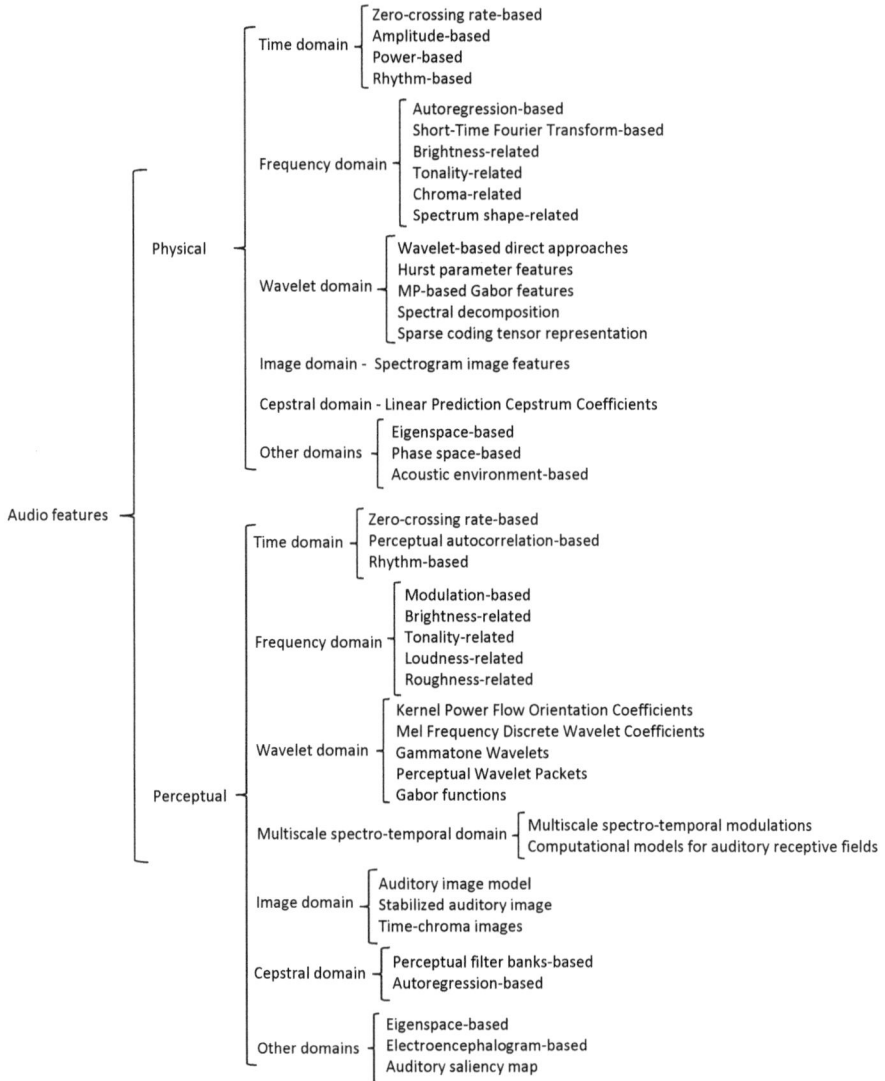

Figure 5. Taxonomy of physical *vs.* perceptual based audio features extraction techniques.

4. Physical Audio Features Extraction Techniques

This section describes the main physical audio features extraction techniques reported in the literature, categorized according to the previously defined taxonomy.

4.1. Time Domain Physical Features

Possibly the most significant trait of time domain features is that they do not require applying any kind of transformation on the original audio signal, and their computation is performed directly on the samples of the signal itself. This approach to audio feature extraction constitutes one of the most elementary and classic, and as such they appear in previous reviews on the topic (e.g., in the work by Mitrović *et al.* [17]).

Time domain physical audio features can be classified into the following categories: zero crossing-based features, amplitude-based features, power-based features and rhythm-based features.

The following paragraphs describe the most commonly used time domain features belonging to these categories.

4.1.1. Zero-Crossing Rate-Based Physical Features

This kind of physical features are based on the analysis of the sing-rate change of the analyzed audio input, which is a simple yet effective parameterization used in several machine hearing applications.

- **Zero-crossing rate (ZCR):** it is defined as the number of times the audio signal waveform crosses the zero amplitude level during a one second interval, which provides a rough estimator of the dominant frequency component of the signal (Kedem [39]). Features based on this criterion have been applied to speech/music discrimination, music classification (Li *et al.* [40], Bergstra *et al.* [41], Morchen *et al.* [42], Tzanetakis and Cook [28], Wang *et al.* [9])), singing voice detection in music and environmental sound recognition (see the works by Mitrović *et al.* [17] and Peltonen *et al.* [18]), musical instrument classification (Benetos *et al.* [10]), voice activity detection in noisy conditions (Ghaemmaghami *et al.* [43]) or for audio-based surveillance systems (as in Rabaoui *et al.* [24]).
- **Linear prediction zero-crossing ratio (LP-ZCR):** this feature is defined as the ratio between the ZCR of the original audio and the ZCR of the prediction error obtained from a linear prediction filter (see El-Maleh *et al.* [44]). Its use is intended for discriminating between signals that show different degree of correlation (e.g., between voiced and unvoiced speech).

4.1.2. Amplitude-Based Features

Amplitude-based features are based on a very simple analysis of the temporal envelope of the signal. The following paragraphs describe the most commonly used amplitude-based temporal features, including the one from the Moving Picture Experts Group (MPEG) [45], (previously reviewed by Mitrović *et al.* [17]), and a feature extraction approach typically used to characterize voice pathologies which has recently found application in music analysis.

- **Amplitude descriptor (AD):** it allows for distinguishing sounds with different signal envelopes, being applied, for instance, for the discrimination of animal sounds (Mitrović *et al.* [46]). It is based on collecting the energy, duration, and variation of duration of signal segments based on their high and low amplitude by means of an adaptive threshold (a level-crossing computation).
- **MPEG-7 audio waveform (AW):** this feature is computed from a downsampled waveform envelope, and it is defined as the maximum and minimum values of a function of a non-overlapping analysis time window [45]. AW has been used as a feature in environmental sound recognition, like in the works of Muhammad and Alghathbar [47], or by Valero and Alías [48].
- **Shimmer:** it computes the cycle-to-cycle variations of the waveform amplitude. This feature has been generally applied to study pathological voices (Klingholz [49], Kreiman and Gerratt [50], Farrús *et al.* [51]). However, it has also been applied to discriminate vocal and non-vocal regions from audio in songs (as in Murthy and Koolagudi [52]), characterize growl and screaming singing styles (Kato and Ito [53]), prototype, classify and create musical sounds (Jenssen [54]) or to improve speaker recognition and verification (Farrús *et al.* [51]) to name a few.

4.1.3. Power-Based Features

The following paragraphs describe the most relevant and classic temporal audio features based on signal power.

- **Short-time energy:** using a frame-based procedure, short-time energy (STE) can be defined as the average energy per signal frame (which is in fact the MPEG-7 audio power descriptor [45]). Nevertheless, there exist also other STE definitions in the literature that compute power in the spectral domain (e.g., see Chu *et al.* [55]). STE can be used to detect the transition from unvoiced to voices speech and *vice versa* (Zhang and Kuo [56]). This feature has also been used in applications like musical onset detection (Smith *et al.* [57]), speech recognition (Liang and Fan [58]), environmental sound recognition (Peltonen *et al.* [18], Muhammad and Alghathbar [47], Valero and Alías [48]) and audio-based surveillance systems (Rabaoui *et al.* [24]).

- **Volume:** according to the work by Liu *et al.* [59], volume is defined as the Root-Mean Square (RMS) of the waveform magnitude within a frame. It has been used for speech segmentation applications, e.g., see Jiang *et al.* [60].

- **MPEG-7 temporal centroid:** it represents the time instant containing the signal largest average energy, and it is computed as the temporal mean over the signal envelope (and measured in seconds) [45]. The temporal centroid has been used as an audio feature in the field of environmental sound recognition, like in the works by Muhammad and Alghathbar [47], and Valero and Alías [48]).

- **MPEG-7 log attack time:** it characterizes the attack of a given sound (e.g., for musical sounds, instruments can generate either smooth or sudden transitions) and it is computed as the logarithm of the elapsed time from the beginning of a sound signal to its first local maximum [45]. Besides being applied to musical onset detection (Smith *et al.* [57]), log attack time (LAT) has been used for environmental sound recognition (see Muhammad and Alghathbar [47], and Valero and Alías [48]).

4.1.4. Rhythm-Based Physical Features

Rhythm represents an relevant aspect of music and speech, but it can also be significant in environmental and human activity related sounds (e.g., the sound of a train, finger tapping, *etc.*), since it characterizes structural organization of sonic events (changes in energy, pitch, timbre, *etc.*) along the time axis. Since the review by Mitrović *et al.* [17], there have been little significant contributions to the derivation of rhythm-based features. Thus, the following paragraphs describe the most relevant and classic rhythm-based features found in the literature.

- **Pulse metric:** this is a measure that uses long-time band-passed autocorrelation to determine how rhythmic a sound is in a 5-second window (as defined by Scheirer and Slaney [61]). Its computation is based on finding the peaks of the output envelopes in six frequency bands and its further comparison, giving a high value when all subbands present a regular pattern. This feature has been used for speech/music discrimination.

- **Pulse clarity:** it is a high-level musical dimension that conveys how easily in a given musical piece, or a particular moment during that piece, listeners can perceive the underlying rhythmic or metrical pulsation (as defined in the work by Lartillot *et al.* [62]). In that work, the authors describe several descriptors to compute pulse clarity based on approaches such as the analysis of the periodicity of the onset curve via autocorrelation, resonance functions, or entropy. This feature has been employed to discover correlations with qualitative measures describing overall properties of the music used in psychology studies in the work by Friberg *et al.* [63].

- **Band periodicity:** this is a measure of the strength of rhythmic or repetitive structures in audio signals (see Lu *et al.* [64]). Band periodicity is defined within a frequency band, and it is obtained as the mean value along all the signal frames of the maximum peak of the subband autocorrelation function.

- **Beat spectrum/spectrogram:** it is a two-dimensional parametrization based on time variations and lag time, thus providing an interpretable representation that reflects temporal changes of tempo (see the work by Foote [22,65]). Beat spectrum shows relevant peaks at rhythm periods that match the rhythmic properties of the signal. Beat spectrum can be used for discriminating between music (or between parts within an entire music signal) with different tempo patterns.
- **Cyclic beat spectrum:** or CBS for short, this is a representation of the tempo of a music signal that groups multiples of the fundamental period of the signal together in a single tempo class (Kurth *et al.* [66]). Thus, CBS gives a more compact representation of the fundamental beat period of a song. This feature has been employed in the field of audio retrieval.
- **Beat tracker:** this a feature is derived following an algorithmic approach based on signal subband decomposition and the application of a comb filter analysis in each subband (see Scheirer [67]). Beat tracker mimics at large extent the human ability to track rhythmic beats in music and allows obtaining not only tempo but also compute beat timing positions.
- **Beat histogram:** it provides a more general tempo perspective and summarizes the beat tempos present in a music signal (Tzanetakis and Cook [28]). In this case, Wavelet transform (see Section 4.3 for further details) is used to decompose the signal in octaves for performing subsequent accumulation of the most salient periodicities in each subband to generate the so-called beat histogram. This feature has been used for music genre classification [28].

4.2. Frequency Domain Physical Features

Audio features on the frequency domain constitute the largest set of audio features reported in the literature (Mitrović *et al.* [17]). They are usually obtained from the Short-Time Fourier Transform (STFT) transform or derived from an autoregression analysis. In general terms, physical frequency domain features describe physical properties of the signal frequency content. Moreover, this type of features can be further decomposed as follows:

- Autoregression-based
- STFT-based
- Brightness-related
- Tonality-related
- Chroma-related
- Spectrum shape-related

The following paragraphs describe these subcategories of physical frequency-based features.

4.2.1. Autoregression-Based Frequency Features

Autoregression-based features are derived from linear prediction analysis of signals, which usually captures typical spectral predominances (e.g., formants) of speech signals.

The most commonly employed physical frequency features based on signal autoregression are described below.

- **Linear prediction coefficients:** or LPC for short, this feature represents an all-pole filter that captures the spectral envelope (SE) of a speech signal (formants or spectral resonances that appear in the vocal tract), and have been extensively used for speech coding and recognition applications. LPC have been applied also in audio segmentation and general purpose audio retrieval, like in the works by Khan *et al.* [68,69].
- **Line spectral frequencies:** also referred to as Line Spectral Pairs (LSP) in the literature, Line Spectral Frequencies (LSF) are a robust representation of LPC parameters for quantization and interpolation purposes. They can be computed as the roots phases of the palindromic and the antipalindromic polynomials that constitute the LPC polynomial representation, which in turns

represent the vocal tract when the glottis is closed and open, respectively (see Itakura [70]). Due to its intrinsic robustness they have been widely applied in a diverse set of classification problems like speaker segmentation (Sarkar and Sreenivas [71]), instrument recognition and in speech/music discrimination (Fu [13]).

- **Code excited linear prediction features:** or CELP for short, this feature was introduced by Schroeder and Atal [72] and has become one of the most important influences in nowadays speech coding standards. This feature comprises spectral features like LSP but also two codebook coefficients related to signal's pitch and prediction residual signal. CELP features have been also applied in the environmental sound recognition framework, like in the work by Tsau *et al.* [73].

4.2.2. STFT-Based Frequency Features

This kind of audio features are generally derived from the signal spectrogram obtained from STFT computation. While some of the features belonging to this category are computed from the analysis of the spectrogram envelope (e.g., subband energy ratio, spectral flux, spectral slope, spectral peaks or MPEG-7 spectral envelope, normalized spectral envelope, and stereo panning spectrum feature), others are obtained from the STFT phase (like group delay functions and/or modified group delay functions).

The following list summarizes the most widely employed STFT-based features.

- **Subband energy ratio:** it is usually defined as a measure of the normalized signal energy along a predefined set of frequency subbands. In a broad sense, it coarsely describes the signal energy distribution of the spectrum (Mitrović *et al.* [17]). There are different approximations as regards the number and characteristics of analyzed subbands (e.g., Mel scale, ad-hoc subbands, *etc.*). It has been used for audio segmentation and music analysis applications (see Jiang *et al.* [60], or Srinivasan *et al.* [74]) and environmental sound recognition (Peltonen *et al.* [18]).
- **Spectral flux:** or SF for short, this feature is defined as the 2-norm of the frame-to-frame spectral amplitude difference vector (see Scheirer and Slaney [61]), and it describes sudden changes in the frequency energy distribution of sounds, which can be applied for detection of musical note onsets or, more generally speaking, detection of significant changes in the spectral distribution. It measures how quickly the power spectrum changes and it can be used to determine the timbre of an audio signal. This feature has been used for speech/music discrimination (like in Jiang *et al.* [60], or in Khan *et al.* [68,69]), musical instrument classification (Benetos *et al.* [10]), music genre classification (Li *et al.* [40], Lu *et al.* [12], Tzanetakis and Cook [28], Wang *et al.* [9]) and environmental sound recognition (see Peltonen *et al.* [18]).
- **Spectral peaks:** this feature was defined by Wang [8] as constellation maps that show the most relevant energy bin components in the time-frequency signal representation. Hence, spectral peaks is an attribute that shows high robustness to possible signal distortions (low signal-to-noise ratio (SNR)–see Klingholz [49], equalization, coders, *etc.*) being suitable for robust recognition applications. This feature has been used for automatic music retrieval (e.g., the well-known Shazam search engine by Wang [8]), but also for robust speech recognition (see Farahani *et al.* [75]).
- **MPEG-7 spectrum envelope and normalized spectrum envelope:** the audio spectrum envelope (ASE) is a log-frequency power spectrum that can be used to generate a reduced spectrogram of the original audio signal, as described by Kim *et al.* [76]. It is obtained by summing the energy of the original power spectrum within a series of frequency bands. Each decibel-scale spectral vector is normalized with the RMS energy envelope, thus yielding a normalized log-power version of the ASE called normalized audio spectrum envelope (NASE) (Kim *et al.* [76]). ASE feature has been used in audio event classification [76], music genre classification (Lee *et al.* [77]) and environmental sound recognition (see Muhammad and Alghathbar [47], or Valero and Alías [48]).
- **Stereo panning spectrum feature:** or SPSF for short, this feature provides a time-frequency representation that is intended to represent the left/right stereo panning of a stereo audio

signal (Tzanetakis *et al.* [78]). Therefore, this feature is conceived with the aim of capturing relevant information of music signals, and more specifically, information that reflects typical postproduction in professional recordings. The additional information obtained through SPSF can be used for enhancing music classification and retrieval system accuracies (Tzanetakis *et al.* [79]).

- **Group delay function:** also known as GDF, it is defined as the negative derivative of the unwrapped phase of the signal Fourier transform (see Yegnanarayana and Murthy [80]) and reveals information about temporal localization of events (*i.e.*, signal peaks). This feature has been used for determining the instants of significant excitation in speech signals (like in Smits and Yegnanarayana [81], or Rao *et al.* [82]) and in beat identification in music performances (Sethares *et al.* [83]).

- **Modified group delay function:** or MGDF for short, it is defined as a smoother version of the GDF, reducing its intrinsic spiky nature by introducing a cepstral smoothing process prior to GDF computation. It has been used in speaker identification (Hegde *et al.* [84]), but also in speech analysis, speech segmentation, speech recognition and language identification frameworks (Murthy and Yegnanarayana [85]).

4.2.3. Brightness-Related Physical Frequency Features

Brightness is an attribute that is closely related to the balance of signal energy in terms of high and low frequencies (a sound is said to be bright when it has more high than low frequency content). The most relevant brightness-related physical features found in the literature are the following:

- **Spectral centroid:** or SC for short, this feature describes the center of gravity of the spectral energy. It can be defined as the first moment (frequency position of the mean value) of the signal frame magnitude spectrum as in the works by Li *et al.* [40], or by Tzanetakis and Cook [28], or obtained from the power spectrum of the entire signal in MPEG-7. SC reveals the predominant frequency of the signal. In the MPEG-7 standard definition [45], the audio spectrum centroid (ASC) is defined by computing SC over the power spectrum obtained from an octave-frequency scale analysis and roughly describes the sharpness of a sound. SC has been applied in musical onset detection (Smith *et al.* [57]), music classification (Bergstra *et al.* [41], Li *et al.* [40], Lu *et al.* [12], Morchen *et al.* [42], Wang *et al.* [9]), environmental sound recognition (like in Peltonen *et al.* [18], Muhammad and Alghathbar [47], Valero and Alías [48]) and, more recently, to music mood classification (Ren *et al.* [86]).

- **Spectral center:** this feature is defined as the median frequency of the signal spectrum, where both lower and higher energies are balanced. Therefore, is a measure close to spectral centroid. It has been shown to be useful for automatic rhythm tracking in musical signals (see Sethares *et al.* [83]).

4.2.4. Tonality-Related Physical Frequency Features

The fundamental frequency is defined as the lowest frequency of an harmonic stationary audio signal, which in turn can be qualified as tonal sound. In music, tonality is a system that organizes the notes of a musical scale according to musical criteria. Moreover, tonality is related to the notion of harmonicity, which describes the structure of sounds that are mainly constituted by a series of harmonically related frequencies (*i.e.*, a fundamental frequency and its multiples), which are typical characteristics of (tonal) musical instruments sounds and voiced speech.

The following paragraphs describe the most widely employed tonality-related features that do not incorporate specific auditory models for their computation.

- **Fundamental frequency:** it is also denoted as F0. The MPEG-7 standard defines audio fundamental frequency feature as the first peak of the local normalized spectro-temporal autocorrelation function [45]. There are several methods in the literature to compute F0, e.g., autocorrelation-based methods, spectral-based methods, cepstral-based methods, and combinations (Hess [87]). This feature has been used in applications like musical onset detection (Smith *et al.* [57]), musical genre classification (Tzanetakis and Cook [28]), audio retrieval (Wold *et al.* [88]) and environmental sound recognition (Muhammad and Alghathbar [47], Valero and Alías [48]). In the literature F0 is sometimes denoted as *pitch* as it may represent a rough estimate of the perceived tonality of the signal (e.g., pitch histogram and pitch profile).

- **Pitch histogram:** instead of using a very specific and local descriptor like fundamental frequency, the pitch histogram describes more compactly the pitch content of a signal. Pitch histogram has been used for musical genre classification by Tzanetakis and Cook [28], as it gives a general perspective of the aggregated notes (frequencies) present in a musical signal along a certain period.

- **Pitch profile:** this feature is a more precise representation of musical pitch, as it takes into account both pitch mistuning effects produced in real instruments and also pitch representation of percussive sounds. It has been shown that use of pitch profile feature outperforms conventional chroma-based features in musical key detection, like in Zhu and Kankanhalli [89].

- **Harmonicity:** this feature is useful for distinguishing between tonal or harmonic (e.g., birds, flute, *etc.*) and noise-like sounds (e.g., dog bark, snare drum, *etc.*). Most traditional harmonicity features either use an impulse train (like in Ishizuka *et al.* [90]) to search for the set of peaks in multiples of F0, or uses the autocorrelation-inspired functions to find the self-repetition of the signal in the time- or frequency-domain (as in Kristjansson *et al.* [91]). Spectral local harmonicity is proposed in the work by Khao [92], a method that uses only the sub-regions of the spectrum that still retain a sufficient harmonic structure. In the MPEG-7 standard, two harmonicity measures are proposed. Harmonic ratio (HR) is a measure of the proportion of harmonic components in the power spectrum. The Upper limit of harmonicity (ULH) is an estimation of the frequency beyond which the spectrum no longer has any harmonic structure. Harmonicity has been used also in the field of environmental sound recognition (Muhammad and Alghathbar [47], Valero and Alías [48]). Some other harmonicity-based features for music genre and instrument family classification have been defined, like harmonic concentration, harmonic energy entropy or harmonic derivative (see Srinivasan and Kankanhalli [93]).

- **Inharmonicity:** this feature measures the extent to which the partials of a sound are separated with respect to its ideal position in a harmonic context (whose frequencies are integers of a fundamental frequency). Some approaches take into account only partial frequencies (like Agostini *et al.* [94,95]), while others also consider partial energies and bandwidths (see Cai *et al.* [96]).

- **Harmonic-to-Noise Ratio:** Harmonic-to-noise Ratio (HNR) is computed as the relation between the energy of the harmonic part and the energy of the rest of the signal in decibels (dB) (Boersma [97]). Although HNR has been generally applied to analyze pathological voices (like in Klingholz [49], or in Lee *et al.* [98]), it has also been applied in some music-related applications such as the characterization of growl and screaming singing styles, as in Kato and Ito [53].

- **MPEG-7 spectral timbral descriptors:** the MPEG-7 standard defines some features that are closely related to the harmonic structure of sounds, and are appropriate for discrimination of musical sounds: MPEG-7 harmonic spectral centroid (HSC) (the amplitude-weighted average of the harmonic frequencies, closely related to brightness and sharpness), MPEG-7 harmonic spectral deviation (HSD) (amplitude deviation of the harmonic peaks from their neighboring harmonic peaks, being minimum if all the harmonic partials have the same amplitude), MPEG-7 harmonic spectral spread (HSS) (the power-weighted root-mean-square deviation of the harmonic peaks from the HSC, related to harmonic bandwidths), and MPEG-7 harmonic spectral variation (HSV) (correlation of harmonic peak amplitudes in two adjacent frames, representing the harmonic variability over time). MPEG-7 spectral timbral descriptors have been employed for environmental sound recognition (Muhammad and Alghathbar [47],Valero and Alías [48]).

- **Jitter:** computes the cycle-to-cycle variations of the fundamental frequency (Klingholz [49]), that is, the average absolute difference between consecutive periods of speech (Farrús *et al.* [51]). Besides typically being applied to analyze pathological voices (like in Klingholz [49], or in Kreiman and Gerratt [50]), it has also been applied to prototyping, classification and creation of musical sounds (Jensen [54]), improve speaker recognition (Farrús *et al.* [51]), characterize growl and screaming singing styles (Kato and Ito [53]) or discriminate vocal and non-vocal regions from audio songs (Murthy and Koolagudi [52]), among others.

4.2.5. Chroma-Related Physical Frequency Features

Chroma is related to perception of pitch, in the sense that it is a complement of the tone height. In a musical context, two notes that are separated one or more octaves have the same chroma (e.g., C4 and C7 notes), and produce a similar effect on the human auditory perception.

The following paragraphs describe chroma-related frequency features, which are basically computed from direct physical approaches:

- **Chromagram:** also known as chroma-based feature, chromagram is a spectrum-based energy representation that takes into account the 12 pitch classes within an octave (corresponding to pitch classes in musical theory) (Shepard [99]), and it can be computed from a logarithmic STFT (Bartsch and Wakefield [100]). Then, it constitutes a very compact representation suited for musical and harmonic signals representation following a perceptual approach.
- **Chroma energy distribution normalized statistics:** or CENS for short, this feature was conceived for music similarity matching and has shown to be robust to tempo and timbre variations (Müller *et al.* [101]). Therefore, it can be used for identifying similarities between different interpretations of a given music piece.

4.2.6. Spectrum Shape-Related Physical Frequency Features

Another relevant set of frequency features are the ones that try to describe the shape of the spectrum of the audio signal. The following paragraphs describe the most widely employed, and some of the newest contributions in this area.

- **Bandwidth:** usually defined as the second-order statistic of the signal spectrum, it helps to discriminate tonal sounds (with low bandwidths) from noise-like sounds (with high bandwidths) (see Peeters [34]). However, it is difficult to distinguish between complex tonal sounds (e.g., music, instruments, *etc.*) from complex noise-like sounds using only this feature. It can be defined over the power spectrum or in its logarithmic version (see Liu *et al.* [59], or Srinivasan and Kankanhalli [93]) and it can be computed over the whole spectrum or within different subbands (like in Ramalingam and Krishnan [102]). MPEG-7 defines audio spectrum spread (ASS) as the standard deviation of the signal spectrum, which constitutes the second moment while (being the ASC the first one). Spectral bandwidth has been used for music classification (Bergstra *et al.* [41], Lu *et al.* [12], Morchen *et al.* [42], Tzanetakis and Cook [28]), and environmental sound recognition (Peltonen *et al.* [18], Muhammad and Alghathbar [47], Valero and Alías [48]).
- **Spectral dispersion:** this is a measure closely related to spectral bandwidth. The only difference is that it takes into account the spectral center (median) instead of the spectral centroid (mean) (see Sethares *et al.* [83]).
- **Spectral rolloff point:** defined as the 95th percentile of the power spectral distribution (see Scheirer and Slaney [61]), spectral rolloff point can be regarded as a measure of the skewness of the spectral shape. It can be used, for example, for distinguishing between voiced from unvoiced speech sounds. It has been used in music genre classification (like in Li and Ogihara [103], Bergstra *et al.* [41], Li *et al.* [40], Lu *et al.* [12], Morchen *et al.* [42], Tzanetakis and Cook [28], Wang *et al.* [9]), speech/music discrimination (Scheirer and Slaney [61]), musical instrument classification (Benetos *et al.* [10]), environmental sound recognition

(Peltonen *et al.* [18]), audio-based surveillance systems (Rabaoui *et al.* [24]) and music mood classification (Ren *et al.* [86]).

- **Spectral flatness:** this is a measure of uniformity in the frequency distribution of the power spectrum, and it can be computed as the ratio between the geometric and the arithmetic mean of a subband (see Ramalingam and Krishnan [102]) (equivalent to the MPEG-7 audio spectrum flatness (ASF) descriptor [45]). This feature allows distinguishing between noise-like sounds (high value of spectral flatness) and more tonal sounds (low value). This feature has been used in audio fingerprinting (see Lancini *et al.* [104]), musical onset detection (Smith *et al.* [57]), music classification (Allamanche *et al.* [105], Cheng *et al.* [106], Tzanetakis and Cook [28]) and environmental sound recognition (Muhammad and Alghathbar [47], Valero and Alías [48]).

- **Spectral crest factor:** in contrast to spectral flatness measure, spectral crest factor measures how peaked the power spectrum is, and it is also useful for differentiation of noise-like (lower spectral crest factor) and tonal sounds (higher spectral crest factor). It can be computed as the ratio between the maximum and the mean of the power spectrum within a subband, and has been used for audio fingerprinting (see Lancini *et al.* [104], Li and Ogihara [103]) and music classification (Allamanche *et al.* [105], Cheng *et al.* [106]).

- **Subband spectral flux:** or SSF for short, this feature is inversely proportional to spectral flatness, being more relevant in subbands with non-uniform frequency content. In fact, SSF measures the proportion of dominant partials in different subbands, and it can be measured accumulating the differences between adjacent frequencies in a subband. It has been used for improving the representation and recognition of environmental sounds (Cai *et al.* [96]) and music mood classification (Ren *et al.* [86]).

- **Entropy:** this is another measure that describes spectrum uniformity (or flatness), and it can be computed following different approaches (Shannon entropy, or its generalization named Renyi entropy) and also in different subbands (see Ramalingam and Krishnan [102]). It has been used for automatic speech recognition, computing the Shannon entropy in different equal size subbands, like in Misra *et al.* [107].

- **Octave-based Spectral Contrast:** also referred to as OSC, it is defined as the difference between peaks (that generally corresponds to harmonic content in music) and valleys (where non-harmonic or noise components are more dominant) measured in subbands by octave-scale filters and using a neighborhood criteria in its computation (Jiang *et al.* [108]). To represent the whole music piece, mean and standard deviation of the spectral contrast and spectral peak of all frames are used as the spectral contrast features. OSC features have been used for music classification (Lee *et al.* [77], Lu *et al.* [12], Yang *et al.* [109]) and music mood classification, as in Ren *et al.* [86].

- **Spectral slope:** this is a measure of the spectral slant by means of a simple linear regression (Morchen *et al.* [42]), and it has been used for classification purposes in speech analysis applications (Shukla *et al.* [110]) and speaker identification problems (Murthy *et al.* [111]).

- **Spectral skewness and kurtosis:** spectral skewness, which is computed as the 3rd order moment of the spectral distribution, is a measure that characterizes the asymmetry of this distribution around its mean value. On the other hand, spectral kurtosis describes the flatness of the spectral distribution around its mean, and its computed as the 4th order moment (see Peeters *et al.* [34]). Both parameters have been applied for music genre classification (Baniya *et al.* [112]) and music mood classification (Ren *et al.* [86]).

4.3. Wavelet-Based Physical Features

A Wavelet is a mathematical function used to divide a given function or continuous-time signal into different scale components. The Wavelet transform (WT) has advantages over the traditional Fourier transform for representing functions that have discontinuities and sharp peaks, and for accurately deconstructing and reconstructing finite, non-periodic and/or non-stationary signals (Mallat [113]). In the work by Benedetto and Teolis [114], a link between auditory functions and

Wavelet analysis was provided, while in the work by Yang *et al.* [115] an analytical framework to model the early stages of auditory processing, based on Wavelet and multiresolution analysis was proposed.

In the following paragraphs, we describe the most commonly used wavelet-based physical frequency features.

- **Wavelet-based direct approaches:** different type or families of wavelets have been used and defined in the literature in the field of audio processing. Daubechies wavelets have been used in blind source speech separation (see the work by Missaoui and Lachiri [116]) and Debechies together with Haar wavelets have been used in music classification (Popescu *et al.* [117]), while Coiflets wavelet have been applied recently to de-noising of audio signals (Vishwakarma *et al.* [118]). Other approaches like Daubechies Wavelet coefficient histogram (DWCH) features, are defined as the first three statistical moments of the coefficient histograms that represent the subbands obtained from Daubechies Wavelet audio signal decomposition (see Li *et al.* [40,119]). They have been applied in the field of speech recognition (Kim *et al.* [120]), music analysis applications such as genre classification, artist style identification and emotion detection (as in Li *et al.* [40,119,121], Mandel and Ellis [122], Yang *et al.* [109]) or mood classification (Ren *et al.* [86]). Also, in the work by Tabinda and Ahire [123], different wavelet families (like Daubechies, symlet, coiflet, biorthogonal, stationary and dmer) are used in audio steganography (an application for hiding data in cover speech which is imperceptible from the original audio).

- **Hurst parameter features:** or pH for short, is a time-frequency statistical representation of the vocal source composed of a vector of Hurst parameters (defined by Hurst [25]), which was computed by applying a wavelet-based multidimensional transformation of the short-time input speech in the work by Sant'Ana [124]. Thanks to its statistical definition, pH is robust to channel distortions as it models the stochastic behavior of input speech signal (see Zao *et al.* [125], or Palo *et al.* [126]). pH was originally applied as a means to improve speech-related problems, such as text-independent speaker recognition [124], speech emotion classification [125,126], or speech enhancement [127]. However, it has also been applied to sound source localization in noisy environments recently, as in the work by Dranka and Coelho [128].

An alternative means to represent signals using a finite dictionary of basis functions (or atoms) is matching pursuit (MP), described by Mallat in [129], an algorithm that provides an efficient way of sparsely decomposing a signal by selecting the "best" subset of basis vectors from a given dictionary. The selection of the "best" elements in the dictionary is based on maximizing the energy removed from the residual signal at each step of the algorithm. This allows obtaining a reasonable approximation of the signal with a few basis functions, which provides an interpretation of the signal structure. The dictionary of basis functions can be composed of Wavelet functions, Wavelet packets, or Gabor functions, to name a few.

The following paragraphs describe some relevant audio features using MP-based signal decompositions.

- **MP-based Gabor features:** Wolfe *et al.* [130] proposed the construction of multiresolution Gabor dictionaries appropriate for audio signal analysis, which is applied for music and speech signals observed in noise, obtaining a more efficient spectro-temporal representation compared to a full multiresolution decomposition. In this work, Gabor atoms are given by time-frequency shifts of distinct window functions. Ezzat *et al.* describe in [131] the use of 2D Gabor filterbank and illustrate its response to different speech phenomena such as harmonicity, formants, vertical onsets/offsets, noise, and overlapping simultaneous speakers. Meyer and Kollmeier propose in [132] the use of spectro-temporal Gabor features to enhance automatic speech recognition performance in adverse conditions, and obtain better results when Hanning-shaped Gabor filters are used in contrast to more classical Gaussian approaches. Chu *et al.* [14] proposed using MP and a dictionary of Gabor functions to represent the time dynamics of environmental sounds, which are typically noise-like with a broad flat spectrum, but may include strong temporal domain signatures. Coupled with Mel Frequency Cepstral Coefficients (see Section 5.6 for more details), the MP-based Gabor features allowed improved environmental sound recognition. In [133], Wang *et al.* proposed a nonuniform scale-frequency map based on Gabor atoms selected via MP, onto which Principal Component Analysis and Linear Discriminate Analysis are subsequently applied to generate the audio feature. The proposed feature was employed for environmental sound classification in home automation.

- **Spectral decomposition:** in [134], Zhang *et al.* proposed an audio feature extraction scheme applied to audio effect classification and based on spectral decomposition by matching-pursuit in the frequency domain. Based on psychoacoustic studies, a set of spectral sinusoid-Gaussian basis vectors are constructed to extract pitch, timbre and residual in-harmonic components from the spectrum, and the audio feature consists of the scales of basis vectors after dimension reduction. Also in [135], Umapathy *et al.* applied an Adaptive Time Frequency Transform (ATFT for short) algorithm for music genre classification as a Wavelet decomposition but using Gaussian-based kernels with different frequencies, translations and scales. The scale parameter, which characterizes the signal envelope, captures information about rhythmic structures, and it has been used for music genre identification (see Fu [13]).

- **Sparse coding tensor representation:** this work presents an evolution of Gabor atom MP-based audio feature extraction of Chu *et al.* [14]. The method proposed in the work by Zhang and He [136] tries to preserve the distinctiveness of the atoms selected by the MP algorithm by using a frequency-time-scale tensor derived from the sparse coding of the audio signal. The three tensor dimensions represent the frequency, time center and scale of transient time-frequency components with different dimensions. This feature was coupled with MFCC and applied to perform sound effects classification.

4.4. Image Domain Physical Features

This approach to feature extraction is based on a joint two-dimensional image-based m of the audio signal. Typically, one of the dimensions corresponds to a frequency vision of the signal, while the other corresponds to a time view (as defined by Walters [137]).

- **Spectrogram image features:** or SIF for short, are features that comprise a set of techniques that focus on applying techniques from the image processing field to the time-frequency representations (using Fourier, cepstral, or other types of frequency mapping techniques) of the sound to be analyzed (Chu *et al.* [14], Dennis *et al.* [138]). Spectrogram image features like subband power distribution (SPD), a two-dimensional representation of the distribution of normalized spectral power over time against frequency, have been shown to be useful for sound event recognition (Dennis [4]). The advantage of the SPD over the spectrogram is that the sparse, high-power elements of the sound event are transformed to a localized region of the SPD, unlike in the spectrogram where they may be scattered over time and frequency. Also, Local Spectrogram features (LS) are introduced by Dennis [4] with the ability to detect an arbitrary combination of overlapping sounds, including two or more different sounds or the same sound overlapping itself. LS features are used to detect keypoints in the spectrogram and then characterize the sound using the Generalized Hough Transform (GHT), a kind of universal transform that can be used to find arbitrarily complex shapes in grey level images, and that it can model the geometrical distribution of speech information over the wider temporal context (Dennis *et al.* [139]).

In Section 5.5 other approaches for image-based audio feature extraction which incorporate perceptual auditory models are reviewed.

4.5. Cepstral Domain Physical Features

Cepstral features are compact representations of the spectrum and provide a smooth approximation based on the logarithmic magnitude. They have been largely used for speaker identification and speech recognition but they have also been employed in the context of audio retrieval.

The main cepstral domain physical features found in the literature are the following:

- **Complex cepstrum:** is defined as the Inverse Fourier transform of the logarithm (with unwrapped phase) of the Fourier transform of the signal (see Oppenheim and Schafer [140]), and has been used for pitch determination of speech signals (Noll [141]) but also for identification of musical instruments (see Brown [142]).
- **Linear Prediction Cepstrum Coefficients:** or LPCC for short. This feature is defined as the inverse Fourier transform of the logarithmic magnitude of the linear prediction spectral complex envelope (Atal [143]), and provide a more robust and compact representation especially useful for automatic speech recognition and speaker identification (Adami and Couto Barone [144]) but also for singer identification (Shen *et al.* [145]), music classification (Xu *et al.* [146], Kim and Whitman [147]) and environmental sound recognition (see Peltonen *et al.* [18], or Chu *et al.* [14]).

4.6. Other Domains

The literature contains other approaches to audio feature extraction that operate on domains different to the ones just reviewed. Some of the most significant physical-based features are the eigenspace domain, the phase space domain, and the acoustic environment domain. The following paragraphs briefly describe these approaches.

- **Eigenspace:** audio features expressed in the eigenspace are usually obtained from sound segments of several seconds of duration, which are postprocessed by dimensionality reduction algorithms in order to obtain a compact representation of the main signal information. This dimensionality reduction is normally performed by means of Principal Component Analysis (PCA) (or alternatively, via Singular Value Decomposition or SVD), which is equivalent to a projection of the original data onto a subspace defined by its eigenvectors (or eigenspace), or Independent Component Analysis (ICA). Some of the most relevant eigendomain physical features found in the literature are: i) MPEG-7 audio spectrum basis/projection feature, which is a combination of two descriptors (audio spectrum basis or ASB–and audio spectrum projection or ASP) conceived for audio retrieval and classification [45,76]. ASB feature is a compact representation of the signal spectrogram obtained through SVD, while ASP is the spectrogram projection against a given audio spectrum basis. ASB and ASP have been used for environmental sound recognition, as in Muhammad and Alghathbar [47]; and ii) Distortion discriminant analysis (DDA) feature, which is a compact time-invariant and noise-robust representation of an audio signal, that is based on applying hierarchical PCA to a time-frequency representation derived from a modulated complex lapped transform (MCLT) (see Burges *et al.* [148], or Malvar [149]). Therefore, this feature serves as a robust audio representation against many signal distortions (time-shifts, compression artifacts and frequency and noise distortions).
- **Phase space:** this type of features emerge as a response to the linear approach that has usually been employed to model speech. However, linear models do not take into account nonlinear effects occurring during speech production, thus constituting a simplification of reality. This is why approaches based on nonlinear dynamics try to bridge this gap. A first example are the nonlinear features for speech recognition presented in the work by Lindgren *et al.* [150], which are based on the so-called reconstructed phase space generated from time-lagged versions of the original time series. The idea is that reconstructed phase spaces have been proven to recover the full dynamics of the generating system, which implies that features extracted from it can potentially contain more and/or different information than a spectral representation. In the works

by Kokkinos and Maragos [151] and by Pitsikalis and Maragos [152], a similar idea is employed to compute for short time series of speech sounds useful features like Lyapunov exponents.

- **Acoustic environment features:** this type of features try to capture information from the acoustic environment where the sound is measured. As an example, in the work by Hu *et al.* [153], the authors propose the use of Direct-to-Reverberant Ratio (DRR), the ratio between the Room Impulse response (RIR) energy of the direct path and the reverberant components, to perform speaker diarization. In this approach, they don't use a direct measure of the RIR, but a Non-intrusive Room Acoustic parameter estimator (NIRA) (see Parada *et al.* [154]). This estimator is a data-driven approach that uses 106 features derived from pitch period importance weighted signal to noise ratio, zero-crossing rate, Hilbert transformation, power spectrum of long term deviation, MFCCs, line spectrum frequency and modulation representation.

5. Perceptual Audio Features Extraction Techniques

The concept of perceptual audio features is based on finding ways to describe general audio properties based on human perception. The literature contains several attempts to derive this type of features, be it through the integration of perception in the very parameterization process, or through the computation of signal features capable of extracting perceptually relevant aspects from the audio signal (see Richard *et al.* [31]).

Interestingly, there exist some works focused on bridging the gap between features and subjective perception, aiming at the discovery of correlations between perceptual audio features and qualitative audio descriptive measures used in psychology studies, such as the work by Friberg *et al.* [63].

This section describes the main perceptual audio features extraction techniques reported in the literature, categorized according to the defined taxonomy (see Figure 5).

5.1. Time Domain Perceptual Features

In the context of time domain perceptual features we can found zero-crossing features, perceptual autocorrelation-based features and rhythm pattern.

5.1.1. Zero-Crossing Rate-Based Perceptual Features

The following zero-crossing-based features which incorporate some bio-inspired auditory model can be found in the literature:

- **Zero-crossing peak amplitudes (ZCPA):** were designed for automatic speech recognition (ASR) in noisy environments by Kim *et al.* [155], showing better results that linear prediction coefficients. This feature is computed from time-domain zero crossings of the signal previously decomposed in several psychoacoustic scaled subbands. The final representation of the feature is obtained on a histogram of the inverse zero-crossings lengths over all the subband signals. Subsequently, each histogram bin is scaled with the peak value of the corresponding zero crossing interval. In [156], Wang and Zhao applied ZCPA to noise-robust speech recognition.
- **Pitch synchronous zero crossing peak amplitudes (PS-ZCPA):** were proposed by Ghulam *et al.* [157] and they were designed for improving robustness of ASR in noisy conditions. The original method is based on an auditory nervous system, as it uses a mel-frequency spaced filterbank as a front-end stage. PS-ZCPA considers only inverse zero-crossings lengths whose peaks have a height above a threshold obtained as a portion of the highest peak within a signal pitch period. PS-ZCPA are only computed in voiced speech segments, being combined with the preceding ZCPA features obtained from unvoiced speech segments. In [158], the same authors presented a new version of the PS-ZCPA feature, using a pitch-synchronous peak-amplitude approach that ignores zero-crossings.

5.1.2. Perceptual Autocorrelation-Based Features

Autocorrelation is a measure of the self-similarity of the signal in the time domain with diverse applications to audio feature extraction. In this section, we revise those features derived from autocorrelation providing a measure of perceptual-based parameters related to acoustic phenomena.

- **Autocorrelation function features:** or ACF for short, this feature introduced by Ando in [159], has been subsequently applied by the same author to environmental sound analysis [29] and recently adapted to speech representation [160]. To compute ACF, the autocorrelation function is firstly computed from the audio signal, and then this function is parameterized by means of a set of perceptual-based parameters related to acoustic phenomena (signal loudness, perceived pitch, strength of perceived pitch and signal periodicity).
- **Narrow-band autocorrelation function features:** also known as NB-ACF, this feature was introduced by Valero and Alías [15], where the ACF concept is reused in the context of a filter bank analysis. Specifically, the features are obtained from the autocorrelation function of audio signals computed after applying a Mel filter bank (which are based on the Mel scale, a perceptual scale of pitches judged by listeners to be equal in distance from one another). These features have been shown to provide good performance for indoor and outdoor environmental sound classification. In [161], the same authors improved this technique by substituting the Mel filter bank employed to obtain the narrow-band signals by a Gammatone filter bank with Equivalent Rectangular Bandwidth bands. In addition, the Autocorrelation Zero Crossing Rate (AZCR) was added, following previous works like the one by Ghaemmaghami *et al.* [43].

5.1.3. Rhythm Pattern

As defined by Mitrović *et al.* [17], this feature is a two-dimensional representation of acoustic versus modulation frequency that is built upon a specific loudness sensation, and it is obtained by Fourier analysis of the critical bands over time and incorporating a weighting stage that is inspired by the human auditory system. This feature has shown to be useful in music similarity retrieval (Pampalk *et al.* [162], Rauber *et al.* [163]).

5.2. Frequency Domain Perceptual Features

Frequency-based features can also be defined on the perceptual frequency domain. This type of features are based in some signal properties measured taking into account the human auditory perception. The main perceptual properties represented by this type of features include:

- Modulation-based
- Brightness-related
- Tonality-related
- Loudness-related
- Roughness-related

The following paragraphs describe these subcategories of frequency-based perceptual features.

5.2.1. Modulation-Based Perceptual Frequency Features

Modulation-based perceptual frequency features represent the low-frequency (e.g., around 20 Hz) modulation content present in audio signals, which produce both amplitude and frequency variations. These variations are easily observed in audio signals that incorporate beats and rhythm (e.g., rhythmic patterns in music, audio signals coming from industrial machineries, speech signals, *etc.*). This modulation information can reflect structural evolution along time of the frequency content of a sound and can be measured separately for each frequency band.

The following paragraphs describe the most relevant modulation frequency features found in the literature based on a perceptual-based approach, including those reviewed by Mitrović *et al.* [17] and some recent contribution to the field:

- **4 Hz modulation energy:** is defined with the aim of capturing the most relevant hearing sensation of fluctuation in terms of amplitude- and frequency-modulated sounds (see Fastl [164]). The authors propose a model of fluctuation strength, based on a psychoacoustical magnitude, namely the temporal masking pattern. This feature can be computed filtering each subband of a signal spectral analysis by a 4 Hz band-pass filter along time and it has been used for music/speech discrimination (see Scheirer and Slaney [61]).

- **Computer model of amplitude-modulation sensitivity of single units in the inferior culliculus:** the work by Hewitt and Meddis [165] introduces a computer model of a neural circuit that replicates amplitude-modulation sensitivity of cells in the central nucleus of the inferior culliculus (ICC) is presented, allowing for the encoding of signal periodicity as a rate-based code.

- **Joint acoustic and modulation frequency features:** these are time-invariant representations that model the non-stationary behavior of an audio signal (Sukittanon and Atlas [166]). Modulation frequencies for each frequency band are extracted from demodulation of the Bark-scaled spectrogram using the Wavelet transform (see Section 4.3). These features have been used for audio fingerprinting by Sukittanon and Atlas [166], and they are similar to rhythm pattern feature (related to rhythm in music).

- **Auditory filter bank temporal envelopes:** or AFTE for short, this is another attempt to capture modulation information related to sound [167]. Modulation information is here obtained through bandpass filtering the output bands of a logarithmic-scale filterbank of 4th-order Gammatone bandpass filters. These features have been used for audio classification and musical genre classification by McKinney and Breebaart [167], and by Fu *et al.* [13].

- **Modulation spectrogram:** also referred to as MS, this feature displays and encodes the signal in terms of the distribution of slow modulations across time and frequency, as defined by Greenberg and Kingsbury [168]. In particular, it was defined to represent modulation frequencies in the speech signal between 0 and 8 Hz, with a peak sensitivity at 4 Hz, corresponding closely to the long-term modulation spectrum of speech. The MS is computed in critical-band-wide channels and incorporates a simple automatic gain control, and emphasizes spectro-temporal peaks. MS has been used for robust speech recognition (see Kingsbury *et al.* [169], or Baby *et al.* [170]), music classification (Lee *et al.* [77]), or content-based audio identification incorporating a Wavelet transform (Sukittanon *et al.* [171]). Recently, the MS features have been separated through a tensor factorization model, which represents each component as modulation spectra being activated across different subbands at each time frame, being applied for monaural speech separation purposes in the work by Barker and Virtanen [172] and for the classification of pathological infant cries (Chittora and Patil [173]).

- **Long-term modulation analysis of short-term timbre features:** in the work by Ren *et al.* [86] the use of a two-dimensional representation of acoustic frequency and modulation frequency to extract joint acoustic frequency and modulation frequency features is proposed, using an approach similar than in the work by Lee *et al.* [77]. Long-term joint frequency features, such as acoustic-modulation spectral contrast/valley (AMSC/AMSV), acoustic-modulation spectral flatness measure (AMSFM), and acoustic-modulation spectral crest measure (AMSCM), are then computed from the spectra of each joint frequency subband. By combining the proposed features, together with the modulation spectral analysis of MFCC and statistical descriptors of short-term timbre features, this new feature set outperforms previous approaches with statistical significance in automatic music mood classification.

5.2.2. Brightness-Related Perceptual Frequency Features

The second subtype of perceptual frequency features are those that aim at describing the brightness of the sound (see Section 4.2.3 for its physical-based counterpart). In this subcategory we can find sharpness, a measure of the signal strength for high frequencies, which is closely related to audio brightness, and it has been used for audio similarity analysis (Herre *et al.* [174], Peeters *et al.* [34]). Sharpness can be computed similarly to SC (see Section 4.2.3), but based on specific loudness instead of the magnitude spectrum (in Zwicker and Fastl [175]), thus, being the perceptual variant of SC (Peeters *et al.* [34], Richard *et al.* [31]).

5.2.3. Tonality-Related Perceptual Frequency Features

Tonality is a sound property that is closely related to the subjective perception of the main frequency of harmonic signals, and it allows distinguishing noise-like sounds from sinusoidal-like sounds, and especially those sinusoidal sounds whose frequencies are harmonically related. Contrary to the previous reviewed pitch-based audio features (see Section 4.2.4), psychoacoustical pitch feature incorporates auditory-based models. This is a measure that models human pitch perception (as defined by Meddis and O'Mard [176]), by incorporating a band pass filtering that emphasizes the most relevant frequency band for pitch perception, the use of specific filter bank model (Gammatone) that mimics the frequency selectivity of the cochlea, and use of inner hair-cell models that allows computing autocorrelation functions of continuous firing probabilities. The final feature is computed summing across channels all these autocorrelation functions. Previous approaches, like the work by Slaney and Lyon [177], combine a cochlear model with a bank of autocorrelators.

5.2.4. Loudness-Related Perceptual Frequency Features

This section summarizes those features that are related to the loudness of the audio signal, a notion that is defined as the subjective impression of the intensity of a sound (see Peeters *et al.* [34]).

*   **Loudness:** in the original work by Olson [178] the loudness measurement procedure of a complex sound (e.g., speech, music, noise) is described as the sum of the loudness index (using equal loudness contours) for each of the several subbands in which the audio us previously divided. In the work by Breebaart and McKinney [179] the authors compute loudness by firstly computing the power spectrum of the input frame and then normalizing by subtracting (in dB) an approximation of the absolute threshold of hearing, and then filtering by a bank of gammatone filters and summing across frequency to yield the power in each auditory filter, which corresponds to the internal excitation as a function of frequency. These excitations are then compressed, scaled and summed across filters to arrive at the loudness estimate.
*   **Specific loudness sensation:** this is a measure of loudness (in Sone units, a perceptual scale for loudness measurement (see Peeters *et al.* [34]) in a specific frequency range. It incorporates both Bark-scale frequency analysis and the spectral masking effect that emulates the human auditory system (Pampalk *et al.* [162]). This feature has been applied to audio retrieval (Morchen *et al.* [42]).
*   **Integral loudness:** this feature closely measures the human sensation of loudness by spectral integration of loudness over several frequency groups (Pfeiffer [180]). This feature has been used for discrimination between foreground and background sounds (see Linehart *et al.* [181], Pfeiffer *et al.* [182]).

5.2.5. Roughness-Related Perceptual Frequency Features

In the work by Daniel and Weber [183], roughness is defined as a basic psychoacoustical sensation for rapid amplitude variations which reduces the sensory pleasantness and the quality of noises. According to psychophysical theories, the roughness of a complex sound (a sound comprising many partials or pure tone components) depends on the distance between the partials measured in critical bandwidths.

In fact, roughness is considered a perceptual or psycho-acoustic feature, but it also captures amplitude modulations. In particular, it is defined as the perception of temporal envelope modulations in the range of about 20–150 Hz and is maximal for modulations near 70 Hz.

To compute the roughness feature, the following pipeline is defined in the work by McKinney and Breeebaart [167]: (i) the temporal (Hilbert) envelope of each filter of a bank of Gammatone filters is computed; (ii) a correlation factor for each filter based on the correlation of its output with that from two filters above and below it in the filter bank is obtained; (iii) the roughness estimate is then calculated by filtering the power in each filter output with a set of bandpass filters (centered near 70 Hz) that pass only those modulation frequencies relevant to the perception of roughness (Zwicker and Fastl [175]), multiplying by the correlation factor and then summing across frequency and across the filter bank.

5.3. Wavelet-Based Perceptual Features

In the following paragraphs, we describe the most commonly used Wavelet-based perceptual frequency features, which represent an extension of the wavelet-based physical features previously reviewed in Section 4.3.

- **Kernel Power Flow Orientation Coefficients (KPFOC):** in the works by Gerazov and Ivanovski [184,185], a bank of 2D kernels is used to estimate the orientation of the power flow at every point in the auditory spectrogram calculated using a Gammatone filter bank (Valero and Alías [161]), obtaining an ASR front-end with increased robustness to both noise and room reverberation with respect to previous approaches, and specially for small vocabulary tasks.
- **Mel Frequency Discrete Wavelet Coefficients:** or MFDWC for short, account for the perceptual response of the ear by applying the discrete WT to the Mel-scaled log filter bank energies obtained from the input signal (see Gowdy and Tufekci [186]). MFDWC, which were initially defined to improve speech recongition problems (Tavenei *et al.* [187]), have been subsequently applied to other machine hearing realed applications such as speaker verification/identification (see Tufekci and Gurbuz [188], Nghia *et al.* [189]), and audio-based surveillance systems (Rabaoui *et al.* [24]).
- **Gammatone wavelet features:** is a subtype of audio features formulated in the Wavelet domain that accounts for perceptual modelling is the Gammatone Wavelet features (GTW) (see Valero and Alías [190], Venkitaraman *et al.* [191]). These features are obtained by replacing typical mother functions, such as Morlet (Burrus *et al.* [192]), Coiflet (Bradie [193]) or Daubechies [194]) by Gammatone functions, which model the auditory system. GTW features show superior classification accuracy both in noiseless and noisy conditions when compared to Daubechies Wavelet features in classification of surveillance-related sounds, as exposed by Valero and Alías [190].
- **Perceptual wavelet packets:** the Wavelet packet transform is an implementation version of the discrete WT, where the filtering process is iterated on both the low frequency and high frequency components (see Jiang *et al.* [195]), which has been optimized perceptually by including the representation of the input audio into critical bands described by Greenwood [196] and Ren *et al.* [197]. Wavelet packet transform has been used in different applications like in the work by Dobson *et al.* [198] for audio coding purposes, or in audio watermarking (Artameeyanant [199]). Perceptual Wavelet Packets (PWP) have been applied to bio-acoustic signal enhancement (Ren *et al.* [197]), speech recognition (Rajeswari *et al.* [200]), and more recently also for baby crying sound events recognition (Ntalampiras [201]).
- **Gabor functions:** the work by Kleinschmidt models the receptive field of cortical neurons also as two-dimensional complex Gabor function [202]. More recently, Heckman *et al.* have studied in [203] the use of Gabor functions against learning the features via Independent Component Analysis technique for the computation of local features in a two-layer hierarchical bio-inspired approach. Wu *et al.* employ two-dimensional Gabor functions with different scales and directions

to analyze the localized patches of the power spectrogram [204], improving the speech recognition performance in noisy environments, compared with other previous speech feature extraction methods. In a similar way, Schröder *et al.* [205] propose an optimization of spectro-temporal Gabor filterbank features for the audio events detection task. In [206,207], Lindeberg and Friberg describe a new way of deriving the Gabor filters as a particular case (using non-causal Gaussian windows) of frequency selective temporal receptive fields, representing the first layer of their scale-space theory for auditory signals.

5.4. Multiscale Spectro-Temporal-Based Perceptual Features

There are different approaches in the bibliography that use the concept of spectro-temporal response field and that incorporate two-stage processes inspired in the auditory system. Those approaches rely on the fact that measurements in the primary auditory cortex of different animals revealed its spectro-temporal organization, *i.e.*, the receptive fields are selective to modulations in the time-frequency domain. In the following paragraphs, proposals that incorporate spectro-temporal analysis of audio signals at different (temporal and/or frequency) scales are reviewed.

- **Multiscale spectro-temporal modulations:** it consists of two basic stages, as defined by Mesgarani *et al.* [208]. An early stage models the transformation of the acoustic signal into an internal neural representation referred to as an auditory spectrogram (using bank of 128 constant-Q bandpass filters with center frequencies equally spaced on a logarithmic frequency axis). Subsequently a central stage analyzes the spectrogram to estimate the content of its spectral and temporal modulations using a bank of modulation-selective filters (equivalent to a two-dimensional affine wavelet transform of the auditory spectrogram, with a spectro-temporal mother wavelet resembling a two-dimensional spectro-temporal Gabor function) mimicking those described in a model of the mammalian primary auditory cortex. In [208], Mesgarani *et al.* use multiscale spectro-temporal modulations to discriminate speech from nonspeech consisting of animal vocalizations, music, and environmental sounds. Moreover, these features have been applied to music genre classification (Panagakis *et al.* [209]) and voice activity detection (Ng *et al.* [210]).

- **Computational models for auditory receptive fields:** in [206,207], Lindeberg and Friberg describe a theoretical and methodological framework to define computational models for auditory receptive fields. The proposal is also based on a two-stage process: (i) a first layer of frequency selective temporal receptive fields where the input signal is represented as multi-scale spectrograms, which can be specifically configured to simulate the physical resonance system in the cochlea spectrogram; (ii) a second layer of spectro-temporal receptive fields which consist of kernel-based 2D processing units in order to capture relevant auditory changes in both time and frequency dimensions (after logarithmic representation in both amplitude and frequency axes, and ignoring phase information), including from separable to non-separable (introducing an specific glissando parameter) spectro-temporal patterns. The presented model is closely related to biological receptive fields (*i.e.*, those that can be physiologically measured from neurons, in the inferior colliculus and the primary auditory cortex, as reported by Qiu *et al.* in [211]). This work gives an interesting perspective unifying in one theory a way to axiomatically derive representations like Gammatone (see Patterson and Holdsworth [212]) or Gabor filterbanks (Wolfe *et al.* [130]), regarding the causality of the filters used in the first audio analysis stage (see Section 5.6). A set of new auditory features are proposed respecting auditory invariances, being the result of the output 2D spectrogram after the kernel-based processing, using different operators like: spectro-temporal smoothings, onset and offset detections, spectral sharpenings, ways for capturing frequency variations over time and glissando estimation.

5.5. Image Domain Perceptual Features

In this section, the image-based audio features introduced in Section 4.4 are extended whenever they include some aspect of psycho-acoustics or perceptual models of hearing. Following a perceptual image-based approach, the main features described in the literature are the following:

- **Spectrogram image features:** as introduced in Section 4.4, spectrogram image features have been also derived from front-end parametrizations which make use of psychoacoustical models. In [213], Dennis *et al.* use GHT to construct a codebook from a Gaussian Mixture Model-Hidden Markov Model based ASR, in order to train an artificial neural network that learns a discriminative weighting for optimizing the classification accuracy in a frame-level phoneme recognition application. In this work MFCC are used as front-end parametrization. The same authors compute in [214] a robust sparse spike coding of the 40-dimension Mel-filtered spectrogram (detection of high energy peaks that correlate with a codebook dictionary) to learn a neural network for sound event classification. The results show a superior reliability when the proposed parameterization is used against the conventional raw spectrogram.

- **Auditory image model:** or AIM for short, this feature extraction technique includes functional and physiological modules to simulate auditory spectral analysis, neural encoding and temporal integration, including new forms of periodicity-sensitive temporal integration that generate stabilized auditory images (Patterson *et al.* [215,216]). The encoding process is based on a three stage system. Briefly, the spectral analysis stage converts the sound wave into the model's representation of basilar membrane motion (BMM). The neural encoding stage stabilizes the BMM in level and sharpens features like vowel formants, to produce a simulation of the neural activity pattern produced by the sound in the auditory nerve. The temporal integration stage stabilizes the repeating structure in the NAP and produces a simulation of our perception, referred to as the auditory image.

- **Stabilized auditory image:** based on the AIM features, the stabilized auditory image (SAI) is defined as a two-dimensional representation of the sound signal (see Walters [137]): the first dimension of a SAI frame is simply the spectral dimension added by a previous filterbank analysis, while the second comes from the strobed temporal integration process by which an SAI is generated. SAI has been applied to speech recognition and audio search [137] and more recently, a low-resolution overlapped SIF has been introduced together with Deep Neural Networks (DNN) to perform robust sound event classification in noisy conditions (McLoughlin *et al.* [217]).

- **Time-chroma images:** this feature is a two dimensional representation for audio signals that plots the chroma distribution of an audio signal over time, as described in the work by Malekesmaeili and Ward [218]. This feature employs a modified definition of the chroma concept called chroma^n, which is defined as the set of all pitches that are apart by n octaves. Coupled with a fingerprinting algorithm that extracts local fingerprints from the time-chroma image, the proposed feature allows improved accuracy in audio copy detection and song identification.

5.6. Cepstral Domain Perceptual Features

Within the perceptual-based cepstral domain features two subtypes of features are found: perceptual filter bank-based and autoregression-based.

The following paragraphs describe the most relevant features belonging to these two categories of cepstral features which incorporate perceptual-based schemes.

5.6.1. Perceptual Filter Banks-Based Cepstral Features

Perceptual filter banks-based cepstral features are based on the computation of cepstral-based parameters following an approach based on, firstly, obtaining the logarithm of the magnitude's Fourier transform, or using an specific filter bank decomposition with some possible perceptual criteria; and secondly, performing a Fourier transform (or Cosine transform) of the previous result.

This type of features comprises the well-known Mel Frequency Cepstral Coefficients and their variants, which are often based on using different frequency scales before the last Fourier-based stage. Some examples include Equivalent Rectangular Bandwidths (ERB) (see Moore *et al.* [219]), Bark (see Zwicker [220]), critical bands (as in the work by Greenwood [196]) and octave-scale (see Maddage *et al.* [221]).

Another aspect connected to these type of features is that they are mostly based on the computation of a *cochleagram* (the resulting time-frequency output of the filterbanks), which in some sense try to model the frequency selectivity of the cochlea (as in the work by Richard *et al.* [31]).

The following paragraphs describe the most relevant features in this area.

- **Mel Frequency Cepstral Coefficients**: also denoted as MFCC, have been largely employed in the speech recognition field but also in the field of audio content classification (see the work by Liang and Fan [58]), due to the fact that their computation is based on perceptual-based frequency scale in the first stage (the human auditory model in which is inspired the frequency Mel-scale). After obtaining the frame-based Fourier transform, outputs of a Mel-scale filter bank are logarithmized and finally they are decorrelated by means of the Discrete Cosine Transform (DCT). Only first DCT coefficients (usually from 8 to 13) are used to gather information that represents the low frequency component of the signal's spectral envelope (mainly related to timbre). MFCC's have been used also for music classification (see the works by Benetos *et al.* [10], Bergstra *et al.* [41], Tzanetakis and Cook [28], or Wang *et al.* [9]), singer identification (as in Shen *et al.* [145]), environmental sound classification (see Beritelli and Grasso [222], or Peltonen *et al.* [18]), audio-based surveillance systems (Rabaoui *et al.* [24]), being also embedded in hearing aids (see Zeng and Liu [223]) and even employed detect breath sound as an indicator of respiratory health and disease (Lei *et al.* [224]). Also, some particular extensions of MFCC have been introduced in the context of speech recognition and speaker verification in the aim of obtaining more robust spectral representation in the presence of noise (e.g., in the works by Shannon and Paliwal [225], Yuo *et al.* [27], or Choi [226]).

- **Greenwood Function cepstral coefficients**: building on the seminal work by Greenwood [196], where it was stated that many mammals have a logarithmic cochlear-frequency response, Clemins *et al.* [32] introduced Greenwood function Cepstral Coefficients (GFCC), extracting the equal loudness curve from species-specific audiogram measurements as an audio feature extraction for the analysis of environmental sound coming from the vocalization of those species. Later, this features were applied also to multichannel speech recognition by Trawicki *et al.* [227].

- **Noise-robust audio features**: or NRAF for short, these features incorporate a specific human auditory model based on a three stage process (a first stage of filtering in the cochlea, transduction of mechanical displacement in electrical activity–log compression in the hair cell stage–, and a reduction stage using decorrelation that mimics the lateral inhibitory network in the cochlear nucleus) (see Ravindran *et al.* [228]).

- **Gammatone cepstral coefficients**: also known as GTCC, Patterson *et al.* in [229,230] proposed a filterbank based on Gammatone function that predicts human masking data accurately, while Hohman proposed in [231] an efficient implementation of a Gammatone filterbank (using the 4th-order linear Gammatone filter) for the frequency analysis and resynthesis of audio signals. Valero and Alías derived the Gammatone cepstral coefficients feature by maintaining the effective computation scheme from MFCC but changing the Mel filter bank by a Gammatone filter bank [232]. Gammatone filters were originally designed to model the human auditory spectral response, given their good approximation in terms of impulse response, magnitude response and filter bandwidth, as described by Patterson and Holdsworth [212]. Gammatone-like features have been used also in audio processing (see the work of Johannesma in [233]), in speech recognition applications (see Shao *et al.* [234], or Schlüter *et al.* [235]), water sound event detection for tele-monitoring applications (Guyot *et al.* [236]), for road noise sources classification (Socoró *et al.* [237]) and computational auditory scene analysis (Shao *et al.* [238]). In [206,207] Lindeberg and Friberg describe an axiomatic way of obtaining Gammatone filters as a particular case of a multi-scale spectrogram when the analysis filters are constrained to be causal. They also define a new family of generalized Gammatone filters that allow for additional degrees of freedom in the trade-off between the temporal dynamics and the spectral selectivity of time-causal spectrograms. This approach represents a part of a unified theory for constructing computational models for auditory receptive fields (see a brief description in Section 5.4).

- **GammaChirp filterbanks:** Irino and Patterson proposed in [239] an extension of the Gammatone filter which was called Gammachirp filter, with the aim of obtaining a more accurate model of the auditory sensitivity, providing an excellent fit to human masking data. Specifically, this approach is able to represent the natural asymmetry of the auditory filter and its dependence on the signal strength. Abdallah and Hajaiej [240] defined GammaChirp Cepstral coefficients (GC-Cept) substituting the typical Mel filterbank in MFCC by a Gammachirp filterbank of 32 filters over speech signals (within the speech frequency band up to 8 KHz). They showed a better performance of the new Gammachirp filterbanks compared with the MFCC in a text independent speaker recognition system for noisy environments.

5.6.2. Autoregression-Based Cepstral Features

A common trait of autoregression-based cepstral features is that linear predictive analysis is incorporated within the cepstral-based framework. This group of features includes perceptual linear prediction, relative spectral-perceptual linear prediction and linear prediction cepstrum coefficients, which are described next.

- **Perceptual Linear Prediction:** or PLP for short, this feature represents a more accurate representation of spectral contour by means of a linear prediction-based approach that incorporates also some specific human hearing inspired properties like use of a frequency Bark-scale and asymmetrical critical-band masking curves, as described by Hermansky [241]. These features, were later revised and improved by Hönig *et al.* [242] for speech recognition purposes and recently applied to baby crying sound events recognition by Ntalampiras [201].
- **Relative Spectral-Perceptual Linear Prediction:** also referred to as RASTA-PLP, this is a noise-robust version of the PLP feature introduced by Hermansky and Morgan [243]. The objective is to incorporate human-like abilities to disregard noise when listening in speech communication by means of filtering each frequency channel with a bandpass filter that mitigate slow time variations due to communication channel disturbances (e.g., steady background noise, convolutional noise) and fast variations due to analysis artifacts. Also, the RASTA-PLP process uses static nonlinear compression and expansion blocks before and after the bandpass processing. There is a close relation between RASTA processing and delta cepstral coefficients (*i.e.*, first derivatives of MFCC), which are broadly used in the contexts of speech recognition and statistical speech synthesis. This features have also been applied for audio-based surveillance systems by Rabaoui *et al.* [24].
- **Generalized Perceptual Linear Prediction:** also denoted as gPLP, is defined as an adaptation of PLP originally developed for human speech processing to represent their vocal production mechanisms of mammals by substituting a species-specific frequency warping and equal loudness curve from humans by those from the analyzed species (see the work by Clemins *et al.* [32,33]).

5.7. Other Domains

The literature contains other approaches to perceptual audio feature extraction that operate on domains different to the ones just reviewed. Some of the most significant are the eigenspace-based features, the electroencephalogramn-based features and the auditory saliency map. The following paragraphs briefly describe these approaches.

- **Eigenspace-based features:** in this category, we find Rate-scale-frequency (RSF) features, which describe modulation components present in certain frequency bands of the auditory spectrum, and they are based in the same human auditory model that incorporates the noise-robust audio features (NRAF), as described in the work by Ravindran *et al.* [228] (see Section 5.6.1). RFS represent a compact and decorrelated representation (they are derived performing a Principal Component Analysis stage) of the two-dimensional Wavelet transform applied to the audio spectrum;

- **Electroencephalogram-based features:** or EEG-based features for short, these find application in human-centered favorite music estimation, as introduced by Sawata *et al.* [244]. In that work, the authors compute features from the EEG signals of a user that is listening to his/her favorite music, while simultaneously computing several features from the audio signal (root mean square, brightness, ZCR or tempo, among others). Subsequently, both types of features are correlated by means of kernel canonical correlation analysis (KCCA), which allows deriving a projection between the audio features space and the EEG-based feature space. By using the obtained projection, the new EEG-based audio features can be derived from audio features, since this projection provides the best correlation between both feature spaces. As a result, it becomes possible to transform original audio features into EEG-based audio features with no need of further EEG signals acquisition.

- **Auditory saliency map:** is a bottom-up auditory attention model which computes an auditory saliency map from the input sound derived by Kalinli *et al.* [245,246], and it has been applied to environmental sounds in the work by De Coensel and Botteldooren [247] (perception of transportation noise). The saliency map holds non-negative values and its maximum defines the most salient location in 2D auditory spectrum. First, auditory spectrum of sound is estimated using an early auditory (EA) system model, consisting of cochlear filtering, inner hair cell (IHC), and lateral inhibitory stages mimicking the process from basilar membrane to the cochlear nucleus in the auditory system (using a set of constant-Q asymmetric band-pass filters uniformly distributed along a logarithmic frequency axis). Next, the auditory spectrum is analyzed by extracting a set of multi-scale features (2D spectro-temporal receptive filters) which consist of intensity, frequency contrast, temporal contrast and orientation feature channels. Subsequently, center-surround differences (point wise differences across different center-based and surrounding-based scales) are calculated from the previous feature channels, resulting in feature maps. From the computed 30 features maps (six for each intensity, frequency contrast, temporal contrast and twelve for orientation) an iterative and nonlinear normalization algorithm (simulating competition between the neighboring salient locations using a large 2D difference of Gaussians filter) is applied to the possible noisy feature maps, obtaining reduced sparse representations of only those locations which strongly stand-out from their surroundings. All normalized maps are then summed to provide bottom-up input to the saliency map.

6. Conclusions

This work has presented an up-to-date review of the most relevant audio feature extraction techniques related to machine hearing which have been developed for the analysis of speech, music and environmental sounds. With the aim of providing a self-contained reference for audio analysis applications practitioners, this review covers the most elementary and classic approaches to audio feature extraction, dating back to the 1970s, through to the most recent contributions for the derivation of audio features based on new domains of computation and bio-inspired paradigms.

To that effect, we revisit classic audio feature extraction techniques taking the complete work by Mitrović *et al.* [17] as a reference, and extend those approaches by accounting for the latest advances in this research field. Besides extending that review with features computed on time, frequency and cepstral domains, we describe feature extraction techniques computed on the wavelet and image domains, obtained from multilinear or non-linear parameterizations, together with those derived from specific representations such as the machine-pursuit algorithm or the Hurst parameterization. Moreover, it is worth noting that a significant number of novel bio-inspired proposals are also described (e.g., including an auditory model such as Mel and Gammatone filter-banks, or derived from the computation of the autocorrelation function or the auditory image model). The described audio features extraction techniques are classified depending on whether they have a physical or a perceptual basis.

It is worth mentioning that the increase of complexity in the field of audio parameterization, specifically as regards the more recent perceptual and bio-inspired approaches, makes it difficult to obtain a clear taxonomy that accommodates all the proposals found in the literature. For instance, a different perspective that goes beyond the proposed taxonomy of audio features could be applied to organize some of the described perceptual features.

Concretely, many perceptual features are based on obtaining a first set of features that try to emulate the physical resonance system in the cochlea using filterbanks with specific frequency positions and bandwidths (e.g., image domain perceptual features, perceptual filter banks-based cepstral features, auditory saliency maps, some of the wavelet-based perceptual features, or most of the modulation-based perceptual frequency features).

Moreover, some of these perceptually-based approaches define a second stage to obtain a set of meaningful features that correlate in some sense with psycho-acoustical responses from the previous spectrogram-based representation. In some cases, this second-stage features are derived from kernel-based 2D processing (like in the wavelet-based perceptual features) while other approaches propose more elaborate processing stages (e.g., auditory saliency maps).

Furthermore, the description of the main concepts and principles behind all reviewed feature extraction techniques has considered the specific particularities of the three main types of audio inputs considered: speech, music and environmental sounds. Furthermore, we have included some classic and recent examples to illustrate the application of these techniques in several specific machine hearing related problems, e.g., for speech: segmentation, recognition, speaker verification/identification or language identification; for music: annotation, recommendation, genre classification, instrument recognition, song identification, or mood classification; for environmental sound: recognition, classification, audio-based surveillance or computational auditory scene analysis, among others.

Finally, we would like to note that this work has been written not as a thorough collection of *all* existing audio features extraction techniques related to audio analysis, but as an attempt to collate up-to-date approaches found in the literature of this dynamic field of research. Furthermore, we expect that the new works proposing innovative approaches in machine hearing require the development of novel audio feature extraction techniques that will extend this work.

Acknowledgments: This research has been partially funded by the European Commission under project LIFE DYNAMAP LIFE13 ENV/IT/001254 and the Secretaria d'Universitats i Recerca del Departament d'Economia i Coneixement (Generalitat de Catalunya) under grant refs. 2014-SGR-0590 and 2015-URL-Proj-046.

Author Contributions: The authors contributed equally to this work. Francesc Alías led the choice of the focus of the review and the initiative of preparing and writing the manuscript, while Joan Claudi Socoró and Xavier Sevillano led the bibliographic search and review of the references and also contributed to the writing of the paper.

Conflicts of Interest: The authors declare no conflict of interest.

Abbreviations

The following abbreviations are used in this manuscript:

ACF	Autocorrelation Function features
AD	Amplitude Descriptor
AFTE	Auditory Filter bank Temporal Envelopes
AMSC/AMSV	Acoustic-Modulation Spectral Contrast/Valley
AMSFM	Acoustic-Modulation Spectral Flatness Measure (AMSFM)
AMSCM	Acoustic-Modulation Spectral Crest Measure (AMSCM)
ASB	Audio Spectrum Basis
ASC	Audio Spectrum Centroid
ASE	audio Spectrum Envelope
ASF	Audio Spectrum Flatness
ASP	Audio Spectrum Projection
ASR	Automatic Speech Recognition
ASS	Audio Spectrum Spread
ATFT	Adaptive Time Frequency Transform
AW	Audio Waveform
AZCR	Autocorrelation Zero Crossing Rate
BMM	Basilar Membrane Motion
CASA	Computational Auditory Scene Analysis
CBS	Cyclic Beat Spectrum
CELP	Code Excited Linear Prediction
CENS	Chroma Energy distribution Normalized Statistics
dB	Decibels
DCT	Discrete Cosine Transform
DDA	Distortion Discriminant Analysis
DNN	Deep Neural Networks
DWCH	Daubechies Wavelet Coefficient Histogram
DRR	Direct-to-Reverberant Ratio
ERB	Equivalent Rectangular Bandwidth
EA	Early Auditory model
F0	Fundamental frequency
GC-Cept	GammaChirp Cepstral coefficients
GDF	Group Delay Functions
GFCC	Greenwood Function Cepstral Coefficients
GHT	Generalised Hough Transform
gPLP	Generalized Perceptual Linear Prediction
GTCC	Gammatone Cepstral Coefficients
GTW	Gammatone Wavelet features
HNR	Harmonic-to-Noise Ratio
HR	Harmonic Ratio
HSC	Harmonic Spectral Centroid
HSD	Harmonic Spectral Deviation
HSS	Harmonic Spectral Spread
HSV	Harmonic Spectral Variation
ICA	Independent Component Analysis
IHC	Inner Hair Cell
KCCA	Kernel Canonical Correlation Analysis
KPFOCs	Kernel Power Flow Orientation Coefficients
LAT	Log Attack Time
LPC	Linear Prediction Coefficient
LPCC	Linear Prediction Cepstrum Coefficients
LP-ZCR	Linear Prediction Zero Crossing Ratio
LS	Local Spectrogram features
LSF	Line Spectral Frequencies
LSP	Line Spectral Pairs

MCLT	Modulated Complex Lapped Transform
MFCC	Mel-Frequency Cepstrum Coefficient
MFDWC	Mel Frequency Discrete Wavelet Coefficients
MGDF	Modified Group Delay Functions
MP	Matching Pursuit
MPEG	Moving Picture Experts Group
MS	Modulation spectrogram
NASE	Normalized Spectral Envelope
NB-ACF	Narrow-Band Autocorrelation Function features
NIRA	Non-intrusive Room Acoustic parameter
NRAF	Noise-Robust Audio Features
OSC	Octave-based Spectral Contrast
PCA	Principal Component Analysis
pH	Hurst parameter features
PLP	Perceptual Linear Prediction
PS-ZCPA	Pitch Synchronous Zero Crossing Peak Amplitudes
PWP	Perceptual Wavelet Packets
RASTA-PLP	Relative Spectral-perceptual Linear Prediction
RIR	Room Impulse Response
RMS	Root Mean Square
RSF	Rate-Scale-Frequency
SAI	Stabilised Auditory Image
SC	Spectral Centroid
SE	Spectral Envelope
SF	Spectral Flux
SIF	Spectrogram Image Features
SNR	Signal-to-Noise Ratio
SPD	Subband Power Distribution
SPSF	Stereo Panning Spectrum Feature
STE	Short-time Energy
STFT	Short Time Fourier Transform
ULH	Upper Limit of Harmonicity
WT	Wavelet Transform
ZCPA	Zero Crossing Peak Amplitudes
ZCR	Zero Crossing Rate

References

1. Lyon, R.F. Machine Hearing: An Emerging Field. *IEEE Signal Process. Mag.* **2010**, *27*, 131–139.
2. Gerhard, D. *Audio Signal Classification: History and Current Techniques;* Technical Report TR-CS 2003-07; Department of Computer Science, University of Regina: Regina, SK, Canada, 2003.
3. Temko, A. Acoustic Event Detection and Classification. Ph.D. Thesis, Universitat Politècnica de Catalunya, Barcelona, Spain, 23 January 2007.
4. Dennis, J. Sound Event Recognition in Unstructured Environments Using Spectrogram Image Processing. Ph.D. Thesis, School of Computer Engineering, Nanyang Technological University, Singapore, 2014.
5. Bach, J.H.; Anemüller, J.; Kollmeier, B. Robust speech detection in real acoustic backgrounds with perceptually motivated features. *Speech Commun.* **2011**, *53*, 690–706.
6. Kinnunen, T.; Li, H. An overview of text-independent speaker recognition: From features to supervectors. *Speech Commun.* **2011**, *52*, 12–40.
7. Pieraccini, R. *The Voice in the Machine. Building Computers That Understand Speech*; MIT Press: Cambridge, MA, USA, 2012.
8. Wang, A.L.C. An industrial-strength audio search algorithm. In Proceedings of the 4th International Conference on Music Information Retrieval (ISMIR), Baltimore, MD, USA, 26–30 October 2003; pp. 7–13.

9. Wang, F.; Wang, X.; Shao, B.; Li, T.; Ogihara, M. Tag Integrated Multi-Label Music Style Classification with Hypergraph. In Proceedings of the 10th International Society for Music Information Retrieval Conference (ISMIR), Kobe, Japan, 26–30 October 2009; pp. 363–368.

10. Benetos, E.; Kotti, M.; Kotropoulos, C. Musical Instrument Classification using Non-Negative Matrix Factorization Algorithms and Subset Feature Selection. In Proceedings of the 2006 IEEE International Conference on Acoustics, Speech, and Signal Processing (ICASSP), Toulouse, France, 14–19 May 2006; Volume 5, pp. V:221–V:225.

11. Liu, M.; Wan, C. Feature selection for automatic classification of musical instrument sounds. In Proceedings of the ACM/IEEE Joint Conference on Digital Libraries (JCDL), Roanoke, VA, USA, 24–28 June 2001; pp. 247–248.

12. Lu, L.; Liu, D.; Zhang, H.J. Automatic Mood Detection and Tracking of Music Audio Signals. *IEEE Trans. Audio Speech Lang. Process.* **2006**, *14*, 5–18.

13. Lyon, R.F. A Survey of Audio-Based Music Classification and Annotation. *IEEE Trans. Multimedia* **2011**, *13*, 303–319.

14. Chu, S.; Narayanan, S.S.; Kuo, C.J. Environmental Sound Recognition With Time-Frequency Audio Features. *IEEE Trans. Audio Speech Lang. Process.* **2009**, *17*, 1142–1158.

15. Valero, X.; Alías, F. Classification of audio scenes using Narrow-Band Autocorrelation features. In Proceedings of the 20th European Signal Processing Conference (EUSIPCO), Bucharest, Romania, 27–31 August 2012; pp. 2012–2019.

16. Schafer, R.M. *The Soundscape: Our Sonic Environment and the Tuning of the World*; Inner Traditions/Bear & Co: Rochester, VT, USA, 1993.

17. Mitrović, D.; Zeppelzauer, M.; Breiteneder, C. Features for content-based audio retrieval. *Adv. Comput.* **2010**, *78*, 71–150.

18. Peltonen, V.; Tuomi, J.; Klapuri, A.; Huopaniemi, J.; Sorsa, T. Computational Auditory Scene Recognition. In Proceedings of the 2002 IEEE International Conference on Acoustics, Speech, and Signal Processing (ICASSP), Orlando, FL, USA, 13–17 May 2002; Volume 2, pp. II:1941 – II:1944.

19. Geiger, J.; Schuller, B.; Rigoll, G. Large-scale audio feature extraction and SVM for acoustic scene classification. In Proceedings of the 2013 IEEE Workshop on Applications of Signal Processing to Audio and Acoustics (WASPAA), New Paltz, NY, USA, 20–23 October 2013, pp. 1–4.

20. Oppenheim, A.V.; Schafer, R.W. *Discrete-Time Signal Processing*; Prentice Hall: Upper Saddle River, NJ, USA, 1989.

21. Gygi, B. Factors in the Identification of Environmental Sounds. Ph.D. Thesis, Indiana University, Bloomington, IN, USA, 12 July 2001.

22. Foote, J.; Uchihashi, S. The Beat Spectrum: A New Approach To Rhythm Analysis. In Proceedings of the 2001 IEEE International Conference on Multimedia and Expo (ICME), Tokyo, Japan, 22–25 August 2001, pp. 881–884.

23. Bellman, R. *Dynamic Programming*; Dover Publications: Mineola, NY, USA, 2003.

24. Rabaoui, A.; Davy, M.; Rossignol, S.; Ellouze, N. Using One-Class SVMs and Wavelets for Audio Surveillance. *IEEE Trans. Inf. Forensics Secur.* **2008**, *3*, 763–775.

25. Hurst, H.E. Long-term storage capacity of reservoirs. *Trans. Amer. Soc. Civ. Eng.* **1951**, *116*, 770–808.

26. Eronen, A.J.; Peltonen, V.T.; Tuomi, J.T.; Klapuri, A.P.; Fagerlund, S.; Sorsa, T.; Lorho, G.; Huopaniemi, J. Audio-based context recognition. *IEEE Trans. Audio Speech Lang. Process.* **2006**, *14*, 321–329.

27. Yuo, K.; Hwang, T.; Wang, H. Combination of autocorrelation-based features and projection measure technique for speaker identification. *IEEE Trans. Speech Audio Process.* **2005**, *13*, 565–574.

28. Tzanetakis, G.; Cook, P.R. Musical genre classification of audio signals. *IEEE Trans. Speech Audio Process.* **2002**, *10*, 293–302.

29. Ando, Y. A theory of primary sensations and spatial sensations measuring environmental noise. *J. Sound Vib.* **2001**, *241*, 3–18.

30. Valero, X.; Alías, F. Hierarchical Classification of Environmental Noise Sources by Considering the Acoustic Signature of Vehicle Pass-bys. *Arch. Acoustics* **2012**, *37*, 423–434.

31. Richard, G.; Sundaram, S.; Narayanan, S. An Overview on Perceptually Motivated Audio Indexing and Classification. *Proc. IEEE* **2013**, *101*, 1939–1954.

32. Clemins, P.J.; Trawicki, M.B.; Adi, K.; Tao, J.; Johnson, M.T. Generalized Perceptual Features for Vocalization Analysis Across Multiple Species. In Proceedings of the IEEE International Conference on Acoustics, Speech and Signal Processing (ICASSP), Toulouse, France, 14–19 May 2006; Volume 1.
33. Clemins, P.J.; Johnson, M.T. Generalized perceptual linear prediction features for animal vocalization analysis. *J. Acoust. Soc. Am.* **2006**, *120*, 527–534.
34. Peeters, G. *A Large Set of Audio Features for Sound Description (Similarity And Classification) in the CUIDADO Project*; Technical Report; IRCAM: Paris, France, 2004.
35. Sharan, R.; Moir, T. An overview of applications and advancements in automatic sound recognition. *Neurocomputing* **2016**, doi:10.1016/j.neucom.2016.03.020.
36. Gubka, R.; Kuba, M. A comparison of audio features for elementary sound based audio classification. In Proceedings of the 2013 International Conference on Digital Technologies (DT), Zilina, Slovak Republic, 29–31 May 2013; pp. 14–17.
37. Boonmatham, P.; Pongpinigpinyo, S.; Soonklang, T. A comparison of audio features of Thai Classical Music Instrument. In Proceedings of the 7th International Conference on Computing and Convergence Technology (ICCCT), Seoul, South Korea, 3–5 December 2012, pp. 213–218.
38. Van Hengel, P.W.J.; Krijnders, J.D. A Comparison of Spectro-Temporal Representations of Audio Signals. *IEEE/ACM Trans. Audio Speech Lang. Process.* **2014**, *22*, 303–313.
39. Kedem, B. Spectral Analysis and Discrimination by Zero-crossings. *Proc. IEEE* **1986**, *74*, 1477–1393.
40. Li, T.; Ogihara, M.; Li, Q. A Comparative Study on Content-based Music Genre Classification. In Proceedings of the 26th Annual International ACM SIGIR Conference on Research and Development in Information Retrieval, Toronto, ON, Canada, 28 July–1 August 2003; pp. 282–289.
41. Bergstra, J.; Casagrande, N.; Erhan, D.; Eck, D.; Kégl, B. Aggregate features and ADABOOST for music classification. *Mach. Learn.* **2006**, *65*, 473–484.
42. Mörchen, F.; Ultsch, A.; Thies, M.; Lohken, I. Modeling timbre distance with temporal statistics from polyphonic music. *IEEE Trans. Audio Speech Lang. Process.* **2006**, *14*, 81–90.
43. Ghaemmaghami, H.; Baker, B.; Vogt, R.; Sridharan, S. Noise robust voice activity detection using features extracted from the time-domain autocorrelation function. In Proceedings of the 11th Annual Conference of the International Speech (InterSpeech), Makuhari, Japan, 26–30 September 2010; Kobayashi, T., Hirose, K., Nakamura, S., Eds.; pp. 3118–3121.
44. El-Maleh, K.; Klein, M.; Petrucci, G.; Kabal, P. Speech/music discrimination for multimedia applications. In Proceedings of the 2000 IEEE International Conference on Acoustics, Speech, and Signal Processing (ICASSP), Istanbul, Turkey, 5–9 June 2000; Volume 4, pp. 2445–2448.
45. International Organization for Standardization (ISO)/International Organization for Standardization (IEC). Information technology—Multimedia content description interface. Part 4 Audio, 2002. Available online: http://mpeg.chiariglione.org/standards/mpeg-7/audio (accessed on 4 May 2016).
46. Mitrović, D.; Zeppelzauer, M.; Breiteneder, C. Discrimination and retrieval of animal sounds. In Proceedings of the 12th International Multi-Media Modelling Conference, Beijing, China, 4–6 January 2006; pp. 339–343.
47. Muhammad, G.; Alghathbar, K. Environment Recognition from Audio Using MPEG-7 Features. In Proceedings of the 4th International Conference on Embedded and Multimedia Computing, Jeju, Korea, 10–12 December 2009; pp. 1–6.
48. Valero, X.; Alías, F. Applicability of MPEG-7 low level descriptors to environmental sound source recognition. In Proceedings of the EAA EUROREGIO 2010, Ljubljana, Slovenia, 15–18 September 2010.
49. Klingholz, F. The measurement of the signal-to-noise ratio (SNR) in continuous speech. *Speech Commun.* **1987**, *6*, 15–26.
50. Kreiman, J.; Gerratt, B.R. Perception of aperiodicity in pathological voice. *J. Acoust. Soc. Am.* **2005**, *117*, 2201–2211.
51. Farrús, M.; Hernando, J.; Ejarque, P. Jitter and shimmer measurements for speaker recognition. In Proceedings of the 8th Annual Conference of the International Speech Communication Association (InterSpeech), Antwerp, Belgium, 27–31 August 2007; pp. 778–781.
52. Murthy, Y.V.S.; Koolagudi, S.G. Classification of vocal and non-vocal regions from audio songs using spectral features and pitch variations. In Proceedings of the 28th Canadian Conference on Electrical and Computer Engineering (CCECE), Halifax, NS, Canada, 3–6 May 2015; pp. 1271–1276.

53. Kato, K.; Ito, A. Acoustic Features and Auditory Impressions of Death Growl and Screaming Voice. In Proceedings of the Ninth International Conference on Intelligent Information Hiding and Multimedia Signal Processing (IIH-MSP), Beijing, China, 16–18 October 2013; Jia, K., Pan, J.S., Zhao, Y., Jain, L.C., Eds.; pp. 460–463.

54. Jensen, K. Pitch independent prototyping of musical sounds. In Proceedings of the IEEE 3rd Workshop on Multimedia Signal Processing, Copenhagen, Denmark, 13–15 September 1999; pp. 215–220.

55. Chu, W.; Cheng, W.; Hsu, J.Y.; Wu, J. Toward semantic indexing and retrieval using hierarchical audio models. *J. Multimedia Syst.* **2005**, *10*, 570–583.

56. Zhang, T.; Kuo, C.C. Audio content analysis for online audiovisual data segmentation and classification. *IEEE Trans. Speech Audio Process.* **2001**, *9*, 441–457.

57. Smith, D.; Cheng, E.; Burnett, I.S. Musical Onset Detection using MPEG-7 Audio Descriptors. In Proceedings of the 20th International Congress on Acoustics (ICA), Sydney, Australia, 23–27 August 2010; pp. 10–14.

58. Liang, S.; Fan, X. Audio Content Classification Method Research Based on Two-step Strategy. *Int. J. Adv. Comput. Sci. Appl. (IJACSA)* **2014**, *5*, 57–62.

59. Liu, Z.; Wang, Y.; Chen, T. Audio Feature Extraction and Analysis for Scene Segmentation and Classification. *J. VLSI Signal Process. Syst. Signal Image Video Technol.* **1998**, *20*, 61–79.

60. Jiang, H.; Bai, J.; Zhang, S.; Xu, B. SVM-based audio scene classification. In Proceedings of the 2005 IEEE International Conference on Natural Language Processing and Knowledge Engineering, Wuhan, China, 30 October–1 November 2005; pp. 131–136.

61. Scheirer, E.D.; Slaney, M. Construction and evaluation of a robust multifeature speech/music discriminator. In Proceedings of the IEEE International Conference on Acoustics, Speech, and Signal Processing (ICASSP), Munich, Germany, 21–24 April 1997; pp. 1331–1334.

62. Lartillot, O.; Eerola, T.; Toiviainen, P.; Fornari, J. Multi-feature modeling of pulse clarity: Design, validation and optimization. In Proceedings of the Ninth International Conference on Music Information Retrieval (ISMIR), Philadelphia, PA, USA, 14–18 September 2008; pp. 521–526.

63. Friberg, A.; Schoonderwaldt, E.; Hedblad, A.; Fabiani, M.; Elowsson, A. Using listener-based perceptual features as intermediate representations in music information retrieval. *J. Acoust. Soc. Am.* **2014**, *136*, 1951–1963.

64. Lu, L.; Jiang, H.; Zhang, H. A robust audio classification and segmentation method. In Proceedings of the 9th ACM International Conference on Multimedia, Ottawa, ON, Canada, 30 September–5 October 2001; Georganas, N.D., Popescu-Zeletin, R., Eds.; pp. 203–211.

65. Foote, J. Automatic Audio Segmentation using a Measure of Audio Novelty. In Proceedings of the IEEE International Conference on Multimedia and Expo (ICME), New York, NY, USA, 30 July–2 August 2000; p. 452.

66. Kurth, F.; Gehrmann, T.; Müller, M. The Cyclic Beat Spectrum: Tempo-Related Audio Features for Time-Scale Invariant Audio Identification. In Proceedings of the 7th International Conference on Music Information Retrieval (ISMIR), Victoria, BC, Canada, 8–12 October 2006; pp. 35–40.

67. Scheirer, E.D. Tempo and beat analysis of acoustic musical signals. *J. Acoust. Soc. Am.* **1998**, *103*, 588–601.

68. Khan, M.K.S.; Al-Khatib, W.G.; Moinuddin, M. Automatic Classification of Speech and Music Using Neural Networks. In Proceedings of the 2nd ACM International Workshop on Multimedia Databases, Washington, DC, USA, 8–13 November 2004; pp. 94–99.

69. Khan, M.K.S.; Al-Khatib, W.G. Machine-learning Based Classification of Speech and Music. *Multimedia Syst.* **2006**, *12*, 55–67.

70. Itakura, F. Line spectrum representation of linear predictor coefficients of speech signals. In Proceedings of the 89th Meeting of the Acoustical Society of America, Austin, TX, USA, 8–11 April 1975.

71. Sarkar, A.; Sreenivas, T.V. Dynamic programming based segmentation approach to LSF matrix reconstruction. In Proceedings of the 9th European Conference on Speech Communication and Technology (EuroSpeech), Lisbon, Portugal, 4–8 September 2005; pp. 649–652.

72. Schroeder, M.; Atal, B. Code-excited linear prediction (CELP): High-quality speech at very low bit rates. In Proceedings of the IEEE International Conference on Acoustics, Speech, and Signal Processing (ICASSP), Tampa, FL, USA, 26–29 April 1985; Volume 10, pp. 937–940.

73. Tsau, E.; Kim, S.H.; Kuo, C.C.J. Environmental sound recognition with CELP-based features. In Proceedings of the 10th International Symposium on Signals, Circuits and Systems, Iasi, Romania, 30 June–1 July 2011; pp. 1–4.

74. Srinivasan, S.; Petkovic, D.; Ponceleon, D. Towards Robust Features for Classifying Audio in the CueVideo System. In Proceedings of the Seventh ACM International Conference on Multimedia (Part 1), Orlando, FL, USA, 30 October–5 November 1999; pp. 393–400.

75. Farahani, G.; Ahadi, S.M.; Homayounpoor, M.M. Use of Spectral Peaks in Autocorrelation and Group Delay Domains for Robust Speech Recognition. In Proceedings of the 2006 IEEE International Conference on Acoustics Speech and Signal Processing, (ICASSP), Toulouse, France, 14–19 May 2006; pp. 517–520.

76. Kim, H.; Moreau, N.; Sikora, T. Audio classification based on MPEG-7 spectral basis representations. *IEEE Trans. Circuits Syst. Video Technol.* **2004**, *14*, 716–725.

77. Lee, C.; Shih, J.; Yu, K.; Lin, H. Automatic Music Genre Classification Based on Modulation Spectral Analysis of Spectral and Cepstral Features. *IEEE Trans. Multimedia* **2009**, *11*, 670–682.

78. Tzanetakis, G.; Jones, R.; McNally, K. Stereo Panning Features for Classifying Recording Production Style. In Proceedings of the 8th International Conference on Music Information Retrieval (ISMIR), Vienna, Austria, 23–27 September 2007; Dixon, S., Bainbridge, D., Typke, R., Eds.; pp. 441–444.

79. Tzanetakis, G.; Martins, L.G.; McNally, K.; Jones, R. Stereo Panning Information for Music Information Retrieval Tasks. *J. Audio Eng. Soc.* **2010**, *58*, 409–417.

80. Yegnanarayana, B.; Murthy, H.A. Significance of group delay functions in spectrum estimation. *IEEE Trans. Signal Process.* **1992**, *40*, 2281–2289.

81. Smits, R.; Yegnanarayana, B. Determination of instants of significant excitation in speech using group delay function. *IEEE Trans. Speech Audio Process.* **1995**, *3*, 325–333.

82. Rao, K.S.; Prasanna, S.R.M.; Yegnanarayana, B. Determination of Instants of Significant Excitation in Speech Using Hilbert Envelope and Group Delay Function. *IEEE Signal Process. Lett.* **2007**, *14*, 762–765.

83. Sethares, W.A.; Morris, R.D.; Sethares, J.C. Beat tracking of musical performances using low-level audio features. *IEEE Trans. Speech Audio Process.* **2005**, *13*, 275–285.

84. Hegde, R.M.; Murthy, H.A.; Rao, G.V.R. Application of the modified group delay function to speaker identification and discrimination. In Proceedings of the 2004 IEEE International Conference on Acoustics, Speech, and Signal Processing, (ICASSP), Montreal, QC, Canada, 17–21 May 2004; pp. 517–520.

85. Murthy, H.A.; Yegnanarayana, B. Group delay functions and its applications in speech technology. *Sadhana* **2011**, *36*, 745–782.

86. Ren, J.M.; Wu, M.J.; Jang, J.S.R. Automatic Music Mood Classification Based on Timbre and Modulation Features. *IEEE Trans. Affect. Comput.* **2015**, *6*, 236–246.

87. Hess, W. *Pitch Determination of Speech Signals: Algorithms and Devices; Springer Series in Information Sciences;* Springer-Verlag: Berlin, Germany, 1983; Volume 3.

88. Wold, E.; Blum, T.; Keislar, D.; Wheaton, J. Content-Based Classification, Search, and Retrieval of Audio. *IEEE MultiMedia* **1996**, *3*, 27–36.

89. Zhu, Y.; Kankanhalli, M.S. Precise pitch profile feature extraction from musical audio for key detection. *IEEE Trans. Multimedia* **2006**, *8*, 575–584.

90. Ishizuka, K.; Nakatani, T.; Fujimoto, M.; Miyazaki, N. Noise robust voice activity detection based on periodic to aperiodic component ratio. *Speech Commun.* **2010**, *52*, 41–60.

91. Kristjansson, T.T.; Deligne, S.; Olsen, P.A. Voicing features for robust speech detection. In Proceedings of the 9th European Conference on Speech Communication and Technology (EuroSpeech), Lisbon, Portugal, 4–8 September 2005; pp. 369–372.

92. Khao, P.C. *Noise Robust Voice Activity Detection;* Technical Report; Nanyang Technological University: Singapore, 2012.

93. Srinivasan, S.H.; Kankanhalli, M.S. Harmonicity and dynamics-based features for audio. In Proceedings of the 2004 IEEE International Conference on Acoustics, Speech, and Signal Processing (ICASSP), Montreal, QC, Canada, 17–21 May 2004; pp. 321–324.

94. Agostini, G.; Longari, M.; Pollastri, E. Musical instrument timbres classification with spectral features. In Proceedings of the Fourth IEEE Workshop on Multimedia Signal Processing (MMSP), Cannes, France, 3–5 October 2001; Dugelay, J., Rose, K., Eds.; pp. 97–102.

95. Agostini, G.; Longari, M.; Pollastri, E. Musical Instrument Timbres Classification with Spectral Features. *EURASIP J. Appl. Signal Process.* **2003**, *2003*, 5–14.
96. Cai, R.; Lu, L.; Hanjalic, A.; Zhang, H.; Cai, L. A flexible framework for key audio effects detection and auditory context inference. *IEEE Trans. Audio Speech Lang. Process.* **2006**, *14*, 1026–1039.
97. Boersma, P. Accurate short-term analysis of the fundamental frequency and the harmonics-to-noise ratio of a sampled sound. *Proc. Instit. Phonet. Sci.* **1993**, *17*, 97–110.
98. Lee, J.W.; Kim, S.; Kang, H.G. Detecting pathological speech using contour modeling of harmonic-to-noise ratio. In Proceedings of the 2014 IEEE International Conference on Acoustics, Speech and Signal Processing (ICASSP), Florence, Italy, 4–9 May 2014; pp. 5969–5973.
99. Shepard, R.N. Circularity in Judgments of Relative Pitch. *J. Acoust. Soc. Am.* **1964**, *36*, 2346–2353.
100. Bartsch, M.A.; Wakefield, G.H. Audio thumbnailing of popular music using chroma-based representations. *IEEE Trans. Multimedia* **2005**, *7*, 96–104.
101. Müller, M.; Kurth, F.; Clausen, M. Audio Matching via Chroma-Based Statistical Features. In Proceedings of the International Conference on Music Information Retrieval (ISMIR), London, UK, 11–15 September 2005; pp. 288–295.
102. Ramalingam, A.; Krishnan, S. Gaussian Mixture Modeling of Short-Time Fourier Transform Features for Audio Fingerprinting. *IEEE Trans. Inf. Forensics Secur.* **2006**, *1*, 457–463.
103. Li, T.; Ogihara, M. Music genre classification with taxonomy. In Proceedings of the 2005 IEEE International Conference on Acoustics, Speech, and Signal Processing (ICASSP), Philadelphia, PA, USA, 18–23 March 2005; pp. 197–200.
104. Lancini, R.; Mapelli, F.; Pezzano, R. Audio content identification by using perceptual hashing. In Proceedings of the 2004 IEEE International Conference on Multimedia and Expo (ICME), Taipei, Taiwan, 27–30 June 2004; pp. 739–742.
105. Allamanche, E.; Herre, J.; Hellmuth, O.; Fröba, B.; Kastner, T.; Cremer, M. Content-based Identification of Audio Material Using MPEG-7 Low Level Description. In Proceedings of the 2nd International Symposium on Music Information Retrieval (ISMIR), Bloomington, IN, USA, 15–17 October 2001.
106. Cheng, H.T.; Yang, Y.; Lin, Y.; Liao, I.; Chen, H.H. Automatic chord recognition for music classification and retrieval. In Proceedings of the 2008 IEEE International Conference on Multimedia and Expo (ICME), Hannover, Germany, 23–26 June 2008; pp. 1505–1508.
107. Misra, H.; Ikbal, S.; Bourlard, H.; Hermansky, H. Spectral entropy based feature for robust ASR. In Proceedings of the 2004 IEEE International Conference on Acoustics, Speech, and Signal Processing (ICASSP), Montreal, QC, Canada, 17–21 May 2004; pp. 193–196.
108. Jiang, D.; Lu, L.; Zhang, H.; Tao, J.; Cai, L. Music type classification by spectral contrast feature. In Proceedings of the 2002 IEEE International Conference on Multimedia and Expo (ICME), Lausanne, Switzerland, 26–29 August 2002; Volume I, pp. 113–116.
109. Yang, Y.; Lin, Y.; Su, Y.; Chen, H.H. A Regression Approach to Music Emotion Recognition. *IEEE Trans. Audio Speech Lang. Process.* **2008**, *16*, 448–457.
110. Shukla, S.; Dandapat, S.; Prasanna, S.R.M. Spectral slope based analysis and classification of stressed speech. *Int. J. Speech Technol.* **2011**, *14*, 245–258.
111. Murthy, H.A.; Beaufays, F.; Heck, L.P.; Weintraub, M. Robust text-independent speaker identification over telephone channels. *IEEE Trans. Speech Audio Process.* **1999**, *7*, 554–568.
112. Baniya, B.K.; Lee, J.; Li, Z.N. Audio feature reduction and analysis for automatic music genre classification. In Proceedings of the 2014 IEEE International Conference on Systems, Man and Cybernetics (SMC), San Diego, CA, USA, 5–8 October 2014; pp. 457–462.
113. Mallat, S.G. A theory for multiresolution signal decomposition: The wavelet representation. *IEEE Trans. Patt. Anal. Mach. Intell.* **1989**, *11*, 674–693.
114. Benedetto, J.; Teolis, A. An auditory motivated time-scale signal representation. In Proceedings of the IEEE-SP International Symposium Time-Frequency and Time-Scale Analysis, Victoria, BC, Canada, 4–6 October 1992; pp. 49–52.
115. Yang, X.; Wang, K.; Shamma, S.A. Auditory representations of acoustic signals. *IEEE Trans. Inf. Theory* **1992**, *38*, 824–839.
116. Missaoui, I.; Lachiri, Z. Blind speech separation based on undecimated wavelet packet-perceptual filterbanks and independent component analysis. *ICSI Int. J. Comput. Sci. Issues* **2011**, *8*, 265–272.

117. Popescu, A.; Gavat, I.; Datcu, M. Wavelet analysis for audio signals with music classification applications. In Proceedings of the 5th Conference on Speech Technology and Human-Computer Dialogue (SpeD), Bucharest, Romaina, 18–21 June 2009; pp. 1–6.
118. Vishwakarma, D.K.; Kapoor, R.; Dhiman, A.; Goyal, A.; Jamil, D. De-noising of Audio Signal using Heavy Tailed Distribution and comparison of wavelets and thresholding techniques. In Proceedings of the 2nd International Conference on Computing for Sustainable Global Development (INDIACom), New Delhi, India, 11–13 March 2015; pp. 755–760.
119. Li, T.; Ogihara, M. Music artist style identification by semisupervised learning from both lyrics and content. In Proceedings of the 12th Annual ACM Interantional Conference on Multimeda, New York, NY, USA, 10–16 October 2004; pp. 364–367.
120. Kim, K.; Youn, D.H.; Lee, C. Evaluation of wavelet filters for speech recognition. In Proceedings of the 2000 IEEE International Conference on Systems, Man, and Cybernetics, Nashville, TN, USA, 8–11 October 2000; Volume 4, pp. 2891–2894.
121. Li, T.; Ogihara, M. Toward intelligent music information retrieval. *IEEE Trans. Multimedia* **2006**, *8*, 564–574.
122. Mandel, M.; Ellis, D. Song-level features and SVMs for music classification. In Proceedings of International Conference on Music Information Retrieval (ISMIR)–MIREX, London, UK, 11–15 September 2005.
123. Mirza Tabinda, V.A. Analysis of wavelet families on Audio steganography using AES. In *International Journal of Advances in Computer Science and Technology (IJACST)*; Research India Publications: Hyderbad, India, 2014; pp. 26–31.
124. Sant'Ana, R.; Coelho, R.; Alcaim, A. Text-independent speaker recognition based on the Hurst parameter and the multidimensional fractional Brownian motion model. *IEEE Trans. Audio Speech Lang. Process.* **2006**, *14*, 931–940.
125. Zão, L.; Cavalcante, D.; Coelho, R. Time-Frequency Feature and AMS-GMM Mask for Acoustic Emotion Classification. *IEEE Signal Process. Lett.* **2014**, *21*, 620–624.
126. Palo, H.K.; Mohanty, M.N.; Chandra, M. Novel feature extraction technique for child emotion recognition. In Proceedings of the 2015 International Conference on Electrical, Electronics, Signals, Communication and Optimization (EESCO), Visakhapatnam, India, 24–25 January 2015; pp. 1–5.
127. Zão, L.; Coelho, R.; Flandrin, P. Speech Enhancement with EMD and Hurst-Based Mode Selection. *IEEE/ACM Trans. Audio Speech Lang. Process.* **2014**, *22*, 899–911.
128. Dranka, E.; Coelho, R.F. Robust Maximum Likelihood Acoustic Energy Based Source Localization in Correlated Noisy Sensing Environments. *J. Sel. Top. Signal Process.* **2015**, *9*, 259–267.
129. Mallat, S.G.; Zhang, Z. Matching pursuits with time-frequency dictionaries. *IEEE Trans. Signal Process.* **1993**, *41*, 3397–3415.
130. Wolfe, P.J.; Godsill, S.J.; Dorfler, M. Multi-Gabor dictionaries for audio time-frequency analysis. In Proceedings of the IEEE Workshop on the Applications of Signal Processing to Audio and Acoustics (WASPAA), New Paltz, NY, USA, 21–24 October 2001; pp. 43–46.
131. Ezzat, T.; Bouvrie, J.V.; Poggio, T.A. Spectro-temporal analysis of speech using 2-d Gabor filters. In Proceedings of the 8th Annual Conference of the International Speech Communication Association (InterSpeech), Antwerp, Belgium, 27–31 August 2007; pp. 506–509.
132. Meyer, B.T.; Kollmeier, B. Optimization and evaluation of Gabor feature sets for ASR. In Proceedings of the 9th Annual Conference of the International Speech Communication Association (InterSpeech), Brisbane, Australia, 22–26 September 2008; pp. 906–909.
133. Wang, J.C.; Lin, C.H.; Chen, B.W.; Tsai, M.K. Gabor-based nonuniform scale-frequency map for environmental sound classification in home automation. *IEEE Trans. Autom. Sci. Eng.* **2014**, *11*, 607–613.
134. Zhang, X.; Su, Z.; Lin, P.; He, Q.; Yang, J. An audio feature extraction scheme based on spectral decomposition. In Proceedings of the 2014 International Conference on Audio, Language and Image Processing (ICALIP), Shanghai, China, 7–9 July 2014; pp. 730–733.
135. Umapathy, K.; Krishnan, S.; Jimaa, S. Multigroup classification of audio signals using time-frequency parameters. *IEEE Trans. Multimedia* **2005**, *7*, 308–315.
136. Zhang, X.Y.; He, Q.H. Time-frequency audio feature extraction based on tensor representation of sparse coding. *Electron. Lett.* **2015**, *51*, 131–132.
137. Walters, T.C. Auditory-Based Processing of Communication Sounds. Ph.D. Thesis, Clare College, University of Cambridge, Cambridge, UK, June 2011.

138. Dennis, J.W.; Dat, T.H.; Li, H. Spectrogram Image Feature for Sound Event Classification in Mismatched Conditions. *IEEE Signal Process. Lett.* **2011**, *18*, 130–133.

139. Dennis, J.; Tran, H.D.; Li, H. Generalized Hough Transform for Speech Pattern Classification. *IEEE/ACM Trans. Audio Speech Lang. Process.* **2015**, *23*, 1963–1972.

140. Oppenheim, A.; Schafer, R. Homomorphic analysis of speech. *IEEE Trans. Audio Electroacoust.* **1968**, *16*, 221–226.

141. Noll, A.M. Cepstrum Pitch Determination. *J. Acoust. Soc. Am.* **1967**, *41*, 293–309.

142. Brown, J.C. Computer identification of musical instruments using pattern recognition with cepstral coefficients as features. *J. Acoust. Soc. Am.* **1999**, *105*, 1933–1941.

143. Atal, B.S. Effectiveness of linear prediction characteristics of the speech wave for automatic speaker identification and verification. *J. Acoust. Soc. Am.* **1974**, *55*, 1304–1312.

144. Adami, A.G.; Couto Barone, D.A. A Speaker Identification System Using a Model of Artificial Neural Networks for an Elevator Application. *Inf. Sci. Inf. Comput. Sci.* **2001**, *138*, 1–5.

145. Shen, J.; Shepherd, J.; Cui, B.; Tan, K. A novel framework for efficient automated singer identification in large music databases. *ACM Trans. Inf. Syst.* **2009**, *27*, 18:1–18:31.

146. Xu, C.; Maddage, N.C.; Shao, X. Automatic music classification and summarization. *IEEE Trans. Speech Audio Process.* **2005**, *13*, 441–450.

147. Kim, Y.E.; Whitman, B. Singer Identification in Popular Music Recordings Using Voice Coding Features. In Proceedings of the 3rd International Conference on Music Information Retrieval (ISMIR), Paris, France, 13–17 October 2002; pp. 164–169.

148. Burges, C.J.C.; Platt, J.C.; Jana, S. Extracting noise-robust features from audio data. In Proceedings of the IEEE International Conference on Acoustics, Speech, and Signal Processing (ICASSP), Orlando, FL, USA, 13–17 May 2002; pp. 1021–1024.

149. Malvar, H.S. A modulated complex lapped transform and its applications to audio processing. In Proceedings of the 1999 IEEE International Conference on Acoustics, Speech, and Signal Processing (ICASSP), Phoenix, AZ, USA, 15–19 March 1999; pp. 1421–1424.

150. Lindgren, A.C.; Johnson, M.T.; Povinelli, R.J. Speech recognition using reconstructed phase space features. In Proceedings of the IEEE International Conference on Acoustics, Speech, and Signal Processing (ICASSP), Hong Kong, China, 5–10 April 2003; pp. 60–63.

151. Kokkinos, I.; Maragos, P. Nonlinear speech analysis using models for chaotic systems. *IEEE Trans. Speech Audio Process.* **2005**, *13*, 1098–1109.

152. Pitsikalis, V.; Maragos, P. Speech analysis and feature extraction using chaotic models. In Proceedings of the IEEE International Conference on Acoustics, Speech, and Signal Processing (ICASSP), Orlando, FL, USA, 13–17 May 2002; pp. 533–536.

153. Hu, M.; Parada, P.; Sharma, D.; Doclo, S.; van Waterschoot, T.; Brookes, M.; Naylor, P. Single-channel speaker diarization based on spatial features. In Proceedings of the IEEE Workshop on Applications of Signal Processing to Audio and Acoustics (WASPAA), New Paltz, NY, USA, 18–21 October 2015; pp. 1–5.

154. Parada, P.P.; Sharma, D.; Lainez, J.; Barreda, D.; van Waterschoot, T.; Naylor, P. A single-channel non-intrusive C50 estimator correlated with speech recognition performance. *IEEE/ACM Trans. Audio Speech Lang. Process.* **2016**, *24*, 719–732.

155. Kim, D.S.; Jeong, J.H.; Kim, J.W.; Lee, S.Y. Feature extraction based on zero-crossings with peak amplitudes for robust speech recognition in noisy environments. In Proceedings of the 1996 IEEE International Conference on Acoustics, Speech, and Signal Processing (ICASSP), Atlanta, GA, USA, 7–10 May 1996; Volume 1, pp. 61–64.

156. Wang, Y.; Zhao, Z. A Noise-robust Speech Recognition System Based on Wavelet Neural Network. In Proceedings of the Third International Conference on Artificial Intelligence and Computational Intelligence (AICI)–Volume Part III, Taiyuan, China, 24–25 September 2011; pp. 392–397.

157. Ghulam, M.; Fukuda, T.; Horikawa, J.; Nitta, T. A noise-robust feature extraction method based on pitch-synchronous ZCPA for ASR. In Proceedings of the 8th International Conference on Spoken Language Processing (ICSLP), Jeju Island, Korea, 4–8 September 2004; pp. 133–136.

158. Ghulam, M.; Fukuda, T.; Horikawa, J.; Nitta, T. A Pitch-Synchronous Peak-Amplitude Based Feature Extraction Method for Noise Robust ASR. In Proceedings of the 2006 International Conference on Acoustics, Speech and Signal Processing (ICASSP), Toulouse, France, 14–19 May 2006; Volume 1, pp. I-505–I-508.

159. Ando, Y. *Architectural Acoustics: Blending Sound Sources, Sound, Fields, and Listeners*; Springer: New York, NY, USA, 1998; pp. 7–19.
160. Ando, Y. Autocorrelation-based features for speech representation. *Acta Acustica Unit. Acustica* **2015**, *101*, 145–154.
161. Valero, X.; Alías, F. Narrow-band autocorrelation function features for the automatic recognition of acoustic environments. *J. Acoust. Soc. Am.* **2013**, *134*, 880–890.
162. Pampalk, E.; Rauber, A.; Merkl, D. Content-based organization and visualization of music archives. In Proceedings of the 10th ACM International Conference on Multimedia (ACM-MM), Juan-les-Pins, France, 1–6 December 2002; Rowe, L.A., Mérialdo, B., Mühlhäuser, M., Ross, K.W., Dimitrova, N., Eds.; pp. 570–579.
163. Rauber, A.; Pampalk, E.; Merkl, D. Using Psycho-Acoustic Models and Self-Organizing Maps to Create a Hierarchical Structuring of Music by Musical Styles. In Proceedings of the 3rd International Conference on Music Information Retrieval (ISMIR), Paris, France, 13–17 October 2002.
164. Fastl, H. Fluctuation strength and temporal masking patterns of amplitude-modulated broadband noise. *Hear. Res.* **1982**, *8*, 441–450.
165. Hewitt, M.J.; Meddis, R. A computer model of amplitude-modulation sensitivity of single units in the inferior colliculus. *J. Acoust. Soc. Am.* **1994**, *95*, 2145–2159.
166. Sukittanon, S.; Atlas, L.E. Modulation frequency features for audio fingerprinting. In Proceedings of the IEEE International Conference on Acoustics, Speech, and Signal Processing (ICASSP), Orlando, FL, USA, 13–17 May 2002; pp. 1773–1776.
167. McKinney, M.F.; Breebaart, J. Features for audio and music classification. In Proceedings of the 4th International Conference on Music Information Retrieval (ISMIR), Baltimore, MD, USA, 26–30 October 2003.
168. Greenberg, S.; Kingsbury, B. The modulation spectrogram: In pursuit of an invariant representation of speech. In Proceedings of the 1997 IEEE International Conference on Acoustics, Speech, and Signal Processing (ICASSP), Munich, Germany, 21–24 April 1997; Volume 3, pp. 1647–1650.
169. Kingsbury, B.; Morgan, N.; Greenberg, S. Robust speech recognition using the modulation spectrogram. *Speech Commun.* **1998**, *25*, 117–132.
170. Baby, D.; Virtanen, T.; Gemmeke, J.F.; Barker, T.; van hamme, H. Exemplar-based noise robust automatic speech recognition using modulation spectrogram features. In Proceedings of the 2014 Spoken Language Technology Workshop (SLT), South Lake Tahoe, NV, USA, 7–10 December 2014; pp. 519–524.
171. Sukittanon, S.; Atlas, L.E.; Pitton, J.W. Modulation-scale analysis for content identification. *IEEE Trans. Signal Process.* **2004**, *52*, 3023–3035.
172. Barker, T.; Virtanen, T. Semi-supervised non-negative tensor factorisation of modulation spectrograms for monaural speech separation. In Proceedings of the 2014 International Joint Conference on Neural Networks (IJCNN), Beijing, China, 6–11 July 2014; pp. 3556–3561.
173. Chittora, A.; Patil, H.A. Classification of pathological infant cries using modulation spectrogram features. In Proceedings of the 9th International Symposium on Chinese Spoken Language Processing (ISCSLP), Singapore, 12–14 September 2014; pp. 541–545.
174. Herre, J.; Allamanche, E.; Ertel, C. How Similar Do Songs Sound? Towards Modeling Human Perception of Musical Similarity. In Proceedings of the IEEE ASSP Workshop on Applications of Signal Processing to Audio and Acoustics (WASPAA), New Paltz, NY, USA, 19–22 October 2003; pp. 83–86.
175. Zwicker, E.; Fastl, H.H. *Psychoacoustics: Facts and Models*; Springer Verlag: New York, NY, USA, 1998.
176. Meddis, R.; O'Mard, L. A unitary model of pitch perception. *J. Acoust. Soc. Am.* **1997**, *102*, 1811–1820.
177. Slaney, M.; Lyon, R.F. A perceptual pitch detector. In Proceedings of the 1990 International Conference on Acoustics, Speech, and Signal Processing (ICASSP), Albuquerque, NM, USA, 3–6 April 1990; Volume 1, pp. 357–360.
178. Olson, H.F. The Measurement of Loudness. *Audio* **1972**, *56*, 18–22.
179. Breebaart, J.; McKinney, M.F. Features for Audio Classification. In *Algorithms in Ambient Intelligence*; Verhaegh, W.F.J., Aarts, E., Korst, J., Eds.; Springer: Dordrecht, The Netherlands, 2004; Volume 2, Phillips Research, Chapter 6, pp. 113–129.
180. Pfeiffer, S. *The Importance of Perceptive Adaptation of Sound Features in Audio Content Processing*; Technical Report 18/98; University of Mannheim: Mannheim, Germany, 1998.

181. Lienhart, R.; Pfeiffer, S.; Effelsberg, W. Scene Determination Based on Video and Audio Features. In Proceedings of the IEEE International Conference on Multimedia Computing and Systems (ICMCS), Florence, Italy, 7–9 June 1999; Volume I, pp. 685–690.

182. Pfeiffer, S.; Lienhart, R.; Effelsberg, W. Scene Determination Based on Video and Audio Features. *Multimedia Tools Appl.* **2001**, *15*, 59–81.

183. Daniel, P.; Weber, R. Psychoacoustical roughness: Implementation of an optimized model. *Acustica* **1997**, *83*, 113–123.

184. Gerazov, B.; Ivanovski, Z. Kernel Power Flow Orientation Coefficients for Noise-Robust Speech Recognition. *IEEE/ACM Trans. Audio Speech Lang. Process.* **2015**, *23*, 407–419.

185. Gerazov, B.; Ivanovski, Z. Gaussian Power flow Orientation Coefficients for noise-robust speech recognition. In Proceedings of the 22nd European Signal Processing Conference (EUSIPCO), Lisbon, Portugal, 1–5 September 2014; pp. 1467–1471.

186. Gowdy, J.N.; Tufekci, Z. Mel-scaled discrete wavelet coefficients for speech recognition. In Proceedings of the IEEE International Conference on Acoustics, Speech, and Signal Processing (ICASSP), Istanbul, Turkey, 5–9 June 2000; Volume 3, pp. 1351–1354.

187. Tavanaei, A.; Manzuri, M.T.; Sameti, H. Mel-scaled Discrete Wavelet Transform and dynamic features for the Persian phoneme recognition. In Proceedings of the 2011 International Symposium on Artificial Intelligence and Signal Processing (AISP), Tehran, Israel, 15–16 June 2011; pp. 138–140.

188. Tufekci, Z.; Gurbuz, S. Noise Robust Speaker Verification Using Mel-Frequency Discrete Wavelet Coefficients and Parallel Model Compensation. In Proceedings of the IEEE International Conference on Acoustics, Speech, and Signal Processing (ICASSP), Philadelphia, PA, USA, 18–23 March 2005; Volume 1, pp. 657–660.

189. Nghia, P.T.; Binh, P.V.; Thai, N.H.; Ha, N.T.; Kumsawat, P. A Robust Wavelet-Based Text-Independent Speaker Identification. In Proceedings of the International Conference on Conference on Computational Intelligence and Multimedia Applications (ICCIMA), Sivakasi, Tamil Nadu, 13–15 December 2007; Volume 2, pp. 219–223.

190. Valero, X.; Alías, F. Gammatone Wavelet features for sound classification in surveillance applications. In Proceedings of the 20th European Signal Processing Conference (EUSIPCO), Bucharest, Romania, 27–31 August 2012; pp. 1658–1662.

191. Venkitaraman, A.; Adiga, A.; Seelamantula, C.S. Auditory-motivated Gammatone wavelet transform. *Signal Process.* **2014**, *94*, 608–619.

192. Burrus, C.S.; Gopinath., R.A.; Guo, H. *Introduction to Wavelets and Wavelet Transforms: A Primer*; Prentice Hall: Upper Saddle River, NJ, USA, 1998.

193. Bradie, B. Wavelet packet-based compression of single lead ECG. *IEEE Trans. Biomed. Eng.* **1996**, *43*, 493–501.

194. Daubechies, I. The wavelet transform, time-frequency localization and signal analysis. *IEEE Trans. Inf. Theory* **1990**, *36*, 961–1005.

195. Jiang, H.; Er, M.J.; Gao, Y. Feature extraction using wavelet packets strategy. In Proceedings of the 42nd IEEE Conference on Decision and Control, Maui, HI, USA, 9–12 December 2003; Volume 5, pp. 4517–4520.

196. Greenwood, D.D. Critical Bandwidth and the Frequency Coordinates of the Basilar Membrane. *J. Acoust. Soc. Am.* **1961**, *33*, 1344–1356.

197. Ren, Y.; Johnson, M.T.; Tao, J. Perceptually motivated wavelet packet transform for bioacoustic signal enhancement. *J. Acoust. Soc. Am.* **2008**, *124*, 316–327.

198. Dobson, W.K.; Yang, J.J.; Smart, K.J.; Guo, F.K. High quality low complexity scalable wavelet audio coding. In Proceedings of the 1997 IEEE International Conference on Acoustics, Speech, and Signal Processing (ICASSP), Munich, Germany, 21–24 April 1997; Volume 1, pp. 327–330.

199. Artameeyanant, P. Wavelet audio watermark robust against MPEG compression. In Proceedings of the 2010 International Conference on Control Automation and Systems (ICCAS), Gyeonggi-do, South Korea, 27–30 October 2010; pp. 1375–1378.

200. Rajeswari; Prasad, N.; Satyanarayana, V. A Noise Robust Speech Recognition System Using Wavelet Front End and Support Vector Machines. In Proceedings of International Conference on Emerging research in Computing, Information, Communication and Applications (ERCICA), Bangalore, India, 2–3 August 2013; pp. 307–312.

201. Ntalampiras, S. Audio Pattern Recognition of Baby Crying Sound Events. *J. Audio Eng. Soc* **2015**, *63*, 358–369.

202. Kleinschmidt, M. Methods for Capturing Spectro-Temporal Modulations in Automatic Speech Recognition. *Acta Acustica Unit. Acustica* **2002**, *88*, 416–422.

203. Heckmann, M.; Domont, X.; Joublin, F.; Goerick, C. A hierarchical framework for spectro-temporal feature extraction. *Speech Commun.* **2011**, *53*, 736–752.

204. Wu, Q.; Zhang, L.; Shi, G. Robust Multifactor Speech Feature Extraction Based on Gabor Analysis. *IEEE Trans. Audio Speech Lang. Process.* **2011**, *19*, 927–936.

205. Schröder, J.; Goetze, S.; Anemüller, J. Spectro-Temporal Gabor Filterbank Features for Acoustic Event Detection. *IEEE/ACM Trans. Audio Speech Lang. Process.* **2015**, *23*, 2198–2208.

206. Lindeberg, T.; Friberg, A. Idealized Computational Models for Auditory Receptive Fields. *PLoS ONE* **2015**, *10*, 1–58.

207. Lindeberg, T.; Friberg, A. Scale-Space Theory for Auditory Signals. In Proceedings of the 5th International Conference on Scale Space and Variational Methods in Computer Vision (SSVM), 31 May–4 June 2015; Lège Cap Ferret, France; pp. 3–15.

208. Mesgarani, N.; Slaney, M.; Shamma, S.A. Discrimination of speech from nonspeech based on multiscale spectro-temporal Modulations. *IEEE Trans. Audio Speech Lang. Process.* **2006**, *14*, 920–930.

209. Panagakis, I.; Benetos, E.; Kotropoulos, C. Music Genre Classification: A Multilinear Approach. In Proceedings of the 9th International Conference on Music Information Retrieval (ISMIR), Philadelphia, PA, USA, 14–18 September 2008; pp. 583–588.

210. Ng, T.; Zhang, B.; Nguyen, L.; Matsoukas, S.; Zhou, X.; Mesgarani, N.; Veselý, K.; Matejka, P. Developing a Speech Activity Detection System for the DARPA RATS Program. In Proceedings of the 13th Annual Conference of the International Speech (InterSpeech), Portland, OR, USA, 9–13 September 2012; pp. 1969–1972.

211. Qiu, A.; Schreiner, C.; Escabi, M. Gabor analysis of auditory midbrain receptive fields: spectro-temporal and binaural composition. *Neurophysiology* **2003**, *90*, 456–476.

212. Patterson, R.D.; Holdsworth, J. A functional model of neural activity patterns and auditory images. *Adv. Speech Hear. Lang. Process.* **1996**, *3*, 547–563.

213. Dennis, J.; Tran, H.D.; Li, H.; Chng, E.S. A discriminatively trained Hough Transform for frame-level phoneme recognition. In Proceedings of the 2014 IEEE Acoustics, Speech and Signal Processing (ICASSP), Florence, Italy, 4–9 May 2014; pp. 2514–2518.

214. Dennis, J.; Tran, H.D.; Li, H. Combining robust spike coding with spiking neural networks for sound event classification. In Proceedings of the 2015 IEEE International Conference on Acoustics, Speech and Signal Processing (ICASSP), Brisbane, Australia, 19–24 April 2015; pp. 176–180.

215. Patterson, R.D.; Allerhand, M.H.; Giguère, C. Time-domain modeling of peripheral auditory processing: A modular architecture and a software platform. *J. Acoust. Soc. Am.* **1995**, *98*, 1890–1894.

216. Patterson, R.D.; Robinson, K.; Holdsworth, J.; Mckeown, D.; Zhang, C.; Allerhand, M. Complex sounds and auditory images. *Audit. Physiol. Percept.* **1992**, *83*, 429–446.

217. McLoughlin, I.; Zhang, H.; Xie, Z.; Song, Y.; Xiao, W. Robust Sound Event Classification Using Deep Neural Networks. *IEEE/ACM Trans. Audio Speech Lang. Process.* **2015**, *23*, 540–552.

218. Malekesmaeili, M.; Ward, R.K. A local fingerprinting approach for audio copy detection. *Signal Process.* **2014**, *98*, 308–321.

219. Moore, B.C.J.; Peters, R.W.; Glasberg, B.R. Auditory filter shapes at low center frequencies. *J. Acoust. Soc. Am.* **1990**, *88*, 132–140.

220. Zwicker., E. Subdivision of the Audible Frequency Range into Critical Bands (Frequenzgruppen). *J. Acoust. Soc. Am.* **1961**, *33*, 248.

221. Maddage, N.C.; Xu, C.; Kankanhalli, M.S.; Shao, X. Content-based music structure analysis with applications to music semantics understanding. In Proceedings of the 12th ACM International Conference on Multimedia, New York, NY, USA, 10–16 October 2004; Schulzrinne, H., Dimitrova, N., Sasse, M.A., Moon, S.B., Lienhart, R., Eds.; pp. 112–119.

222. Beritelli, F.; Grasso, R. A pattern recognition system for environmental sound classification based on MFCCs and neural networks. In Proceedings of the International Conference on Signal Processing and Communication Systems (ICSPCS), Gold Coast, Australia, 15–17 December 2008.

223. Zeng, W.; Liu, M. Hearing environment recognition in hearing aids. In Proceedings of the 12th International Conference on Fuzzy Systems and Knowledge Discovery (FSKD), Zhangjiajie, China, 15–17 August 2015; pp. 1556–1560.

224. Lei, B.; Rahman, S.A.; Song, I. Content-based classification of breath sound with enhanced features. *Neurocomputing* **2014**, *141*, 139–147.

225. Shannon, B.J.; Paliwal, K.K. MFCC computation from magnitude spectrum of higher lag autocorrelation coefficients for robust speech recognition. In Proceedings of the 8th International Conference on Spoken Language Processing (ICSLP), Jeju Island, Korea, 4–8 October 2004.

226. Choi, E.H.C. On compensating the Mel-frequency cepstral coefficients for noisy speech recognition. In Proceedings of the 29th Australasian Computer Science Conference (ACSC2006), Hobart, Tasmania, Australia, 16–19 January 2006; Estivill-Castro, V., Dobbie, G., Eds.; Volume 48, pp. 49–54.

227. Trawicki, M.B.; Johnson, M.T.; Ji, A.; Osiejuk, T.S. Multichannel speech recognition using distributed microphone signal fusion strategies. In Proceedings of the 2012 International Conference on Audio, Language and Image Processing (ICALIP), Shanghai, China, 16–18 July 2012; pp. 1146–1150.

228. Ravindran, S.; Schlemmer, K.; Anderson, D.V. A Physiologically Inspired Method for Audio Classification. *EURASIP J. Appl. Signal Process.* **2005**, *2005*, 1374–1381.

229. Patterson, R.D.; Moore, B.C.J. Auditory filters and excitation patterns as representations of frequency resolution. *Freq. Sel. Hear.* **1986**, 123–177.

230. Patterson, R.D.; Nimmo-Smith, I.; Holdsworth, J.; Rice, P. An Efficient Auditory Filterbank Based on the Gammatone Function. In Proceedings of the IOC Speech Group on Auditory Modelling at RSRE, Malvern, UK, 14–15 December 1987; pp. 1–34.

231. Hohmann, V. Frequency analysis and synthesis using a Gammatone filterbank. *Acta Acustica Unit. Acustica* **2002**, *88*, 433–442.

232. Valero, X.; Alías, F. Gammatone Cepstral Coefficients: Biologically Inspired Features for Non-Speech Audio Classification. *IEEE Trans. Multimedia* **2012**, *14*, 1684–1689.

233. Johannesma, P.I.M. The pre-response stimulus ensemble of neurons in the cochlear nucleus. In Proceedings of the Symposium on Hearing Theory, Eindhoven, The Netherlands, 22–23 June 1972; pp. 58–69.

234. Shao, Y.; Srinivasan, S.; Wang, D. Incorporating Auditory Feature Uncertainties in Robust Speaker Identification. In Proceedings of the 2007 IEEE International Conference on Acoustics, Speech and Signal Processing (ICASSP), Honolulu, HI, USA, 15–20 April 2007; Volume 4, pp. IV-277–IV-280.

235. Schlüter, R.; Bezrukov, I.; Wagner, H.; Ney, H. Gammatone Features and Feature Combination for Large Vocabulary Speech Recognition. In Proceedings of the IEEE International Conference on Acoustics, Speech, and Signal Processing (ICASSP), Honolulu, HI, USA, 15–20 April 2007; pp. 649–652.

236. Guyot, P.; Pinquier, J.; Valero, X.; Alías, F. Two-step detection of water sound events for the diagnostic and monitoring of dementia. In Proceedings of the 2013 IEEE International Conference on Multimedia and Expo (ICME), San Jose, CA, USA, 15–19 July 2013; pp. 1–6.

237. Socoró, J.C.; Ribera, G.; Sevillano, X.; Alías, F. Development of an Anomalous Noise Event Detection Algorithm for dynamic road traffic noise mapping. In Proceedings of the 22nd International Congress on Sound and Vibration (ICSV22), Florence, Italy, 12–16 July 2015.

238. Shao, Y.; Srinivasan, S.; Jin, Z.; Wang, D. A computational auditory scene analysis system for speech segregation and robust speech recognition. *Comput. Speech Lang.* **2010**, *24*, 77–93.

239. Irino, T.; Patterson, R.D. A time-domain, level-dependent auditory filter: The gammachirp. *J. Acoust. Soc. Am.* **1997**, *101*, 412–419.

240. Abdallah, A.B.; Hajaiej, Z. Improved closed set text independent speaker identification system using Gammachirp Filterbank in noisy environments. In Proceedings of the 11th International MultiConference on Systems, Signals Devices (SSD), Castelldefels-Barcelona, Spain, 11–14 February 2014; pp. 1–5.

241. Hermansky, H. Perceptual linear predictive (PLP) analysis of speech. *J. Acoust. Soc. Am.* **1990**, *87*, 1738–1752.

242. Hönig, F.; Stemmer, G.; Hacker, C.; Brugnara, F. Revising Perceptual Linear Prediction (PLP). In Proceedings of the 9th European Conference on Speech Communication and Technology (EuroSpech), Lisbon, Portugal, 4–8 September 2005; pp. 2997–3000.

243. Hermansky, H.; Morgan, N. RASTA processing of speech. *IEEE Trans. Speech Audio Process.* **1994**, *2*, 578–589.

244. Sawata, R.; Ogawa, T.; Haseyama, M. Human-centered favorite music estimation: EEG-based extraction of audio features reflecting individual preference. In Proceedings of the 2015 IEEE International Conference on Digital Signal Processing (DSP), Signapore, 21–24 July 2015; pp. 818–822.
245. Kalinli, O.; Narayanan, S.S. A saliency-based auditory attention model with applications to unsupervised prominent syllable detection in speech. In Proceedings of the 8th Annual Conference of the International Speech Communication Association (InterSpeech), Antwerp, Belgium, 27–31 August 2007; pp. 1941–1944.
246. Kalinli, O.; Sundaram, S.; Narayanan, S. Saliency-driven unstructured acoustic scene classification using latent perceptual indexing. In Proceedings of the IEEE International Workshop on Multimedia Signal Processing (MMSP), Rio De Janeiro, Brazil, 5–7 October 2009; pp. 1–6.
247. De Coensel, B.; Botteldooren, D. A model of saliency-based auditory attention to environmental sound. In Proceedings of the 20th International Congress on Acoustics (ICA), Sydney, Australia, 23–27 August 2010; pp. 1–8.

applied sciences

MDPI

Article

Auralization of Accelerating Passenger Cars Using Spectral Modeling Synthesis

Reto Pieren *, Thomas Bütler and Kurt Heutschi

Empa, Swiss Federal Laboratories for Material Science and Technology, CH-8600 Duebendorf, Switzerland;
thomas.buetler@empa.ch (T.B.); kurt.heutschi@empa.ch (K.H.)
* Correspondence: reto.pieren@empa.ch; Tel.: +41-58-765-60-31

Academic Editor: Vesa Valimaki
Received: 28 September 2015 / Accepted: 15 December 2015 / Published: 24 December 2015

Abstract: While the technique of auralization has been in use for quite some time in architectural acoustics, the application to environmental noise has been discovered only recently. With road traffic noise being the dominant noise source in most countries, particular interest lies in the synthesis of realistic pass-by sounds. This article describes an auralizator for pass-bys of accelerating passenger cars. The key element is a synthesizer that simulates the acoustical emission of different vehicles, driving on different surfaces, under different operating conditions. Audio signals for the emitted tire noise, as well as the propulsion noise are generated using spectral modeling synthesis, which gives complete control of the signal characteristics. The sound of propulsion is synthesized as a function of instantaneous engine speed, engine load and emission angle, whereas the sound of tires is created in dependence of vehicle speed and emission angle. The sound propagation is simulated by applying a series of time-variant digital filters. To obtain the corresponding steering parameters of the synthesizer, controlled experiments were carried out. The tire noise parameters were determined from coast-by measurements of passenger cars with idling engines. To obtain the propulsion noise parameters, measurements at different engine speeds, engine loads and emission angles were performed using a chassis dynamometer. The article shows how, from the measured data, the synthesizer parameters are calculated using audio signal processing.

Keywords: auralization; road traffic noise; passenger cars; sound synthesis

PACS: 43.50.Lj; 43.50.Rq; 43.28.Js; 43.60.Cg; 43.58.Jq

---

1. Introduction

Noise caused by traffic is a relevant health factor in urban environments, along major transport routes and in the vicinity of airports. Noise, in contrast to sound, can principally not be measured, but has to be assessed. For the most relevant noise sources, objective quantities have been derived that correlate with the annoyance as reported by people. However, these correlations are usually weak. One reason for this is the fact that the describing quantities used so far represent the acoustic situation only in a very simplified manner. A method to further investigate the signal properties relevant to noise is to conduct listening experiments where different stimuli are presented to test persons. Relying on audio recordings allows for little variation of different signal aspects only. A more versatile method with a much higher degree of freedom, as well as full control of the influencing signal parameters is to synthesize the stimuli and, thus, to auralize an acoustical environment.

Auralization has been in use for quite some time in architectural acoustics, namely in the fields of room and building acoustics [1–3], but it has only recently been discovered for environmental acoustical applications. Today, most auralizations are generated based on computer models and digital signal processing. However, between applications, the individual simulation steps may vary significantly.

In room and building acoustical auralizations, it is common to utilize (anechoic) recordings as the source material, whereas in environmental acoustics, it is often required to artificially synthesize the source signals [4–8]. Furthermore, the sound propagation simulation substantially differs. In room acoustical auralizations, the focus lies on the simulation of many room reflections, in particular specular *vs.* diffuse reflections, and diffraction [2,3]. In building acoustical auralizations, sound transmission through structures is simulated using sound insulation prediction models [2]. For environmental acoustical auralizations, however, particular emphasis is placed on a detailed simulation of direct sound and ground reflection. Leastwise, in terms of sound reproduction, the same techniques are tapplicable.

Early attempts in auralizations for environmental noise applications have been made by a group at NASA Langley Research Center, where aircraft flyovers have been binaurally rendered based on monaural recordings [9]. In the same period, a synthesis model for the traction noise of electric rail-bound vehicles was developed at the RWTH Aachen University and used to study sound quality [4]. Newer auralization models also try to synthesize the sounds of aircraft [7,8,10]. One of these models has already been combined with 3D visualizations to make aircraft noise both heard and seen in immersive virtual reality environments [8]. In the recently-completed Swiss project "VisAsim" [11,12], outdoor auralizations of wind farms were linked to synchronous GIS-based 3D visualizations. Within the Swedish project "LISTEN" [13,14] and the European project "HOSANNA" [5,15], tools for the auralization of road traffic noise were developed. The main motivation is to provide more intuitive information about traffic noise scenarios for city planners, noise consultants and decision makers. There is common agreement in the point that information about noise in the form of dB values is difficult to communicate to the public [16]. On the other hand, there is growing interest in the perceptual aspects of noise abatement measures. While in the past, the quality of noise mitigation measures was understood as the A-weighted sound pressure level reduction, the focus shifts to perceptual efficiency [17]. In this respect the optimal measure is not necessarily the one with the highest dB(A) drop, but the one with the highest annoyance reduction. The idea of describing and subsequently improving the acoustical environment with respect to human perception corresponds to the soundscape concept.

Generally, the auralization process consists of three modules [2]. The first module provides the signal emitted at the source. In its simplest form, this emission module makes use of suitable audio recordings. Obviously, the disadvantage of this strategy is its inflexibility, the limitations with respect to possible emission signal variations and the fundamental difficulty to obtain recordings that are not contaminated by unwanted sound. In addition, due to the Doppler effect, recordings of moving sources can be used for the auralization of situations with similar geometries only. It is therefore promising to use a digital emission synthesizer that is capable of generating the audio signal as radiated by the source. In [5], a granular synthesis using an enhanced pitch-synchronous overlap-and-add (PSOLA) method is presented for engine noise. This method features low computational costs and is therefore well suited for real-time applications. It is however limited in terms of flexibility, as it does not allow for inter- and extrapolations between measured signals. This limitation can be overcome by the synthesizer presented in this article, at the expense of computational costs.

The second module needed in the auralization process is a filter that simulates the sound propagation effects of the wave traveling from the source to the receiver. These effects involve geometrical spreading, propagation delay and atmospheric absorption. In outdoor situations, along with the direct sound path, also reflections occur, particularly ground reflections, leading to the ground effect. When either the source or the receiver moves, additionally, the Doppler effect arises. However, also, due to the time-varying propagation situation, all of the other effects change over time, which means that the propagation filter becomes time variant [18].

The third module is a reproduction system, which renders the received signals to headphones or a multi-channel loudspeaker system. In this process, for every received signal, its corresponding emission angle with respect to the observer is required. For the rendering, a variety of different methods exists [19]. Further, it has to be assured that the reproduction system has a linear frequency response and that it is correctly calibrated.

In the ongoing research project "TAURA", a traffic noise auralizator is developed that covers road traffic and railway noise. It will form the basis for future experiments to refine the characterization of noise. The key element is a synthesizer that simulates the acoustical emission of a great many different vehicles, operating on a wide variety of surfaces and under different operating conditions. In the TAURA model, road traffic noise is created by the superposition of individual vehicle pass-by sounds. The objective of this article is to describe how these single pass-by sounds can be generated in the case of passenger cars.

This paper extends the work presented in [20] and is structured in two main parts: In Section 2, the auralization model of accelerating passenger cars is developed step by step and presented. Thereby, an emission synthesizer, which is based on spectral modeling synthesis, and propagation filtering algorithms are elucidated. Section 3 shows how the model parameters can be estimated based on controlled measurements. On that account, a series of signal analysis steps to obtain the steering parameters of the synthesizer is proposed. The article ends with conclusions in Section 4.

2. Model Development

2.1. Overview

This section presents an overview of the model to auralize accelerating passenger cars. Further, the key assumptions and motivations used for the model development are presented. In the model, each car is represented by two moving point sources. The geometrical situation is depicted in Figure 1, in which the distance of the straight driving lane to the receiver is D, the emission angle φ, the angle of inclination α and the point source positions S1 and S2, respectively. Describing the kinematics of the vehicle, its speed $v(t)$ in km/h as a function of time t (at the source) is used throughout this paper. In correspondence with the Harmonoise model [21], the point sources are vertically stacked and located at heights of 0.01 and 0.3 m above ground. By not attributing separate point sources to each vehicle axle, we limit the applicability of the model to situations with source–receiver distances clearly larger than the axial distance, while in return, saving computational costs.

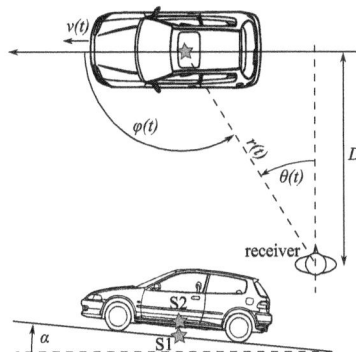

Figure 1. Sketch of the geometrical situation showing the two source positions S1 and S2, the inclination angle of the road α, the distance D, the instantaneous vehicle speed v, emission angle φ, immmission angle θ and source–receiver distance r.

Road traffic noise is mainly composed of propulsion noise and tire noise [18,21–23]. Both contributions differ in their relevance for the total noise, depending on the vehicle, its operating conditions and the pavement type. This motivates that in the presented model, the contributions of propulsion noise and tire noise are simulated individually.

In accordance with Vorländer's definition of the "principle of auralization" [2], the presented model to auralize accelerating passenger cars comprises a separate emission, a propagation and a reproduction module. The emission module is described in Section 2.2, the propagation module in Section 2.3 and the reproduction module in Section 2.4. Figure 2 shows the block diagram of the model. The input variables describe the vehicle, driver, road surface, geometry, ground type and the weather; the input variables marked by * are time dependent.

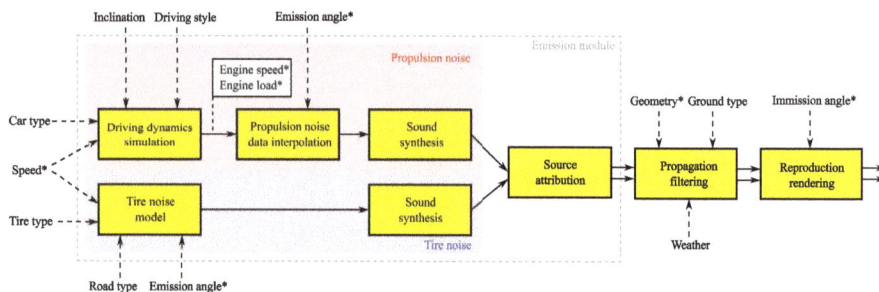

Figure 2. Simulation flowchart of the auralization of accelerating passenger cars. The input variables marked by a * are time dependent.

The emission module describes the emitted sound of an individual passenger car pass-by, *i.e.*, its outside sound. In Figure 2, the emission module is edged by a gray dashed line. It contains several simulation blocks creating two separate signal paths. The upper path (red box) represents the propulsion noise and the lower path (blue box) the tire noise simulation. Key elements are two sound synthesizers that artificially generate audible emission signals for propulsion and tire noise, respectively. The synthesizer parameters can be obtained from controlled measurements, as elucidated in Section 3. Section 2.2 describes how the synthesized emission signals are attributed to the two point sources. As a proper interface to the propagation module, the point source signals are defined at a virtual reference distance of $r_0 = 1$ m [6]. They feature time-varying sound characteristics and already include source directivity.

For both moving source positions, S1 and S2, the sound propagation to the (static) receiver position is simulated in the propagation module. To generate the receiver signals, both corresponding source signals, s_1 and s_2, are filtered by a series of time-varying digital filters, as described in Section 2.3. These filters depend on the instantaneous propagation geometry, the ground type and the weather conditions (*cf.* Figure 2). Finally, in the reproduction module, the receiver signals are summed up and rendered for multi-channel reproduction using the instantaneous immission angle. Section 2.4 exemplifies a possible stereo rendering procedure.

2.2. Emission Module

The emission module describes the emitted sounds of an individual passenger car. Its structure is depicted in Figure 2. As described above, the acoustical emission of the passenger car is assumed to consist of the two contributions: tire noise and propulsion noise. Their corresponding emission signals are denoted as s_{tire} and s_{prop}, respectively.

Tire noise strongly depends on tire type [22,24], road surface type [18,21,22,25] and vehicle speed [18,21–23]. Further, the horn effect mainly determines the horizontal directivity of tire noise [21,23,26]. To model these effects, the tire noise contribution is assumed to depend on the road and tire type, as well as on vehicle speed v and the emission angle φ. Section 2.2.1 shows how the signals s_{tire} are calculated based on these input parameters.

In current noise prediction models, propulsion noise is commonly calculated as a function of vehicle speed, acceleration and road inclination [18,21,27]. This is due to the fact that these models are developed and used in cases for which the engaged gear is not known. The gear, however, strongly influences the sound of propulsion [23]. For a given speed, acceleration and road inclination, by changing the gear, the engine speed, as well as the engine load change. From an engine's viewpoint, it is these two parameters that are sufficient to fully describe the engine condition. Section 2.2.2 explains how in the auralization model, engine speed n and engine load Γ are calculated by simulating the driving dynamics of the vehicle. These simulations require information on the vehicle and the driving style, the road inclination α and the vehicle speed $v(t)$ as a function of time t. Further, propulsion noise features a directivity [23], which is also taken into account in the auralization model. Section 2.2.3 shows how, based on n, Γ and the emission angle φ, the signal s_{prop} is calculated.

The audible emission signals s_{tire} and s_{prop} are generated artificially by two digital sound synthesizers. The synthesizers are based on a combination of additive and subtractive synthesis. In additive synthesis, the signal is constructed by the sum of sinusoids, each having a time-varying amplitude and phase [28–30]. On the other hand, subtractive synthesis uses filters to shape a more complex source signal, e.g., a sawtooth wave or white noise [28,29]. The combination of both techniques is known as spectral modeling synthesis [30–32]. However, in contrast to the applications presented in [31] and [32], in the presented model, the sounds are not synthesized using the short-time Fourier transform (STFT), but directly in the time domain. The structure of the synthesizer is similar to the one recently published for wind turbine sounds [6].

The signal of propulsion noise, s_{prop}, is fully attributed to the upper point source S2. However, the sound power of the tire noise contribution is attributed to the point sources by 80%/20% [21]. This translates to a ratio of 2:1 of their respective sound pressure signals. The conditions of incoherent signals and energy conservation yield a normalization factor of $1/\sqrt{5}$. Thus, the sound pressure source signals are:

$$s_2(t) = \frac{1}{\sqrt{5}} \cdot s_{\text{tire},2}(t) + s_{\text{prop}}(t) \tag{1}$$

$$s_1(t) = \frac{2}{\sqrt{5}} \cdot s_{\text{tire},1}(t) \tag{2}$$

at reference distance $r_0 = 1$ m for source positions S2 and S1, respectively. Indices 2 and 1 indicate that different, uncorrelated signals for the sound of tires are generated for the two source positions.

2.2.1. Sound of Tires

The emission signal of the sound of tires is assumed to consist of broadband noise only, *i.e.*, discrete tones due to, e.g., tire tread resonances or discrete vibrational tire resonances are not taken into account.

The spectral shaping of the broadband noise components is performed in 1/3 octave bands. For each 1/3 octave band i, white noise is generated and filtered by a digital pink filter. This pre-shaping helps to produce a smoother spectrum of the resulting signal [6]. The output of the pink filter is bandpass filtered by an eighth order Butterworth filter (Class 0 according to the standard IEC 1260:1995 [33]) and normalized to unit signal power to obtain the signal $\xi_i(t)$. For stability reasons, also at low frequencies, the filters are implemented as cascaded second-order sections (SOS).

The sound pressure emission signals of the sound of tires component are thus calculated by [6]:

$$s_{\text{tire}}(t) = \sum_{i=1}^{N_b} p_0 10^{L_{\text{tire},i}(v(t),\varphi(t))/20} \cdot \xi_i(t) \tag{3}$$

with N_b being the number of considered 1/3 octave bands, the reference pressure $p_0 = 20\ \mu$Pa and normalized bandpass filtered pink noise signals $\xi_i(t)$. A total of $N_b = 29$ bands from 20 Hz to 12.5 kHz are used. For the sound pressure level $L_{\text{tire},i}$ of band i, a common logarithmic speed relationship [22] with additive correction terms is assumed:

$$L_{\text{tire},i}(t) = A_i + B_i \cdot \log\left(\frac{v(t)}{v_0}\right) + \Delta L_{\text{road},i}(v(t)) + \Delta L_{\text{dir},i}(\varphi(t)) \tag{4}$$

with reference speed $v_0 = 70$ km/h, regression parameters A_i and B_i, the road surface correction $\Delta L_{\text{road},i}$ and a horizontal directivity $\Delta L_{\text{dir},i}$. For the road surface correction, the Swiss "sonRoad" model [18] offers the parameter Δ_{BG} for 10 surface types. However, Δ_{BG} does not depend on frequency or vehicle speed. The recently-published EU directive on establishing common noise assessment methods (CNOSSOS-EU) [27] contains spectral corrections in octave bands in the form of:

$$\Delta L_{\text{road},i}(v(t)) = \alpha_i + \beta \cdot \log\left(\frac{v(t)}{v_0}\right) \tag{5}$$

with experimental regression parameters α_i and β, which are tabulated for 15 different road surface types. The horizontal directivity simulates the horn effect [26] and only applies for signal $s_1(t)$, i.e., for the lower source position (S1). The empirically-obtained relationship [21]:

$$\Delta L_{\text{dir},i}(\varphi(t)) = \begin{cases} -2.5 + 4\,|\cos\varphi(t)| + C, & 800\text{ Hz} \le f_{c,i} \le 6.3\text{ kHz} \\ 0, & \text{otherwise} \end{cases} \tag{6}$$

with the 1/3 octave band center frequencies $f_{c,i}$ and the correction C is employed. C accounts for a limited emission angle range during emission measurements, e.g., it amounts to 0.9 dB for an angle range $45° < \varphi < 135°$.

2.2.2. Driving Dynamics

Figure 2 shows that as a first step of the propulsion noise simulation, the driving dynamics of the car are calculated in order to obtain the instantaneous engine speed $n(t)$ and engine load $M(t)$. The engine speed in engaged mode reads [34]:

$$n(t) = 60 \cdot i_{\text{gear}}(t) \cdot i_{\text{ax}} \cdot \frac{v(t)/3.6}{2\pi r_{\text{tire,dyn}}} \quad \text{[rpm]} \tag{7}$$

with the instantaneous vehicle speed v given in km/h, the gear ratio i_{gear}, the axle ratio i_{ax} and the dynamic tire radius $r_{\text{tire,dyn}} \approx 0.3$ m. The traction F_T is modeled by: [34,35]

$$F_T(t) = F_B(t) + \bar{e}m \cdot a(t) + mg\sin(\alpha) \quad \text{[N]} \tag{8}$$
$$F_B(t) = F_0 + F_1 \cdot v(t) + F_2 \cdot v^2(t) \tag{9}$$
$$a(t) = \frac{dv(t)/3.6}{dt} \tag{10}$$

with the vehicle mass m, gravity g, the inclination angle of the road α, the translational acceleration a of the car and a mean equivalent mass factor $\bar{e} = 1.15$ for the rotational accelerations for each individual gear. The basic driving resistance F_B (consisting of rolling resistance and aerodynamic drag) is modeled by the coast-down parameters F_0, F_1 and F_2 with units N, N/(km/h) and N/(km/h)^2, respectively. These parameters have to be provided by the manufacturer during the type approval procedure. The engine load (torque) is formulated by [34]:

$$M(t) = \frac{r_{\text{tire,dyn}} \cdot F_T(t)}{\eta \cdot i_{\text{gear}} \cdot i_{\text{ax}}} \quad \text{[Nm]} \tag{11}$$

with a globally-set efficiency factor $\eta = 0.9$ for the power transmission from the engine to the wheels. The engine load in percent is defined by [36]:

$$\Gamma(t) \equiv \frac{M(t)}{M_{\max}(n(t))} \cdot 100 \quad \text{[\%]} \tag{12}$$

with $\Gamma = 100\%$ at full load. At idling engine, $M = 0$ Nm and $\Gamma = 0\%$. In engine overrun operation (e.g., while engine braking), the engine delivers a negative torque to the crankshaft, which means that the engine load M becomes negative. In the model, this state is approximated as idle, *i.e.*, M is set to zero.

Gearbox shifts are modeled by three consecutive processes: the clutch is disengaged; a new gear is put in (at idling engine); and the clutch is engaged again. In dependence of the driving style, these processes vary in their respective durations. In the model, for a sporty driving style, the total gear change takes 0.6 s, whereas for a cozy, economic driving style, the gear change takes 1.3 s. Furthermore, the moments of a gear change strongly depend on the driving style and can be formulated as a function of engine speed and engine load, which is also the basic working principle of an automatic gearbox.

Figure 3a and 3b show two simulated engine condition courses within an engine load *vs.* engine speed diagram. The black lines show the simulated temporal progression of the engine condition during a virtual pass-by. Both simulations start at the same initial engine condition of 900 rpm and 3 Nm, marked by green stars. In both cases, the passenger car starts in first gear and accelerates from $v = 7$ km/h to 50 km/h, however with differing accelerations a and driving styles. For the medium driving style and an acceleration of $a = 1$ m/s^2 (Figure 3a), three gear changes occur at around 2000 rpm (Sample 1); whereas for the sporty driving style and an acceleration of $a = 2$ m/s^2 (Figure 3b), only two gear changes happen, but at higher engine speeds of 3000 to 4000 rpm (Sample 2). The temporal behaviors of the engine states for these two examples are published as supplementary data (see the videos in Supplementary File).

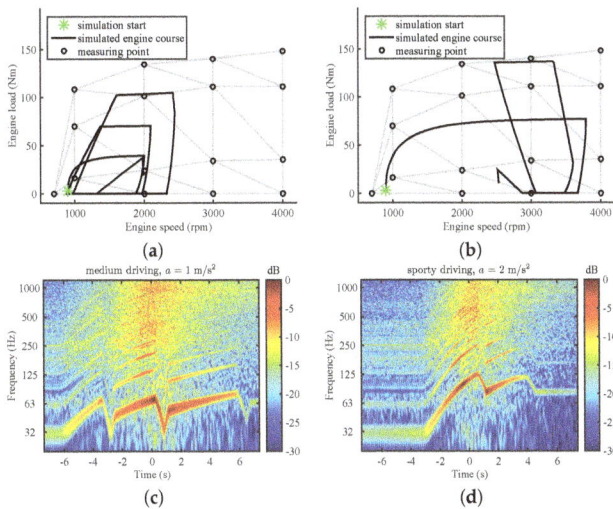

Figure 3. Simulation results: The upper graphs (**a**,**b**) show two simulated engine condition courses of an accelerating Ford Focus 1.8i with different accelerations and driving styles. The gray triangles show the interpolation grid spanned by the measuring points marked as circles, as introduced in Section 2.2.3. The lower two graphs (**c**,**d**) show the spectrograms of the corresponding synthesized pass-by sounds (normalized to 0 dB). Their calculation is elucidated in Section 2.3.

2.2.3. Sound of Propulsion

The structure of the emission synthesizer for the sound of propulsion is depicted in Figure 4. The sound pressure emission signal of the sound of propulsion is assumed to consist of a deterministic signal representing the most important engine orders and a quasi-stochastic signal:

$$s_{\text{prop}}(t) = s_{\text{prop,ord}}(t) + s_{\text{prop,noise}}(t) \tag{13}$$

Engine order ν corresponds to an event taking place ν times per engine revolution. The engine order signal is composed of the sum of the engine orders ν, which are generated using additive synthesis [29,30]. The engine order signal is thus calculated by [6]:

$$s_{\text{prop,ord}}(t) = \sum_{\nu} p_0 10^{L^{\dagger}_{\text{prop,ord},\nu}(t)/20} \cdot \sqrt{2} \cos(\beta_{\nu}(t)) \tag{14}$$

A proper selection of the essential orders ν strongly depends on the specific vehicle type and its condition. In the context of sound design, it is known that at least orders up to $\nu = 18$ are relevant [23]. Further, the sound characteristics can be influenced by half-orders [23]. In this model, it was decided to synthesize orders $\nu = 1$ to 30 in half-order steps, resulting in a total of 59 orders. This somewhat arbitrary, but safe choice leaves room for optimization. In Equation (14), $L^{\dagger}_{\text{prop,ord},\nu}$ denotes the order level and the instantaneous order phase:

$$\beta_{\nu}(t) = \phi^{\dagger}_{\nu}(t) + 2\pi \int_{-\infty}^{t} F_{\nu}(\tau)d\tau \tag{15}$$

with the order phase $\phi^{\dagger}_{\nu}$ and the order frequency:

$$F_{\nu}(t) = \nu \cdot n(t)/60. \tag{16}$$

Listening tests revealed that in this application, the order phase is a relevant synthesizer parameter. For a four-stroke engine with N_{cyl} cylinders, the engine order corresponding to the ignition, and mostly the predominant order, is $\nu_{\text{ign}} = N_{\text{cyl}}/2$ [23]. Thus, the ignition frequency reads [5]:

$$F_{\text{ign}}(t) = N_{\text{cyl}}/2 \cdot n(t)/60 \tag{17}$$

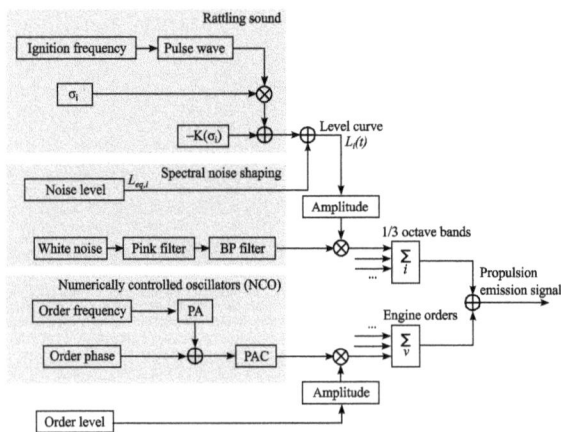

Figure 4. Signal flow chart of the synthesizer for the sound of propulsion.

For time-discrete signals, Equations (14) and (15) can be interpreted as a modified numerically-controlled oscillator (NCO) [28], whereas Equation (15) corresponds to the phase accumulator (PA), and the phase-to-amplitude converter (PAC) is realized in Equation (14). This formulation concurrently implements a frequency modulation by F and a phase modulation by ϕ.

The noise signal component of the sound of propulsion is synthesized similarly as the sound of tires (Equation (3)) by:

$$s_{\text{prop,noise}}(t) = \sum_{i=1}^{N_b} p_0 10^{L_{\text{prop,noise},i}(t)/20} \cdot \zeta'_i(t) \tag{18}$$

A total of $N_b = 29$ bands from 20 Hz to 12.5 kHz are used. The 1/3 octave band level function is formulated as:

$$L_{\text{prop,noise},i}(t) = L^\dagger_{\text{eq,prop,noise},i} + \sigma^\dagger_i \cdot R(t) - K(\sigma^\dagger_i) \tag{19}$$

with the level $L^\dagger_{\text{eq,prop,noise},i}$, a level standard deviation $\sigma^\dagger_i$ and a level fluctuation function $R(t)$ with zero mean and unit power. The constant K ensures that despite the level fluctuations, the equivalent continuous level (Leq) is not altered. This level modulation simulates the rattling sound component that elicits a roughness sensation, which is particularly characteristic for low engine speeds and diesel engines. Motivated by measurement data that showed the strongest level fluctuations at the ignition frequency, R is modeled by a quasi-periodic function with period $1/F_{\text{ign}}(t)$. The first half-period of R is composed of a Hann window, whereas the second half-period is held constant.

In summary, the presented synthesizer needs about 180 input parameters to generate a stationary signal for the sound of propulsion. However, during a pass-by, the sound of propulsion may considerably vary, and so do these parameters. These parameters, which are marked by † in the above equations, simultaneously depend on the engine speed n, the engine load M and the emission angle φ and are hence time dependent. They are calculated by a triangulation-based linear 3D interpolation of measurement data. Measurements were taken on a discrete grid, typically $n \approx \{1000, 2000, 3000, 4000\}$ rpm, $\Gamma \approx \{0, 40, 70, 100\}\%$ and $\varphi = \{0, 60, 120, 180\}°$. The measuring point pairs $\langle n, M \rangle$ of a measurement performed on a Ford Focus 1.8i are depicted as circles in Figure 3. The topmost points at 1000 to 4000 rpm are at full load, *i.e.*, $\Gamma = 100\%$. Furthermore, the adopted Delaunay triangulation, which is used for the interpolation, is shown with gray lines. The synthesizer parameters are evaluated with a temporal resolution of 20 ms and linearly interpolated to the audio sampling rate, f_s. For the interpolation of the order phase, ϕ, its cyclic behavior has to be considered in order to avoid spurious phase fluctuations.

2.3. Propagation Filtering

The sound propagation model described in this section incorporates the following effects:

- Propagation delay
- Doppler effect (frequency shift and amplification)
- Convective amplification
- Geometrical spreading
- Ground reflection
- Air absorption

Other outdoor sound propagation effects that may be relevant in certain situations are screening [37–39], foliage attenuation [38], meteorological effects due to an inhomogeneous atmosphere [38–42], as well as reflections at artificial [38,43] and natural surfaces [42,44,45]. Most published environmental noise auralization models simulate some of the above listed effects by applying a 1/3 octave filter bank and adjusting the filter gains [8,14,15]. In this model, however, all of these effects are applied in the time domain, *i.e.*, by time-variant digital filters. Sound propagation is modeled by two paths, namely for direct sound and a single ground reflection (in the following account indicated by subscripts "dir" and "gr", respectively). The sound pressure of a point source has a $1/r$ distance dependency. Thus, to model geometrical spreading, the emitted sound pressure signals, x, are divided by their path length r_{dir} or r_{gr}, respectively. The interaction of a sound wave with the ground influences its amplitude and phase as a function of frequency. This effect can be modeled by convolution of the ground-reflected signal with a time-variant filter [46]. Furthermore, the attenuation due to air absorption can be efficiently modeled using a filter [46]. Considering these aspects, the receiver signal y is calculated by:

$$y(t') = h_{\text{air},t'}(t') * \left(\frac{x_{\text{dir}}(t')}{r_{\text{dir}}(t')} + h_{\text{gr},t'}(t') * \frac{x_{\text{gr}}(t')}{r_{\text{gr}}(t')} \right) \tag{20}$$

where $*$ denotes linear convolution, t' the receiver time axis, r_{dir} is the source–receiver distance, r_{gr} is the distance source–ground reflection point–receiver, x_{dir} and x_{gr} are delayed versions of the emitted sounds and $h_{\text{air},t'}$ and $h_{\text{gr},t'}$ denote impulse responses of time-dependent filters described in Sections 2.3.2 and 2.3.3. The modeling of effects due to source motion and the propagation delay are explained in Section 2.3.1. Note that the immission angle $\theta(t')$, which is needed for surround reproduction, has to be evaluated on the receiver axis, as well.

Figure 3 shows normalized spectrograms of two synthesized pass-by sounds. For the synthesis of the sound of propulsion, the respective engine condition courses depicted in a and b of Figure 3 were used. They are described in Section 2.2.2. The temporal behavior of the engine states, the spectrograms, as well as the auralizations are published as supplementary data (see the videos in the Supplementary File). In both simulations, the car passes the receiver at Time 0 at 30 km/h. The receiver is located 1.2 m above a hard ground at a distance $D = 7.5$ m. As at the pass-by, the engine speed still increases, the Doppler frequency shift is not directly observable in the course of the order frequencies. However, the gear change moments can be well observed as local decreases of order frequencies. As a consequence of the used engine condition courses, these frequency drops occur at higher frequencies for the sporty driving style (d).

2.3.1. Effects Due to Source Motion and Propagation Delay

Due to the travel time of sound and the movement of the source, the source and the receiver have differing time axis. By neglecting wind and turbulence, the warped time axis at the receiver is given by:

$$t' = t + \Delta t(t) = t + \frac{r_{\text{dir/gr}}(t)}{c_0} \tag{21}$$

where $r_{\text{dir/gr}}$ denotes the sound propagation distance of the direct sound or the ground reflected sound, respectively, and c_0 is sound speed in still air. A constant sound speed of $c_0 = 340$ m/s is assumed. Since the receiver signal is supposed to have a constant sampling rate of f_s, the corresponding times, t_s, on the emission time axis have to be found. This is achieved by linear interpolation of Equation (21). The emission signals x for the direct and the ground-reflected path, respectively, with respect to the receiver time t' are:

$$x_{\text{dir/gr}}(t') = s(t) \cdot \mathscr{D}^2(t) \tag{22}$$

with the Doppler factor:

$$\mathscr{D} \equiv \frac{f'}{f} = \frac{dt}{dt'} \tag{23}$$

Equation (22) describes the kinematic and the aerodynamic effect of source motion. The former is known as the Doppler effect, *i.e.*, the Doppler frequency shift and amplification. The latter is known as convective amplification. The exponent two of the Doppler factor indicates that a Lighthill [47] monopole [48] and/or dipole source [49] is assumed.

The change of the time axis in Equation (22) realizes the propagation delay, as well as the Doppler frequency shift. For digital signals, this change corresponds to an asynchronous resampling process. It can by implemented using a variable delay-line with delay Δt [50]. If Δt is just rounded to the nearest sample, audible artifacts occur, so-called "zipper noise". Therefore, an interpolation strategy has to be used. As we are only interested in sequential access to the emission signal, a fractional delay filter can be used [51]. In [18] and [50], a linear interpolator is proposed. This however produces high frequency attenuation, as well as strong nonlinear distortions due to aliasing. Therefore, here, we introduce a band-limited interpolation or, respectively, a windowed sinc interpolation [51]:

$$s(k_s) = \sum_{k=\lfloor k_s \rfloor - b + 1}^{\lfloor k_s \rfloor + b} s[k] K(|k_s - k|) \tag{24}$$

with the floor function $\lfloor . \rfloor$, the integer sample index k, the non-integer sample index $k_s = t_s f_s$ and the Hamming kernel:

$$K(m) = \begin{cases} [0.54 + 0.46 \cos(\pi m / b)] \operatorname{sinc}(m) & \text{if } 0 \leq m < b \\ 0 & \text{otherwise} \end{cases} \tag{25}$$

with an integer b describing the filter length. To keep the computational effort low, in the implementation of Equation (24), values of the kernel K are stored in a look-up table. The Doppler factor $\mathscr{D}$ in Equation (22) is implemented by approximating the derivative in Equation (23) by finite differences as:

$$\mathscr{D}(k_{s,i}) \cong k_{s,i+1} - k_{s,i} \tag{26}$$

with index i.

In order to validate different implementations of Equation (22), numerical simulations were performed. Figure 5 compares the signal attenuation introduced by different interpolation schemes. The high frequency attenuation of the linear interpolation can be improved by a windowed sinc interpolation and controlled by parameter b. Figure 6 shows spectrograms of receiver signals calculated by the same three interpolation schemes. As an extreme case, a virtual source emitting a 1 kHz pure tone travels at constant speed $v = 150$ km/h and passes a static receiver at a distance of $D = 7.5$ m. Figure 6 shows that by introducing a windowed sinc interpolation of sufficient filter length, artifacts due to aliasing can be significantly reduced compared to a linear interpolation (a). The minimal kernel size b required for a decent sound quality cannot be stated in general, as it strongly depends on the application, *i.e.*, the source signal, the propagation situation and, not least, the sampling frequency. However, in the example of Figure 6, already, $b = 10$ reaches a good sound quality, without audible artifacts. Nevertheless, to be on the safe side, a value of $b = 100$ was adopted. The careful choice of b, however, provides the potential for optimization in terms of sound quality and computational cost.

For sound speed $c_0 = 340$ m/s, the Mach number $M \equiv v/c_0 \approx 0.12$. At times $t = \pm\infty$, the received frequencies f' due to the Doppler shift are given by [2,52]:

$$f'_{t=\pm\infty} = f \cdot \mathscr{D}_{t=\pm\infty} = \frac{f}{1 \pm M} \tag{27}$$

with f being the emitted frequency. In our example, according to Equation (27), the received frequency changes by a factor of 1.28 across the pass-by, which corresponds to a musical interval that is larger than a major third. The sum of the Doppler and the convective amplification amounts to [48]:

$$G_{t=\pm\infty} = -40 \cdot \log 10 \left(1 \pm M\right) \quad [\text{dB}] \tag{28}$$

Equation (28) yields an amplification of 2.3 dB at $t = -\infty$ and an attenuation by 2.0 dB at $t = \infty$, resulting in a level difference of 4.3 dB across the pass-by. The numerical implementations of Equation (22) corresponded well with these theoretical values.

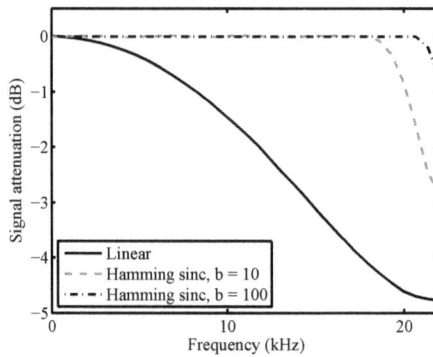

Figure 5. Spectral attenuation due to different resampling strategies.

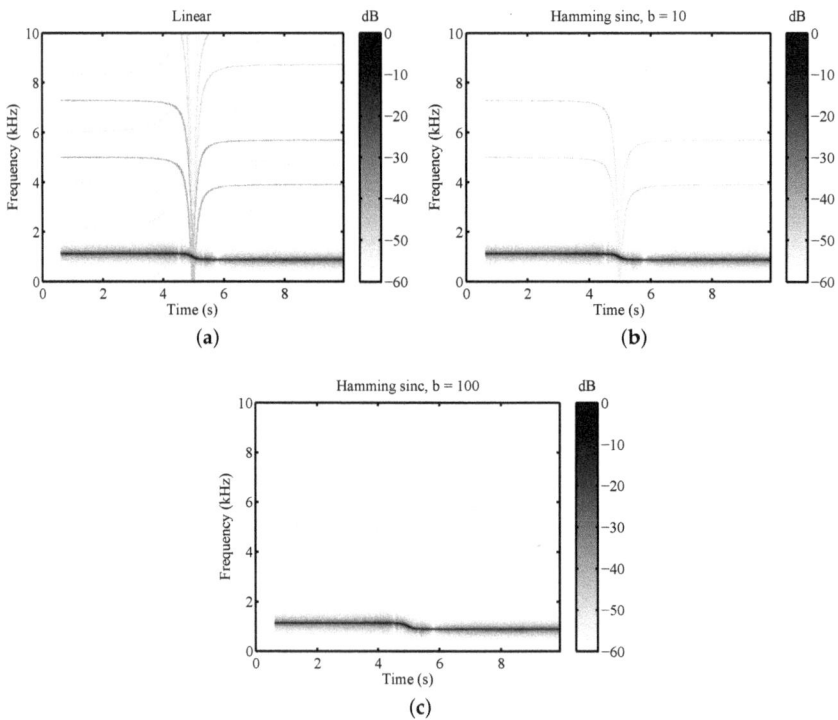

(a)

(b)

(c)

Figure 6. Non-linear distortions (aliasing) due to different resampling strategies: linear interpolation (a) and Hamming sinc interpolations with different filter lengths, $b = 10$ for (b) and $b = 100$ for (c), respectively. The simulation was performed for a source emitting a 1 kHz pure tone that travels at constant speed $v = 150$ km/h and passes a static receiver at a distance of $D = 7.5$ m.

2.3.2. Ground Effect

In Equation (20), the ground effect is modeled in a physical way as the interference between direct and ground reflected sound. A flat topography is assumed, *i.e.*, only one ground-reflected path is modeled, which is implemented by adding a second signal path. The ground-reflected sound differs from direct sound by scaling with its propagation distance and a complex reflection factor, as well as an additional delay. The complex reflection factor depends on frequency, geometry and ground surface type and is realized by the filter $h_{gr,t'}$. $h_{gr,t'}$ is the impulse response of the spherical wave reflection coefficient at an infinite locally-reacting surface. The ground surface is acoustically described by a frequency-depending surface impedance, for which the widely-used empirical model of Delany and Bazley [53] was used.

In [46], the additional delay of the ground-reflected sound was modeled by a digital delay of integer length. However, in this application, due to the higher relative source speed and short delays, audible artifacts ("zipper noise") occur. Therefore, a separate resampling is performed in Equation (22) for the ground-reflected sound. Furthermore, this type of processing eliminates the spectrally-fluctuating errors (see Figure 1 in [46]).

The spherical wave reflection coefficient filter $h_{gr,t'}$ is implemented by an FIR filter designed using the inverse FFT, as described in [46]. It has to be made sure that the filter lag is compensated. However, compared to [46] for this application, substantially more filter taps are required to reproduce the correct interference pattern. Figure 7 shows simulation results for the standard configuration of road traffic noise emission measurements (a) and a receiver point at distance $D = 100$ m and height 2 m with sound propagation over grassy ground (b). For the former case a filter with 40 taps is sufficient, as the difference to the simulation with a filter with 400 taps stays well below 1 dB for all frequencies (nearly perfect coincidence of curves in Figure 7a). For the latter case, however, Figure 7b shows that such a short filter is not able to correctly reproduce the interference pattern and creates large errors at mid and low frequencies. A filter length of 400 taps allows simulations that are in good agreement with the exact solution for both cases. Large errors only occur near the Nyquist frequency. An update interval of the filter coefficients of 200 ms is used.

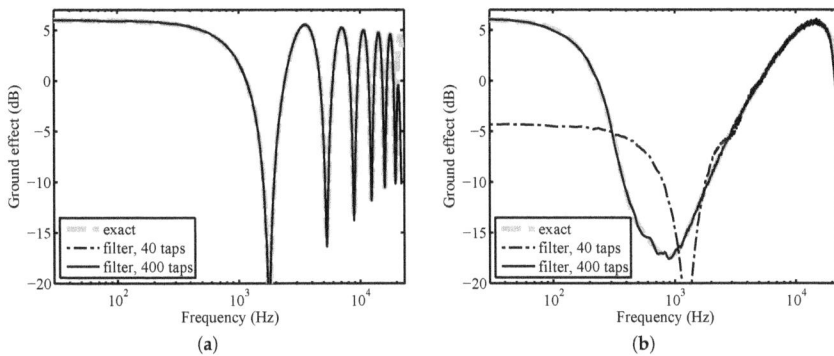

Figure 7. Simulated ground effect spectra for a point source at a height of 0.3 m in the reference situation (**a**) with a receiver at a height of 1.2 m at a horizontal distance of $D = 7.5$ m and propagation over hard ground (flow resistivity 20,000 kPa·s·m^{-2}); and a distant situation (**b**) with a receiver at a height of 2 m at horizontal distance $D = 100$ m and propagation over grassy ground (flow resistivity 200 kPa·s·m^{-2}).

2.3.3. Air Absorption

For performance reasons, the identical air absorption filter $h_{air,t'}$ is applied to the direct and ground reflected path in Equation (20). $h_{air,t'}$ are linear-phase FIR filters designed using the inverse FFT, as described in [46]. The frequency-dependent sound attenuation coefficients for atmospheric absorption as a function of relative humidity and temperature are calculated according to the standard ISO 9613-1 [54]. A filter length of 30 taps is used with an update interval of the filter coefficients of 200 ms.

2.4. Reproduction Rendering

The rendering of the immission signals for reproduction strongly depends on the type of reproduction system. For surround reproduction via multiple loudspeakers, techniques, such as Ambisonics [55,56] or amplitude panning (e.g., Vector Base Amplitude Panning (VBAP) [57] or Multiple-Direction Amplitude Panning (MDAP) [58]), are possible candidates. For binaural reproduction over headphones, generally, head-related transfer functions (HRTF) should be applied. In this paper, for simplicity, a simulation of the "ORTF" stereo technique [59,60] is used. If the listener is facing the road, this allows for a reproduction with sufficient accuracy via headphones and a stereo speaker set-up. The cardioid microphone pattern and the time difference between the left and right channel are modeled by:

$$L(t') = 0.5\,(1 + \cos(\theta - 55°)) \cdot y(t' + u) \tag{29}$$
$$R(t') = 0.5\,(1 + \cos(\theta + 55°)) \cdot y(t') \tag{30}$$

with the time-varying time difference:

$$u = \frac{0.17 \cdot \sin(\theta(t'))}{c_0} \tag{31}$$

The immission angle $\theta(t')$ has to be evaluated on the receiver axis (see Equation (21)). $y(t' + u)$ is calculated using a windowed sinc interpolation strategy according to Equation (24). As a consequence of this interpolation, high-frequency attenuation, as shown in Figure 5, and nonlinear distortions are introduced to channel L.

3. Model Parameter Estimation

This section presents procedures to obtain the model parameters of the emission synthesizer described in Section 2.2. The procedures are based on controlled measurements. The following sections describe the measurements, as well as the signal processing that is applied to the acquired data.

3.1. Tire Noise

The emission parameters for tire noise were obtained from pass-by measurements with idling engine. For an individual tire type, pass-bys by the same passenger car at different speeds were recorded at a sampling frequency of f_s = 44.1 kHz with a calibrated measurement microphone in a set-up referring to the standard ISO 11819-1 [25] and depicted in Figure 8a. The pass-by speed was measured by radar, and the pass-by time was determined from synchronous video. Under the assumption of constant speed, a time-dependent backpropagation to the source was performed. Thereby, two equal incoherent point sources at the nearby wheels were assumed, *i.e.*, placed at the side of the car, horizontally separated by the wheelbase and set on the ground. For the temporal accordance, the sound propagation delay, as well as the filter group delays of the 1/3 octave band filters have to be taken into account. Consequently, emission levels at reference distances r_0 were obtained by integration over an emission angle range of 90°. Applying a logarithmic transformation to the measured pass-by speeds, the linear regression parameters A_i and B_i of Equation (4) were fitted in a least-squares sense. Despite the idling engine, some low 1/3 octave bands were contaminated by the engine sound. To correct for this, in the first step, for each band, a quality criterion based on the correlation coefficient and the slope of the regression line was deployed. Adverse bands were identified and imputed based on the values of adjacent valid bands. In the second step, low-frequency peaks of A_i were smoothed by a nonlinear method. Figure 9 contains measured parameters A and B in 1/3 octave bands of 13 tires.

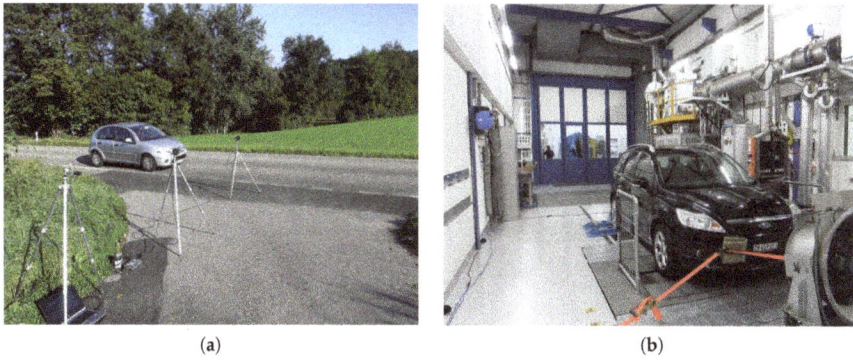

(a)   (b)

Figure 8. Photographs showing the measurement set-ups for tire noise (**a**) and propulsion noise (**b**). In (**a**), the coast-by situation is depicted with two measurement microphones placed at different distances and a camera connected to a laptop; (**b**) shows the lab with a passenger car on the chassis dynamometer, the airstream fan in front of the car and two microphones on the floor at the left-hand room edge (emission angles $\varphi = 60°$ and $120°$).

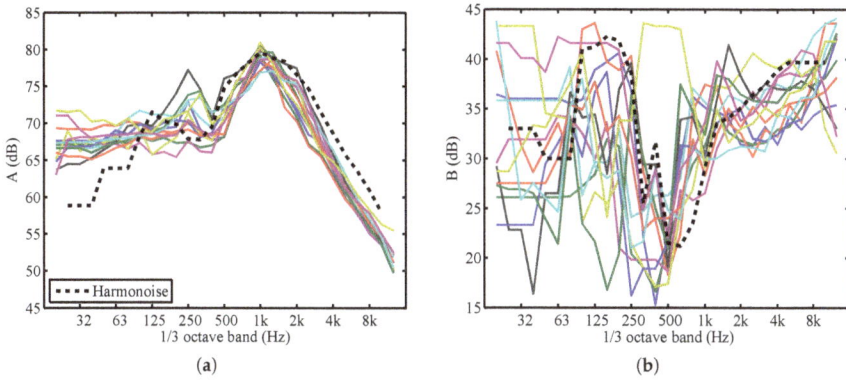

(a)   (b)

Figure 9. Measured tire noise regression parameters A_i (**a**) and B_i (**b**) of 13 tires and the values according to the Harmonoise model [21] (dotted lines).

3.2. Propulsion Noise

To obtain the emission synthesizer parameters of the propulsion noise, controlled measurements on a chassis dynamometer (see Figure 8b) and at idling engine under free field conditions were performed. Calibrated audio recordings at a sampling frequency of $f_s = 44.1$ kHz at different microphone positions around the vehicle and at different engine conditions were taken. During the measurements on the chassis dynamometer, four microphones were placed on the ground at emission angles $\varphi \approx \{0, 60, 120, 180\}°$ at distances $r = 1$ to 2 m from the vehicle. During the free field measurements, four additional microphones were placed on the ground at the identical emission angles, but at larger distances of $r' = 4.5$ to 7 m. The free field measurements were used to correct for the room influences of the lab, as explained in Section 3.2.5. On the chassis dynamometer, measurements were typically taken at engine speeds of $n \approx \{1000, 2000, 3000, 4000\}$ rpm and engine loads of $\Gamma \approx \{0, 40, 70, 100\}\%$. The measuring point pairs $\langle n, M \rangle$ of a measurement performed on a Ford Focus 1.8i are depicted as circles in Figure 3. The topmost points at 1000 to 4000 rpm are at full load, *i.e.*, $\Gamma = 100\%$. To confine tire noise, low vehicle speeds were aimed for by choosing low driving gears. Mostly, it was the second gear, which resulted in vehicle speeds <50 km/h.

To these recordings, a series of signal analysis steps were applied, which are outlined in a signal flowchart in Figure 10. These steps are further explained in the following sections. For the signal processing, a signal length of 4 s is used.

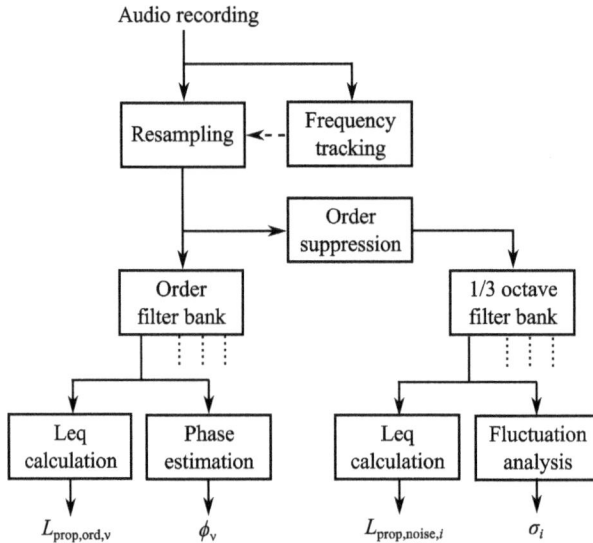

Audio recording

```
                   Audio recording
                          │
            ┌─────────────┼──────────────┐
            ▼                             ▼
      ┌──────────┐  ◄--  ┌──────────────┐
      │Resampling│       │  Frequency   │
      └──────────┘       │   tracking   │
            │            └──────────────┘
            │            ┌──────────────┐
            ├───────────►│    Order     │
            │            │ suppression  │──────────┐
            ▼            └──────────────┘          ▼
      ┌──────────┐                          ┌──────────┐
      │  Order   │                          │1/3 octave│
      │filter bank│                         │filter bank│
      └──────────┘                          └──────────┘
        ┌────┴────┐                          ┌────┴────┐
        ▼         ▼                          ▼         ▼
    ┌───────┐ ┌────────┐              ┌───────┐ ┌──────────┐
    │  Leq  │ │ Phase  │              │  Leq  │ │Fluctuation│
    │ calc  │ │estimate│              │ calc  │ │ analysis │
    └───────┘ └────────┘              └───────┘ └──────────┘
```

Figure 10. Signal analysis flowchart to obtain the synthesizer parameters of propulsion noise as described in Section 2.2.3 from audio recordings.

3.2.1. Resampling

The emission synthesizer uses detailed information about the engine orders. These parameters are obtained by a narrowband analysis, which is described in the following section. Although during the measurements, the engine speed was kept fairly constant, the instantaneous order frequencies slightly fluctuate as exemplarily shown in Figure 11. This figure shows the spectrogram of a recording made at the rear of a car with an inline, four cylinder engine idling at 1100 rpm. To be able to separate engine orders and broadband noise by the narrowband analysis, a preceding resampling of the slightly non-stationary signals is performed. In order to actuate the resampling process, the instantaneous ignition frequency, $F_{\text{ign}}(t)$, of the engine is required. This data are extracted from the audio recordings.

In the first step, the average ignition frequency is estimated. From the signal taken closest to the exhaust, the power spectral density (PSD) with a frequency resolution <1 Hz is calculated. Based on the rough indication of the engine speed taken from the car's tachometer, a first estimate of the ignition frequency, F_{ign}, is obtained using Equation (17). The location of the maximum value of the PSD within a search range around this frequency yields a better, second estimate. Particularly, for low engine speeds, at which the ignition frequency can be as low as 20 Hz, this estimate is still not precise enough due to the low relative resolution at low frequencies. Thus, this estimate is further enhanced by considering the double ignition frequency, $2F_{\text{ign}}$, (*i.e.*, engine order $v = 4$ in Equation (16) for a four-cylinder engine) within a smaller range of the PSD.

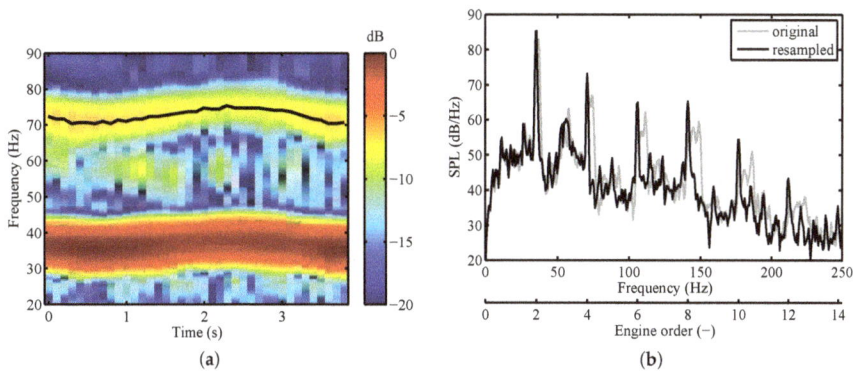

Figure 11. Normalized spectrogram (**a**) of the measured sound pressure signal with tracked double ignition frequency (drawn as a black line) and power spectral density (**b**) of the original and asynchronously resampled sound pressure signal, respectively. The recording was conducted at the rear of the BMW with an inline, four-cylinder engine idling at 1100 rpm.

In the second step, this information is used to track the course of the ignition frequency, $F_{ign}(t)$. This task is generally known as pitch detection [61–63]. A wide variety of algorithms exist that work in the time or frequency domain or a combination of them. In our application, a spectral method was established, in which the course of one discrete frequency component (*i.e.*, an engine order) is tracked in a spectrogram. The spectrogram $S(t, f)$ expressed in decibels is computed by the short-time Fourier transform (STFT):

$$S(t,f) = 10 \cdot \log \left(|STFT(t,f)|^2 \right) \tag{32}$$

STFT is calculated using the FFT. Windows of 200 ms with a 50% overlap, *i.e.*, a temporal resolution of $\Delta t = 100$ ms, are multiplied by a Hann window function. To obtain a high frequency resolution of $\Delta f < 0.5$ Hz, the signals are zero padded. A section of such a spectrogram is depicted in Figure 11. Within the spectrogram, the "highest cost" path between time $t = 0$ and $t = T$, within a certain frequency range around a reference frequency, F_r, is sought. F_r is chosen to be the first multiple of the mean ignition frequency above 55 Hz. This is a compromise between signal power and frequency localization: typically, the power decreases for increasing even orders (see Figure 11b), but higher orders exhibit larger absolute frequency variations (see Figure 11a). In the example of Figure 11, F_r is 71 Hz (corresponding to the fourth engine order), as the mean ignition frequency lies at 35 Hz.

The optimization task is solved by dynamic programming, which breaks the complex problem down into many simple subproblems. This method prevents taking possible wrong local decisions and guarantees that the best solution is found. A well-known algorithm that uses dynamic programming is dynamic time warping (DTW), which is often applied in, e.g., automatic speech recognition (ASR). Additionally, we make use of the *a priori* knowledge that the engine speed does not change rapidly over time. This is introduced as a requirement on the slope of the optimal path, F_{opt}. The algorithm described below is based on an algorithm developed for object tracking in video data [64]. Within the search section of the discrete spectrogram, the local score q is calculated by:

$$q(m,l) = S(m,l) - \min_{m,l} (S(m,l)) \tag{33}$$

for which holds $q \geq 0$. From q for each time step $m = \{1...M\}$ and frequency bin l, the global score $Q(m,l)$ is recursively computed by:

$$Q(m,l) = q(m,l) + \max_{l' \in \{l-c \leq l \leq l+c\}} Q(m-1,l') \tag{34}$$

with a positive integer c realizing the requirement:

$$|C| \leq c \frac{\Delta f}{\Delta t} \tag{35}$$

on the absolute value of the slope C of F_{opt} given in Hertz per second. For the forward processing described by Equation (34), the starting condition is that the initial global score, $Q(1,l)$, is set to the local score, *i.e.*, $Q(1,l) = q(1,l)$. During the evaluation of Equation (34), it is essential that the back pointers:

$$B(m,l) = \underset{l' \in \{l-c \leq l \leq l+c\}}{\arg \max} \; Q(m-1,l') \tag{36}$$

to the optimal predecessors are stored. From the global score Q, the end point of the optimal path is found by:

$$F_{opt}(M) = \arg \max_l Q(M,l) \tag{37}$$

Using the back pointers B, the optimal path can be traced back using a recursive procedure known as backtracking:

$$F_{opt}(m-1) = B\left(m, F_{opt}(m)\right) \tag{38}$$

In Appendix, we provide a simple MATLAB code, which solves Equations (34) to (38). Figure 11 shows the optimal path drawn as a black line following the frequency component around 70 Hz.

In the third step, the sound pressure signal is asynchronously resampled based on the course of the tracked engine order. The warped time axis is calculated by:

$$t_{warp} = t + \frac{F_{opt}(t)}{F_r} \tag{39}$$

where $F_{opt}(t)$ is the linearly-interpolated version of $F_{opt}(m)$. For the resampling of the sound pressure signals, a windowed sinc interpolation, as described by Equation (24), is adopted. Figure 11 illustrates the effect of the asynchronous resampling on the power spectral density. In contrast to the original signal, for the resampled signal, all even engine orders from two to 12 can be clearly identified as equidistant, narrow peaks.

3.2.2. Order Analysis

From the resampled signals, information about the engine orders is extracted. Therefore, a filter bank consisting of one bandpass filter per considered engine order is generated and applied to the signal. Eighth order Butterworth filters centered around the engine order frequency F_v with a 6-Hz bandwidth are employed. Figure 12 shows the magnitude frequency response of the filter bank. At the output of each filter, the corresponding order level, $L_{prop,ord,v}$ in Equation (14), is calculated as an equivalent continuous level (Leq). Figure 13 exemplifies measured order levels at idling engine and full load recorded at the rear of a VW Touran running at 1000 and 3000 rpm.

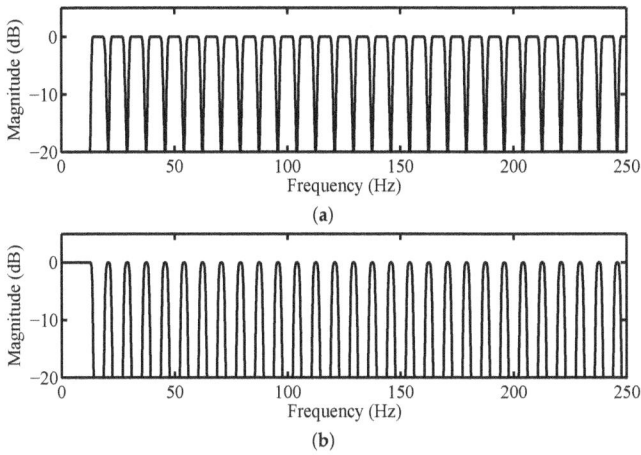

Figure 12. Magnitude frequency response of the engine order analysis filter bank (**a**) and engine order suppression filter (**b**) for engine speed $n = 1000$ rpm, engine orders $\nu = 1, 1.5, 2, ..., 15$ and $N_{cyl} = 4$.

The order phases are detected using the cross-correlation function. Since the above-described infinite impulse response (IIR) filter bank introduces phase shifts, the outputs of the filter bank are time reversed and sent once again through the same filter bank and, finally, time reversed. In doing so, a zero-phase forward and reverse digital IIR filtering is implemented. This signal, $g_\nu(t)$, is cross-correlated with a prototype function $\cos(2\pi F_\nu t)$ to obtain the time shift:

$$\kappa_\nu = \arg\max_\tau \left\{ \int g_\nu(t + \tau) \cos(2\pi F_\nu t) dt \right\} \tag{40}$$

from which the phase shift of Equation (15) can be derived as:

$$\phi_\nu = -2\pi F_\nu \kappa_\nu \tag{41}$$

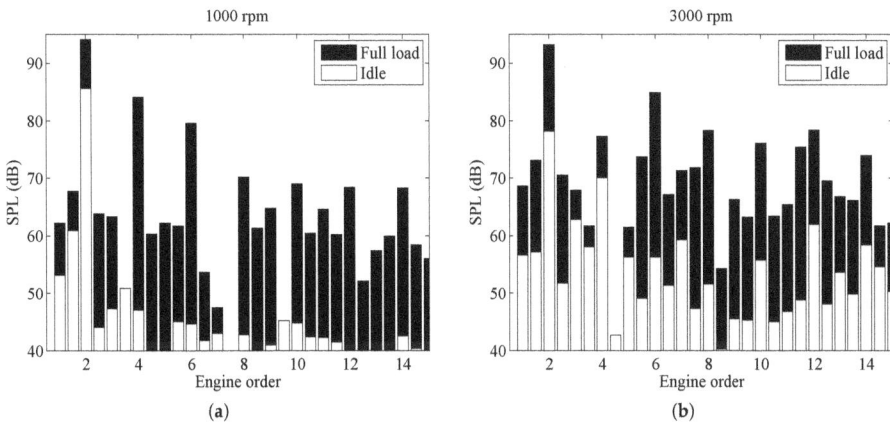

Figure 13. Comparison of engine order levels with idling engine (white) and full load (black) at 1000 (**a**) and 3000 rpm (**b**). Recorded at the rear of a VW Touran 1.6 FSI.

Figure 14 compares the sound pressure signals of a recording and the corresponding synthesis consisting of engine orders with constant phases, which were estimated by Equation (41).

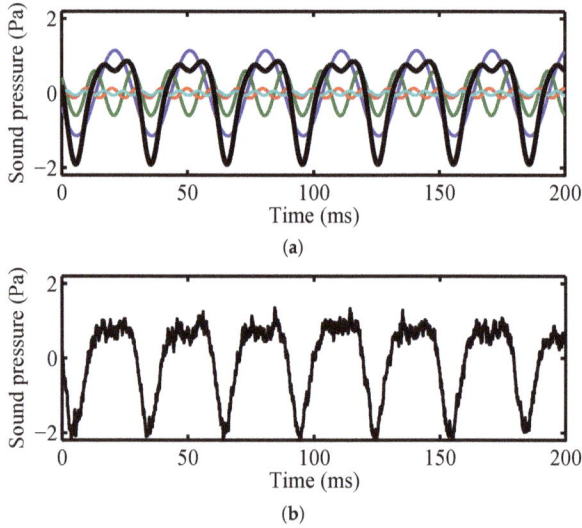

(a)

(b)

Figure 14. Comparison of sound pressure signals of a recording (**a**) and the corresponding synthesis consisting of engine orders with estimated phases (**b**). For the purpose of illustration only, the four dominant engine orders (colored lines) are used. The recording was conducted at the rear of a Ford Focus 1.8i at 1000 rpm and full load.

3.2.3. Noise Analysis

The noise levels and their short-term level fluctuations are obtained by a series of filtering operations. Starting with the resampled signal, in a first attempt, the engine orders are suppressed using cascaded notch filters. These filters are designed analogously to the engine order filter bank from the previous section, except that instead of bandpass filters, band-stop filters are generated (see Figure 12). Figure 15 shows two power spectral densities, which illustrate the effect of the order suppression filter. After this operation, the signal is split into sub-bands for further analysis. The signal is therefore decomposed into 1/3 octave bands using a 1/3 octave band filter bank. Each of the N_b filters yields a signal $q_i(t)$.

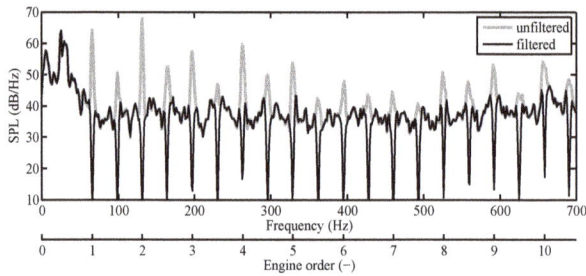

Figure 15. Power spectral densities illustrating the effect of the order suppression filter, which is applied to a recording of an inline, four-cylinder engine idling at 4000 rpm.

From $q_i(t)$, the noise levels, $L_{eq,prop,noise,i}$ in Equation (19), are calculated as Leqs. Moreover, from $q_i(t)$, using a moving average filter, smoothed level-time curves:

$$L_{q,i}(t) = 10 \cdot \log \left(\frac{1}{K} \int_0^K \frac{q_i^2(t+\tau)}{p_0^2} d\tau \right) \tag{42}$$

are calculated using a window length of $K = 4$ ms. Subsequently, from $L_{q,i}$, the mean value is subtracted to obtain a DC-free level fluctuation signal:

$$\Psi_{q,i}(t) = L_{q,i}(t) - \overline{L_{q,i}(t)} \tag{43}$$

Figure 16a exemplifies such a fluctuation signal for the 2.5-kHz band recorded at the front of a diesel engine car. The periodic structure is clearly visible. Following [6], the autocorrelation function (ACF) is used to estimate the standard deviations σ_i (used in Equation (19)) of the level fluctuations with period $1/F_{ign}$ by:

$$\sigma_i^2 = ACF_{\Psi_{q,i}}(1/F_{ign}) \tag{44}$$

Figure 16b shows the square root of the ACF of the level fluctuation signal depicted in Figure 16a. Clear peaks can be observed at lag zero and multiples of the ignition period of 34 ms. The standard deviation σ amounts to about 5 dB. The fact that a higher peak appears at the double ignition period, at 68 ms, indicates that the signal contains an additional level modulation with a modulation frequency equal to the half ignition frequency. This can also be observed in Figure 16a in which every second peak is about 5 dB higher than the previous one.

Figure 17 shows measured spectra of the standard deviations σ_i. The measurements were performed in front of five cars idling at low engine speeds. It can be seen that diesel cars feature higher values compared to gasoline engine cars. This finding corresponds to the increased rattling sound noticed in the field.

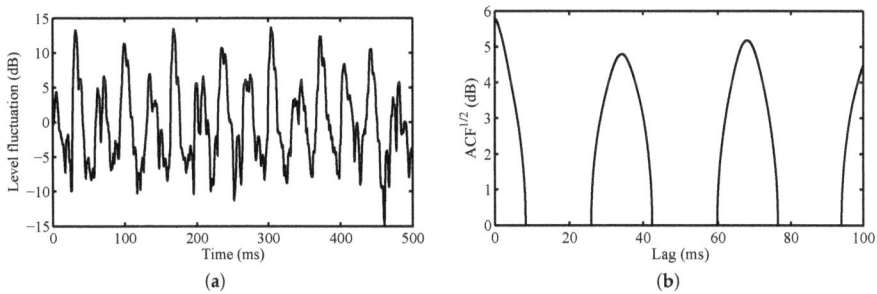

(a)   (b)

Figure 16. Level fluctuation signal (**a**) of the 2.5-kHz 1/3 octave band and its square root of the autocorrelation function (**b**) from a recording taken at the front of an idling four-cylinder diesel engine at 870 rpm, corresponding to an ignition period of 34 ms. The right plot indicates that for this band, the level standard deviation, σ, amounts to about 5 dB.

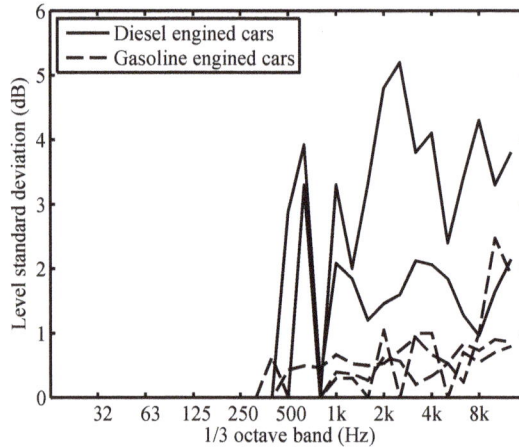

Figure 17. Level fluctuation standard deviations, σ, in 1/3 octave bands measured at the front of two diesel engine and three gasoline engine cars idling at 900 rpm.

3.2.4. Background Noise Corrections

On the test rig, the main background noise sources were the tire noise, the airstream fan, the room ventilation and the dynamometer itself. Firstly, to confine the tire noise, the measurements were performed at low vehicle speeds, *i.e.*, low gears. Secondly, during the measurements, the airstream fan (depicted in Figure 8) was briefly switched off for periods of about 10 s. However, the dropping tonal components of the fan still strongly interfered with the propulsion noise of the car (see Figure 18). Therefore, several shifted analysis time windows were deployed, and the minimal levels and the maximum level standard deviations, σ, across these windows were exploited. Thirdly, background noise measurements with a switched off engine at different vehicle speeds were performed. For each ordinary measurement, the corresponding background noise was identically analyzed and used for level corrections.

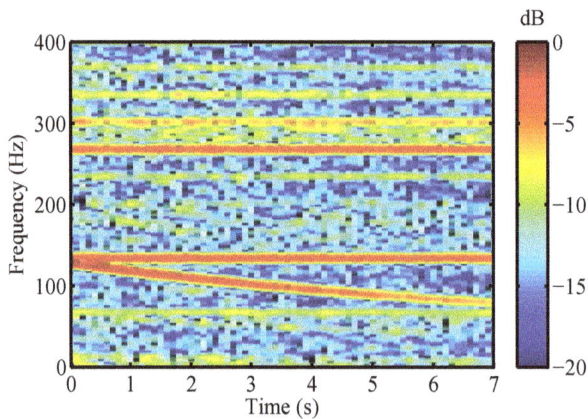

Figure 18. Normalized spectrogram of a microphone position in front of a Ford Focus 1.8i at 4000 rpm and full load on the dynamometer. The dropping tonal component around 100 Hz stems from the briefly switched off airstream fan of the lab.

3.2.5. Backpropagation

As the interface to the propagation model, the emission signals are defined at a (virtual) reference distance of $r_0 = 1$ meter from the source position. For the measured levels L_{lab}, the following inverse sound propagation model is used:

$$L_{Em,1\,m} = L_{lab} + 20\log\left(\frac{r_{Ac}}{r_0}\right) + A_{room} + A_{gr} \tag{45}$$

with the ground effect $A_{gr} = -6$ dB, as all microphones were mounted on the ground. For the microphones placed close to the room edge (emission angles $\varphi = 60°$ and $120°$), the room correction A_{room} was set to -6 dB for frequency bands below 1 kHz and to -3 dB otherwise. As for the microphones placed in front and at the back of the car, the distance to the closest wall was about three meters; A_{room} was set to 0 dB for these signals. r_{Ac} is the distance to the acoustical center, which, by assuming geometrical spreading of a point source, is obtained by simultaneous free field measurements at two points at distances r and r' ($r' > r$) with:

$$r_{Ac} = \frac{r' - r}{1 - 10^{(L_{ff,r'} - L_{ff,r})/20}} \tag{46}$$

Parameter r_{Ac} is evaluated separately for each emission angle, engine speed and frequency band (or engine order, respectively).

4. Conclusions

In the proposed auralization model, emission sounds of accelerating passenger cars are artificially generated based on spectral modeling synthesis. Whereas the sound of tires is synthesized as stationary noise, which is time-dependently shaped in third octave bands, the realistic synthesis of sounds of propulsion requires more subtlety.

It is synthesized as the superposition of a noise component and tones. Frequency-dependent periodic short-term modulations are applied to the noise component in order to create a rattling sound eliciting a roughness sensation. The tones are related to the engine orders. It was found that a large number of engine orders are needed (50...100) to convincingly represent different engine speeds and loads. Moreover auralizations revealed that the order phases have to be included as synthesizer parameters. In conclusion, the presented emission synthesizer gives complete control over the signal characteristics, but is computationally much more demanding than a synthesizer based on granular synthesis [5] with limited flexibility. However, a hybrid approach could profit from both advantages, *i.e.*, by pre-creating relevant signal grains using spectral modeling synthesis and usage for granular synthesis in real-time applications.

Analysis of the propagation filtering algorithms yielded two main insights. Aliasing, arising from the simulation of the Doppler effect, can be reduced by incorporating a band-limited resampling strategy, such as the windowed sinc interpolation. Furthermore, due to the low source height, a significantly higher number of filter taps is needed to correctly simulate the ground effect in relevant situations, as compared to elevated sources, such as airplanes or wind turbines [46].

We conclude that with the presented synthesizer structure, audio signals from vehicle pass-bys can be represented in a compact and elegant manner. To give the reader an impression of the subjective quality of the proposed model, auralizations of two examples are published as supplementary data (see the videos in the Supplementary File and Section 2.3 for details).

Acknowledgments: This work was supported by the Swiss National Science Foundation (SNSF) (# 200021-149618). In addition to the four anonymous reviewers, we would like to thank Kurt Eggenschwiler, Jean Marc Wunderli, Beat Schäffer, Christoph Zellmann, Stefan Plüss, Markus Studer, Ioannis Karipidis, Simon Holdener and Markus Haselbach for their help in the measurement campaigns. The first author also thanks Albert Pieren for his valuable explanations on automotive engineering.

Author Contributions: Reto Pieren drafted the main manuscript. Thomas Bütler conducted the measurements on the chassis dynamometer and assisted in the development of the driving dynamics simulation. Kurt Heutschi supervised the research and helped in the preparation of the manuscript.

Conflicts of Interest: The authors declare no conflict of interest.

Author Contributions: Appendix Code of the Frequency Tracking Algorithm

The frequency tracking algorithm described in Section 3.2.1 uses dynamic programming in order to find the optimal solution. This task can be easily implemented in, e.g., MATLAB. For the convenience of the reader, in this appendix, we provide a MATLAB code that solves Equations (34) to (38).

```
% initialization
M = size(q,1); %total time steps
N = size(q,2); %total frequency bins
Q(1,:) = q(1,:); %init of total score at first time step
% forward processing
for m = 2:M
  for k = 1:N
    kb = max(k-c,1);
    kt = min(k+c,N);
    [maxVal, maxIdx] = max(Q(m-1,kb:kt));
    Q(m,k) = q(m,k) + maxVal;
    B(m,k) = kb + maxIdx-1;
  end
end
% end point of optimal path
[~, P(M)] = max(Q(M,:));
% backtracking
for m = M:-1:2
  P(m-1) = B(m,P(m));
end
```

References

1. Kleiner, M.; Dalenbäck, B.I.; Svensson, P. Auralization—An Overview. *J. Audio Eng. Soc.* **1993**, *41*, 861–875.
2. Vorländer, M. *Auralization: Fundamentals of Acoustics, Modelling, Simulation, Algorithms and Acoustic Virtual Reality*; Springer: Berlin, Germany, 2008.
3. Savioja, L.; Svensson, U. Overview of geometrical room acoustic modeling techniques. *J. Acoust. Soc. Am.* **2015**, *138*, 708–730.
4. Klemenz, M. Sound Synthesis of Starting Electric Railbound Vehicles and the Influence of Consonance on Sound Quality. *Acta Acust. United Acust.* **2005**, *91*, 779–788.
5. Jagla, J.; Maillard, J.; Martin, N. Sample-based engine noise synthesis using an enhanced pitch-synchronous overlap-and-add method. *J. Acoust. Soc. Am.* **2012**, *132*, 3098–3108.
6. Pieren, R.; Heutschi, K.; Müller, M.; Manyoky, M.; Eggenschwiler, K. Auralization of Wind Turbine Noise: Emission Synthesis. *Acta Acust. United Acust.* **2014**, *100*, 25–33.
7. Arntzen, M.; Simons, D. Modeling and synthesis of aircraft flyover noise. *Appl. Acoust.* **2014**, *84*, 99–106.
8. Sahai, A.; Wefers, F.; Pick, P.; Stumpf, E.; Vorländer, M.; Kuhlen, T. Interactive simulation of aircraft noise in aural and visual virtual environments. *Appl. Acoust.* **2016**, *101*, 24–38.
9. Rizzi, S.; Sullivan, B.; Sondridge, C. A three-dimensional virtual simulator for aircraft flyover presentation. In Proceedings of the 2003 International Conference on Auditory Display, Boston, MA, USA, 6–9 July 2003; pp. 87–90.

10. Rietdijk, F.; Heutschi, K.; Zellmann, C. Determining an empirical emission model for the auralization of jet aircraft. In Proceedings of the 10th European Conference on Noise Control, Maastricht, The Netherlands, 31 May–3 June 2015; pp. 781–784.

11. Manyoky, M.; Hayek Wissen, U.; Klein, T.; Pieren, R.; Heutschi, K.; Grêt-Regamey, A. Concept for collaborative design of wind farms facilitated by an interactive GIS-based visual-acoustic 3D simulation. In Proceedings of the Digital Landscape Architecture, Bernburg, Germany, 31 May 2012; pp. 297–306.

12. Manyoky, M.; Hayek Wissen, U.; Heutschi, K.; Pieren, R.; Grêt-Regamey, A. Developing a GIS-Based Visual-Acoustic 3D Simulation for Wind Farm Assessment. *ISPRS Int. J. Geo-Inf.* **2014**, *3*, 29–48.

13. Forssén, J.; Kaczmarek, T.; Alvarsson, J.; Lundén, P.; Nilsson, M.E. Auralization of traffic noise within the LISTEN project—Preliminary results for passenger car pass-by. In Proceedings of the Eighth European Conference on Noise Control, Edinburgh, UK, 26–28 October 2009.

14. Peplow, A.; Forssén, J.; Lundén, P.; Nilsson, M.E. Exterior Auralization of Traffic Noise within the LISTEN project. In Proceedings of the European Conference on Acoustics (Forum Acusticum 2011), Aalborg, Denmark, 27 June–1 July 2011; pp. 665–669.

15. Maillard, J.; Jagla, J. Real Time Auralization of Non-Stationary Traffic Noise—Quantitative and Perceptual Validation in an Urban Street. In Proceedings of the AIA-DAGA Conference on Acoustics, Merano, Italy, 18–21 March 2013.

16. McDonald, P.; Rice, H.; Dobbyn, S. Auralisation and Dissemination of Noise Map Data Using Virtual Audio. In Proceedings of the Eighth European Conference on Noise Control, Edinburgh, UK, 26–28 October 2009.

17. Fiebig, A.; Genuit, K. Development of a synthesis tool for soundscape design. In Proceedings of the Eighth European Conference on Noise Control, Edinburgh, UK, 26–28 October 2009.

18. Heutschi, K. SonRoad: New Swiss Road Traffic Noise Model. *Acta Acust. United Acust.* **2004**, *90*, 548–554.

19. Havelock, D.; Kuwano, S.; Vorländer, M. *Handbook of Signal Processing in Acoustics*; Springer: New York, NY, USA, 2008.

20. Pieren, R.; Bütler, T.; Heutschi, K. Auralisation of accelerating passenger cars. In Proceedings of the 10th European Conference on Noise Control, Maastricht, The Netherlands, 31 May–3 June 2015; pp. 757–762.

21. Jonasson, H.G. Acoustical Source Modelling of Road Vehicles. *Acta Acust. United Acust.* **2007**, *93*, 173–184.

22. Sandberg, U.; Ejsmont, J. *Tyre/Road Noise Reference Book*; INFORMEX, Harg: Kisa, Sweden, 2002.

23. Zeller, P. *Handbuch Fahrzeugakustik—Grundlagen, Auslegung, Berechnung, Versuch*, 2nd ed.; Vieweg Teubner Verlag: Wiesbaden, Geramny, 2012.

24. United Nations. UN/ECE Regulation No. 117: Uniform Provisions Concerning the Approval of Tyres with Regard to Rolling Sound Emissions and/or to Adhesion on Wet Surfaces and/or to Rolling Resistance. *Off. J. Eur. Union* **2011**, *54*, 3–63.

25. International Organization of Standardization. ISO 11819-1: Acoustics—Measurement of the Influence of Road Surfaces on Traffic Noise—Part 1: Statistical Pass-By Method; ISO: Geneva, Switzerland, 1997.

26. Schutte, J.; Wijnant, Y.; de Boer, A. The Influence of the Horn Effect in Tyre/Road Noise. *Acta Acust. United Acust.* **2015**, *101*, 690–700.

27. European Commission. Commission Directive (EU) 2015/996 of 19 May 2015 establishing common noise assessment methods according to Directive 2002/49/EC of the European Parliament and of the Council. *Off. J. Eur. Union* **2015**, *58*, 1–823.

28. Roads, C. *The Computer Music Tutorial*; MIT Press: Cambridge, MA, USA, 1996.

29. Rossing, T. *Springer Handbook of Acoustics*; Springer: New York, NY, USA, 2007.

30. Smith, J., III. *Spectral Audio Signal Processing*; W3K Publishing: Standford, UK, 2011.

31. Serra, X.; Smith, J., III. Spectral Modeling Synthesis: A Sound Analysis/Synthesis System Based on a Deterministic Plus Stochastic Decomposition. *Comput. Music J.* **1990**, *14*, 12–24.

32. Serra, X. Musical Sound Modeling with Sinusoids plus Noise. In *Musical Signal Processing*; Swets & Zeitlinger: Lisse, The Netherlands, 1997.

33. International Electrotechnical Commission. *IEC 1260:1995: Electroacoustics—Octave-Band and Franctional-Octave-Band Filters*; IEC: Geneva, Switzerland, 1995.

34. Reif, K. *Bosch Automotive Handbook*, 8th ed.; Wiley: Chichester, UK, 2011.

35. United Nations. *UN/ECE/TRANS/WP.29/2014/27: Proposal for a New Global Technical Regulation on the Worldwide harmonized Light vehicles Test Procedure (WLTP)*; UNECE: Geneva, Switzerland, 2014.

36.   International Organization of Standardization. *ISO 15031-5: Road Vehicles—Communication between Vehicle and External Equipment for Emissions-Related Diagnostics—Part 5: Emissions-related Diagnostic Services*; ISO: Geneva, Switzerland, 2015.

37.   Maekawa, Z. Noise reduction by screens. *Appl. Acoust.* **1968**, *1*, 157–173.

38.   International Organization of Standardization. *ISO 9613-2: Acoustics—Attenuation of Sound during Propagation Outdoors—Part 2: General Method of Calculation*; ISO: Geneva, Switzerland, 1996.

39.   Van Maercke, D.; Defrance, J. Development of an Analytical Model for Outdoor Sound Propagation within the Harmonoise Project. *Acta Acust. United Acust.* **2007**, *93*, 201–212.

40.   Daigle, G.; Embleton, T.; Piercy, J. Propagation of sound in the presence of gradients and turbulence near the ground. *J. Acoust. Soc. Am.* **1986**, *79*, 613–627.

41.   Hofmann, J.; Heutschi, K. An engineering model for sound pressure in shadow zones based on numerical simulations. *Acta Acust. United Acust.* **2005**, *91*, 661–670.

42.   Wunderli, J.; Pieren, R.; Heutschi, K. The Swiss shooting sound calculation model sonARMS. *Noise Control Eng. J.* **2012**, *90*, 224–235.

43.   Heutschi, K. Calculation of Reflections in an Urban Environment. *Acta Acust. United Acust.* **2009**, *95*, 644–652.

44.   Wunderli, J. An Extended Model to Predict Reflections from Forests. *Acta Acust. United Acust.* **2012**, *98*, 263–278.

45.   Pieren, R.; Wunderli, J. A Model to Predict Sound Reflections from Cliffs. *Acta Acust. United Acust.* **2011**, *97*, 243–253.

46.   Heutschi, K.; Pieren, R.; Müller, M.; Manyoky, M.; Hayek Wissen, U.; Eggenschwiler, K. Auralization of Wind Turbine Noise: Propagation Filtering and Vegetation Noise Synthesis. *Acta Acust. United Acust.* **2014**, *100*, 13–24.

47.   Lighthill, M. On sound generated aerodynamically. I. General Theory. *Proc. R. Soc. Lond. Ser. A* **1952**, *211*, 564–587.

48.   Morse, P.; Ingard, K. *Theoretical Acoustics*; Mc Gray-Hill Book Company: New York, NY, USA, 1968.

49.   Lighthill, M. The Bakerian Lecture, 1961: Sound Generated Aerodynamically. *Proc. R. Soc. Lond. Ser. A* **1962**, *267*, 147–182.

50.   Smith, J.; Serafin, S.; Abel, J.; Berners, D. Doppler Simulation and the Leslie. In Proceedings of the 5th International Conference on Digital Audio Effects, Hamburg, Germany, 26–28 September 2002; pp. 932–937.

51.   Laakso, T.I.; Välimäki, V.; Karjalainen, M.; Laine, U.K. Splitting the unit delay—Tools for fractional delay filter design. *IEEE Signal Process. Mag.* **1996**, *13*, 30–60.

52.   Bruneau, M. *Fundamentals of Acoustics*; ISTE Ltd: London, UK, 2006.

53.   Delany, M.E.; Bazley, E.N. Acoustical properties of fibrous absorbent materials. *Appl. Acoust.* **1970**, *3*, 105–116.

54.   International Organization of Standardization. *ISO 9613-1: Acoustics—Attenuation of Sound During Propagation Outdoors—Part 1: Calculation of the Absorption of Sound by the Atmosphere*; ISO: Geneva, Switzerland, 1993.

55.   Gerzon, M. Periophony: With-Height Sound Reproduction. *J. Audio Eng. Soc.* **1973**, *21*, 2–10.

56.   Gerzon, M. Ambisonics in Multichannel Broadcasting and Video. *J. Audio Eng. Soc.* **1985**, *33*, 859–871.

57.   Pulkki, V. Virtual Sound Source Positioning Using Vector Base Amplitude Panning. *J. Audio Eng. Soc.* **1997**, *45*, 456–466.

58.   Pulkki, V. Uniform spreading of amplitude panned virtual sources. In Proceedings of the IEEE Workshop on Applications of Signal Processing to Audio and Acoustics, New Paltz, NY, USA, 17–20 October 1999.

59.   Ballou, G. *Handbook for Sound Engineers*. 4th ed.; Elsevier: Amsterdam, The Netherlands, 2008.

60.   Dickreiter, M.; Dittel, V.; Hoeg, W.; Wöhr, M. *Handbuch der Tonstudiotechnik—Band 1*, 7th ed.; ARD.ZDF medienakademie: Nürnberg, Geramny, 2008.

61.   Rabiner, L.; Cheng, M.; Rosenberg, A.; McGonegal, C. A comparative performance study of several pitch detection algorithms. *IEEE Trans. Acoust. Speech Signal Process.* **1976**, *24*, 399–418.

62.   De la Cuadra, P.; Master, A. Efficient pitch detection techniques for interactive music. In Proceedings of the International Computer Music Conference, Havana, Cuba, 17–23 September 2001; pp. 87–90.

63.   Gerhard, D. *Pitch Extraction and Fundamental Frequency: History and Current Techniques, Technical Report TR-CS 2003-06*; Department of Computer Scinece, University of Regina: Regina, SK, Canada, 2003.

64. Dreuw, P.; Deselaers, T.; Rybach, D.; Keysers, D.; Ney, H. Tracking Using Dynamic Programming for Appearance-Based Sign Language Recognition. In Proceedings of the IEEE International Conference on Automatic Face and Gesture Recognition, Southampton, UK, 2–6 April 2006; pp. 293–298.

![applied sciences logo] *applied sciences*

MDPI

Article

Frequency-Dependent Amplitude Panning for the Stereophonic Image Enhancement of Audio Recorded Using Two Closely Spaced Microphones

Chan Jun Chun and Hong Kook Kim *

School of Information and Communications, Gwangju Institute of Science and Technology (GIST), Gwangju 61005, Korea; cjchun@gist.ac.kr

* Correspondence: hongkook@gist.ac.kr; Tel.: +82-62-715-2228; Fax: +82-62-715-2204

Academic Editor: Vesa Valimaki

Received: 19 November 2015; Accepted: 21 January 2016; Published: 1 February 2016

Abstract: In this paper, we propose a new frequency-dependent amplitude panning method for stereophonic image enhancement applied to a sound source recorded using two closely spaced omni-directional microphones. The ability to detect the direction of such a sound source is limited due to weak spatial information, such as the inter-channel time difference (ICTD) and inter-channel level difference (ICLD). Moreover, when sound sources are recorded in a convolutive or a real room environment, the detection of sources is affected by reverberation effects. Thus, the proposed method first tries to estimate the source direction depending on the frequency using azimuth-frequency analysis. Then, a frequency-dependent amplitude panning technique is proposed to enhance the stereophonic image by modifying the stereophonic law of sines. To demonstrate the effectiveness of the proposed method, we compare its performance with that of a conventional method based on the beamforming technique in terms of directivity pattern, perceived direction, and quality degradation under three different recording conditions (anechoic, convolutive, and real reverberant). The comparison shows that the proposed method gives us better stereophonic images in a stereo loudspeaker reproduction than the conventional method without any annoying effects.

Keywords: stereophonic image for stereo loudspeakers; frequency-dependent amplitude panning; azimuth-frequency analysis; two closely spaced omni-directional microphones

---

1. Introduction

Stereo loudspeaker reproduction is widely used to provide a more natural listening experience because of the distinguished relative positions of objects and events in the horizontal plane [1]. In fact, a stereo audio system can deliver a more immersive illusion than a mono system, because spatial information (e.g., inter-channel time difference (ICTD) and inter-channel level difference (ICLD) [2]) helps listeners perceive a horizontal direction [3]. In addition, according to duplex theory [2], the ICTD and ICLD are dominant for horizontal sound localization at low frequencies (below 1–2 kHz) and high frequencies (above 1–2 kHz), respectively.

Stereo audio recording techniques can be classified into three different categories depending on the placement and characteristics of microphones: coincident, near-coincident, and spaced recording techniques [4,5]. Coincident recording techniques such as the XY and mid-side (MS) techniques [5] place stereo microphones as close together as possible at different angles to capture a stereophonic image, where the stereophonic image is about sound recording and reproduction concerning the perceived spatial locations of the sound source. Thus, a good stereophonic image means that the location of the sound source can be clearly perceived, while a poor one means that the location of the source is difficult to be perceived [6]. In addition, near-coincident recording techniques such as the Office de Radiodiffusion Télévision Française (ORTF) and Nederlandse Omroep Stichting (NOS) techniques [4,7] place microphones slightly apart. In the ORTF technique, the microphone spacing is similar to the human ear spacing, while the spacing for the NOS technique is approximately 30 cm. In general, both coincident and near-coincident recording techniques utilize directional microphones to realize good directional characteristics [4]. However, with spaced recording techniques, including AB techniques [4,8], stereophonic images can be obtained by the ICTD between stereo microphones, because omni-directional microphones are often used in such techniques [4,9].

To date, numerous portable video and audio capture devices have been released to the market. These devices usually capture stereo audio as well as high-quality video. Unfortunately, because most portable devices are limited in size, and coincident or near-coincident recording techniques are most appropriate for such devices. However, when the audio signals captured by these recording techniques are reproduced, the stereophonic images often do not feel sufficient [10–13]. This is because the body of a portable device equipped with directional microphones acts as wall reflection in a recording, which is referred to as shadowed directivity [10]. As an alternative, a spaced recording technique can be applied to capture stereo audio from such portable devices [11–13]. In this case, the width or length of the portable device body such as a smart phone or digital camera is approximately 10 cm, thus the allowable distance between two microphones is less than 3 cm. This hardware limitation makes it difficult to match the perceived azimuth angle of the reproduced sound source to the actual that of the original sound source, when a spaced recording technique is applied [8].

To mitigate this, a stereophonic image enhancement method using head-related transfer functions (HRTFs) was proposed in [11]. This method was more successful at enhancing stereophonic images than original stereo signals, but it had a somewhat limited sweet spot, and it was difficult to deliver reliable stereo quality to a listener located beyond this sweet spot. If the original stereo signals were nearly monaural signals, it is difficult to enhance a stereophonic image properly, even though the illusion of wider stereo loudspeaker spacing is created. Umayahara *et al.* proposed a stereophonic image control method that linearly interpolated the spectra of the left and right channels in the frequency domain [12]. This method used the same interpolation factor for all frequencies; thus, its performance for enhancing a stereophonic image might have been limited when the direction of the input stereo signals was dependent on the frequency [14]. In another study, a delay-and-sum (DS) beamformer was utilized to convert AB stereo signals into XY stereo signals [13]. However, in real reverberant recording environments, the DS beamformer changed the direction of stereophonic images due to the reverberant effects [15]. This was because the reverberation time changed according to the frequency, and the DS beamforming weights could not be adapted to this reverberation time change [14]. These results suggest that frequency-dependent amplitude panning for stereophonic image enhancement is necessary for real reverberant environments.

Accordingly, we propose frequency-dependent amplitude panning for stereophonic image enhancement when two omni-directional microphones are closely spaced, as deployed in portable devices. In [16,17], frequency-dependent amplitude panning was also used to enhance the stereophonic image for portable devices equipped with closely spaced stereo microphones, where the ratio of spectral magnitudes between the left- and right-channel signal was used for the panning. On the other hand, the proposed method first introduces azimuth-frequency (A-F) analysis [18] to estimate the direction of the input audio according to frequency. In other words, we first apply short-time Fourier transform (STFT) to the input stereo audio signal, and then project the spectral component of each frequency bin on an azimuth plane that is generated by converting the time difference between two microphones into their level difference. Next, the direction at each frequency bin is assigned as the azimuth at which the projected magnitude is minimized. Finally, a frequency-dependent amplitude panning technique is proposed to enhance the stereophonic image by modifying the stereophonic law of sines [19].

To evaluate the performance of the proposed method, three different recording environments are considered: anechoic, convolutive, and real reverberant. First, the directivity pattern of the proposed method is compared with that of a conventional method based on the DS beamformer [13] in the three recording environments. Second, the directional accuracy of the stereo audio processed by the proposed method is compared to that of the audio processed by the conventional method by measuring the subjective directions of listeners depending on the horizontal direction of the sources. Finally, a degraded mean opinion score (DMOS) assessment [20] is carried out to evaluate the quality as an aspect of the audio distortion after the proposed method has been applied.

The remainder of this paper is organized as follows: a conventional method based on the DS beamformer [13] is described in Section 2. Section 3 describes the proposed stereophonic image enhancement method based on A-F analysis and a frequency-dependent amplitude panning technique. Section 4 then evaluates the performance of the proposed method applied to audio signals recorded in anechoic, convolutive, and real reverberant environments. Finally, Section 5 concludes this paper.

2. Conventional Stereophonic Image Enhancement

In this section, we describe a conventional stereophonic image enhancement method applied to closely spaced omni-directional microphones based on a DS beamforming technique [13,21]. Figure 1 shows a block diagram of the conventional method. As shown in the figure, the DS beamformer of the conventional method compensates for the delay between the two channels, where the delay, n_d, is determined depending on the distance, l, between the two microphones. After that, a free-field response filter, $h(n)$, [13] is applied to the beamformed signals to obtain the enhanced stereo signals, $y_L(n)$ and $y_R(n)$, respectively by:

$$y_L(n) = h(n) * (x_R(n) - x_L(n - n_d)) \tag{1}$$

$$y_R(n) = h(n) * (x_L(n) - x_R(n - n_d)) \tag{2}$$

where $*$ indicates the linear convolution operator, and $x_L(n)$ and $x_R(n)$ are the stereo audio sequences obtained by the omni-directional microphones.

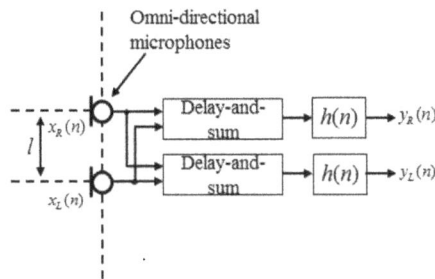

Figure 1. Block diagram of a conventional stereophonic image enhancement method.

The directivity patterns for the beamformed signal of the left channel at 2.5 kHz and 4 kHz are depicted in Figure 2a,b, respectively. Note that the directivity patterns for the right channel are exactly the opposite of those for the left channel. As illustrated in the figure, the conventional method provides different directional responses depending on the frequency. That is, it has a cardioid and a super-cardioid directivity pattern for 2.5 and 4 kHz, respectively. It is expected that the stereophonic image of the audio signal at 2.5 kHz should be enhanced but that at 4 kHz could be distorted due to the negative rear lobe of the super-cardioid pattern [13]. To remedy this problem, a Wiener filter has been applied to the beamformed signal to reduce the negative rear-lobe effects [13]. Nevertheless, it was reported that the DS beamformer with a Wiener filter could not change the direction of a stereophonic image when audio recording was performed in a reverberant environment [13]. This is because the reverberation time differs from the frequency and the DS beamformer cannot be adapted to such reverberation time changes [14].

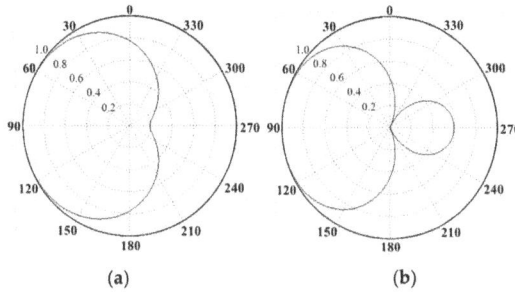

Figure 2. Directivity patterns of the DS beamformer (l = 3 cm, direction = 90°) at (**a**) 2.5 kHz and (**b**) 4 kHz.

Therefore, we propose a method that operates differently depending on the frequency, referred to as the frequency-dependent amplitude panning for stereophonic image enhancement (FDAP-SIE) method and compare the performance of our proposed method with that of the DS beamformer described in this section.

3. Proposed Frequency-Dependent Amplitude Panning for Stereophonic Image Enhancement

3.1. Overview

In this section, we propose an FDAP-SIE method applied to audio recording with two closely spaced omni-directional microphones and illustrate the block diagram of the proposed method in Figure 3. First, the left- and right-channel input signals, respectively designated $x_L(n)$ and $x_R(n)$, are each segmented into a sequence of frames of 2048 samples by applying a Hanning window, and each frame is overlapped with 1024 samples of the previous frame. Then, a 2048-point STFT is applied to each segment to obtain $X_L(k)$ and $X_R(k)$. Next, A-F analysis is carried out to estimate the direction of the sound sources in each frequency bin. After that, frequency-dependent amplitude panning is applied to $X_L(k)$ and $X_R(k)$ according to the estimated direction for the k-th frequency bin. Finally, an inverse STFT followed by an overlap-add method is applied to obtain the enhanced stereophonic signal.

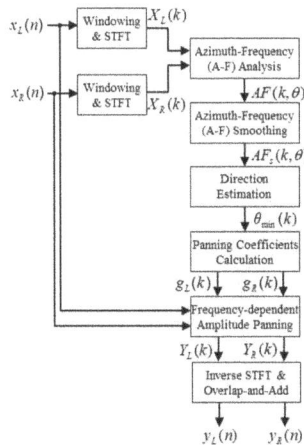

Figure 3. Block diagram of the proposed frequency-dependent amplitude panning for stereophonic image enhancement.

3.2. Azimuth-Frequency Analysis Using Time Delay

A stereo signal recorded using a stereo omni-directional microphone array, $x_L(n)$ and $x_R(n)$, can be represented as a delayed and attenuated version of the desired signal, $s(n)$, such as [21]

$$\begin{bmatrix} x_L(n) \\ x_R(n) \end{bmatrix} = \begin{bmatrix} a_L s(n) \\ a_R s(n-\tau) \end{bmatrix} + \begin{bmatrix} v_L(n) \\ v_R(n) \end{bmatrix}, \tag{3}$$

where $v_L(n)$ and $v_R(n)$ are ambient noise recorded by the left and right microphones, respectively. In addition, a_L and a_R are the respective attenuation factors, and τ is the relative time delay measured between the left and right microphones. Note here that Equation (3) is designed using the far-field model [18,19], because the spacing between the stereo omni-directional microphones is small. Moreover, we can assume $a_L = a_R \approx 1$ [22]. Applying an N-point STFT to Equation (3) provides the following relationship:

$$\mathbf{X} = \mathbf{d}S(k) + \mathbf{V}, \tag{4}$$

where $\mathbf{X}^T = \begin{bmatrix} X_L(k) & X_R(k) \end{bmatrix}$ and $\mathbf{V}^T = \begin{bmatrix} V_L(k) & V_R(k) \end{bmatrix}$. In addition, $S(k)$ is the k-th spectral component of $s(n)$, and $\mathbf{d}$ is a steering vector of

$$\mathbf{d}^T = \begin{bmatrix} 1 & \exp\left(-j\frac{2\pi k\tau}{N}\right) \end{bmatrix}, \tag{5}$$

where τ can be determined by the speed of sound c, the spacing between the microphones l, and the direction of the source θ, as $\tau = (f_s/c)l\sin\theta$, where f_s is the sampling rate. Thus, we have the following equation:

$$\mathbf{d}^T = \begin{bmatrix} 1 & \exp\left(-j\frac{2\pi k}{N}\frac{f_s}{c}l\sin\theta\right) \end{bmatrix}. \tag{6}$$

If θ is known, we can separate $S(k)$ and $\mathbf{d}$ from Equation (4). Then, we can modify $\mathbf{d}$ by replacing θ with another value to improve the obtained stereophonic images. This is because the listener cannot feel the actual direction of $S(k)$ when two stereo microphones are placed very close together. In practice, it is difficult to separate the direction $\mathbf{d}$ and source $S(k)$, and it is even more difficult to do so under ambient noise conditions and/or with multiple sound sources [23]. Therefore, instead of separating the sound source and its steering vector in this paper, we apply a panning law to the recorded signal $\mathbf{X}$, with the estimated direction. To estimate the source direction, we consider the time delay τ in Equation (3) using the stereo signal $x_L(n)$ and $x_R(n)$, where we have assumed that $v_L(n)$ and $v_R(n)$ are negligible under high signal-to-noise ratio (SNR) conditions. In other words, the time delay is estimated as $\hat{\tau} = \operatorname{argmin}_\tau |x_L(n) - x_R(n-\tau)|$. We can then extend this concept in the frequency domain, as:

$$\hat{\tau}(k) = \operatorname*{argmin}_\tau \left| X_L(k) - e^{-j\frac{2\pi}{N}k\tau} X_R(k) \right|. \tag{7}$$

In this paper, τ in Equation (7) can be considered as a function of the direction θ. Therefore, the right-hand side of Equation (7), which is a function of the frequency k, and the direction θ, is referred to as an A-F plane and defined as [18].

$$AF(k,\theta) = \left| X_L(k) - e^{-j\frac{2\pi}{N}k\tau(\theta)} X_R(k) \right|. \tag{8}$$

We can estimate the direction $\hat{\theta}(k)$ so that $AF(k,\theta)$ is minimized at the k-th frequency bin. However, when $AF(k,\theta)$ is used for estimating $\hat{\theta}(k)$, many local minima exist. To mitigate this problem, a smoothing window is applied to $AF(k,\theta)$ prior to estimating $\hat{\theta}(k)$, such that:

$$AF_s(k,\theta) = \frac{1}{B(k)+1} \sum_{m=k-B(k)/2}^{k+B(k)/2} AF(m,\theta), \tag{9}$$

where $B(k)$ corresponds to a critical bandwidth of the auditory filter [2]. For example, $B(k) = 6\,(150\,\text{Hz})$ when $k = 43\,(1\,\text{kHz})$. Thus, the direction at each frequency bin is estimated so that $AF_s(k,\theta)$ is minimized, such that:

$$\hat{\theta}(k) = \underset{\theta}{\operatorname{argmin}} AF_s(k,\theta). \tag{10}$$

Figure 4 illustrates an A-F plane, $AF(k,\theta)$, and a smoothed A-F plane, $AF_s(k,\theta)$, computed for a stereo signal that is recorded from a stereo microphone array in an anechoic room, where a white noise source is angled at 15° and placed 1.5 m from the center of the microphone array. In the figure, a 2048-point STFT is applied to each frame of white noise, and θ is changed from $-90°$ to $90°$ at $1°$ steps. In addition, the distance between the two microphones is $l = 3$ cm and $f_s = 48$ kHz. As shown in the figure, the direction of the white noise is easily estimated at low frequencies, but there are multiple minima at mid-to-high frequencies. As shown in Figure 4c, the estimated direction of the white noise is 15°, which is identical to the direction at which the white noise is located for recording.

Next, we repeat the experiment above by recording white noise in a reverberant room whose reverberation time (RT_{60}) is measured as 230 ms, and the A-F planes and estimated direction are shown in Figure 4. Comparing Figure 5a with Figure 4a, the A-F plane in the reverberant room is more blurred than that in the anechoic room. This is because the reverberation muddles the direction of the sound source, making it seem as though multiple sound sources are being recorded by the stereo microphones. Owing to the smoothing window, the smoothed A-F plane shown in Figure 5b becomes similar to that in Figure 4b. Therefore, as shown in Figure 5c, the direction of white noise can be estimated correctly, especially at mid-to-high frequencies, while there are some errors at low frequencies. Since it is known that stereophonic images are mostly affected by mid-to-high frequencies, the quality of stereophonic images is not significantly affected by such errors at low frequencies [24].

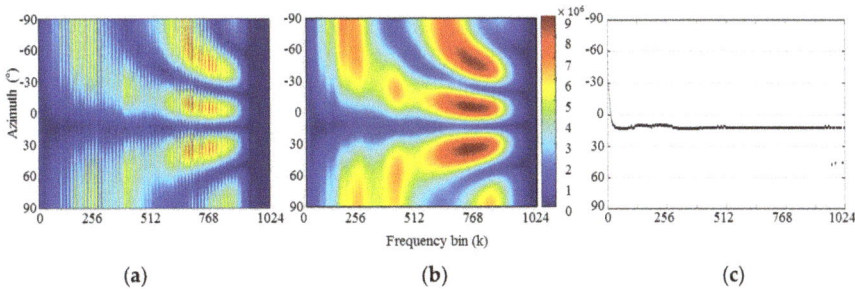

(a)	(b)	(c)

Figure 4. A-F planes and estimated direction for white noise in an anechoic room: (a) $AF(k,\theta)$; (b) $AF_s(k,\theta)$; (c) estimated direction using $AF_s(k,\theta)$.

3.3. Frequency-Dependent Amplitude Panning

This subsection describes how the estimated direction in each frequency bin is used for stereophonic image enhancement. Figure 6 illustrates the concept of the process described in this subsection. As shown in Figure 6a, a sound source is located at an angle of θ. However, the close spacing between the stereo microphones could mean that it is perceived as being at a lesser angle—*i.e.*,

$\theta_p \ll \theta$. Thus, we have to increase the perceived angle by applying frequency-dependent amplitude panning such that $\theta_0 \approx \theta \gg \theta_p$.

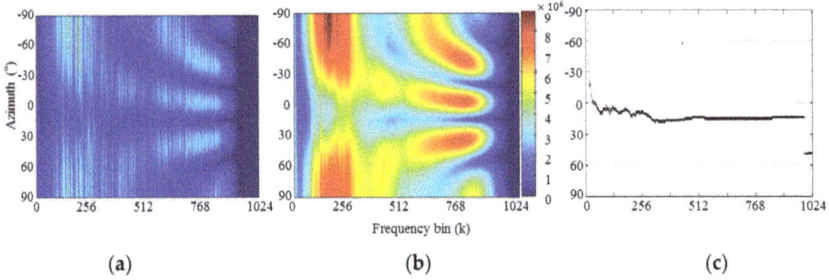

| (a) | (b) | (c) |

Figure 5. A-F planes and estimated direction for white noise in a reverberant room with $RT_{60} = 230$ ms: (a) $AF(k, \theta)$; (b) $AF_s(k, \theta)$; (c) estimated direction using $AF_s(k, \theta)$.

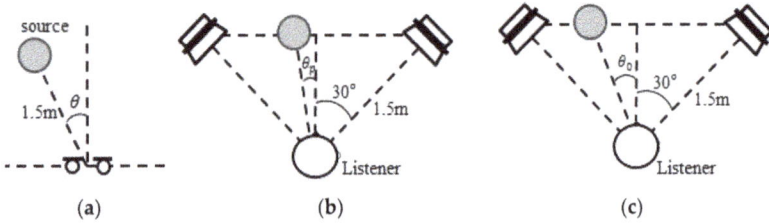

| (a) | (b) | (c) |

Figure 6. Illustrations of stereophonic image enhancement: (a) Original sound source; (b) perceived sound source without any enhancement technique; and (c) perceived sound source after applying the proposed method.

Many panning methods have been reported [19,25,26]. Among them, the stereophonic law of sines [19] has been popularly used to reproduce a source using two loudspeakers, and it is realized as:

$$\frac{\sin\theta}{\sin\theta_0} = \frac{g_L - g_R}{g_L + g_R}, \tag{11}$$

where θ_0 is the physical angle between stereo loudspeakers and θ is the desired angle at which the sound source should be located in terms of perception. Thus, g_L and g_R become the respective scale factors that are multiplied with the sound source according to the desired angle, as:

$$y_L(n) = g_L s(n), \tag{12}$$

and

$$y_R(n) = g_R s(n), \tag{13}$$

where $s(n)$ is the sound source, and $y_L(n)$ and $y_R(n)$ are respectively the panned signals of the left and right channel.

In this paper, we extend the stereophonic law of sines so that it is applied in the frequency domain. For a given direction at the k-th frequency bin $\hat{\theta}(k)$, as described in Section 3.2, the frequency-dependent scale factors, $g_L(k)$ and $g_R(k)$, are obtained using the following equation:

$$\frac{\sin(\hat{\theta}(k))}{\sin\theta_0} = \frac{g_L(k) - g_R(k)}{g_L(k) + g_R(k)}, \tag{14}$$

where θ_0 is also the physical angle between stereo loudspeakers, as described in Equation (10). As in Equations (11) and (12), the scale factors to Equation (13) are multiplied to the k-th spectral magnitude of the sound source as:

$$Y_L(k) = g_L(k)S(k),\qquad(15)$$

and

$$Y_R(k) = g_R(k)S(k).\qquad(16)$$

Here, while $S(k)$ should be separated from $\mathbf{X}$ according to Equation (4), the spectral magnitude of the sound source is approximated as the mid signal of the recorded sound. That is $S(k) \approx (X_L(k) + X_R(k))/2$. Finally, by applying an inverse STFT followed by the overlap-add method, the output signal with an enhanced stereophonic image is obtained.

4. Performance Evaluation

To demonstrate the effectiveness of the proposed method, three different recording environments were considered: anechoic, convolutive, and real reverberant. Figure 7 illustrates the configuration for the room impulse response (RIR) filter design. The dimensions of the room were 4.5 m × 7.5 m × 2.5 m, and a stereo microphone array with 3-cm spacing was located in the room at the coordinates denoted in the figure. To simulate the convolutive environment, a RIR filter was designed based on the image method [27], and the response of the left channel is shown in Figure 8. As shown in the figure, RT_{60} of this RIR was measured as 230 ms.

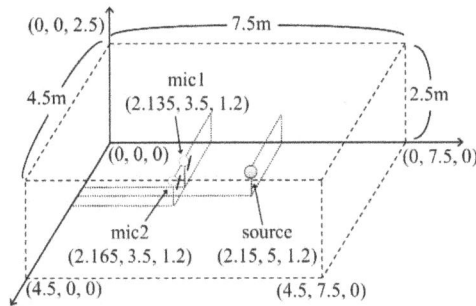

Figure 7. Experimental setup for simulating the room impulse response to simulate a reverberant environment.

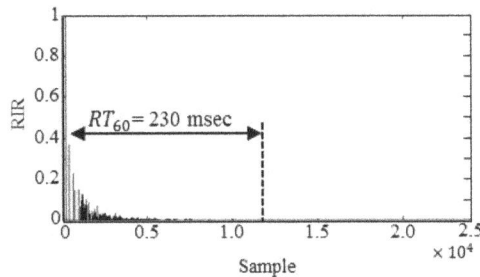

Figure 8. Simulated room impulse response of the left channel based on the image method, where RT_{60} was measured as 230 ms.

The performance of the proposed method was evaluated in terms of three different measurements. First, the directivity pattern of the proposed method was compared with that of a conventional method based on a DS beamformer [13] in the three recording environments. Second, the accuracy of direction estimates for the stereo audio processed by the proposed method was compared with that processed by the conventional method by measuring the perceived directions of listeners depending on the horizontal directions of sources. Third, a DMOS assessment [20] was carried out to evaluate the audio quality degradation after the proposed method had been applied.

4.1. Directivity Pattern Performance

Figure 9 shows an experimental setup for evaluating the performance of the directivity patterns. A stereo microphone array was placed with 3-cm spacing, and one loudspeaker was located at 60° from the center of the microphone array at a distance of 1.5 m. The white noise at a sampling rate of 48 kHz was played out via loudspeaker and recorded by the microphone array. The recorded signal was processed by both the DS beamformer and the proposed method. After that, the recorded and processed white noise signals were all played through stereo loudspeakers that were configured according to International Telecommunication Union Radiocommunication Sector (ITU-R) Recommendation BS.775-1 [28]. Then, a dummy head [29] was rotated from 0° to 350° at 10° steps to measure the directivity patterns.

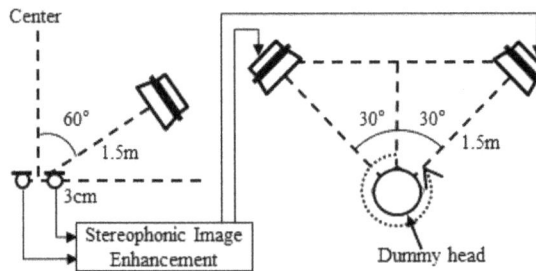

Figure 9. Experimental setup for evaluating the directivity patterns.

Figure 10 compares the directivity patterns of the original source with those obtained by the DS beamformer and the proposed method in three different environments (anechoic, convolutive, and real reverberant room). As shown in Figure 10a, the directivity for the original signal was towards 0° in the anechoic environment, while the actual directivity was set to 60°. However, the directivities of the signals processed by the conventional and proposed methods were approximately 30°, which was the same angle of the loudspeakers against the dummy head. Consequently, we concluded that the proposed and conventional methods significantly enhanced the originally recorded signal.

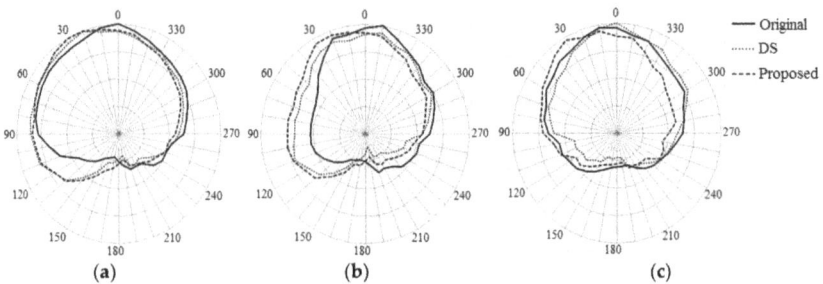

Figure 10. Comparison of directivity pattern of the original source with those obtained by the DS beamformer and the proposed method in three different environments: (**a**) anechoic; (**b**) convolutive; and (**c**) real reverberant room.

Next, when comparing the directivity in convolutive and real reverberant environments, it was clear that the proposed method could locate the sound source to approximately 30°, while the DS beamformer failed to do so. This was because the simulated and real reverberation limited the stereophonic image enhancement of the DS beamformer. However, the proposed method was not affected by the reverberation, due to the frequency-dependent direction estimation and panning.

4.2. Perceived Direction Performance

Figure 11 shows the experimental setup for evaluating listeners' directional perception. To record stereo signals, a stereo microphone array was placed with 3-cm spacing, and a sound source was played through one loudspeaker from the center of the microphone array at a distance of 1.5 m. For the evaluation, each listener was sitting in an anechoic room of dimensions 2130 mm × 3370 mm × 3000 mm, in which two loudspeakers had been placed as shown in the figure. Note that here the model of all the loudspeakers was Genelec 6010A. In this experiment, we prepared five audio clips that were excerpted from the sound quality assessment material (SQAM) [30]; Table 1 describes the genre and musician of each audio clip. Note that since all audio clips were sampled at 44.1 kHz, we upsampled the audio clips to 48 kHz to ensure a consistent experimental environment. Then, we recorded five audio clips in three different environments where the loudspeaker was rotated from 0° to 90° at a 15° step towards the right direction, resulting in seven different directions. After that, the recorded signals were processed by the DS beamformer and the proposed method (some audio samples can be found at [31]). After the processed clips were played at sound pressure level (SPL) 90 dB, eight participants (four males and four females) with no auditory diseases were asked to indicate their perceived directions for the original and processed signals. Note that the participants were allowed head movement.

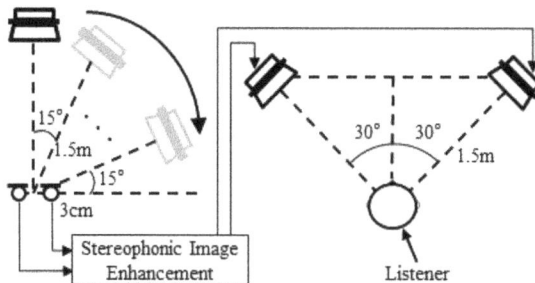

Figure 11. Experimental setup for evaluating the perceived directions.

Table 1. Detailed information on five audio clips used for the evaluation of perceived direction.

Track	Genre	Description
49	Speech	Female English speech
66	Orchestra	Wind ensemble, Stravinsky
67	Orchestra	Wind ensemble, Mozart
69	Pop music	Abba
70	Pop music	Eddie Rabbit

Figure 12 compares the perceived azimuths averaged over five audio clips and eight participants for each source direction in three different recording environments, where the dashed line indicates the target direction, and the vertical bar on each bar chart corresponds to the standard deviation. As shown in the figure, the originally recorded signals were all perceived at around 0°–10° for all environments, even though the actual source angles were above 15°. By applying the DS beamformer to enhance the stereophonic image, the perceived angles increased. However, errors between the actual angle (dashed straight line) and the perceived angle increased as the actual angle increased, especially in anechoic and real reverberant environments. The proposed method provided smaller perceived errors than the DS beamformer, which implies that it could enhance stereophonic images for all recording environments compared to the conventional method.

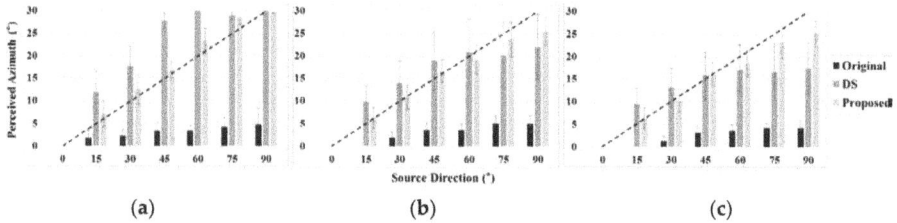

Figure 12. Comparison of the perceived azimuth depending on the direction of the source in three different environments: (**a**) anechoic; (**b**) convolutive; and (**c**) real reverberant room.

4.3. Audio Quality Degradation

To evaluate the quality degradation of audio signals processed by the proposed method, we performed a DMOS assessment test according to ITU Telecommunication Standardization Sector (ITU-T) Recommendation P.800 [20]. The experimental conditions such as audio clips, participants, and listening room are identical to those of the experiment described in Section 4.2. Each participant listened to a pair of audio clips composed of an original and processed version by either the DS beamformer or the proposed method. Then, each was asked to rate the degree of quality degradation from five to one. Table 2 describes the scores and their meanings for DMOS assessment.

Table 2. Score and description of degraded mean opinion score (DMOS) assessment.

Score	Description
5	Degradation is inaudible
4	Degradation is audible but not annoying
3	Degradation is slightly annoying
2	Degradation is annoying
1	Degradation is very annoying

Table 3 compares the results of DMOS assessment between the conventional DS beamformer-based method and the proposed method in three different recording environments. We conducted a statistical analysis and indicated the 95% confidence intervals (CIs) as numbers in parentheses in Table 3. As shown in the table, the proposed method provided average DMOS scores of approximately four for all environments, which implied that there were no annoying effects [32]. However, there was significant quality degradation in the conventional method, especially in the real reverberant environment. It was revealed from statistical analysis that the quality degradation of the audio signals enhanced by the proposed method was statistically less than those enhanced by the DS beamformer.

Table 3. Comparison of DMOS assessment results of the conventional and proposed methods for three different recording environments where the numbers in parentheses indicate 95% CIs.

Method Environment	DS	Proposed
Anechoic	3.61 (0.2898)	4.04 (0.2622)
Convolutive	3.31 (0.3147)	3.97 (0.2581)
Real Reverberant	3.46 (0.2950)	3.90 (0.2711)

5. Conclusions

In this paper, we proposed a frequency-dependent stereophonic image enhancement method that could be applied to two closely spaced omni-directional microphones available for portable audio recording devices. First, the A-F plane was obtained from the spectral magnitudes of stereo audio signals. Next, the direction at each frequency bin was estimated as the azimuth at which the A-F plane was minimized. Finally, a frequency-dependent amplitude panning technique was also proposed to enhance the stereophonic image from the stereophonic law of sines. The performance of the proposed method was evaluated in three different recording environments: anechoic, convolutive, and real reverberant. First, the directivity pattern of the proposed method was compared to that of a conventional method based on a DS beamformer. Second, the directional accuracy of the stereo audio processed by the proposed method was compared to that processed by a conventional method by the measurement of listeners' perceived directions. Finally, a DMOS assessment test was carried out to evaluate quality degradation after the proposed method had been applied. Consequently, it was revealed that the proposed method gave better directivity, higher directional accuracy, and less quality degradation than the conventional method. It was argued here that, compared to the conventional method, the proposed method could improve performance with the help of frequency-dependent processing.

We have only experimented with a single source throughout this study, so we are planning to examine what happens when multiple sources are available. One possible approach will be to detect multiple directions from the A-F analysis and propose an appropriate panning method that can treat multiple angles.

Acknowledgments: This work was supported in part by the National Research Foundation of Korea (NRF) grant funded by the government of Korea (MSIP) (No. 2015R1A2A1A05001687), and by the ICT R&D program of MSIP/IITP (R01261510340002003, Development of hybrid audio contents production and representation technology for supporting channel and object based audio).

Author Contributions: All authors discussed the contents of the manuscript. Hong Kook Kim contributed to the research idea and the framework of this study. Chan Jun Chun performed the experimental work.

Conflicts of Interest: The authors declare no conflict of interest.

References

1. Breebaart, J.; Faller, C. *Spatial Audio Processing: MPEG Surround and Other Applications*; John Wiley & Sons, Ltd.: Chichester, UK, 2007.
2. Blauert, J. *Spatial Hearing: The Psychophysics of Human Sound Localization*; MIT Press: Cambridge, MA, USA, 1997.
3. Rumsey, F. Spatial quality evaluation for reproduced sound: Terminology, meaning, and a scene-based paradigm. *J. Audio Eng. Soc.* **2002**, *50*, 651–666.
4. Rumsey, F. *Spatial Audio*; Focal Press: Woburn, MA, USA, 2001.
5. Hibbing, M. XY and MS microphone techniques in comparison. *J. Audio Eng. Soc.* **1989**, *37*, 823–831.
6. Bennett, J.C.; Barker, K.; Edeko, F.O. A new approach to the assessment of stereophonic sound system performance. *J. Audio Eng. Soc.* **1985**, *33*, 314–321.
7. Kim, J.K.; Chun, C.J.; Kim, H.K. Design of a coincident microphone array for 5.1-channel audio recording using the mid-side recording technique. *Adv. Sci. Technol. Lett.* **2012**, *14*, 61–64.
8. Dooley, W.; Streicher, T. MS stereo: A powerful technique for working in stereo. *J. Audio Eng. Soc.* **1982**, *30*, 707–718.
9. Eargle, J. *The Microphone Book*; Focal Press: Oxford, UK, 2004.
10. Menounou, P.; Papaefthymio, E.S. Shadowing of directional noise sources by finite noise barriers. *Appl. Acoust.* **2000**, *71*, 351–367. [CrossRef]
11. Aarts, R.M. Phantom sources applied to stereo-base widening. *J. Audio Eng. Soc.* **2000**, *48*, 181–189.
12. Umayahara, T.; Hokari, H.; Shimada, S. Stereo width control using interpolation and extrapolation of time-frequency representation. *Audio Speech Lang. Process. IEEE Trans.* **2006**, *14*, 1364–1377. [CrossRef]

13. Faller, C. Conversion of two closely spaced omnidirectional microphone signals to an XY stereo signal. In Proceedings of the 129th AES Convention, San Francisco, CA, USA, 4–7 November 2010; p. 8188.

14. Marsch, J.; Porschmann, C. Frequency dependent control of reverberation time for auditory virtual environments. *Appl. Acoust.* **2000**, *61*, 189–198. [CrossRef]

15. Usher, J.; Woszczyk, W. Interaction of source and reverberance spatial imagery in multichannel loudspeaker audio. In Proceedings of the 118th AES Convention, Barcelona, Spain, 28–31 May 2005; p. 6370.

16. Cobos, M.; Lopez, J.J. Method and Apparatus for Stereo Enhancement in Audio Recordings. PCT Patent PCT/ES2009/000409, 31 July 2009.

17. Cobos, M.; Lopez, J.J. Interactive enhancement of stereo recordings using time-frequency selective panning. In Proceedings of the 40th Audio Engineering Society Conference, Tokyo, Japan, 8–10 October 2010; pp. 2–10.

18. Barry, D.; Coyle, E.; Lawlor, B. Real-time sound source separation: Azimuth discrimination and resynthesis. In Proceedings of the 117th AES Convention, San Francisco, CA, USA, 28–31 October 2004; p. 6258.

19. Bauer, B.B. Phasor analysis of some stereophonic phenomena. *J. Acoust. Soc. Am.* **1961**, *33*, 1536–1539. [CrossRef]

20. P.800: Methods for Subjective Determination of Transmission Quality. Available online: https://www.itu.int/rec/T-REC-P.800-199608-I/en (accessed on 27 January 2016).

21. Brandstein, M.; Ward, D.B. *Microphone Arrays: Signal. Processing Techniques and Applications*; Springer-Heidelberg: New York, NY, USA, 2001.

22. Kennedy, R.A.; Abhayapala, P.T.D.; Ward, D.B.; Williamson, R.C. Nearfield broadband frequency invariant beamforming. In Proceedings of the IEEE International Conference on Acoustics, Speech and Signal Processing (ICASSP), Atlanta, GA, USA, 7–10 May 1996; pp. 905–908.

23. Duong, N.Q.K.; Vincent, E.; Gribonval, R. Under-determined reverberant audio source separation using local observed covariance and auditory-motivated time-frequency representation. *Audio Speech Lang. Process. IEEE Trans.* **2010**, *18*, 1830–1840. [CrossRef]

24. Pulkki, V.; Karjalainen, M. Localization of amplitude-panned virtual sources I: Stereophonic panning. *J. Audio Eng. Soc.* **2001**, *49*, 739–752.

25. Choi, T.S.; Park, Y.C.; Youn, D.H.; Lee, S.P. Virtual sound rendering in a stereophonic loudspeaker setup. *Audio Speech Lang. Process. IEEE Trans.* **2011**, *19*, 1962–1974.

26. Pulkki, V. Virtual source positioning using vector base amplitude panning. *J. Audio Eng. Soc.* **1977**, *45*, 456–466.

27. Allen, J.B.; Berkley, D.A. Image method for efficiently simulating small-room acoustics. *J. Acoust. Soc. Am.* **1979**, *65*, 943–951. [CrossRef]

28. BS.775: Multichannel Stereophonic Sound System with and without Accompanying Picture. Available online: https://www.itu.int/rec/R-REC-BS.775/en (accessed on 27 January 2016).

29. Product Information KU 100. Available online: http://www.coutant.org/ku100/ku100.pdf (accessed on 27 January 2016).

30. Sound Quality Assessment Material Recordings for Subjective Tests—Users' Handbook for the EBU-SQAM Compact Disc. Available online: https://tech.ebu.ch/docs/tech/tech3253.pdf (accessed on 27 January 2016).

31. Chun, C.J.; Kim., H.K. Some Audio Samples Processed by Frequency-Dependent Amplitude Panning for the Stereophonic Image Enhancement. Available online: http://hucom.gist.ac.kr/2016ApplSci/sample.html (accessed on 27 January 2016).

32. Spanias, A.; Painter, T.; Atti, V. *Audio Signal. Processing and Coding*; John Wiley & Sons, Inc.: Hoboken, NJ, USA, 2007.

applied
sciences

MDPI

Article

Semantically Controlled Adaptive Equalisation in Reduced Dimensionality Parameter Space †

Spyridon Stasis *, Ryan Stables * and Jason Hockman *

Digital Media Technology Lab, Birmingham City University, Birmimgham B42 2SU, UK
* Correspondence: spyridon.stasis@bcu.ac.uk (S.S.); ryan.stables@bcu.ac.uk (R.S.);
  jason.hockman@bcu.ac.uk (J.H.); Tel.: +4412-1331-7957 (R.S.); +4412-1202-2386 (J.H.)
† This paper is an extended version of our paper published in the 18th International Conference on Digital
  Audio Effects, Trondheim, Norway, 30 November–3 December 2015.

Academic Editor: Vesa Valimaki
Received: 24 February 2016; Accepted: 5 April 2016; Published: 20 April 2016

Abstract: Equalisation is one of the most commonly-used tools in sound production, allowing users to control the gains of different frequency components in an audio signal. In this paper we present a model for mapping a set of equalisation parameters to a reduced dimensionality space. The purpose of this approach is to allow a user to interact with the system in an intuitive way through both the reduction of the number of parameters and the elimination of technical knowledge required to creatively equalise the input audio. The proposed model represents 13 equaliser parameters on a two-dimensional plane, which is trained with data extracted from a semantic equalisation plug-in, using the timbral adjectives *warm* and *bright*. We also include a parameter weighting stage in order to scale the input parameters to spectral features of the audio signal, making the system adaptive. To maximise the efficacy of the model, we evaluate a variety of dimensionality reduction and regression techniques, assessing the performance of both parameter reconstruction and structural preservation in low-dimensional space. After selecting an appropriate model based on the evaluation criteria, we conclude by subjectively evaluating the system using listening tests.

Keywords: equalisation; adaptive audio effects; semantics; dimensionality reduction; intelligent music production

1. Introduction

Equalisation, as described in [1], is an integral part of the music production workflow, with applications in live sound engineering, recording, music production, and mastering, in which multiple frequency dependent gains are imposed upon an audio signal. Generally, the process of equalisation can be categorised under one of the following headings as described in [2], corrective equalisation: in which problematic frequencies are often attenuated in order to prevent issues such as acoustic feedback, and creative equalisation: in which the audio spectrum is modified to achieve a desired timbral aesthetic. Whilst the former is primarily based on adapting the effect parameters to the changes in the audio signal, the latter often involves a process of translation between a perceived timbral adjective such as *bright*, *flat*, or *sibilant* and an audio effect input space, by which a music producer must reappropriate a perceptual representation of a timbral transformation as a configuration of multiple parameters in an audio processing module. As music production is an inherently technical process, this mapping procedure is not necessarily trivial, and is made more complex by the source-dependent nature of the task.

2. Background

2.1. Semantically-Controlled Audio Effects

Engineers and producers generally use a wide variety of timbral adjectives to describe sound, each with varying levels of agreement. By modelling these adjectives, we are able to provide perceptually meaningful abstractions, which lead to a deeper understanding of musical timbre and systems that facilitate the process of audio manipulation. The extent to which timbral adjectives can be accurately modelled is defined by the level of exhibited agreement, a concept investigated in [3], in which terms such as *bright, resonant,* and *harsh* all exhibit strong agreement scores, and terms such as *open, hard,* and *heavy* all show low subjective agreement scores. It is common for timbral descriptors to be represented in low-dimensional space; *brightness,* for example, is shown to exhibit a strong correlation with spectral centroid [4,5] and has further dependency on the fundamental frequency of the signal [6]. Similarly, studies such as [7,8] demonstrate the ability to reduce complex data to lower-dimensional spaces using dimensionality reduction.

Recent studies have also focused on modification of the audio signal using specific timbral adjectives, where techniques such as spectral morphing [9] and additive synthesis [10] have been applied. For the purposes of equalisation, timbral modification has also been implemented via psychoacoustic measurements such as loudness [11], spectral masking [12], and semantically-meaningful controls and intuitive parameter spaces. SocialEQ [13], for example, collects timbral adjective data via a web interface and approximates the configuration of a graphic equaliser curve using multiple linear regression. Similarly, subjEQt [14] provides a two-dimensional interface, created using a Self-Organising Map, in which users can navigate between presets such as *boomy, warm,* and *edgy* using natural neighbour interpolation. This is a similar model to 2DEQ [15], in which timbral descriptors are projected onto a two-dimensional space using Principal Component Analysis (PCA). The Semantic Audio Feature Extraction (SAFE) project provides a similar non-parametric interface for semantically controlling a suite of audio plug-ins, in which semantics data is collected within a given Digital Audio Workstation (DAW). Adaptive presets can then be selectively derived based on audio features, parameter data, and music production metadata.

2.2. Aims

In this study, we propose a system that projects the controls of a parametric equaliser comprising five biquad filters, as detailed in [16], arranged in series onto an editable two-dimensional space, allowing the user to manipulate the timbre of an audio signal using an intuitive interface. Whilst the axes of the two-dimensional space are somewhat arbitrary, underlying timbral characteristics are projected onto the space via a training stage using two-term musical semantics data. In addition to this, we propose a signal processing method of adapting the parameter modulation process to the incoming audio data based on feature extraction applied to the long-term average spectrum (LTAS), as detailed in [17–19], capable of running in near-real-time. The model is implemented using the SAFE architecture (detailed in [20]), and is provided as an extension of the current Semantic Audio Parametric Equaliser (available for download at [21]), as shown in Figure 1a.

3. Methodology

In order to model the desired relationship between the two parameter spaces, a number of problems must be addressed. Firstly, the data reduction process should account for maximal variance in high-dimensional space without bias towards a smaller subset of the equaliser parameters. Similarly, we should be able to map to the high-dimensional space with minimal reconstruction error, given a new set of (x, y) coordinates. This process of mapping between spaces is nontrivial, due to loss of information in the reconstruction process. Furthermore, the low-dimensional parameter space should be configured in a way that preserves an underlying timbral characteristic in the data, thus allowing a user to transform the incoming audio signal in a musically meaningful way. Finally, the process of parameter space modification should not be agnostic of the incoming audio signal, meaning that any mapping between the two-dimensional plane and the equaliser parameters should be expressed as a function of the (x, y) coordinates and some representation of the signal spectral energy. In addition to this, the system should be capable of running in near-real time, enabling its use in a DAW environment.

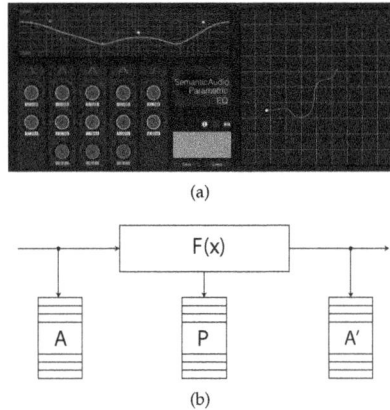

(a)

(b)

Figure 1. An overview of the Semantic Audio Feature Extraction (SAFE) equaliser and its feature extraction architecture. (**a**) The extended Semantic Audio Equalisation plug-in with the two-dimensional interface. To modify the *brightness/warmth* of an audio signal, a point is positioned in two-dimensional space; (**b**) The SAFE feature extraction process, where A represents the audio features captured before the effect is applied, A' represents the features captured after the effect is applied, and P represents the parameter vector.

To address these problems, we develop a model that consists of two phases. The first is a training phase, in which a map is derived from a corpus of semantically-labelled parameter data, and the second is an implementation phase in which a user can present (x, y) coordinates and an audio spectrum, resulting in a 13-dimensional vector of parameter state variables. To optimise the mapping process, we experiment with a combination of 6 dimensionality reduction techniques and 5 reconstruction methods, followed by a stacked-autoencoder (sAE) model that encapsulates both the dimensionality reduction and reconstruction processes. The techniques were chosen to represent a variable range of complexity and nonlinearity, and were intended to provide a selection of possible solutions to the problem, in which the highest performing section would be used for implementation. With the intention of scaling the parameters to the incoming audio signal, we derive a series of weights based on a selection of features, extracted from the signal LTAS coefficients. To evaluate the model performance under a range of conditions, we train it with binary musical semantics data and measure both objective and subjective performance based on the reconstruction of the input space and the structural preservation in reduced dimensionality space.

3.1. Dataset

For the training of the model, we compile a dataset of 800 semantically-annotated equaliser parameter space settings, comprising 40 participants equalising 10 musical instrument samples using two descriptive terms: *warm* and *bright*. To do this, participants were presented with the musical instrument samples in a DAW and asked to use a parametric equaliser to achieve the two timbral settings. After each setting was recorded, the data were recorded and the equaliser was reset to unity gain. During the test, samples were presented to the participants in a random order across separate DAW channels. Furthermore, the musical instrument samples were all performed unaccompanied, were Root Mean Square (RMS) normalised and ranged from 20 to 30 s in length. All of the participants had normal hearing, were aged 18–40, and all had at least 3 years' music production experience.

The descriptive terms (*warm* and *bright*) were selected for a number of reasons; firstly, the agreement levels exhibited by participants tend to be high (as suggested by [3]), meaning there should be less intra-class variance when subjectively assigning parameter settings. When measured using an agreement metric, defined by [13] as the log number of terms over the trace of the covariance matrix, *warm* and *bright* were the two highest ranked terms in a dataset of 210 unique adjectives. Secondly, the two terms are deemed to be sufficiently different enough to form an audible timbral variation in low dimensional space. While the two terms do not necessarily exhibit orthogonality (for example, *brightness* can be modified with constant *warmth* [9]), they have relatively dissimilar timbral profiles, with *brightness* widely accepted to be highly correlated with the signal's spectral centroid, and *warmth* often attributed to the ratio of the first three harmonics to the remaining harmonic partials in the magnitude spectrum [22].

The parameter settings were collected using a modified build of the SAFE data collection architecture, in which descriptive terms, audio feature data, parameter data, and metadata can be collected remotely within the DAW environment and uploaded to a server. As illustrated in Figure 1b, the SAFE architecture allows for the capture of audio feature data before and after processing has been applied. Similarly, the interface parameters P are captured and stored in a linked database. For the purpose of this experiment, the architecture was modified by adding the functionality to capture LTAS coefficients, with a window size of 1024 samples and a hop size of 256.

While the SAFE project comprises a number of DAW plug-ins, we focus solely on the parametric equaliser, which utilises five biquad filters arranged in series, consisting of a low-shelving filter (LS), three peaking filters (Pf_n), and a high-shelving filter (HS), where the LS and HS filters each have two parameters and the (Pf_n) filters each have three, as described in Table 1.

Table 1. A list of the parameter space variables and their ranges of possible values, taken from the Semantic Audio Feature Extraction (SAFE) parametric equaliser interface.

n	Assignment	Range	n	Assignment	Range
0	LS gain	−12–12 dB	7	Pf_1 Q	0.1–10 Hz
1	LS Freq	22–1000 Hz	8	Pf_2 Gain	−12–12 dB
2	Pf_0 Gain	−12–12 dB	9	Pf_2 Freq	220–10,000 Hz
3	Pf_0 Freq	82–3900 Hz	10	Pf_2 Q	0.1–10 Hz
4	Pf_0 Q	0.1–10 Hz	11	HS Gain	−12–12 dB
5	Pf_1 Gain	−12–12 dB	12	HS Freq	580–20,000 Hz
6	Pf_1 Freq	180–4700 Hz			

3.2. Evaluation Criteria

To evaluate the model under various conditions and to select an appropriate mapping topology, we apply objective metrics to the data during the dimensionality reduction and reconstruction processes. These allow us to evaluate the extent to which (1) the dimensionality reduction technique retains the structure of the high-dimensional data (*trustworthiness, continuity, K-Nearest Neighbours (K-NN)*), (2) the classes are separable in low-dimensional space (*Jeffries–Matusita Distance*), and (3) the system accurately reconstructs the high-dimensional parameter space (*reconstruction error*).

3.2.1. Trustworthiness and Continuity

To evaluate the structural preservation of each technique, the metrics *trustworthiness* and *continuity* [23] are applied to the dataset. Here, the distance of point i in high-dimensional space is measured against its k closest neighbours using rank order, and the extent to which each rank changes in low-dimensional space is measured. For n samples, let $r(i, j)$ be the rank in distance of sample i to sample j in the high-dimensional space U_i^k. Similarly, let $\hat{r}(i,j)$ be the rank of the distance between sample i and sample j in low-dimensional space V_i^k. Using the k-nearest neighbours, the map is considered *trustworthy* if these k neighbours are also placed close to point i in the low-dimensional space, as shown in Equation (1).

$$T(k) = 1 - \frac{2}{nk(2n-3k-1)} \sum_{i=1}^{n} \sum_{j \in U_i^{(k)}} (r(i,j) - k) \tag{1}$$

Similarly, *continuity* (shown in Equation 2) measures the extent to which original clusters of datapoints are preserved, and can be considered the inverse to *trustworthiness*, finding sample points that are close to point i in low-dimensional space, but not in the high-dimensional plane.

$$C(k) = 1 - \frac{2}{nk(2n-3k-1)} \sum_{i=1}^{n} \sum_{j \in V_i^{(k)}} (\hat{r}(i,j) - k) \tag{2}$$

In both of these equations, a normalising factor is used to bound the *trustworthiness* and *continuity* scores between 0 and 1. These measures evaluate the extent to which the local structure of the original dataset is preserved in a low-dimensional map; this is described in [24], where it is shown that the local structure of the dataset needs to be retained for a successful map of the datapoints.

3.2.2. K-NN

In order to measure the similarities in inter-class structures within the high and low dimensional space, we apply a K-NN classifier with $k = 1$, as described in [25], and then measure the differences in classification accuracies. The nearest neighbours are found using Euclidean distances with 13 and 2 dimensions, respectively. The accuracies are derived using K-fold cross validation with $K = 20$, where 20% of the data is partitioned for testing. This allows us to measure the extent to which the between-class structures have been preserved in the reduction process, and effectively acts as a supervised structural preservation metric.

3.2.3. Jeffries–Matusita Distance

In order to evaluate the extent to which timbral descriptors lie at opposing ends of the mapped parameter space, we can measure the extent to which the timbre classes are separable using a distance metric. Typically, this can be done by finding the divergence between class distributions using a technique such as Kullback–Leibler Divergence (KLD), as we proposed in [26]; however, as explained in [27], this does not satisfy the triangle inequality based on the measurement's asymmetry. While two-sided KLD addresses this, as explained in [28], [29] proposes Jeffries–Matusita Distance (JMD) as a more appropriate alternative. JMD (as shown in Equation 4) is a metric derived from the Bhattacharya (BH) distance, as in Equation (3), which bounds the output of the measure from 0 (no separability) to 2 (perfect separability).

$$BH_{i,j} = \frac{1}{8}(m_i - m_j)^T \left(\frac{S_i + S_j}{2}\right)^{-1}(m_i - m_j) + 0.5\ln\left(\frac{0.5(|S_i + S_j|)}{\sqrt{|S_i||S_j|}}\right) \tag{3}$$

$$JMD_{1,2} = \sqrt{2(1 - e^{-BH_{i,j}})} \tag{4}$$

Here m represents the mean and S represents the covariance of classes i and j, respectively.

3.2.4. Reconstruction Error

To measure the reconstruction accuracy (low-to-high-dimensionality mapping) of the model, we measure the input/output error for each pair-wise combination of dimensionality reduction and reconstruction techniques by computing the mean absolute error between predicted and actual parameter values. This is done using *K*-fold cross validation with $k = 20$ iterations, and a test partition size of 20% (160 training examples). As some of the dimensionality reduction techniques are unable to embed new information into the reduced-dimensionality space, the first part of the test process (*i.e.*, the prediction of new low-dimensional values as implemented in [26]) was omitted, and only regression and interpolation techniques were evaluated.

3.3. Subjective Evaluation

Using the metrics defined in Section 3.2, we are able to select an appropriate model which is capable of accurately reducing the dataset while preserving the data structure and accurately reconstructing the input parameters with minimal error. To validate this, we implement subjective user tests in which participants are asked to equalise a series of audio samples using the reduced-dimensionality interface. To do this, 10 participants were asked to apply the process to 10 input sounds using only the two-dimensional interface. Each participant was asked to achieve a *warm* or *bright* output sound for each stimuli. During the test, samples were presented to participants in a random order across separate DAW channels, and the equaliser parameters remained hidden. No indication was given as to the underlying distribution of datapoints. The stimuli comprised unaccompanied musical instrument samples and ranged from 20 to 30 s in length. The samples were primarily taken from electric guitars and included a variety of genres, taken from the Mixing Secrets Multitrack Audio Dataset [30]. All of the participants had normal hearing, were aged 18–35, and had varied music production experience, from 0 to 5 years.

4. Model

The proposed system maps between the equaliser parameter space, consisting of 13 filter parameters and a two-dimensional plane, while preserving the context-dependent nature of the audio effect. After an initial training phase, the user can then submit (x, y) coordinates to the system using a track-pad interface, resulting in a timbral modification via the corresponding filter parameters. To demonstrate this, we train the model with two class (*bright*, *warm*) musical semantics data taken from the SAFE equaliser database, thus resulting in an underlying transition between opposing timbral descriptors in two-dimensional space. By training the model in this manner, we intend to retain the high-dimensional structure of the dataset in the two-dimensional space while minimising the reconstruction error inherent to dimensionality reduction methods.

The model (illustrated in Figure 2) has two key operations. The first involves weighting the parameters by computing the vector $\alpha_n(A)$ from the input signal long-term spectral energy (A). We can then modify the parameter vector (P) to obtain a weighted vector (P'). The second component scales the dimensionality of (P'), resulting in a compact audio-dependent representation. During the model implementation phase, we apply an unweighting procedure based on the (x, y) coordinates and the signal modified spectrum. This is done by multiplying the estimated parameters with the inverse weight vector, resulting in an approximation of the original parameters. In addition to the weighting and dimensionality reduction stages, a scale-normalisation procedure is applied, aiming to convert the ranges of each parameter (given in Table 1), to $(0 < p_n < 1)$. This converts the data into a suitable format for dimensionality reduction.

4.1. Parameter Scaling

As the configuration of the filter parameters assigned to each descriptor by the user during equalisation is likely to vary based on the audio signal being processed, the first requirement of the model is to apply weights to the parameters based on knowledge of the audio data at the time of processing. To do this, we selectively extract features from the signal LTAS before and after the filter is applied. This is possible due to the configuration of the data collection architecture, highlighted in Figure 1b. The weights (α_m) can then be expressed as a function of the LTAS, where the function's definition varies based on the parameter representation (*i.e.*, gain, centre frequency, or bandwidth of the corresponding filter). We use the LTAS to prevent the parameters from adapting each time a new frame is read. In practice, we are able to do this by presenting users with means to store the audio data, rather than continually extracting it from the audio stream. Each weighting is defined as the ratio between a spectral feature taken from the filtered audio signal (A'_k) and the signal filtered by an enclosing rectangular window (R_k). Here, the rectangular window is bounded by the minimum and maximum frequency values attainable by the observed filter $f_k(A)$.

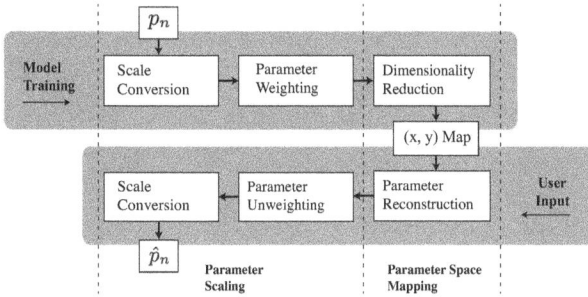

Figure 2. An overview of the proposed model. The grey horizontal paths represent training and implementation (user input) phases.

We can define the equaliser as an array of biquad functions arranged in series, as depicted in Equation (5).

$$f_k = f_{k-1}(A, \vec{P}_{k-1})$$
$$k = 1, \dots, K-1$$
(5)

Here, $K = 5$ represents the number of filters used by the equaliser and f_k represents the kth biquad function, which we can define by its transfer function, given in Equation (6).

$$H_k(z) = c \cdot \frac{1 + b_1 z^{-1} + b_2 z^{-2}}{1 + a_1 z^{-1} + a_2 z^{-2}}$$
(6)

The LTAS is then modified by the filter as in Equation (7) and the weighted parameter vector can be derived using the function expressed in Equation (8).

$$A'_k = |H_k(e^{j\omega})| A_k$$
(7)

$$p'_n = \alpha_m(k) \cdot p_n$$
(8)

where p_n is the nth parameter in the vector P. The weighting function is then defined by the parameter type (m), where $m = 0$ represents gain, $m = 1$ represents centre-frequency, and $m = 2$ represents bandwidth. For gain parameters, the weights are expressed as a ratio of the spectral energy in the filtered spectrum (A') to the spectral energy in the enclosing rectangular window (R_k), derived in Equation (9) and illustrated in Figure 3.

$$\alpha_0(k) = \frac{\sum_i (A'_k)_i}{\sum_i (R_k)_i}$$
(9)

For frequency parameters ($m = 1$), the weights are expressed as a ratio of the respective spectral centroids of A' and R_k, as demonstrated in Equation (10), where bn_i are the corresponding frequency bins.

$$\alpha_1(k) = \left(\frac{\sum_i (A'_k)_i bn_i}{\sum_i (A'_k)_i} \right) \Big/ \left(\frac{\sum_i (R_k)_i bn_i}{\sum_i (R_k)_i} \right)$$
(10)

Finally, the weights for bandwidth parameters ($m = 2$) are defined as the ratio of spectral spread exhibited by both A' and R_k. This is demonstrated in Equation (11), where $(x)_{sc}$ represents the spectral centroid of x.

$$\alpha_2(k) = \left(\frac{\sum_i (bn_i - (A'_k)_{sc})^2 (A'_k)_i}{\sum_i (A'_k)_i} \right) \bigg/ \left(\frac{\sum_i (bn_i - (R_k)_{sc})^2 (R_k)_i}{\sum_i (R_k)_i} \right) \tag{11}$$

During the implementation phase, retrieval of the unweighted parameters, given a weighted vector, can be achieved by simply multiplying the weighted parameters with the inverse weights vector, as in Equation (12).

$$\hat{p}_n = \alpha_m^{-1}(k) \cdot p'_n \tag{12}$$

where $\hat{p}$ is a reconstructed version of p, after dimensionality reduction has been applied.

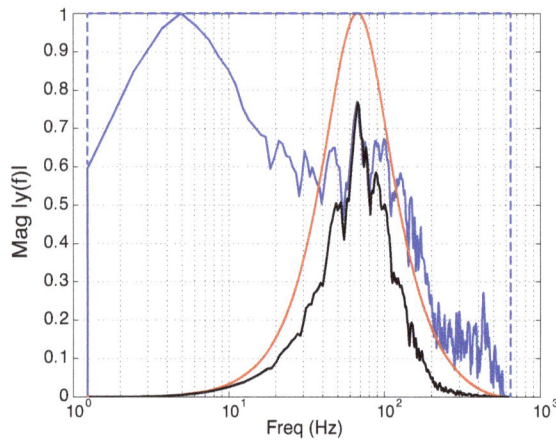

Figure 3. An example spectrum taken from an input example, weighted by the biquad coefficients, where the red line represents a peaking filter, the black line represents the biquad-filtered spectrum, and the blue line represents the spectral energy in the rectangular window (R_k).

To ensure the parameters are in a consistent format for each of the dimensionality scaling algorithms, a scale normalisation procedure is applied using Equation (13), where during the training process, the p_{min} and p_{max} represent the minimum and maximum values for each parameter (given in Table 1), and q_{min} and q_{max} represent 0 and 1. During the implementation process, these values are exchanged such that q_{min} and q_{max} represent the minimum and maximum values for each parameter and p_{min} and p_{max} represent 0 and 1.

$$\rho_n = \frac{(p_n - q_{min})(p_{max} - p_{min})}{q_{max} - q_{min}} + p_{min} \tag{13}$$

Additionally, a sorting algorithm was used to place the three mid-band filters in ascending order based on their centre frequency. This prevents normalisation errors due to the frequency ranges, allowing filters to be rearranged by the user.

4.2. Parameter Space Mapping

Once the filters have been weighted by the audio signal, the mapping from 13 equaliser variables to a two-dimensional subspace can be accomplished using a range of dimensionality reduction techniques. In this study, we expand on [26] and evaluate the performance of six dimensionality reduction techniques. Here, the algorithms that were used for the dimensionality reduction are available as part of the dimensionality reduction toolbox in [31]. In addition to this, parameter space mapping is evaluated by measuring the quality of reduction with rank-based measures and nearest neighbour classification algorithms. In dimensionality reduction, the reconstruction process is often less common due to the nature of the task (e.g., feature optimisation, data reduction). We evaluate the efficacy of two regression-based techniques and three interpolation techniques at mapping two-dimensional interface variables to a vector of equaliser parameters. This is done by approximating functions using the weighted parameter data and measuring the reconstruction error. Finally, we evaluate an sAE model of data reduction, in which the parameter space is both reduced and reconstructed in the same algorithm; we are then able to isolate the reconstruction (decoder) stage for the implementation process.

Dimensionality reduction is implemented using the following techniques: PCA, a widely used method of embedding data into a linear subspace of reduced dimensionality by finding the eigenvectors of the covariance matrix, originally proposed by [32]; *Kernel PCA* (kPCA), a non-linear manifold mapping technique in which the eigenvectors are computed from a kernel matrix as opposed to the covariance matrix, as defined by [33]; *probabilistic PCA* (pPCA), a method that considers standard PCA as a latent variable model and makes use of an Expectation Maximisation (EM) algorithm, a method for finding the maximum-likelihood estimate of the parameters in an underlying distribution from a given data set, depending on unobserved latent variables [34] as described in [35]; *Factor Analysis* (FA), a statistical analysis technique that identifies the relationship between different variables of a dataset and groups those variables by the correlation of the underlying factors [36]; *Diffusion Maps* (DM), a technique inspired by the field of dynamical systems, reducing the dimensionality of data by embedding the original dataset in a low-dimensional space by retrieving the eigenvectors of Markov random walks [37]; *Linear Discriminant Analysis* (LDA), a supervised projection technique that maps to a linear subspace while maximising the separability between data points that belong to different classes (see [38]). As LDA projects the data-points onto the dimensions that maximise inter-class variance for C classes, the dimensionality of the subspace is set to $C - 1$. This means that in a binary classification problem such as ours, we need to reconstruct the second dimension arbitrarily. For each of the other algorithms, we select the first two variables for mapping, and for the kPCA algorithm, the feature distances are computed using a Gaussian kernel.

The parameter reconstruction process was implemented using the following techniques: *Linear Regression* (LR), a process by which a linear function is used to estimate latent variables; *Natural Neighbour Interpolation* (NaNI), a method for interpolating between scattered data points using Voronoi tessellation, as used by [14] for a similar application; *Nearest Neighbour Interpolation* (NeNI), an interpolation method where the query point takes the value of the nearest neighbour [39]; *Linear Interpolation* (LI), an interpolation technique that assumes a linear relationship between the existing points in a dataset; *Support Vector Regression* (SVR), a non-linear kernel-based regression technique (see [40]), for which we choose a Gaussian kernel function.

An autoencoder is an Artificial Neural Network (ANN) with a topology capable of learning a compact representation of a dataset by optimising a matrix of weights, such that a loss function representing the difference between the output and input vectors is minimised. Autoencoders can then be stacked together using the output of the prior layer as the input for the next in order to construct a deep network architecture. Each autoencoder is then trained individually, learning to minimise the reconstruction error between its input and the predicted output. This approach has been used for data compression [41], and by extension, dimensionality reduction. This type of ANN is often used in order to improve the classification accuracy of logistic regression [42]; however, since our problem involves data reconstruction as opposed to classification, a logistic layer is not implemented.

Network Topology

The autoencoder was built using the Theano Python library [43], where we observed an error of 0.086 using a single hidden layer with N (in this case $N = 2$) units. To reduce the error, a mirrored

$[13 - 9 - 2]$ architecture was selected empirically, resulting in an error measurement of 0.08. To improve reconstruction accuracy further, noise was introduced at each stage in the network, as demonstrated by [44]. Here, the first autoencoder was corrupted with 0.6 magnitude noise, and the second with 0.5. This approach is able to further reduce the reconstruction error to 0.0784. Additionally, we replace the previously-used stochastic gradient descent algorithm with an *RMSprop* method [45] with a batch size of 10 as the pre-training and fine-tuning methods of optimization, and a learning rate of 0.01 and 0.001, respectively. This approach allows for faster optimization, as shown in [46]. For the weighted parameters, we found that a three-layer denoising autoencoder with an architecture of $[13 - 9 - 6 - 2]$ and noise of magnitude $(0.5, 0.4, 0.3)$ is able to outperform our two-layer denoising autoencoder model.

5. Results

5.1. Parameter Space Evaluation

To evaluate the extent to which structures in the parameter space are preserved in the reduced dimensionality map, we report the *trustworthiness*, *continuity*, and class-wise similarity (k-NN). This is applied to data shown in Figure 4a–g, in which a two-dimensional projection of the 13 equaliser parameters is given for both *warm* and *bright* samples in the dataset.

5.1.1. Low-Dimensional Mapping Accuracy

From Table 2 we show that for *trustworthiness*, pPCA achieves the highest rating (0.8426), with the sAE also performing similarly (0.842). The rest of the techniques are also able to achieve a high score, ranging from 0.81 for kPCA to 0.839 for standard PCA. The only technique that does not perform to the same standard is LDA, as the algorithm maximizes the separability of classes in the data instead of preserving the structure of the original dataset unrelated to its classes. For *continuity*, we can see that the majority of the techniques perform similarly, with scores ranging from 0.943 for the sAE to 0.958 for kPCA. However, as was the case with *trustworthiness*, LDA does not perform as well (0.868), due to the map reduction process.

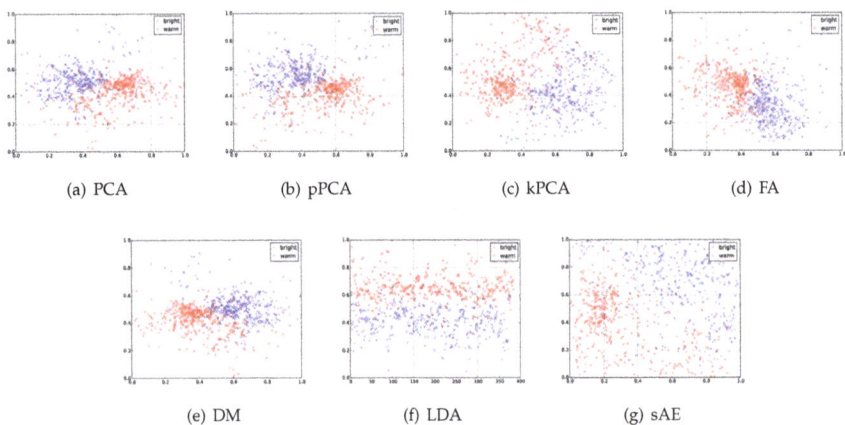

(a) PCA (b) pPCA (c) kPCA (d) FA

(e) DM (f) LDA (g) sAE

Figure 4. Two-dimensional parameter-space representations using seven data reduction techniques, where the red data points are taken from parameter spaces described as *bright* and the blue points are described as *warm*. (**a**), (**c**), (**f**), and (**g**). PCA: Principal Components Analysis; pPCA: Probabilistic PCA; kPCA: Kernel PCA; FA: Factor Analysis; DM: Diffusion Maps; LDA: Linear Discriminant Analysis; sAE: stacked-autoencoder.

Table 2. *Trustworthiness* and *continuity* scores for the different dimensionality reduction techniques (higher values are better) and classification accuracy of 1-NN classification

Technique	Trustworthiness	Continuity	1-NN Classification
Original	-	-	91.21%
PCA	0.8398	0.9541	87.61%
pPCA	0.8426	0.9567	87.92%
kPCA	0.8102	0.9583	86.14%
FA	0.8337	0.9490	86.19%
DM	0.8395	0.9533	87.89%
LDA	0.7292	0.8684	85.40%
sAE	0.8420	0.9439	84.01%

Trustworthiness and *continuity* metrics were used with a varying number of neighbours, ranging from 1 to 250. Here, the sAE exhibits higher scores for a lower number of neighbours ($<$120), as shown in Figure 5a—a result that suggests the system is better at retaining the local structure of the data, which is a necessary goal for a successful mapping technique. Furthermore, while the *continuity* score of the autoencoder is lower than the remaining dimensionality reduction techniques (Table 2), its error from the best performing technique in terms of *continuity* (kPCA) is only 0.015, which is deemed negligible.

5.1.2. Class Preservation

The classification of 1-NN in the original dataset achieves an average of 91.21% for 100 iterations of the algorithm. None of the dimensionality reduction techniques are able to replicate this response, with pPCA achieving the highest score (87.92%), as seen in Table 2. On the other hand, the sAE achieved an accuracy of 84.01%, the lowest among the techniques being tested, 7.2% worse than the classification accuracy of the algorithm in the high-dimensional dataset. This result reveals that sAE is not as capable as other reduction techniques in preserving the classes on the low-dimensional space; however, as sAE is able to achieve better results than the other techniques for *trustworthiness* for a lower number of neighbours, and its performance in 1-NN is not drastically worse (3.91%) than the best technique in pPCA, it can be considered a minor problem.

5.1.3. Class Separation

By applying JMD (Equation 4) to the dimensionality reduction techniques, we find that kPCA outperforms the rest of the techniques used, achieving 0.607, whereas the optimised autoencoder model performs slightly less favourably with a score 0.558, as shown in Table 3. The only technique that was excluded from this process was LDA, for two reasons: (1) it is a supervised technique that specifically maximizes the separability between the different classes in the low-dimensional space, and (2) in the context of our study, LDA has reduced the dataset to a single dimension, while all the other techniques have reduced the dimensionality to two dimensions. While class-separability is not necessarily correlated with accurate preservation of structure, high separability will allow users to effectively modulate between contrasting timbral descriptors.

Table 3. Jeffries–Matusita Distance (JMD) scores showing separation across different dimensionality reduction techniques.

Separability Measure	PCA	pPCA	kPCA	FA	DM	sAE
JMD	0.5142	0.5152	0.6076	0.4862	0.5125	0.5581

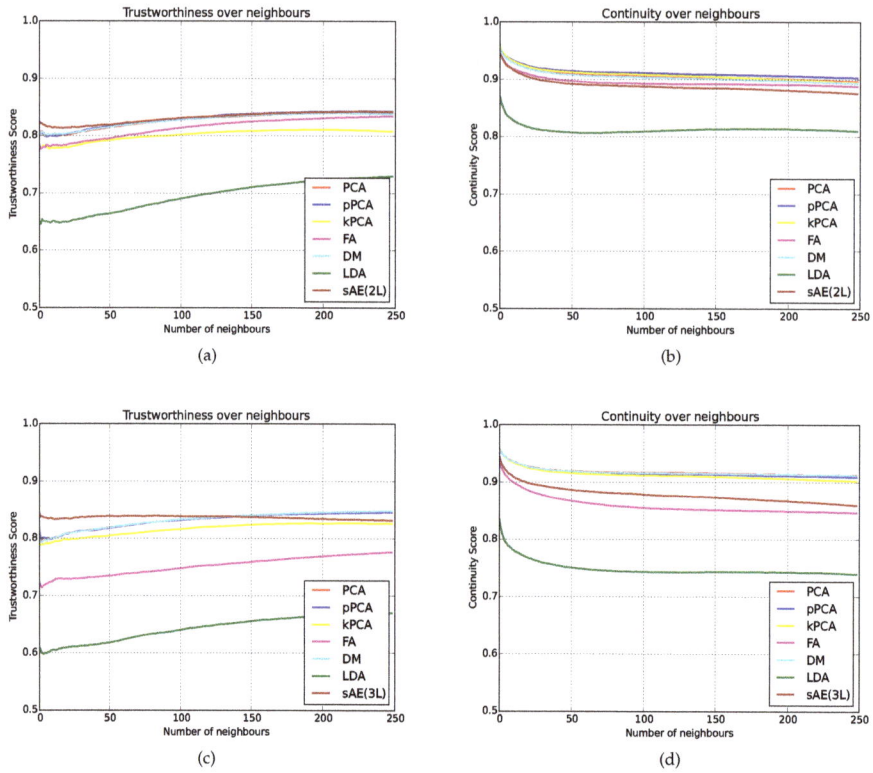

Figure 5. *Trustworthiness* and *continuity* plots across the different dimensionality reduction techniques for number of neighbors (1 : 250). (**a**) Trustworthiness; (**b**) Continuity; (**c**) Trustworthiness (Weighted Parameters); (**d**) Continuity (Weighted Parameters).

5.2. Parameter Reconstruction Error

In [26], the sAE was able to achieve the lowest reconstruction error, 0.086, while the technique that came the closest to its accuracy was kPCA with support vector regression, achieving an error of 0.09. The sAE technique still outperforms all the other combinations of techniques, as can be seen in Table 4, achieving an overall error 0.074. It should also be noted that the sAE is able to reconstruct the most parameters of the equaliser (6) more accurately than any other combination of techniques.

Table 4. Mean reconstruction error per parameter using combinations of dimensionality reduction and reconstruction techniques, with the lowest reconstruction error highlighted in grey. The final column shows the mean (μ) error across all techniques, while the model with the lowest mean reconstruction error (Stacked Autoencoder, sAE) is highlighted in green. LR: Linear Regression; SVR: Support Vector Regression; NaNI: Natural Neighbour Interpolation; NeNI: Nearest Neighbour Interpolation; LI: Linear Interpolation.

P:	0	1	2	3	4	5	6	7	8	9	10	11	12	μ
PCA-LR	0.099	0.070	0.142	0.047	0.041	0.139	0.079	0.028	0.124	0.090	0.029	0.102	0.109	0.084
LDA-LR	0.194	0.070	0.150	0.047	0.041	0.171	0.082	0.028	0.116	0.090	0.030	0.123	0.106	0.096
kPCA-LR	0.081	0.070	0.136	0.047	0.040	0.150	0.082	0.027	0.130	0.084	0.029	0.120	0.107	0.085
pPCA-LR	0.099	0.069	0.138	0.046	0.039	0.142	0.078	0.027	0.126	0.092	0.030	0.104	0.108	0.084
DM-LR	0.104	0.070	0.138	0.047	0.040	0.139	0.081	0.027	0.126	0.091	0.031	0.102	0.106	0.085
FA-LR	0.151	0.068	0.156	0.042	0.040	0.143	0.068	0.029	0.144	0.084	0.030	0.103	0.094	0.089
PCA-SVR	0.086	0.064	0.123	0.046	0.040	0.137	0.079	0.028	0.125	0.089	0.031	0.097	0.095	0.080
LDA-SVR	0.196	0.068	0.152	0.048	0.040	0.171	0.081	0.028	0.116	0.087	0.031	0.123	0.105	0.096
kPCA-SVR	0.077	0.069	0.136	0.045	0.039	0.144	0.079	0.026	0.130	0.088	0.032	0.111	0.099	0.083
pPCA-SVR	0.089	0.066	0.128	0.047	0.040	0.136	0.077	0.027	0.128	0.088	0.031	0.096	0.097	0.081
DM-SVR	0.088	0.067	0.121	0.047	0.040	0.133	0.078	0.026	0.124	0.089	0.031	0.096	0.095	0.080
FA-SVR	0.144	0.062	0.137	0.041	0.039	0.144	0.066	0.026	0.144	0.085	0.030	0.098	0.082	0.084
PCA-NaNI	0.091	0.080	0.137	0.054	0.045	0.149	0.092	0.029	0.144	0.107	0.032	0.104	0.107	0.090
LDA-NaNI	0.263	0.098	0.209	0.071	0.046	0.216	0.117	0.031	0.149	0.124	0.033	0.158	0.128	0.126
kPCA-NaNI	0.083	0.082	0.159	0.056	0.042	0.154	0.095	0.029	0.160	0.116	0.033	0.125	0.108	0.096
pPCA-NaNI	0.092	0.078	0.139	0.050	0.041	0.148	0.090	0.028	0.139	0.106	0.034	0.105	0.106	0.089
DM-NaNI	0.094	0.080	0.139	0.052	0.043	0.146	0.091	0.026	0.143	0.107	0.030	0.107	0.103	0.089
FA-NaNI	0.152	0.070	0.157	0.046	0.041	0.164	0.075	0.028	0.159	0.098	0.033	0.102	0.087	0.093
PCA-NeNI	0.099	0.093	0.163	0.060	0.047	0.177	0.106	0.030	0.162	0.123	0.035	0.121	0.121	0.103
LDA-NeNI	0.252	0.100	0.194	0.060	0.042	0.217	0.109	0.031	0.151	0.120	0.037	0.158	0.115	0.122
kPCA-NeNI	0.092	0.096	0.187	0.060	0.042	0.175	0.110	0.025	0.180	0.128	0.029	0.135	0.124	0.106
pPCA-NeNI	0.103	0.088	0.162	0.059	0.042	0.170	0.107	0.027	0.160	0.123	0.034	0.120	0.117	0.101
DM-NeNI	0.110	0.090	0.161	0.059	0.046	0.175	0.101	0.025	0.159	0.124	0.034	0.122	0.116	0.102
FA-NeNI	0.176	0.082	0.171	0.054	0.041	0.193	0.087	0.028	0.205	0.114	0.034	0.138	0.096	0.109
PCA-LI	0.092	0.078	0.141	0.055	0.042	0.149	0.095	0.026	0.143	0.114	0.033	0.108	0.108	0.091
LDA-LI	0.254	0.097	0.195	0.062	0.043	0.209	0.107	0.032	0.153	0.115	0.037	0.155	0.113	0.121
kPCA-LI	0.083	0.082	0.159	0.058	0.039	0.159	0.102	0.028	0.160	0.114	0.030	0.127	0.115	0.096
pPCA-LI	0.091	0.080	0.138	0.053	0.047	0.148	0.095	0.029	0.146	0.112	0.034	0.108	0.107	0.091
DM-LI	0.098	0.076	0.142	0.051	0.045	0.149	0.089	0.030	0.146	0.112	0.033	0.108	0.105	0.091
FA-LI	0.160	0.070	0.153	0.046	0.041	0.172	0.078	0.028	0.176	0.102	0.032	0.119	0.087	0.097
sAE(2-Layer)	0.073	0.046	0.126	0.039	0.027	0.149	0.067	0.014	0.123	0.091	0.017	0.099	0.096	0.074

5.3. Parameter Weighting

In order to evaluate the effectiveness of the signal specific weights, we measure the reconstruction accuracy of each system after the weights have been applied (see Table 5). Overall, the systems exhibit a general improvement in the reconstruction accuracy of the gain and Q parameters. All the systems have improved accuracy measurements, with the highest performing pair being PCA with SVR, achieving an error of 0.059. Similarly, the sAE with the same architecture, with hidden layer sizes [9, 2], is able to achieve a reconstruction accuracy of 0.06—a further improvement from the 0.0748 error observed with unweighted parameters. For the weighted parameters we found that a three-layer denoising autoencoder was able to outperform our two-layer autoencoder, improving the reconstruction accuracy by 0.02.

Finally, the parameter weighting stage improves the *trustworthiness* of the low-dimensional mapping when using PCA, pPCA, kPCA, DM, and sAE, whilst FA and LDA exhibited significantly lower scores, as presented in Table 6. On the other hand, the *continuity* of the systems had very little change, with pPCA, kPCA, DM, FA, and sAE showing very minor reductions, LDA showing significant reduction, and PCA showing an improvement. In this case, sAE with parameter weighting still outperforms the other techniques in terms of *trustworthiness* for a lower number of neighbours, as in Figure 5c, and the performance in terms of continuity sees the sAE performing better than FA (Figure 5d).

Table 5. Mean reconstruction error per parameter using combinations of dimensionality reduction and reconstruction techniques for the weighted parameterers, with the lowest reconstruction error highlighted in grey. The final column shows the mean (μ) error across all techniques, while the model with the lowest mean reconstruction error (Stacked Autoencoder) is highlighted in green.

P:	0	1	2	3	4	5	6	7	8	9	10	11	12	μ
PCA-LR	0.052	0.059	0.062	0.040	0.023	0.114	0.075	0.018	0.107	0.088	0.020	0.034	0.106	0.061
LDA-LR	0.149	0.068	0.116	0.047	0.022	0.118	0.083	0.017	0.101	0.088	0.020	0.028	0.105	0.074
kPCA-LR	0.039	0.066	0.056	0.043	0.021	0.113	0.084	0.016	0.112	0.089	0.021	0.035	0.105	0.062
pPCA-LR	0.054	0.066	0.062	0.042	0.022	0.111	0.074	0.017	0.108	0.090	0.022	0.036	0.110	0.063
DM-LR	0.058	0.068	0.066	0.041	0.023	0.111	0.074	0.016	0.110	0.091	0.020	0.036	0.107	0.063
FA-LR	0.149	0.062	0.141	0.035	0.021	0.111	0.063	0.015	0.066	0.075	0.022	0.024	0.091	0.067
PCA-SVR	0.046	0.059	0.059	0.041	0.021	0.111	0.071	0.015	0.103	0.087	0.021	0.035	0.099	0.059
LDA-SVR	0.155	0.070	0.120	0.047	0.023	0.121	0.081	0.016	0.109	0.094	0.020	0.027	0.104	0.076
kPCA-SVR	0.036	0.068	0.052	0.044	0.023	0.111	0.080	0.016	0.106	0.090	0.022	0.035	0.108	0.061
pPCA-SVR	0.047	0.061	0.058	0.041	0.023	0.113	0.074	0.016	0.106	0.094	0.021	0.035	0.101	0.061
DM-SVR	0.050	0.063	0.060	0.042	0.024	0.110	0.074	0.016	0.103	0.089	0.020	0.035	0.100	0.060
FA-SVR	0.141	0.050	0.136	0.036	0.023	0.108	0.058	0.017	0.064	0.075	0.019	0.024	0.092	0.065
PCA-NaNI	0.048	0.066	0.064	0.047	0.026	0.127	0.081	0.019	0.116	0.096	0.024	0.038	0.111	0.066
LDA-NaNI	0.195	0.092	0.152	0.062	0.025	0.160	0.106	0.020	0.135	0.123	0.026	0.033	0.123	0.096
kPCA-NaNI	0.038	0.075	0.061	0.051	0.026	0.137	0.098	0.020	0.120	0.102	0.024	0.039	0.110	0.069
pPCA-NaNI	0.046	0.065	0.064	0.045	0.027	0.128	0.080	0.022	0.117	0.094	0.021	0.036	0.110	0.066
DM-NaNI	0.054	0.070	0.069	0.046	0.028	0.128	0.084	0.019	0.118	0.100	0.024	0.038	0.109	0.068
FA-NaNI	0.164	0.055	0.163	0.040	0.023	0.124	0.069	0.019	0.077	0.090	0.025	0.029	0.104	0.076
PCA-NeNI	0.057	0.077	0.080	0.057	0.029	0.157	0.100	0.022	0.140	0.119	0.022	0.043	0.126	0.079
LDA-NeNI	0.195	0.096	0.157	0.063	0.027	0.157	0.105	0.023	0.132	0.122	0.027	0.032	0.123	0.097
kPCA-NeNI	0.042	0.081	0.072	0.058	0.030	0.154	0.108	0.024	0.145	0.112	0.025	0.045	0.125	0.079
pPCA-NeNI	0.054	0.072	0.076	0.055	0.027	0.155	0.097	0.022	0.137	0.110	0.022	0.042	0.130	0.077
DM-NeNI	0.059	0.075	0.084	0.053	0.030	0.158	0.095	0.022	0.143	0.114	0.025	0.045	0.129	0.079
FA-NeNI	0.185	0.064	0.190	0.047	0.029	0.144	0.085	0.020	0.091	0.109	0.025	0.033	0.117	0.088
PCA-LI	0.052	0.070	0.069	0.050	0.027	0.136	0.087	0.021	0.127	0.102	0.026	0.038	0.119	0.071
LDA-LI	0.192	0.103	0.154	0.062	0.027	0.161	0.110	0.018	0.140	0.135	0.025	0.035	0.124	0.099
kPCA-LI	0.037	0.069	0.064	0.049	0.027	0.138	0.094	0.020	0.122	0.106	0.024	0.040	0.113	0.069
pPCA-LI	0.052	0.071	0.069	0.049	0.026	0.137	0.084	0.020	0.125	0.102	0.024	0.039	0.116	0.070
DM-LI	0.054	0.070	0.070	0.046	0.029	0.132	0.085	0.020	0.121	0.099	0.024	0.037	0.113	0.069
FA-LI	0.170	0.056	0.162	0.040	0.026	0.124	0.070	0.021	0.077	0.093	0.025	0.030	0.103	0.077
sAE(3-Layer)	0.065	0.053	0.081	0.040	0.021	0.106	0.075	0.015	0.077	0.081	0.017	0.028	0.096	0.058

Table 6. *Trustworthiness* and *continuity* scores (including weighting) for the different dimensionality reduction techniques (higher values are better), and classification accuracy of 1-nn classification

Technique	Trustworthiness	Continuity	1-NN Classification
Original	-	-	84.9%
PCA	0.8463	0.9562	67.85%
pPCA	0.8454	0.9552	67.39%
kPCA	0.8263	0.9566	69.40%
FA	0.7761	0.9359	59.52%
DM	0.8477	0.9561	66.03%
LDA	0.6702	0.8340	73.92%
sAE(3-Layer)	0.8440	0.9431	73.51%

5.4. User Evaluation

We evaluate the performance of the selected mode (sAE) using subjective tests in which we present the user with various samples and ask them to equalise it using the low-dimensional space (shown in Figure 6). We then measure the class separability using the JMD metric presented in Section 3.2. In Table 7 we present the degree of separation between user inputs using high-dimensional and low-dimensional responses from the subjective data. From this we can deduce that the overlap between *warm* and *bright* descriptors has decreased, with a value of 0.8527. This is higher than the high-dimensional dataset instances (0.5581). Furthermore, we see an increase in separation between the high-dimensional classes and the opposing low-dimensional classes. For instance, the high-dimensional *warm* examples and the low-dimensional *bright* examples achieve a separation of 0.7719, again higher than the original separation between the high-dimensional classes. Similarly, a strong positive correlation between high-dimensional and low-dimensional equalisation is exhibited by examples in the same class, a desired effect that displays the ability of the users to choose the corresponding regions for the two descriptors.

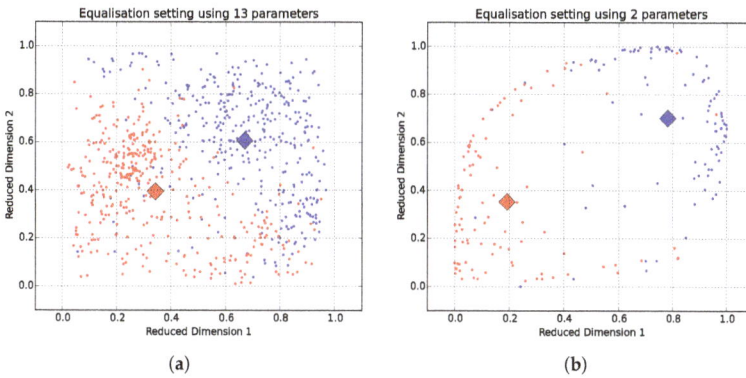

Figure 6. Equalisation settings shown in reduced dimensionality space where the figure (**a**) shows the results of users recording warm and bright samples using 13 parameters; (**b**) the results of users producing the same descriptors using a sAE-based two-dimensional equaliser. Here, diamonds represent the class centroids.

Table 7. Jeffries-Matusita Distance (JMD) scores showing separation for data gathered from 13-dimensional parameters and a two-dimensional interface using *warm*(W) and *bright*(B) examples. Higher scores are desirable for the first four measurements, while lower scores are better for the last two columns.

Separability	W(13-d)/B(13-d)	W(2-d)/B(2-d)	W(13-d)/B(2-d)	B(13-d)/W(2-d)	W(13-d)/W(2-d)	B(13-d)/B(2-d)
JMD	0.5581	0.8527	0.7719	0.6988	0.0846	0.1439

Table 8. Pearson correlation between the reconstructed equaliser curves.

Metric	B(13-d)/B(2-d)	W(13-d)/W(2-d)	W(13-d)/B(13-d)	W(2-d)/B(2-d)
Pearson correlation	0.9346	0.9247	-0.7594	-0.9121

This is reinforced by the low Euclidean distances between class centroids (shown in Figure 6) and strong positive coherence (spectral correlation) between the equaliser curves achieved using the 13- and 2-dimensional interfaces (shown in Figure 7a,b). These results are provided through the Pearson correlation measures in Table 8, revealing a positive correlation between the high-dimensional and low-dimensional datasets for the same descriptor: 0.9346 for warm and 0.9247 for bright, and a negative correlation between opposite high-dimensional and low dimensional descriptors: −0.7594 and −0.9121, respectively.

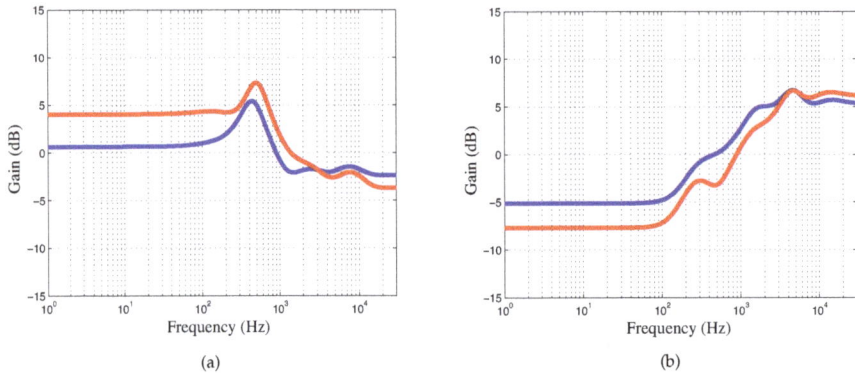

(a)

(b)

Figure 7. The reconstructed equaliser curves for the centroid of the *warm* and *bright* descriptors for both the high-dimensional (red) and low-dimensional (blue) datasets. (**a**) Reconstructed *warm* equaliser curve; (**b**) Reconstructed *bright* equaliser curve.

6. Discussion

For reconstruction accuracy we find that the sAE is able to outperform all pairwise combinations of dimensionality reduction and reconstruction techniques, whether the system includes parameter weighting or not (Table 4 and Table 5). Furthermore, the sAE is able to achieve the second highest *trustworthiness* score (see Table 2, Figure 5a,c) in low-dimensional space, and performs to a high standard in the preservation of high-dimensional clusters (*continuity*), as in Table 2 and Figure 5b,d. Using a sAE however, the class-separability in low-dimensional space is reduced when parameter weighting is applied. Furthermore, the system is able to reconstruct the most parameters of the equaliser accurately (six for the unweighted parameters and five for the weighted parameters), while FA with SVR is the only combination able to accurately reconstruct five parameters for the weighted reconstruction. It achieves lower results for overall reconstruction accuracy (0.065), trustworthiness (0.7761), and classification (59.52%), and marginally lower for continuity (0.9359).

Whilst the parameter reconstruction of the autoencoder is sufficiently accurate for our application, it is bound by the intrinsic dimensionality of the data, defined as the minimum number of variables required to accurately represent the variance in lower dimensional space. For the *bright/warm* parameter-space data used in this experiment, we can show that the intrinsic dimensionality requires three variables when computed using Maximum Likelihood Estimation [47]. As our application requires a two-dimensional interface, this means the reconstruction accuracy is inherently limited.

Additionally, the user tests revealed that the two-dimensional slider using a sAE is able to accurately reconstruct the equaliser curve, retaining the characteristics associated with *warm* (boost on low-mid and cut on high-end) and *bright* (cut on low-end and boost on high-end), as displayed in Figure 7a,b. Participants of the experiment also commented that the underlying two-dimensional map is easy to quickly learn and provides an intuitive tool for controlling an audio equaliser. Taking into account that the final audio effect should be incorporated alongside the equaliser, with the high-dimensional parameters also available to the users, and with indications as to where the semantic regions are placed, it can be expected that the resulting effect will feature a quick way of achieving the different descriptors (using the two-dimensional slider) and a further fine-tuning stage (via changing the high-dimensional equaliser parameters) if that is necessary.

Providing the model training is applied offline, mapping techniques such as PCA, LDA, DM, pPCA, kPCA, and FA are all capable of running in real-time given the lower degree of computational complexity, as do reconstruction methods such as the interpolation techniques (LI, NaNI, NeNI) and the sAE. Similarly, while the sAE requires iterative training, which will have variable training times based on the number of iterations, the learning rate and the number of neurons and hidden layers, it still offers a fast implementation as the user-input process is relatively lightweight.

7. Conclusions

We have presented a model for the modulation of equalisation parameters using a two-dimensional control interface. The model utilises a sAE to modify the dimensionality of the input data and a weighting process that adapts the parameters to the LTAS of the input audio signal. We train the model with semantics data in order to get the appropriate decoder weights and bias units, which can then be applied to any new input data. This data is given by a user as the position of the cursor changes in an (x,y) Cartesian space. This new information will compute high-dimensional values, which will be rescaled and unweighted, and consequently passed to the equaliser parameters. We show that the sAE model achieves better reconstruction accuracy than other regression and interpolation techniques, achieving an error as low as 0.058. Similarly, the *trustworthiness* and *continuity* of the system perform similarly to (and in some cases outperform) the rest of the dimensionality reduction techniques. Through subjective testing, we can show that the 2D equaliser provides users with an intuitive tool to recreate the high-dimensional equaliser settings extracted from the original dataset. This is demonstrated by comparing the centroids taken from the high and low-dimensional maps and by comparing the equalisation curves when applied to *warm* and *bright* samples.

Acknowledgments: The work of the first author is supported by The Alexander S. Onassis Public Benefit Foundation.

Author Contributions: The work was done in close collaboration. Spyridon Stasis conducted the experiments, derived results and contributed to the manuscript. Ryan Stables defined the mathematical models and drafted sections of the manuscript. Jason Hockman co-developed the models and contributed to the manuscript.

Conflicts of Interest: The authors declare no conflict of interest.

References

1. Valimaki, V.; Reiss, J. All About Audio Equalization: Solutions and Frontiers. *Appl. Sci.*, unpublished work, 2016.
2. Bazil, E. *Sound Equalization Tips and Tricks*; PC Publishing, Norfolk, UK, 2009.
3. Sarkar, M.; Vercoe, B.; Yang, Y. Words that describe timbre: A study of auditory perception through language. In Proceedings of the 2007 Language and Music as Cognitive Systems Conference, Cambridge, UK, 11–13 May 2007; pp. 11–13.
4. Beauchamp, J.W. Synthesis by spectral amplitude and "Brightness" matching of analyzed musical instrument tones. *J. Audio Eng. Soc.* **1982**, *30*, 396–406.
5. Schubert, E.; Wolfe, J. Does timbral brightness scale with frequency and spectral centroid? *Acta Acust. United Acust.* **2006**, *92*, 820–825.
6. Marozeau, J.; de Cheveigné, A. The effect of fundamental frequency on the brightness dimension of timbre. *J. Acoust. Soc. Am.* **2007**, *121*, 383–387.
7. Grey, J.M. Multidimensional perceptual scaling of musical timbres. *J. Acoust. Soc. Am.* **1977**, *61*, 1270–1277.
8. Zacharakis, A.; Pastiadis, K.; Reiss, J.D.; Papadelis, G. Analysis of musical timbre semantics through metric and non-metric data reduction techniques. In Proceedings of the 12th International Conference on Music Perception and Cognition, Thessaloniki, Greece, 23–28 July 2012; pp. 1177–1182.
9. Brookes, T.; Williams, D. Perceptually-motivated audio morphing: Brightness. In Proceedings of the 122nd Convention of the Audio Engineering Society, Vienna, Austria, 5–8 May 2007.

10. Zacharakis, A.; Reiss, J. An additive synthesis technique for independent modification of the auditory perceptions of brightness and warmth. In Proceedings of the 130th Convention of the Audio Engineering Society, London, UK, 13–16 May 2011.

11. Hafezi, S.; Reiss, J.D. Autonomous multitrack equalization based on masking reduction. *J. Audio Eng. Soc.* **2015**, *63*, 312–323.

12. Perez-Gonzalez, E.; Reiss, J. Automatic equalization of multichannel audio using cross-adaptive methods. In Proceedings of the 127th Convention of the Audio Engineering Society, New York, USA, 9–12 October 2009.

13. Cartwright, M.; Pardo, B. Social-EQ: Crowdsourcing an equalization descriptor map. In Proceedings of the 14th ISMIR Conference, Curitiba, Brazil, 4–8 November 2013; pp. 395–400.

14. Mecklenburg, S.; Loviscach, J. SubjEQt: Controlling an equalizer through subjective terms. In Proceedings of the CHI-06, Montreal, QC, Canada, 22–27 April 2006; pp. 1109–1114.

15. Sabin, A.T.; Pardo, B. 2DEQ: An intuitive audio equalizer. In Proceedings of the 7th ACM Conference on Creativity and Cognition, Berkeley, CA, USA, 27–30 October 2009; pp. 435–436.

16. Bristow-Johnson, R. Cookbook formulae for audio EQ biquad filter coefficients. Available online: http://www.musicdsp.org/files/Audio-EQ-Cookbook.txt (accessed on 25 February 2016).

17. Verfaille, V.; Arfib, D. A-DAFx: Adaptive digital audio effects. In Proceedings of the COST G-6 Conference on Digital Audio Effects (DAFX-01), Limerick, Ireland, 6–8 December 2001.

18. Verfaille, V.; Zölzer, U.; Arfib, D. Adaptive digital audio effects (A-DAFx): A new class of sound transformations. *IEEE Trans. Audio Speech Lang. Process.* **2006**, *14*, 1817–1831.

19. Zölzer, U.; Amatriain, X.; Arfib, D. *DAFX: Digital Audio Effects*; Wiley Online Library: New York, NY, USA, 2011.

20. Stables, R.; Enderby, S.; de Man, B.; Fazekas, G.; Reiss, J.D. SAFE: A system for the extraction and retrieval of semantic audio descriptors. In Proceedings of the 15th ISMIR Conference, Taipei, Taiwan, 27–31 October 2014.

21. Semantic Audio: The SAFE Project. Available online: http://www.semanticaudio.co.uk/ (accessed on 20 January 2016).

22. Brookes, T.; Williams, D. Perceptually-motivated audio morphing: Warmth. In Proceedings of the 128th Convention of the Audio Engineering Society, London, UK, 22–25 May 2010.

23. Venna, J.; Kaski, S. Local multidimensional scaling with controlled tradeoff between trustworthiness and continuity. In Proceedings of 5th Workshop on Self-Organizing Maps, Paris, France, 5–8 September 2005; pp. 695–702.

24. Van der Maaten, L.J.P.; Postma, E.O.; van den Herik, H.J. Dimensionality reduction: A comparative review. *J. Mach. Learn. Res.* **2009**, *10*, 66–71.

25. Sanguinetti, G. Dimensionality reduction of clustered data sets. *IEEE Trans. Pattern Anal. Mach. Intell.* **2008**, *30*, 535–540.

26. Stasis, S.; Stables, R.; Hockman, J. A model for adaptive reduced-dimensionality equalisation. In Proceedings of the 18th International Conference on Digital Audio Effects, Trondheim, Norway, 30 November–3 December 2015.

27. Chaudhuri, K.; McGregor, A. Finding metric structure in information theoretic clustering. *COLT Citeseer* **2008**, *8*. Available online: https://people.cs.umass.edu/ mcgregor/papers/08-colt.pdf (accessed on 20 April 2016).

28. Johnson, D.; Sinanovic, S. Symmetrizing the kullback-leibler distance, Computer and Information Technology Institute, Department of Electrical and Computer Engineering, Rice University, Houston, Texas, USA, 2001. Available online: http://www.ece.rice.edu/~dhj/resistor.pdf (accessed on 10 March 2016).

29. Bruzzone, L.; Roli, F.; Serpico, S.B. An extension of the Jeffreys-Matusita distance to multiclass cases for feature selection. *IEEE Trans. Geosci. Remote Sens.* **1995**, *33*, 1318–1321.

30. Senior,M. Mixing secrets for the small studio additional resources. Available online: http://www.cambridge-mt.com/ms-mtk.htm (accessed on 20 January 2016).

31. Van der Maaten, L.J.P. An introduction to dimensionality reduction using Matlab. *Report* **2007**, *1201*. Available online: http://citeseerx.ist.psu.edu/viewdoc/download?doi=10.1.1.107.1327&rep=rep1&type=pdf (accessed on 20 April 2016).

32. Hotelling, H. Analysis of a complex of statistical variables into principal components. *J. Educ. Psychol.* **1933**, *24*, 417–441.

33. Schölkopf, B.; Smola, A.; Müller, K.R. Nonlinear component analysis as a kernel eigenvalue problem. *Neural Comput.* **1998**, *10*, 1299–1319.

34. Bilmes, J.A. A gentle tutorial of the EM algorithm and its application to parameter estimation for Gaussian mixture and hidden Markov models. *Int. Comput. Sci. Inst.* **1998**, *4*. Available online: http://lasa.epfl.ch/teaching/lectures/ML_Phd/Notes/GP-GMM.pdf (accessed on 20 April 2016).

35. Roweis, S. EM algorithms for PCA and SPCA. *Adv. Neural Inf. Process. Syst.* **1998**, 626–632.

36. Khosla, N. Dimensionality Reduction Using Factor Analysis. Ph.D. Thesis, Griffith University, Brisbane, Queensland, Australia, December 2004.

37. Nadler, B.; Lafon, S.; Coifman, R.R.; Kevrekidis, I.G. Diffusion maps, spectral clustering and reaction coordinates of dynamical systems. *Appl. Comput. Harmonic Anal.* **2006**, *21*, 113–127.

38. Fisher, R.A. The use of multiple measurements in taxonomic problems. *Ann. Eugen.* **1936**, *7*, 179–188.

39. Bobach, T.; Umlauf, G. Natural neighbor interpolation and order of continuity. University of Kaiserslautern, Computer Science Department/IRTG, Kaiserslautern, Germany, 2006. Available online: http://www-umlauf.informatik.uni-kl.de/~bobach/work/publications/dagstuhl06.pdf (accesed on 20 January 2016).

40. Drucker, H.; Burges, C.J.C.; Kaufman, L.; Smola, A.; Vapnik, V. Support vector regression machines. *Adv. Neural Inf. Process. Syst.* **1997**, *9*, 155–161.

41. Hinton, G.E.; Salakhutdinov, R.R. Reducing the dimensionality of data with neural networks. *Science* **2006**, *313*, 504–507.

42. Bengio, Y. Learning deep architectures for AI. *Found. Trends Mach. Learn.* **2009**, *2*, 1–127.

43. Bergstra, J.; Breuleux, O.; Bastien, F.; Lamblin, P.; Pascanu, R.; Desjardins, G.; Turian, J.; Warde-Farley, D.; Bengio, Y. Theano: A CPU and GPU math compiler in Python. In Proceedings of the 9th Python in Science Conference (SciPy), Austin, Texas, USA, 28 June–3 July 2010.

44. Vincent, P.; Larochelle, H.; Lajoie, I.; Bengio, Y.; Manzagol, P.A. Stacked denoising autoencoders: Learning useful representations in a deep network with a local denoising criterion. *J. Mach. Learn. Res.* **2010**, *11*, 3371–3408.

45. Tieleman, T.; Hinton, G. Lecture 6e - rmsprop: Divide the gradient by a running average of its recent magnitude, 2012. Available online: http://www.cs.toronto.edu/~tijmen/csc321/slides/lecture_slides_lec6.pdf (accessed on 20 January 2016).

46. Dauphin, Y.; de Vries, H.; Bengio, Y. Equilibrated adaptive learning rates for non-convex optimization. In Proceedings of Advances in Neural Information Processing Systems, Montreal, QC, Canada, 7–12 December 2015, pp. 1504–1512.

47. Levina, E.; Bickel, P.J. Maximum likelihood estimation of intrinsic dimension. In Proceeding of Advances in Neural Information Processing Systems, Vancouver, BC, Canada, 13–18 December 2004.

![applied sciences logo]

applied
sciences

MDPI

Article

Influence of the Quality of Consumer Headphones in the Perception of Spatial Audio

Pablo Gutierrez-Parera * and Jose J. Lopez

Institute of Telecommunications and Multimedia Applications (ITEAM), Universitat Politècnica de València,
Valencia 46022, Spain; jjlopez@dcom.upv.es
* Correspondence: pabgupa@iteam.upv.es; Tel.: +34-963-877-007

Academic Editor: Vesa Valimaki
Received: 29 February 2016; Accepted: 12 April 2016; Published: 22 April 2016

Abstract: High quality headphones can generate a realistic sound immersion reproducing binaural recordings. However, most people commonly use consumer headphones of inferior quality, as the ones provided with smartphones or music players. Factors, such as weak frequency response, distortion and the sensitivity disparity between the left and right transducers could be some of the degrading factors. In this work, we are studying how these factors affect spatial perception. To this purpose, a series or perceptual tests have been carried out with a virtual headphone listening test methodology. The first experiment focuses on the analysis of how the disparity of sensitivity between the two transducers affects the final result. The second test studies the influence of the frequency response relating quality and spatial impression. The third test analyzes the effects of distortion using a Volterra kernels scheme for the simulation of the distortion using convolutions. Finally, the fourth tries to relate the quality of the frequency response with the accuracy on azimuth localization. The conclusions of the experiments are: the disparity between both transducers can affect the localization of the source; the perception of quality and spatial impression has a high correlation; the distortion produced by the range of headphones tested at a fixed level does not affect the perception of binaural sound; and that some frequency bands have an important role in the front-back confusions.

Keywords: headphones; spatial sound; quality; perception; binaural; subjective test; distortion; frequency response; front-back confusion

PACS: 43.66.Pn, 43.66.Yw

1. Introduction

With the advent of high definition TV, 3D video and mobile devices, spatial audio technologies have gained great popularity in recent years. Speaker sets have evolved from the classic stereo systems into many channels, not only considering 2D configurations, but also height speakers. The variety of formats (from 5.1 to 22.2 and also headphone systems) and the reproduction techniques (Vector Base Amplitude Panning (VBAP) [1], Wave Field Synthesis (WFS) [2], Ambisonics, *etc.*) open up many possibilities for the recreation of acoustic environments and especially the creation of new musical experiences. Audio reproduction systems based on loudspeakers are the most popular, but the headphone-based systems are increasing in popularity because of the private hearing they provide in any type of environment, as well as the widespread use of mobile devices nowadays. Headphones are commonly employed to reproduce stereo recordings, but binaural material represents a step forward.

The reproduction of binaural sound over headphones uses the principles of the human auditory system [3]. It assumes that, if we are able to reproduce in the listener's ears with headphones the same pressures that the listener experiences in a natural environment, a realistic acoustic immersion can be simulated [4].

To have a correct sense of spatial immersion, high quality microphones should be employed in conjunction with acoustic mannequins. In addition, high quality headphones should be used for playback. However, low end headphones are widely used in most cases, either for economic reasons or simply because they are included with mobile devices. It is generally known that low cost headphones usually provide a poorer sense of immersion, but the degrading factors that cause such a loss in quality have not been sufficiently studied, as well as the level of their effects. In this research, we are laying the groundwork for a strategy to study the factors that affect the spatial sensation on listening with headphones and their relationship to perceived quality.

Hypothesis and Planning of the Study

Different factors can affect the perception of the spatial sound image. Our hypothesis states that three main factors are responsible for this degradation. Some of these degrading factors could be the frequency response, the distortion and the disparity between the left-right transducers, especially in low cost headphones. To determine this, we propose a series of perceptual tests [5] to particularly study these factors.

Section 2 describes the methodology, the headphones employed in the study, as well as the technique used to measure and simulate them. Sections 3–6 explain a series of perceptual tests that constitute the bulk of this research. Firstly, Section 3 presents a perceptual test carried out to study the influence of the sensitivity disparity between left and right transducers and to establish the degree to which perception of the sound source position in the azimuth is affected. Although in high quality headphones, manufacturers match transducers with similar sensibilities, these low cost headphones have different sensibilities due to broader manufacturing tolerances. Another second subjective perceptual test described in Section 4 was conducted to evaluate the effect of the frequency response in the perception of quality and spatial impression with headphones. As frequency response is the factor that varies most among different headphones due to their quality, this test is of particular interest to better understand how frequency response affects the spatial sound impression. Section 5 outlines the third perceptual test planned to evaluate the effect of harmonic distortion in listening with headphones. Distortion can be considerable if high dynamic sound and high reproduction levels are employed. Section 6 explains the fourth and last test, which studies the relation of the frequency response with the accuracy of localization in the horizontal plane. The capacity of a headphone to generate a good spatial immersion can be different from its capacity to generate precise locations. To explore this point, azimuth localization is tested here for different kinds of headphones. The discussion and conclusions of these experiments are presented in Section 7.

2. Headphones Measurements and Virtual Headphone Simulation

It is well known in loudspeaker testing that visual cues play an undesirable role in the results provided by test subjects. Similarly, when testing headphones, tactile cues can also influence results. Consequently, it can be challenging to conduct a double-blind comparative listening test for headphones. It is difficult to hide the possible influencing variables, such as brand, design or price. In addition, the manual substitution of different headphones on the subject's head can be disruptive and introduces useless fatigue on the subject [6]. Moreover, the fitting and tactile sensations are impossible to remove, making them an important bias factor [7].

In order to avoid these effects, it is appropriate to use a virtual headphone simulation to perform the listening tests [8,9]. This method employs one reference headphone to simulate the different headphones under test. In this way, listeners can evaluate the simulated versions of the different headphones wearing just the reference headphone, therefore avoiding the manual change of headphones and removing the visual and tactile biases. Some other advantages are obtained with this virtual method: listeners can have immediate access to the different headphones, and the procedure test becomes more flexible, transparent, controlled and repeatable.

The reliability of this virtual simulation method has been previously studied, finding good correlation between standard listening tests using real headphones and the virtual simulation method. However, in some cases, some discrepancy related to a specific model or sound signal [8] has been found due to the visual and tactile bias present in the standard test [10].

Due to the great advantages of a virtual test over a standard one, this study used a virtual headphone listening test methodology. This will remove the strong bias that would appear in this study due to the great difference in appearance and fitting characteristics among the consumer headphones and high quality ones used in this test.

2.1. Headphone Selection

Different headphones were selected in order to represent a range of commercial and readily-available headphones. According to this principle and the scope of the study described in previous sections, seven different headphones were selected plus a high quality reference one. A Sennheiser HD800 (Sennheiser, Wedemark, Germany) was chosen as the reference headphone (REF). The reason for this selection is due to its great fidelity, response, low distortion and accurate timbral reproduction. The other seven headphones were selected to cover a wide range of possible common uses. The brands and models of the rest of the headphones will be omitted, as they are not necessary for the result analysis.

The headphones used in the study were classified as:

(a) REF, Reference, Sennheiser HD800 headphone (open and circumaural)
(b) HQop, High Quality open headphone (circumaural)
(c) MQcl, Medium Quality closed headphone (circumaural)
(d) BDso, Big Diaphragm semi-open headphone (circumaural)
(e) LCmul, Low Cost multimedia headphone (supra-aural)
(f) AirL, Airline headphone (supra-concha)
(g) Woh, Wireless open headphone (circumaural)
(h) LCmul2, Low Cost multimedia headphone 2 (supra-aural)

The reference headphone was the only one that participants used, saw and had contact with during the tests. The rest of the headphones were simulated through the reference one. Then, all of the participants performed the test using the same high quality reference headphone (REF, Sennheiser HD800). The resulting signals for the rest of the headphones (used in Tests 2, 3 and 4 and described in their sections) were simulated by means of proper signal processing algorithms and heard through the reference headphone.

2.2. Frequency Responses Measures

To measure the response of the different headphones, a swept-sine method was employed [11] using a Head and Torso Simulator (HATS) Model B & K Type 4100 (Brüel & Kjær, Nærum, Denmark) (Figure 1). This technique gave us both the frequency response, as well as the first and second distortion harmonics needed for the simulation of the different headphones.

Figure 1. Set-up for measuring the headphones with the Head and Torso Simulator (HATS).

To avoid differences in the amplitude level of the measures, the selected criterion was to achieve the same equivalent power between 100 Hz to 10 kHz for all of the headphones (for calibration, we employed band pass pink noise between 100 Hz to 10 kHz instead of 20 Hz to 20 kHz in order to minimize the influence of roll-off in low and high frequencies in low quality headphones). This decision allowed us to measure all of the headphones in the same reproduction conditions and to achieve the same reproduction level in this band of frequencies. The reproduced pressure level for all of the headphones was equivalent to 69 Sound Pressure Level dB (dBSPL) of pink noise in the reference headphones. This level was selected in informal tests as a pleasant listening level. Besides, this level allowed the measurement of the different headphone models without any saturation distortion in equivalent conditions.

Each of the headphones, including the reference one, were measured with the mentioned swept-sine method. The resulting impulse responses ($h_i[n]$) were truncated to 50 ms (2205 samples for a 44,100-Hz sampling frequency) and windowed with a half Hamming window. This length provides good resolution in low frequencies until 20 Hz. To minimize errors related to headphone positioning on the ear of the HATS simulator, five resets of the headphones were done and measured. The curves shown in Figure 2 are based on the average of those measures.

The first curve corresponds to the reference headphone (a)-REF, which shows a smooth response and flat below 3 kHz. The next three, (b)-HQop, (c)-MQcl, (d)-BDso headphones, were chosen as good mid-quality range with different characteristics: open, closed and semi-open. Their frequency responses below 6 kHz are quite flat, with the exception of some irregularities in the (c)-MQcl curve and a peak down at 4.5 kHz that decreases to −14 dB. There is another peak up in the curve (d)-BDso at 6 kHz of 15 dB. The next curves (e to h) represent the frequency responses of the multimedia (e)-LCmul, airline (f)-AirL, wireless (g)-Woh and another multimedia (h)-LCmul2 headphones, that were chosen to be an example of mid- and poor quality headphones. Their frequency responses have important peaks and valleys that affect the sound. Curve (e)-LCmul has a reinforcement in frequencies around 1.5 kHz and a big dip in 3.5 kHz, and curve (f)-AirL has a strong peak in 140 Hz, as well as other distortions up to 4.5 kHz. Curve (g)-Woh is flatter in the mid frequencies with a small reinforcement in 1.5 kHz and a decay around 4.5 kHz. In the case of curve (h)-LCmul2, it is important to note the rapid decline above 3 kHz and the lack of proper high frequency beyond 5 kHz. All of these headphones are intended to be a small representation of quality range in commercial headphones.

2.3. Headphones Frequency Response Simulation

The seven headphones under study were simulated to be reproduced with the reference headphones ((a)-REF-Sennheiser HD800). The simulation of each headphone was done by filtering with its frequency response, but compensating the effect of the reference headphone using its inverted frequency response. Equation (1) shows the process for the simulation, where $H_i(\omega)$ is the measured response of the headphone to simulate, $H_{HD800}(\omega)$ is the measured response of the reference headphone and $H_{i\,corrected}(\omega)$ is the response of the simulated headphone, which is applied to the corresponding stimulus.

$$H_{i\,corrected}(\omega) = \frac{H_i(\omega)}{H_{HD800}(\omega)} \tag{1}$$

These virtual headphone equalizations include not only the magnitude response, but also the phase of the headphone measured. Although it is generally accepted that phase does not seem to affect the perceived accuracy of the simulations [12], especially if the stimuli material is a typical music program, it can be noticed with pink noise stimuli. All of the impulse responses of the headphones measured, the correction of the reference headphone and its application convolving with the stimulus, respect and keep the original phases. Moreover, accurate phase processing guaranties that our filtering will not alter in any way the Interaural Time Difference (ITD) between left and right transducers.

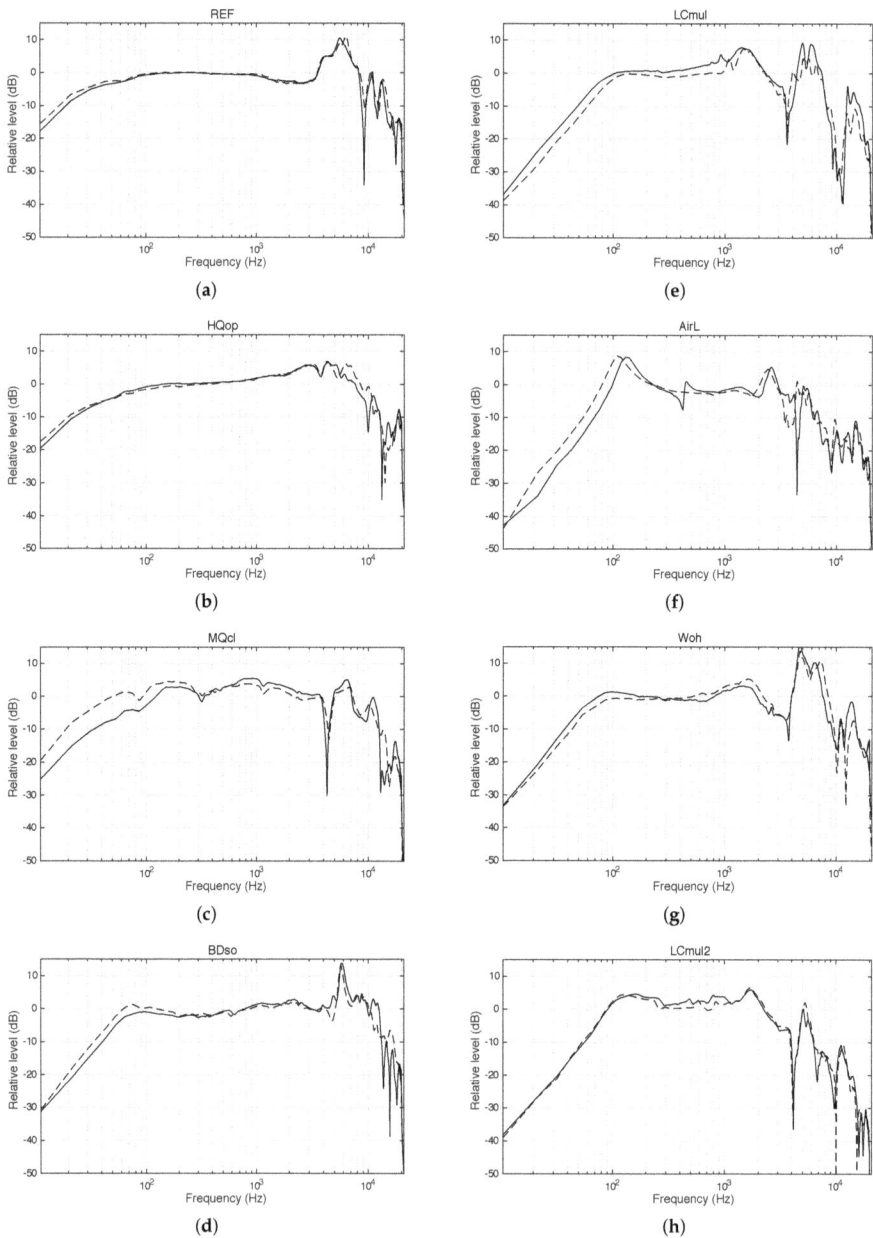

Figure 2. Frequency response of the headphones used in the study. Solid curve, left channel; dashed curve, right channel. (**a**) REF, Reference headphone; (**b–h**) headphones under study. (**b**) HQop, High Quality open headphone; (**c**) MQcl, Medium Quality closed headphone; (**d**) BDso, Big Diaphragm semi-open headphone; (**e**) LCmul, Low Cost multimedia headphone; (**f**) AirL, Airline headphone; (**g**) Woh, Wireless open headphone; (**h**) LCmul2, Low Cost multimedia headphone 2.

The filter implementation of Equation (1) was carried out in MATLAB (Matrix Laboratory, R2015a, MathWorks Inc., Natick, MA, USA, 2015) in the time domain, using Equation (2); where $h_{i\,corrected}[n]$ is the response for the simulation of the virtual headphone, $h_i[n]$ is the impulse response of the headphone to simulate and $h_{HD800}^l[n]$ is the inverted impulse response of the reference headphone.

$$h_{i\,corrected}[n] = h_i[n] * h_{HD800}^l[n] \tag{2}$$

To obtain $h_{HD800}^l[n]$, we firstly recorded the impulse response of the reference headphone h_{HD800} with 2205 sample points (50 ms, fs = 44,100 Hz). Secondly, the Fast Fourier Transform (FFT) of the response was computed, with zero padding up to a size of 4096, which guaranties a spectral resolution of 10 Hz. This is low enough to see details of the frequency response. Thirdly, the resulting FFT was inverted, taking into account a boost limitation of +15 dB. This limitation was included to avoid an excess of boost at a couple of very narrow notches of the h_{HD800} response (see Figure 2a), assuring that final signals are inside the reproducible dynamic margin and free from artifacts. Lastly, the inverted and limited response was then used to properly compute the inverse FFT and next Hamming windowed to obtain the $h_{HD800}^l[n]$. This process guaranties the avoidance of undesirable effects, such as circular convolution or others.

Finally, the different headphones were simulated applying the simulation filter $h_{i\,corrected}[n]$ to the sound materials for each test, obtaining the different stimuli. This was the procedure used for Tests 2 (Section 4) and 4 (Section 6).

2.4. Non-Linear Distortion Simulations

As commented on before, the swept-sine method employed to measure the frequency response of the headphones provides, apart from the frequency response, distortion harmonics simultaneously. Figure 3 shows the frequency response and the second and third distortion harmonic of the reference ((a)-REF) and the airline ((b)-AirL) headphones. Both of these headphones are a good example of low (a) and high distortion (b).

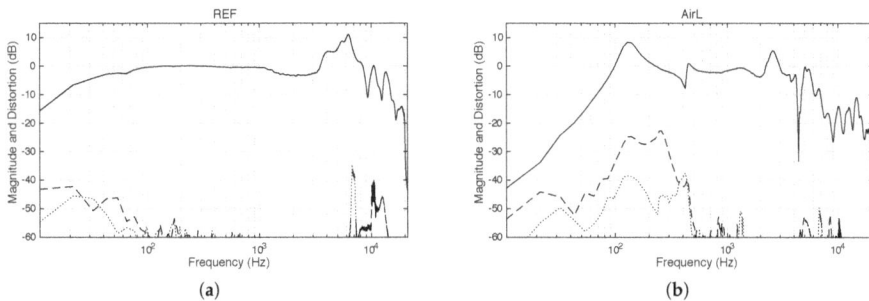

Figure 3. Frequency response with distortion of two headphones (left channel). Solid curve, magnitude; dashed curve, second order distortion harmonic; dotted curve, third order distortion harmonic. (**a**) REF, Reference headphone; (**b**) AirL, Airline headphone.

To simulate the non-linear distortion of each headphone, the method described in [13], which uses Volterra kernels and a series of linear convolutions, was chosen. With this method, the transfer function of a system is described by means of a Volterra series expansion. The output signal can be represented as the sum of the linear convolution of the measured impulse responses with the input signal and the corresponding frequency-shifted version. Applying Fourier transforms to these series results in a linear equation system. The solution of this system allows the computation of the diagonal Volterra kernels obtaining the impulse response terms for the main response and the first two distortion orders; Equation (3).

$$\begin{cases} H_1 = H_1' + H_3' \\ H_2 = -2\hat{H}_2' \\ H_3 = -4H_3' \end{cases} \tag{3}$$

where H_1', H_2', H_3' are the measured frequency responses and H_1, H_2, H_3 are the Volterra kernels (^ represents the Hilbert transform).

Using these equations, the second and third distortion orders were simulated by convolution, applying them to Equation (4), where $x(n)$ is the input signal and M is the number of samples of the kernel:

$$y(n) = \sum_{i=0}^{M-1} h_1(i) \cdot x(n-i) + \sum_{i=0}^{M-1} h_2(i) \cdot x^2(n-i) + \sum_{i=0}^{M-1} h_3(i) \cdot x^3(n-i) \tag{4}$$

More details of this technique can be found in [13]. This procedure was followed for Test 3 (Section 5).

2.5. Binaural Room Impulse Responses Measurements

In order to generate the spatiality of sound sources, some Binaural Room Impulse Responses (BRIR) [14] were measured with a HATS B & K Type 4100.

Reverberation is an influential factor for spatial localization [3,15], and because of this, we decided to record our own BRIR with natural reverberation instead of using dry responses from a library. The impulse responses were recorded in a rectangular room with a volume of 132 m³ and a reverberation time of about 0.7 s. Nine different azimuth angles were recorded (0°, 30°, 60°, 90°, 135°, 225°, 270°, 300°, 330°) in the horizontal plane at 1.5 m of distance.

These measures were used to simulate binaural sound source positions in Test 4 (Section 6).

3. Test 1. Sensitivity Disparity between Left-Right Transducers

3.1. Test Description

The idea of this test is to evaluate how sensitivity disparity between the left and right transducers affects the perception of the source azimuth. To do that, a subjective perceptual test was carried out applying some volume level variations to different binaural sounds and checking how this affects the accuracy of horizontal localization.

In this test, participants had to listen, wearing headphones, to some binaural recordings obtained with a HATS on specific angles in the horizontal plane. Different variations of the original level between left and right transducers were applied to these sounds and then presented to the listeners. Participants should then indicate the direction of arrival, marking the angle in a Graphical User Interface (GUI).

The volume level variations applied were 0 (no modification), 1, 2 or 4 dB more on the left channel than the right one. Four different angles of direction of arrival were chosen, $-30°$, $0°$, $65°$ and $90°$ of azimuth in the horizontal plane. Besides, the influence of different types of sounds was also studied.

These sounds were specifically recorded for this test using a binaural mannequin (B & K Model 4100) at the specific angles under study. A 44,100-Hz sampling frequency was employed, obtaining full audio band recordings. The mannequin was in a semianecoic room, and sources were placed around it at 1 m apart. Four different sounds were recorded: a timbal drum hit, voice, a whistle and pink noise. The impulsivity of the timbal hit is an interesting characteristic regarding sound localization, also interesting for its low frequency content. Both voice and whistle are easily recognizable common sounds, which make them useful for the test. Moreover, the reduced spectral content of the whistle can be an interesting feature that can affect the test. The voice signal was the syllables "ba-be-bi-bo-bu", pronounced by a male voice. This sound has diverse vocalic contents and bilabial consonantal phoneme /b/, which produces impulsive sound. Pink noise was employed to evaluate a wide spectrum signal. All of these sounds were reproduced by the Sennheiser HD800 reference headphones.

According to the different types of sounds described above, the total number of stimuli presented to each participant in this test was: 4 angles × 4 types of sounds × 4 level variations = 64 stimuli. These stimuli were randomly presented, and the participant could listen to each of them as many times as he or she wanted.

During the test, participants also had the possibility of hearing a reference stimulus at any time, choosing between −90°, −45°, 0°, 45° and 90° of azimuth.

To perform the test, a simple Graphical User Interface (GUI) was developed in MATLAB that brings the user full control of the test. The participant could select the perceived sound source direction angle in an arc of −90° to 90° of azimuth (with a 5° resolution). It was also possible for the subject to freely control and listen to the reference stimulus.

The test was performed by 20 people, 10 men and 10 women (21 to 45 years, with an average age of 32). The average runtime of the test was 9 min. Every participant did a training session before taking the test, so all could listen to all of the stimuli and become familiar with the GUI and the assigned task. Some preliminary results of this test were previously published by the authors in [16].

3.2. Results

Figure 4a shows the average of the answered angles (for all of the level variation cases) according to the reproduced angle. The average of the answers has a deviation to the left-hand side. This is expected since the variations (0, 1, 2, 4 dB) were always more in favor of the left channel than the right.

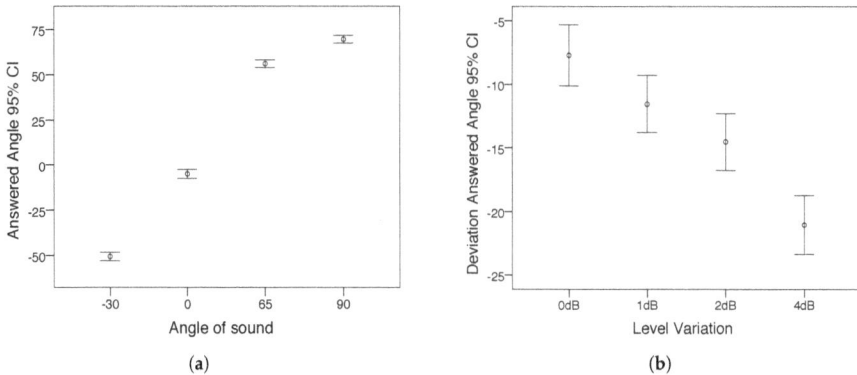

Figure 4. (a) Average of the answered angles *versus* reproduced angles (degrees); (b) average of the deviation of the answered angles (degrees) *versus* level variation (dB).

The tendency of this angle deviation to the left can be seen in Figure 4b, considering the level variation applied (0, 1, 2, 4 dB).

An Analysis of Variance (ANOVA) indicates that the level variation has a very significant influence ($F = 27.338$, df = 3, $p < 0.001$) over the deviation in the answers.

If we consider just the central angles used in the experiment (0° and 65°), a smaller average deviation can be seen (Figure 5). This leads us to believe that listeners tended to divert the location of the sounds perceived on the sides more, which means that the introduced level variations made the lateral angles disperse more than the central ones.

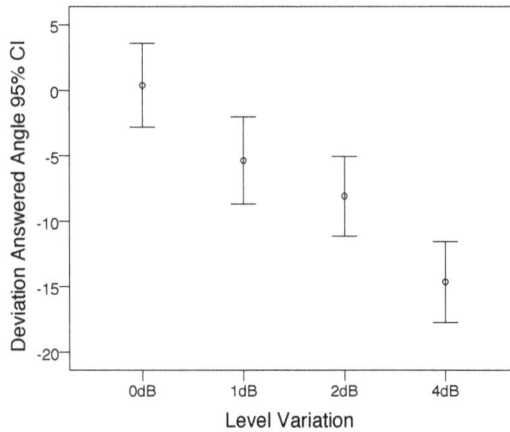

Figure 5. Average deviation of the answered angles (degrees) *versus* level variation (dB), considering only the angles 0° and 65°.

On the other hand, the influence of the type of sound (timbal, voice, whistle or pink noise) on the deviation in responses can be seen in Figure 6a. Voice and pink noise have lower deviation than timbal and whistle sounds, especially in cases of 0 and 1 dB of deviation. Besides, voice stimuli and pink noise manifest a more separate and clear deviation at varying levels.

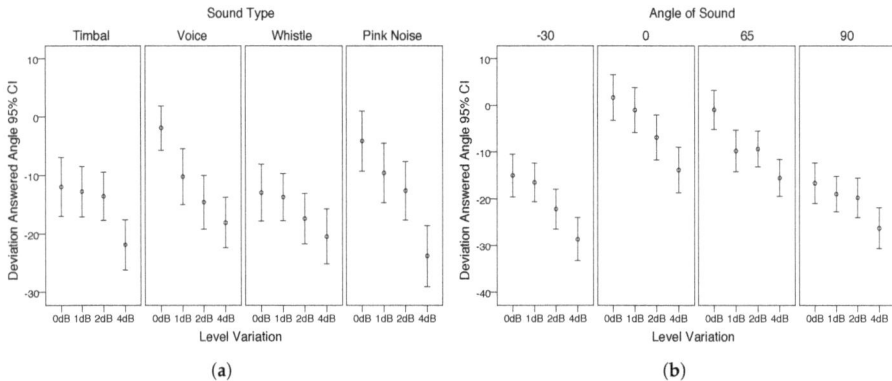

Figure 6. Average deviation of the answered angles (degrees) *versus* the level variation (dB): (a) considering the type of sound; (b) considering the angle reproduction of sound.

The influence of the type of sound over the deviation of answers is significant ($F = 4.409$, df = 3, $p = 0.004$) according to an analysis of variance. The sound angle reproduction has a very significant influence ($F = 54.932$, df = 3, $p < 0.001$) over the deviation of the answers. In Figure 6b, the deviation of the answers for each sound angle reproduction is represented. Angles 0° and 65° present less deviation to the left. The biggest deviation of the answers corresponds to the angle −30°, and it could be due to the fact that it was the only angle on the left side.

4. Test 2. Frequency Response about Quality and Spatial Impressions

4.1. Test Description

In this test, participants listened to some excerpts of sound with headphones and rated their quality and their sound spatial image. These different headphones were simulated as described in Section 2.3 by means of the convolution of their frequency responses with the stimuli sounds, and all of them were reproduced with the reference headphones.

Due to the fact that different frequency responses produce noticeable effects, the perceptual test was designed according to the recommendation International Telecommunication Union, recommendation by Radiocommunication sector (ITU-R) 1534-2 [17], which describes the MUltiple Stimuli with Hidden Reference and Anchor (MUSHRA) perceptual test. This kind of test describes a method to assess intermediate quality audio systems and also all of the requirements needed to accomplish the test with rigor. Besides, this test sets a zero to 100 continuous scale (zero–bad; 100–excellent) to evaluate quality and other parameters of sounds and systems, always using a reference sound. All systems are compared to a reference of maximum quality, and the different systems are also compared between them.

Two different tasks were evaluated during the test by the participants. The first task was to indicate the quality of the sound with respect to the reference. The second task was to evaluate the spatial impression (locations, sensations of depth, immersion, reality of the audio event) [18] with respect to the reference.

Five different excerpts of audio (12 to 14 s) were employed as source material (see Table 1), and all of them were reproduced simulating the different headphones under study. All of these sound fragments were chosen by their spatial, stereophonic and timbral attributes.

Table 1. Music program used for listening Tests 2 and 3.

Artist	Track	CD	Description
Bettina Flater	*Haugebonden*	Women en Mi	female voice and guitar
Paco de Lucía	*Zambra Gitana*	Canción Andaluza	male voice and guitar
Jerry Glodsmith	*Night Boarders*	OST The Mummy	high dynamic orchestral
The Chad Fisher Group	*Basin Street Blues*	live	jazz (binaural)
Smashing Pumpkins	audience sound	live	audience and drums (binaural)

In this test, five headphones simulations were done, corresponding to headphones (b)-HQop, (c)-MQcl, (d)-BDso, (e)-LCmul and (f)-AirL (described in Section 2.1, with frequency responses in Figure 2). Each of the five sound excerpts previously mentioned were reproduced by the virtual headphone simulation described in Section 2.3. A virtual headphone simulation for each sound was presented randomly in series to the listeners, as well as a hidden reference ((a)-REF) and also two anchor signals. The first Anchor signal (ANC1) was a 7-kHz low pass filtered version of the sound (according to the mid-quality anchor of the ITU recommendation 1534-2 [17]), and the second Anchor signal (ANC2) was a monaural version of the sound. This second anchor was determined to set a reference for the spatial impression question.

To perform the test, a GUI was developed in MATLAB according to the recommendation [17], which allowed participants to freely listen to each of the sounds and to the reference, as many times as they wanted. The different sound fragments were presented randomly as a series with all of the different headphone simulations, to compare to the reference sound. Once the participant had scored all of the simulations of a series, a new sound excerpt was presented to be evaluated. This process was repeated twice, once for each question of the test (the first about quality and the second about spatial impression), with a pause in between.

The number of stimuli of this test was: (5 headphones simulations + 1 hidden reference + 2 anchor signals) × 5 sound excerpts = 40 stimuli, presented in five series of eight stimuli plus the reference. As commented before, these 40 stimuli were presented twice in a different random order, to answer the two different questions.

The test was performed by 11 people, seven men and four women (21 to 37 years, with an average age of 30). As the test had two different questions, they were separated into two parts with a rest pause in the middle. The average runtime of the test was 22 min for the first part and 16 min for the second. Every participant did a training session before preforming the actual test, so all of them could listen to all of the stimuli and become familiar with the GUI and the assigned tasks.

4.2. Results

Figure 7a shows the average of the normalized (zero to 100) quality answers for the hidden reference, all five headphones simulated and the two anchors. As shown, the reference has been properly identified in most cases. The three supposedly good quality headphones have high scores; meanwhile, the two supposedly poor quality ones have the lowest scores. Both anchors remain in the middle of the scores of these two groups.

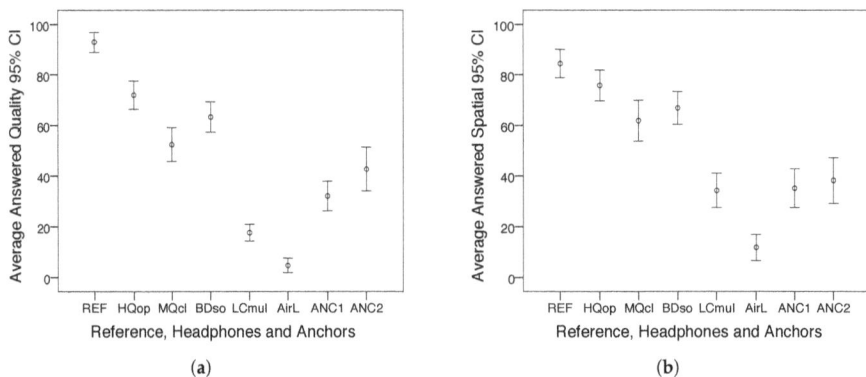

Figure 7. (a) Average answered quality *versus* reference, headphones and anchors (minutes); (b) average answered spatial impression *versus* reference, headphones and anchors.

An analysis of variance confirms that the headphones have a very significant influence ($F = 58.33$, df = 7, $p < 0.001$) over the quality perceived.

Figure 7b shows the average of the normalized (zero to 100) spatial impression answers for the hidden reference, the five headphones simulated and the two anchors. The results seem to be similar to the answers about quality, with a high correlation of $r^2 = 0.648$. Nevertheless, in this case, the confidence intervals are a bit wider, and the scores have some differences. The three supposedly good quality headphones have high scores again, but the confidence intervals do not separate them very much. There is a bigger difference between the two supposedly poor quality headphones, and the low cost multimedia ((e)-LCmul) ones are in the same range as both anchor signals. It is also noticeable that the Anchor Signal 2 (ANC2) as a monaural signal does not have a lower score.

In any case, an ANOVA confirms that the headphones have a very significant influence ($F = 58.33$, df = 7, $p < 0.001$) over the perceived spatial impression. No significant influence of the type of sound has been detected, even though some of them were binaural recordings.

5. Test 3. Non-Linear Distortion

5.1. Test Description

The objective of this test is to evaluate how the effect of harmonic distortion in headphones affects the spatial impression.

Several stimuli with and without the simulation of their harmonic distortion were presented to the participants that had to score their perception.

The effect of these distortions is very subtle. For that reason, the perceptual test was designed according to the recommendation ITU-R 1116-2 [19], which describes a method to assess small impairments in audio systems. This recommendation also establishes rigorous requirements of room, equipment and other arrangements. A continuous scale from one to five (1 – very annoying; 5 – imperceptible) is used to evaluate degradations with respect to a reference signal. The recommendation proposes an ABC test in which two stimuli, A and B, are presented to be compared against a known reference. One of these two stimuli, A or B, is always a hidden reference, and the other a degraded signal.

One single question was presented to the participants: "What degradation of quality and spatial impression do you hear with respect to the reference?"

The same five audio excerpts previously described in Test 2 were used here (see Table 1), as well as the same five virtual headphone simulations (b)-HQop, (c)-MQcl, (d)-BDso, (e)-LCmul and (f)-AirL (described in Section 2.1, with frequency responses in Figure 2). No anchors beyond the proposed scale were used this time.

Two different versions of the headphones simulations were presented in this test. One without and the other with the distortion simulated with the method described in Section 2.4. These two versions of the same stimulus were presented each time to the participants. They have then to rate the distorted against the not distorted version of the same sound in a double-blind manner (A *vs.* B). In each trial, there was always a non-distorted version sound that acted as the known reference (C sound), which according to the recommendation [19] has to be compared to the A and B sounds.

The number of stimuli of this test was then: 5 headphones simulations × 2 versions (with and without distortion) × 5 sound excerpts = 50 stimuli, presented in twenty five series of two stimuli plus the reference. All of these pairs were presented randomly to each participant.

To perform the test, a GUI was developed according to the recommendation, which allowed participants to freely listen to each of the sounds to evaluate and the reference, as many times as they wanted.

The five headphones under study were simulated (including distortion) to be reproduced with the reference headphones ((a)-REF, frequency response in Figure 2).

This test was performed by the same 11 people of the previous Test 2; seven men and four women (21 to 37 years, with an average age of 30). The average runtime of the test was 16 min. Every participant did a training session before preforming this test, so all of them could listen to all of the stimuli and become familiar with the GUI and the assigned task.

5.2. Results

According to the recommendation [19], the difference between the score of the hidden reference and the score of the degraded signal is analyzed. Figure 8 shows these differences for each of the headphones simulated.

No significance has been found. Then, distortion can be considered as imperceptible. Therefore, it has no effect in spatial perception, at least with the fixed level used to simulate all headphones (69 dBSPL).

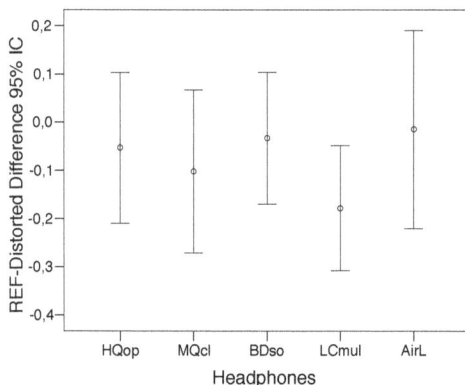

Figure 8. Difference between hidden reference and distorted signals *versus* headphones.

6. Test 4. Frequency Response about Azimuth Localization

6.1. Test Description

The results obtained in Test 2 are significant, but do not provide information about the accuracy in the localization of sources. For that reason, a test to evaluate the influence of frequency response on this accuracy was carried out.

Attempts to describe different spatial attributes have been a constant pursuit in the field of spatial audio [18,20,21]. The diffuse term employed in Test 2 to ask about spatial characteristics (spatial impression) was intended to relate in a simple way the perception of quality with the feeling of spaciousness. A more specific study of spatial attributes is then necessary to better evaluate the performing of the different headphones. In this direction, the localization accuracy in azimuth is one of the most studied spatial attributes [22–25] and therefore a good anchor point to contrast the previous Test 2 with a localization experiment. Therefore, this test tries to establish a relation of the influence of the frequency response on the azimuth localization in the horizontal plane.

As commented on in Section 2.5, to simulate the position of the sound sources in the horizontal plane, recordings of BRIRs in a medium-sized room were done. Nine different azimuth angles, 0°, 30°, 60°, 90°, 135°, 225°, 270°, 300° and 330°, were used.

Four types of sound were employed: door, voice (female), guitar and pink noise. A closing door is an impulsive sound with quite low frequency content, which can be useful for sound localization. The guitar sound was composed by various impulsive sounds in different main frequencies, one for each chord. Voice is an easily-recognizable common sound, and female was chosen to have some energy in high frequencies. The words "*estímulo sonoro*" (*sound stimulus* in Spanish) were employed. They present the repeated fricative phoneme /s/ with high frequency content and the phoneme /t/, a occlusive articulation that generates impulsive sound. Pink noise was employed to evaluate a wide spectrum signal.

For this test, seven different headphones plus a hidden reference were simulated (Section 2.1). Besides these, an additional anchor auralization (low pass filtered (LPF) sounds at 7 kHz) for each angle was employed (ANC1).

Therefore, the number of stimuli in this test was: 9 angles × 4 types of sound × (7 headphones simulation + 1 hidden reference + 1 anchor auralization) = 324 stimuli. These stimuli were presented in random order in two parts of 162 stimuli, with a rest in between.

To perform the test, a GUI was developed in MATLAB, which allowed participants to freely listen to the stimuli from a random list as many times as they wanted. Participants should indicate the perceived angle of the sound source. The GUI consists of a circle of points, which represents the top view of the listener, with a 5° resolution. Additionally, it included a parallel control to freely listen to a reference sound (pink noise) in the angles of 0, 45, 90, 135, 180, 225, 270 and 315 degrees.

The test was performed by 16 people, 10 men and 6 women (21 to 36 years, average age of 30). The average runtime was of 21 and 17 min for each part.

6.2. Results

A Cronbach's alpha analysis over the answers has been performed giving a value of $\alpha = 0.982$, which shows a high internal consistency.

A one-way ANOVA showed a significant influence between the headphones and the deviation of the answered angle (deviation = answered angle–real angle) ($F = 2.399$; df = 8; $p = 0.014$).

A first exploration of the participants' answers reveals that several front-back confusions [26,27] occur. For this reason, an evaluation of the amount of front-back confusions was performed for each of the headphones simulated. An ANOVA showed that there is a very significant influence of the type of headphones on the number of front-back confusions ($F = 46.307$; df = 8; $p < 0.001$). In Figure 9, we can see that headphones (f)-AirL and (h)-LCmul2 produce an average of nearly 50% of front-back confusions. This can be logical, as both headphones are supposed to be in the low quality range. However, the (c)-MQcl headphone stands out in the group of high quality ones, as it has 30.2% of front-back confusions, more confusions than the (e)-LCmul headphone, with a significant difference. A comparison of the frequency response of the headphones that produce more front-back confusions ((f)-AirL, (h)-LCmul2 and (c)-MQcl) reveals that they share in common strong irregularities in the band of 100 to 1600 Hz. On the other side, other headphones of medium and low quality ranges that have less front-back confusions do not present these strong irregularities in that four-octave band. Because of that, we suspect this can be an affecting factor disturbing the front-back discrimination.

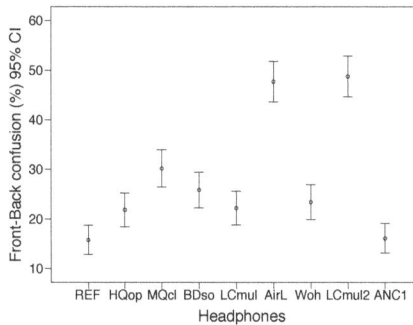

Figure 9. Percentage of front-back confusions for the reference, headphones and the anchor.

There is no significant influence of the type of sound crossed with the headphones. The sound guitar is the only one that produces slightly less front-back confusions for all of the headphones.

Due to the strong front-back confusion, the analysis of the deviation of the perceived sound with respect to the reproduced sound will produce large angle errors with complicated analysis of the results. A front-back confusion produces a bigger error for sources in the median plane than lateral sources, avoiding an analysis of the deviation angle (perceived angle–reproduced angle) with respect to the source position.

To overcome this setback, we propose a modified analysis of the error consisting of a preprocessing of the listener responses based on reflecting to the correct semi-plane the ones that have front-back confusion, leaving untouched the ones that do not. This correction eliminates big jumps in the deviation, focusing the experiment in the performance analysis of the headphones reproducing correctly the main spatial cues as ITD and the low frequency part of Interaural Level Difference (ILD). The high frequency part is more related to the pinna effect that is not considered with the reflection applied.

Taking into account the strong front-back confusion, the analysis of the answer deviation from the reproduction angle of the sound was performed introducing the correction of the front-back confusion. Therefore, a symmetric image of the responses in the back (90° to 270°) is brought to the front.

Figure 10 shows the deviation angle of the answers for the reproduction angle of the sounds, both of them front-back corrected. We can see that the deviations are quite uniform across the different headphones, except for the angles 90° and 270° in the cases of (f)-AirL and (h)-LCmul2. Looking at Figure 2, it is easy to see that the frequency responses of these two headphones present irregularities and deep level drops between 4 and 7 kHz. It is noticeable that the anchor LPF 7-kHz sounds auralized in the different angles (ANC1) are not affected by this problem, supporting the suspicion that the commented band is important for sources located in lateral positions.

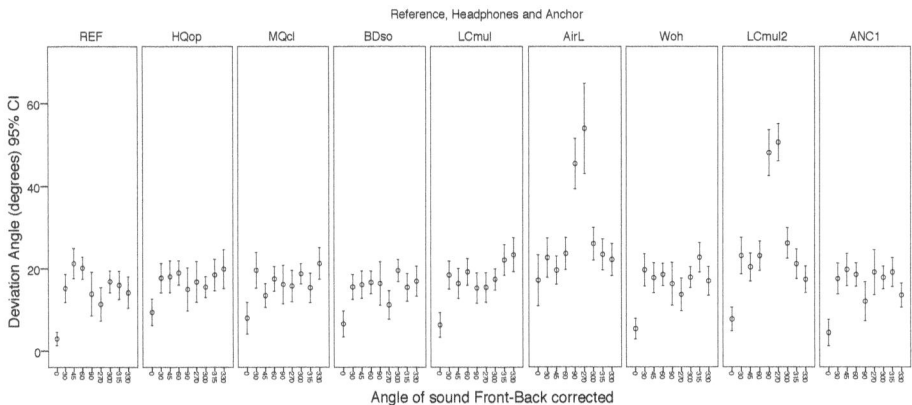

Figure 10. Deviation in degrees of the answers for every reproduced angle of sound. The reference, headphones under testing and anchor are represented.

7. Conclusions

This study outlines the influence of different quality parameters in headphones in the context of spatial sound reproduction. Four different perceptual tests have been done to analyze: (1) the effects of the sensitivity disparity between the transducers; (2) the influence of the frequency response over the perception of quality and the spatial impression; (3) the effects of non-linear distortion; and (4) the influence of the frequency response over azimuth localization.

The following main conclusions can be drawn:

1. The sensitivity disparities between left and right transducers affect the localization of sound sources, starting from level differences of 1 dB.
2. The quality and uniformity of the frequency response have an important influence in the *spatial impression*.
3. Additionally, the *spatial impression* has a high correlation with the subjective *perceived quality*.
4. The binaural recordings do not obtain significant better results for the parameter *spatial impression* compared to two-channel stereo mixes.
5. The distortion introduced by consumer level low quality headphones does not affect the perception of the spatial sound image.
6. It has been ratified that much front-back confusion is produced, both for high and low quality headphones.
7. We found that irregularities of the frequency response in the band of 100 to 1600 Hz seem to especially affect the front-back discrimination.
8. We also found that a poor response in the band of 4 to 7 kHz degrades the accuracy in lateral position localization.

All of these conclusions have been supported with statistical and ANOVA analysis. Some other interesting comments and clarifications about these conclusions can be added:

In addition to Conclusion 1, the angles chosen in the disparity test are a determining factor, whereby the more lateralized the angle, the larger the deviation. An increased number of angular positions may be of interest in later studies.

In relation to Conclusions 2 and 3, it is worth remarking that the mono anchor signal (ANC2) has obtained equal or even better results for *spatial impression* than some headphones ((e)-LCmul, (f)-AirL) and the stereo LPF anchor (ANC1). This fact seems to be in relation to a deficient high frequency reproduction and the general listening sensation, as evidenced by the high correlation statistics obtained with the parameter *perceived quality*.

In relation to Conclusion 5, other works, such as [28], have not found significant perception of the distortion. However, this earlier study used high quality headphones, while ours does so also with low quality consumer headphones, and we have also analyzed the influence on spatial reproduction.

Finally, taking into account these three characteristics, *perceived quality*, *spatial impression* and accuracy in *azimuth localization*, we have concluded that the first two are highly correlated. Surprisingly, and contrary to how it might seem *a priori*, there is virtually no correlation between spatial impression and accuracy in localization, because the strong influence that the subjective perceived quality has over the spatial image perception. An illustrating example can be seen with the (f)-LCmul headphone. It would be interesting to deepen this relationship in future work.

Based on the results of this study, some general guidelines for the design of headphones suitable for spatial sound reproduction can be suggested. A sensitivity difference between left-right transducers less than 1 dB should be assured in the manufacturing process to avoid azimuth localization errors. A flat frequency response between 100 to 1600 Hz is desirable to reduce front-back confusion. Finally, a good frequency response in the band 4 to 7 kHz would guarantee a good accuracy in the localization of lateral sources.

Acknowledgments: The Spanish Ministry of Economy and Competitiveness supported this work under the projects TEC2012-37945-C02-01, TEC2015-68076-R and the grant BES-2013-065034.

Author Contributions: All authors discussed the contents of the manuscript. Jose Javier Lopez contributed to the research idea and the framework of this study. Pablo Gutierrez-Parera performed the experimental work and analyzed the data.

Conflicts of Interest: The authors declare no conflict of interest.

Abbreviations

The following abbreviations are used in this manuscript:

ANOVA: Analysis of Variance
BRIR: Binaural Room Impulse Responses
dBSPL: Sound Pressure Level dB
FFT: Fast Fourier Transform
GUI: Graphical User Interface
HATS: Head and Torso Simulator
ILD: Interaural Level Difference
ITD: Interaural Time Difference
ITU-R: International Telecommunication Union, recommendation by Radiocommunication sector
LPF: Low Pass Filter
MATLAB: Matrix Laboratory, MathWorks software
MUSHRA: MUltiple Stimuli with Hidden Reference and Anchor
OST: Original film Soundtrack
VBAP: Vector Base Amplitude Panning
WFS: Wave Field Synthesis

References

1. Pulkki, V. Virtual sound source positioning using vector base amplitude panning. *J. Audio Eng. Soc.* **1997**, *45*, 456–466.
2. Berkhout, A.; de Vries, D. Acoustic control by wave field synthesis. *J. Acoust. Soc. Am.* **1993**, *93*, 2764–2778.
3. Blauert, J. *Spatial Hearing: The Psychophysics of Human Sound Localization*; MIT Press: Cambridge, MA, USA, 1997.
4. Begault, D.R. *3D Sound for Virtual Reality and Multimedia Applications*; Academic Press Professional Inc.: San Diego, CA, USA, 1994.
5. Bech, S.; Zacharov, N. *Perceptual Audio Evaluation—Theory, Method and Application*; John Wiley & Sons Ltd.: Sussex, UK, 2006.
6. Olive, S.E.; Welti, T. The relationship between perception and measurement of headphone sound quality. In Proceedings of the 133rd AES Convention, San Francisco, CA, USA, 26–29 October 2012.
7. Opitz, M. Headphones listening tests. In Proceedings of the 121st AES Convention, San Francisco, CA, USA, 5–8 October 2006.
8. Hirvonen, T.; Vaalgamaa, M.; Backman, J.; Karjalainen, M. Listening test methodology for headphone evaluation. In Proceedings of the 114th AES Convention, Amsterdam, The Netherlands, 22–25 March 2003.
9. Briolle, F.; Voinier, T. Transfer function and subjective quality of headphones: Part 2, subjective quality evaluations. In Proceedings of the 11th AES International Conference, Portland, OR, USA, 29–31 May 1992.
10. Olive, S.E.; Welti, T.; McMullin, E. A virtual headphone listening test methodology. In Proceedings of the 51st AES International Conference, Helsinki, Finland, 22–24 August 2013.
11. Farina, A. Simultaneous measurement of impulse response and distortion with a swept-sine technique. In Proceedings of the 108th AES Convention, Paris, France, 18–22 February 2000.
12. Lindau, A.; Brinkmann, F. Perceptual evaluation of headphone compensation in binaural synthesis based on non-individual recordings. *J. Audio Eng. Soc.* **2012**, *60*, 54–62.
13. Farina, A.; Armelloni, E. Emulation of not-linear, time-variant device by the convolution technique. In Proceedings of the Congresso AES Italy 2005, Como, Italy, 3–5 November 2005.
14. Pulkki, V.; Karjalainen, M. *Communication Acoustics: An Introduction to Speech, Audio and Psychoacoustics*; John Wiley & Sons Ltd: Sussex, UK, 2015.
15. Rumsey, F. *Spatial Audio*; Focal Press: Oxford, UK, 2001.
16. Gutierrez-Parera, P.; Lopez, J.J.; Aguilera, E. On the influence of headphones quality in the spatial immersion produced by binaural recordings. In Proceedings of the 138th AES Convention, Warsaw, Poland, 7–10 May 2015.
17. Rec. ITU-R BS. 1534-2. *Method for the Subjective Assessment of Intermediate Quality Level of Audio Systems*; International Telecommunication Union (ITU): Geneva, Switzerland, 2014.
18. Rumsey, F. Spatial quality evaluation for reproduced sound: Terminology, meaning and a scene-based paradigm. *J. Audio Eng. Soc.* **2002**, *50*, 651–666.
19. Rec. ITU-R BS. 1116-2. *Methods for the Subjective Assessment of Small Impairments in Audio Systems*; International Telecommunication Union (ITU): Geneva, Switzerland, 2014.
20. Letowski, T. Sound quality assessment: Cardinal concepts. In Proceedings of the 87th AES Convention, New York, NY, USA, 18–21 October 1989.
21. Zacharov, N.; Koivuniemi, K. Unravelling the perception of spatial sound reproduction. In Proceedings of the 19th AES International Conference, Bavaria, Germany, 21–24 June 2001.
22. Shinn-Cunningham, B. Learning reverberation: Considerations for spatial auditory displays. In Proceedings of the International Conference on Auditory Display (ICAD), Atlanta, GA, USA, 2–5 April 2000.
23. Minnair, P.; Olesen, S.K.; Christensen, F.; Møller, H. Localization with binaural recordings from artificial and human heads. *J. Audio Eng. Soc.* **2001**, *49*, 323–336.
24. Santala, O.; Pulkki, V. Directional perception of distributed sound sources. *J. Acoust. Soc. Am.* **2011**, *129*, 1522–1530.
25. Mendoça, C.; Campos, G.; Dias, P.; Santos, J.A. Learning auditory space: Generalization and long-term effects. *PLoS ONE* **2013**, *8*, e77900.

26. So, R.H.Y.; Ngan, B.; Horner, A.; Braasch, J.; Blauert, J.; Leung, K.L. Toward orthogonal non-individualised head-related transfer functions for forward and backward directional sound: Cluster analysis and an experimental study. *Ergonomics* **2010**, *53*, 767–781.
27. Zhang, P.X.; Hartmann, W.M. On the ability of human listeners to distinguish between front and back. *Hear. Res.* **2010**, *260*, 30–46.
28. Temme, S.; Olive, S.E.; Tatarunis, S.; Welti, T.; McMullin, E. The correlation between distortion audibility and listener preference in headphones. In Proceedings of the 137th AES Convention, Los Angeles, CA, USA, 9–12 October 2014.

applied
sciences

MDPI

Article

Blockwise Frequency Domain Active Noise Controller Over Distributed Networks

Christian Antoñanzas *, Miguel Ferrer, Maria de Diego and Alberto Gonzalez

Institute of Telecommunication and Multimedia Applications, Universitat Politecnica de Valencia,
Camino de Vera s/n; 46022 Valencia, Spain; mferrer@dcom.upv.es (M.F.); mdediego@dcom.upv.es (M.d.D.);
agonzal@dcom.upv.es (A.G.)
* Correspondence: chanma@iteam.upv.es; Tel.: +34-963-879-580

Academic Editor: Vesa Valimaki
Received: 2 March 2016; Accepted: 20 April 2016; Published: 28 April 2016

Abstract: This work presents a practical active noise control system composed of distributed and collaborative acoustic nodes. To this end, experimental tests have been carried out in a listening room with acoustic nodes equipped with loudspeakers and microphones. The communication among the nodes is simulated by software. We have considered a distributed algorithm based on the Filtered-x Least Mean Square (FxLMS) method that introduces collaboration between nodes following an incremental strategy. For improving the processing efficiency in practical scenarios where data acquisition systems work by blocks of samples, the frequency-domain partitioned block technique has been used. Implementation aspects such as computational complexity, processing time of the network and convergence of the algorithm have been analyzed. Experimental results show that, without constraints in the network communications, the proposed distributed algorithm achieves the same performance as the centralized version. The performance of the proposed algorithm over a network with a given communication delay is also included.

Keywords: active noise control; distributed networks; acoustic sensor networks; filtered-x least mean square

---

1. Introduction

Active Noise Control (ANC) systems try to cancel, or at least minimize, some undesired noise by generating some sound signals specifically designed to cancel the first [1]. The global noise reduction is virtually impossible in an entire enclosure. Alternatively, we can attempt to control the noise field within a certain area to create local zones of quiet [2]. In particular, the system is intended to reduce the disturbance signal, called primary noise, at specific spatial points monitored by microphones, called error sensors. Generally speaking, the use of a large number of microphones strategically located produces larger zones of quiet. Similarly, multiple transducers are commonly used to improve the system performance, resulting in a multichannel ANC system. Typically, multichannel ANC systems use a single centralized processor managed by a control algorithm that has access to all the signals involved in the system. However, distributed systems offer a good solution to satisfy both high computational requirements and multiple signals capture, management and generation. Distributed systems are characterized by their flexibility, versatility and scalability. Flexibility allows the system to select the suitable strategy depending on the objective application. In addition, they have the versatility to adapt quickly and easily to different situations. Moreover, these systems can increase the number of controllers without redesigning the system. Therefore, a distributed system can be understood as a set of centralized systems that distribute the computational burden as well as the acquisition and signal generation to reach a common target. For example, a multichannel centralized controller can be distributed into several single-channel controllers (see Figure 1). In both systems, the error signals $e_k(t)$ are the signals recorded at the N microphones, the anti-noise signals $y_j(t)$ are the filter output signals reproduced by the N loudspeakers, the acoustic channels $h_{j,k}$ are the impulse responses between the jth loudspeaker and the kth microphone, and the reference signal $x(t)$ is the noise signal recorded at the reference sensor used by the adaptive controller to design the output signals $y_j(t)$ (where $k = 1, 2, \ldots, N$ and $j = 1, 2, \ldots, N$).

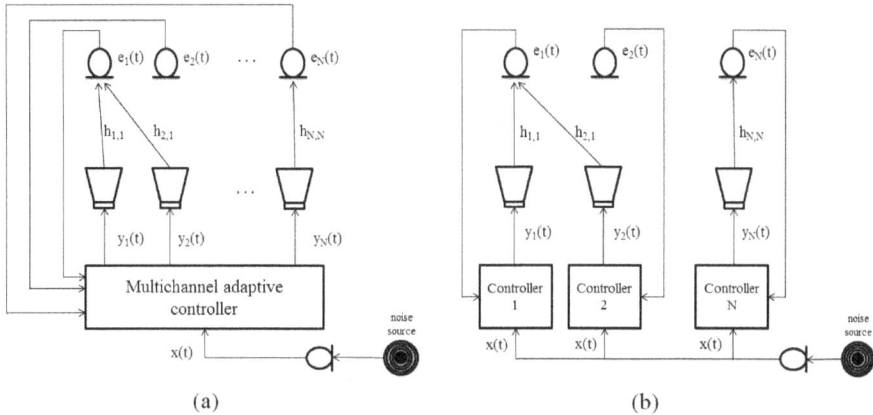

Figure 1. Scheme of (**a**) centralized Active Noise Control (ANC) system with a multichannel controller and (**b**) distributed ANC system with single-channel controllers.

While centralized systems work with all the signals generated by the loudspeakers and captured by the microphones, distributed systems employ independent processors which control a subset of loudspeakers from the signals picked up by a subset of microphones. The concept of distributed ANC systems for sound control applications was first introduced in [3], which considers processors that work independently and do not interchange local information. Thus, the computational burden is distributed among the processors and, in the case that there was no acoustic interaction among loudspeakers and microphones, it is possible to reach the centralized cancelation solution. As Figure 1b shows and it is stated in [3], each error signal $e_k(t)$ is used only for its corresponding controller, and the reference signal $x(t)$ is common to all of them. Therefore, the use of a network that allows communication among the controllers would be beneficial for the distributed ANC system to achieve results equivalent to those of the centralized method. Recently, the miniaturization of electronic components is enabling a low-cost implementation of many types of electronic devices with a high performance as well. These devices are usually equipped with sensors and actuators, and they also contain increasingly powerful and efficient processors with communication capability. Moreover, despite their small size, they allow for a great autonomy as a result of the low power requirements. In the last decade, the cooperation and communications among these kind of devices to perform a specific task through the so-called Wireless Sensor Networks (WSN) has been addressed [4]. Such networks have many advantages and applications compared to the wired networks [5,6] such as scalability and low computational cost. For the purpose of monitoring and transmission of multimedia content, Wireless Multimedia Sensor Networks (WMSN) [7] are available, which require increased computational cost, synchronization, data transmission and energy consumption because of certain characteristics of the multimedia signals [8]. A subclass of WMSN are the Wireless Acoustic Sensor Networks (WASN) [9,10]. WASNs are a popular and efficient solution for different applications in multiple acoustic areas, such as environmental audio monitoring, binaural hearing aids, audio surveillance [11–13] as well as industrial monitoring and control [10]. WASNs offer many possibilities for developing new applications and strategies for audio processing due to the expansion of the area of interest and the low power, low cost and small size of the sensors [14]. These networks are usually composed of wireless devices, called nodes, randomly distributed in the environment and formed by a single microphone plus a processing unit with communication capability [15]. These passive nodes are dedicated to estimate parameters or signals common to all of them [16] or to a particular node [17]. However, for sound field control applications, such as ANC, elements or devices capable of measuring and generating signals are necessary. Furthermore, the network should focus on the estimate of the signals that will feed the loudspeakers in order to control and modify the sound field. Therefore, it is necessary to redefine the concept of acoustic node in order to use it in ANC applications. Thus, we define acoustic nodes as devices capable of obtaining information from one or more microphones and capable of generating signals via one or more loudspeakers. Moreover, every node has the ability to individually process signals as well as to interchange the necessary information with the other nodes using a suitable communication network. Therefore, a distributed network use a set of nodes, placed strategically to reach a common objective. An output signal at each node is generated as a result of processing the signal captured by the node as well as the information received from other nodes, when there exists communication among the nodes. Every node processes signals independently and all the nodes are relevant for the proper performance of the global system. It should be noted that the selection of the network topology will affect how data is processed by each node. Some common topology architectures are mesh, star, tree and ring topologies [18]. The proper election of the topology depends on the communication constraints dictated by the network such as the amount and frequency of the transmitted data, transmission distance, battery life of the node, *etc*. Note that, in real-time applications, some of these topologies introduce delays that could seriously affect the system performance [10]. Therefore, the use of synchronization mechanisms among nodes is necessary. Consequently, our target is to take advantage of the benefits of wireless acoustic sensor networks to implement active noise control systems in real environments. This work deals with the practical implementation of a distributed and collaborative active noise controller. To our knowledge, no other system of this type has been already reported.

The paper is organized as follows: in Section 2, we present a distributed solution that minimizes the power of the sum of the measured signals at the sensors' locations over an Acoustic Sensor Network (ASN) without communication constraints and justifies the requirement of the distributed processing and the block-data processing in the frequency domain. In Section 3, the prototype description is given. The experimental results to compare the performance of both the distributed solution proposed and the centralized algorithm are shown in Section 4, including a discussion considering communication constraints. Finally, Section 5 outlines the main conclusions of the present work.

Notation: For the sake of clarity, the following notation has been used throughout this work: boldface lower-case denote vectors and matrices that contain information of signals in the time-domain (e.g., **e**) and boldface upper-case letters denote vectors and matrices that contain information of the Fast Fourier Transform (FFT) of the previous signals (e.g., **E**).

2. Collaborative Distributed Algorithm for an *N*-Nodes ASN Based on an Incremental Strategy

Consider a network of N nodes that supports an ANC system composed by N sensors and N actuators, as shown in Figure 2. For the sake of simplicity, we consider one disturbance noise and a distributed network of single-channel acoustic nodes in a homogeneous network. This implies that all the nodes have the same computation and communication capabilities, execute the same algorithm and are composed of a single sensor and a single actuator. The signals recorded at the sensors are called error signals and denoted by $e_k(t)$ (where $k = 1, 2, \ldots, N$), the actuators emit the filter output signals $y_j(t)$ (where $j = 1, 2, \ldots, N$), the acoustic channel impulse response between actuator j and sensor k is estimated as $\mathbf{s}_{jk}$, which is defined as the Finite Impulse Response (FIR) filter that models the estimation of the real acoustic channel $\mathbf{h}_{jk}$. In feedforward systems, the reference signal $x(t)$, is captured by a reference sensor employed to detect the acoustic noise far away from the area of interest. There exits only one noise source and all the nodes share the same reference signal $x(t)$. We aim at canceling the acoustic noise signal in the sensors location, $d_k(t)$, designing an adaptive filter $\mathbf{w}_k(t)$ at every node. Assuming that $\mathbf{w}_k(t)$ varies slowly, it can be written:

$$\mathbf{e}(t) = \mathbf{d}(t) + \mathbf{w}^T(t)\mathbf{x}_f(t), \qquad (1)$$

where the vectors $\mathbf{e}(t) = [e_1(t) \ e_2(t) \cdots e_N(t)]^T$ and $\mathbf{d}(t) = [d_1(t) \ d_2(t) \cdots d_N(t)]^T$ contain the error signals and desired signals, respectively, of the N nodes of the network, vector $\mathbf{w}(t) = [\mathbf{w}_1(t) \ \mathbf{w}_2(t) \cdots \mathbf{w}_N(t)]^T$ of size $LN \times 1$, concatenates the N adaptive filters $\mathbf{w}_k(t)$ that contain the L filter coefficients of the kth node at the time instant t. Matrix $\mathbf{x}_f(t) = [\mathbf{x}_{f,1}(t) \ \mathbf{x}_{f,2}(t) \cdots \mathbf{x}_{f,N}(t)]$ is the concatenation of N vectors of size $LN \times 1$ defined as $\mathbf{x}_{f,k}(t) = [x_{f,1k}(t) \ x_{f,2k}(t) \cdots x_{f,Nk}(t)]^T$ that contain the last L samples of the reference signal $x(t)$ filtered through the acoustic channel $\mathbf{h}_{jk}$ that links the actuator at the jth node with the sensor at the kth node. The objective is to estimate the coefficients vector $\mathbf{w}(t)$ that minimizes a cost function $J(t)$ that depends on the error signals $e_k(t)$. To do this, we use a gradient-descent method to estimate the coefficients in an iterative manner,

$$\mathbf{w}(t) = \mathbf{w}(t-1) - \mu \nabla_{\mathbf{w}} E\{J(t)\}, \qquad (2)$$

where μ is the step-size parameter, $E\{.\}$ is the expectation operator and $\nabla_{\mathbf{w}} = [\nabla_{\mathbf{w}_0} \ \nabla_{\mathbf{w}_1} \cdots \ \nabla_{\mathbf{w}_N}]$ being $\nabla_{\mathbf{w}_k}$ the gradient operator defined as the partial derivatives with respect to the coefficients vector $\mathbf{w}(t)$. The cost function is approximated by its instantaneous value by using the Least Mean Square (LMS) method [19], $(\nabla_{\mathbf{w}} E\{J(t)\} \approx \nabla_{\mathbf{w}} (J(t)))$. Moreover and as stated in [20], we consider the sum of the power of the N instantaneous error signals as cost function,

$$J(t) = \sum_{k=1}^{N} e_k^2(t) = \mathbf{e}(t)\mathbf{e}^T(t). \qquad (3)$$

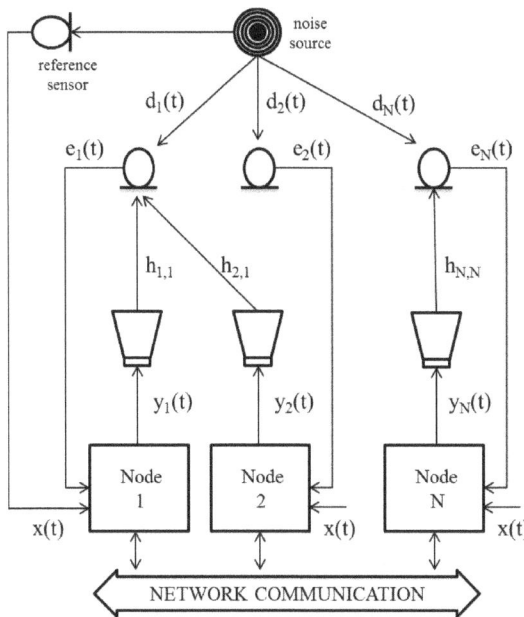

Figure 2. Acoustic Sensor Network (ASN) of N nodes for an ANC system.

From Equation (3) and applying the gradient operator in Equation (1), we obtain

$$\mathbf{w}(t) = \mathbf{w}(t-1) - \mu \mathbf{x}_f(t)\mathbf{e}(t), \tag{4}$$

that can be rewritten as

$$\mathbf{w}(t) = \mathbf{w}(t-1) - \mu \sum_{k=1}^{N} \mathbf{x}_{f,k}(t)\, e_k(t). \tag{5}$$

Note that the computation of $\mathbf{x}_{f,k}(t)$ involves the real acoustic path $\mathbf{h}_{jk}$ which, in practical environments, can be estimated as $\mathbf{s}_{jk}$. Thus, the vector of size $LN \times 1$ that contains the last L samples of reference signal $x(t)$ filtered through $\mathbf{s}_{jk}$ is named as $\mathbf{v}_k(t)$ and substituting in Equation (5) leads to

$$\mathbf{w}(t) = \mathbf{w}(t-1) - \mu \sum_{k=1}^{N} \mathbf{v}_k(t)\, e_k(t). \tag{6}$$

It can be seen in Equation (6) that all the error signals are necessary to calculate the coefficients of each filter, so that a central unit that receives and transmits all the information through the network is required (see Figure 3a). The problem is that, if the number of nodes increases or if multichannel nodes are used, it is not straightforward to transmit information between the central unit and each node due to an increase in the bandwidth required for communication. Moreover, any failure in the central unit will cause no information to be processed. Therefore, a distributed network (see Figure 3b) with the computational burden shared among the nodes becomes necessary to solve these problems. Since the number of signals processed at every node is low, this type of processing provides a more efficient computational performance. In addition, depending on the strategy used to exchange information among the nodes, the bandwidth in data transmission can be reduced. It should be noted that a distributed ASN in the context of this paper means that, not only are the nodes physically distributed in the area of interest, but also the processing (or computation) is divided among the nodes. Previous works [16,21] showed that the implementation of the LMS algorithm over distributed networks using collaborative strategies achieves good results. However, those works do not consider acoustic nodes that acoustically interact with the environment both controlling and modifying it, as it was introduced in [20]. A distributed ANC system based on the Multiple Error Filtered-x Least Mean Square (MEFxLMS) algorithm [20] and using incremental communication strategies with sample-by-sample data acquisition was presented in [22].

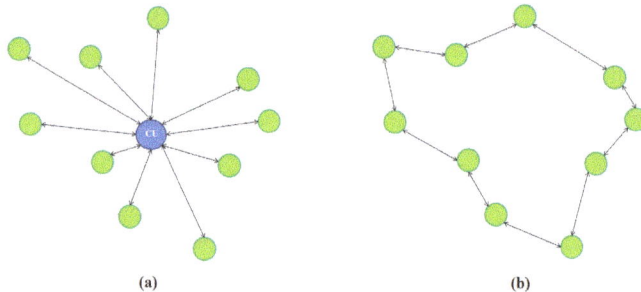

(a)   (b)

Figure 3. A centralized ASN (**a**) and distributed ASN (**b**). CU is the central unit in the centralized case.

In the case of a ring network based on an incremental strategy, the coefficients of the adaptive filters are calculated by distributing the calculation among different nodes by transmitting information to an adjacent node in a consecutive order. Thus, every node can calculate a portion of the sum of the filter updating equation and supply to the next node the partial result to update the coefficients with its respective information. If this step is performed with an incremental strategy, the last node will have the complete updated coefficients. Finally, these coefficients are disseminated to the rest of the nodes to allow the system to generate the appropriate cancelation signals before the next iteration begins. This means $2(N-1)$ interchanges of the filter coefficients among the nodes (see Figure 4). Every kth node can share information with its adjacent $k+1$th node. Each node should calculate the adaptive filters of all the nodes of the network so, in a network of N nodes, the network state would be defined by N adaptive filters one of each node. If we define the global state of the network $\mathbf{w}(t)$ as the adaptive filter coefficients of each node of the network at the time instant t and considering $\mathbf{w}^k(t)$ a local version of $\mathbf{w}(t)$ at the kth node, Equation (6) may be written as:

$$\mathbf{w}^k(t) = \mathbf{w}^{k-1}(t) - \mu \mathbf{v}_k(t)\, e_k(t). \tag{7}$$

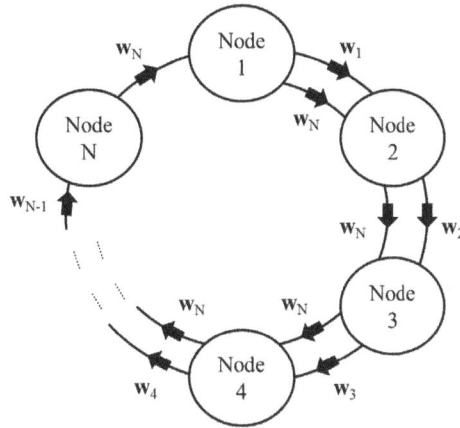

Figure 4. Ring network with incremental communication.

Equation (7) is the updating rule of the state of the kth node by using an iterative algorithm that minimizes Equation (3). Moreover, note that $\mathbf{w}^0(t) = \mathbf{w}^N(t-1) = \mathbf{w}(t-1)$.

However, in real-time applications, the obtained equations have to be considered using block processing in the frequency domain in order to ensure a more efficient computational performance. The reason is because most common audio cards work with block data buffers. Moreover, hardware platforms that work with blocks of samples such as Digital Signal Processors (DSPs) or Graphics Processing Units (GPUs), use libraries of frequency-domain operations for a more efficient processing [23]. Furthermore, if the adaptive filters and the estimated secondary paths are longer than the sample block, they have to be split up into partitions [24]. For all these reasons, we consider the Frequency-domain Partitioned Block technique for the adaptive filtering operation [25,26] based on the conventional filtered-x scheme (FPBFxLMS). Notation in Table 1 will be used to describe the following equations. Samples are processed by blocks of size B. L is the length of the adaptive filters, and M is the length of the FIR filters that model the estimated secondary path. F and P are the number of partitions of both the adaptive filters and the estimated secondary paths, respectively. Furthermore, the index n between brackets denotes block iteration and the super-indexes f and p denotes the number of the partition.

The vector $\mathbf{x}_B[n] = [x(Bn)\ x(Bn-1) \cdots x(Bn-B+1)]^T$ contains the last B samples of $x(t)$ at discrete time instant $t=Bn$, and the error vector $\mathbf{e}_{k,B}[n]$ contains the last block of size B of the error signal $e_k(t)$ in the node k at discrete time instant $t=Bn$, $\mathbf{e}_{k,B}[n] = [e_k(Bn)\ e_k(Bn-1) \cdots e_k(Bn-B+1)]^T$. The vector $\mathbf{X}[n]$ contain the FFT of size $2B$ of the vector $\mathbf{x}_B[n]$ in the actual block iteration and the same vector in the previous block iteration, $\mathbf{X}[n] = \text{FFT}[\mathbf{x}_B[n-1]\ \mathbf{x}_B[n]]$ and the vector $\mathbf{E}_k[n]$ is the FFT of size $2B$ of the error vector $\mathbf{e}_{k,B}[n]$ preceded by a vector of zeros of size B, $\mathbf{0}_B$. Note that we only consider the last B samples of the $2B$-Inverse Fast Fourier Transform (IFFT) operation, IFFT$\{\mathbf{Y}[n]\}$, as the valid samples of the adaptive filter output $\mathbf{y}_B[n]$ since the first B samples suffer the effects of circular convolution due both to data management and the FFT sizes. Now, we define

$$\mathbf{W}[n] = \left[\ \mathbf{W}_1[n], \mathbf{W}_2[n], \dots, \mathbf{W}_N[n], \ \right] \tag{8}$$

being $\mathbf{W}_k[n] = \left[\ \mathbf{W}_k^1[n], \mathbf{W}_k^2[n], \dots, \mathbf{W}_k^F[n] \ \right]$ a matrix of size $[2B \times F]$ where $\mathbf{W}_k^f[n]$ contains the $2B$-FFT of the fth partition of the adaptive filter of the kth node. The N nodes collaborate with each other by updating their part of $\mathbf{W}[n]$ and transferring $\mathbf{W}[n]$ to the next node. Therefore, every node will use a local version of the global state of the network (denoted by $\hat{\mathbf{W}}_k[n]$) at the kth node at the nth block iteration. Notice that only the F partitions of $2B$ coefficients of their adaptive filter are needed to generate the kth node output signal:

$$\mathbf{W}_k[n-1] = \hat{\mathbf{W}}_N[n-1]_{(:,1+F(k-1):Fk)}. \tag{9}$$

Table 1. Notation of the description of the FPBFxLMS algorithm.

B	Block size
L	Length of the adaptive filters
M	Length of the Finite Impulse Response (FIR) filters that model the estimated secondary paths
F	L/B, number of partitions of the adaptive filters
P	M/B, Number of partitions of the estimated secondary paths
n	index that denotes block iteration
f, p	super-indexes that denote partition number.
$\mathbf{s}_{jk}$	M-length estimation of the acoustic path that links the actuator at the jth node with the sensor at the kth node.
$\mathbf{S}_{jk}^p$	Fast Fourier Transform (FFT) of size $2B$ of the pth partition of the acoustic path $\mathbf{s}_{jk}$.

Moreover, we define

$$\mathbf{V}_k[n] = \left[\ \mathbf{V}_{1k}[n], \mathbf{V}_{2k}[n], \dots \mathbf{V}_{Nk}[n], \ \right] \tag{10}$$

being $\mathbf{V}_{jk}[n] = \left[\ \mathbf{V}_{jk}^1[n], \mathbf{V}_{jk}^2[n], \dots, \mathbf{V}_{jk}^F[n] \ \right]$ a matrix of size $[2B \times F]$ where vector $\mathbf{V}_{jk}^f[n]$ contains the $2B$-FFT of the reference signal filtered with the fth partition of the estimated acoustic channel $\mathbf{s}_{jk}$. Each node estimates the coefficients of the rest of the nodes to achieve a global solution using the information of the previous node and the adaptation matrix calculated using the signals that each node own. Derived from the FPBFxLMS algorithm in [26], we can rewrite Equation (7) in the frequency domain with blocks of B samples. Thus, the update equation of the adaptive filter coefficients of the kth node at the nth block iteration is given by

$$\hat{\mathbf{W}}_k[n] = \hat{\mathbf{W}}_{k-1}[n] - \mu \, \text{FFT}\{[\, [\, \text{IFFT}\{\underline{\mathbf{E}}_k[n] \circ \mathbf{V}_k[n]^*\} \,]_{[1:B,:]} \quad \mathbf{0}_{[B \times FN]} \,]\}, \tag{11}$$

where $\underline{\mathbf{E}}_k[n]$ is the multiplication of vector $\mathbf{E}_k[n]$ by $\mathbf{1}_{[1 \times FN]}$, a row vector of ones of size FN. Constant μ is the step-size parameter and the operators FFT and IFFT perform the direct and inverse fast Fourier transform of size $2B$ of each column of the matrixes involved. $\circ$ denotes the element-wise product of two matrices, $*$ denotes complex conjugation and $\mathbf{0}_{[B \times FN]}$ is a matrix of zeros of size $[B \times FN]$. Note that we only consider the first B samples of the $2B$-IFFT operation. Once all the nodes have finished the filter coefficient updates, the global vector $\hat{\mathbf{W}}_N[n]$ is disseminated to the rest of the nodes for the $(n+1)$th iteration. Note that in (11), $\hat{\mathbf{W}}_0[n] = \hat{\mathbf{W}}_N[n-1]$, as we stated in Equation (7). **Algorithm 1** illustrates the summary of the algorithm instructions, which are executed per block iteration at each node.

Algorithm 1 Distributed FPBP x LMS Algorithm for *N*-nodes ASN

1: **for all** *node* $1 \leq k \leq N$ **do**

2: $\quad \mathbf{W}_k[n-1] = \hat{\mathbf{W}}_k[n-1]_{(:,1+F(k-1):Fk)}$

3: $\quad \mathbf{Y}_k[n] = \sum_{f=1}^{F} \mathbf{W}_k[n-1] \circ \mathbf{X}[n-f+1]$

4: $\quad$ **for all** $1 \leq j \leq N$ **do**

5: $\quad\quad \mathbf{V}_{jk}[n] = \sum_{p=1}^{P} \mathbf{S}_{jk}^{p} \circ \mathbf{X}[n-p+1]$

6: $\quad$ **end for**

7: $\quad \mathbf{V}_k[n] = [\, \mathbf{V}_{1k}[n], \mathbf{V}_{2k}[n], \ldots \mathbf{V}_{Nk}[n] \,]$

9: $\quad \mathbf{E}_k[n] = \text{FFT}\{[\, \mathbf{0}_B \quad \mathbf{e}_B[n] \,]\}$

10: $\quad \underline{\mathbf{E}}_k[n] = \mathbf{E}_k[n] \cdot \mathbf{1}_{[1 \times FN]}$

11: $\quad \hat{\mathbf{W}}_k[n] = \hat{\mathbf{W}}_{k-1}[n] - \mu \, \text{FFT}\{[\, [\, \text{IFFT}\{\underline{\mathbf{E}}_k[n] \circ \mathbf{V}_k[n]^*\} \,]_{[1:B,:]} \quad \mathbf{0}_{[B \times FN]} \,]\}$

12: **end for**

13: **for all** *node* $0 \leq k \leq N$ **do**

14: $\quad \hat{\mathbf{W}}_k[n] = \hat{\mathbf{W}}_N[n]$

15: **end for**

---

3. Prototype Description

The distributed ANC prototype is depicted in Figure 5. The addition of the recent DSP System Toolbox [27] in the computing environment MATLAB® provides multichannel real-time audio recording, processing and reproduction at low latency. Audio objects based on Object Oriented Programming (OOP) have been optimized for iterative computations that process large streams of audio data. Moreover, Audio Stream Input/Output (ASIO) drivers [28] have been incorporated to this software providing a low-latency and high fidelity interface between MATLAB® and audio card. The hardware implementation is composed by a CPU (Intel Core i7 3.07 GHz) and an audio card (MOTU 24 I/O). The communication between both components is performed by ASIO drivers and it is controlled by using the MATLAB® System objects provided by the DSP System Toolbox of MATLAB® software. The audio card stores the input data from the sensor of each node in buffers of size B and send them to the CPU through the ASIO drivers. The CPU, with MATLAB® support, executes the audio processing, saves the output data in buffers and sends them back to the audio card through the ASIO drivers, to be reproduced by the loudspeaker of each node.

The selection of the block size B in the audio card is critical to determine the latency of the algorithm. The latency is the sum of the time spent on storing data in the input buffers, on processing these data and on sending them to the output buffers as well. A real-time application must satisfy that the time spent to fill up the input buffers (buffering time) was higher than the time spent in data processing (processing time). The buffering time is defined as B/fs, fs being the sampling rate. In a centralized ANC system, the processing time is the time that the algorithm takes to process data. However, in a distributed ANC system, it also includes the time in updating its own global state of the network ($\hat{\mathbf{W}}$) and in delivering this information among the nodes. As we consider an incremental network composed of N nodes, every node must transfer $2L \times N$ coefficients (size of $\hat{\mathbf{W}}$) to the following node $2(N-1)$ times in each block iteration (see Figure 4). Therefore, the processing time of the whole network at each block iteration (algorithm processing time + N times the updating time of $\hat{\mathbf{W}}$ + $2(N-1)$ times the transmission time of $\hat{\mathbf{W}}$) has to be less than the buffering time (see Figure 6). As an example, a transfer rate at least of 16.5 megabytes per second (MBps) would be necessary with an incremental network composed of four nodes, a general filter length of 4096 taps and single-precision floating-point data format. Therefore, using a standard Ethernet network of 1 Gbps ($\approx$ 125 MBps), we would have enough rate capacity to perform the required data transfer among the nodes. It should be noted that, if the processing time of the network increases due to the addition of more nodes, the buffering time must be increased in order to satisfy the real-time condition. Hence, if we assume a fixed sampling rate, the block size B must be increased.

Another important aspect that should be guaranteed is the causality of the system. The algorithm has to satisfy that the sum of the buffering delay and the maximum delay of the acoustic paths that join the actuators with the error sensors should be less than or equal to the minimum delay of the acoustic paths that join the noise source with the error sensors [29]. Causality constraint can be relaxed when a harmonic excitation is considered, but it is important in broadband noise control. As we perform an ANC system in a acoustic enclosure, where the wavelength is relatively low in comparison with the physical dimensions of the system, the causality condition is fulfilled by carefully choosing the distances between the noise source, actuators and error sensors. However, if this condition is not met, a decrease in buffering time is required. When these requirements are fulfilled and assuming a network of synchronized nodes, the proposed distributed algorithm can achieve the same performance as the centralized version.

The sampling rate of the audio card was fixed at 44.1 kHz, the lowest possible rate, because of audio signals are typically sampled at that value (CD quality). The audio card offers different values of B between 16 and 2048 samples. Due to the real-time condition explained previously, we have selected $B = 2048$ as a tradeoff between the processing time of the network and the buffering time.

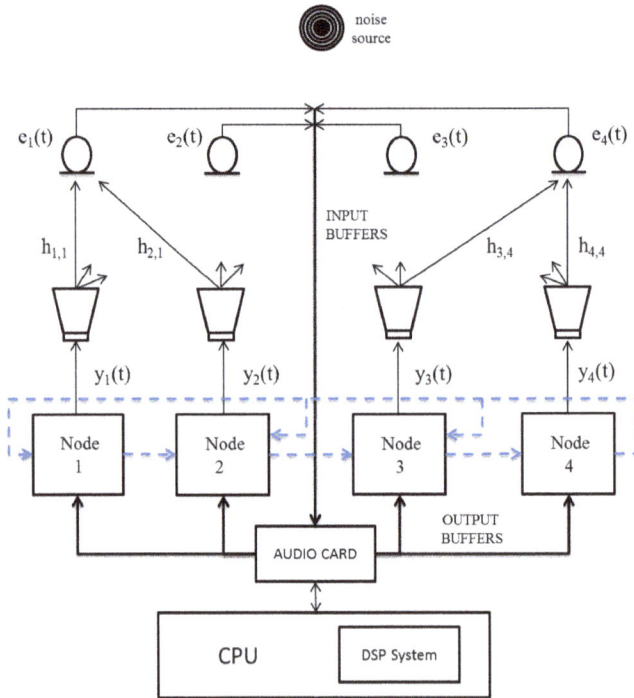

Figure 5. Scheme of the ANC prototype. The incremental communication strategy is represented by dashed lines.

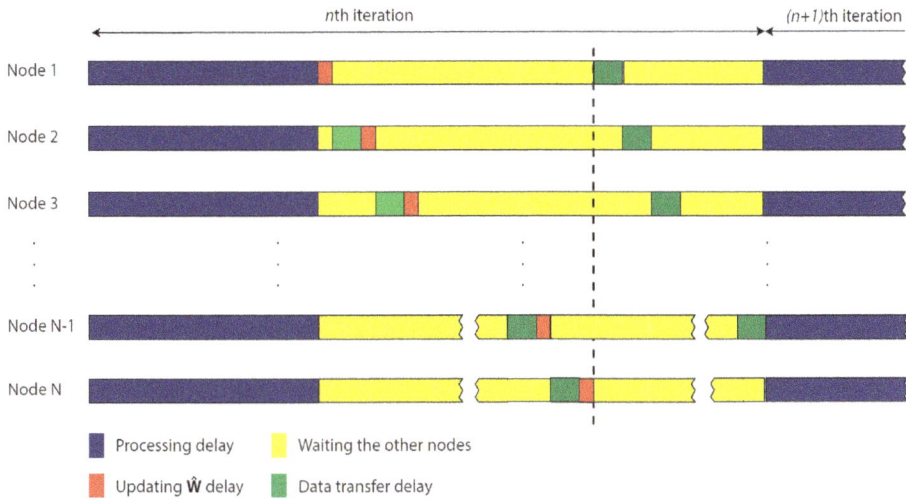

Figure 6. Timing diagram of the processes carried out by each node of the network at each block iteration.

It should be noted that the algorithms execute their processing in real time but, for simplicity, the communication network and the network distributed processing are emulated. All the nodes share the same central processing unit (CPU) but the implemented code allows a distributed and independent processing, simulating the processing carried out in a real distributed ASNs. Therefore, a basic networking hardware dedicated to the information exchange (a physical layer of the network) is not considered. The communication among the nodes is virtual, thanks to the code designed in MATLAB® software.

4. Experimental Results

In this section, we show the experiments carried out to validate the performance of the distributed algorithm for unconstrained and constrained networks. For this purpose, in the first stage, we have evaluated and compared the convergence behavior, the noise reduction and the computational complexity of both distributed and centralized ANC systems in a ideal network. In the second stage, a performance analysis of the distributed algorithm over non-ideal networks has been carried out. The experiments have used the prototype described in Section 3 inside a listened room of 9.36 meters long, 4.78 meters wide and 2.63 meters high, located at the Audio Processing Laboratory of the Polytechnic University of Valencia [30]. This room has an array of 96 loudspeakers mounted in an orthogonal structure. A photograph of the listening room and the different settings can be seen in Figure 7. For all the designed ASNs, the real acoustic responses between all the loudspeakers and all microphones are identified off-line using adaptive methods and modeled as FIR filters of $M = 4096$ coefficients at a sampling rate of 44.1 kHz. We have considered a zero-mean Gaussian white noise with unit variance as disturbance noise, which is emitted by a loudspeaker located in front of the loudspeakers and microphones that compose the ASN, as it is shown in Figure 7. Since practical ANC systems are suitable to cancel low-frequency noise signals, we have considered a broadband white noise limited to 200 Hz. Furthermore, we have considered a block size of $B = 2048$ and an adaptive filter length of $L = 4096$ coefficients. A constant step size parameter of $\mu = 8 \times 10^{-5}$, as the highest value that ensures the stability of the algorithms, has been used.

Figure 7. Photograph of the ANC prototype in the listening room at the Audio Processing Laboratory of the Polytechnic University of Valencia.

The configuration of the ASN is depicted in Figure 5. Four nodes composed by one loudspeaker and one microphone each of them were considered. The loudspeakers were selected from the array with an equal separation of 20 cm between adjacent loudspeakers. The microphones were placed opposite to the loudspeakers and separated 40 cm away from them. The separation between the microphones was 20 cm. Future applications closely related to the tested distribution would be related to the creation of local quiet zones in enclosures using moderate size networks of sound nodes with likely acoustic interaction between at least a few of them—for example, a cabin of a public transport (train, plane, bus, *etc.*) where the separation between actuators and sensors would be similar than the detailed above. For the experimental tests, two different scenarios have been evaluated:

- ANC system over a four-node ideal network controlled by the FPBFxLMS distributed algorithm and its centralized version.
- ANC system over a four-node non-ideal network controlled by the FPBFxLMS distributed algorithm, comparing the results with the same algorithm but introducing constant delays in the data exchanges through the network. Now we assume that the nodes interchange information every Np block iterations, being p a constant positive integer and N the number of single-channel nodes. During the remainder Np-1 block iterations, we assume two cases. In case 1, the nodes just wait for the arrival of new network information trying to simulate a network with limited power, which saves as much energy as possible. In case 2, the nodes will use its local information to update their filter coefficients. Moreover, we assume that the diffusion of the global state of the network $\hat{\mathbf{W}}_N[n]$ from the last node N to the rest of the nodes (see Figure 4) is not considered in this scenario.

In order to evaluate the performance of the different algorithms, we define the instantaneous Noise Reduction at node k, $NR_k(t)$, as the ratio in dB between the estimated error power with and without the application of the active noise controller:

$$NR_k(t) = 10 \times log_{10} \left[\frac{e_k^2(t)}{d_k^2(t)} \right], \tag{12}$$

where $d_k^2(t)$ is the signal power picked up at the kth microphone when the ANC system is inactive and $e_k^2(t)$ is the error signal power measured at the kth microphone when the ANC system works. Moreover, these signals powers have been estimated using an exponential windowing from the instantaneous signals.

In the first scenario, we have evaluated the performance of both the distributed and the centralized FPBFxLMS algorithm in a network composed of four single-channel nodes with no communications constraints. Figure 8 shows the $NR_k(t)$ by using Equation (12) for both algorithms. As expected, the distributed implementation exhibits the same performance than the centralized implementation for the four nodes. Both algorithms show a robust and stable performance providing an attenuation up to 11 dB for the worst node and almost 20 dB for the best node.

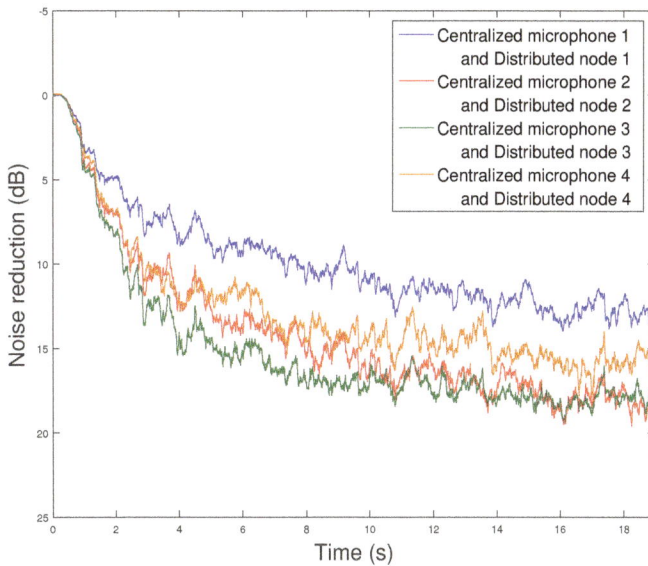

Figure 8. Noise reduction obtained by both the four-node distributed system and the centralized system with a 1:4:4 configuration.

Table 2 compares the computational complexity in terms of multiplications, additions, and FFTs per iteration of the FPBFxLMS algorithm, implemented for a centralized and a distributed ANC systems. For the centralized implementation, we consider a multichannel ANC system with one disturbance noise and the same number of microphones and loudspeakers (1:N:N configuration). For the distributed implementation, we consider a network of N single-channel nodes. It is important to note that the complexity of the network as a whole is at least as high as the centralized algorithm. However, note that each node processes the algorithm simultaneously except the last addition of the global network state of the previous node (see line 11 in Algorithm 1). Therefore, in Table 2, we have only computed the operations of one single-channel node. Since we use a value of $M = L$, and $B = L/2$ (two partitions) the computational complexity only depends on L and N. First, the third column of Table 2 shows the computational complexity of both algorithms related to values of L and N. Then, this computational complexity is particularized for $N = 1$, $N = 4$ and $N = 8$. As expected, when $N = 1$, both implementations need the same number of operations. This is because both the centralized and the distributed ANC systems become a single-channel system. Moreover, for $N = 4$, we compare the operations of a centralized ANC system with a 1:4:4 configuration (16 channels) with the operations of a single-channel node of a network of four nodes. Finally, the same is done for $N = 8$. Results show that in a centralized ANC system, the computational complexity increases significatively with the number of channels. This fact represents a bottleneck in massive multichannel ANC systems. Otherwise, the increase of computational complexity in a distributed ANC system is not so significant.

Table 2. Total number of multiplications (MUX), additions (ADD) and Fast Fourier Transforms (FFTs) per blockwise iteration of the FPBFxLMS algorithm in both: (1) centralized and (2) distributed ANC systems. L: length of the adaptive filters; N: number of nodes.

	Operations	Generic	$N = 1$	$N = 4$	$N = 8$
	MUX	$4LN + 4LN^2$	$8L$	$80L$	$288L$
(1)	ADD	$LN + 3LN^2$	$4L$	$52L$	$200L$
	FFTs	$2 + 6N$	8	26	50
	MUX	$2L + 6LN$	$8L$	$26L$	$50L$
(2)	ADD	$L + 3LN$	$4L$	$13L$	$25L$
	FFTs	$4 + 4N$	8	20	36

In the second scenario, the influence of a constant delay between information exchanges within a four-node network on the behavior of the distributed FPBFxLMS algorithm has been evaluated. The $NR_k(t)$ curves of the best node in the network with different delay values of p are presented in Figure 9a for case 1 and Figure 9c for case 2. In both cases, as p increases, the performance of the algorithm gets worse. However, when the nodes performs their local updating while waiting for the network information (*case 2*), the influence of the increased delay is lower than in case 1, obtaining an $NR_k(t)$ in all cases only 5 dB lower than the case of an ideal network. Similar results are obtained for the node with the worst performance. Figure 9 shows the $NR_k(t)$ curves of the worst node in the network with different values of p, (**b**) for the case 1 and (**d**) for the case 2. Therefore, for networks with communication constraints, the behavior of the system can be improved if the nodes are allowed to update their filter coefficients while waiting for the network information. It is important to take into account that the final behavior of both cases depends on the degree of acoustic interaction between nodes, and, therefore, it should be studied in future works.

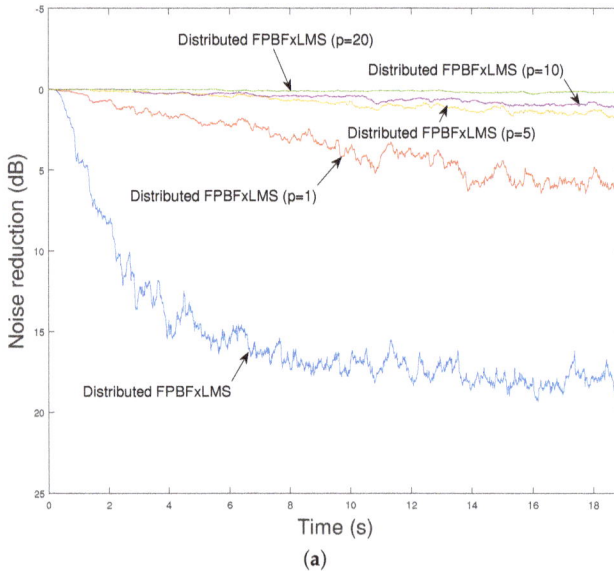

(a)

Figure 9. *Cont.*

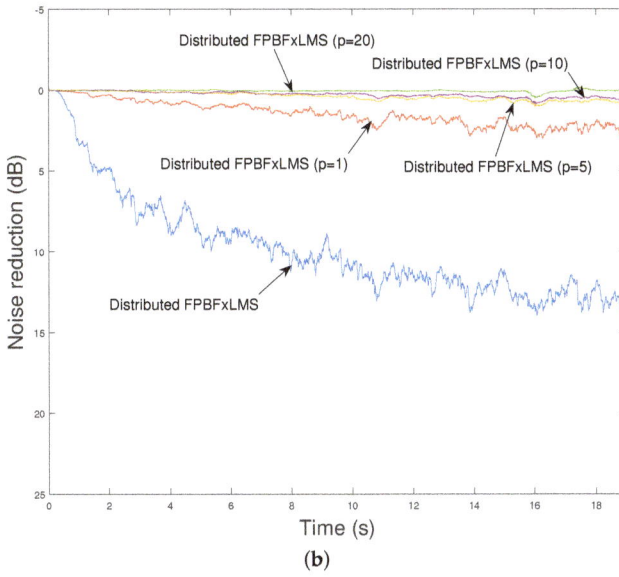

(b)

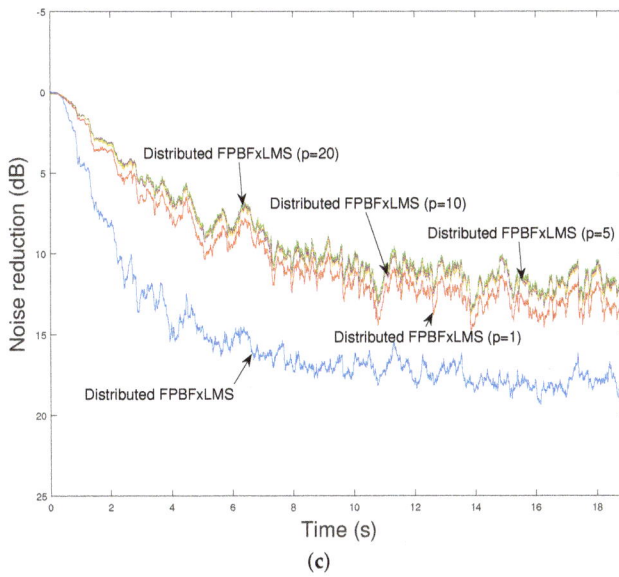

(c)

Figure 9. *Cont.*

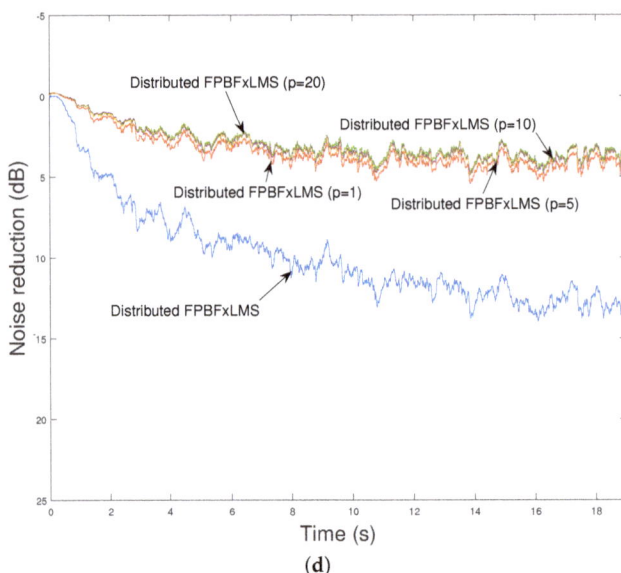

Figure 9. Noise reduction obtained for the distributed FPBPxLMS algorithm using a four-node ASN for different latency values at the node with the best performance (**a**) (case 1) (**c**) (case 2) and with the worst performance (**b**) (case 1) (**d**) (case 2).

Finally, note that all the results accomplished in this work depend on particular settings, but their behavior can be easily extrapolated to other configurations.

5. Conclusions

In this paper, an experimental validation of an ANC system has developed over a network of distributed acoustic nodes using a collaborative incremental strategy. For this purpose, a distributed version of the FPBFxLMS algorithm has been introduced. In order to evaluate the performance of the proposed algorithms in a practical environment, we have implemented the adaptive control in a real-time audio toolbox provided by MATLAB®. Results show that the distributed implementation of the algorithm exhibits the same performance as its centralized version when there are no communication constraints in the network.

Moreover, the computational complexity of the distributed algorithm has been studied and compared with the centralized version. Since each node of the distributed algorithm can perform almost all the operations simultaneously, the computational complexity is split among the nodes and the processing time of the algorithm is significantly reduced. However, when we consider the use of physical nodes in a real network, we will have to analyze some aspects with regards to the communication capabilities among the nodes in the network. In this sense, we can conclude that the distributed ANC system performs well if the network has enough bandwidth to transfer the network information without violating the real-time condition.

The performance of the proposed distributed FPBFxLMS algorithm in an ideal ASN and an ASN with a constant delay in the communications has been also shown. The behavior of the system worsens when delay increases, but, in this experiment, it shows an acceptable and stable performance despite data loss in the information exchange when the node uses its local information while waiting for network information. However, the influence of the degree of acoustic interaction between nodes on the network communication strategies must be considered in future works.

Acknowledgments: This work has been supported by the European Union (European Regional Development Fund) together with Spanish Government through TEC2015-67387-C4-1-R project, the grant BES-2013-063783 and Generalitat Valenciana through the PROMETEOII/2014/003 project.

Author Contributions: All authors discussed the results and implications and commented on the manuscript at all stages. Alberto Gonzalez, Maria de Diego and Miguel Ferrer conceived the study, supervised and edited the work. All authors contributed to the formulation development. Christian Antoñanzas performed the experimental work, analyzed the data and wrote the manuscript.

Conflicts of Interest: The authors declare no conflict of interest.

References

1. Elliott, S. *Signal Processing for Active Control*; Academic Press: Cambridge, MA, USA, 2000.
2. Pawełczyk, M. Active noise control-a review of control-related problems. *Arch. Acoust.* **2008**, *33*, 509–520.
3. Elliott, S.; Boucher, C. Interaction between multiple feedforward active control systems. *IEEE Trans. Speech Audio Process.* **1994**, *2*, 521–530.
4. Akyildiz, I.; Su, W.; Sankarasubramaniam, Y.; Cayirci, E. A survey on sensor networks. *IEEE Commun. Mag.* **2002**, *40*, 102–114.
5. Yick, J.; Mukherjee, B.; Ghosal, D. Wireless sensor network survey. *Comput. Netw.* **2008**, *52*, 2292–2330.
6. Puccinelli, D.; Haenggi, M. Wireless sensor networks: Applications and challenges of ubiquitous sensing. *IEEE Circuits Syst. Mag.* **2005**, *5*, 19–31.
7. Harjito, B.; Han, S. Wireless multimedia sensor networks applications and security challenges. In Proceedings of the 2010 International Conference on Broadband, Wireless Computing, Communication and Applications, Fukuoka, Japan, 4–6 November 2010; Institute of Electrical & Electronics Engineers (IEEE): Piscataway, NJ, USA, 2010.
8. Akyildiz, I.; Melodia, T.; Chowdury, K. Wireless multimedia sensor networks: A survey. *IEEE Wirel. Commun.* **2007**, *14*, 32–39.
9. Bertrand, A. Applications and trends in wireless acoustic sensor networks: A signal processing perspective. In Proceedings of the 2011 18th IEEE Symposium on Communications and Vehicular Technology in the Benelux (SCVT), Ghent, Belgium, 22–23 November 2011; Institute of Electrical & Electronics Engineers (IEEE): Piscataway, NJ, USA, 2011.
10. Flammini, A.; Ferrari, P.; Marioli, D.; Sisinni, E.; Taroni, A. Wired and wireless sensor networks for industrial applications. *Microelectron. J.* **2009**, *40*, 1322–1336.
11. Bertrand, A.; Moonen, M. Robust distributed noise reduction in hearing aids with external acoustic sensor nodes. *EURASIP J. Adv. Signal Process.* **2009**, *2009*, 530435.
12. Guo, Y.; Hazas, M. Acoustic source localization of everyday sounds using wireless sensor networks. In Proceedings of the 12th ACM International Conference Adjunct Papers on Ubiquitous Computing–Adjunct, Copenhagen, Denmark, 26–29 September 2010; Association for Computing Machinery (ACM): New York, NY, USA, 2010.
13. Wang, H. Wireless Sensor Networks for Acoustic Monitoring. Ph.D. Thesis, University of California, Los Angeles, CA, USA, 2006.
14. Bertrand, A. Signal Processing Algorithms for Wireless Acoustic Sensor Networks. Ph.D. Thesis, University of Leuven, Leuven, Belgium, 2011.
15. Kwon, H.; Berisha, V.; Spanias, A. Real-time sensing and acoustic scene characterization for security applications. In Proceedings of the 2008 3rd International Symposium on Wireless Pervasive Computing, Santorini, Greece, 7–9 May 2008; Institute of Electrical & Electronics Engineers (IEEE): Piscataway, NJ, USA, 2008.
16. Lopes, C.; Sayed, A. Incremental adaptive strategies over distributed networks. *IEEE Trans. Signal Process.* **2007**, *55*, 4064–4077.
17. Bertrand, A.; Moonen, M. Distributed adaptive node-specific signal estimation in fully connected sensor networks—Part I: Sequential node updating. *IEEE Trans. Signal Process.* **2010**, *58*, 5277–5291.
18. Andrew, S.; Tanenbaum, D.J.W. *Computer Networks*; Prentice Hall: Upper Saddle River, NJ, USA, 2012.
19. Widrow, B.; Stearns, S.D. *Adaptive Signal Processing*; Prentice Hall: Englewood Cliffs, NJ, USA, 1985; Volume 1.
20. Elliott, S.; Stothers, I.; Nelson, P. A multiple error LMS algorithm and its application to the active control of sound and vibration. *IEEE Trans. Acoust. Speech Signal Process.* **1987**, *35*, 1423–1434.

21. Lopes, C.G.; Sayed, A.H. Diffusion least-mean squares over adaptive networks: Formulation and performance analysis. *IEEE Trans. Signal Process.* **2008**, *56*, 3122–3136.

22. Ferrer, M.; de Diego, M.; Piñero, G.; Gonzalez, A. Active noise control over adaptive distributed networks. *Signal Process.* **2015**, *107*, 82–95.

23. Lorente, J.; Ferrer, M.; de Diego, M.; Gonzalez, A. GPU implementation of multichannel adaptive algorithms for local active noise control. *IEEE/ACM Trans. Audio Speech Language Process.* **2014**, *22*, 1624–1635.

24. Borrallo, J.P.; Otero, M.G. On the implementation of a partitioned block frequency domain adaptive filter (PBFDAF) for long acoustic echo cancellation. *Signal Process.* **1992**, *27*, 301–315.

25. Farhang-Boroujeny, B. *Adaptive Filters: Theory and Applications*; John Wiley & Sons: Hoboken, NJ, USA, 2013.

26. Shynk, J. Frequency-domain and multirate adaptive filtering. *IEEE Signal Process. Mag.* **1992**, *9*, 14–37.

27. Mathworks. DSP System Toolbox. Available online: http://mathworks.com/help/dsp/ (accessed on 11 November 2015).

28. Steinberg, ASIO. Available online: http://www.steinberg.net/nc/en/company/developers/sdk-download-portal.html (accessed on 27 July 2015).

29. Burdisso, R.A. Causality analysis of feedforward-controlled systems with broadband inputs. *J. Acoust. Soc. Am.* **1993**, *94*, 234–242.

30. Audio and communications signal processing group (GTAC). Available online: http://www.gtac.upv.es (accessed on 31 March 2016).

applied
sciences

MDPI

Article

Augmenting Environmental Interaction in Audio Feedback Systems†

Seunghun Kim [1], Graham Wakefield [2] and Juhan Nam [1,*]

[1]   Graduate School of Culture Technology, KAIST, Daejeon 34141, Korea; seunghun.kim@kaist.ac.kr
[2]   Arts, Media, Performance & Design, York University, Toronto, ON M3J 1P3, Canada; grrrwaaa@yorku.ca
*   Correspondence: juhannam@kaist.ac.kr; Tel.: +82-42-350-2926
†   This paper is an extended version of paper published in the 41st International Computer Music Conference, Denton, TX, USA, 25 September–1 October 2015

Academic Editor: Vesa Valimaki
Received: 2 March 2016; Accepted: 18 April 2016; Published: 28 April 2016

Abstract: Audio feedback is defined as a positive feedback of acoustic signals where an audio input and output form a loop, and may be utilized artistically. This article presents new context-based controls over audio feedback, leading to the generation of desired sonic behaviors by enriching the influence of existing acoustic information such as room response and ambient noise. This ecological approach to audio feedback emphasizes mutual sonic interaction between signal processing and the acoustic environment. Mappings from analyses of the received signal to signal-processing parameters are designed to emphasize this specificity as an aesthetic goal. Our feedback system presents four types of mappings: approximate analyses of room reverberation to tempo-scale characteristics, ambient noise to amplitude and two different approximations of resonances to timbre. These mappings are validated computationally and evaluated experimentally in different acoustic conditions.

Keywords: audio feedback; digital filters; digital signal processing; music; reverberation

PACS: J0101

---

1. Introduction

Audio feedback is an acoustic phenomenon that occurs when sound played by a loudspeaker is received by a microphone to create a persistent loop through a sound system. While audio feedback is generally regarded as an undesired situation, for example when a public address system manifests an unpleasant howling tone, there have been numerous artistic examples and compositions that make use of its tone-generating nature. Jimi Hendrix is an oft-cited example of how electric guitar players create feedback-based tones by holding their instruments close to the amplifiers, and Steve Reich's *Pendulum Music* (1968) [1] features phasing feedback tones generated by suspending microphones above loudspeakers.

A modern approach to audio feedback in experimental improvisation and compositional works utilizes computer-based control over sound generation and organization: Sanfilippo introduced various examples [2]. Di Scipio's *Audio Ecosystems* [3] is a prominent example in which a self-feeding feedback loop interconnects a digital system with its acoustic environment.

By inserting a network of low-level components, represented by a chain of transducers and other acoustic components inserted into the loop, these audio feedback systems can lead to nonlinear behaviours [3], since specific performances cannot be accurately predicted in advance. Many audio feedback systems mostly have focused on design of the low-level relations to generate and organize the feedback sounds while paying less attention to control over the overall sonic shape [4].

We previously proposed a new concept of audio feedback systems [5] that supports intentional control through tendency design, while preserving other attractive nonlinear features of feedback systems, which could open up new possibilities of musical applications combining nonlinearity and interactivity. In this paper, we explore this concept further by taking account of the relation between system and room acoustics.

Room acoustics are an essential yet under-examined factor in the shaping of audio feedback. Our work is designed to augment the interaction between system and room acoustics. This context-based control supports the intentional control of audio feedback through the generation of long-term sonic behaviours that respond appropriately to the acoustics of the environment. Our prototypes map signal-inferred properties of room reverberation, ambient noise level and resonances of the acoustic environment to tempo, amplitude and timbre characteristics of the acoustic feedback, respectively. In this paper, these mappings are validated through simulations and evaluated experimentally in different acoustic conditions.

2. Related Work

2.1. Audio Feedback in Computer Music

Figure 1 shows the common structure of audio feedback. The incoming signal is connected to the output via transformation, and output re-enters the system again after a certain delay. In the digital domain, acoustic feedback is generally emulated using a delay line and filter. In some cases feedback sounds may occur suddenly due to slight increases of gain or changes of distance between a microphone and a loudspeaker, leading to the magnitude of the open-loop transfer function exceeding unity in a particular frequency region [6]. It is also defined as the violation of the Barkhausen stability criterion [7].

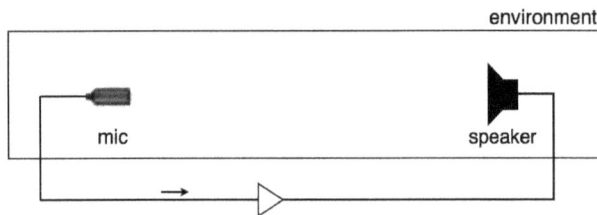

Figure 1. Basic structure for audio feedback. Signals from a microphone are amplified, emitted by a loudspeaker, then received by the microphone again to be endlessly re-amplified.

The Karplus-Strong algorithm and digital waveguide synthesis [8–10] are also based on audio feedback mechanisms, in which the fundamental resonant frequency is determined by the internal delay. This is also evident in audio feedback occurring in physical environment: the fundamental frequency of the Larsen tone is mainly determined by the delay formed as a combination of the system-internal delay and the delay caused by the physical distance between microphone and loudspeaker. Acoustic feedback effects of the electric guitar have been emulated through feedback loop models [11,12]. Gabrielli et, al. presented control of audio feedback simulation over harmonic content using a nonlinear digital oscillator consisting of a bandpass filter and a nonlinear component [7].

Because of its unique sound-generating nature, performers have deliberately generated positive feedback tones to be included in music, and designers of new musical instruments have created interfaces incorporating feedback tones. Examples of acoustic feedback used for new musical instruments include the *hybrid Virtual/Physical Feedback Instruments* (VPFI) [13], in which a physical instrument (e.g., a pipe) and virtual components (e.g., audio digital signal processing (DSP) processors such as a low-pass filter) constitute a feedback cycle. Overholt et al. documents the role of feedback in actuated musical instruments, which uses the audio signal as a control signal for the instrument again and thereby is co-manipulated by performer and machine [14].

Sanfilippo presented a common structure for feedback-based audio/music systems (Figure 2) [2]. While only a single connection exists between a microphone and a loudspeaker (via signal processing components) in the basic audio feedback model, a subsystem is added for sonic control based on signal analysis of the real-time microphone input; this can be used to trigger the Larsen tone and to control internal signal-processing states.

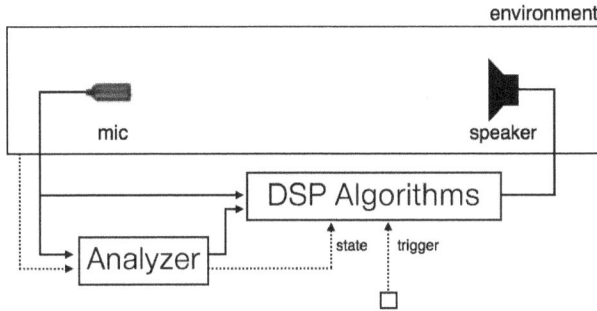

Figure 2. Structure of feedback-based audio/music systems (redesigned diagram of Figure 2 in [2]). The analyzer extracts the environmental information from the audio signal and controls the state in digital signal processing (DSP) algorithms. Other agents may exist to trigger the audio feedback.

2.2. The Feature of Openness in Audio Feedback

In these feedback-based music systems, a complex of computational signal-processing components and a physical space are naturally connected through sound, typically mediated by transducers (microphones and loudspeakers). The generated sound diffuses in the room, reflected by walls and other objects, and re-enters the computer via the microphones. No emitted sound will re-enter the system unmodified, and in addition environmental noise will be included in the input signal to further stimulate the system. The physical, acoustic part of the system can be regarded as a medium for relationships between sonic agents (including circular links), or it can be considered as another agent in the network.

The role of the environment is thereby essential to audio feedback. The openness to ambient noise, the sensitivity to the acoustic properties of the shared physical environment, and the sonic richness of its combination with signal-processing, are attractive aspects of the audio feedback-based music systems. The acoustic characteristics of a room, such as resonance and reverberation, influence the resulting sound by changing resonant frequencies and timbre (energy distribution over frequency). Although the momentary dynamics of the system may be unpredictable, the long-term dynamics of Larsen tones (e.g., stable frequencies) are known to be strongly determined by the resonant modes of the chamber in which they are placed as well as the placement of microphones and speakers.

Di Scipio's *Audible Ecosystemic Interface* (AESI) is a compositional work that interacts with its acoustic environment through audio feedback, depending on ambient noise as its energy source [3]. In AESI, the machine/environment relationship is the primary site of design. Features extracted from the received sound are used as parameters for sound synthesis.

However Kollias observed that with AESI, the composer has lost control over the overall sonic shape, as it only determines microstructural sonic design [4,15]. With *Ephemeron*, Kollias demonstrated a network of several systems recognizing and expressing sound respectively. In addition to designing low-level relationships, a performer can control the sound at a high level interactively changing operating states and interrupting stabilities.

Syntxis is also a feedback system sensitive to the acoustic environment [16]. It uses a genetic algorithm to gradually evolve bandpass filter banks toward the resonant peaks in acoustic feedback, and thus the total system adapts to the acoustic characteristics of a physical space. Di Scipio's *Background Noise Study* also extracts information from a microphone signal to control a delay line and amplitude gain followed by three algorithms including bandpass filter, resampling and granular synthesis, to create a rich sound in a space through the sounds from each of these components [17].

2.3. Context-Based Control of Audio Feedback

Although acoustic characteristics of an environment inevitably affect any audio feedback system, the combined effect of these influences with signal-processing in the feedback loop is not easily predicted [18]. The influences might lead to sonic results that conflict with sonic intentions, or the influences might be so weak as to diminish the role of the environment in the sonic result. A deeper understanding of the relation between software system and acoustic environment is required to better support compositional intentions and affirm the specificity of the role of the environment in the result.

Our system extends such dependency in a system by teasing out specificities in sonic feedback systems and mapping them to control parameters, generating long-term sonic behaviours that significantly respond to the acoustics of the environment. The structure is similar to that of the *Audible Ecosystems* (and also the "Control" Information Rate in [2]): features extracted from the received sound are used as parameters for sound synthesis. In our system, however, the selected feature is explicitly designed to be sensitive to information regarding the acoustic environment, in order to augment the specific effects of any particular physical space.

We may consider the relation of environmental characteristics to control parameters of a system well-defined and composed when we can observe desired long-term tendencies in the sound. Furthermore, if these tendencies show greater differentiation according to the physical environment in which they are placed, the system can be said to have augmented the specificity of its interactions with an acoustic environment.

3. System Design and Validation

Based on the above notions, we designed an audio feedback system in which control of the signal-processing parameters depends on recognized features of the feedback sound, aiming in particular for sensitivity to information it carries regarding the acoustic environment. These dependencies are used to shape musical tendencies in the sonic output, with analogical approximation to the following composed interactions:

1. The amount of reverberation in the space determines a tempo-scale characteristic
2. The system will change its output level depending on the volume level of ambient noise
3. A prominent resonance determines a timbre characteristic
4. The distribution of room modes determines a timbre characteristic, energy of higher frequency partials in particular

Figure 3 shows the overall diagram for simulating the effects (tempo, volume and timbre) of audio feedback and validating the mapping. The amplifier is adaptively controlled based on the peak amplitude of the input signal. Convolution of an impulse response simulates the room acoustic response, which is shaped by propagation and reflections in physical space. We used impulse response data from Fokke van Saane [19] and Aleksey Vaneev [20]. A noise function simulates room ambient noise, used as an excitation energy source in the feedback loop. We used a biquad high-pass filter with a cutoff frequency below 80Hz to remove excessive amplification of low frequency feedback, which is uncommon in a real acoustic environment. A low-pass filter, similar to the damping loop filter of a waveguide model, is also used to reduce the howling that can be introduced by excessive amplification.

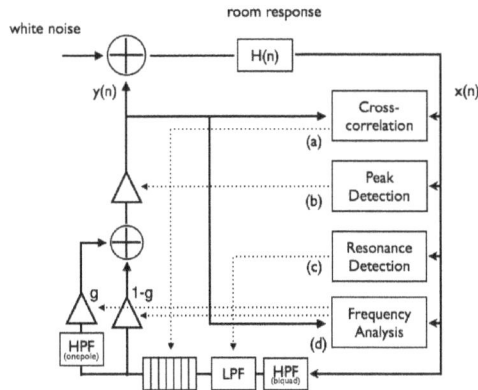

Figure 3. Overview of the system for simulating (**a**) tempo control by delay line length depending on room reverberation (Section 3.1), (**b**) volume control by gain threshold depending on volume of ambient noise (Section 3.2), (**c**) timbre control by cutoff frequency of a low-pass filter depending on a prominent resonance (Section 3.3) and (**d**) timbre control by gain of a one-pole high-pass filter according to the distribution of room modes (Section 3.4).

Table 1 compares the approximation methods and control parameters to implement the proposed mappings. It is inevitable that we can attain only approximate information about the acoustic environment from the real-time audio input, since complete real-time segregation of ambient noise and acoustic reflection information from the received feedback sound is practically impossible. Accepting this limitation, we have designed our system to infer what properties it can have, and use them to augment the specificity of the result.

Table 1. Comparison of input information, approximation methods, control targets and methods for composition of musical tendencies depending on information carried in the feedback signals regarding the acoustic environment.

Input Information	Approximation Methods	Control Targets	Control Methods
Reverberation	Cross-correlation of input/output	Tempo	Delay line length
Ambient noise volume	Peak amplitude	Output level	Gain threshold
Acoustic resonance	Freq. of maximum energy	Timbre	LPF cutoff freq.
Distribution of room modes	Variance of magnitudes from the transfer function	Timbre	HPF gain

3.1. Reverberation Level with Tempo

Tempo effects in the feedback sounds emerge from a combination of the adaptive gain control and the long delay line in the feedback loop. The synthesis of Larsen tones requires positive amplification of feedback, and once established, negative feedback to prevent saturation. This is achieved in our system by gradually increasing/decreasing the signal amplitude using a ramp function. The choice of amplification or de-amplification is determined by comparison of the input signal to high/low threshold values. When the peak amplitude of the signal over a three millisecond window exceeds the high threshold value (0.7), de-amplification is applied, otherwise the signal is amplified when the peak amplitude does not exceed the low threshold value (0.3). When the length of a delay line set to more than approximately 5000 samples (e.g., 100 ms), this range is more relevant to tempo-scale periodic occurrences of the Larsen tones (see Figure 4).

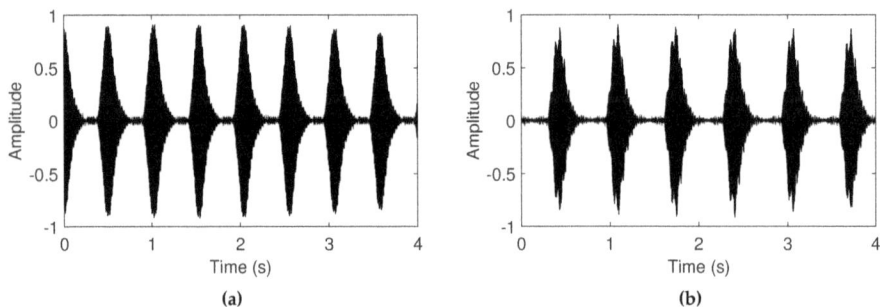

Figure 4. Simulated tempo-scale effects of feedback sounds when length of the delay line is (a) 22,000 samples (2 Hz) and (b) 28,500 samples (1.55 Hz). The signals are normalized to compare the tempo-scale periodic occurrences of the Larsen tones, depending on delay line length.

The cross-correlation of input x[n] with delayed output y[n] is used to derive an approximate measure of the amount of reverberation in the physical space. Cross-correlation measures the similarity between two signals as a function of time-lag, and is defined as:

$$\hat{r}_{xy}(l) = \frac{1}{N} \sum_{n=0}^{N-1} \tilde{x}(n)y(n+l)$$

$$l = D_1, D_1 + 1, D_1 + 2, ..., D_2$$

(1)

If a sound causes strong reverberation, the reflected sound is also large and cross-correlation is expected to be high. In order to exclude direct (non-reflected) sound propagation, the cross-correlation value is taken as the mean over time-lags from 5000 (D1) to 18,000 (D2) samples (*i.e.*, from 113 to 408 ms). This value is then divided by the maximum value over time-lags from zero to 18000 samples to minimize effect from amplitude of the feedback sound from the system. (Without this division, a louder feedback sound would also increase the cross-correlation value).

The resulting approximation of reverberation level is mapped to the length of a delay line in the feedback loop, over a range of approximately 20,000 to 55,000 samples (450 to 1250 ms). In this experiment, two opposite mappings were investigated: proportional mapping tends to generate slow tempo by a longer delay line in strongly reverberant spaces (fast tempo in weakly reverberant spaces), and reflected mapping tends to generate fast tempo by a shorter delay line in strongly reverberant spaces (slow tempo in weakly reverberant spaces). In terms of preserving the feedback signal energy, these mappings reduce or intensify the effect of the reverberant characteristics of the acoustic environment, defined as:

$$L_1 = a_1 \times r + c_1$$

$$L_2 = -a_2 \times r + c_2$$

(2)

where L_1 and L_2 indicate delay line length using the proportional (L_1) and reflected (L_2) mappings, respectively, and r means cross-correlation value. The constants were chosen to achieve delay times of approximately 20,000 to 55,000 samples, such as $a_1 = 80,000$, $c_1 = 20,000$, $a_2 = 80,000$, $c_2 = 60,000$. Figure 5 evaluates the two mappings and Table 2 compares the average delay line lengths and the reverberation characteristics (RT60 and the cross-correlation values), using the impulse response data in several locations. These results affirm that the mapping results in intentional control of tendencies in audio feedback according to inferences of room reverberation.

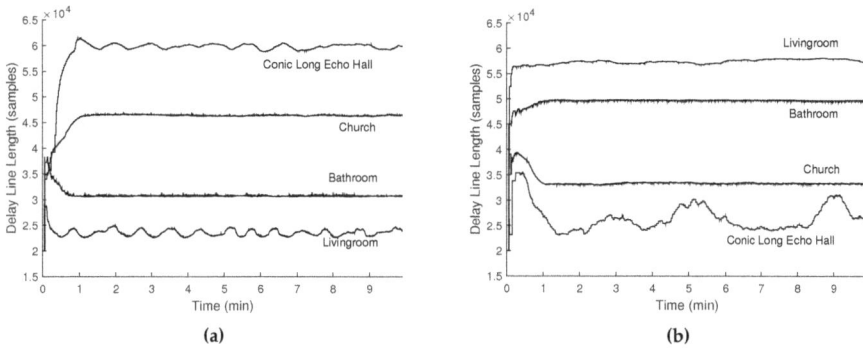

(a) (b)

Figure 5. Delay line length curves simulated using impulse response data of several locations with different reverberant characteristics (measured in Table 2), using the (**a**) proportional (L_1) and (**b**) reflected (L_2) mappings.

Table 2. Comparison of the reverberant characteristics measured by RT60 (the reverberation time over a 60 dB decay range), the cross-correlation values when the delay line is set to 28,000 samples, and average delay line lengths (in sample) using the proportional (L_1) and reflected (L_2) mappings.

Room Types	RT60 (s)	Xcorr	L_1	L_2
Livingroom	0.28	0.0407	23496	46991
Bathroom	0.58	0.1365	30674	42633
Church	0.97	0.4205	46311	30691
Long Echo Hall	3.07	0.4904	59604	25302

3.2. Ambient Noise Level with Amplitude Control

As mentioned in the previous Section 3.1, the adaptive gain control determines amplification or de-amplification by the high/low threshold values. In the previous tempo control, the threshold values were fixed to 0.3 and 0.7. However, use of different thresholds generates different output levels: a large threshold value tends to generate loud feedback sounds (Figure 6). The range of the thresholds are constrained between 0.2 and 1.0 to ensure the Larsen tone generation is possible.

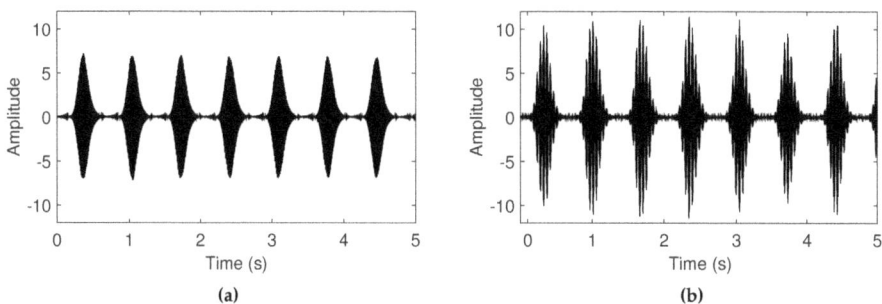

(a) (b)

Figure 6. Simulated amplitude control of feedback sounds when the low and high threshold values are (**a**) small (0.3 and 0.6) and (**b**) large (0.5 and 1.0). The length of the delay line is set to 30,000 samples (1.47 Hz).

In order to control the system's output level depending on the volume level of ambient noise, the threshold values themselves are determined by measuring the amplitude of ambient noise. The threshold is updated when the amplitude of the output signal, which is measured as the maximum amplitude value in the audio frame, is below the current low threshold, in order to exclude the feedback itself from the adaptation: if the amplitude is below threshold, it is assumed that the ambient sound dominates in the input, and the threshold value is updated according to the peak amplitude of the input signal. Figure 7 compares the average threshold and peak amplitude values of the input signals according to the volume level of ambient noise, which is simulated using white noises with different amplitudes.

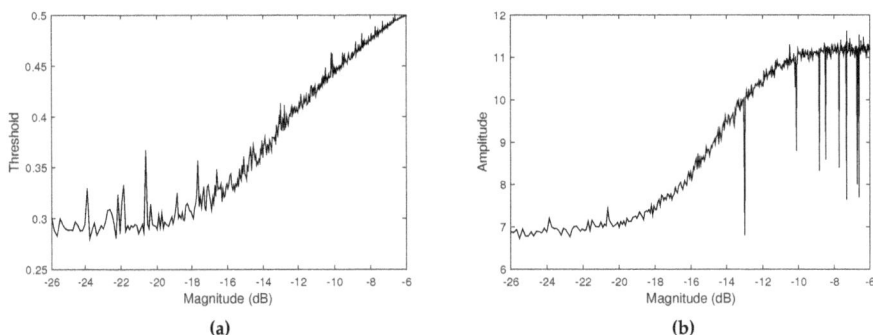

Figure 7. (a) Low threshold and (b) average peak amplitude curves of the feedback signals according to volume level of ambient noise. Ambient noise is simulated using different volumes of white noises.

3.3. Acoustic Resonance with Timbre

The length of the prevailing path in a feedback loop principally determines the acoustic resonance of the audio feedback, and this path is predominantly influenced by the distance between a microphone and a loudspeaker. We chose the resonant frequency as the simplest characteristic representation of acoustic resonance, and approximated it by the position of maximum energy in the spectrum. The distance between a microphone and a loudspeaker is simulated by appending a silence to the beginning of the impulse response, in order to delay the direct sound and the following reflections. Figure 8 compares the distances and the approximations of resonant frequencies.

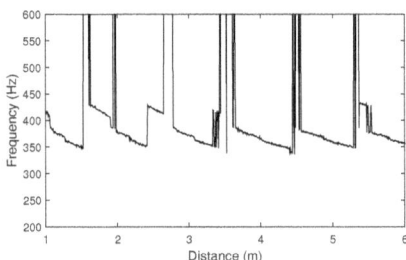

Figure 8. Resonant frequency approximations by the position of maximum energy in the spectra of the simulated feedback signals. Distance is simulated as the length of a silence appended to the beginning of the impulse response.

Timbre, a target control feature, is the quality of sound that distinguishes different sounds of the same frequency and amplitude. Our system purposes to control timbre of the feedback sounds in terms of brightness, overall distribution of energy over frequency, by the cutoff frequency parameter of the low-pass filter. Spectral centroid is used as a measure of brightness, and Figure 9 shows the controlled different timbres through the comparison of spectra and spectral centroids: since the cutoff frequency is the attenuating position, higher cutoff frequency generates brighter sounds and lower cutoff frequency generates darker sounds.

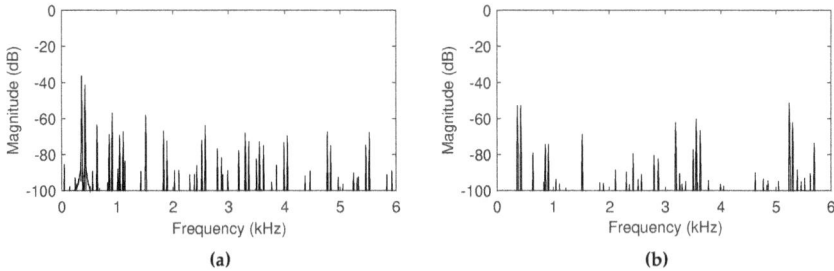

Figure 9. Plots of the magnitude spectrum of the normalized feedback signals when the cutoff frequency of the low-pass filter is (**a**) 300 Hz and (**b**) 6000 Hz. Spectral centroids are 3327 Hz and 3965 Hz respectively. The distance is set to 4 meters.

The approximation of resonant frequency is mapped to the cutoff frequency parameter of the low-pass filter. An intriguing point is that the change in timbre (brightness) mostly influences the position of maximum energy in a spectrum, which is used as the input information. This enables evaluation of this mapping by observing changes in the approximations of resonant frequencies. However, the resonant frequency does not always behave in proportion to cutoff frequency change: timbre (brightness) may show abrupt changes by only a slight change in cutoff frequency (the *nonlinearity* explained in [2]). Thus, we designed the cutoff frequency to adaptively change depending on variation of the frequency approximation.

Similarly to the previous tempo control designs (Section 3.1), two opposite mappings were also investigated in accordance with the changing direction. For example, if the estimated frequency is bigger than a threshold, the cutoff frequency parameter gradually increases (proportional mapping) or decreases (reflected mapping) using a ramp function; on the other hand, when the estimated frequency is lower than a threshold, the parameter decreases (proportional mapping) or increases (reflected mapping). The effect of changing the distance is thereby intensified by the proportional mapping and reduced by the reflected mapping: Figure 10 evaluates the two mappings.

3.4. Distribution of Room Modes with Timbre

In Section 3.1, we approximated a measure of the amount of reverberation according to reverberation time. In this section, another approximation regarding reverberation is investigated, emphasizing related to spectral rather than of temporal characteristics. High reverberation diffusion means that reflections are scattered along various paths, producing a smooth magnitude response; on the other hand, strong reflections exist at specific frequencies in low reverberation diffusion, producing an uneven magnitude response. While in the previous Section 3.3 we measure the strongest path of audio feedback, this section is concerned with diffusion of the reflection paths.

In order to discern such distribution of room modes, we used the variance of magnitudes from the transfer function, derived from the frequency responses of the input and output signals. Table 3 compares actual measures of the distribution through spectral flatness [21] and variances from the impulse response data in several locations, and approximate measures by the variances from the derived transfer functions. Figure 11 also details these approximate measures.

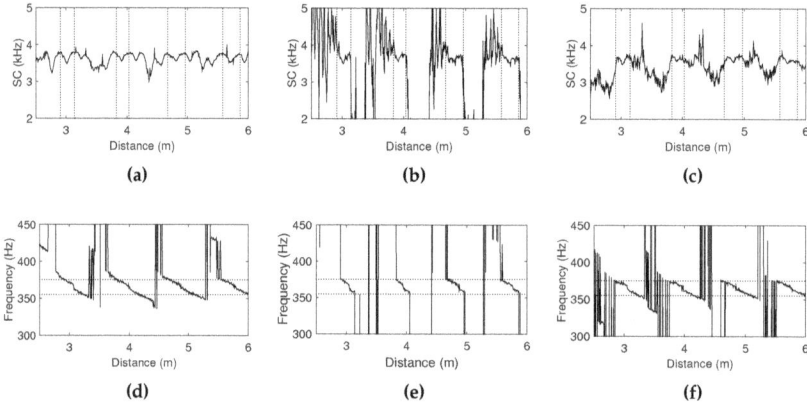

Figure 10. Approximations of spectral centroids and resonant frequencies from the simulated feedback signals when the cutoff frequency of the low-pass filter is (**a,d**) constant and (**b,e**) controlled with the proportional mapping and (**c,f**) reflected mappings. Vertical dot lines in the upper figures represent the positions where the approximated resonant frequencies deviate from the thresholds. These thresholds are depicted as the horizontal dot lines in the lower figures.

Table 3. Comparison of the distributions of room modes measured by spectral flatness and variance of the magnitudes of the impulse response data ($H(f)$), average variance of the frequency magnitudes of the estimated transfer function ($H'(f)$), and average gain values mapped by $H'(f)$ (g).

Room Types	Spectral Flatness	Var($|H(f)|$)	Var($|H'(f)|$)	g
Long Echo Hall	0.85	29.71	52.90	0.56
Livingroom	0.83	31.52	61.08	0.74
Church	0.73	52.68	70.37	0.95
Bathroom	0.64	79.15	90.00	1.0

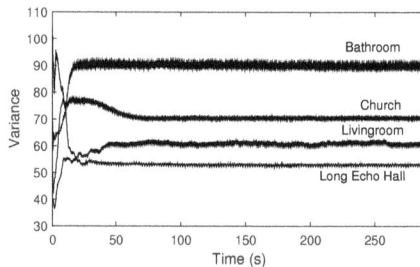

Figure 11. Variances of magnitudes of transfer functions simulated using impulse response data of several locations with different distributions of room modes (measured in Table 3).

The resulting approximation of the distribution is mapped to the gains of the unmodified output signal and a one-pole high-pass filter having a cutoff frequency of 4000 Hz, which are mixed thereafter: high variance increases the gain of the high-pass filter (decreases the gain of the unmodified signal) and the sum of the gains is always unity in linear scale. The last column in Table 3 shows the average gain values mapped by the variance measurements. This gain control changes spectral centroid as well as distribution width of spectral energy, which is associated with softness of a sound [22]. Figure 12 evaluates the mappings by comparing the changes in spectral centroids.

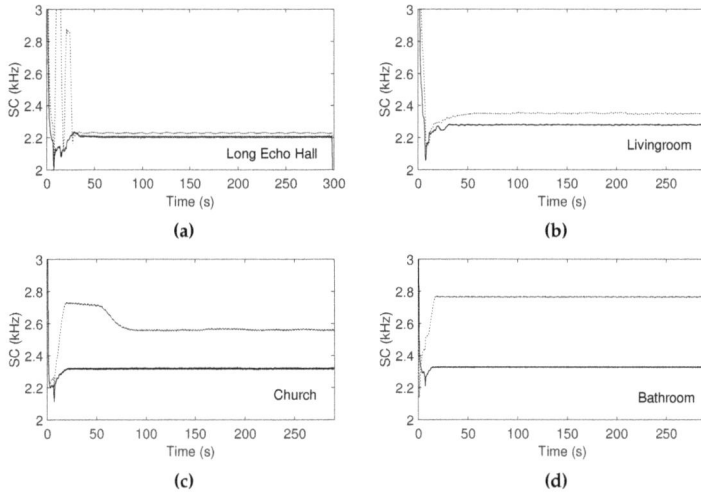

Figure 12. Spectral centroids measured using the impulse response data of (**a**) long echo hall, (**b**) living room, (**c**) church, and (**d**) bathroom. Solid curves represent unmodified signals, and dot curves represent output signals mixed with the output of the one-pole high-pass filter, where the mix ratio between the two signals is controlled depending on the estimated variances.

4. Experiments in a Real Room

The designed dependencies were implemented as software authored using Max/MSP 7, developed by Cycling '74 (Santa Cruz, CA, USA) and openFrameworks, a C++ open source toolkit (Figure 13), and investigated by experiments in a 3 m × 5 m small living room with artificially controlled acoustic characteristics. Tests were performed with external audio devices 8030A loudspeaker and SM58 microphone, manufactured by Genelec (Iisalmi, Finland) and Shure (Niles, IL, USA), respectively.

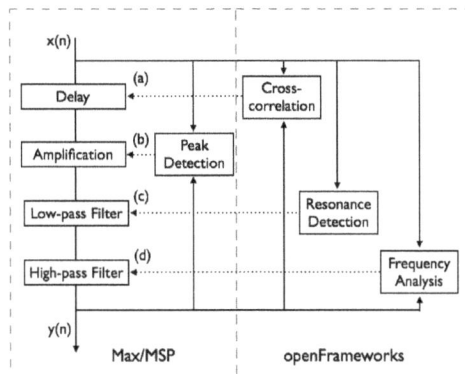

Figure 13. Overview of the system implemented as software (Max/MSP and openFrameworks), for (**a**) tempo control depending on room reverberation (Section 4.1), (**b**) volume control depending on ambient noise volume (Section 4.2) and (**c**) timbre control depending on frequency response of the acoustic environment (Section 4.3) and (**d**) timbre control depending on distribution of room modes (Section 4.4).

4.1. Observations in Different Room Reverberations

In the previous work, we evaluated in rooms with different reverberant acoustics, while in this paper we used a reverberation model named *yafr2* that is provided within the Max/MSP software to gain more parametric control over room properties while observing changes. We experimented with different reverberation time parameters while holding other properties constant. Figure 14 and Table 4 evaluate the two mappings of delay line length from cross-correlation value. The constants in Equation (2) were chosen as $a_1 = 33,000$, $c_1 = 1000$ and $a_2 = 33,000$, $c_2 = 20,000$. Although the differences are not as clearly differentiated than those from the simulated rooms, we still observe differences in the long-term averages as expected, both when using the proportional mapping, and also with the reflected mapping.

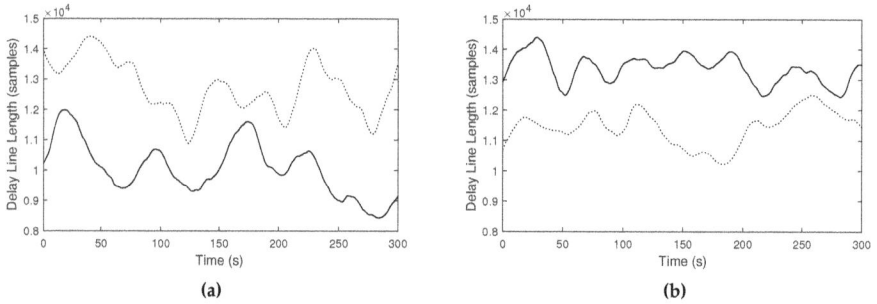

(a)　　　　　　　　　　　　　　　　(b)

Figure 14. Delay line length curves measured in the strongly reverberant (RT30 of 8.50 seconds, solid curves) and weakly reverberant environment (RT30 of 0.65 seconds, dot curves), using the (a) proportional (L_1) and (b) reflected (L_2) mappings.

Table 4. Comparison of the reverberant characteristics (RT30) and the average delay line lengths during the first 5 minutes, using the proportional (L_1) and reflected (L_2) mappings.

RT30 (s)	L_1	L_2
0.65	12720	11450
1.33	11690	11940
8.50	10070	13360

4.2. Observations in Different Ambient Noise Levels

This experiment was conducted under different ambient noise amplitudes. Figure 15 evaluates the mapping. We observed different threshold values under different noisy conditions: they increase in noisy conditions and decrease in quiet conditions, as expected.

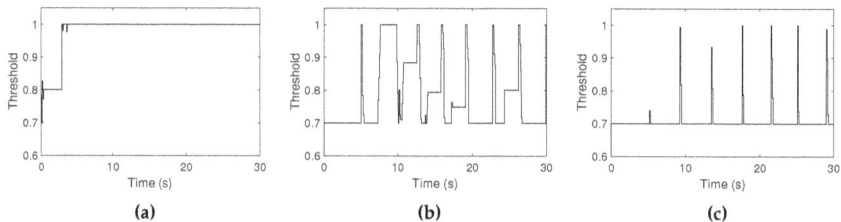

(a)　　　　　　　　　(b)　　　　　　　　　(c)

Figure 15. High threshold curves measured when the volume level of ambient noise is measured as (a) −25 dB, (b) −30 dB, and (c) −47 dB. The range of the high threshold is constrained between 0.7 and 1.0.

4.3. Observations with Different Acoustic Resonances

This experiment evaluated timbre changes with different distances between the microphone and the loudspeaker. Figure 16a compares the approximations of resonant frequencies using a constant or variable cutoff frequency of the low-pass filter, the latter driven by the estimated resonant frequency. We also observed the tendencies in the long-term averages as expected: the frequency change tends to be intensified by the proportional mapping and reduced by the reflected mapping. Even when the frequency is not noticeably changed, we still observed the small change in terms of brightness in comparison of the magnitude spectra (Figure 16b).

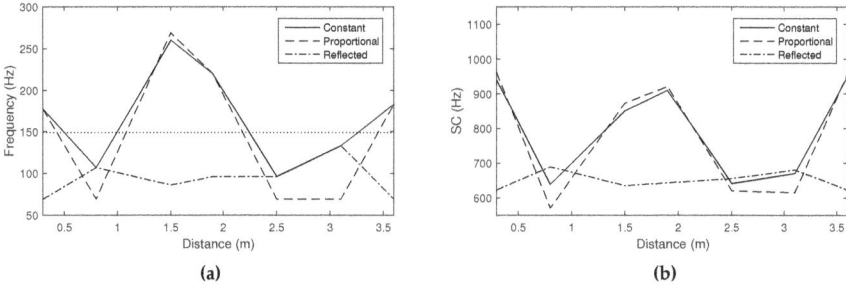

(a)   (b)

Figure 16. Approximated (**a**) resonant frequencies and (**b**) spectral centroids according to the distance between the loudspeaker and the microphone, with different cutoff frequency settings of the low-pass filter: constant (straight lines), determined by the measured resonant frequency, using the proportional (L_1, dash lines) and reflected mappings (L_2, dash-dot lines). The threshold was set to 140 Hz (dot line in (**a**)).

4.4. Observations with Different Distributions of Room Modes

Like the Section 4.1, different room properties are generated through diffusion parameters in the reverberation model while holding other properties constant. Figure 17 shows the measured variances and Figure 18 evaluates the mappings when the gain is constant and depends on the measurements. We also observed different variance curves and deviations of spectral centroids, as expected.

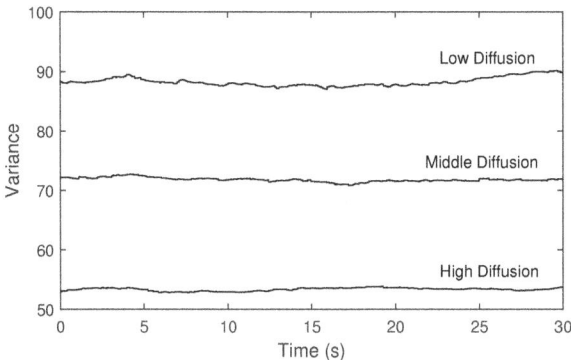

Figure 17. Variances of magnitudes of the measured transfer functions when the diffusion parameter is set to low, middle and high.

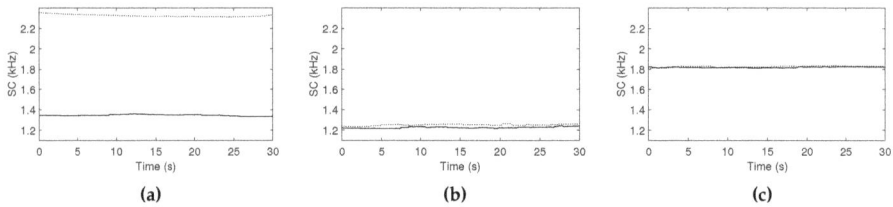

Figure 18. Spectral centroids measured when the diffusion parameter is set to (**a**) low, (**b**) middle and (**c**) high. Solid curves represent unmodified signals, and dot curves represent output signals mixed with the output of the one-pole high-pass filter, where the mix ratio between the two signals is controlled depending on the estimated variances.

5. Conclusions

In this paper we presented several methods for augmenting composed audio feedback interactions within acoustic environments, including room response, ambient noise and placement of transducers; supporting sound generation that strikes a balance between unpredictable short-term behavior and intentional long-term tendencies. We designed long-term tendencies in terms of tempo, amplitude and timbre characteristics depending upon reverberant properties, ambient noise level, and acoustic resonance respectively, each inferred indirectly from the environment. We measured these mappings through software simulations as well as acoustic experiments.

Beyond regarding the acoustic environment as a filter and source of disturbance, it can also be considered a site of discovery. The composed system attempts to differentiate and affirm itself through reflections, yet by doing so it also augments or exaggerates the specificity of the environment. This duality is also evident in the analysis: comparing input and output signals cannot fully segregate external and internal influence, as the feedback sounds depend upon parameters within the system, which in turn depend on the analysis. Through this paper, we are satisfied that affirming specificity was achieved, but we believe this is only an initial step in developing truly adaptive, self-augmenting responsive sonic environments.

Acknowledgments: This work was supported by KAIST (Project No. G04140049).

Author Contributions: The work was done in close collaboration. Seunghun Kim developed the models, conducted the experiments, derived results and drafted the main manuscript. Graham Wakefield contributed to the manuscript and advised in the refinement of the system. Juhan Nam co-developed the models, supervised the research and contributed to the manuscript.

Conflicts of Interest: The authors declare no conflict of interest.

References

1. Reich, S. *Pendulum Music*; Universal Editor: London, UK, 1968.
2. Sanfilippo, D.; Valle, A.; Elettronica, M. Feedback Systems: An Analytical Framework. *Comput. Music J.* **2013**, *37*, 12–27.
3. Di Scipio, A. "Sound is the interface": From Interactive to Ecosystemic Signal Processing. *Organ. Sound* **2003**, *8*, 269–277.
4. Kollias, P.A. Ephemeron : Control over Self-Organised Music. In Proceedings of the 5th International Conference of Sound and Music Computing, Berlin, Germany, 31 July–3 August 2008; pp. 138–146.
5. Kim, S.; Nam, J.; Wakefield, G. Toward Certain Sonic Properties of an Audio Feedback System by Evolutionary Control of Second-Order Structures. In Proceedings of the 4th International Conference (and 12th European event) on Evolutionary and Biologically Inspired Music, Sound, Art and Design (Part of Evostar 2015), Copenhagen, Denmark, 8–10 April 2015.
6. Berdahl, E.; Harris, D. Frequency Shifting for Acoustic Howling Suppression. In Proceedings of the 13th International Conference on Digital Audio Effects, Graz, Austria, 6–10 September 2010.

7.   Gabrielli, L.; Giobbi, M.; Squartini, S.; Valimaki, V. A Nonlinear Second-Order Digital Oscillator for Virtual Acoustic Feedback. In Processing of the 2014 IEEE International Conference on Acoustics, Speech and Signal Processing (ICASSP), Florence, Italy, 4–9 May 2014; pp. 7485–7489.

8.   Karjalainen, M.; Välimäki, V.; Tolonen, T. Plucked-String Models: From the Karplus-Strong Algorithm to Digital Waveguides and Beyond. *Comput. Music J.* **1998**, 17–32.

9.   Kim, S.; Kim, M.; Yeo, W.S. Digital Waveguide Synthesis of the Geomungo with a Time-varying Loss Filter. *J. Audio Eng. Soc.* **2013**, *61*, 50–61.

10.   Smith, J.O. Physical Modeling using Digital Waveguides. *Comput. Music J.* **1992**, *16*, 74–91.

11.   Sullivan, C.R. Extending the Karplus-Strong Algorithm to Synthesize Electric Guitar Timbres with Distortion and Feedback. *Comput. Music J.* **1990**, *14*, 26–37.

12.   Gustafsson, F. System and Method for Simulation of Acoustic Feedback. US Patent 7,572,972, 2009.

13.   Waters, S. Performance Ecosystems: Ecological Approaches to Musical Interaction. EMS: Electroacoustic Music Studies Network, 2007; pp. 1–20.

14.   Overholt, D.; Berdahl, E.; Hamilton, R. Advancements in Actuated Musical Instruments. *Organ. Sound* **2011**, *16*, 154–165.

15.   Kollias, P.A. The Self-Organising Work of Music. *Organ. Sound* **2011**, *16*, 192–199.

16.   Scamarcio, M. Space as an Evolution Strategy. Sketch of a Generative Ecosystemic Structure of Sound. In Proceedings of the Sound and Music Computing Conference, Berlin, Germany, 31 July–3 August 2008; pp. 95–99.

17.   Di Scipio, A. Listening to Yourself through the Otherself: On Background Noise Study and Other Works. *Organ. Sound* **2011**, *16*, 97–108.

18.   Aufermann, K. Feedback and Music: You Provide the Noise, the Order Comes by Itself. *Kybernetes* **2005**, *34*, 490–496.

19.   Impulse Responses Made by Fokke van Saane. Available online: http://fokkie.home.xs4all.nl/IR.htm (accessed on 23 April 2016).

20.   Free Reverb Impulse Responses. Available online: http://www.voxengo.com/impulses (accessed on 23 April 2016).

21.   Lerch, A. *An Introduction to Audio Content Analysis: Applications in Signal Processing and Music Informatics*; John Wiley & Sons: New York, NY, USA, 2012.

22.   Fritz, C.; Blackwell, A.F.; Cross, I.; Woodhouse, J.; Moore, B.C. Exploring Violin Sound Quality: Investigating English Timbre Descriptors and Correlating Resynthesized Acoustical Modifications with Perceptual Properties. *J. Acoust. Soc. Am.* **2012**, *131*, 783–794.

applied sciences

MDPI

Article

Full-Band Quasi-Harmonic Analysis and Synthesis of Musical Instrument Sounds with Adaptive Sinusoids

Marcelo Caetano [1,*], George P. Kafentzis [2], Athanasios Mouchtaris [2,3] and Yannis Stylianou [2]

[1]   Sound and Music Computing Group, Institute for Systems and Computer Engineering, Technology and Science (INESC TEC), 4200-465 Porto, Portugal

[2]   Multimedia Informatics Lab, Department of Computer Science, University of Crete, 700-13 Heraklion, Greece; kafentz@csd.uoc.gr (G.P.K.); mouchtar@ics.forth.gr (A.M.); yannis@csd.uoc.gr (Y.S.)

[3]   Signal Processing Laboratory, Institute of Computer Science, Foundation for Technology & Research-Hellas (FORTH), 700-13 Heraklion, Greece

*   Correspondence: mcaetano@inesctec.pt; Tel.: +351-22-209-4217

Academic Editor: Vesa Valimaki
Received: 16 February 2016; Accepted: 19 April 2016; Published: 2 May 2016

Abstract: Sinusoids are widely used to represent the oscillatory modes of musical instrument sounds in both analysis and synthesis. However, musical instrument sounds feature transients and instrumental noise that are poorly modeled with quasi-stationary sinusoids, requiring spectral decomposition and further dedicated modeling. In this work, we propose a full-band representation that fits sinusoids across the entire spectrum. We use the extended adaptive Quasi-Harmonic Model (eaQHM) to iteratively estimate amplitude- and frequency-modulated (AM–FM) sinusoids able to capture challenging features such as sharp attacks, transients, and instrumental noise. We use the signal-to-reconstruction-error ratio (SRER) as the objective measure for the analysis and synthesis of 89 musical instrument sounds from different instrumental families. We compare against quasi-stationary sinusoids and exponentially damped sinusoids. First, we show that the SRER increases with adaptation in eaQHM. Then, we show that full-band modeling with eaQHM captures partials at the higher frequency end of the spectrum that are neglected by spectral decomposition. Finally, we demonstrate that a frame size equal to three periods of the fundamental frequency results in the highest SRER with AM–FM sinusoids from eaQHM. A listening test confirmed that the musical instrument sounds resynthesized from full-band analysis with eaQHM are virtually perceptually indistinguishable from the original recordings.

Keywords: musical instruments; analysis and synthesis; sinusoidal modeling; AM–FM sinusoids; adaptive modeling; nonstationary sinusoids; full-band modeling

PACS: 43.75.Zz; 43.75.De; 43.75.Ef; 43.75.Fg; 43.75.Gh; 43.75.Kk; 43.75.Mn; 43.75.Pq; 43.75.Qr

1. Introduction

Sinusoidal models are widely used in the analysis [1,2], synthesis [2,3], and transformation [4,5] of musical instrument sounds. The musical instrument sound is modeled by a waveform consisting of a sum of time-varying sinusoids parameterized by their amplitudes, frequencies, and phases [1–3]. Sinusoidal analysis consists of the estimation of parameters, synthesis comprises techniques to retrieve a waveform from the analysis parameters, and transformations are performed as changes of the parameter values. The time-varying sinusoids, called partials, represent how the oscillatory modes of the musical instrument change with time, resulting in a flexible representation with perceptually meaningful parameters. The parameters completely describe each partial, which can be manipulated independently.

Several important features can be directly estimated from the analysis parameters, such as fundamental frequency, spectral centroid, inharmonicity, spectral flux, onset asynchrony, among many others [2]. The model parameters can also be used in musical instrument classification, recognition, and identification [6], vibrato detection [7], onset detection [8], source separation [9], audio restoration [10], and audio coding [11]. Typical transformations are pitch shifting, time scaling [12], and musical instrument sound morphing [5]. Additionally, the parameters from sinusoidal models can be used to estimate alternative representations of musical instrument sounds, such as spectral envelopes [13] and the source-filter model [14,15].

The quality of the representation is critical and can impact the results for the above applications. In general, sinusoidal models render a close representation of musical instrument sounds because most pitched musical instruments are designed to present very clear modes of vibration [16]. However, sinusoidal models do not result in perfect reconstruction upon resynthesis, leaving a modeling residual that contains whatever was not captured by the sinusoids [17]. Musical instrument sounds have particularly challenging features to represent with sinusoids, such as sharp attacks, transients, inharmonicity, and instrumental noise [16]. Percussive sounds produced by plucking strings (such as harpsichords, harps, and the *pizzicato* playing technique) or striking percussion instruments (such as drums, idiophones, or the piano) feature sharp onsets with highly nonstationary oscillations that die out very quickly, called transients [18]. Flute sounds characteristically comprise partials on top of breathing noise [16]. The reed in woodwind instruments presents a highly nonlinear behavior that also results in attack transients [19], while the stiffness of piano strings results in a slightly inharmonic spectrum [18]. The residual from most sinusoidal representations of musical instrument sounds contains perceptually important information [17]. However, the extent of this information ultimately depends on what the sinusoids are able to capture.

The standard sinusoidal model (SM) [1,20] was developed as a parametric extension of the short-time Fourier transform (STFT) so both analysis and synthesis present the same time-frequency limitations as the Discrete Fourier Transform (DFT) [21]. The parameters are estimated with well-known techniques, such as peak-picking and parabolic interpolation [20,22], and then connected across overlapping frames (partial tracking [23]). Peak-picking is known to bias the estimation of parameters because errors in the estimation of frequencies can bias the estimation of amplitudes [22,24]. Additionally, the inherent time-frequency uncertainty of the DFT further limits the estimation because long analysis windows blur the temporal resolution to improve the frequency resolution and *vice-versa* [21]. The SM uses quasi-stationary sinusoids (QSS) under the assuption that the partials are relatively stable inside each frame. QSS can accurately capture the lower frequencies because these have fewer periods inside each frame and thus less temporal variation. However, higher frequencies have more periods inside each frame with potentially more temporal variation lost by QSS. Additionally, the parameters of QSS are estimated using the center of the frame as the reference and the values are less accurate towards the edges because the DFT has a stationary basis [25]. This results in the loss of sharpness of attack known as pre-echo.

The lack of transients and noise is perceptually noticeable in musical instrument sounds represented with QSS [17,26]. Serra and Smith [1] proposed to decompose the musical instrument sound into a sinusoidal component represented with QSS and a residual component obtained by subtraction of the sinusoidal component from the original recording. This residual is assumed to be noise not captured by the sinusoids and commonly modeled by filtering white noise with a time-varying filter that emulates the spectral characteristics of the residual component [1,17]. However, the residual contains both errors in parameter estimation and transients plus noise missed by the QSS [27].

The time-frequency resolution trade-off imposes severe limits on the detection of transients with the DFT. Transients are essentially localized in time and usually require shorter frames which blur the peaks in the spectrum. Daudet [28] reviews several techniques to detect and extract transients with sinusoidal models. Multi-resolution techniques [29,30] use multiple frame sizes to circumvent the time-frequency uncertainty and to detect modulations at different time scales. Transient modeling synthesis (TMS) [26,27,31] decomposes sounds into sinusoids plus transients plus noise and models each separately. TMS performs sinusoidal plus residual decomposition with QSS and then extracts the transients from the residual.

An alternative to multiresolution techniques is the use of high-resolution techniques based on total least squares [32] such as ESPRIT [33], MUSIC [34], and RELAX [35] to fit exponentially damped sinusoids (EDS). EDS are widely used to represent musical instrument sounds [11,36,37]. EDS are sinusoids with stationary (*i.e.*, constant) frequencies modulated in amplitude by an exponential function. The exponentially decaying amplitude envelope from EDS is considered suitable to represent percussive sounds when the beginning of the frame is synchronized with the onsets [38]. However, EDS requires additional partials when there is no synchronization, which increases the complexity of the representation. ESPRIT decomposes the signal space into sinusoidal and residual, further ranking the sinusoids by decreasing magnitude of eigenvalue (*i.e.*, spectral energy). Therefore, the first K sinusoids maximize the energy upon resynthesis regardless of their frequencies.

Both the SM and EDS rely on sinusoids with stationary frequencies, which are not appropriate to represent nonstationary oscillations [21]. Time-frequency reassignment [39–41] was developed to estimate nonstationary sinusoids. Polynomial phase signals [20,25] such as splines [21] are commonly used as an alternative to stationary sinusoids. McAulay and Quatieri [20] were among the first to interpolate the phase values estimated at the center of the analysis window across frames with cubic polynomials to obtain nonstationary sinusoids inside each frame. Girin *et al.* [42] investigated the impact of the order of the polynomial used to represent the phase and concluded that order five does not improve the modeling performance sufficiently to justify the increased complexity. However, even nonstationary sinusoids leave a residual with perceptually important information that requires further modeling [25].

Sinusoidal models rely on spectral decomposition assuming that the lower end of the spectrum can be modeled with sinusoids while the higher end essentially contains noise. The estimation of the separation between the sinusoidal and residual components has proved difficult [27]. Ultimately, spectral decomposition misses partials on the higher end of the spectrum because the separation is artificial, depending on the spectrum estimation method rather than the spectral characteristics of musical instrument sounds. We consider spectral decomposition to be a consequence of artifacts from previous sinusoidal models instead of an acoustic property of musical instruments. Therefore, we propose the full-band modeling of musical instrument sounds with adaptive sinusoids as an alternative to spectral decomposition.

Adaptive sinusoids (AS) are nonstationary sinusoids estimated to fit the signal being analyzed usually via an iterative parameter re-estimation process. AS have been used to model speech [43–46] and musical instrument sounds [25,47]. Pantazis [45,48] developed the adaptive Quasi-Harmonic Model (aQHM), which iteratively adapts the frequency trajectories of all sinusoids at the same time based on the Quasi-Harmonic Model (QHM). Adaptation improves the fit of a spectral template via an iterative least-squares (LS) parameter estimation followed by frequency correction. Later, Kafentzis [43] devised the extended adaptive Quasi-Harmonic Model (eaQHM), capable of adapting both amplitude and frequency trajectories of all sinusoids iteratively. In eaQHM, adaptation is equivalent to the iterative projection of the original waveform onto nonstationary basis functions that are locally adapted to the time-varying characteristics of the sound, capable of modeling sudden changes such as sharp attacks, transients, and instrumental noise. In a previous work [47], we showed that eaQHM is capable of retaining the sharpness of the attack of percussive sounds.

In this work, we propose full-band modeling with eaQHM for a high-quality analysis and synthesis of isolated musical instrument sounds with a single component. We compare our method to QSS estimated with the standard SM [20] and EDS estimated with ESPRIT [36]. In the next section, we discuss the differences in full-band spectral modeling and traditional decomposition for musical instrument sounds. Next, we describe the full-band quasi-harmonic adaptive sinusoidal modeling behind eaQHM. Then, we present the experimental setup, describe the musical instrument sound database used in this work and the analysis parameters. We proceed to the experiments, present the results, and evaluate the performance of QSS, EDS, and eaQHM in modeling musical instrument sounds. Finally, we discuss the results and present conclusions and perspectives for future work.

2. Full-Band Modeling

Spectrum decomposition splits the spectrum of musical instrument sounds into a sinusoidal component and a residual as illustrated in Figure 1a. Spectrum decomposition assumes that there are partials only up to a certain cutoff frequency f_c, above which there is only noise. Figure 1a represents the spectral peaks as spikes on top of colored noise (wide light grey frequency bands) and f_c as the separation between the sinusoidal and residual components. Therefore, f_c determines the number of sinusoids because only the peaks at the lower frequency end of the spectrum are represented with sinusoids (narrow dark grey bars) and the rest is considered wide-band and stochastic noise existing across the whole range of the spectrum. There is noise between the spectral peaks and at the higher end of the spectrum. In a previous study [17], we showed that the residual from the SM is perceptually different from filtered (colored) white noise. Figure 1a shows that there are spectral peaks left in the residual because the spectral peaks above f_c are buried under the estimation noise floor (and sidelobes). Consequently, the residual from sinusoidal models that rely on spectral decomposition such as the SM is perceptually different from filtered white noise.

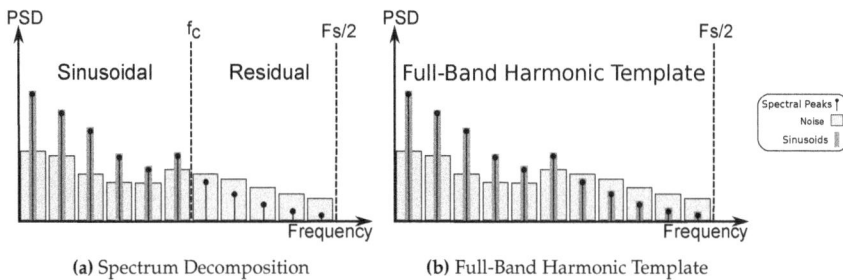

(a) Spectrum Decomposition   (b) Full-Band Harmonic Template

Figure 1. Illustration of the spectral decomposition and full-band modeling paradigms.

From an acoustic point of view, the physical behavior of musical instruments can be modeled as the interaction between an excitation and a resonator (the body of the instrument) [16]. This excitation is responsible for the oscillatory modes whose amplitudes are shaped by the frequency response of the resonator. The excitation signal commonly contains discontinuities, resulting in wide-band spectra. For instance, the vibration of the reed in woodwinds can be approximated by a square wave [49], the friction between the bow and the strings results in an excitation similar to a sawtooth wave [16], the strike in percussion instruments can be approximated by a pulse [2], while the vibration of the lips in brass instruments results in a sequence of pulses [50] (somewhat similar to the glottal excitation, which is also wide band [46]).

Figure 1b illustrates a full-band harmonic template spanning the entire frequency range, fitting sinusoids to spectral peaks in the vicinity of harmonics of the fundamental frequency f_0. The spectrum of musical instruments is known to present deviations from perfect harmonicity [16], but quasi-harmonicity is supported by previous studies [51] that found deviations as small as 1%. In this work, the full-band harmonic template becomes quasi-harmonic after the estimation of parameters via least-squares followed by a frequency correction mechanism (see details in Section 3.1). Therefore, full-band spectral modeling assumes that both the excitation and the instrumental noise are wide band.

3. Adaptive Sinusoidal Modeling with eaQHM

In what follows, $x(n)$ is the original sound waveform and $\hat{x}(n)$ is the sinusoidal model with sample index n. Then, the following relation holds:

$$x(n) = \hat{x}(n) + e(n), \tag{1}$$

where $e(n)$ is the modeling error or residual. Each frame of $x(n)$ is

$$x(n, m) = x(n) w(n - mH), \quad m = 0, \cdots, M - 1, \tag{2}$$

where m is the frame number, M is the number of frames, and H is the hop size. The analysis window $w(n)$ has L samples and it defines the frame size. Typically, $H < L$ such that the frames m overlap.

Figure 2 presents an overview of the modeling steps in eaQHM. The feedback loop illustrates the adaptation cycle, where $\hat{x}(n)$ gets closer to $x(n)$ with each iteration. The iterative process stops when the fit improves by less than a threshold ε. The dark blocks represent parameter estimation based on the quasi-harmonic model (QHM), followed by interpolation of the parameters across frames before additive [1] resynthesis (instead of overlap add (OLA) [52]). The resulting time-varying sinusoids are used as nonstationary basis functions for the next iteration, so the adaptation procedure illustrated in Figure 3 iteratively projects $x(n)$ onto $\hat{x}(n)$. Next, QHM is summarized, followed by parameter interpolation and then eaQHM.

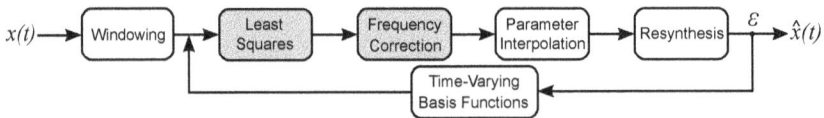

Figure 2. Block diagram depicting the modeling steps in the extended adaptive Quasi-Harmonic Model (eaQHM). The blocks with a dark background correspond to parameter estimation, while the feedback loop illustrates adaptation as iteration cycles around the loop. See text for the explanation of the symbols.

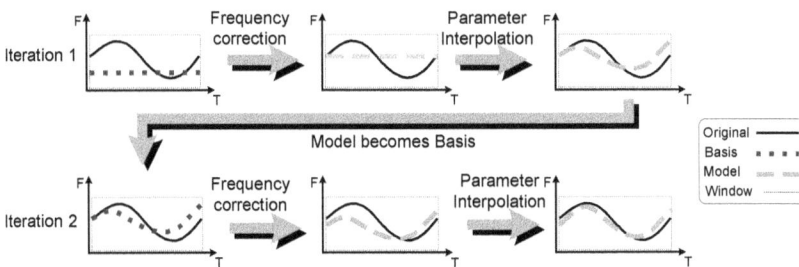

Figure 3. Illustration of the adaptation of the frequency trajectory of a sinusoidal partial inside the analysis window in eaQHM. The figure depicts the first and second iterations of eaQHM around the loop in Figure 2, showing local adaptation as the iterative projection of the original waveform onto the model.

3.1. The Quasi-Harmonic Model (QHM)

QHM [48] projects $x(n, m)$ onto a template of sinusoids $e^{j2\pi n \hat{f}_k / f_s}$ with constant frequencies $\hat{f}_k$ and sampling frequency f_s. QHM estimates the parameters of $\hat{x}(n, m)$ using

$$\hat{x}\left(n,m\right) = \sum_{k=-K}^{K} \left(\mathbf{a}_k + n\mathbf{b}_k\right) e^{j2\pi \hat{f}_k n / f_s}, \tag{3}$$

where k is the partial number, K is the number of real sinusoids, $\mathbf{a}_k$ the complex amplitude and $\mathbf{b}_k$ is the complex slope of the kth sinusoid. The term $n\mathbf{b}_k$ arises from the derivative of $e^{j2\pi n \hat{f}_k / f_s}$ with respect to frequency. The constant frequencies $\hat{f}_k$ define the spectral template used by QHM to fit the analysis parameters $\mathbf{a}_k$ and $\mathbf{b}_k$ by least-squares (LS) [44,45]. In principle, any set of frequencies $\hat{f}_k$ can be used because the estimation of $\mathbf{a}_k$ and $\mathbf{b}_k$ also provides a means of correcting the initial frequency values $\hat{f}_k$ by making $\hat{f}_k$ converge to nearby frequencies f_k present in the signal frame. The mismatch between f_k and $\hat{f}_k$ leads to an estimation error $\eta_k = f_k - \hat{f}_k$. Pantazis *et al.* [48] showed that QHM provides an estimate of η_k given by

$$\hat{\eta}_k = \frac{f_s}{2\pi} \frac{\mathrm{Re}\left\{\mathbf{a}_k\right\} \mathrm{Im}\left\{\mathbf{b}_k\right\} - \mathrm{Im}\left\{\mathbf{a}_k\right\} \mathrm{Re}\left\{\mathbf{b}_k\right\}}{\left|\mathbf{a}_k\right|^2}, \tag{4}$$

which corresponds to the frequency correction block in Figure 2. Then $\hat{x}\left(n,m\right)$ is locally synthesized as

$$\hat{x}\left(n,m\right) = \sum_{k=-K}^{K} \hat{a}_k e^{j\left(2\pi \hat{F}_k n / f_s + \hat{\phi}_k\right)}, \tag{5}$$

where $\hat{a}_k = \left|\mathbf{a}_k\right|$, $\hat{F}_k = \hat{f}_k + \hat{\eta}_k$, and $\hat{\phi}_k = \angle \mathbf{a}_k$ are constant inside the frame m.

The full-band harmonic spectral template shown in Figure 1b is obtained by setting $\hat{f}_k = k f_0$ with k an integer and $1 \leq k \leq f_s / 2 f_0$. The f_0 is not necessary to estimate the parameters, but it improves the fit because the initial full-band harmonic template approximates better the spectrum of isolated quasi-harmonic sounds. QHM assumes that the sound being analyzed contains a single source, so, for isolated notes from pitched musical instruments, a constant f_0 is used across all frames m.

3.2. Parameter Interpolation across Frames

The model parameters $\hat{a}_k$, $\hat{F}_k$, and $\hat{\phi}_k$ from Equation (5) are estimated as samples at the frame rate $1/H$ of the amplitude- and frequency-modulation (AM–FM) functions $\hat{a}_k\left(n\right)$ and $\hat{\phi}_k\left(n\right) = 2\pi / f_s \hat{F}_k\left(n\right) + \hat{\phi}_k$, which describe, respectively, the long-term amplitude and frequency temporal variation of each sinusoid k. For each frame m, $\hat{a}_k\left(\tau,m\right)$ and $\hat{F}_k\left(\tau,m\right)$ are estimated using the sample index at the center of the frame $n = \tau$ as reference. Resynthesis of $\hat{x}\left(n,m\right)$ requires $\hat{a}_k\left(n,m\right)$ and $\hat{F}_k\left(n,m\right)$ at the signal sampling rate f_s. Equation (5) uses constant values, resulting in locally stationary sinusoids with constant amplitudes and frequencies inside each frame m.

However, the parameter values might vary across frames, resulting in discontinuities such as $\hat{a}_k\left(\tau,m\right) \neq \hat{a}_k\left(\tau,m+1\right)$ due to temporal variations happening at the frame rate $1/H$. OLA resynthesis [52] uses the analysis window $w\left(n\right)$ to taper discontinuities at the frame boundaries by resynthesizing $\hat{x}\left(n,m\right) = \hat{x}\left(n\right) w\left(n\right)$ for each m similarly to Equation (2) and then overlap-adding $\hat{x}\left(n,m\right)$ across m to obtain $\hat{x}\left(n\right)$.

Additive synthesis is an alternative to OLA that results in smoother temporal variation [20] by first interpolating $\hat{a}_k\left(\tau,m\right)$ and $\hat{\phi}_k\left(\tau,m\right)$ across m and then summing over k. In this case, $\hat{a}_k\left(n\right)$ is obtained by linear interpolation of $\hat{a}_k\left(\tau,m\right)$ and $\hat{a}_k\left(\tau,m+1\right)$. Recursive calculation across m results in a piece-wise linear approximation of $\hat{a}_k\left(n\right)$. $\hat{F}_k\left(n\right)$ is estimated via piece-wise polynomial interpolation of $\hat{F}_k\left(\tau,m\right)$ across m with quadratic splines, and $\hat{\phi}_k\left(n\right)$ is obtained integrating $\hat{F}_k\left(n\right)$ in two steps because $\hat{\phi}_k\left(\tau,m\right)$ is wrapped around 2π across m. First, $\bar{\phi}_k\left(n\right)$ is calculated as

$$\bar{\phi}_k\left(n\right) = \hat{\phi}_k\left(\tau,m\right) + \frac{2\pi}{f_s} \sum_{u=m}^{m+1} \hat{F}_k\left(u\right). \tag{6}$$

The calculation of $\bar{\phi}_k(n)$ using Equation (6) does not guarantee that $\bar{\phi}_k(\tau, m+1) = \hat{\phi}_k(\tau, m+1) + 2\pi P$, with P the closest integer to unwrap the phase (see details in [45]). Thus, $\hat{\phi}_k(n)$ is calculated as

$$\hat{\phi}_k(n) = \hat{\phi}_k(\tau, m) + \frac{2\pi}{f_s} \sum_{u=m}^{m+1} \left[\hat{F}_k(u) + \gamma \sin\left(\frac{\pi(u - m\tau)}{(m+1)\tau - m\tau} \right) \right], \tag{7}$$

where the term given by the sine function ensures continuity with $\hat{\phi}_k(\tau, m+1)$ when γ is

$$\gamma = \frac{\pi}{2} \left[\frac{\hat{\phi}_k(\tau, m+1) + P - \bar{\phi}_k(\tau, m+1)}{(m+1)\tau - m\tau} \right], \tag{8}$$

with P given by $|\hat{\phi}_k(\tau, m+1) - \bar{\phi}_k(\tau, m+1)|$ (see [45]).

3.3. The Extended Adaptive Quasi-Harmonic Model (eaQHM)

Pantazis *et al.* [45] proposed adapting the phase of the sinusoids. The adaptive procedure applies LS, frequency correction, and frequency interpolation iteratively (see Figure 2), projecting $x(n, m)$ onto $\hat{x}(n, m)$. Figure 3 shows the first and second iterations to illustrate adaptation of one sinusoid. Kafentzis *et al.* [43] adapted both the instantaneous amplitude and the instantaneous phase of $\hat{x}(n, m)$ with a similar iterative procedure in eaQHM. The analysis stage uses

$$\hat{x}(n, m) = \sum_{k=-K}^{K} (\mathbf{a}_k + n\mathbf{b}_k)\, \hat{A}_k(n, m)\, e^{j\Phi_k(n,m)}, \tag{9}$$

where $\hat{A}_k(n, m)$ and $\Phi_k(n, m)$ are functions of the time-varying instantaneous amplitude and phase of each sinusoid, respectively [43,45], obtained from the parameter interpolation step and defined as

$$\hat{A}_k(n, m) = \frac{\hat{a}_k(n)}{\hat{a}_k(\tau, m)}, \tag{10a}$$

$$\hat{\Phi}_k(n, m) = \hat{\phi}_k(n) - \hat{\phi}_k(\tau, m), \tag{10b}$$

where $\hat{a}_k(n)$ is the piece-wise linear amplitude and $\hat{\phi}_k(n)$ is estimated using Equation (7). Finally, eaQHM models $x(n)$ as a set of amplitude and frequency modulated nonstationary sinusoids given by

$$\hat{x}_i(n) = \sum_{k=-K}^{K} \hat{a}_{k,i-1}(n)\, e^{j\hat{\phi}_{k,i-1}(n)}, \tag{11}$$

where $\hat{a}_{k,i-1}(n)$ and $\hat{\phi}_{k,i-1}(n)$ are the instantaneous amplitude and phase from the previous iteration $i-1$. Adaptation results from the iterative projection of $x(n)$ onto $\hat{x}(n)$ from $i-1$ as the model $\hat{x}(n)$ are used as nonstationary basis functions locally adapted to the time-varying behavior of $x(n)$. Note that Equation (9) is simply Equation (3) with a nonstationary basis $\hat{A}_k(n, m)\, e^{j\Phi_k(n,m)}$. In fact, Equation (9) represents the next parameter estimation step, which will be again followed by frequency correction as in Figure 2. The convergence criterion for eaQHM is either a maximum number of iterations i or an adaptation threshold ε calculated as

$$\frac{\text{SRER}^{i-1} - \text{SRER}^i}{\text{SRER}^{i-1}} < \varepsilon, \tag{12}$$

where the signal-to-reconstruction-error ratio (SRER) is calculated as

$$\text{SRER} = 20\log_{10} \frac{\text{RMS}(x)}{\text{RMS}(x - \hat{x})} = 20\log_{10} \frac{\text{RMS}(x)}{\text{RMS}(e)}. \tag{13}$$

The SRER measures the fit between the model $\hat{x}(n)$ and the original recording $x(n)$ by dividing the total energy in $x(n)$ by the energy in the residual $e(n)$. The higher the SRER, the better the fit. Note that ε stops adaptation whenever the fit does not improve from iteration $i-1$ to i regardless of the absolute SRER value. Thus, even sounds from the same instruments can reach different SRER.

4. Experimental Setup

We now investigate the full-band representation of musical instrument sounds and the nonstationarity of the adaptive AM–FM sinusoids from eaQHM. We aim to show that spectral decomposition fails to capture partials at the higher end of the spectrum so full-band quasi-harmonic modeling increases the quality of analysis and resynthesis by capturing sinusoids across the full range of the spectrum. Additionally, we aim to show that adaptive AM–FM sinusoids from eaQHM capture nonstationary partials inside the frame. We compare full-band modeling with eaQHM against the SM [1,20] and EDS estimated with ESPRIT [36] using the same number of partials K. We assume that the musical instrument sounds under investigation can be well represented as quasi-harmonic. Thus, we set K_{max} to the highest harmonic number k below Nyquist frequency $f_s/2$ or equivalently the highest integer K that satisfies $Kf_0 \leq f_s/2$. The fundamental frequency f_0 of all sounds was estimated using the sawtooth waveform inspired pitch estimator (SWIPE) [53] because in the experiments the frame size L, the maximum number of partials K_{max}, and the full-band harmonic template depend on f_0. In the SM, K is the number of spectral peaks modeled by sinusoids. For EDS, ESPRIT uses K to determine the separation between the dimension of the signal space (sinusoidal component) and of the residual.

The SM is considered the baseline for comparison due to the quasi-stationary nature of the sinusoids and the need for spectral decomposition. EDS estimated with ESPRIT is considered the state-of-the-art due to the accurate analysis and synthesis and constant frequency of EDS inside the frame m. We present a comparison of the local and global SRER as a function of K and L for the SM and EDS against eaQHM in two experiments. In experiment 1, we vary K from 2 to K_{max} and record the SRER. In experiment 2, we vary L from $3T_0 f_s$ to $8T_0 f_s$ samples and record the SRER, where $T_0 = 1/f_0$ is the fundamental period. The local SRER is calculated within the first frame $m = 0$, where we expect the attack transients to be. The first frame is centered at the onset with $\tau = 0$ (and the first half is zero-padded), so artifacts such as pre-echo (in the first half of the frame) are also expected to be captured by the local SRER. The global SRER is calculated across all frames, thus considering the whole sound signal $\hat{x}(n)$. Next, we describe the musical instrument sounds modeled and the selection of parameter values for the algorithms.

4.1. The Musical Instrument Sound Dataset

In total, 92 musical instrument sounds were selected. "Popular" and "Keyboard" musical instruments are from the RWC Music Database: Musical Instrument Sound [54]. All other sounds are from the Vienna Symphonic Library [55] database of musical instrument samples. Table 1 lists the musical instrument sounds used. The recordings were chosen to represent the range of musical instruments commonly found in traditional Western orchestras and in popular recordings. Some instruments feature different registers (alto, baritone, bass, *etc*). All sounds used belong to the same pitch class (C), ranging in pitch height from C2 ($f0 \approx 65$ Hz) to C6 ($f0 \approx 1046$ Hz). The dynamics is indicated as *forte* ("f") or *fortissimo* ("ff"), and the duration of most sounds is less than 2 s. Normal attack ("na") and no vibrato ("nv") were chosen whenever available. Presence of vibrato ("vib"), progressive attack ("pa"), and slow attack ("sa") are indicated, as well as different playing modes such as *staccato* ("stacc"), *sforzando* ("sforz"), and *pizzicato* ("pz"), achieved by plucking string instruments. Extended techniques were also included, such as *tongue ram* ("tr") for the flute, *près de la table* ("pdlt") for the harp, muted ("mu") strings, and bowed idiophones (vibraphone, xylophone, *etc*.) for short ("sh") and long ("lg") sounds.

Different mallet materials such as metal ("met"), plastic ("pl"), and wood ("wo") and hardness such as soft ("so"), medium ("med"), and hard ("ha") are indicated.

Table 1. Musical instrument sounds used in all experiments. See text in Section 4.1 for a description of the terms in brackets. Sounds **in bold** were used in the listening test described in Section 6. The quasi-harmonic model (QHM) failed for the sounds *in italics* marked *.

Family	Musical Instrument Sounds
Brass	Bass Trombone (C3 f nv na), Bass Trombone (C3 f stac), Bass Trumpet (C3 f na vib), Cimbasso (C3 f nv na), Cimbasso (C3 f stac), *Contrabass Trombone** (C2♯ f stac), Contrabass Tuba (C3 f na), Contrabass Tuba (C3 f stac), Cornet (C4 f), French Horn (C3 f nv na), **French Horn** (C3 f stac), Piccolo Trumpet (C5 f nv na), Piccolo Trumpet (C5 f stac), Tenor Trombone (C3 f na vib), **Tenor Trombone** (C3 f nv sa), Tenor Trombone (C3 f stac), **C Trumpet** (C4 f nv na), C Trumpet (C4 f stac), Tuba (C3 f vib na), Tuba (C3 f stac), Wagner Tuba (C3 f na), Wagner Tuba (C3 f stac)
Woodwinds	Alto Flute (C4 f vib na), Bass Clarinet (C3 f na), **Bass Clarinet** (C3 f sforz), Bass Clarinet (C3 f stac), Bassoon (C3 f na), Bassoon (C3 f stac), Clarinet (C4 f na), Clarinet (C4 f stac), *Contra Bassoon** (C2 f stac), *Contra Bassoon** (C2 f sforz), English Horn (C4 f na), English Horn (C4 f stac), **Flute** (C4 f nv na), Flute (C4 f stac), Flute (C4 f tr), Flute (C4 f vib na), **Oboe 1** (C4 f stac), **Oboe 2** (C4 f nv na), Oboe (C4 f pa), Piccolo Flute (C6♯ f vib sforz), Piccolo Flute (C6 f nv ha ff)
Plucked Strings	Cello (C3 f pz vib), **Harp** (C3 f), Harp (C3 f pdlt), Harp (C3 f mu), Viola (C3 f pz vib), Violin (C4 f pz mu)
Bowed Strings	**Cello** (C3 f vib), Cello (C3 f stac), **Viola** (C3 f vib), Viola (C4 f stac), Violin (C4 f), Violin (C4♯ ff vib), Violin (C4 f stac)
Struck Percussion	**Glockenspiel** (C4 f), Glockenspiel (C6 f wo), Glockenspiel (C6 f pl), Glockenspiel (C6 f met), Marimba (C4 f), Vibraphone (C4 f ha 0), Vibraphone (C4 f ha fa), Vibraphone (C4 f ha sl), Vibraphone (C4 f med 0), Vibraphone (C4 f med fa), Vibraphone (C4 f med 0 mu), **Vibraphone** (C4 f med sl), Vibraphone (C4 f so 0), Vibraphone (C4 f so fa), Xylophone (C5 f GA L), Xylophone (C5 met), Xylophone (C5 f HO L), Xylophone (C5 f mP L), **Xylophone** (C5 f wP L)
Bowed Percussion	Vibraphone (C4 f sh vib), Vibraphone (C4 f sh nv), Vibraphone (C4 f lg nv)
Popular	**Accordion** (C3♯ f), **Acoustic Guitar** (C3 f), Baritone Sax (C3 f), **Bass Harmonica** (C3♯ f), Chromatic Harmonica (C4 f), Classic Guitar (C3 f), Mandolin (C4 f), **Pan Flute** (C5 f), **Tenor Sax** (C3♯ f), **Ukulele** (C4 f)
Keyboard	**Celesta** (C3 f na nv), Celesta (C3 f stac), Clavinet (C3 f), **Piano** (C3 f)

In what follows, we will present the results for 89 sounds because QHM failed to adapt for the three sounds marked * in Table 1. The estimation of parameters for QHM uses LS [45]. The matrix inversion fails numerically when the matrix is close to singular (see [44]). The fundamental frequency (C2 ≈ 65 Hz) of these sounds determines a full-band harmonic spectral template whose frequencies are separated by C2, which results in singular matrices.

4.2. Analysis Parameters

The parameter estimation for the SM follows [20] with a Hann window for analysis, and phase interpolation across frames via cubic splines followed by additive resynthesis. The estimation of parameters for EDS uses ESPRIT with a rectangular window for analysis and OLA resynthesis [36]. Parameter estimation in eaQHM used a Hann window for analysis and additive resynthesis following Equation (11). In all experiments, ε in Equation (12) is set to 0.01 and $f_s = 16$ kHz for all sounds. The step size for analysis (and OLA synthesis) was $H = 16$ samples for all algorithms, corresponding to 1 ms. The frame size is $L = qT_0f_s$ samples with q an integer. The size of the FFT for the SM is kept constant at $N = 4096$ samples with zero padding.

5. Results and Discussion

5.1. Adaptation Cycles in eaQHM

Figure 4 shows the global and local SRER as a function of the number of adaptation cycles (iterations). Each plot was averaged across the sounds indicated, while the plot "all instruments" is an average of the previously shown. The SRER increases quickly after a few iterations, slowly converging to a final value considerably higher than before adaptation. Iteration 0 corresponds to QHM initialized with the full-band harmonic template, thus Figure 4 demonstrates that the adaptation of the sinusoids by eaQHM increases the SRER when compared to QHM.

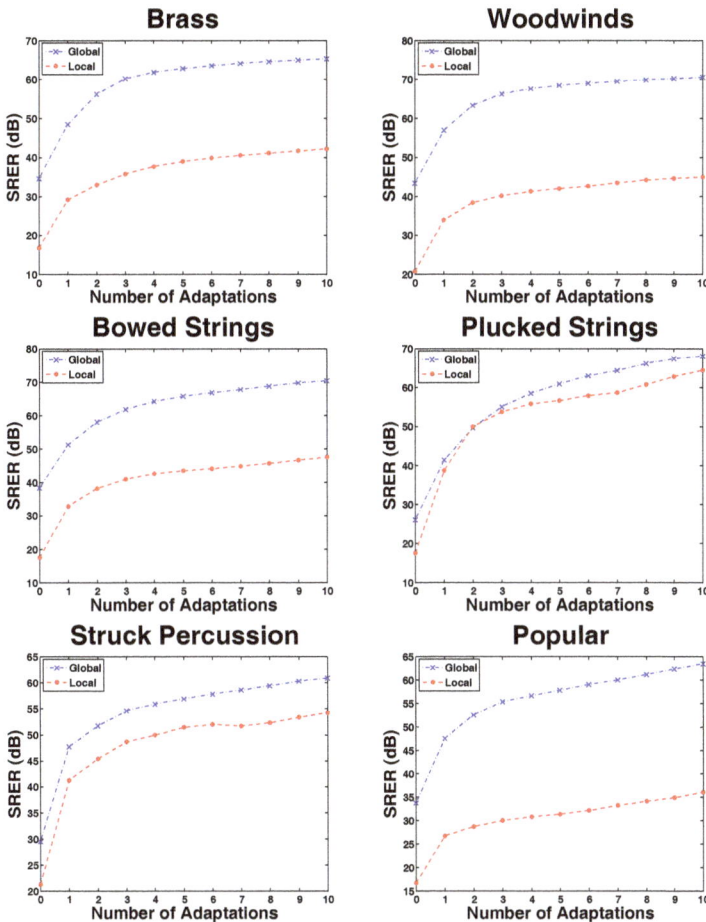

Figure 4. *Cont.*

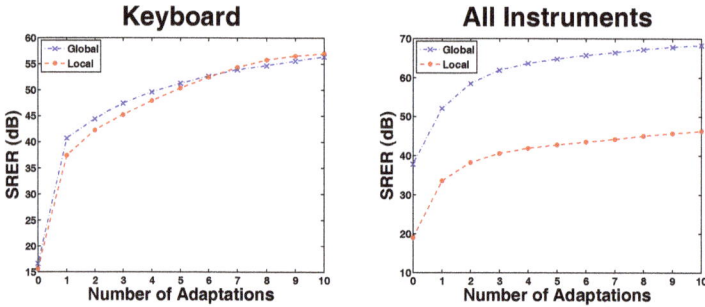

Figure 4. Plot of the signal-to-reconstruction-error ratio (SRER) as a function of number of adaptations to illustrate how adaptation increases the SRER in eaQHM. Iteration 0 corresponds to QHM initialized with the full-band harmonic spectral template.

5.2. Experiment 1: Variation Across K (Constant $L = 3T_0 f_s$)

We ran each algorithm varying K (the frame size was kept at $L = 3T_0 f_s$) and recorded the resulting local and global SRER values. We started from $K = 2$ and increased K by two partials up to K_{max}. Figure 5 shows the local and global SRER (averaged across sounds) as a function of K for the SM, EDS, and eaQHM. Sounds with different f_0 values have different K_{max}. Figure 5 shows that the addition of partials for the SM does not result in an increase in SRER after a certain K. EDS tends to continuously increase the SRER with more partials that capture more spectral energy. Finally, eaQHM increases the SRER up to K_{max}.

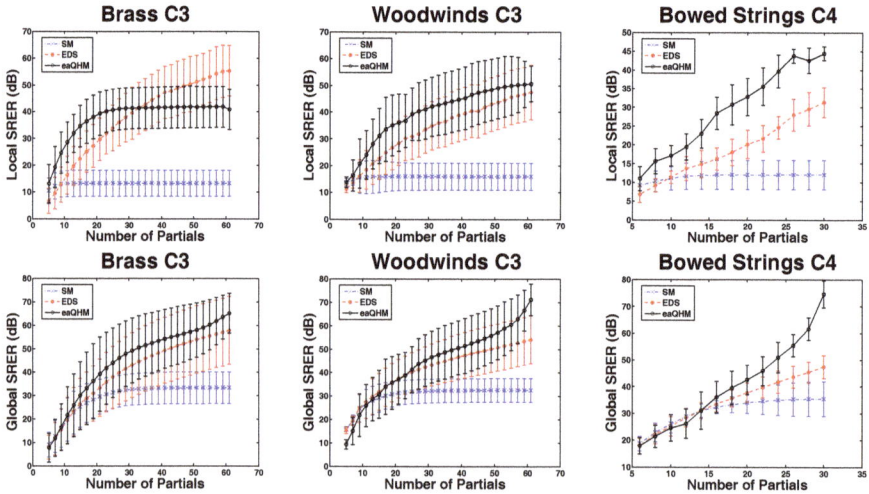

Figure 5. *Cont.*

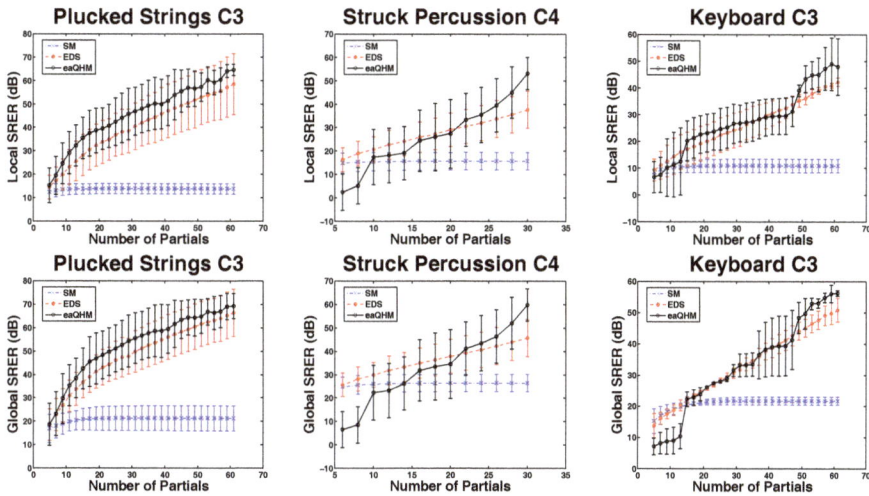

Figure 5. Comparison between local and global SRER as a function of the number of partials for the three models (the standard sinusoidal model (SM), exponentially damped sinusoids (EDS), and eaQHM). The bars around the mean are the standard deviation across different sounds from the family indicated. The distributions are not symmetrical as suggested by the bars.

The SM, EDS, and eaQHM use different analysis and different synthesis methods, which partially explains the different behavior under variation of K. More importantly, the addition of partials for each algorithm uses different criteria. Both the SM and EDS use spectral energy as a criterion, while eaQHM uses the frequencies of the sinusoids assuming quasi-harmonicity. In the SM, a new sinusoid is selected as the next spectral peak (increasing frequency) with spectral energy above a selected threshold regardless of the frequency of the peak. In fact, the frequency is estimated from the peak afterwards. For EDS, K determines the number of sinusoids used upon resynthesis. However, ESPRIT ranks the sinusoids by decreasing eigenvalue rather than the frequency, adding partials with high spectral energy that will increase the fit of the reconstruction. The frequencies of the new partials are not constrained by harmonicity. Finally, eaQHM uses the spectral template to search for nearby spectral peaks with LS and frequency correction. The sinusoids will converge to spectral peaks in the neighborhood of the harmonic template with K harmonically related partials starting from f_0. Therefore, $K_{\max}$ in eaQHM corresponds to full-band analysis and synthesis but not necessarily for the SM or EDS.

5.3. Experiment 2: Variation Across L (Constant $K = K_{\max}$)

We ran each algorithm varying L from $3T_0 f_s$ to $8T_0 f_s$ with a constant number of partials $K_{\max}$ and measured the resulting local and global SRER. In the literature [46], $L = 3T_0 f_s$ is considered a reasonable value for speech and audio signals when using the SM. We are unaware of a systematic investigation of how L affects modeling accuracy for EDS. Figure 6 shows the local and global SRER (averaged across sounds) as a function of L expressed as q times $T_0 f_s$, so sounds with different f_0 values have different frame size L in samples.

Figure 6 shows that the SRER decreases with L for all algorithms. The SM seldom outperforms EDS or eaQHM, but it is more robust against variations of L. For the SM, L affects both spectral estimation and temporal representation. In the FFT, L determines the trade-off between temporal and spectral resolution, which affects the performance of the peak picking algorithm for parameter estimation. The temporal representation is affected because the parameters are an average across L referenced to the center of the frame. In turn, ESPRIT estimates EDS with constant frequency inside the frames referenced to the beginning of the frame, thus L affects the temporal modeling accuracy more than the spectral estimation. However, the addition of sinusoids might compensate for the stationary frequency of EDS inside the frame. Finally, the SRER for eaQHM decreases considerably when L increases because L adversely affects the frequency correction and interpolation mechanisms. Frequency correction is applied at the center of the analysis frame and eaQHM uses spline interpolation to capture frequency modulations across frames. Thus, adaptation improves the fit more slowly for longer L, generally reaching a lower absolute SRER value.

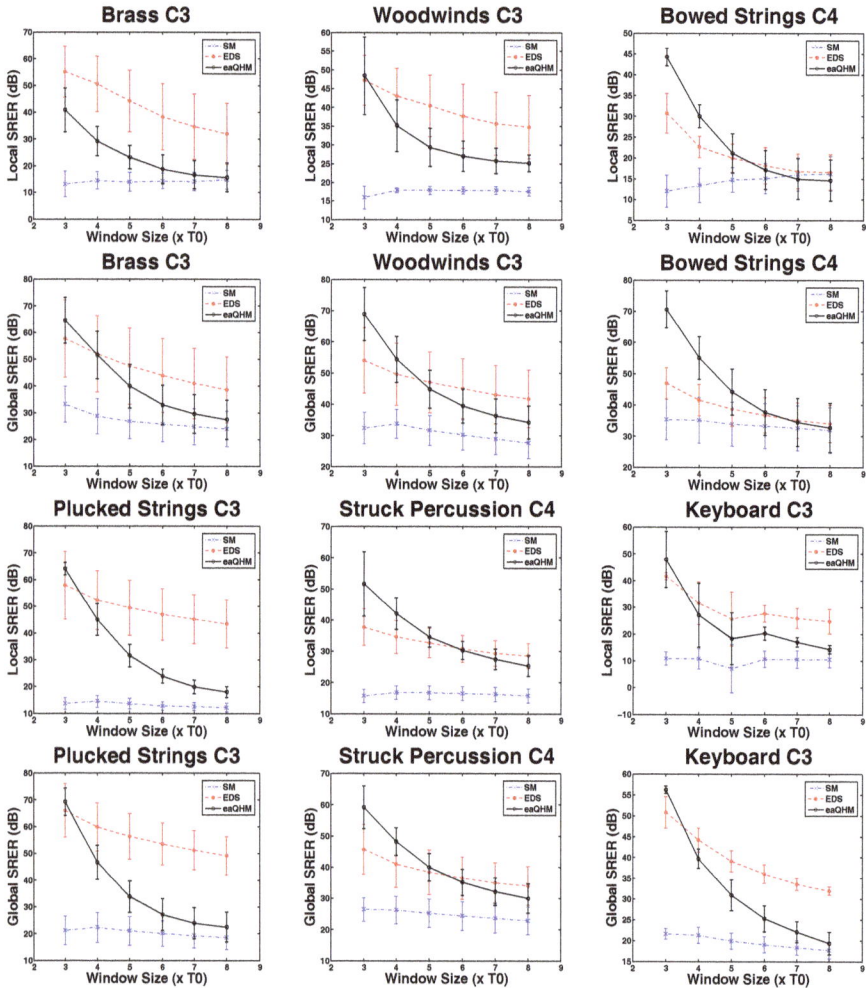

Figure 6. Comparison between local and global SRER as a function of the size of the frame for the three models (SM, EDS, and eaQHM). The bars around the mean are the standard deviation across different sounds from the family indicated. The distributions are not symmetrical as suggested by the bars.

5.4. Full-Band Quasi-Harmonic Analysis with AM–FM Sinusoids

To simplify the comparison and reduce the information, we present the differences of SRER instead of absolute SRER values. For each sound, we subtract the absolute SRER values (in dB) for the SM and EDS from that of eaQHM to obtain the differences of SRER. The local value measures the fit for the attack and the global value measures the overall fit. Positive values indicate that eaQHM results in higher SRER than the other method for that particular sound, while a negative value means the opposite. The different SRER values are averaged across all musical instruments that belong to the family indicated. Table 2 shows the comparison of eaQHM against EDS and the SM with $K = K_{max}$ and $L = 3T_0 f_s$ clustered by instrumental family. The distributions are not symmetrical around the mean as suggested by the standard deviation.

Table 2. Local and global difference of signal-to-reconstruction-error ratio (SRER) comparing eaQHM with exponentially damped sinusoids (EDS) and eaQHM with the standard sinusoidal model (SM) for the frame size $L = 3T_0 f_s$ and number of partials $K = K_{max}$. The three C2 sounds are not included.

Family	SRER (eaQHM-EDS)		SRER (eaQHM-SM)	
	Local (dB)	Global (dB)	Local (dB)	Global (dB)
Brass	-9.4 ± 7.0	12.5 ± 6.8	27.3 ± 5.8	31.9 ± 4.0
Woodwinds	7.8 ± 3.9	22.0 ± 5.9	30.9 ± 7.5	36.1 ± 4.7
Bowed Strings	12.2 ± 4.2	24.1 ± 6.7	35.0 ± 4.7	40.0 ± 4.7
Plucked Strings	8.3 ± 5.0	4.7 ± 3.4	49.5 ± 4.3	46.6 ± 5.1
Bowed Percussion	-2.7 ± 2.5	16.3 ± 2.2	12.7 ± 2.6	37.6 ± 3.6
Struck Percussion	10.5 ± 4.8	10.1 ± 2.6	28.6 ± 13.3	26.0 ± 11.3
Popular	6.3 ± 3.3	11.9 ± 7.0	26.5 ± 10.8	27.5 ± 11.6
Keyboard	5.7 ± 3.4	5.4 ± 4.3	37.0 ± 8.0	34.6 ± 2.0
Total	5.3 ± 2.4	13.2 ± 3.3	31.0 ± 7.1	35.0 ± 5.9

Thus, Table 2 summarizes the result of full-band quasi-harmonic analysis with adaptive AM–FM sinusoids from eaQHM comparing with the SM and EDS under the same conditions, namely the same number of sinusoids $K = K_{max}$ and frame size $L = 3T_0 f_s$. When eaQHM is compared to the SM, both local and global difference SRER are positive for all families. This means that full-band quasi-harmonic modeling with eaQHM results in a better fit for the analysis and synthesis of musical instrument sounds.

When eaQHM is compared to EDS, all global difference SRER are positive and all local difference SRER are positive except for *Brass* and *Bowed Percussion*. Thus, EDS can fit the attack of *Brass* and *Bowed Percussion* better than eaQHM. The exponential amplitude envelope of EDS is considered suitable to model percussive sounds with sharp attacks such as harps, pianos, and marimbas [36,37]. The musical instrument families that contain percussive sounds are *Plucked strings*, *Struck percussion*, and *Keyboard*. Table 2 shows that eaQHM outperformed EDS locally and globally for all percussive sounds. The ability to adapt the amplitude of the sinusoidal partials to the local characteristics of the waveform makes eaQHM extremely flexible to fit both percussive and nonpercussive musical instrument sounds. On the other hand, both *Brass* and *Bowed Percussion* present slow attacks typically lasting longer than one frame L. Note that $L/f_s = 3T_0 \approx 22$ ms for C3 ($f_0 \approx 131$ Hz) while *Bowed Percussion* can have attacks longer than 100 ms. Therefore, one frame $L = 3T_0 f_s$ does not measure the fit for the entire duration of the attack.

Note that the local SRER is important because the global SRER measures the overall fit without indication of *where* the differences lie in the waveform. For musical instrument sounds, differences in the attack impact the results differently than elsewhere because the attack is among the most important perceptual features in dissimilarity judgment [56–58]. Consequently, when comparing two models with the global SRER, it is only safe to say that a higher SRER indicates that resynthesis results in a waveform that is closer to the original recording.

5.5. Full-Band Modeling and Quasi-Harmonicity

Time-frequency transforms such as the STFT represent L samples in a frame with N DFT coefficients provided that $N \geq L$. Note that $N \in \mathbb{C}$, corresponding to $p = 2N$ real numbers. There is signal expansion whenever the representation uses p parameters to represent L samples and $p > L$. Sinusoidal models represent L samples in a frame with K sinusoids. In turn, each sinusoid is described by p parameters, requiring pK parameters to represent L samples. Therefore, there is a maximum number of sinusoids to represent a frame without signal expansion. For example, white noise has a flat spectrum across that would take a large number of sinusoids close together in frequency resulting in signal expansion.

The pK parameters to represent L samples can be interpreted as the degrees of freedom of the fit. As a general rule, more parameters mean greater flexibility of representation (hence potentially a better fit), but with the risk of over-fitting. Table 3 shows a comparison of the number of real parameters p (per sinusoid k per frame m) for the analysis and synthesis stages of the SM, EDS, and eaQHM. Note that eaQHM and EDS require more parameters than the SM at the *analysis* stage, but eaQHM and the SM require fewer parameters than EDS for the *synthesis* stage. The difference is due to the resynthesis strategy used by each algorithm. EDS uses OLA resynthesis, which requires all analysis parameters for resynthesis, while both eaQHM and the SM use additive resynthesis.

Table 3. Comparison of the number of real parameters p per sinusoid k per frame m for the analysis and synthesis stages of the SM, EDS, and eaQHM. The table presents the number of real parameters p to estimate and to resynthesize each sinusoid inside a frame.

	Number of Real Parameters p Per Sinusoid k Per Frame m		
	SM	**EDS**	**eaQHM**
Analysis	$p = 3$	$p = 4$	$p = 4$
Synthesis	$p = 3$	$p = 4$	$p = 3$

Harmonicity of the partials guarantees that there are no signal expansions in full-band modeling with sinusoids. Consider $L = qT_0 f_s$ with q an integer and $T_0 = 1/f_0$. Using $K_{max} \approx f_s/2f_0$ quasi-harmonic partials and p parameters per partial, it takes at most $pK_{max} = (pf_s)/2f_0$ numbers to represent $L = qT_0 f_s = (qf_s)/f_0$ samples, which gives the ratio $r = (pK_{max})/L = p/2q$. Table 3 shows that analysis with eaQHM requires $p = 4$ real parameters. Thus, a frame size with $q > 2$ is enough to guarantee no signal expansion. This result is due to the full-band paradigm using K_{max} harmonically related partials, not a particular model. The advantage of full-band modeling results from the use of one single component instead of decomposition.

Table 4 compares the complexity of SM, EDS, and eaQHM in Big-O notation. The complexity of SM is $\mathcal{O}(N \log N)$, which is the complexity of the FFT algorithm for size N inputs. ESPRIT estimates the parameters of EDS with singular value decomposition (SVD), whose algorithmic complexity is $\mathcal{O}(L^2 + K^3)$ for an L by K matrix (frame size *versus* the number of sinusoids). Adaptation in eaQHM is an iterative fit where each iteration i requires running the model again as described in Section 3. For each iteration i, eaQHM estimates the parameters with least squares (LS) via calculation of the pseudoinverse matrix using QR decomposition. The algorithmic complexity of QR decomposition is $\mathcal{O}(K^3)$ for a square matrix of size K (the number of sinusoids).

Adaptation of the sinusoids in eaQHM can result in over-fitting. The amplitude and frequency modulations capture temporal variations inside the frame such as transients and instrumental noise around the partials. However, adaptation must not capture noise resulting from sources such as quantization, which is extraneous to the sound. Ideally, the residual should contain only external additive noise without any perceptually important information from the sound [17].

Table 4. Comparison of algorithmic complexity in Big-O notation. The table presents the complexity as a function of the size of the input N, L, and K and the number of iterations i. See text for details.

	Algorithmic Complexity		
	SM	**EDS**	**eaQHM**
Complexity	$\mathcal{O}(N \log N)$	$\mathcal{O}(L^2 + K^3)$	$\mathcal{O}(iK^3)$

6. Evaluation of Perceptual Transparency with a Listening Test

We performed a listening test to validate the full-band representation of musical instrument sounds with eaQHM. The aim of the test was to evaluate whether full-band modeling with eaQHM resulted in resynthesized musical instrument sounds that are perceptually indistinguishable from the original recordings. The 21 sounds **in bold** in Table 1 were selected for the listening test, which presented pairs *original* and *resynthesis*. The participants were instructed to listen to each pair as many times as necessary and to answer the question "Can you tell the difference between the two sounds in each pair?" Full-band (FB) resynthesis with eaQHM (using a harmonic template with $K = K_{max}$ sinusoids) was used for all 21 musical instrument sounds. For nine of these sounds, half-band (HB) resynthesis with eaQHM (using a harmonic template with $K = K_{max}/2$ sinusoids) was also included as control group to test the aptitude of the listeners and compare against the FB version. All HB versions were placed at random positions among the FB, so the test presented 30 pairs overall. The listening test can be accessed at [59].

In total, 20 people aged between 26 and 40 took the test. The participants declared themselves as experienced with listening tests and familiar with signal processing techniques. Figure 7 shows the result of the listening test as the percentage of the people who answered "no" to the question, indicating that they cannot tell the difference between the original recording and the resynthesis. In general, the result of the listening test shows that full-band modeling with eaQHM results in perceptually indistinguishable resynthesis for most musical instrument sounds tested. The figure indicates that 10 out of the 21 FB sounds tested were rated perceptually identical to the original by 100% of the listeners. As expected, most HB sounds fall under 30% (except *Tenor Trombone*) and most FB sounds lie above 70% (except *Pan Flute*). Table 1 shows that *Tenor Trombone* is played at C3 and *Pan Flute* at C5. The Tenor Trombone sound is not bright, which indicates that there is little spectral energy at the higher frequency end of the spectrum. Thus, the HB version synthesized with fewer partials than K_{max} was perceived as identical to the original by some listeners. The Pan Flute sound contains a characteristic breathing noise captured as AM–FM elements in eaQHM. However, the breathing noise in the full-band version sounds brighter than the original recording and most listeners were able to tell the difference.

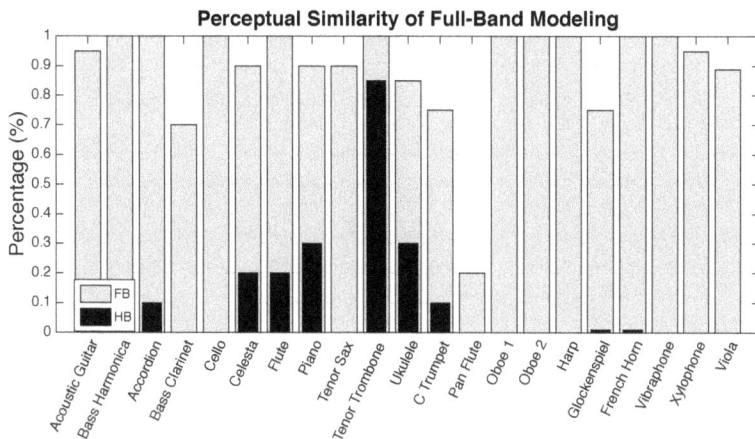

Figure 7. Result of the listening test on perceptual similarity of full-band (FB) and half-band (HB) resynthesis with eaQHM compared to the original recording. The sounds used in the listening test appear **in bold** in Table 1.

7. Conclusions

We proposed the full-band quasi-harmonic modeling of musical instrument sounds with adaptive AM–FM sinusoids from eaQHM as an alternative to spectrum decomposition. We used the SRER to measure the fit of the sinusoidal model to the original recording of 89 percussive and nonpercussive musical instruments sounds from different families. We showed that full-band modeling with eaQHM results in higher global SRER values when compared to the standard SM and to EDS estimated with ESPRIT for K_{max} sinusoids and frame size $L = 3T_0f_s$. EDS resulted in higher local SRER than eaQHM for two of nine instrumental families, namely *Brass* and *Bowed Percussion*. A listening test confirmed that full-band modeling with eaQHM resulted in perceptually indistinguishable resynthesis for most musical instrument sounds tested.

Future work should investigate a method to prevent over-fitting with eaQHM. Additionally, the use of least-squares to estimate the parameters leads to matrices that are badly conditioned for sounds with low fundamental frequencies. A more robust estimation method to prevent bad-conditioning would improve the stability of eaQHM. Currently, eaQHM can only estimate the parameters of isolated sounds. We intend to develop a method for polyphonic instruments and music. Future work also involves using eaQHM in musical instrument sound transformation, estimation of musical expressivity features such as vibrato, and solo instrumental music. The companion webpage [60] contains sound examples. Finally, the proposal of a full-band representation of musical instrument sounds with adaptive sinusoids motivates further investigation on full-band extensions of other sinusoidal methods, such as SM and EDS used here.

Acknowledgments: This work was partly supported by project "NORTE-01-0145-FEDER-000020" financed by the North Portugal Regional Operational Programme (NORTE 2020), under the PORTUGAL 2020 Partnership Agreement, and through the European Regional Development Fund (ERDF) and by the European Union's Horizon 2020 research and innovation programme under the Marie Skłodowska-Curie Grant 644283. The latter project also supplied funds for covering the costs to publish in open access.

Author Contributions: Marcelo Caetano conceived and designed the experiments, analyzed the data, and wrote the manuscript. George P. Kafentzis performed the experiments, helped analyze the results, and revised the manuscript. Athanasios Mouchtaris supervised the research and revised the manuscript. Yannis Stylianou supervised the research.

Conflicts of Interest: The authors declare no conflict of interest.

References

1. Serra, X.; Smith, J.O. Spectral modeling synthesis: A sound analysis/synthesis system based on a deterministic plus stochastic decomposition. *Comput. Music J.* **1990**, *14*, 49–56.
2. Beauchamp, J.W. Analysis and synthesis of musical instrument sounds. In *Analysis, Synthesis, and Perception of Musical Sounds*; Beauchamp, J.W., Ed.; Modern Acoustics and Signal Processing; Springer: New York, NY, USA, 2007; pp. 1–89.
3. Quatieri, T.; McAuley, R. Audio signal processing based on sinusoidal analysis/synthesis. In *Applications of Digital Signal Processing to Audio and Acoustics*; Kahrs, M., Brandenburg, K., Eds.; Kluwer Academic Publishers: Berlin/Heidelberg, Germany, 2002; Chapter 9, pp. 343–416.
4. Serra, X.; Bonada, J. Sound Transformations based on the SMS high level attributes. *Proc. Digit. Audio Eff. Workshop* **1998**, *5*. Available online: http://mtg.upf.edu/files/publications/dafx98-1.pdf (accessed on 26 April 2016).
5. Caetano, M.; Rodet, X. Musical Instrument sound morphing guided by perceptually motivated features. *IEEE Trans. Audio Speech Lang. Process.* **2013**, *21*, 1666–1675.
6. Barbedo, J.; Tzanetakis, G. Musical instrument classification using individual partials. *IEEE Trans. Audio Speech Lang. Process.* **2011**, *19*, 111–122.
7. Herrera, P.; Bonada, J. Vibrato Extraction and parameterization in the spectral modeling synthesis framework. *Proc. Digit. Audio Eff. Workshop* **1998**, *99*. Available online: http://www.mtg.upf.edu/files/publications/dafx98-perfe.pdf (accessed on 26 April 2016).

8. Glover, J.; Lazzarini, V.; Timoney, J. Real-time detection of musical onsets with linear prediction and sinusoidal modeling. *EURASIP J. Adv. Signal Process.* **2011**, doi:10.1186/1687-6180-2011-68.
9. Virtanen, T.; Klapuri, A. Separation of harmonic sound sources using sinusoidal modeling. In Proceedings of the 2000 IEEE International Conference on Acoustics, Speech, and Signal Processing (ICASSP), Istanbul, Turkey, 5–9 June 2000; Volume 2, pp. II765–II768.
10. Lagrange, M.; Marchand, S.; Rault, J.B. Long interpolation of audio signals using linear prediction in sinusoidal modeling. *J. Audio Eng. Soc.* **2005**, *53*, 891–905.
11. Hermus, K.; Verhelst, W.; Lemmerling, P.; Wambacq, P.; Huffel, S.V. Perceptual audio modeling with exponentially damped sinusoids. *Signal Process.* **2005**, *85*, 163–176.
12. Nsabimana, F.; Zolzer, U. Audio signal decomposition for pitch and time scaling. In Proceedings of the International Symposium on Communications, Control, and Signal Processing (ISCCSP), St Julians, Malta, 12–14 March 2008; pp. 1285–1290.
13. El-Jaroudi, A.; Makhoul, J. Discrete all-pole modeling. *IEEE Trans. Commun. Technol.* **1969**, *39*, 481–488.
14. Caetano, M.; Rodet, X. A source-filter model for musical instrument sound transformation. In Proceedings of the 2012 IEEE International Conference on Acoustics, Speech, and Signal Processing (ICASSP), Kyoto, Japan, 25–30 March 2012; pp. 137–140.
15. Wen, X.; Sandler, M. Source-Filter Modeling in the Sinusoidal Domain. *J. Audio Eng. Soc.* **2010**, *58*, 795–808.
16. Fletcher, N.H.; Rossing, T.D. *The Physics of Musical Instruments*, 2nd ed.; Springer: New York, NY, USA, 1998.
17. Caetano, M.; Kafentzis, G.P.; Degottex, G.; Mouchtaris, A.; Stylianou, Y. Evaluating how well filtered white noise models the residual from sinusoidal modeling of musical instrument sounds. In Proceedings of the IEEE Workshop on Applications of Signal Processing to Audio and Acoustics (WASPAA), New Paltz, NY, USA, 20–23 October 2013; pp. 1–4.
18. Bader, R.; Hansen, U. Modeling of musical instruments. In *Handbook of Signal Processing in Acoustics*; Havelock, D., Kuwano, S., Vorländer, M., Eds.; Springer: New York, NY, USA, 2009; pp. 419–446.
19. Fletcher, N.H. The nonlinear physics of musical instruments. *Rep. Prog. Phys.* **1999**, *62*, 723–764.
20. McAulay, R.J.; Quatieri, T.F. Speech analysis/synthesis based on a sinusoidal representation. *IEEE Trans. Acoust. Speech Signal Process.* **1986**, *34*, 744–754.
21. Green, R.A.; Haq, A. B-spline enhanced time-spectrum analysis. *Signal Process.* **2005**, *85*, 681–692.
22. Belega, D.; Petri, D. Frequency estimation by two- or three-point interpolated Fourier algorithms based on cosine windows. *Signal Process.* **2015**, *117*, 115–125.
23. Prudat, Y.; Vesin, J.M. Multi-signal extension of adaptive frequency tracking algorithms. *Signal Process.* **2009**, *89*, 96–973.
24. Candan, Ç. Fine resolution frequency estimation from three DFT samples: Case of windowed data. *Signal Process.* **2015**, *114*, 245–250.
25. Röbel, A. Adaptive additive modeling with continuous parameter trajectories. *IEEE Trans. Audio Speech Lang. Process.* **2006**, *14*, 1440–1453.
26. Verma, T.S.; Meng, T.H.Y. Extending spectral modeling synthesis with transient modeling synthesis. *Comput. Music J.* **2000**, *24*, 47–59.
27. Laurenti, N.; De Poli, G.; Montagner, D. A nonlinear method for stochastic spectrum estimation in the modeling of musical sounds. *IEEE Trans. Audio Speech Lang. Process.* **2007**, *15*, 531–541.
28. Daudet, L. A review on techniques for the extraction of transients in musical signals. *Proc. Int. Symp. Comput. Music Model. Retr.* **2006**, *3902*, 219–232.
29. Jang, H.; Park, J.S. Multiresolution sinusoidal model with dynamic segmentation for timescale modification of polyphonic audio signals. *IEEE Trans. Speech Audio Process.* **2005**, *13*, 254–262.
30. Beltrán, J.R.; de León, J.P. Estimation of the instantaneous amplitude and the instantaneous frequency of audio signals using complex wavelets. *Signal Process.* **2010**, *90*, 3093–3109.
31. Levine, S.N.; Smith, J.O. A compact and malleable sines+transients+noise model for sound. In *Analysis, Synthesis, and Perception of Musical Sounds*; Beauchamp, J.W., Ed.; Modern Acoustics and Signal Processing; Springer: New York, NY, USA, 2007; pp. 145–174.
32. Markovsky, I.; Huffel, S.V. Overview of total least-squares methods. *Signal Process.* **2007**, *87*, 2283–2302.
33. Roy, R.; Kailath, T. ESPRIT-estimation of signal parameters via rotational invariance techniques. *IEEE Trans. Acoust. Speech Signal Process* **1989**, *37*, 984–995.

34. Van Huffel, S.; Park, H.; Rosen, J. Formulation and solution of structured total least norm problems for parameter estimation. *IEEE Trans. Signal Process.* **1996**, *44*, 2464–2474.

35. Liu, Z.S.; Li, J.; Stoica, P. RELAX-based estimation of damped sinusoidal signal parameters. *Signal Process.* **1997**, *62*, 311–321.

36. Nieuwenhuijse, J.; Heusens, R.; Deprettere, E.F. Robust exponential modeling of audio signals. In Proceedings of the 1998 IEEE International Conference on Acoustics, Speech, and Signal Processing (ICASSP), Seattle, WA, USA, 12–15 May 1998; Volume 6, pp. 3581–3584.

37. Badeau, R.; Boyer, R.; David, B. EDS Parametric Modeling And Tracking of Audio Signals. In Proceedings of the 5th International Conference on Digital Audio Effects (DAFx), Hambourg, Germany, 26–28 September 2002; pp. 26–28.

38. Jensen, J.; Heusdens, R. A comparison of sinusoidal model variants for speech and audio representation. In Proceedings of the 2002 11th European Signal Processing Conference (EUSIPCO), Toulouse, France, 3–6 September 2002; pp. 1–4.

39. Auger, F.; Flandrin, P. Improving the readability of time-frequency and time-scale representations by the reassignment method. *IEEE Trans. Signal Process.* **1995**, *43*, 1068–1089.

40. Fulop, S.A.; Fitz, K. Algorithms for computing the time-corrected instantaneous frequency (reassigned) spectrogram, with applications. *J. Acoust. Soc. Am.* **2006**, *119*, 360–371.

41. Li, X.; Bi, G. The reassigned local polynomial periodogram and its properties. *Signal Process.* **2009**, *89*, 206–217.

42. Girin, L.; Marchand, S.; Di Martino, J.; Röbel, A.; Peeters, G. Comparing the order of a polynomial phase model for the synthesis of quasi-harmonic audio signals. In Proceedings of the IEEE Workshop on Applications of Signal Processing to Audio and Acoustics (WASPAA), New Paltz, NY, USA, 19–22 October 2003; pp. 193–196.

43. Kafentzis, G.P.; Pantazis, Y.; Rosec, O.; Stylianou, Y. An extension of the adaptive quasi-harmonic model. In Proceedings of the 2012 IEEE International Conference on Acoustics, Speech, and Signal Processing, Kyoto, Japan, 25–30 March 2012; pp. 4605–4608.

44. Kafentzis, G.P.; Rosec, O.; Stylianou, Y. On the modeling of voiceless stop sounds of speech using adaptive quasi-harmonic models. In Proceedings of the Annual Conference of the International Speech Communication Association (INTERSPEECH), Portland, OR, USA, 9–13 September 2012.

45. Pantazis, Y.; Rosec, O.; Stylianou, Y. Adaptive AM–FM signal decomposition with application to speech analysis. *IEEE Trans. Audio Speech Lang. Process.* **2011**, *19*, 290–300.

46. Degottex, G.; Stylianou, Y. Analysis and synthesis of speech using an adaptive full-band harmonic model. *IEEE Trans. Audio Speech Lang. Process.* **2013**, *21*, 2085–2095.

47. Caetano, M.; Kafentzis, G.P.; Mouchtaris, A.; Stylianou, Y. Adaptive sinusoidal modeling of percussive musical instrument sounds. In Proceedings of the European Signal Processing Conference (EUSIPCO), Marrakech, Morocco, 9–13 September 2013; pp. 1–5.

48. Pantazis, Y.; Rosec, O.; Stylianou, Y. On the Properties of a time-varying quasi-harmonic model of speech. In Proceedings of the Annual Conference of the International Speech Communication Association (INTERSPEECH), Brisbane, Australia, 22–26 September 2008; pp. 1044–1047.

49. Smyth, T.; Abel, J.S. Toward an estimation of the clarinet reed pulse from instrument performance. *J. Acoust. Soc. Am.* **2012**, *131*, 4799–4810.

50. Smyth, T.; Scott, F. Trombone synthesis by model and measurement. *EURASIP J. Adv. Signal Process.* **2011**, doi:10.1155/2011/151436.

51. Brown, J.C. Frequency ratios of spectral components of musical sounds. *J. Acoust. Soc. Am.* **1996**, *99*, 1210–1218.

52. Borss, C.; Martin, R. On the construction of window functions with constant overlap-add constraint for arbitrary window shifts. In Proceedings of the 2012 IEEE International Conference on Acoustics, Speech and Signal Processing (ICASSP), Kyoto, Japan, 25–30 March 2012; pp. 337–340.

53. Camacho, A.; Flory, H.Y. A sawtooth waveform inspired pitch estimator for speech and music. *J. Acoust. Soc. Am.* **2008**, *124*, 1638–1652.

54. Goto, M.; Hashiguchi, H.; Nishimura, T.; Oka, R. RWC Music Database: Music Genre Database and Musical Instrument Sound Database. In Proceedings of the International Conference on Music Information Retrieval (ISMIR), Baltimore, MD, USA, 26–30 October 2003; pp. 229–230. Available online: http://staff.aist.go.jp/m.goto/RWC-MDB/ (accessed on 26 April 2016).
55. Vienna Symphonic Library–GmbH. Available online: http://www.vsl.co.at/ (accessed on 26 April 2016).
56. Grey, J.M.; Gordon, J.W. Multidimensional perceptual scaling of musical timbre. *J. Acoust. Soc. Am.* **1977**, *61*, 1270–1277.
57. Krumhansl, C.L. Why is musical timbre so hard to understand? In *Structure and Perception of Electroacoustic Sound and Music*; Nielzén, S., Olsson, O., Eds.; Excerpta Medica: New York, NY, USA, 1989; pp. 43–54.
58. McAdams, S.; Giordano, B.L. The perception of musical timbre. In *The Oxford Handbook of Music Psychology*; Hallam, S., Cross, I., Thaut, M., Eds.; Oxford University Press: New York, NY, USA, 2009; pp. 72–80.
59. Listening Test. Webpage for the Listening Test. Available online: http://ixion.csd.uoc.gr/kafentz/listest/pmwiki.php?n=Main.JMusLT (accessed on 26 April 2016).
60. AdaptiveSinMus. Companion webpage with sound examples. Available online: http://www.csd.uoc.gr/kafentz/listest/pmwiki.php?n=Main.AdaptiveSinMus (accessed on 26 April 2016).

applied
sciences

MDPI

Article

Psychoacoustic Approaches for Harmonic Music Mixing [†]

Roman B. Gebhardt [1,*], Matthew E. P. Davies [2] and Bernhard U. Seeber [1]

[1] Audio Information Processing, Technische Universität München, Arcisstraße 21, 80333 Munich, Germany; seeber@tum.de
[2] Sound and Music Computing Group, Instituto de Engenharia de Sistemas e Computadores, Tecnologia e Ciência - INESC TEC, Rua Dr. Roberto Frias, 4200-465 Porto, Portugal; mdavies@inesctec.pt
* Correspondence: roman.gebhardt@tum.de
† This paper is an extended version of our paper "Harmonic Mixing Based on Roughness and Pitch Commonality" published in the Proceedings of the 18th International Conference on Digital Audio Effects (DAFx-15), Trondheim, Norway, 30 November–3 December 2015; pp. 185–192.

Academic Editor: Vesa Valimaki
Received: 29 February 2016; Accepted: 25 April 2016; Published: 3 May 2016

Abstract: The practice of harmonic mixing is a technique used by DJs for the beat-synchronous and harmonic alignment of two or more pieces of music. In this paper, we present a new harmonic mixing method based on psychoacoustic principles. Unlike existing commercial DJ-mixing software, which determines compatible matches between songs via key estimation and harmonic relationships in the circle of fifths, our approach is built around the measurement of musical consonance. Given two tracks, we first extract a set of partials using a sinusoidal model and average this information over sixteenth note temporal frames. By scaling the partials of one track over ±6 semitones (in 1/8th semitone steps), we determine the pitch-shift that maximizes the consonance of the resulting mix. For this, we measure the consonance between all combinations of dyads within each frame according to psychoacoustic models of roughness and pitch commonality. To evaluate our method, we conducted a listening test where short musical excerpts were mixed together under different pitch shifts and rated according to consonance and pleasantness. Results demonstrate that sensory roughness computed from a small number of partials in each of the musical audio signals constitutes a reliable indicator to yield maximum perceptual consonance and pleasantness ratings by musically-trained listeners.

Keywords: audio content analysis; audio signal processing; digital DJ interfaces; music information retrieval; music technology; musical consonance; psychoacoustics; sound and music computing; spectral analysis

1. Introduction

The digital era of DJ-mixing has opened up DJing to a huge range of users and has also enabled new technical possibilities in music creation and remixing. The industry-leading DJ-software tools now offer users of all technical abilities the opportunity to rapidly and easily create DJ mixes out of their personal music collections or those stored online. Central to these DJ-software tools is the ability to robustly identify tempo and beat locations, which, when combined with high quality audio time-stretching, allow for automatic "beat-matching" (*i.e.*, temporal synchronization) of music [1].

In addition to leveraging knowledge of the beat structure, these tools also extract harmonic information, typically in the form of an estimated key. Knowing the key of different pieces of music allows users to engage in so-called "harmonic mixing", where the aim is not only to align music in time, but also in key. Different pieces of music are deemed to be harmonically compatible if their keys exactly match or adhere to well-known relationships within the circle of fifths, e.g., those in relative keys (major and relative minor) or those separated by a perfect fourth or perfect fifth occupying adjacent positions [2]. When this information is combined with audio pitch-shifting (*i.e.*, the ability to transpose a piece of music by some number of semitones independently of its temporal structure), it provides a seemingly powerful means to "force" the harmonic alignment between two pieces of otherwise harmonically incompatible music in the same way beat matching works for the temporal dimension [3]. To illustrate this process by example, consider two musical excerpts, one in D minor and the other in F minor. Since both excerpts are in a minor key, the key-based match can be made by simply transposing the second down by three semitones. Alternatively, if one excerpt is in A major and the other is in G# minor, this would require pitch shifting the second excerpt down by two semitones to F# minor, which is the relative minor of A major.

While the use of tempo and key detection along with high quality music signal processing techniques is certainly effective within specific musical contexts, in particular for harmonically- and temporally-stable house music (and other related genres), we believe the key-based matching approach has several important limitations. Perhaps the most immediate of these limitations is that the underlying key estimation might be error-prone, and any errors would then propagate into the harmonic mixing. In addition to this, a global property, such as musical key, provides almost no information regarding what is in the signal itself and, in turn, how this might affect perceptual harmonic compatibility for listeners when two pieces are mixed. Similarly, music matching based on key alone provides no obvious means for ranking the compatibility, and hence, choosing, among several different pieces of the same key [3]. Likewise, assigning one key for the duration of a piece of music cannot indicate where in time the best possible mixes (or mashups) between different pieces of music might occur. Even with the ability to use pitch-shifting to transpose the musical key, it is important to consider the quantization effect of only comparing whole semitone shifts. The failure to consider fine-scale tuning could lead to highly dissonant mistuned mixes between songs that still share the same key.

Towards overcoming some of the limitations of key-based mixing, beat-synchronous chromagrams [3,4] have been used as the basis for harmonic alignment between pieces of music. However, while the chromagram provides a richer representation of the input signal than using key alone, it nevertheless relies on the quantization into discrete pitch classes and the folding of all harmonic information into a single octave to faithfully represent the input. In addition, harmonic similarity is used as a proxy for harmonic compatibility.

Therefore, to fully address the limitations of key-based harmonic mixing, we propose a new approach based on the analysis of consonance. We base our approach on the well-established psychoacoustic principles of sensory consonance and harmony as defined by Ernst Terhardt [5,6], where our goal is to discover the optimal, consonance-maximizing alignment between two music excerpts. In this way, we avoid looking for harmonic similarity and seek to move towards a direct measurement of harmonic compatibility. To this end, we first extract a set of frequencies and amplitudes using a sinusoidal model and average this information over short temporal frames. We fix the partials of one excerpt and apply a logarithmic scaling to the partials of the other over a range of one full octave in 1/8th semitone steps. Through an exhaustive search, we can identify the frequency shift that maximizes the consonance between the two excerpts and then apply the appropriate pitch-shifting factor prior to mixing the two excerpts together. A graphical overview of our approach is given in Figure 1.

Searching across a wide frequency range in small steps allows both for multiple possible harmonic alignments and the ability to compensate for differences in tuning. In comparison with an existing commercial DJ-mixing system, we demonstrate that our approach is able to provide mixes that were considered significantly both more consonant and more pleasant by musically-trained listeners.

In comparison to our previous work [7], the main contribution of this paper relates to an extended evaluation. To this end, we largely maintain the original description of our original method, but we provide the results of a new listening test, a more detailed statistical analysis and an examination of the effect of the parameterization of our model.

The remainder of this paper is structured as follows. In Section 2, we review existing approaches for the measurement of consonance based on roughness and pitch commonality. In Section 3, we describe our approach for consonance-based music mixing driven by these models. We then address the evaluation of our approach in Section 4 via a listening test and explore the effect of the parameterization of our model. Finally, in Section 5, we present conclusions and areas for future work.

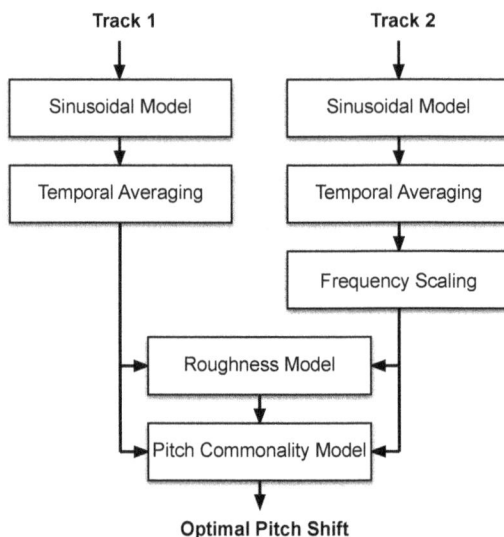

Figure 1. An overview of the proposed approach for consonance-based mixing. Each input track is analyzed by a sinusoidal model with 90-ms frames (with a 6-ms hop size). These are median-averaged into sixteenth note temporal frames. The frequencies of Track 2 are scaled over a single octave range and the sensory roughness calculated between the two tracks per frequency shift. The frequency shifts leading to the lowest roughness are used to determine the harmonic consonance via a model of pitch commonality.

2. Consonance Models

In this section, we present the theoretical approaches for the computational estimation of consonance that will form the core of the overall implementation described in Section 3 for estimating the most consonant combination of two tracks. To avoid misunderstandings due to ambiguous terminology, we define consonance by means of Terhardt's psychoacoustic model [5,6], which is divided into two categories: The first, sensory consonance, combines roughness (and fluctuations, standing for slow beatings and therefore equated with roughness throughout), sharpness (referring to high energy in high registers of a sound's timbre) and tonalness (the degree of tonal components a sound holds). The second, harmony, is mostly built upon Terhardt's virtual pitch theory, which describes the effect of perceiving an imaginary root pitch of a sonority's harmonic pattern. This, in terms of musical consonance, he calls the root relationship, whereas he describes pitch commonality as the degree of how similar the harmonic patterns of two sonorities are. We take these categories as the basis for our approach. To estimate the degree of sensory consonance, we use a modified version of Hutchinson and Knopoff's [8] roughness model. For calculating the pitch commonality of a combination of sonorities, we propose a model that combines Parncutt's [9] pitch categorization procedure with Hofmann-Engl's [10] virtual pitch and chord similarity model. Both models take a sequence of sinusoids, expressed as frequencies, f_i in Hz, and amplitudes, M_i in dBSPL (sound pressure level), as input.

2.1. Roughness Model

As stated above, the category of sensory consonance can be divided into three parts: roughness, tonalness and sharpness. While sharpness is closely connected to the timbral properties of musical audio [6], we do not attempt to model or modify this aspect, since it can be considered independent of the interaction of two pieces of music, which is the object of our investigation in this paper. Parncutt and Strasburger [11] discuss the strong relationship between roughness and tonalness as a sufficient reason to only analyze one of the two properties. The fact that roughness has been more extensively explored than tonalness and that most sensory consonance models build exclusively upon it motivates the use of roughness as our sole descriptor for sensory consonance in this work. For each of the partials of a spectrum, the roughness that is evoked by the co-occurrence with other partials is computed, then weighted by the dyads' amplitudes and, finally, summed for every sinusoid.

The basic structure of this procedure is a modified version of Hutchinson and Knopoff's [12] roughness model for complex sonorities that builds on the roughness curve for pure tone sonorities proposed by Plomp and Levelt [13] (this approach also forms the basis of work by Sethares [14] and Bañuelos [15] on the analysis of consonance in tuning systems and musical performance, respectively). A function that approximates the graph estimated by Plomp and Levelt is proposed by Parncutt [16]:

$$g(y) = \begin{cases} (\exp(1)\frac{y}{0.25}\exp(-\frac{y}{0.25}))^2 & y < 1.2 \\ 0 & \text{otherwise} \end{cases} \tag{1}$$

where $g(y)$ is the degree of roughness of a dyad and y the frequency interval between two partials (f_i and f_j) expressed in the critical bandwidth (CBW) of the mean frequency $\bar{f}$, such that:

$$y = \frac{|f_j - f_i|}{\text{CBW}(\bar{f})} \tag{2}$$

and:

$$\bar{f} = \frac{f_i + f_j}{2}. \tag{3}$$

Since pitch perception is based on ratios, we substitute CBW($\bar{f}$) with Moore and Glasberg's [17] equation for the equivalent rectangular bandwidth ERB($\bar{f}$) in Equation (2).

$$\text{ERB}(\bar{f}) = 6.23(10^{-3}\bar{f})^2 + 93.39(10^{-3}\bar{f}) + 28.52 \tag{4}$$

which Parncutt [16] also cites as offering "possible minor improvements." The roughness values $g(y)$ for every dyad are then weighted by the dyad's amplitudes (M_i and M_j) to obtain a value of the overall roughness D of a complex sonority with N partials:

$$D = \frac{\sum_{i=1}^{N} \sum_{j=i+1}^{N} M_i M_j g_{ij}}{\sum_{i=1}^{N} M_i^2}. \tag{5}$$

2.2. Pitch Commonality Model

As opposed to sensory consonance, which can be applied to any arbitrary sound, the second category of Terhardt's consonance model [5,6] is largely specified on musical sounds. This is why the incorporation of an aspect based on harmony should be of critical importance in a system that aligns music according to consonance. Nevertheless, the analysis of audio with a harmonic model of consonance is currently under-explored in the literature. Existing consonance-based tools for music typically focus on roughness alone [14,18,19]. Relevant approaches that include harmonic analysis perform note extraction, categorization in an octave-ranged chromagram and, as a consequence of this, key detection, but the psychoacoustic aspect of harmony is rarely applied. One of our main aims in this work is therefore to use the existing theoretical background to develop a model that estimates the consonance in terms of root relationship and pitch commonality and ultimately to combine this with a roughness model.

The fundament of the approach lies in harmonic patterns in the spectrum. The extraction of these patterns is taken from the pre-processing stage of the pitch categorization procedure of Parncutt's model for the computational analysis of harmonic structure [9,11].

For a given set of partials, the audibilities of pitch categories in semitone intervals are produced. Since this corresponds directly to the notes of the chromatic scale, the degree of audibility for different pitch categories can be attributed to a chord. Hofmann-Engl's [10] virtual pitch model then will be used to compute the "Hofmann-Engl pitch sets" of these chords, which will be subsequently compared for their commonality.

2.2.1. Pitch Categorization

Parncutt's algorithm detects the particular audibilities for each pure tone, considering the frequency-specific threshold of hearing, masking effects and the theory of virtual pitch. Following Terhardt [20], the threshold in quiet L_{TH} is formulated as:

$$L_{TH} = 3.64 f_i^{-0.8} - 6.5 \exp\left(-0.6(f_i - 3.3)^2\right) + 10^{-3} f_i^4. \tag{6}$$

Next, the auditory level YL of a pure tone with its specific frequency f_i is defined as its level in dB above its threshold in quiet,

$$YL(f_i) = \max(0, M_i - L_{TH}(f_i)) \tag{7}$$

Masking depends on the distance of pure tones in critical bandwidths. To simulate the effects of masking in the model, the pitch of the pure tone is examined on a scale that corresponds to critical bandwidths. To this end, the pure tone height, $H_p(f_i)$, for every pitch category, f_i, in the spectrum is computed, using the analytic formula by Moore and Glasberg [17] that expresses the critical band rate in ERB (equivalent rectangular bandwidth):

$$H_p(f_i) = H_1 \log_e\left(\frac{f_i + f_1}{f_i + f_2}\right) + H_0. \tag{8}$$

As parameters, Moore and Glasberg propose H_1 = 11.17 ERB, H_0 = 43.0 ERB, f_1 = 312 Hz and f_2 = 14,675 Hz.

The partial masking level $ml(f_i, f_j)$, which is the degree of how much every pure tone in the sonority with the frequency f_i is masked by an adjacent pure tone with its specific frequency f_j and auditory level $YL(f_j)$, is estimated as:

$$ml(f_i, f_j) = YL(f_j) - k_m |H_p(f_j) - H_p(f_i)| \tag{9}$$

where k_m can take values between 12 and 18 dB (chosen value: 12 dB). The partial masking level is specified in dB. The overall masking level, $ML(f_i)$, of every pure tone is obtained by summing its partial masking levels, which are converted first to amplitudes and, then, after the addition, back to dB levels:

$$ML(f_i) = \max(0, (20 \log_{10} \sum_{P \neq P'} 10^{(ml(f_i, f_j)/20)})). \tag{10}$$

In the case of a pure tone with frequency f_i that is not masked, $ml(f_i, f_j)$ will take a large negative value. This negative value for $ML(f_i)$ is avoided by the use of the max operator when comparing the calculated value to zero.

Following this procedure for each component, we can now obtain its audible level $AL(f_i)$ by subtracting its overall masking level from its auditory level $YL(f_i)$:

$$AL(f_i) = \max(0, (YL(f_i) - ML(f_i))). \tag{11}$$

To incorporate the saturation of each pure tone with increasing audible level, the audibility $A_p(f_i)$ is estimated for each pure tone component:

$$A_p(f_i) = 1 - \exp(\frac{-AL(f_i)}{AL_0}). \tag{12}$$

where, following Hesse [21], AL_0 is set to 15 dB. Due to the need to extract harmonic patterns and to consider virtual pitches, the still audible partials are now assigned discrete semitone values. To this end, frequency values that fall into a certain interval are assigned to so-called pitch categories, P, which are defined by their center frequencies in Hz:

$$P(f_i) = 12\log_2(\frac{f_i}{440}) + 57 \tag{13}$$

where the standard pitch of 440 Hz (musical note A_4) is represented by Pitch Category 57.

For the detection of harmonic patterns in the sonority, a template is used to detect partials of harmonic complex tones shifted over the spectrum in a step size of one pitch category. One pattern's element is given by the formula:

$$P_n = P_1 + \lfloor 12\log_2(n) + 0.5 \rfloor \tag{14}$$

where P_1 represents the pitch category of the lowest element (corresponding to the fundamental) and P_n the pitch category of the n-th harmonic.

Wherever there is a match between the template and the spectrum for each semitone-shift, a complex-tone audibility $A_c(P_1)$ is assigned to the template's fundamental. To take the lower audibility of higher harmonics into account, they are weighted by their harmonic number, n:

$$A_c(P_1) = \frac{1}{k_T} \left(\sum_n \sqrt{\frac{A_p(P_n)}{n}} \right)^2. \tag{15}$$

where the free parameter k_T is set to three. To estimate the audibility, $A(P)$, of a component that considers both the spectral- and complex-tone audibility of every category, the overall maximum of the two is taken as the general audibility. This choice is supported by Terhardt *et al.* [20], who state that only either a pure or a complex tone can be perceived at once:

$$A(P) = \max(A_p(P), A_c(P)). \tag{16}$$

2.2.2. Pitch-Set Commonality and Harmonic Consonance

The resulting set of pitch categories can be interpreted as a chord with each pitch category's note sounding according to its audibility $A(P)$. With the focus on music and given the importance of the triad in Western culture [22], we extract the three notes of the sonority with the highest audibility.

To compare two chords according to their pitch commonality, Hofmann-Engl proposes to estimate their similarity by the aid of the pitch sets that are produced by his virtual pitch model [23]. The obtained triad is first inserted into a table similar to the one Terhardt uses to analyze a chord for its root note (see [6]), with the exception that Hofmann-Engl's table contains one additional subharmonic. The notes are ordered from low to high along with their corresponding different subharmonics. A major difference to Terhardt's model is the introduction of two weights w_1 and w_2 to estimate the strength β_{note} for a specific note to be the root of the chord with $Q = 3$ tones for all 12 notes of an octave:

$$\beta_{note} = \frac{\sum_{q=1}^{Q} w_{1,note} \, w_{2,q}}{Q} \tag{17}$$

where the result is a set of 12 strengths of notes or so-called "Hofmann-Engl pitches" [23]. As an example, the pitch set deriving from a C major triad is shown in Figure 2. The fusion weight, $w_{1,note}$, is based on note similarity and gives the subharmonics more impact in decreasing order. This implies that the unison and the octave have the highest weight, then the fifth, the major third, and so on. The maximum value of $w_{1,note}$ is $c = 6$ Hh (Helmholtz; unit set by Hofmann-Engl). The fusion weight is decreased by the variable b, which is $b = 1$ Hh for the fifth, $b = 2$ Hh for the major third, $b = 3$ Hh for the minor seventh, $b = 4$ Hh for the major second and $b = 5$ Hh for the major seventh. All other intervals take the value $b = 6$ and are therefore weighted zero, according to the formula:

$$w_{1,note} = \frac{c^2 - b^2}{c}. \tag{18}$$

The weight according to pitch order, w_2, adds greater importance to lower notes, assuming that a lower note is more likely to be perceived as the root of the chord than a higher one and is calculated as:

$$w_{2,q} = \sqrt{\frac{1}{q}} \tag{19}$$

where q represents the position of the note in the chord. For the comparison between two sonorities (e.g., from different tracks), the Pearson correlation $r_{set_1 set_2}$ is calculated for the pair of Hofmann-Engl pitch sets, as Hofmann-Engl [23] proposes to determine chord similarity and, therefore, consonance, C, in the sense of harmony as:

$$C = r_{set_1 set_2}. \tag{20}$$

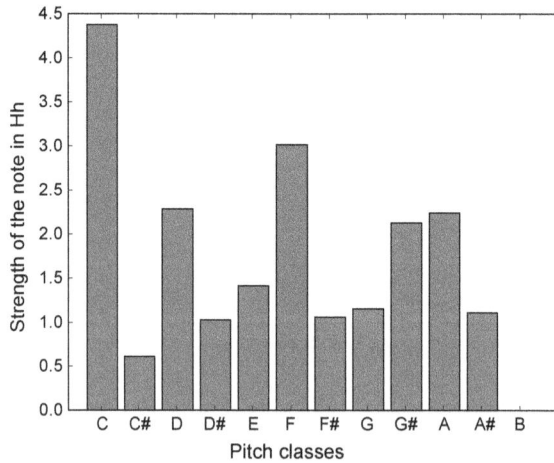

Figure 2. Hofmann-Engl pitch set for a C major triad, for which each pitch class of the chromatic scale has a strength (*i.e.*, likelihood) of being perceived as the root of the C major chord, which is measured in Helmholtz (Hh).

A graphical example showing the harmonic consonance for different triads compared to the C major triad is shown in Figure 3.

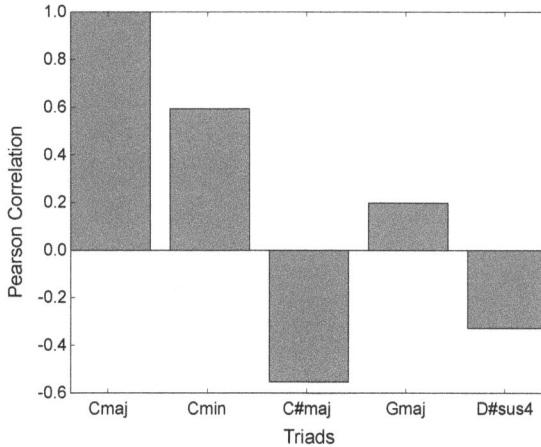

Figure 3. Harmonic consonance C, from Equation (20), measured as the correlation of two different pitch sets of different triads with a C major triad as the reference.

3. Consonance-Based Mixing

Based on the models of roughness and pitch commonality presented in the previous section, we now describe our approach for consonance-based mixing between two pieces of music.

3.1. Data Collection and Pre-Processing

We first explain the necessary pre-processing steps that allow the subsequent measurement of consonance between two pieces of music. For the purpose of this paper, we make several simplifications concerning the properties of the musical audio content we intend to mix.

Given that one of our aims is to compare consonance-based mixing to key-based matching methods in DJ-mixing software (see Section 4), we currently only consider electronic music (e.g., house music), which is both harmonically stable and typically has a fixed tempo. We collected a set of 30 tracks of recent electronic music for which we manually annotated the tempo and beat locations and isolated short regions within each track lasting precisely 16 beats (*i.e.*, four complete bars). In order to focus entirely on the issue of harmonic alignment without the need to address temporal alignment, we force the tempo of each excerpt to be exactly 120 beats per minute. For this beat quantization process, we use the open source pitch-shifting and time-stretching library, Rubber Band [24], to implement any necessary tempo changes. Accordingly, our database of musical excerpts consists of a set of 8 s (*i.e.*, 500 ms per beat) mono .wav files sampled at 44.1 kHz with 16-bit resolution. Further details concerning this dataset are in Section 4.1.

To provide an initial set of frequencies and amplitudes, we use a sinusoidal model, namely the "Spectral Modeling Synthesis Tools" Python software package by Serra [25,26], with which we extract sinusoids using the default window size and hop sizes of 4001 and 256 samples, respectively. In order to focus on the harmonic structure present in the musical input, we extract the $I = 20$ partials with the highest amplitude under 5 kHz.

For our chosen genre of electronic music and our assembled dataset, we observed that the harmonic structure remained largely constant over the duration of each 1/16th note (*i.e.*, 125 ms). Therefore, to strike a balance between temporal resolution and computational complexity, we summarize the frequencies and amplitudes by taking the frame-wise median over the duration of each 1/16th note. Thus, for each excerpt, we obtain a set of frequencies and amplitudes, $f_{\gamma,i}$ and $M_{\gamma,i}$, where i indicates the partial number (up to $I = 20$) and γ each 1/16th note frame (up to $\Gamma = 64$). An overview of the extraction of sinusoids and temporal averaging is shown in Figure 4. In Section 4.2, we examine the effect of this choice of parameters.

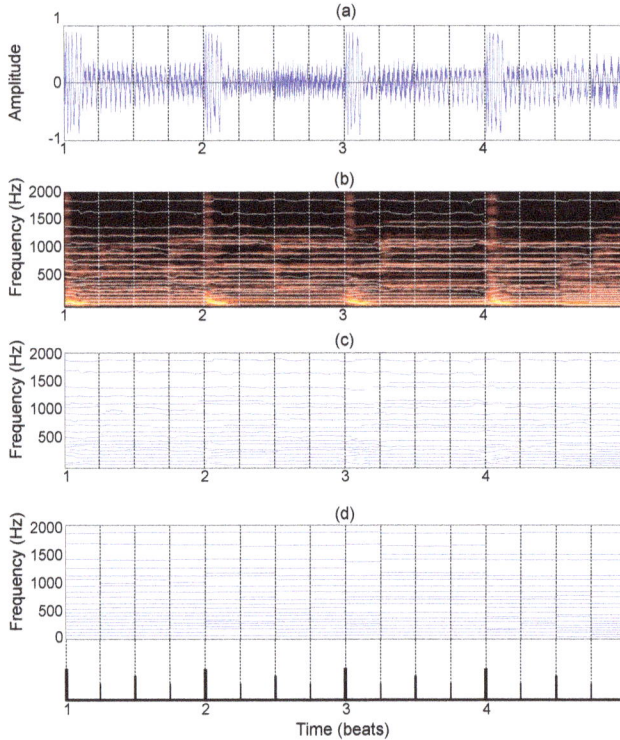

Figure 4. Overview of sinusoidal modeling and temporal averaging. (**a**) A one-bar (*i.e.*, 2 s) excerpt of an input audio signal sampled at 44.1 kHz at 120 beats per minute. Sixteenth notes are overlaid as vertical dotted lines. (**b**) The spectrogram (frame size = 4001 samples, hop size = 256 samples, Fast Fourier Transform (FFT) size = 4096), which is the input to the sinusoidal model (with overlaid solid grey lines showing the raw tracks of the sinusoidal model). (**c**) The raw tracks of the sinusoidal model. (**d**) The sinusoidal tracks averaged over sixteenth note temporal frames, each of a duration of 125 ms.

3.2. Consonance-Based Alignment

For two input musical excerpts, T^1 and T^2, with corresponding frequencies and amplitudes $f^1_{\gamma,i}, M^1_{\gamma,i}$ and $f^2_{\gamma,i}, M^2_{\gamma,i}$, respectively, we seek to find the optimal consonance-based alignment between them. At this stage, we could attempt to modify (*i.e.*, pitch shift) both excerpts, T^1 and T^2, so as to minimize the overall stretch factor between them. However, we conceptualize the harmonic mixing problem as one in which there is a user-selected query, T^1, to which we will mix T^2. In this sense, we can retain the possibility to rank multiple different excerpts in terms of how well they match T^1. To this end, we fix all information regarding T^1 and modify only T^2. This setup offers the additional advantage that only one excerpt will contain artifacts resulting from pitch shifting.

Our approach centers on the calculation of consonance as a function of a frequency shift, s, and is based on the hypothesis that under some frequency shift applied to T^2, the consonance between T^1 and T^2 will be maximized, and this, in turn, will lead to the optimal mix between the two excerpts.

In total, we create $S = 97$ shifts, which cover the range of ± 6 semitones in 1/8th semitone steps (*i.e.*, 48 downward and 48 upward shifts around a single "no shift" option). We scale the frequencies of the partials $f_{\gamma,i}^2$ as follows:

$$f_{\gamma,i}^2[s] = 2^{\log_2(f_{\gamma,i}^2) + \frac{s-48}{96}} \quad s = 0, \ldots, S-1. \tag{21}$$

For each 1/16th note temporal frame, γ, and per shift, s, we then merge the corresponding frequencies and amplitudes between both tracks (as shown in Figures 5 and 6), such that:

$$f_\gamma[s] = \left[f_\gamma^1 \; f_\gamma^2[s] \right] \tag{22}$$

and:

$$M_\gamma[s] = \left[M_\gamma^1 \; M_\gamma^2[s] \right]. \tag{23}$$

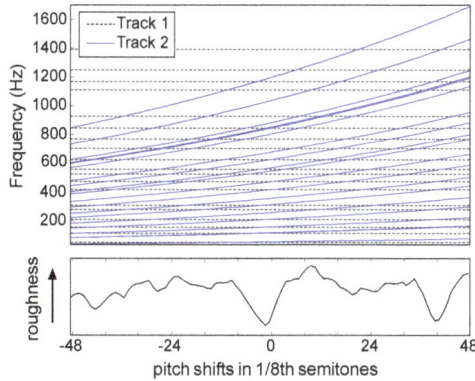

Figure 5. (Upper plot) Frequency scaling applied to the partials of one track (solid lines) compared to the fixed partials of the other (dotted lines) for a single temporal frame. (Lower plot) The corresponding roughness as a function of frequency scaling over that frame.

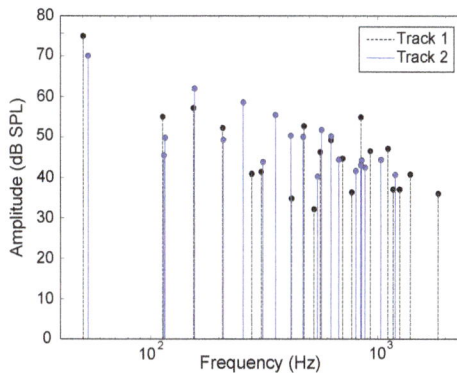

Figure 6. The partials of two excerpts for one temporal frame, γ.

We then calculate the roughness, $D_\gamma[s]$ according to Equation (5) in Section 2.1 with the merged partials and amplitudes as input. Figure 7 illustrates the interaction between the partials for a single frame within two equivalent visualizations, first with the partials between the two tracks separated and, then, once they have been merged. In this way, we can observe the interactions between roughness-creating partials between the two tracks in a given frame or, alternatively, examine a visualization that corresponds to their mixture.

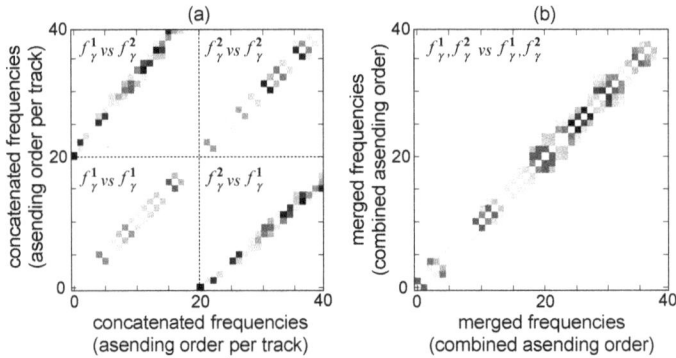

Figure 7. Visualization of the roughness matrix g_{ij} from Equation (1) for the frequencies f_γ^1 for one temporal frame of T^1 and f_γ^2 for the same frame of T^2. Darker shades indicate higher roughness. (**a**) The frequencies are sorted in ascending order per track to illustrate the internal roughness of T^1 and T^2, as well as the "cross-roughness" between them. (**b**) Here, the full set of frequencies is merged and then sorted to show the roughness of the mixture.

Then, to calculate the overall roughness, $\bar{D}[s]$, as a function of frequency shift, s, we take the mean of the roughness values $D_\gamma[s]$ across the $\Gamma = 64$ temporal frames of the excerpt:

$$\bar{D}[s] = \frac{1}{\Gamma} \sum_{\gamma=0}^{\Gamma-1} D_\gamma[s], \tag{24}$$

for which a graphical example is shown in Figure 8.

Having calculated the roughness across all possible frequency shifts, we now turn our focus towards the measurement of pitch commonality as described in Section 2.2. Due both to the high computational demands of the pitch commonality model and the rounding that occurs due to the allocation of discrete pitch categories, we do not calculate the harmonic consonance as a function of all possible frequency shifts. Instead, we extract all local minima from $\bar{D}[s]$, label these frequency shifts, s^*, and then proceed with this subset. In this way, we use the harmonic consonance, C, as a means to filter and further rank the set of possible alignments (*i.e.*, minima) arising from the roughness model.

While the calculation of $D_\gamma[s]$ relies on the merged set of frequencies and amplitudes from Equations (22) and (23), the harmonic consonance compares two individually-calculated Hofman-Engl pitch sets. To this end, we calculate Equations (8) to (17) independently for f_γ^1 and $f_\gamma^2[s^*]$ to create set_γ^1 and $\text{set}_\gamma^2[s^*]$ and, hence, $C_\gamma[s^*]$ from Equation (20). As with the roughness, the overall harmonic consonance $\bar{C}[s^*]$ is then calculated by taking the mean across the temporal frames:

$$\bar{C}[s^*] = \frac{1}{\Gamma} \sum_{\gamma=0}^{\Gamma-1} C_\gamma[s^*]. \tag{25}$$

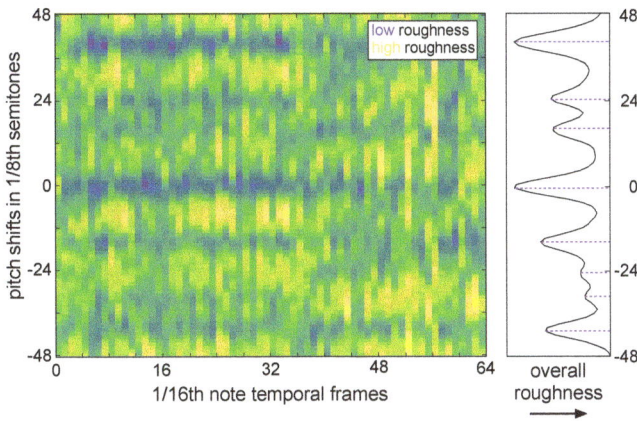

Figure 8. Visualization of roughness, $D_\gamma[s]$, over 64 frames for the full range of pitch shifts. Purple regions indicate lower roughness, while yellow indicates higher roughness. The subplot on the right shows the average roughness curve, $\bar{D}[s]$, as a function of pitch shift, where the roughness minima point to the left and are shown with purple dashed lines.

Since no prior method exists for combining the roughness and harmonic consonance, we adopt a simple approach to equally weight their contributions to give an overall measure of consonance based on roughness and pitch commonality:

$$\rho[s^*] = \hat{D}[s^*] + \hat{C}[s^*] \tag{26}$$

where $\hat{D}[s^*]$ corresponds to the raw roughness values $\bar{D}[s^*]$, which have been inverted (to reflect sensory consonance as opposed to roughness) and then normalized to the range [0,1], and $\hat{C}[s^*]$ similarly represents the [0,1] normalized version of $\bar{C}[s^*]$. The overall consonance $\rho[s^*]$ takes values that range from zero (minimum consonance) to two (maximum consonance), as shown in Figure 9. The maximum score of two is achieved only when the roughness and harmonic consonance detect the same pitch shift index as most consonant.

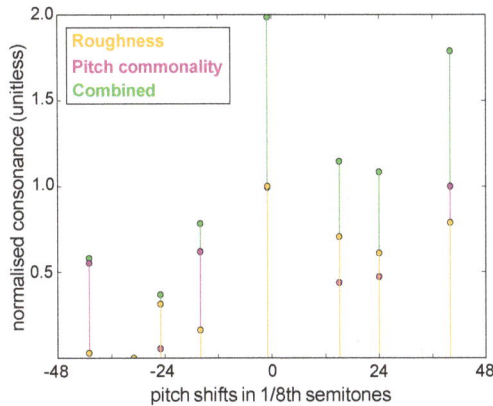

Figure 9. Values of consonance from the sensory consonance model, $\hat{D}[s^*]$, the harmonic consonance, $\hat{C}[s^*]$, and the resulting overall consonance, $\rho[s^*]$. Pitch shift index -1 (*i.e.*, -0.125 semitones) holds the highest consonance value and is the system's choice for the most consonant shift.

3.3. Post-Processing

The final stage of the consonance-based mixing is to implement the mix between tracks T^1 and T^2 under the consonance-maximizing pitch shift, *i.e.*, $\arg\max_{s^*}(\rho[s^*])$. As in Section 3.1, we again use the Rubber Band Library [24] to perform the pitch shifting on T^2, as this was found to give better audio quality than implementing the pitch shift directly using the output of the sinusoidal model. To avoid loudness differences between the two tracks prior to mixing, we normalize each audio excerpt to a reference loudness level (pink noise at 83 dB SPL) using the replay gain method [27].

4. Evaluation

The primary purpose of our evaluation is to determine whether the roughness curve can provide a robust means for identifying consonant harmonic alignments between two musical excerpts. If this is the case, then pitch shifting and mixing according to the minima of the roughness curve should lead to consonant (and hence, pleasant) musical results, where as mixing according to the maxima should yield dissonant musical combinations. To explore the relationship between roughness and consonance, we designed and conducted a listening test to obtain consonance and pleasantness ratings for a set of musical excerpts mixed according to different pitch shifts. Following this, we then investigated the effect of varying the main parameters of the pre-processing stage (*i.e.*, the number of partials I and the number of temporal frames Γ), as described in Section 3.1, by examining the correlation between roughness values and listener ratings under different parameterizations.

4.1. Listening Test

To evaluate the ability of our model to provide consonant mixes between different pieces of music, we conducted a listening test using excerpts from our dataset of 30 short musical excerpts of recent house music (each 8 s in duration and lasting exactly 16 beats). While our main concern is in evaluating the properties of the roughness curve, we also included a comparison against a key-based matching method using the key estimation from the well-known DJ software Traktor 2 (version 6.1) from Native Instruments [28]. In total, we created five conditions for the mix of two individual excerpts, which are summarized as follows:

- **A** No shift: no attempt to harmonically align the excerpts; instead, the excerpts were only aligned in time by beat-matching.
- **B** Key match (Traktor): each excerpt was analyzed by Traktor 2 and the automatically-detected key recorded. The key-based mix was created by finding the smallest pitch shift necessary to create a harmonically-compatible mix according to the circle of fifths, as per the description in the Introduction.
- **C** Max roughness: the roughness curve was analyzed for local maxima, and the pitch shift with the highest roughness (*i.e.*, most dissonant) was chosen to mix the excerpts.
- **D** Min roughness: the roughness curve was analyzed for local minima, and the pitch shift with the lowest roughness (*i.e.*, most consonant) was chosen to mix the excerpts.
- **E** Min roughness and harmony: from the set of extracted minima in Condition **D**, the combined harmonic consonance and roughness was calculated, and the pitch shift yielding the maximum overall consonance $\rho[s^*]$ was selected to mix the excerpts.

We selected the set of stimuli for use in the listening experiment according to two conditions. First and foremost, we required a set of unique pitch shifts across the five conditions per mix, and second, we chose not to have any repeated excerpts either as input nor the track to be pitch-shifted. To this end, we calculated the pitch shifts for each of the five conditions for all possible combinations of the 30 excerpts in the dataset (introduced in Section 3.1) compared to one other. In total, this provided 900 possible combinations of tracks (including the trivial comparison of each excerpt with itself). A breakdown of the number of matching shifts among the conditions is shown in Table 1. By definition, there were no matching pitch shifts between Conditions **C** (max roughness) and **D** (min roughness) or **E** (min roughness and harmony). By contrast, Conditions **D** and **E** matched 385 times.

Table 1. Number of identical shifts (from a maximum of 900) across each of the conditions resulting from the exhaustive combination of all pairs within the 30 excerpt dataset.

Condition	A	B	C	D	E
A	x	92	7	56	45
B		x	2	100	81
C			x	0	0
D				x	385
E					x

Out of 900, a total of 409 combinations gave unique pitch shifts across all five conditions. From this subset of 409, we discarded all cases where the smallest pitch shift between any pair of combinations was lower than 0.25 semitones. Next, we removed all mixes containing duplicate excerpts to avoid single tracks in more than one mix. From this final subset, we kept the 10 mixes (listed in Table 2) with the lowest maximum pitch shift across the conditions. In total, this provided 50 stimuli (10 mixes × 5 conditions) to be rated. A graphical overview of the pitch shifts per condition is shown in Figure 10, for which sound examples are available in the Supplementary Material. All of the stimuli were rendered as mono .wav files at a sampling rate of 44.1 kHz and with 16-bit resolution.

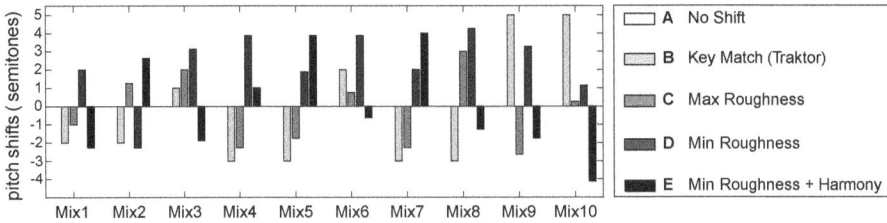

Figure 10. Comparison of suggested pitch shifts for each condition of the listening experiment. Note, the "no shift" condition is always zero.

In total, 34 normal hearing listeners (according to a self-report) participated in the experiments. Their musical training was self-rated as being either: music students, practicing musicians or active in DJing. Eleven of the participants were female, and 23 were male; their ages ranged between 23 and 57. When listening to each mix, the participants were asked to rate two properties: first, how consonant and, second, how pleasant the mixes sounded to them. The question for pleasantness was introduced both to emphasize the distinction between personal taste and musical consonance to the listener and also to consider the fact that a higher level of consonance might not lead to a more pleasant listening experience [9]. Both conditions were rated on a discrete six-point scale (zero to five) using a custom patch developed in Max/MSP. The order of the 50 stimuli was randomized for each participant. After every sound example, the ratings had to be entered before proceeding to the next. To guarantee familiarity with the experimental procedure and stimuli, a training phase preceded the main experiment. This was also used to ensure all participants understood the concept of consonance and to set the playback volume to a comfortable level. All participants took the experiment in a quiet listening environment using high quality headphones.

Regarding our hypotheses on the proposed conditions, we expected Condition **C** (max roughness) to be the least consonant, followed by **A** (no shift). However, without any harmonic alignment, its behavior was not easily predictable, save for the fact that it would be at least 0.25 semitones from any other condition. Of the remaining conditions, which attempted to find a good harmonic alignment, we expected **B** (Traktor) to be less consonant than both **D** (min roughness) and **E** (min roughness and harmony).

4.2. Results

4.2.1. Statistical Analysis

To examine the data collected in the listening experiment, we separately analyzed the consonance and pleasantness ratings using the non-parametric Friedman test where we treated participants and mixes as random effects. For both the consonance and pleasantness ratings, the main effect of the conditions was highly significant (consonance: chi-square = 181.60, $p < 0.00001$; pleasantness: chi-square = 240.73, $p < 0.00001$).

With regard to the interaction across conditions, we performed a *post hoc* analysis via a multiple comparison of means with Bonferroni correction for which the mean rankings and 95% confidence intervals are shown in Figure 11a,b for consonance and pleasantness ratings, respectively.

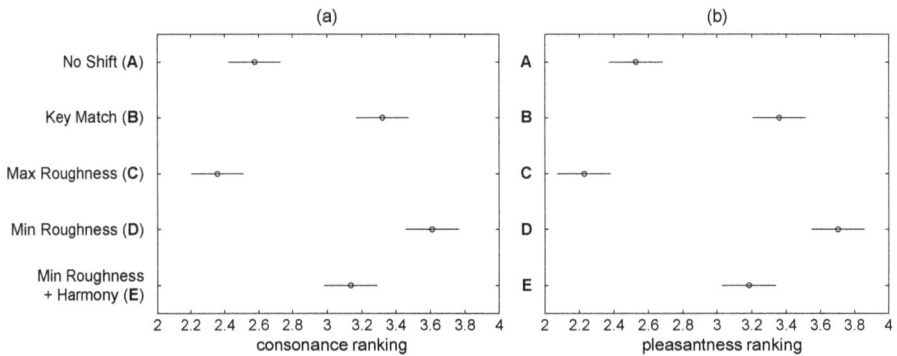

Figure 11. Summary of multiple comparisons of mean rankings (with Bonferroni correction) between conditions for (**a**) consonance and (**b**) pleasantness ratings. Both mixes and participants are treated as random effects. Error bars (95% confidence intervals) without overlap indicate statistically-significant differences in the mean rankings.

There is a very large separation between Conditions **B**, **D** and **E**, *i.e.*, those conditions that attempt to find a good harmonic alignment, and Conditions **A** and **C**, which do not. No significant difference was found between Conditions **A** and **C** (consonance: $p > 0.45$; pleasantness: $p > 0.07$) and likewise for conditions **B** and **E** (consonance: $p > 0.90$; pleasantness: $p = 1.00$). For consonance, the difference between Conditions **D** and **B** is not significant, $p > 0.08$; however it is significant for pleasantness $p < 0.05$.

Inspection of Figure 11 reveals similar patterns regarding consonance and pleasantness ratings, which are generally consistent with our hypotheses stated in Section 4.1. Ratings for Condition **C** (max roughness) are significantly smaller (worse) than for all other conditions, except Condition **A** (no shift). Pitch shifts in Condition **D** (min roughness) are rated significantly highest (best) in terms of pleasantness ratings.

While there is a large separation between the ratings for Conditions **D** and **E**, *i.e.*, our two proposed methods for consonance, such a result should be examined within the context of the experimental design and additional inspection of Table 1. Here, we find that close to 43% of the 900 combinations resulted in an identical choice of pitch shift, implying that both methods often converged on the same result and to a far greater degree than any of the other condition pairs. Since there is no significant difference between the ratings of Conditions **E** and **B** and because these were rated towards the higher end of the (zero to five) scale, we could consider any of three methods to be a valid means of harmonically mixing music signals, nevertheless with Condition **D** the preferred choice.

Looking again at the key-based approach (Condition **B**), it is useful to consider the impact of any misestimation of the key made by Traktor. To this end, we asked a musical expert to annotate the ground truth keys for each of the 20 excerpts used to make the listening test. These annotated keys are shown in Table 2. Despite the apparent simplicity of this type of music from a harmonic perspective, our music expert was unable to precisely label the key in six out of the 20 cases. This was due to the short duration of the excerpts and an insufficient number of different notes to unambiguously choose between a major or minor key. In these cases, the most predominant pitch class was annotated instead. Traktor, on the other hand, always selects a major or minor key (irrespective of the tonality of the music), and in fact, it only agreed with our expert in six of the cases. In addition, we used Traktor to extract the key for the full-length recordings, and in these cases, the key matched between the excerpt and full-length recording only eight out of 20 times. While the inability of Traktor to extract the correct key should lead us to expect poor performance in creating harmonic mixes, this is not especially evident in the results. In fact, it may be that the harmonic simplicity (*i.e.*, the weak sense of any one predominant key) in the excerpts of our chosen dataset naturally lends itself to multiple different harmonic alignments; an observation supported by the results, which show more than one possible option for harmonic alignment being rated towards the higher end of the scales for consonance and pleasantness. A graphical example comparing the output of the key-based matching using Traktor (Condition **B**) and min roughness (Condition **D**) between two excerpts is shown in Figure 12.

Table 2. Track titles and artists for the stimuli used in the listening test along with ground truth annotations made by a musical expert. Those excerpts labeled 'a' were the inputs to the mixes, whereas those labeled 'b' were subject to pitch shifting. In some cases, the harmonic information was too sparse (*i.e.*, too few notes) to make an unambiguous decision between major and minor. In these cases, the predominant root note is indicated. Note, the artist ##### (Mix 4a) is an alias of Aroy Dee (Mix 3b).

Mix No.	Artist	Track Title	Annotated Key
1a	Person Of Interest	Plotting With A Double Deuce	(E)
1b	Locked Groove	Dream Within A Dream	A maj
2a	Stephen Lopkin	The Haggis Trap	(A)
2b	KWC 92	Night Drive	D# min
3a	Legowelt	Elementz Of Houz Music (Actress Mix 1)	B min
3b	Aroy Dee	Blossom	D# min
4a	#####	#####.1	A min
4b	Barnt	Under His Own Name But Also Sir	C min
5a	Julius Steinhoff	The Cloud Song	D min
5b	Donato Dozzy & Tin Man	Test 7	F min
6a	R-A-G	Black Rain (Analogue Mix)	(E)
6b	Lauer	Highdimes	(F)
7a	Massimiliano Pagliari	JP4-808-P5-106-DEP5	(C)
7b	Levon Vincent	The Beginning	D# min
8a	Roman Flügel	Wilkie	C min
8b	Liit	Islando	D# min
9a	Tin Man	No New Violence	C min
9b	Luke Hess	Break Through	A min
10a	Anton Pieete	Waiting	A min
10b	Voiski	Wax Fashion	(E)

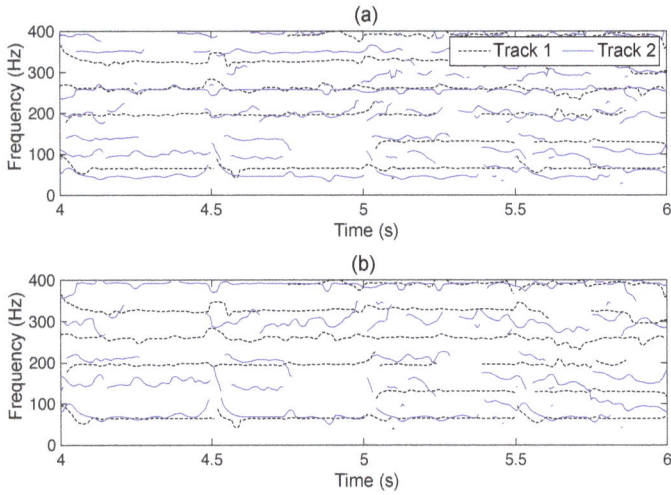

Figure 12. Comparison of extracted sinusoids after pitch shifting using Traktor (**a**) and min roughness (**b**) on Mix 8 from Table 2. Traktor applies a pitch shift of −2.0 semitones to Track 2 (solid blue lines), while the min roughness applies a pitch shift of +2.25 semitones to Track 2. In comparison to Track 1 (dotted black lines), we see that the min roughness approach (**b**) has primarily aligned the bass frequencies (under 100 Hz), whereas Traktor (**a**) has aligned higher partials around 270 Hz.

4.2.2. Effect of Parameterization

Having looked into detail at the interactions between the difference conditions in terms of the ratings, we now revisit the properties of the roughness curve towards understanding the extent to which it provides a meaningful indicator of consonance for harmonic mixing.

To this end, we now investigate the correlation between the ratings obtained from consonance and pleasantness compared to the corresponding points in the roughness curve for each associated pitch shift. While only three of the five conditions (**C**, **D** and **E**) were derived directly from each roughness curve, for completeness, we use the full set of 50 points (*i.e.*, five conditions across 10 mixes).

To gain a deeper insight into the design of our model, which is highly dependent on the extraction of partials using a sinusoidal model, we generate multiple roughness curves under different parameterizations and measure the correlation with the listener ratings for each. We focus on what we consider to be the two most important parameters: I, the number of sinusoids, and Γ, the number of temporal frames after averaging. In this way, we can examine the relationship from a harmonic and temporal perspective. To span the parameter space, we vary I from five up to 80 (default value = 20), and for the temporal averaging, we consider three cases: (i) beat level averaging ($\Gamma = 16$ across four-bar excerpts); (ii) 16th note averaging ($\Gamma = 64$ and our default condition); and (iii) using all frames from the sinusoidal model without any averaging. The corresponding plots for both consonance and pleasantness ratings are shown in Figure 13.

From inspection of the figure, we can immediately see that the number of sinusoids plays a more critical role than the extent/use of temporal averaging. Using more than 25 sinusoids (per frame of each track) has an increasingly negative impact on the Pearson correlation value. Likewise, using too few sinusoids also appears to have a negative impact. Considering the roughness model, having too few observations of the harmonic structure will very likely fail to capture all of the main roughness creating partials. While on the other hand, over-populating the roughness model with sinusoids (many of which may result from percussive or noise-like content) will also obscure the interaction of the "true" harmonic partials in each track. Within the context of our (harmonically simple) dataset, a range of between 15 and 25 partials provides the strongest relationship between roughness values and consonance ratings.

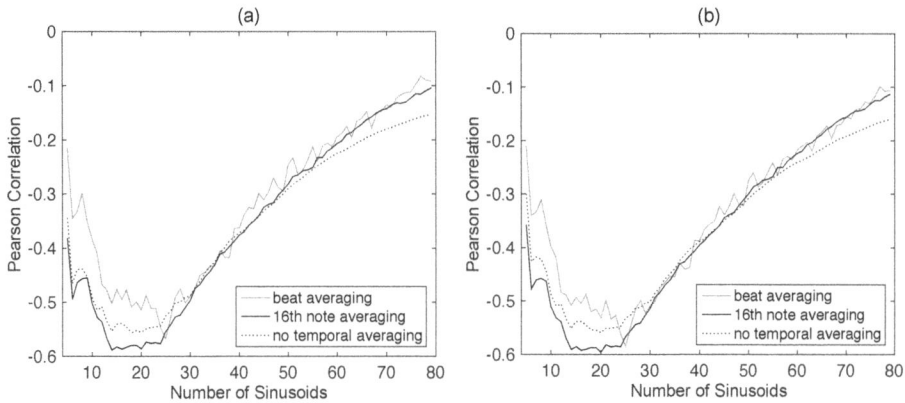

Figure 13. Pearson correlation between (**a**) consonance and (**b**) pleasantness ratings and sensory roughness values under different parameterizations of the model. The number of sinusoids vary from five to 80, and the temporal averaging is shown for beat length frames, 16th note frames and using all frames from the sinusoidal model without averaging. The negative correlation indicates the negative impact of roughness towards consonance ratings.

Looking next at the effect of the temporal averaging, we can see a much noisier relationship when using beat averaging compared to our chosen summarization at the 16th note level. In contrast, the plot is smoothest without any temporal averaging, yet it is moderately less correlated with the data. As with the harmonic dimension, the 16th note segmentation adequately captures the rate at which harmonic content changes in the signal, without losing too much fine detail through the temporal averaging process.

Finally, comparing the plots side by side, we see a near identical pattern for consonance and pleasantness. This behavior is to be expected given the very high correlation between the consonance and pleasantness ratings themselves ($r = 0.76$, $p < 1\times10^{-6}$). In the context of our dataset, this implies that the participants of the listening test considered consonance and pleasantness to be highly inter-dependent and, thus, that the measurement of roughness is a reliable indicator of listener preference for harmonic mixing.

5. Conclusions

In this paper, we have presented a new method for harmonic mixing ultimately targeted towards addressing some of the limitations of key-based DJ-mixing systems. Our approach centers on the use of psychoacoustic models of roughness and pitch commonality to identify an optimal harmonic alignment between different pieces of music across a wide range of possible fine-scaled pitch shifts applied to one of them. Via a listening experiment with musically-trained participants, we demonstrated that, within the context of the musical stimuli used, mixes based on a minimum degree of roughness were perceived as significantly more pleasant than those aligned according to musical key. Furthermore, including a harmonic consonance model in addition to the roughness model provided alternative pitch shifts, which were rated as consonant and pleasant as those from a commercial DJ-mixing system.

Concerning areas for future work, our model has thus far only been tested on very short and harmonically-simple musical excerpts, and therefore, we intend to test it under a wider variety of musical stimuli, including excerpts with more harmonic complexity. In addition, we plan to focus on the adaptation of our model towards longer musical excerpts, perhaps through the use of some structural segmentation into harmonically-stable regions.

We have also yet to consider the role of music with vocals and how to examine the potentially unnatural results that arise from pitch shift singing. To this end, we will explore both singing voice detection and voice suppression. Along similar lines, our roughness-based model can reveal not only which temporal frames give rise to the most roughness, but also precisely which partials contribute within these frames. Hence, we plan to explore methods for the suppression of dissonant partials, towards more consonant mixes.

Lastly, in relation to the interaction between the harmonic consonance and roughness, we will reexamine the rather simplistic combination of these two sources of information, towards a more sophisticated two-dimensional model of sensory roughness and harmony.

Supplementary Materials: Sound examples are available online at www.mdpi.com/2076-3417/6/5/123/s1.

Acknowledgments: M.D. is supported by National Funds through the FCT—Fundação para a Ciência e a Tecnologia within post-doctoral Grant SFRH/BPD/88722/2012 and by Project "NORTE-01-0145-FEDER-000020", which is financed by the North Portugal Regional Operational Programme (NORTE 2020), under the Portugal 2020 Partnership Agreement, and through the European Regional Development Fund (ERDF). B.S. is supported by the Bundesministerium für Bildung und Forschung (BMBF) 01 GQ 1004B (Bernstein Center for Computational Neuroscience Munich).

Author Contributions: All authors conceived and designed the experiments; R.G. and M.D. performed the experiments; M.D. and B.S. analysed the data; R.G., M.D. and B.S. wrote the paper.

Conflicts of Interest: The authors declare no conflict of interest.

References

1. Ishizaki, H.; Hoashi, K.; Takishima, Y. Full-automatic DJ mixing with optimal tempo adjustment based on measurement function of user discomfort. In Proceedings of the International Society for Music Information Retrieval Conference, Kobe, Japan, 26–30 October 2009; pp. 135–140.
2. Sha'ath, I. Estimation of Key in Digital Music Recordings. Master's Thesis, Birkbeck College, University of London, London, UK, 2011.
3. Davies, M.E.P.; Hamel, P.; Yoshii, K.; Goto, M. AutoMashUpper: Automatic creation of multi-song mashups. *IEEE/ACM Trans. Audio Speech Lang. Process.* **2014**, *22*, 1726–1737.
4. Lee, C.L.; Lin, Y.T.; Yao, Z.R.; Li, F.Y.; Wu, J.L. Automatic Mashup Creation By Considering Both Vertical and Horizontal Mashabilities. In Proceedings of the International Society for Music Information Retrieval Conference, Malaga, Spain, 26–30 October 2015; pp. 399–405.
5. Terhardt, E. The concept of musical consonance: A link between music and psychoacoustics. *Music Percept.* **1984**, *1*, 276–295.
6. Terhardt, E. *Akustische Kommunikation (Acoustic Communication)*; Springer: Berlin, Germany, 1998. (In German)
7. Gebhardt, R.; Davies, M.E.P.; Seeber, B. Harmonic Mixing Based on Roughness and Pitch Commonality. In Proceedings of the 18th International Conference on Digital Audio Effects (DAFx-15), Trondheim, Norway, 30 November–3 December 2015; pp. 185–192.
8. Hutchinson, W.; Knopoff, L. The significance of the acoustic component of consonance of Western triads. *J. Musicol. Res.* **1979**, *3*, 5–22.
9. Parncutt, R. *Harmony: A Psychoacoustical Approach*; Springer: Berlin, Germany, 1989.
10. Hofman-Engl, L. Virtual Pitch and Pitch Salience in Contemporary Composing. In Proceedings of the VI Brazilian Symposium on Computer Music, Rio de Janeiro, Brazil, 19–22 July 1999.
11. Parncutt, R.; Strasburger, H. Applying psychoacoustics in composition: "Harmonic" progressions of "non-harmonic" sonorities. *Perspect. New Music* **1994**, *32*, 1–42.
12. Hutchinson, W.; Knopoff, L. The acoustic component of western consonance. *Interface* **1978**, *7*, 1–29.
13. Plomp, R.; Levelt, W.J.M. Tonal consonance and critical bandwidth. *J. Acoust. Soc. Am.* **1965**, *38*, 548–560.
14. Sethares, W. *Tuning, Tibre, Spectrum, Scale*, 2nd ed.; Springer: London, UK, 2004.
15. Bañuelos, D. *Beyond the Spectrum of Music: An Exploration through Spectral Analysis of SoundColor in the Alban Berg Violin Concerto*; VDM: Saarbrücken, Germany, 2008.

16. Parncutt, R. Parncutt's Implementation of Hutchinson & Knopoff, 1978. Available online: http://uni-graz.at/parncutt/rough1doc.html (accessed on 28 January 2016).

17. Moore, B.; Glassberg, B. Suggested formulae for calculating auditory-filter bandwidths and excitation patterns. *J. Acoust. Soc. Am.* **1983**, *74*, 750–753.

18. MacCallum, J.; Einbond, A. Real-Time Analysis of Sensory Dissonance. In *Computer Music Modeling and Retrieval. Sense of Sounds*; Kronland-Martinet, R., Ystad, S., Jensen, K., Eds.; Springer: Berlin, Germany, 2008; Volume 4969, pp. 203–211.

19. Vassilakis, P.N. SRA: A Web-based Research Tool for Spectral and Roughness Analysis of Sound Signals. In Proceedings of the Sound and Music Computing Conference, Lefkada, Greece, 11–13 July 2007; pp. 319–325.

20. Terhardt, E.; Seewan, M.; Stoll, G. Algorithm for Extraction of Pitch and Pitch Salience from Complex Tonal Signals. *J. Acoust. Soc. Am.* **1982**, *71*, 671–678.

21. Hesse, A. Zur Ausgeprägtheit der Tonhöhe gedrosselter Sinustöne (Pitch Strength of Partially Masked Pure Tones). In *Fortschritte der Akustik*; DPG-Verlag: Bad-Honnef, Germany, 1985; pp. 535–538. (In German)

22. Apel, W. *The Harvard Dictionary of Music*, 2nd ed.; Harvard University Press: Cambridge, UK, 1970.

23. Hofman-Engl, L. Virtual Pitch and the Classification of Chords in Minor and Major Keys. In Proceedings of the ICMPC10, Sapporo, Japan, 25–29 August 2008.

24. Rubber Band Library. Available online: http://breakfastquay.com/rubberband/ (accessed on 19 January 2016).

25. Serra, X. SMS-tools. Available online: https://github.com/MTG/sms-tools (accessed on 19 January 2016).

26. Serra, X.; Smith, J. Spectral modeling synthesis: A sound analysis/synthesis based on a deterministic plus stochastic decomposition. *Comput. Music J.* **1990**, *14*, 12–24.

27. Robinson, D. Perceptual Model for Assessment of Coded Audio. Ph.D. Thesis, University of Essex, Colchester, UK, March 2002.

28. Native Instruments Traktor Pro 2 (version 6.1). Available online: http://www.native-instruments.com/en/products/traktor/dj-software/traktor-pro-2/ (accessed on 28 January 2016).

applied
sciences

MDPI

Article

Two-Polarisation Physical Model of Bowed Strings with Nonlinear Contact and Friction Forces, and Application to Gesture-Based Sound Synthesis[†]

Charlotte Desvages * and Stefan Bilbao

Acoustics and Audio Group, University of Edinburgh, Edinburgh EH9 3FD, Scotland, UK; s.bilbao@ed.ac.uk
* Correspondence: charlotte.desvages@ed.ac.uk; Tel.: +44-0-7598-448379
† This paper is an extended version of paper published in the International Conference on Digital Audio Effects (DAFx-15), Trondheim, Norway, 30 November–3 December 2015.

Academic Editor: Vesa Valimaki
Received: 15 March 2016; Accepted: 26 April 2016; Published: 10 May 2016

Abstract: Recent bowed string sound synthesis has relied on physical modelling techniques; the achievable realism and flexibility of gestural control are appealing, and the heavier computational cost becomes less significant as technology improves. A bowed string sound synthesis algorithm is designed, by simulating two-polarisation string motion, discretising the partial differential equations governing the string's behaviour with the finite difference method. A globally energy balanced scheme is used, as a guarantee of numerical stability under highly nonlinear conditions. In one polarisation, a nonlinear contact model is used for the normal forces exerted by the dynamic bow hair, left hand fingers, and fingerboard. In the other polarisation, a force-velocity friction curve is used for the resulting tangential forces. The scheme update requires the solution of two nonlinear vector equations. The dynamic input parameters allow for simulating a wide range of gestures; some typical bow and left hand gestures are presented, along with synthetic sound and video demonstrations.

Keywords: computer generated music; finite difference; musical acoustics; signal synthesis; nonlinear systems; energy balanced scheme; instrument simulation

---

1. Introduction

Sound synthesis techniques for string instruments have evolved, in the past few decades, from abstract synthesis [1] (wavetables, frequency-modulation (FM) synthesis …) towards sampling synthesis, based on a library of pre-recorded sounds, and physical models, emulating the instruments themselves. To this day, the best synthesised sound quality is achieved by sampling techniques; however, the potentially very large storage requirements for these sound libraries are a major argument for using physical models. Beyond the storage concerns, the use of a physical model allows for great flexibility for input parameters (typically, the instrument's shape and material properties, together with the player's controls), as well as output parameters, usually the "listening conditions", that can be changed freely and dynamically along a simulation, as opposed to the case of statically recorded samples.

Physical modelling synthesis for strings debuted in the 1970s, with time stepping methods to discretise and directly solve the 1D wave equation [2–4]. However, the very limited computational power at the time ruled out simulation at an audio sample rate in any reasonable amount of time. The next generation of models therefore focussed on algorithmic simplification, through physically plausible assumptions. The physics-based Karplus-Strong string synthesis algorithm [5,6] was generalised by the digital waveguide framework [7,8]; Karjalainen *et al.* [9] review the use of these models for string synthesis. Their fast execution and realistic sound output found efficient applications in bowed string modelling, and are still widely used to this day [10–14]. Digital waveguides model the forward and backward travelling waves along a string using delay lines—a simple and efficient strategy for certain linear time invariant systems. In particular, they are well suited for systems in one dimension, well described by the wave equation. Another more general class of physical models relies on the modal solutions of the string equation, and has been successfully adapted for bowed strings [15,16].

However, the very assumptions that underlie the efficiency of these methods can lead to difficulties when extensions to more realistic settings are desired—the bowed string and its complex, nonlinear, time-varying interaction with the environment being an excellent example. Time-stepping methods, and more specifically finite difference methods [17], though computationally costly, have regained appeal in musical sound synthesis [18] with the great increase in computing power during the last two decades. String simulation in one dimension is particularly suited for these kind of methods [19,20]. The interactions of, say, a bowed string with its environment can be included in a straightforward manner, as long as they can be described with a system of partial differential equations (PDEs). These methods also allow a greater flexibility for modelling the musician's gestures, and more generally to deal with the dynamic nature of the input and output parameters.

While the number of parameters is small compared to other methods, navigating the space they describe is somewhat of a challenge. Indeed, as opposed to a struck or plucked string, the continuous excitation mechanism of a bowed string makes playability a major issue, for real or virtual instruments. The player shapes the sound and behaviour of his instrument throughout the whole production of a note. Their gestures can be described with a handful of parameters, which must be perfectly coordinated at all times to allow a tone to be created and sustained; indeed, Schelleng [21], following the work of Raman [22] some decades earlier, was the first to analytically show that, under simplifying assumptions and for a certain bow velocity, only a relatively narrow triangular area (in logarithmic scale) in the downwards bow force versus bow-bridge distance parameter space gave rise to the characteristic stable Helmholtz motion desired by most musicians (his work was revised by Schoonderwaldt *et al.* [23], introducing more refined elements of bowed string motion). This area is tied to the concept of playability, and the so-called Schelleng diagrams are widely used in bowed string playability studies [24–27]. Transient quality also constitutes a major part of playability: Guettler [28] investigated the relation between bow downwards force and bow acceleration regarding the quality of initial transients, producing triangular diagrams resembling those of Schelleng, again under simplifying assumptions; Woodhouse *et al.* [29] produced Guettler diagrams with more refined numerical models as well as experimental data, showing the predicted wedge-shaped region. A detailed review of the published literature on bowed string mechanics (and, indeed, violin acoustics in general) was recently written by Woodhouse [30].

After the studies of Schelleng, an experimental study by Askenfelt [31,32] yielded measured values for these control parameters, on a violin; he was the first to develop a measuring rig able to record them all simultaneously during performance. More recently, the variation of these parameters along various bowed string gestures was observed and analysed in detail [15,33,34]. The obtained signals can be mathematically reconstructed [35], or directly used, to be fed into a physical modelling algorithm such as the one presented in this work. The aforementioned studies are mainly interested in bowing gestures; to handle those of the left hand fingers (vibrato, legato, glissando, etc.), one can include a finger model, along with a fingerboard.

This work is concerned with a detailed model of bowed string vibration, emphasising the interactions of the string with the player. A linear bowed string is simulated in two polarisations. The model includes the following features:

- in one polarisation, distributed nonlinear contact interactions between the string and the dynamic left hand fingers, dynamic bow, and fingerboard. A stable finite difference scheme for modelling distributed contact/collisions has recently been established [36,37], that can be used in this stopped string-fingerboard setup [38,39];
- in the other polarisation, orthogonally to the first, distributed nonlinear friction forces between the string and the same three objects. The friction force nonlinearity is modelled with a force/velocity friction curve for the bow [40]. Tangential Coulomb friction also keeps the string captured between the fingers and fingerboard during note production;
- full control over the physical parameters of the system, as well as dynamic variations of the playing parameters. This time domain model is therefore able to reproduce most bowed string gestures;
- introduction of an inertial term in the bow model, in the tangential direction. This leads to a horizontal force bow control, instead of the bow velocity input signal used in most of the aforementioned models;
- in the same direction, introduction of an inertial term and a restoring term in the finger model. The fingertip is massive, absorbs some of the tangential string vibrations, and oscillates about a fixed horizontal finger position.

The main challenges associated with this model are computational. On one hand, the resulting numerical method must be stable; energy methods are used to derive a stability condition from the model system, and establish a power balance to keep track of the energy exchanges between the various parts of the system. On the other hand, the distributed nonlinearities induce heavy computational costs, with the need to resort to iterative nonlinear system solving methods; extensive optimisation is necessary in order to bring the algorithm closer to real-time operation. Finally, the gestural control is obviously limited by the aforementioned playability issues; such a model is able to produce waveforms outside of the generally desired Helmholtz sawtooth, and the absence of direct feedback for the player in the present algorithm indeed makes parameter control a rather fastidious task.

In Section 2, the model equations for the bow/string system are presented, with an elaborate description of finger/string interaction in the case of stopped notes, and the string/fingerboard collision interaction. A globally energy balanced finite difference scheme is presented in Section 3. Finally, bowed string simulation results, with the reproduction of several typical gestures, are presented in Section 4. Some sound and video examples from the computed simulations are available online [41].

2. Model Description

2.1. Context

The choice of time-stepping methods is in line with the larger aim of building full physical models of musical instruments, embedded in virtual acoustic spaces. The work presented here is a step towards designing such a model for a bowed string instrument; this longer term aim guided the choice of whether or not to include certain features of bowed string playing in this work.

As a consequence, although the bowed string model described here uses sensible, physical assumptions, some of its features may not be as refined as one can find in some of the recent literature. This is the result of compromising between physical accuracy, computational complexity, and (subjective) synthetic sound quality. Some features, for instance, have been neglected for dramatically increasing the algorithmic complexity and computational load, while having very little influence to both the output sound and the control quality; this is the case for e.g., the nonlinear intrinsic coupling between the two polarisations of the string, or the thermal properties of the rosin layer coating the bow [42]. Torsion waves are also excluded; although, when present, their synchronisation with transverse waves has been experimentally shown to strongly contribute to tone quality and stability [43], their presence (or absence) may not have a strong bearing on the playability and synthetic sound quality of a simplified physical model [25].

2.2. A Linear, Two-Polarisation String Model

A dual-polarisation string model serves two main purposes in this work. First, treated in this paper, is the ability to simulate realistic, nonlinear impact interactions between the string and any external objects in one direction, translating directly into tangential friction forces in the other direction. This is an intuitive way to include not only a bow that is able to naturally bounce, but also dynamic left hand fingers to stop the string against a fingerboard, while absorbing some of the string vibrations.

Second is the potential to use this algorithm towards a model of a full instrument. The coupling of the two polarisations at the bridge boundary, through a model of the bridge itself coupled to the instrument body, would be straightforwardly achieved with this string model as a starting point. While the incorporation of the bridge and wooden cavity is undoubtedly crucial to the final sound of the virtual instrument, let us first describe the proposed string model.

2.3. The Isolated String

Consider a linear, stiff, and lossy string, of length L. The string displacement in the *vertical* or *normal* polarisation is denoted by $w(x,t)$, while $u(x,t)$ is the string displacement in the *horizontal* or *tangential* polarisation. Both are defined for position $x \in \mathcal{D} = [0, L]$ and time $t \in \mathbb{R}^+ = [0, +\infty]$ (see Figure 1).

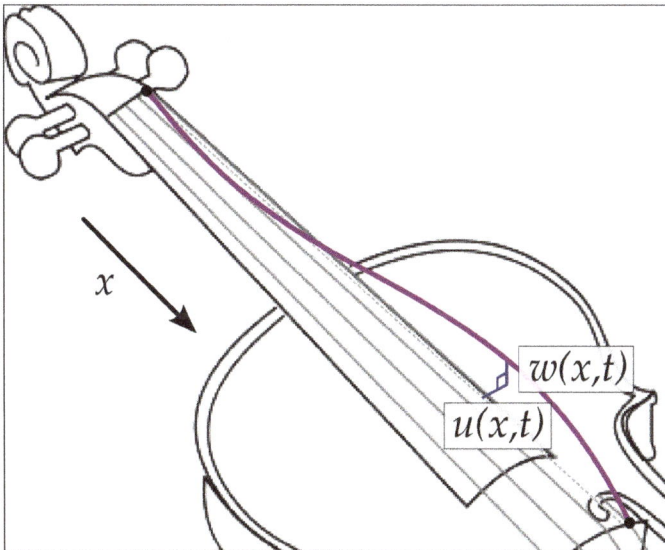

Figure 1. String displacement in two polarisations. The horizontal polarisation $(u(x,t))$ corresponds to the plane containing the bow and the string; the vertical polarisation $(w(x,t))$ is orthogonal to this plane.

The partial differential equations governing the time evolution of $u(x,t)$ and $w(x,t)$ can be written as:

$$\mathcal{L}w = 0 \quad \text{(1a)} \qquad \mathcal{L}u = 0 \quad \text{(1b)}$$

$\mathcal{L}$ is the partial differential operator defined as [20]:

$$\mathcal{L} = \rho\partial_t^2 - T\partial_x^2 + EI_0\partial_x^4 + \lambda_1\rho\partial_t - \lambda_2\rho\partial_t\partial_x^2 \qquad (2)$$

where ∂_t^i is equivalent to $\frac{\partial^i}{\partial \cdot t^i}$. ρ is the linear mass density of the string, in kg/m; T is the tension of the string, in N; EI_0 is the bending stiffness, where E is Young's modulus in Pa, and $I_0 = \frac{\pi r^4}{4}$ is the area moment of inertia of the circular cross-section of the string, with r the string radius in m. Some typical parameters for a violin, a viola, and a cello, can be found in the literature, notably those measured by Pickering [44] and Percival [45].

λ_1 (1/s) and λ_2 (m²/s) are positive damping coefficients, that empirically account for frequency independent and dependent losses in the string, respectively. With more refinement, they could be replaced by a full model accounting for energy dissipation through air viscosity, acoustic radiation, and internal friction, each of these three mechanisms having a more or less pronounced impact across the spectrum [46]. In accordance with the statements in Section 2.1 however, this empirical, simplified loss model is deemed sufficient for the purposes of this work.

The system described by Equations (1) and (2) is accompanied by a set of boundary conditions (four of them for each polarisation of the stiff string). Standard energy conserving conditions of the simply supported type are chosen, assuming an isolated string, with no interaction with the instrument body (note that for the musical string, the effect of bending stiffness being very small with respect to that of tension, there is little difference between the simply supported and clamped boundary conditions):

$$w(0,t) = w(L,t) = 0 \qquad (3a) \qquad\qquad u(0,t) = u(L,t) = 0 \qquad (4a)$$
$$\partial_x^2 w(0,t) = \partial_x^2 w(L,t) = 0 \qquad (3b) \qquad\qquad \partial_x^2 u(0,t) = \partial_x^2 u(L,t) = 0 \qquad (4b)$$

It is worth noting that although definitely worth investigating, intrinsic and/or boundary coupling between the two polarisations is not included in the present model, although, as mentioned in Section 2.2, boundary coupling could be introduced through modelling the bridge and body. As will be clarified in Sections 2.4 and 2.5, physical coupling between the two directions of vibration occurs where the string is in contact with the bow, finger, or fingerboard.

2.4. Vertical Polarisation

2.4.1. The Collision Interaction

The collision model used in this paper was formalised in 1975 by Hunt and Crossley [47], as a means to write a new law governing the mechanics of nonlinear damped impacts. The undamped power-law model was adopted by the musical acoustics community as a means to describe lumped collisions, especially hammer-string collisions in the piano [19,48,49], and mallet-membrane impacts in drums [50,51]. A similar model than that of Hunt and Crossley, with hysteretic damping, was used for the modelling of the interaction between the piano hammer felt and the piano string, with good concordance with experimental results, by Stulov [52]. A numerical time domain framework for this particular model has been developed throughout the recent few years [36,37], and has proven to give rise to stable schemes; this aspect will be further developed in Section 3.3.

This type of contact interaction is chosen for two modelling aspects of bowed string playing. First and foremost, a bow having the ability to naturally bounce is a necessary feature for a wide range of the musician's gestural palette, such as *spiccato* or *ricochet* bowing. Introducing a collision mechanism in the bow model itself allows for this bouncing under realistic playing parameters. The natural frequency of the bouncing bow varies between 6 and 30 Hz, depending on whether the string is bowed closer to the tip or the frog of the bow [53]; this can be tuned by changing the stiffness and mass parameters of the bow model. It is worth noting that the entire dynamics of the bow hair have purposely been excluded from this work; as the primary aim is sound synthesis, it is found here that a dynamic, nonlinear, albeit lumped bow is a satisfying compromise between computational cost and gestural versatility.

The second aspect of the contact interaction lies in the capture of the string between a left hand finger and the fingerboard. The use of the damped impact law allows for realistic simulation of the fingertip reacting against the tension of the string, while significantly absorbing vibrations. The distributed fingerboard acts as a continuous barrier for fingers to slide along, allowing for *glissando* and *vibrato* gestures on the fly. Again, the built-in impact model allows for string rattling effects [39], used, for instance, in jazz double bass playing.

2.4.2. Bow and Finger

At a physical level, in the vertical polarisation, the bow and finger are essentially modelled the same way—that is, a lumped, flexible body pushing down on the string. The distinction lies in the values of the various parameters which define the collision force. For instance, the bow hair may have different stiffness properties than that of the fingertip, and the latter definitely exhibits higher damping than the former.

Adding the finger and bow forces to the string model described in Equation (1a) gives:

$$\mathcal{L}w = -J_F f_F - J_B f_B \tag{5}$$

where the downward forces exerted by the finger and the bow onto the string are respectively denoted by $f_F(t)$ and $f_B(t)$. Their action on the string is localised as defined by the continuous distributions $J_F(x,t)$ and $J_B(x,t)$, possibly time-varying (one can use, e.g., a Dirac delta function to model a pointwise interaction). $f_F(t)$ and $f_B(t)$ can be written using the Hunt and Crossley collision model mentioned in Section 2.4.1:

$$f_F(\Delta_F) = \frac{\dot{\Phi}_F}{\dot{\Delta}_F} + \dot{\Delta}_F \Psi_F \qquad (6a) \qquad\qquad f_B(\Delta_B) = \frac{\dot{\Phi}_B}{\dot{\Delta}_B} + \dot{\Delta}_B \Psi_B \qquad (6b)$$

where the dot notation is used for total time differentiation ($\frac{d}{dt}$).

These contact forces are nonlinear functions of the penetration $\Delta_F(t)$ and $\Delta_B(t)$, corresponding to the distance by which the colliding object (here, the fingertip and the bow hair, respectively) would deform from its resting shape. Figures 2a,b provide a visual interpretation of the finger and the bow penetrations. $\Delta_F(t)$ and $\Delta_B(t)$ are defined as:

$$\Delta_F(t) = \int_{\mathcal{D}} J_F(x,t)w(x,t)dx - w_F(t) \qquad (7a) \qquad\qquad \Delta_B(t) = \int_{\mathcal{D}} J_B(x,t)w(x,t)dx - w_B(t) \qquad (7b)$$

where $w_F(t)$ and $w_B(t)$ are respectively the vertical positions of the finger and bow at time t (see Figure 2).

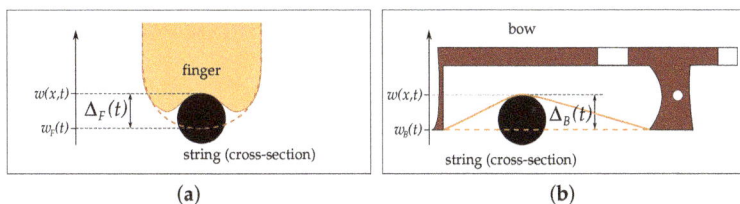

Figure 2. Visualisation of the penetration variables Δ_F and Δ_B. (**a**) $\Delta_F(t)$ represents the distance by which the finger deforms under the surface of the string; (**b**) $\Delta_B(t)$ can be interpreted as the deformation of the flexible bow hair against the string.

$\Phi_F(\Delta_F)$ and $\Phi_B(\Delta_B)$ are nonlinear potential functions, related to the stored collision energy. $\Psi_F(\Delta_F)$ and $\Psi_B(\Delta_B)$ are nonlinear damping coefficients. They can be written as:

$$\Phi_F = \frac{K_F}{\alpha_F + 1} [\Delta_F]_+^{\alpha_F + 1} \qquad (8a) \qquad\qquad \Phi_B = \frac{K_B}{\alpha_B + 1} [\Delta_B]_+^{\alpha_B + 1} \qquad (9a)$$

$$\Psi_F = K_F \beta_F [\Delta_F]_+^{\alpha_F} \qquad (8b) \qquad\qquad \Psi_B = K_B \beta_B [\Delta_B]_+^{\alpha_B} \qquad (9b)$$

where $[\cdot]_+$ means $\max(\cdot, 0)$.

The parameters K_F and K_B, both strictly positive, define the respective stiffnesses of the finger and the bow. α_F and α_B are power law exponents, both larger than 1. β_F and β_B are positive damping factors.

$w_F(t)$ and $w_B(t)$ are respectively the vertical positions of the finger and bow at time t (see Figure 2). Their behaviour is governed by:

$$M_F \ddot{w}_F = f_F + f_{\text{ext}\,w,F} \qquad (10a) \qquad\qquad M_B \ddot{w}_B = f_B + f_{\text{ext}\,w,B} \qquad (10b)$$

where M_F, M_B are the finger and bow masses, respectively (in kg), and $f_{\text{ext}\,w,F}(t)$, $f_{\text{ext}\,w,B}(t)$ are the external forces applied vertically on the finger and bow, respectively (in N).

2.4.3. Fingerboard

The fingerboard forms a distributed barrier under the string. Equation (4) therefore generalises to:

$$\mathcal{L}w = \mathcal{F}_N - J_F f_F - J_B f_B \qquad (11)$$

where the index $\cdot_N$ indicates "neck", to avoid confusion.

$\mathcal{F}_N$ is the contact force density exerted by the neck onto the string, along its length (in N/m). Here, the Hunt and Crossley collision model is used again, this time as a smooth approximation to a rigid collision:

$$\mathcal{F}_N(\Delta_N) = \frac{\partial_t \Phi_N}{\partial_t \Delta_N} + \partial_t \Delta_N \Psi_N \qquad (12)$$

Once again, the contact force is a function of the fingerboard penetration $\Delta_N(x, t)$, defined over $\mathcal{D}$:

$$\Delta_N(x, t) = \varepsilon(x) - w(x, t) \qquad (13)$$

where $\varepsilon(x)$ is the position of the fingerboard with respect to the string at rest (*i.e.*, the action of the instrument). Figure 3 summarises the forces at play in the vertical polarisation.

$\Phi_N(\Delta_N)$ and $\Psi_N(\Delta_N)$ are defined analogously to the finger and bow functions in Equations (8) and (9):

$$\Phi_N = \frac{K_N}{\alpha_N + 1} [\Delta_N]_+^{\alpha_N + 1} \qquad (14a) \qquad\qquad \Psi_N = K_N \beta_N [\Delta_N]_+^{\alpha_N} \qquad (14b)$$

where K_N is chosen very large to approach an ideally rigid collision. Indeed, such a choice ensures that Δ_N stays very small, as should be for fingerboard-like structures [39].

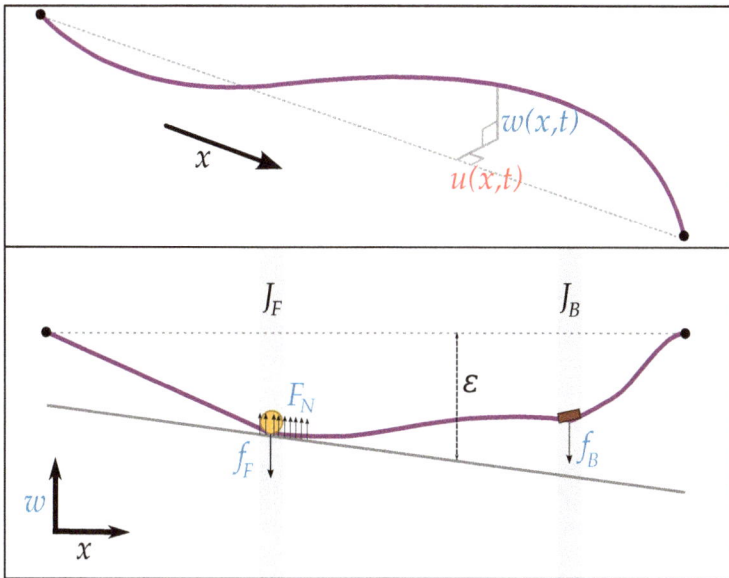

Figure 3. Diagram summarising the interactions of the string with the finger, bow, and fingerboard, in the vertical polarisation.

2.5. Horizontal Polarisation

2.5.1. The Friction Interaction

The vertical contact forces described in Section 2.4 give rise to corresponding tangential friction forces onto the string, in the horizontal polarisation. A classic dry friction model is used, where the friction force is directly proportional to the normal force:

$$F_F = [F_C]_+ \, \varphi \tag{15}$$

where F_F is the tangential friction force (in N), φ is a dimensionless friction coefficient, and F_C is the normal contact force, applied downwards on the string. The neck, finger and bow are not considered adhesive, therefore friction exists only for positive normal forces.

The friction force therefore arises from a friction coefficient, modulated by the normal force applied on the string. As a result, the interactions in the vertical polarisation feed into the horizontal polarisation. It is important to note that, for this particular framework, this is the coupling point between the two modelled directions; as mentioned earlier in Section 2.3, there is no other form of polarisation coupling in this model.

The friction coefficient can be defined as dependent on the relative velocity between the string and the external object. One can therefore establish a force-velocity friction curve, mapping the friction coefficient to a particular value of the relative velocity v_{rel}.

2.5.2. Bow

The most straightforward way to introduce the tangential friction forces is probably to come back to the simple bowed string, where the friction phenomenon is the most appreciable. The bowed string equation in the horizontal polarisation is:

$$\mathcal{L}u = -J_B \, [f_B]_+ \, \varphi_B \tag{16}$$

where $J_B(x,t)$ and $f_B(t)$ are the distribution and the normal collision force defined in Section 2.4.2. Note the negative sign in front of the friction term, reflecting the friction force opposing the motion of the string. $\varphi_B(v_{\mathrm{rel},B})$ is a dimensionless friction coefficient, depending on $v_{\mathrm{rel},B}$, the relative velocity between the bow hair and the string. These are defined as [42]:

$$
\begin{cases}
|\varphi_B| \leqslant 1.2 & \text{if } (v_{\mathrm{rel},B}) = 0 \\
\varphi_B = \mathrm{sign}\,(v_{\mathrm{rel},B})\left(0.4e^{\frac{-|v_{\mathrm{rel},B}|}{0.01}} + 0.45e^{\frac{-|v_{\mathrm{rel},B}|}{0.1}} + 0.35\right) & \text{if } v_{\mathrm{rel},B} \neq 0 \text{ (kinetic)}
\end{cases}
\tag{17a}
$$

$$
v_{\mathrm{rel},B}(t) = \frac{d}{dt}\left(\int_{\mathcal{D}} J_B(x,t)u(x,t) - u_B(t)\right) dx
\tag{17b}
$$

where $u_B(t)$ is the bow transverse displacement. As the bow is pushed across the string, the equation governing the transverse motion of the bow is:

$$
M_B \ddot{u}_B = -\lambda_B \dot{u}_B + [f_B]_+ \varphi_B + f_{\mathrm{ext}\,u,B}
\tag{18}
$$

λ_B is a positive coefficient quantifying the linear energy absorption by the bow in the horizontal direction. The linear damping term is negligible compared to the friction term, when the string is in contact with the bow; indeed, this linear term mainly has a practical purpose in the numerical simulations, that is to avoid the bow drifting away when it is lifted from the string at the end of a note.

$f_{\mathrm{ext}\,u,B}(t)$ is the force with which the player pushes the bow tangentially, in order to establish the desired bow velocity. Note the slight difference with the usual control parameter in most bowed string studies; instead of directly imposing a bow velocity $v_B(t)$, the force applied by the player on the bow is used, resulting in a bow velocity $\dot{u}_B$.

The choice of a friction coefficient depending on relative velocity, $\varphi_B(v_{\mathrm{rel},B})$, while already quite refined and fairly costly to model, is nonetheless somewhat of a trade-off between computational simplification and physical realism. More elaborate models for the bowed string friction interaction, involving viscothermal effects in the rosin layer coating the bow hair, can be used [13,14]; however, they require significantly more advanced implementations. Satisfying results and synthetic sound are obtained with the simple friction curve, although it has been shown that using a temperature-based friction model improves playability [25]. The friction curve employed here for the bow is deduced from experimental measurements in the steady sliding case (e.g., at constant velocity) [42]; it is illustrated in Figure 4a.

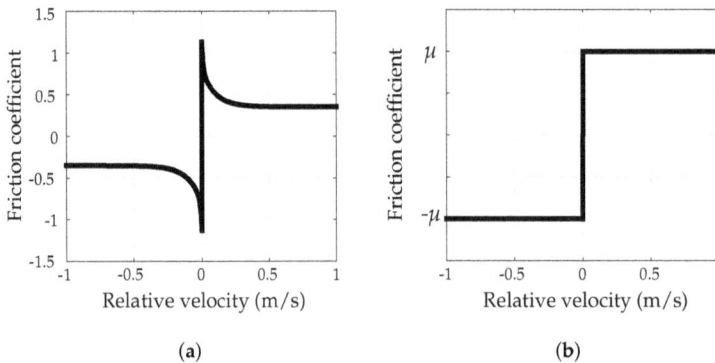

Figure 4. Friction curves for (**a**) bow, and (**b**) neck and finger. (**a**) Friction characteristic for the bow, from Smith *et al.* [42]; (**b**) Coulomb friction characteristic.

2.5.3. Finger

Adding the left hand finger in the horizontal direction yields:

$$\mathcal{L}u = -J_F \, [f_F]_+ \, \varphi_F - J_B \, [f_B]_+ \, \varphi_B \tag{19}$$

where $J_F(x,t)$ and $f_F(t)$ are defined in Section 2.4.2, and $\varphi_F \, (v_{\mathrm{rel},F})$ is a dimensionless friction coefficient.

To the authors' knowledge, there is no experimental data allowing for the calibration of the finger (or the fingerboard) friction curve. The fingers have the joint function, along with the fingerboard, of capturing the string to reduce its speaking length, to a crude approximation. In the absence of such data, a Coulomb-like step characteristic can therefore be assumed (illustrated in Figure 4b, where the static friction case occurs in most playing situations, but the string is capable of slipping under the finger if the left hand grip is too loose:

$$\begin{cases} |\varphi_F \, (v_{\mathrm{rel},F})| \leqslant \mu_F & \text{if } v_{\mathrm{rel},F} = 0 \text{ (static)} \\ \varphi_F \, (v_{\mathrm{rel},F}) = \mu_F \, \mathrm{sign} \, (v_{\mathrm{rel},F}) & \text{if } v_{\mathrm{rel},F} \neq 0 \text{ (kinetic)} \end{cases} \tag{20a}$$

$$v_{\mathrm{rel},F}(t) = \frac{d}{dt} \left(\int_D J_F(x,t) u(x,t) - u_F(t) \right) dx \tag{20b}$$

μ_F is a positive kinetic friction coefficient, quantifying how "sticky" the fingertip is. $u_F(t)$ is the horizontal position of the fingertip, with respect to the resting string axis. It is hypothesised that the fingertip oscillates about the top finger joint, while simultaneously damping the horizontal vibrations of the string. The temporal evolution of $u_F(t)$ can therefore be written as:

$$M_F \ddot{u}_F = -\mathcal{K}_F u_F - \lambda_F \dot{u}_F + [f_F]_+ \, \varphi_F \tag{21}$$

where $\mathcal{K}_F \geqslant 0$ is a spring constant, and $\lambda_F \geqslant 0$ is a damping coefficient. A linear damped oscillator model for the finger is chosen in the horizontal polarisation. Indeed, the choice of a more elaborate contact model such as the one used in the vertical polarisation seems unjustified; while impacts are dominant in the vertical polarisation, e.g., when hammering the string for changing notes, it is clear that collisions only have an auxiliary effect in the tangential polarisation.

Note that this version of the model, before even introducing the fingerboard, could be used to simulate the bowing of natural harmonics of the string.

2.5.4. Fingerboard

The fingerboard friction force is distributed along the whole string; if plucked particularly hard, the string's impact on the fingerboard can tangentially translate into friction, and the string will also slide against the neck of the instrument, adding to the audible rattling effect. The string equation becomes:

$$\mathcal{L}u = - \, [\mathcal{F}_N]_+ \, \varphi_N - J_F \, [f_F]_+ \, \varphi_F - J_B \, [f_B]_+ \, \varphi_B \tag{22}$$

where, again, $\mathcal{F}_N(x,t)$ is the normal fingerboard force defined in Section 2.4.3, and $\varphi_N \, (v_{\mathrm{rel},N})$ is the friction coefficient for the fingerboard, depending on the relative velocity between the fingerboard and the string (that is, the velocity of the string itself, as the fingerboard is not moving):

$$\begin{cases} |\varphi_N \, (v_{\mathrm{rel},N})| \leqslant \mu_N & \text{if } v_{\mathrm{rel},N} = 0 \text{ (static)} \\ \varphi_N \, (v_{\mathrm{rel},N}) = \mu_N \, \mathrm{sign} \, (v_{\mathrm{rel},N}) & \text{if } v_{\mathrm{rel},N} \neq 0 \text{ (kinetic)} \end{cases} \tag{23a}$$

$$v_{\mathrm{rel},N}(x,t) = \partial_t u \tag{23b}$$

Figure 5 summarises the forces at play in the horizontal polarisation.

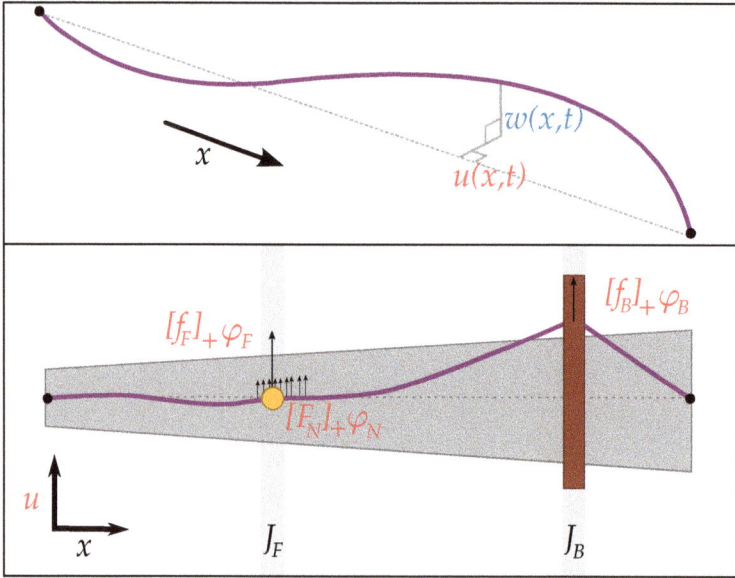

Figure 5. Diagram summarising the interactions of the string with the finger, bow, and fingerboard, in the horizontal polarisation.

2.6. Energy Analysis

One can derive a power balance equation for both polarisations. The transfer of this equation to discrete time provides a tool to help ensure numerical stability.

Multiplying Equation (11) by $\partial_t w$ and integrating over the length of the string, and multiplying Equations (10a) and (10b) by $\dot{w}_F$ and $\dot{w}_B$ respectively, yield the following power balance (for energy-conserving boundary conditions, such as those given in Equation (3)):

$$\dot{H}_w = P_w - Q_w \tag{24}$$

The variation of the total kinetic and potential energy $H_w(t) = H_{w,s}(t) + H_{w,N}(t) + H_{w,F}(t) + H_{w,B}(t)$ is equal to the total power $P_w(t)$ supplied to the system through external excitation (which can be negative), minus the power $Q_w(t) \geqslant 0$ withdrawn from the system through damping. The system is therefore globally energy balanced. The energy is defined as:

$$H_w = H_{w,s} + H_{w,N} + H_{w,F} + H_{w,B} \tag{25a}$$

$$H_{w,s} = \int_{\mathcal{D}_S} \left[\frac{\rho}{2} (\partial_t w)^2 + \frac{T}{2} (\partial_x w)^2 + \frac{EI_0}{2} \left(\partial_x^2 w \right)^2 \right] dx \tag{25b}$$

$$H_{w,N} = \int_{\mathcal{D}_S} \Phi_N \, dx \tag{25c}$$

$$H_{w,F,B} = \Phi_{F,B} + \frac{M_{F,B}}{2} \dot{w}_{F,B}^2 \tag{25d}$$

The power supplied through external excitation is:

$$P_w = \dot{w}_F f_{\text{ext}\,w,F} + \dot{w}_B f_{\text{ext}\,w,B} + \int_{\mathcal{D}_S} \left(f_F w \partial_t J_F + f_B w \partial_t J_B \right) dx \tag{26}$$

The power lost through damping within the string and through collision with the neck, finger and bow is given by:

$$Q_w = Q_{w,s} + Q_\Psi \tag{27a}$$

$$Q_{w,s} = \rho \int_{\mathcal{D}_S} \left[\lambda_1 \left(\partial_t w \right)^2 + \lambda_2 \left(\partial_t \partial_x w \right)^2 \right] dx \tag{27b}$$

$$Q_\Psi = \int_{\mathcal{D}_S} \left(\partial_t \Delta_N \right)^2 \Psi_N \, dx + \dot{\Delta}_F^2 \Psi_F + \dot{\Delta}_B^2 \Psi_B \tag{27c}$$

In the absence of excitation, the energy H_w strictly decreases.

For the horizontal polarisation, multiplying Equation (22) by $\partial_t u$ and integrating over $\mathcal{D}_S$, and multiplying Equations (18) and (21) by $\dot{u}_B$ and $\dot{u}_F$ respectively, yield the power balance:

$$\dot{H}_u = P_u - Q_u \tag{28}$$

Again, the variation of $H_u(t) = H_{u,s}(t) + H_{u,F}(t) + H_{u,B}(t)$ is equal to the total power $P_u(t)$ supplied to the system in the horizontal polarisation through external excitation, minus power losses $Q_u(t) \geqslant 0$ from damping. The energy is defined as:

$$H_u = H_{u,s} + H_{u,F} + H_{u,B} \tag{29a}$$

$$H_{u,s} = \int_{\mathcal{D}_S} \left[\frac{\rho}{2} \left(\partial_t u \right)^2 + \frac{T}{2} \left(\partial_x u \right)^2 + \frac{EI_0}{2} \left(\partial_x^2 u \right)^2 \right] dx \tag{29b}$$

$$H_{u,F} = \frac{M_F}{2} \dot{u}_F^2 + \frac{K_F}{2} u_F^2 \tag{29c}$$

$$H_{u,B} = \frac{M_B}{2} \dot{u}_B^2 \tag{29d}$$

The power supplied or withdrawn by external excitation is:

$$P_u = \left[f_F \right]_+ \varphi_F \int_{\mathcal{D}_S} u \partial_t J_F dx \; + \; \left[f_B \right]_+ \varphi_B \int_{\mathcal{D}_S} u \partial_t J_B dx \; + \; \dot{u}_B f_{\text{ext}\,u,B} \tag{30}$$

The power lost through string damping and friction is:

$$Q_u = Q_{u,s} + Q_\varphi + Q_{u,F} + Q_{u,B} \tag{31a}$$

$$Q_{u,s} = \rho \int_{\mathcal{D}_S} \left[\lambda_1 \left(\partial_t u \right)^2 + \lambda_2 \left(\partial_t \partial_x u \right)^2 \right] dx \tag{31b}$$

$$Q_\varphi = \int_{\mathcal{D}_S} v_{\text{rel},N} \left[\mathcal{F}_N \right]_+ \varphi_N dx \; + \; v_{\text{rel},F} \left[f_F \right]_+ \varphi_F \; + \; v_{\text{rel},B} \left[f_B \right]_+ \varphi_B \tag{31c}$$

$$Q_{u,F,B} = \lambda_{F,B} \dot{u}_{F,B}^2 \tag{31d}$$

Note that $Q_u \geqslant 0$ if $v_{\text{rel}} \varphi \left(v_{\text{rel}} \right) \geqslant 0$, which is true for the friction characteristics of the three objects. The total power of the full system is therefore balanced by:

$$\dot{H} = P - Q \quad \text{(32a)} \qquad H = H_u + H_w \quad \text{(32b)} \qquad P = P_u + P_w \quad \text{(32c)} \qquad Q = Q_u + Q_w \quad \text{(32d)}$$

A power balance of the same type as Equation (32a) is generally used as the base for another class of modelling methods, based on so-called port-Hamiltonian systems [54]. Discretisation of such systems has recently been successfully implemented for time domain physical modelling of acoustic and electroacoustic systems [55].

3. Numerical Scheme

3.1. Framework

3.1.1. Discretising the Equations of Motion

The equations of motion can now be discretised by approximating the partial derivatives with their finite difference [17] counterparts. This method allows a full system simulation, and therefore great flexibility of control for the input parameters and gesture reproduction, at the cost of increased computational requirements. This method has seen a myriad of applications in physical modelling sound synthesis, and more generally musical acoustics simulations [3,18]. In this section, the numerical scheme is defined, the discrete energy balance is detailed, and the scheme update is described.

3.1.2. Grid Functions and Finite Difference (FD) Operators

All the time varying quantities defined in Section 2 are now discretised into series of integer multiples of the time step, and defined at times $t = nk, n \in \mathbb{N}$. k is the time step in s; $F_s = k^{-1}$ is the desired sample rate in Hz. Typically, audio sample rates are used, such as $F_s = 44.1$ kHz.

The space dependent variables are discretised into grid functions, defined at positions $x = lh, l \in \mathcal{D} = [0, \ldots, N_{\mathcal{D}}]$, where h is the grid spacing, in m.

For an arbitrary continuous function $g(x, t)$ defined for $x \in \mathcal{D}$ and $t \in \mathbb{R}^+$, g_l^n is a grid function approximating $g(lh, nk)$, at grid point l and time step n. Introduce the forward and backward unit time and space shift operators, applied to g_l^n:

$$e_{t-}g_l^n = g_l^{n-1} \quad \text{(33a)} \qquad\qquad e_{x-}g_l^n = g_{l-1}^n \quad \text{(34a)}$$
$$e_{t+}g_l^n = g_l^{n+1} \quad \text{(33b)} \qquad\qquad e_{x+}g_l^n = g_{l+1}^n \quad \text{(34b)}$$

Partial differentiation with respect to time and space can be approximated with a number of first order Finite Difference (FD) operators:

$$\delta_{t-} = \frac{1 - e_{t-}}{k} \quad \text{(35a)} \qquad\qquad \delta_{x-} = \frac{1 - e_{x-}}{h} \quad \text{(36a)}$$
$$\delta_{t+} = \frac{e_{t+} - 1}{k} \quad \text{(35b)} \qquad\qquad \delta_{x+} = \frac{e_{x+} - 1}{h} \quad \text{(36b)}$$
$$\delta_{t\cdot} = \frac{e_{t+} - e_{t-}}{2k} \quad \text{(35c)}$$

Higher order partial derivation operators are approximated with:

$$\partial_t^2 \approx \delta_{tt} = \delta_{t-}\delta_{t+} \quad \text{(37a)} \qquad\qquad \partial_x^2 \approx \delta_{xx} = \delta_{x-}\delta_{x+} \quad \text{(37b)}$$

Finally, the averaging FD operators approximate identity:

$$\mu_{t-} = \frac{1 + e_{t-}}{2} \quad \text{(38a)} \qquad \mu_{t+} = \frac{e_{t+} + 1}{2} \quad \text{(38b)} \qquad \mu_{t\cdot} = \frac{e_{t+} + e_{t-}}{2} \quad \text{(38c)}$$

Note that $\delta_{t-}\mu_{t+} = \delta_{t+}\mu_{t-} = \delta_{t\cdot}$.

3.2. The Isolated String

3.2.1. Finite Difference Scheme

The discrete counterparts of $w(x, t)$ and $u(x, t)$ are the grid functions w_l^n and u_l^n, as described in Section 3.1.2. System Equation (1) is discretised as:

$$Lw_l^n = 0 \qquad \text{(39a)} \qquad\qquad Lu_l^n = 0 \qquad \text{(39b)}$$

where L is a finite difference discretisation of the differential operator $\mathcal{L}$ defined in Equation (2), using the finite difference operators described in Section 3.1.2:

$$L = \rho\delta_{tt} - T\delta_{xx} + EI_0\delta_{xx}\delta_{xx} + \lambda_1\rho\delta_{t\cdot} - \lambda_2\rho\delta_{t-}\delta_{xx} \qquad \text{(40)}$$

The use of the backward first order time FD operator in the second damping term allows for the isolated string scheme to stay explicit; this results in simplified computations, at the cost of a slightly altered stability condition.

The discrete simply supported boundary conditions are given as:

$$w_0^n = w_{N_{\mathfrak{D}}}^n = 0 \qquad \text{(41a)} \qquad\qquad u_0^n = u_{N_{\mathfrak{D}}}^n = 0 \qquad \text{(42a)}$$

$$\delta_{xx}w_0^n = \delta_{xx}w_{N_{\mathfrak{D}}}^n = 0 \qquad \text{(41b)} \qquad\qquad \delta_{xx}u_0^n = \delta_{xx}u_{N_{\mathfrak{D}}}^n = 0 \qquad \text{(42b)}$$

3.2.2. Vector-Matrix Notation

The grid functions w_l^n and u_l^n are defined over the discrete domain $\mathfrak{D}$, containing $N_{\mathfrak{D}} + 1$ grid points. The discrete position of the whole string can therefore be described with vectors of length $N_{\mathfrak{D}} + 1$. The first set of boundary conditions, Equations (41a) and (42a), ensure that the two extreme values of such vectors (*i.e.*, the string displacement at the bridge and nut boundary) are 0 at all times. One now only needs to store the state of the string in a vector of size $(N_{\mathfrak{D}} - 1)$, omitting these two extreme values:

$$\mathbf{w}^n = \left[w_1^n, \ldots, w_{N-1}^n\right]^\mathsf{T} \qquad \text{(43a)} \qquad\qquad \mathbf{u}^n = \left[u_1^n, \ldots, u_{N-1}^n\right]^\mathsf{T} \qquad \text{(43b)}$$

The action of spatial FD operators on the grid functions is then equivalent to a matrix-vector multiplication. For simply supported boundary conditions, the notation of spatial FD operators in matrix form naturally follows as:

$$\mathbf{D}_{x-} = \frac{1}{h}\begin{bmatrix} 1 & & & & \\ -1 & 1 & & & \\ & \ddots & \ddots & & \\ & & -1 & 1 & \\ & & & -1 & \end{bmatrix} \qquad \text{(44a)}$$

$$\mathbf{D}_{x+} = -\mathbf{D}_{x-}^\mathsf{T} \qquad \text{(44b)}$$

$$\mathbf{D}_{xx} = \mathbf{D}_{x+}\mathbf{D}_{x-} \qquad \text{(44c)}$$

of size $N_{\mathfrak{D}} \times (N_{\mathfrak{D}} - 1)$, $(N_{\mathfrak{D}} - 1) \times N_{\mathfrak{D}}$, $(N_{\mathfrak{D}} - 1) \times (N_{\mathfrak{D}} - 1)$, and $(N_{\mathfrak{D}} - 1) \times (N_{\mathfrak{D}} - 1)$, respectively. System Equation (39) can now be written as a pair of vector equations:

$$\mathbf{L}\mathbf{w}^n = 0 \qquad \text{(45a)} \qquad\qquad \mathbf{L}\mathbf{u}^n = 0 \qquad \text{(45b)}$$

where $\mathbf{L}$ is the matrix form of the difference operator L defined in Equation (40):

$$\mathbf{L} = \rho\delta_{tt} - T\mathbf{D}_{xx} + EI_0\mathbf{D}_{xx}\mathbf{D}_{xx} + \lambda_1\rho\delta_{t\cdot} - \lambda_2\rho\delta_{t-}\mathbf{D}_{xx} \qquad \text{(46)}$$

3.3. Vertical Polarisation

3.3.1. Discrete System

A discretisation for Equation (11) can now be written in vector-matrix form:

$$\mathbf{L}\mathbf{w}^n = \mu_t . \mathbf{J}_\mathbf{w}^n\, \mathbf{f}_\mathbf{w}^n \tag{47}$$

where $\mathbf{J}_\mathbf{w}^n$ is the $(N_{\mathfrak{D}} - 1) \times (N_{\mathfrak{D}} + 1)$ distribution matrix, and $\mathbf{f}_\mathbf{w}^n$ is a column vector containing all the contact force information:

$$\mathbf{J}_\mathbf{w}^n = \left[\ \mathbf{I}_{N-1} \ \middle| \ -\mathbf{j}_F^n \ \middle| \ -\mathbf{j}_B^n \ \right] \tag{48a} \qquad\qquad \mathbf{f}_\mathbf{w}^n = \left[\ (\mathbf{f}_N^n)^\mathrm{T} \ \middle| \ f_F^n \ \middle| \ f_B^n \ \right]^\mathrm{T} \tag{48b}$$

where $\mathbf{I}_{N-1}$ is the $(N_{\mathfrak{D}} - 1) \times (N_{\mathfrak{D}} - 1)$ identity matrix, and $\mathbf{j}_F^n$ and $\mathbf{j}_B^n$ are discrete spreading operators in column vector form, accounting for the continuous distributions described in Section 2.4.2. If the contact regions are sufficiently wide, these can be sampled from the true shape of the concerned object; otherwise, if contact is localised in a small region, or pointwise, interpolation is needed between grid points, as truncation to the nearest grid point would result in audible artefacts when the finger or bow move along the length of the string.

The spreading vectors are normalised over the grid: for instance, in the simplified case of a fixed position object, located on a grid point, if $J_F(x, t)$ and $J_B(x, t)$ are Dirac delta functions, then $\mathbf{j}_F^n$ and $\mathbf{j}_B^n$ will be all-zero vectors, except for one element of value $\frac{1}{h}$, at the position where the force is applied.

Note the use of the averaging operator $\mu_t .$ in Equation (47), necessary to show an energy balance for this scheme in the case of moving finger or bow (see Section 3.5).

$\mathbf{f}_N^n$, f_F^n and f_B^n are the discrete counterparts of those defined in Sections 2.4.2 and 2.4.3.

3.3.2. Collision Normal Forces

The force vector $\mathbf{f}_\mathbf{w}^n$ can be written as:

$$\mathbf{f}_\mathbf{w}^n = \frac{\delta_t . \boldsymbol{\Phi}^n}{\delta_t . \boldsymbol{\Delta}^n} + (\delta_t . \boldsymbol{\Delta}^n) \odot \boldsymbol{\Psi}^n \tag{49}$$

where the division is pointwise, and $\odot$ is the pointwise product. This expression for $\mathbf{f}_\mathbf{w}^n$ is written as a function of the vector penetration $\boldsymbol{\Delta}^n$, defined as:

$$\boldsymbol{\Delta}^n = \begin{bmatrix} \vdots \\ \Delta_N^n \\ \vdots \\ \hline \Delta_F^n \\ \hline \Delta_B^n \end{bmatrix} \tag{50a}$$

$$\Delta_N^n = \varepsilon - \mathbf{w}^n \tag{50b}$$

$$\Delta_F^n = h\mathbf{j}_F^{n\,\mathrm{T}}\mathbf{w}^n - w_F^n \tag{50c}$$

$$\Delta_B^n = h\mathbf{j}_B^{n\,\mathrm{T}}\mathbf{w}^n - w_B^n \tag{50d}$$

where the elements of vector ε are $\varepsilon_l = \varepsilon(lh)$, and w_F^n and w_B^n are the respective vertical positions of the finger and bow.

$\boldsymbol{\Phi}^n(\boldsymbol{\Delta}^n)$, $\boldsymbol{\Psi}^n(\boldsymbol{\Delta}^n)$ are now vector functions of $\boldsymbol{\Delta}^n$:

$$\boldsymbol{\Phi}^n(\boldsymbol{\Delta}^n) = \frac{K}{\alpha + 1} \odot [\boldsymbol{\Delta}^n]_+^{\alpha+1} \tag{51a} \qquad\qquad \boldsymbol{\Psi}^n(\boldsymbol{\Delta}^n) = K \odot \beta \odot [\boldsymbol{\Delta}^n]_+^{\alpha} \tag{51b}$$

where the exponentiation operation is also element-wise, and the nonlinearity parameters K, α, and β are themselves in vector form:

$$K = \begin{bmatrix} \vdots \\ K_N \\ \vdots \\ \hline K_F \\ \hline K_B \end{bmatrix} \qquad (52a) \qquad \alpha = \begin{bmatrix} \vdots \\ \alpha_N \\ \vdots \\ \hline \alpha_F \\ \hline \alpha_B \end{bmatrix} \qquad (52b) \qquad \beta = \begin{bmatrix} \vdots \\ \beta_N \\ \vdots \\ \hline \beta_F \\ \hline \beta_B \end{bmatrix} \qquad (52c)$$

Finally, the discrete equations for the evolution of the vertical positions of the finger and bow can be written as:

$$\mathbf{M}_{FB}\delta_{tt}\mathbf{w}_{FB}^n = \mathbf{f}_{\mathbf{w}FB}^n + \mathbf{f}_{\mathrm{ext}\,w,FB}^n \qquad (53a)$$

$$\mathbf{M}_{FB} = \begin{bmatrix} M_F & 0 \\ 0 & M_B \end{bmatrix} \quad (53b) \qquad \mathbf{w}_{FB}^n = \begin{bmatrix} w_F^n \\ w_B^n \end{bmatrix} \quad (53c) \qquad \mathbf{f}_{\mathbf{w}FB}^n = \begin{bmatrix} f_F^n \\ f_B^n \end{bmatrix} \quad (53d) \qquad \mathbf{f}_{\mathrm{ext}\,w,FB}^n = \begin{bmatrix} f_{\mathrm{ext}\,w,F}^n \\ f_{\mathrm{ext}\,w,B}^n \end{bmatrix} \quad (53e)$$

3.4. Horizontal Polarisation

Equation (22) is now discretised as:

$$\mathbf{L}\mathbf{u}^n = -\mu_t.\mathbf{J}_{\mathbf{u}}^n\mathbf{f}_{\mathbf{u}}^n \quad (54a) \qquad \mathbf{J}_{\mathbf{u}}^n = \begin{bmatrix} \mathbf{I}_{N-1} \mid \mathbf{j}_F^n \mid \mathbf{j}_B^n \end{bmatrix} \quad (54b)$$

$\mathbf{L}$ is defined in Equation (46). $\mathbf{f}_{\mathbf{u}}^n$ is a column vector containing the friction force information:

$$\mathbf{f}_{\mathbf{u}}^n = [\mathbf{f}_{\mathbf{w}}^n]_+ \odot \begin{bmatrix} \vdots \\ \varphi_N\left(v_{\mathrm{rel},N}^n\right) \\ \vdots \\ \hline \varphi_F\left(v_{\mathrm{rel},F}^n\right) \\ \varphi_B\left(v_{\mathrm{rel},B}^n\right) \end{bmatrix} \qquad (55a)$$

$$v_{\mathrm{rel},N}^n = \delta_t.\mathbf{u}^n \qquad (55b)$$

$$v_{\mathrm{rel},F}^n = h\delta_t.\left({\mathbf{j}_F^n}^{\mathrm{T}}\mathbf{u}^n\right) - \delta_t.u_F^n \qquad (55c)$$

$$v_{\mathrm{rel},B}^n = h\delta_t.\left({\mathbf{j}_B^n}^{\mathrm{T}}\mathbf{u}^n\right) - \delta_t.u_B^n \qquad (55d)$$

where φ_N, φ_F and φ_B are defined in Section 2.5. One can define a vector relative velocity:

$$\mathbf{v}_{\mathrm{rel}}^n = \left[\left(\mathbf{v}_{\mathrm{rel},N}^n\right)^{\mathrm{T}} \mid v_{\mathrm{rel},F}^n \mid v_{\mathrm{rel},B}^n\right]^{\mathrm{T}} \qquad (56)$$

Finally, a matrix equation describes the evolution of the horizontal displacements u_F^n and u_B^n of the finger and bow, respectively:

$$\mathbf{M}_{FB}\delta_{tt}\mathbf{u}_{FB}^n = \mathbf{K}_{FB}\mu_t.\mathbf{u}_{FB}^n - \lambda_{FB}\delta_t.\mathbf{u}_{FB}^n + \mathbf{f}_{\mathbf{u}FB}^n + \mathbf{f}_{\mathrm{ext}\,u,FB}^n \qquad (57a)$$

$$\mathbf{K}_{FB} = \begin{bmatrix} K_F & 0 \\ 0 & 0 \end{bmatrix} \quad (57b) \qquad \lambda_{FB} = \begin{bmatrix} \lambda_F & 0 \\ 0 & \lambda_B \end{bmatrix} \quad (57c) \qquad \mathbf{u}_{FB}^n = \begin{bmatrix} u_F^n \\ u_B^n \end{bmatrix} \quad (57d)$$

$$\mathbf{f}_{\mathbf{u}FB}^n = [\mathbf{f}_{\mathbf{w}FB}^n]_+ \odot \begin{bmatrix} \varphi_F\left(v_{\mathrm{rel},F}^n\right) \\ \varphi_B\left(v_{\mathrm{rel},B}^n\right) \end{bmatrix} \quad (57e) \qquad \mathbf{f}_{\mathrm{ext}\,u,FB}^n = \begin{bmatrix} 0 \\ f_{\mathrm{ext}\,u,B}^n \end{bmatrix} \quad (57f)$$

where the 2-point averaging operator $\mu_{t.}$ is used as a means to avoid restricting the stability limit of this scheme any further; this aspect will be developed in Section 3.5.4.

3.5. Energy Analysis

The results of Section 2.6 can be transferred to discrete time, and the energy exchanges going on in the system can be monitored at all times during the simulation. An energy balance equation is derived between the energy of the closed system (H^n) and the power brought in and out, by external excitation (P^n) and damping Q^n. The maintenance of this energy balance is a means to ensure a stable algorithm.

3.5.1. Vertical Polarisation

For the vertical polarisation, multiplying Equation (47) by $h\,(\delta_{t.}\mathbf{w}^n)^{\mathsf{T}}$ and Equation (53a) by $(\delta_{t.}\mathbf{w}_{FB}^n)^{\mathsf{T}}$ gives the power balance:

$$\delta_{t-}H_w^n = P_w^n - Q_w^n \tag{58}$$

The numerical energy H_w^n is defined as:

$$H_w^n = H_{w,s}^n + H_\Phi^n \tag{59a}$$

$$H_{w,s}^n = \frac{\rho h}{2}|\delta_{t+}\mathbf{w}^n|^2 + \frac{Th}{2}(\mathbf{D}_{x-}\mathbf{w}^n)^{\mathsf{T}}\mathbf{D}_{x-}\mathbf{w}^{n+1} + \frac{EI_0h}{2}(\mathbf{D}_{xx}\mathbf{w}^n)^{\mathsf{T}}\mathbf{D}_{xx}\mathbf{w}^{n+1}$$
$$- \frac{\lambda_2\rho kh}{4}|\delta_{t+}\mathbf{D}_{x-}\mathbf{w}^n|^2 \tag{59b}$$

$$H_\Phi^n = \mathbf{h}^{\mathsf{T}}\mu_{t+}\Phi^n + \frac{1}{2}(\mathbf{M}_{FB}\delta_{t+}\mathbf{w}_{FB}^n)^{\mathsf{T}}\delta_{t+}\mathbf{w}_{FB}^n \tag{59c}$$

where $\mathbf{h} = [\ldots h \ldots |1|1]^{\mathsf{T}}$.

The power P_w^n supplied or withdrawn through excitation is:

$$P_w^n = (\delta_{t.}\mathbf{w}_{FB}^n)^{\mathsf{T}}\mathbf{f}_{ext\,w,FB}^n - h\left((\mu_{t.}\mathbf{w}^n)^{\mathsf{T}}(\delta_{t.}\mathbf{J}^n)\right)\mathbf{f}_{\mathbf{w}}^n \tag{60}$$

The power $Q_w^n \geqslant 0$ dissipated through damping is:

$$Q_w^n = Q_{w,s}^n + Q_\Psi^n \tag{61a}$$

$$Q_{w,s}^n = \lambda_1\rho h|\delta_{t.}\mathbf{w}^n|^2 + \lambda_2\rho h|\delta_{t.}\mathbf{D}_{x-}\mathbf{w}^n|^2 \tag{61b}$$

$$Q_\Psi^n = (\mathbf{h}\odot\delta_{t.}\Delta^n)^{\mathsf{T}}((\delta_{t.}\Delta^n)\odot\Psi^n) \tag{61c}$$

In the absence of external excitation, the numerical energy H_w^n is strictly decreasing.

3.5.2. Horizontal Polarisation

On the other hand, the product of Equation (54a) by $h\,(\delta_{t.}\mathbf{u}^n)^{\mathsf{T}}$, and that of Equation (57a) by $(\delta_{t.}\mathbf{u}_{FB}^n)^{\mathsf{T}}$, yield a numerical power balance for the horizontal polarisation:

$$\delta_{t-}H_u^n = P_u^n - Q_u^n \tag{62}$$

where the numerical energy H_u^n is defined as:

$$H_u^n = H_{u,s}^n + H_{u,FB}^n \tag{63a}$$

$$H_{u,s}^n = \frac{\rho h}{2}|\delta_{t+}\mathbf{u}^n|^2 + \frac{Th}{2}(\mathbf{D}_{x-}\mathbf{u}^n)^\mathsf{T}\mathbf{D}_{x-}\mathbf{u}^{n+1} + \frac{EI_0 h}{2}(\mathbf{D}_{xx}\mathbf{u}^n)^\mathsf{T}\mathbf{D}_{xx}\mathbf{u}^{n+1}$$
$$- \frac{\lambda_2\rho kh}{4}|\delta_{t+}\mathbf{D}_{x-}\mathbf{u}^n|^2 \tag{63b}$$

$$H_{u,FB}^n = \frac{1}{2}(\mathbf{M}_{FB}\delta_{t+}\mathbf{u}_{FB}^n)^\mathsf{T}\delta_{t+}\mathbf{u}_{FB}^n + \frac{1}{2}\mu_{t+}\left((\mathbf{K}_{FB}\mathbf{u}_{FB}^n)^\mathsf{T}\mathbf{u}_{FB}^n\right) \tag{63c}$$

The power P_u^n brought in or out by the player excitation is:

$$P_u^n = h\left((\mu_t.\mathbf{u}^n)^\mathsf{T}(\delta_t.\mathbf{J}^n)\right)\mathbf{f}_u^n + (\delta_t.\mathbf{u}_{FB}^n)^\mathsf{T}\mathbf{f}_{ext\,u,FB}^n \tag{64}$$

The power $Q_u^n \geqslant 0$ dissipated by friction and damping is:

$$Q_u^n = Q_{u,s}^n + Q_\varphi^n + Q_{u,FB}^n \tag{65a}$$

$$Q_{u,s}^n = \lambda_1\rho h|\delta_t.\mathbf{u}^n|^2 + \lambda_2\rho h|\delta_t.\mathbf{D}_{x-}\mathbf{u}^n|^2 \tag{65b}$$

$$Q_\varphi^n = (\mathbf{h}\odot\mathbf{v}_{rel}^n)^\mathsf{T}\mathbf{f}_u^n \tag{65c}$$

$$Q_{u,FB}^n = (\mathbf{\Lambda}_{FB}\delta_t.\mathbf{u}_{FB}^n)^\mathsf{T}\delta_t.\mathbf{u}_{FB}^n \tag{65d}$$

3.5.3. Global Power Balance

The total numerical energy H^n of the system is balanced by:

$$\delta_{t-}H^n = P^n - Q^n \tag{66a}$$

$$H^n = H_u^n + H_w^n \quad (66b) \qquad P^n = P_u^n + P_w^n \quad (66c) \qquad Q^n = Q_u^n + Q_w^n \quad (66d)$$

3.5.4. Non-Negativity of the Numerical Energy and Stability Condition

The stability of this scheme now reduces to a condition on the non-negativity of $H^n = H_w^n + H_u^n$ at all times. For the vertical polarisation, as $\Phi^n \geqslant 0$ by construction, it is straightforward to see from Equation (59c) that $H_\Phi^n \geqslant 0$; the modelling of collision interactions does not have an effect on numerical stability. In the horizontal direction, $H_{u,FB}^n \geqslant 0$ follows directly from Equation (63c); the energy of the finger and bow is unconditionally strictly positive, thanks to the use of the averaging operator in Equation (57).

$H^n \geqslant 0$ is then equivalent to the isolated string energy being non-negative, that is $H_{w,s}^n + H_{u,s}^n \geqslant 0$. This is only ensured at all times if both $H_{w,s}^n \geqslant 0$ and $H_{u,s}^n \geqslant 0$, which are verified for this particular scheme under the same condition, linking the time step k and grid spacing h [18]:

$$h \geqslant \sqrt{\frac{1}{2}\left(\frac{Tk^2}{\rho} + 2\lambda_2 k + \sqrt{\left(\frac{Tk^2}{\rho} + 2\lambda_2 k\right)^2 + 16k^2\frac{EI_0}{\rho}}\right)} \tag{67}$$

3.5.5. Invariant Quantity

Following Equation (66), the quantity E^n can be derived, that should remain constant throughout the simulation:

$$E^n = H^n - k\sum_{i=0}^{n}\left(P^i - Q^i\right) = H^0 \tag{68}$$

The variations of E^n should remain within machine accuracy, and can therefore be monitored as a means of ensuring that the algorithm is running properly. Examining its components separately provides a practical visualisation of the energy exchanges between the various parts of the system along time.

3.6. Scheme Update

As the horizontal friction forces depend on the normal contact forces, the vertical polarisation is updated first. The resulting impact forces are then fed into the horizontal polarisation system, which is subsequently updated.

3.6.1. Vertical Polarisation

Expanding the operators in Equations (47) and (53a), and combining Equations (49) and (51), leads to a two-step recursion algorithm in vector-matrix form, to be updated at each time step n:

$$\mathbf{w}^{n+1} = \mathbf{B}\mathbf{w}^n + \mathbf{C}\mathbf{w}^{n-1} + A\mu_t.\mathbf{J}_w^n \mathbf{f}_w^n \tag{69a}$$

$$\mathbf{w}_{FB}^{n+1} = 2\mathbf{w}_{FB}^n - \mathbf{w}_{FB}^{n-1} + k^2\mathbf{M}_{FB}^{-1}\left(\mathbf{f}_{wFB}^n + \mathbf{f}_{ext\,w,FB}^n\right) \tag{69b}$$

where A is a constant, and $\mathbf{B}$, $\mathbf{C}$ are matrices, written in terms of the physical parameters of the string, the time step k, and the grid spacing h, through the spatial difference matrices:

$$A = \frac{2k^2}{\rho(2+\lambda_1 k)} \tag{70a}$$

$$\mathbf{B} = \frac{2}{2+\lambda_1 k}\left(2 + \left(\frac{Tk^2}{\rho} + \lambda_2 k\right)\mathbf{D}_{xx} - \frac{EI_0 k^2}{\rho}\mathbf{D}_{xx}\mathbf{D}_{xx}\right) \tag{70b}$$

$$\mathbf{C} = \frac{2}{2+\lambda_1 k}\left(\frac{\lambda_1 k}{2} - 1 - \lambda_2 k\mathbf{D}_{xx}\right) \tag{70c}$$

However, the nonlinearity of the contact model doesn't allow for a simple explicit update. Combining Equations (69a) and (69b), and rewriting in terms of $\boldsymbol{\Delta}^n$, leads to a nonlinear equation in matrix form, in terms of the unknown vector $\mathbf{r}^n = \boldsymbol{\Delta}^{n+1} - \boldsymbol{\Delta}^{n-1}$:

$$\boldsymbol{\Lambda}_1^n \mathbf{r}^n + \boldsymbol{\Lambda}_2^n \mathbf{f}_\Phi^n + \mathbf{b}_w^n = 0 \tag{71}$$

where the matrices $\boldsymbol{\Lambda}_1^n$, $\boldsymbol{\Lambda}_2^n$, and the vectors $\mathbf{f}_\Phi^n$, $\mathbf{b}^n$ are given by:

$$\boldsymbol{\Lambda}_2^n = A\,\text{diag}(\overline{h})\left(\mathbf{J}_w^{n+1}\right)^{\mathsf{T}}\mu_t.\mathbf{J}_w^n + k^2\mathbf{M}_{\text{inv}} \tag{72a}$$

$$\boldsymbol{\Lambda}_1^n = \mathbf{I}_{N+1} + \frac{1}{2k}\boldsymbol{\Lambda}_2^n\,\text{diag}(\boldsymbol{\Psi}^n) \tag{72b}$$

$$\mathbf{f}_\Phi^n = \frac{\delta_t.\boldsymbol{\Phi}^n}{\delta_t.\boldsymbol{\Delta}^n} = \frac{\boldsymbol{\Phi}(\mathbf{r}^n + \boldsymbol{\Delta}^{n-1}) - \boldsymbol{\Phi}(\boldsymbol{\Delta}^{n-1})}{\mathbf{r}^n} \tag{72c}$$

$$\mathbf{b}_w^n = \left[\mathbf{0}_{N-1}\,|\,2\left(\mathbf{w}_{FB}^n - \mathbf{w}_{FB}^{n-1}\right)^{\mathsf{T}} + k^2\left(\mathbf{M}_{FB}^{-1}\mathbf{f}_{ext,FB}^n\right)^{\mathsf{T}}\right]^{\mathsf{T}}$$
$$+ \text{diag}\left(\overline{h}\right)\left(\left(\mathbf{J}_w^{n+1}\right)^{\mathsf{T}}\left(\mathbf{B}\mathbf{w}^n + \mathbf{C}\mathbf{w}^{n-1}\right) - \left(\mathbf{J}_w^{n-1}\right)^{\mathsf{T}}\mathbf{w}^{n-1}\right) \tag{72d}$$

where $\mathbf{M}_{\text{inv}}$ is a $(N_\mathfrak{D}+1) \times (N_\mathfrak{D}+1)$ matrix with $\mathbf{M}_{FB}^{-1}$ at its bottom-right corner, and all zeros elsewhere; $\mathbf{0}_{N-1}$ is an all-zero row vector of length $(N_\mathfrak{D}-1)$; and $\overline{h} = [\ldots 1 \ldots |h|h]^{\mathsf{T}}$.

Equation (71) is resolved with an iterative nonlinear system solver.

3.6.2. Horizontal Polarisation

Similarly to the vertical polarisation, a two-step recursion is derived from Schemes Equations (54a) and (57a):

$$\mathbf{u}^{n+1} = \mathbf{B}\mathbf{u}^n + \mathbf{C}\mathbf{u}^{n-1} - A\mu_t.\mathbf{J}_u^n\mathbf{f}_u^n \tag{73a}$$

$$\mathbf{u}_{FB}^{n+1} = \mathbf{B}_{FB}\mathbf{u}_{FB}^n + \mathbf{C}_{FB}^{n-1} + k^2\mathbf{M}_{FB}^{-1}\mathbf{A}_{FB}\left(\mathbf{f}_{uFB}^n + \mathbf{f}_{\text{ext}\,u,FB}^n\right) \tag{73b}$$

where A, $\mathbf{B}$, and $\mathbf{C}$ are the same as defined in Equation (70). $\mathbf{A}_{FB}$, $\mathbf{B}_{FB}$, and $\mathbf{C}_{FB}$ are 2×2 matrices, defined as:

$$\mathbf{A}_{FB} = 2\left(2\mathbf{M}_{FB} + k^2\mathbf{K}_{FB} + k\lambda_{FB}\right)^{-1}\mathbf{M}_{FB} \tag{74a}$$

$$\mathbf{B}_{FB} = 2\mathbf{A}_{FB} \tag{74b}$$

$$\mathbf{C}_{FB} = \frac{1}{2}\mathbf{A}_{FB}\left(-2\mathbf{M}_{FB} - k^2\mathbf{K}_{FB} + k\lambda_{FB}\right)\mathbf{M}_{FB}^{-1} \tag{74c}$$

Due to the friction linearity, the horizontal update is also implicit. Equations (73a) and (73b) can be written as a nonlinear system in terms of the vector variable $\mathbf{v}_{\text{rel}}^n$:

$$\mathbf{v}_{\text{rel}}^n + \mathbf{\Lambda}_3^n\mathbf{f}_u^n + \mathbf{b}_u^n = 0 \tag{75}$$

where the matrix $\mathbf{\Lambda}_3^n$, and the vector $\mathbf{b}_u^n$ are defined as:

$$\mathbf{\Lambda}_3^n = \frac{1}{2k}\left(A\text{diag}\left(\overline{\mathbf{h}}\right)\left(\mathbf{J}_u^{n+1}\right)^{\mathsf{T}}\mu_t.\mathbf{J}_u^n + \mathbf{A}_{\text{obj}}\right) \tag{76a}$$

$$\mathbf{b}_u^n = \frac{1}{2k}\text{diag}\left(\overline{\mathbf{h}}\right)\left(\left(\mathbf{J}_u^{n-1}\right)^{\mathsf{T}}\mathbf{u}^{n-1} - \left(\mathbf{J}_u^{n+1}\right)^{\mathsf{T}}\left(\mathbf{B}\mathbf{u}^n + \mathbf{C}\mathbf{u}^{n-1}\right)\right) + \left[\mathbf{0}_{N-1}|\left(\mathbf{b}_{uFB}^n\right)^{\mathsf{T}}\right]^{\mathsf{T}} \tag{76b}$$

$$\mathbf{b}_{uFB}^n = \frac{1}{2k}\left(\mathbf{B}_{FB}\mathbf{u}_{FB}^n + (\mathbf{C}_{FB} - \mathbf{I}_2)\,\mathbf{u}_{FB}^{n-1} + \mathbf{A}_{FB}\mathbf{f}_{\text{ext}\,u,FB}^n\right) \tag{76c}$$

where $\mathbf{A}_{\text{obj}}$ is a $(N_{\mathfrak{D}} + 1) \times (N_{\mathfrak{D}} + 1)$ matrix with $\mathbf{A}_{FB}$ at its bottom-right corner and zeros elsewhere; $\mathbf{I}_2$ is the 2×2 identity matrix.

3.6.3. Friedlander's Construction and Pitch Flattening

Update Equation (75) is valid for the general case, for any friction curves obeying the condition $v\varphi(v) \geqslant 0$. However, it is important to note that whenever the bow does not share any contact points with the fingers or the backboard (which is the case in most realistic playing scenarios), the last equation of System Equation (75) (corresponding to the bow friction force) is decoupled from the rest of the system. As a result, the bow friction nonlinearity, different in nature from that of the finger and fingerboard, can be treated separately, by the means of solving the nonlinear scalar equation:

Given the known values b_B^n and σ^n, Equation (77) is a scalar equation in terms of $v_{\text{rel},B}^n$ only, and can be solved with the help of Friedlander's graphical construction [56]. Indeed, the solutions to Equation (77) are the abscissas of the intersections of the friction curve $\varphi_B(v_{\text{rel},B}^n)$ with the line defined by $\eta_B\left(v_{\text{rel},B}^n\right) = -\frac{1}{\sigma^n}\left(v_{\text{rel},B}^n + b_B^n\right)$, as shown in Figure 6, with φ_B in black and η_B in blue, for different values of b_B^n. A solution on the vertical part of the curve corresponds to the sticking state ($v_{\text{rel},B}^n = 0$); otherwise, the string is slipping against the bow ($|v_{\text{rel},B}| > 0$).

$$v_{\text{rel},B}^n + \sigma^n\varphi_B(v_{\text{rel},B}^n) + b_B^n = 0 \tag{77}$$

where b_B^n is the last element of $\mathbf{b}^n$, and σ^n is defined as:

$$\sigma^n = \frac{hA}{2k}\left(\mathbf{j}_B^{n+1}\right)^{\mathsf{T}}\mu_t.\mathbf{j}_B^n\left[f_B^n\right]_+ \tag{78}$$

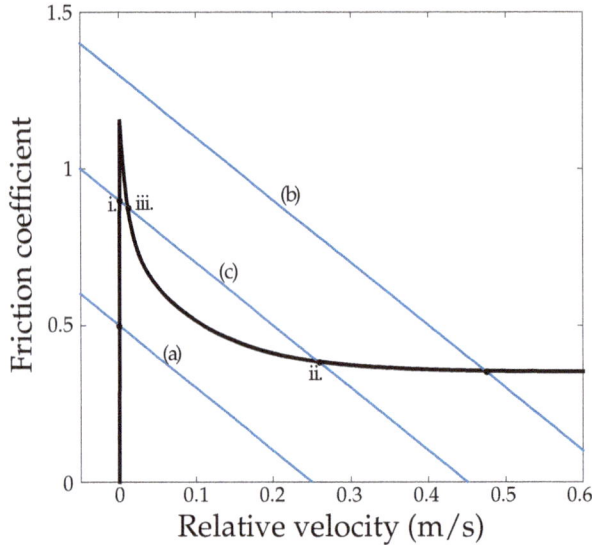

Figure 6. Friedlander construction to solve for the bow relative velocity. The solutions for $v_{rel,B}^n$ lie at the intersections of φ_B (black) and η_B (blue). Case (a): sticking; case (b): slipping. Under certain conditions (case (c)), there are potentially multiple solutions; this ambiguity is resolved by choosing the solution on the same branch of φ_B as the previous time step [40]. i. sticking; ii. slipping; iii. is always unstable, and is not seen in bowed string motion.

As Figure 6 shows (case (c)), one must beware of the possibility of multiple solutions. With this force-velocity curve model, this possibility, although seemingly brought about by the particular discretisation chosen for the bowed string equations, does indeed resemble Friedlander's so-called *ambiguity*, giving rise to pitch flattening phenomena, as McIntyre *et al.* have deduced [40]. They showed that the "correct" solution (Figure 6, i. and iii.) is the one that preserves the current oscillation phase (sticking (Figure 6, i.) or slipping (Figure 6, ii.)) for the longest time, and that the "middle" solution (Figure 6, iii.) is always physically unstable, and never seen in practice. This gives rise to hysteretic behaviour: the relative velocity solution jumps back and forth between the two branches of φ_B at different points depending on whether the transition is from stick to slip, or the other way around. This shows as a lengthening of the stick/slip cycles, and therefore a flattening of the pitch. This effect has later been found to be due to the naturally hysteretic thermal behaviour of the melting rosin, which is somewhat approximated by the hysteresis rule on the simpler friction curve model [42].

As can be seen in Figure 6, multiple solutions can only arise if the (negative) slope of η_B is small enough, provided by the condition:

$$-\frac{1}{\sigma^n} \leqslant \min\left(\varphi_B'\right) \tag{79}$$

From Equation (78), it can be seen that for a given virtual string, as j_B^n is normalised at all times, the variations of σ^n are directly proportional to those of the normal bow force: as $\left[f_B^n\right]_+$ increases, $\frac{-1}{\sigma^n}$ increases. The pitch flattening effect therefore is, for this scheme, a result of pushing down too strongly on the bow; this is indeed what can be observed in real bowed string playing. An expression can be derived for the minimal bow force beyond which this effect can occur; Appendix demonstrates that, for an *ideal* bowed string, the condition on this minimal force derived analytically by Friedlander is found to be the same as that obtained with a similar finite difference scheme as that used in this work, at the stability limit. This is an encouraging indicator of the validity of using a physical hysteresis rule such as that used by McIntyre *et al.* to resolve this seemingly numerical ambiguity, albeit not a definite proof that the minimal bow force expression will be the same, in practice, for different choices of discretisation.

The implementation of the various cases is greatly beneficial for reducing computational time; indeed, the slipping phase only lasts for a fraction of a complete cycle, therefore the equation becomes linear (and trivial to solve) in most cases, no longer requiring a computationally expensive iterative solver.

4. Simulation Results

4.1. Control Parameters

The first aspect of user control for this model lies in the physical parameters of the system. Provided that algorithmic stability is ensured, the end user has total and independent control over all of the physical parameters defined in this work. The measurements mentioned in Section 2.3 can be used as a sort of safeguard, if one is concerned with synthesising realistic bowed string instrument sounds; the more adventurous composer can experiment with different parameters, not limited by physical constraints. It is important to note that because of the nature of bowed string playing, unusual combinations of physical parameters may drastically limit playability.

The second aspect is gesture control, defined by two sets of dynamically varying parameters (three for the bow, two for the finger):

- Bow parameters:
    - bow position along the string, closer to or further away from the bridge
    - bow downwards force (denoted by "bowing pressure" in most papers)
    - bow tangential force, which will determine the bowing velocity (and, in turn, the amplitude of bowed string motion)
- Finger parameters:
    - finger position along the string, determining the (possibly varying) pitch of the resulting note
    - finger downwards force, which can be used to capture the string against the fingerboard, or to lightly push the finger against the string for playing natural harmonics

These five gestural parameters are fed into the algorithm as time series. Their value at a certain time is determined with the help of a score file, defining breakpoint functions for every one of them. The resulting piecewise linear time series can be modified at will; a representative example is the addition of an oscillatory term in the finger position function, so as to simulate a vibrato sound (see Section 4.3).

All the simulations presented here are run at audio sample rate ($F_s = 44.1$ kHz, with the exception of those for which spectrograms are computed, where the sampling rate has been increased in order to enhance the spectrogram resolution), as close as possible to the stability limit. The output waveform is read as the displacement of the last mobile point of the string before the bridge termination. Accompanying simulation videos and synthetic sounds are available online [41].

4.2. Bowed String Motion

As the bow is driven by the external force $f^n_{\text{ext}\,u,B}$, and not an imposed velocity, the amplitude and shape of the force signal to send into the bow is at first less intuitive to gauge. However, while a full parameter exploration study is definitely worth considering (with regards to playability and transient quality; see e.g., [24]), minimal trial and error allowed to successfully reproduce the standard, periodic Helmholtz motion [57] of the bowed string, as well as other typical oscillation states under realistic bowing conditions.

For a given bow-bridge distance and bow velocity, if the player presses the bow too strongly for the returning Helmholtz corner to detach it from the string, the model produces raucous motion, as predicted by Schelleng [21]. If the string is bowed too lightly for it to stick to it for a whole nominal period, the synthesised bowed string motion has the characteristics of multiple slipping. Figure 7 shows the simulated typical sawtooth waveform associated with the Helmholtz motion, the split sawtooth associated with multiple slipping, and the rough, aperiodic waveform resulting from raucous motion of the string.

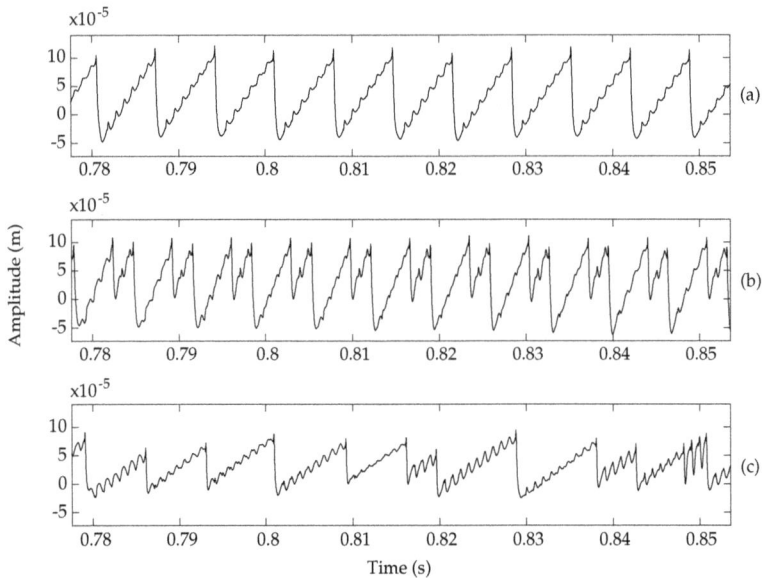

Figure 7. Different simulated waveforms on a cello G string, with fixed bow position $x_B = 0.851L$ m, bow tangential force $f_{ext\,u,B} = 4.4$ N, and bow downwards force $f_{ext\,w,B} = -3$ N (Helmholtz motion, (a)), $f_{ext\,w,B} = -0.5$ N (multiple slipping, (b)), and $f_{ext\,w,B} = -3.8$ N (raucous motion, (c)).

4.3. Gesture-Based Sound Synthesis

Modelling of the fingerboard, as well as that of the two-polarisation dynamics of the left hand finger and the bow, allows the simulation of a broad range of the bowed string player's gestures. Figure 8 illustrates the start of a typical simulated gesture.

4.3.1. Varying Bow Forces and Position

As long as the combination of bow force and bow position along the string stays within the Schelleng-like triangular area allowing to produce a stable tone, the bow control parameters can be varied across the simulation, while keeping the Helmholtz motion sustained. This allows for dynamic variations of a note's timbre and loudness, directly mapped to the same parameters that a player would handle on a real instrument (as opposed to a more abstract representation of their influence over the output sound), greatly enhancing the expressiveness of the synthetic sound. For instance, a sharper, more brilliant sound can be achieved by bowing the string closer to the bridge; loudness is controlled by adjusting the bow tangential force, determining the bow transverse velocity. Figure 9 shows the complex spectral variations associated with dynamic changes in the bow downwards force, tangential force, and position along the string, during a simple synthesised bowed string gesture.

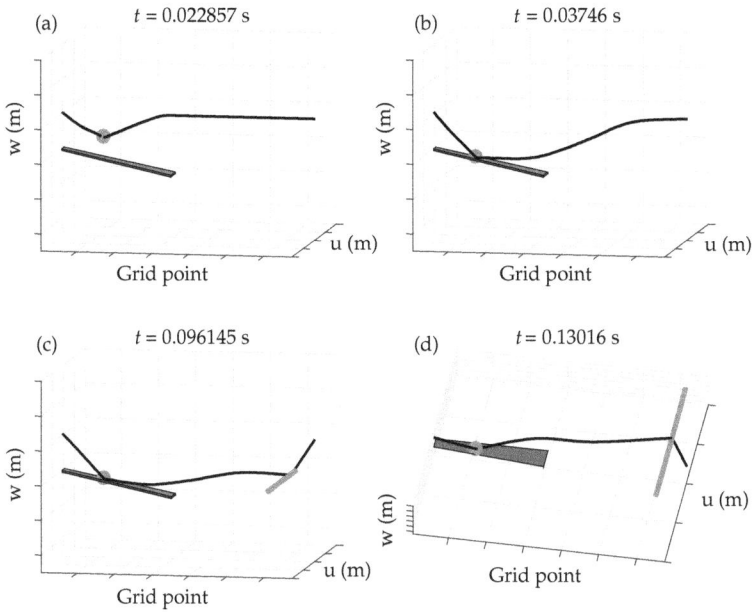

Figure 8. Typical gesture simulation. A finger (**a**) captures the string against the fingerboard (**b**); the bow then comes down (**c**), and starts bowing across (**d**). The string is excited along its speaking length, and some residual oscillations are observed between the nut and the finger.

4.3.2. Moving Finger

Like the bow, the left hand finger can be moved along the string while pressed down against the fingerboard. It is therefore straightforward to introduce gestures such as *glissando*, where the finger slides up or down along a significant portion of the string, and *vibrato*, by oscillating the finger along a central position, at a sub-audio rate. Figure 10 shows the spectrogram for a *glissando* gesture, followed by a *vibrato*; the associated finger position is displayed underneath the spectrogram for reference.

4.3.3. Natural Harmonics

If the finger downwards force is set small enough so that the string does not touch the fingerboard, but large enough so that the finger has a grip on the string, it is possible to bow natural harmonics by placing the finger at an integer ratio of the string length.

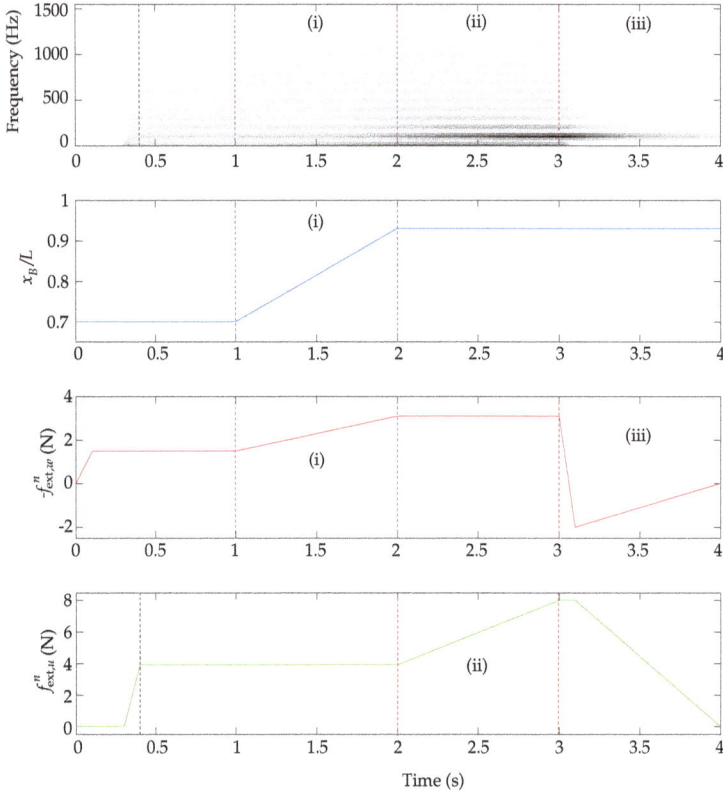

Figure 9. Spectrogram of the synthetic sound produced by bowing a cello string with varying downwards force $f^n_{\text{ext}\,w}$, tangential force $f^n_{\text{ext}\,u}$, and position along the string x^n_B. Higher harmonics appear when increasing the bow force and bowing closer to the bridge ((i), between the two purple dashed lines). A bow tangential force increase does not influence the spectral content of the sound, but increases its global amplitude ((ii), between the two red dashed lines). The free string oscillations decay as soon as the bow is lifted up (iii). The green dashed line marks the start of steady-state bowing.

4.3.4. Bouncing Bow

The bow spontaneously bounces against the string, if the initial *vertical* bow velocity is high enough (that is if the bow is pushed down on the string too fast, typically with a downwards initial velocity of 1–10 m/s), and the downwards force $f^n_{\text{ext}\,w,B}$ is too small to immediately compensate the restoring force due to the tension of the string. Figure 11 shows the resulting waveform, with alternation of sustained oscillations when the bow is in contact with the string, and decaying free string oscillations. With more refined control over the gesture parameters, it is indeed possible to simulate *spiccato* playing.

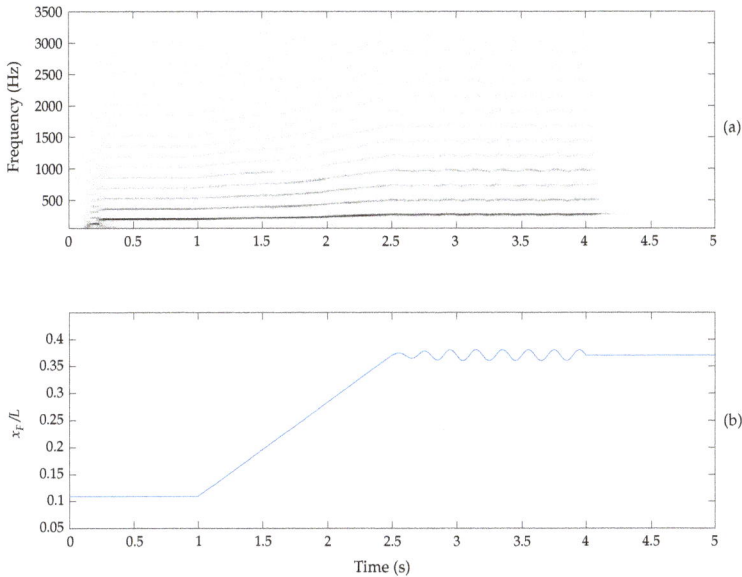

Figure 10. Spectrogram (**a**) of a synthesised gesture, showing a glissando along a cello string followed by a vibrato, accompanied the by variations of the finger position (**b**).

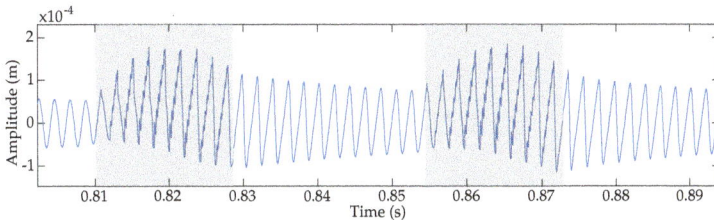

Figure 11. Bow bouncing against a violin string. The shaded areas are those where the bow is in contact with the string; elsewhere, the free string oscillations decay. Note the rounding of the waveform moments after the bow detaches itself, due to frequency-dependent loss modelling. The bow bouncing frequency is related to the bow parameters (M_B, K_B, α_B, β_B).

4.3.5. Rattling

In the absence of a bow, the string can be "plucked" by either initialising the string displacement away from its resting position, or feeding an external force signal in the form of a raised half-cosine into the string at the desired plucking position [39]. If the string is plucked hard enough in both polarisations at the same point, that is equivalent to pulling the string both aside and away from the fingerboard, it will both collide and rub against the fingerboard; the resulting sound is that of the string rattling against the fingerboard, as can be heard in some double bass techniques. *Pizzicato* playing is achieved with a lighter pluck.

4.4. Energy Balance

To demonstrate the balanced numerical energy of the system, we monitor the variations of the quantity E^n defined in Equation (68) along a bowed string simulation, where the bow and finger positions, forces, and the bow tangential force are all time-varying. We normalise E^n with respect to the mean energy $\bar{H}^n$, averaged over the duration of the simulation. As seen in Figure 12b, E^n is invariant until the 10th significant digit. The finite error tolerance for the nonlinear system solvers, as well as the accumulation of round-off error, seem to prevent reaching true floating point accuracy.

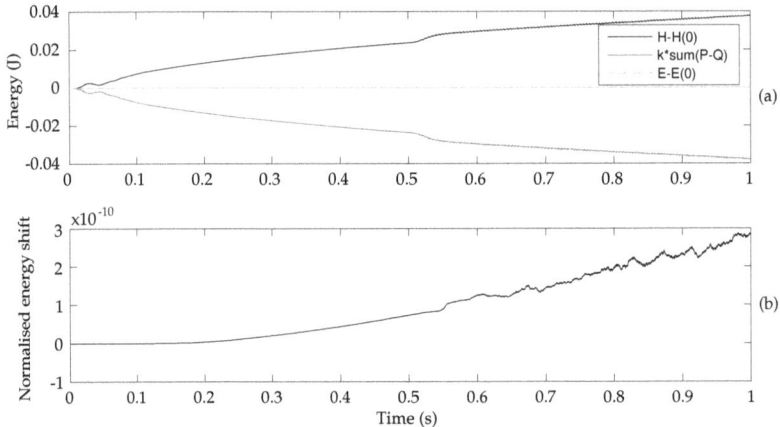

Figure 12. Numerical energy balance for the whole system. (**a**) in both polarisations, the energy is balanced at all times by the cumulative supplied and withdrawn power; (**b**) the total energy is conserved to the 10th significant digit, when normalised with respect to the mean energy. The apparent trend is due to accumulated round-off error.

5. Discussion

This work introduces a novel two polarisation bowed string physical model, including nonlinear damped contact and friction interactions with one bow, one stopping finger, and the distributed fingerboard. An energy-balanced finite difference scheme was presented, resulting in a two-step time recursion.

The inclusion of lumped and distributed interactions with the player (through the bow and finger) and fingerboard allows for simulating full articulated gestures in a relatively instinctive and concrete way, without having to rely on somewhat abstract hypotheses — an eloquent example being the finger model, that accounts for several important phenomena that would be difficult (impossible in fact, for some) to model with a simple absorbing string termination. Here, the simple action of pushing a finger down onto the string results in damped dynamic behaviour in both polarisations, variations of the string's speaking length, possible slipping of the string while captured, while the portion of the string between the nut and finger is still realistically oscillating, and responding to the excitation.

However, an important aspect of gestural control in bowed string playing resides in real-time adjustments of playing parameters during note production. The musician relies on immediate feedback from his instrument, adapting its playing accordingly. Our model, even with the aforementioned possible optimisations, does not run in real-time, making gesture design rather difficult. An interesting study could make use of recorded data from sensors during various gestures, feeding them as time series into the model, rather than our current breakpoint functions. This would help calibrate the model, on the string side as well as for the gestural functions [15,35].

The adaptation of this work to the more realistic case of multiple fingers (and, why not, multiple bows) is trivial, as well as the design of a multiple string environment. The mutual coupling of such strings is the obvious next step, moving towards the design of a full instrument, where strings communicate with a flexible body and with each other through a bridge. The simulated body will eventually take a great part in both the virtual instrument's playability, introducing vibrations feeding back into the strings, and the realism of the synthetic sound; to address the latter, and get a glimpse at the potential of a full instrument model, we have convolved a dry output signal from this string model with the impulse response of a cello body, a principle that is still used to this day for high quality sound synthesis [58]. The resulting sound example can be found online, amongst other relevant samples obtained from the model [41].

Acknowledgments: This work was supported by the Edinburgh College of Art, the Audio Engineering Society, and the European Research Council under grant number StG-2011-279068-NESS.

Author Contributions: Charlotte Desvages and Stefan Bilbao conceived and designed the model; Charlotte Desvages implemented the numerical schemes, obtained simulation data, and wrote the paper, with many improvements, rewritings, and corrections from Stefan Bilbao.

Conflicts of Interest: The authors declare no conflict of interest. The founding sponsors had no role in the design of the study; in the collection, analyses, or interpretation of data; in the writing of the manuscript, and in the decision to publish the results.

Appendix

Consider the simplifiedsystem of an ideal string, coupled to a pointwise bow at constant position x_B, with constant velocity v_B, pushing down with a constant force f_B. Using the same notations as established in Section 2, the transverse displacement of the string $u(x, t)$ obeys:

$$\rho \partial_t^2 u = T \partial_x^2 u - J_B f_B \varphi_B(v_{\text{rel},B}) \tag{A1}$$

where $J_B(x) = \delta(x - x_B)$ is a Dirac delta function, and $v_{\text{rel},B}(t)$ is defined as:

$$v_{\text{rel},B}(t) = \int_{\mathcal{D}} J_B(x, t)\dot{u}(x, t)dx - v_B(t) = \dot{u}(x_B, t) - v_B(t) \tag{A2}$$

System Equation (A1) can be discretised as:

$$\rho \delta_{tt} \mathbf{u}^n = T \mathbf{D}_{xx} \mathbf{u}^n - \mathbf{j}_B f_B \varphi_B(v_{\text{rel},B}^n) \tag{A3a} \qquad\qquad v_{\text{rel},B}^n = h \mathbf{j}_B^{\mathsf{T}} \delta_t.\mathbf{u}^n - v_B \tag{A3b}$$

The stability condition for this scheme can straightforwardly be deduced from that of the full system Equation (67):

$$h \leqslant ck \Leftrightarrow \lambda \leqslant 1 \tag{A4a} \qquad\qquad \lambda = \frac{h}{ck} \tag{A4b} \qquad\qquad c = \sqrt{\frac{T}{\rho}} \tag{A4c}$$

Equation (A3a) can be rearranged to derive an update equation for $\mathbf{u}^{n+1}$:

$$\mathbf{u}^{n+1} = \left(2 + \frac{Tk^2}{\rho}\mathbf{D}_{xx}\right)\mathbf{u}^n - \mathbf{u}^{n-1} - \frac{k^2}{\rho}\mathbf{j}_B f_B \varphi_B(v_{\text{rel},B}^n) \tag{A5}$$

Rewriting Equation (A5) in terms of $v_{\text{rel},B}^n$ yields the following nonlinear equation:

$$v_{\text{rel},B}^n + \sigma \varphi_B(v_{\text{rel},B}^n) + b = 0 \tag{A6}$$

where σ and b are constants given by:

$$\sigma = \frac{k}{2\rho}\|\mathbf{j}_B\|^2 f_B \tag{A7a}$$

$$b = v_B - \frac{h}{2k}\mathbf{j}_B^{\mathsf{T}}\left(\left(2 + \frac{Tk^2}{\rho}\mathbf{D}_{xx}\right)\mathbf{u}^n - 2\mathbf{u}^{n-1}\right) \tag{A7b}$$

The Friedlander construction, detailed in Section 3.6.3, can be used again, giving a condition on σ so that the solution of Equation (A6) is unique:

$$-\frac{1}{\sigma} \leqslant \min(\varphi'_B) \quad \Rightarrow \quad k \leqslant -\frac{2\rho}{\min(\varphi'_B)\|\mathbf{j}_B\|^2 f_B} \tag{A8}$$

Suppose that $\mathbf{j}_B$ is normalised as usual, so that $h\|\mathbf{j}_B\|^2 = 1$. Bound Equation (A8) now becomes:

$$f_B \leqslant \frac{2c\rho\lambda}{-\min(\varphi'_B)} \tag{A9}$$

which is independent of the bow velocity. With the friction curve defined in Equation (17a), $-\min(\varphi'_B) = 44.5$. The minimum bow force beyond which pitch flattening occurs therefore only depends on the string's physical parameters, and the Courant number λ.

Scheme Equation (A3a), without the bow force term, can be proved to provide the exact solution for the 1D wave equation, when the grid spacing and time step are set at the stability limit, that is $\lambda = 1$ ([18], pp. 133–136). In this limit, adding the bow model yields a bound on f_B that is indeed the same as that found analytically by Friedlander [56], and such a bound does not seem to vanish, even when the time step and grid spacing are made very small (keeping λ constant).

References

1. Kleimola, J. Nonlinear Abstract Sound Synthesis Algorithms. Ph.D. Thesis, Aalto University, Espoo, Finland, 22 February 2013.
2. Ruiz, P.M. A Technique for Simulating the Vibration of Strings with a Digital Computer. Ph.D. Thesis, University of Illinois at Urbana-Champaign, Champaign, IL, USA, 1970.
3. Hiller, L.; Ruiz, P. Synthesizing musical sounds by solving the wave equation for vibrating objects: Part 1. *J. Audio Eng. Soc.* **1971**, *19*, 462–470.
4. Bacon, R.A.; Bowsher, J.M. A discrete model of a struck string. *Acustica* **1978**, *41*, 21–27.
5. Karplus, K.; Strong, A. Digital synthesis of plucked-string and drum timbres. *Comput. Music J.* **1983**, *7*, 43–55.
6. Jaffe, D.; Smith, J.O., III. Extensions of the Karplus-Strong plucked-string algorithm. *Comput. Music J.* **1983**, *7*, 56–69.
7. Smith, J.O., III. A new approach to digital reverberation using closed waveguide networks. In Proceedings of the International Computer Music Conference (ICMC), Burnaby, BC, Canada, 19–22 August 1985; pp. 47–53.
8. Smith, J.O. Efficient simulation of the reed-bore and bow-string mechanisms. In Proceedings of the International Computer Music Conference (ICMC), The Hague, The Netherlands, 20–24 October 1986; pp. 275–280.
9. Karjalainen, M.; Välimäki, V.; Tolonen, T. Plucked-string models: From the Karplus-Strong algorithm to digital waveguides and beyond. *Comput. Music J.* **1998**, *22*, 17–32.
10. Woodhouse, J. Physical modeling of bowed strings. *Comput. Music J.* **1992**, *16*, 43–56.
11. Takala, T.; Hiipakka, J.; Laurson, M.; Välimäki, V. An expressive synthesis model for bowed string instruments. In Proceedings of the International Computer Music Conference (ICMC), Berlin, Germany, 27 August–1 September 2000 .
12. Serafin, S.; Avanzini, F.; Ing, D.; Rocchesso, D. Bowed string simulation using an elasto-plastic friction model. In Proceedings of the Stockholm Music Acoustics Conference (SMAC), Stockholm, Sweden, 6–9 August 2003; pp. 1–4.

13. Woodhouse, J. Bowed string simulation using a thermal friction model. *Acta Acust. United Acust.* **2003**, *89*, 355–368.
14. Maestre, E.; Spa, C.; Smith, J.O., III. A bowed string physical model including finite-width thermal friction and hair dynamics. In Proceedings of the International Computer Music Conference (ICMC), Athens, Greece, 22–26 October 2014 .
15. Demoucron, M. On the Control of Virtual Violins—Physical Modelling and Control of Bowed String Instruments. Ph.D. Thesis, Université Pierre et Marie Curie-Paris VI, Paris, France, 24 November 2008.
16. Debut, V.; Delaune, X.; Antunes, J. Identification of the nonlinear excitation force acting on a bowed string using the dynamical responses at remote locations. *Int. J. Mech. Sci.* **2010**, *52*, 1419–1436.
17. Strikwerda, J.C. *Finite Difference Schemes and Partial Differential Equations*; Siam: Philadelphia, PA, USA, 2004.
18. Bilbao, S. *Numerical Sound Synthesis*; John Wiley & Sons, Ltd.: Chichester, UK, 2009.
19. Chaigne, A.; Askenfelt, A. Numerical simulations of piano strings. I. A physical model for a struck string using finite difference methods. *J. Acoust. Soc. Am.* **1994**, *95*, 1112–1118.
20. Bensa, J.; Bilbao, S.; Kronland-Martinet, R.; Smith, J.O., III. The simulation of piano string vibration: From physical models to finite difference schemes and digital waveguides. *J. Acoust. Soc. Am.* **2003**, *114*, 1095–1107.
21. Schelleng, J.C. The bowed string and the player. *J. Acoust. Soc. Am.* **1973**, *53*, 26–41.
22. Raman, C. On the mechanical theory of the vibrations of bowed strings and of musical instruments of the violin family, with experimental verification of the results. *Bull. Indian Assoc. Cultiv. Sci.* **1918**, *15*, 1–158.
23. Schoonderwaldt, E.; Guettler, K.; Askenfelt, A. Schelleng in retrospect: A systematic study of bow-force limits for bowed violin strings. In Proceedings of the International Symposium on Musical Acoustics (ISMA), Barcelona, Spain, 9–12 September 2007.
24. Schumacher, R.T.; Woodhouse, J. The transient behaviour of models of bowed-string motion. *Chaos* **1995**, *5*, 509–523.
25. Serafin, S.; Smith , J.O., III; Woodhouse, J. An investigation of the impact of torsion waves and friction characteristics on the playability of virtual bowed strings. In Proceedings of the IEEE Workshop on Applications of Signal Processing to Audio and Acoustics, New Paltz, NY, USA, 17–20 October 1999; pp. 1–4.
26. Young, D.; Serafin, S. Playability evaluation of a virtual bowed string instrument. In Proceedings of the Conference on New Interfaces for Musical Expression (NIME), Montreal, QC, Canada, 22–24 May 2003; pp. 104–108.
27. Woodhouse, J. Playability of bowed-string instruments. In Proceedings of the Stockholm Music Acoustics Conference (SMAC), Stockholm, Sweden, 30 July–3 August 2013; pp. 3–8.
28. Guettler, K. On the creation of the Helmholtz motion in bowed strings. *Acta Acust. United Acust.* **2002**, *88*, 970–985.
29. Woodhouse, J.; Galluzzo, P.M. The bowed string as we know it today. *Acta Acust. United Acust.* **2004**, *90*, 579–589.
30. Woodhouse, J. The acoustics of the violin: A review. *Rep. Prog. Phys.* **2014**, *77*, doi:10.1088/0034-4885/77/11/115901.
31. Askenfelt, A. Measurement of bow motion and bow force in violin playing. *J. Acoust. Soc. Am.* **1986**, *80*, 1007–1015.
32. Askenfelt, A. Measurement of the bowing parameters in violin playing. II: Bow-bridge distance, dynamic range, and limits of bow force. *J. Acoust. Soc. Am.* **1989**, *86*, 503–516.
33. Maestre, E. Modeling Instrumental Gesture: An Analysis/Synthesis Framework for Violin Bowing. Ph.D. Thesis, University Pompeu Fabra, Barcelona, Spain, 12 November 2009.
34. Schoonderwaldt, E. Mechanics and Acoustics of Violin Bowing. Ph.D. Thesis, KTH Royal Institute of Technology, Stockholm, Sweden, 30 January 2009.
35. Maestre, E. Analysis/synthesis of bowing control applied to violin sound rendering via physical models. In Proceedings of the Meetings on Acoustics, Montreal, QC, Canada, 2–7 June 2013; Volume 19, doi:/10.1121/1.4801073.
36. Bilbao, S.; Torin, A.; Chatziioannou, V. Numerical modeling of collisions in musical instruments. *Acta Acust. United Acust.* **2015**, *101*, 155–173.

37. Chatziioannou, V.; van Walstijn, M. Energy conserving schemes for the simulation of musical instrument contact dynamics. *J. Sound Vib.* **2015**, *339*, 262–279.
38. Desvages, C.; Bilbao, S. Physical modeling of nonlinear player-string interactions in bowed string sound synthesis using finite difference methods. In Proceedings of the International Symposium on Musical Acoustics (ISMA), Le Mans, France, 7–12 July 2014.
39. Bilbao, S.; Torin, A. Numerical simulation of string/barrier collisions: The fretboard. In Proceedings of the International Conference on Digital Audio Effects (DAFx), Erlangen, Germany, 1–5 September 2014.
40. McIntyre, M.E.; Woodhouse, J. On the fundamentals of bowed-string dynamics. *Acustica* **1979**, *43*, 93–108.
41. Bowed String Simulations and Gesture-Based Sound Synthesis. Available online: http://www.charlottedesvages.com/companion/appl-sci-16 (accessed on 4 May 2016).
42. Smith, J.H.; Woodhouse, J. The tribology of rosin. *J. Mech. Phys. Solids* **2000**, *48*, 1633–1681.
43. Bavu, E.; Smith, J.; Wolfe, J. Torsional waves in a bowed string. *Acta Acust. United Acust.* **2005**, *91*, 241–246.
44. Pickering, N.C. Physical properties of violin strings. *Catgut Acoust. Soc. J.* **1985**, *44*, 6–8.
45. Percival, G.K. Physical Modelling Meets Machine Learning: Performing Music with a Virtual String Ensemble. Ph.D. Thesis, University of Glasgow, Glasgow, Scotland, 8 May 2013.
46. Cuesta, C.; Valette, C. Evolution temporelle de la vibration des cordes de clavecin. *Acustica* **1988**, *66*, 37–45. (In French)
47. Hunt, K.H.; Crossley, F.R.E. Coefficient of restitution interpreted as damping in vibroimpact. *J. Appl. Mech.* **1975**, *42*, 440–445.
48. Boutillon, X. Model for piano hammers: Experimental determination and digital simulation. *J. Acoust. Soc. Am.* **1988**, *83*, doi:10.1121/1.396117.
49. Chabassier, J.; Chaigne, A.; Joly, P. Time domain simulation of a piano. Part 1 : Model description. *ESAIM: Math. Model. Numer. Anal.* **2013**, *48*, 1241–1278.
50. Rhaouti, L.; Chaigne, A.; Joly, P. Time-domain modeling and numerical simulation of a kettledrum. *J. Acoust. Soc. Am.* **1999**, *105*, 3545–3562.
51. Chaigne, A.; Joly, P.; Rhaouti, L. Numerical modeling of the timpani. In Proceedings of the European Congress on Computational Methods in Applied Sciences and Engineering, Barcelona, Spain, 11–14 September 2000.
52. Stulov, A. Dynamic behavior and mechanical features of wool felt. *Acta Mech.* **2004**, *169*, 13–21.
53. Guettler, K. A closer look at the string player's bowing gestures. *Catgut Acoust. Soc. J.* **2003**, *4*, 12–16.
54. Van der Schaft, A.J. Port-Hamiltonian systems: An introductory survey. In Proceedings of the International Congress of Mathematicians, Madrid, Spain, 22–30 August 2006; pp. 1339–1365.
55. Falaize, A.; Lopes, N.; Hélie, T.; Matignon, D.; Maschke, B. Energy-balanced models for acoustic and audio systems: A port-Hamiltonian approach. In Proceedings of the Unfold Mechanics for Sounds and Music, Paris, France, 11–12 September 2014.
56. Friedlander, F.G. On the oscillations of a bowed string. In Proceedings of the Mathematical Proceedings of the Cambridge Philosophical Society; Cambridge University Press: Cambridge, UK, 1953; Volume 49, pp. 516–530.
57. Von Helmholtz, H. *On the Sensations of Tone (English Translation A.J. Ellis, 1885, 1954)*; 3rd ed.; Cambridge University Press: Cambridge, UK, 2009.
58. Pérez Carrillo, A.; Bonada, J.; Patynen, J.; Välimäki, V. Method for measuring violin sound radiation based on bowed glissandi and its application to sound synthesis. *J. Acoust. Soc. Am.* **2011**, *130*, 1020–1029.

applied
sciences

MDPI

Article

Dynamical Systems for Audio Synthesis: Embracing Nonlinearities and Delay-Free Loops †

David Medine

Department of Music, University of California, San Diego, CA 92093, USA; dmedine@ucsd.edu;
Tel.: +1-858-822-7547

† This paper is an extended version of our paper published in the 41st International Computer Music
  Conference (ICMC), Denton, TX, USA, 25 September–1 October 2015.

Academic Editor: Vesa Valimaki
Received: 1 February 2016; Accepted: 27 April 2016; Published: 10 May 2016

Abstract: Many systems featuring nonlinearities and delay-free loops are of interest in digital audio,
particularly in virtual analog and physical modeling applications. Many of these systems can be posed
as systems of implicitly related ordinary differential equations. Provided each equation in the network
is itself an explicit one, straightforward numerical solvers may be employed to compute the output
of such systems without resorting to linearization or matrix inversions for every parameter change.
This is a cheap and effective means for synthesizing delay-free, nonlinear systems without resorting
to large lookup tables, iterative methods, or the insertion of fictitious delay and is therefor suitable
for real-time applications. Several examples are shown to illustrate the efficacy of this approach.

Keywords: digital signal processing; sound synthesis and modeling; dynamical systems; virtual
analog; physical modeling

PACS: J0101

---

1. Introduction

There are certain continuous-time systems that are of interest to computer musicians, that
contain a delay-free loop—instantaneous feedback from the output to the input. These are said
to be 'non-computable loops' because in order to process the input, one needs to know the output,
which hasn't been computed yet.

The most straightforward way to deal with this problem is to insert a small amount of delay in
the feedback loop. This was done, for example, in one of the earliest considerations of digitizing the
Moog ladder filter [1]. In that case, the analog circuitry under consideration contains a delay-free path
from output to input in its continuous-time block diagram. The authors add unit delay (z^{-1}) to the
structure in order to make it computable (Figure 1). This has the unwanted side-effect, however, of
coupling the resonance parameter of the filter to the parameter governing cutoff frequency.

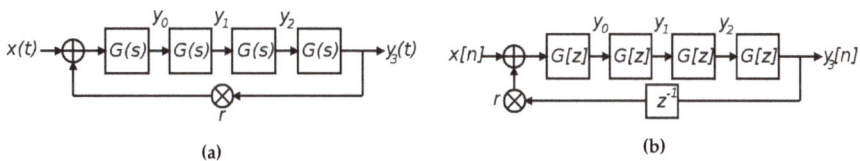

Figure 1. Continuous-time (**a**) and discrete time (**b**) block diagrams of the Moog ladder filter after
Stilson and Smith, 1996 [1].

The circuit emulation software SPICE (Simulation Program with INtegrated Circuit Emphasis) [2] can compute systems with delay-free loops through the use of iterative solvers. SPICE is very robust and is still the gold standard against which to demonstrate accuracy in virtual analog literature. It is not, however, well suited for virtual instrument design since it is not meant to be run in real-time. It is also unsuitable for describing systems that are not represent-able as electrical circuits.

A class of structures that can also emulate circuits (and other things) in real-time are wave digital filters (WDFs). WDFs were first introduced in the early 1970's [3] and have been the subject of a steady stream of study since then. More recently, there has been increased development of WDF theory in the digital audio community, particularly in circuit simulation applications [4–6], but in virtual acoustics as well [7–9]. Until very recently, however, WDF theory was not sufficient for representing systems involving delay-free loops. To this end, the K-method [10] has been suggested as a means of solving coupled, multivariate nonlinearities in WDFs [11,12].

Interestingly, the K-method was originally proposed in [10] as a means of eliminating delay-free loops in virtual acoustic models. Nevertheless, it and its variants are frequently used in circuit simulation applications [13–15]. The K-method operates on dynamical state-space systems that are of the form:

$$\dot{\mathbf{x}} = \mathbf{A}\mathbf{x} + \mathbf{B}\mathbf{u} + \mathbf{C}\mathbf{i}(\mathbf{v}) \tag{1}$$

$$\mathbf{y} = \mathbf{D}\mathbf{x} + \mathbf{E}\mathbf{u} + \mathbf{F}\mathbf{i}(\mathbf{v}) \tag{2}$$

$$\mathbf{v} = \mathbf{L}\mathbf{x} + \mathbf{M}\mathbf{u} + \mathbf{N}\mathbf{i}(\mathbf{v}) \tag{3}$$

where $\mathbf{x}$ is the 'state', $\mathbf{u}$ is an input vector, $\mathbf{y}$ is the output and $\mathbf{i}$ is a nonlinear vector function of $\mathbf{v}$.

After discretizing the system, the nonlinearity in Equation (3) can be solved with an iterative solver such as Newton-Raphson method. Furthermore multidimensional lookup tables can be pre-computed to avoid convergence issues, rendering the solution more suitable to real-time computation. The method also requires discretizing Equation (1) (e.g., with Backward Euler or the trapezoidal rule). A consequence of this is that if an adjustable parameter presents itself in the discretized version of matrix $\mathbf{A}$ it may be the case that multiple matrix inversions need to be performed in order to find the correct discretized coefficients that correspond to $\mathbf{B}$, $\mathbf{C}$, *etc.* There are times when analysis can resolve this issue (as in [15]), but this can potentially lead to significant slowdown if controllable parameters are desired.

If $\mathbf{i}$ is linear or if it is a low order polynomial, an exact solution be found to the system and resorting to such methods may not be necessary. However, this is not always the case. We present several examples from a class of nonlinear dynamical systems with delay-free loops that can be solved explicitly without resorting to iteration, tabulation or frequent re-calculation of filter coefficients. Namely, we examine systems in which each node of the network is expressed as an ordinary differential equation (ODE). Consider, for example:

$$\dot{x} = f_1(x, y, u) \tag{4}$$

$$\dot{y} = f_2(x, y, u), \tag{5}$$

where x and y are states, f_i are functions (perhaps nonlinear) and u is some (perhaps 0 valued) input.

Equations (4) and (5) differ from Equations (1)–(3) in that in Equations (4) and (5), each equation is an explicit ODE whereas only Equation (1) in the other system is a differential one. Most importantly, $\mathbf{i}$ is not an ODE and Equation (3) is implicit (v is a function of itself). On the other hand, since Equations (4) and (5) are both explicit ODEs, they can be directly solved using explicit numerical methods. Since numerical solvers to ODEs can compute both x and y, the fact that both variables are fed back to the nonlinear functions f_1 and f_2 in a delay-free loop does not doom us to non-computability.

2. Methods

The solver used in each example below is explicit fourth order Runge-Kutta (RK4):

$$x_i(n+1) = x_i(n) + \frac{h}{6}(k_{i1} + 2k_{i2} + 2k_{i3} + k_{i4}) \tag{6}$$

$$k_{i1} = v_i(n) \tag{7}$$

$$k_{i2} = v_i(n) + \frac{h}{2}k_{i1} \tag{8}$$

$$k_{i3} = v_i(n) + \frac{h}{2}k_{i2} \tag{9}$$

$$k_{i4} = v_i(n) + hk_{i3}. \tag{10}$$

The subscript i in Equations (6)–(10) is to indicate that there can be N functions (f_i) with inputs v_i—which is some computable value. Each node (each ODE that we solve for) has only one output value (the updated state at time $x_i(n+1)$, but the value v_i, which we take to be an 'input' to the ODE, can be any parameterized combination, linear or not, of all the x_i states in the network; and, if desired, external input.

The time-step h used here is simply the inverse of the sampling frequency: $h = 1/48,000$ for most of the examples shown below. RK4 is chosen because it is explicit, commonly used, and fourth order accurate. It is beyond the scope of this paper to describe the accuracy and numerical stability of RK4 and compare it to other, similar solvers. The curious reader may be referred to chapter 3 in [16] for a thorough investigation of this topic as it pertains to audio synthesis. It is worth noting, however, that RK4 does not always guarantee stability.

Using a C language application programming interface (API) written by the author, a system of interconnected ODEs can be declared and (through the use of helper functions) passed to an implementation of RK4 that will solve each network node $x_i(n+1)$ at every time-step. The API also allows for control objects (for adjusting function parameters) to be declared. These can then be manipulated in real-time while the audio engine is running. This API was used to create a Pure Data [17] (Pd) extern to produce the data shown in the examples below. As a creation argument, the extern takes the path to a shared object (a compiled C file with loadable functions) which specifies the functions to be solved, the network topology, and any required control inputs to adjust function parameters. The extern automatically creates the correct number of inlets and outlets and control values can be passed via Pd's messaging system. Having such an infrastructure for designing, controlling and solving systems such as that given in Equations (4) and (5) is a potentially powerful tool for computer musicians.

This technique of declaring a system of inter-connected ODEs to a software application which will then compute the solution to each node on the network is referred to as 'unsampled digital synthesis' (UDS) in a recent paper given at the 2015 International Computer Music Conference [18]. The application of numerical solvers to ODEs is not new in digital audio. The technique has been used, for example, to model diodes in guitar distortion circuitry [19,20]. Furthermore, such a scheme is essentially a subset of the more general family of techniques known as finite difference methods, whose use in digital audio applications is well studied [21].

3. Results

Below are a number of examples of sound synthesis routines that are networks of coupled ODEs solved using RK4.

3.1. Basic Example: Simple Harmonic Motion

Simple harmonic motion can be modeled as a spring-mass mechanism. In this model, the acceleration of the mass is proportional to its position. The proportion depends on the value of the mass and the stiffness of the spring:

$$m\frac{d^2x}{dt} = -kx(t) \tag{11}$$

here, m is the mass, and k is the spring constant. Since we will consistently be relating rates of change over time, we will drop the time variable t when speaking of these dynamical models. We will use the conventional 'dot' notation to denote the time derivative.

Equation (11) is a second order differential equation, but we may render it a first order system in the usual way:

$$\begin{aligned} \dot{x} &= y_v \\ \dot{y}_v &= -\frac{k}{m}x \end{aligned} \tag{12}$$

In this formulation the introduced variable y_v is the velocity of the system. By scaling this variable by the resonant angular frequency, we arrive at a more symmetrical form in which both x and the variable y are now in the same units (position):

$$\dot{x} = \omega y \tag{13}$$

$$\dot{y} = -\omega x \tag{14}$$

The frequency of this system is $f = \omega/2\pi$ Hz. We note that Equations (13) and (14) is an equivalent system to the the the one given in Equation (11) when $\omega = \sqrt{k/m}$. Differentiating Equation (13) yields $\ddot{x} = \omega\dot{y}$ and substituting Equation (14) for $\dot{y}$ yields $\ddot{x} = -\omega^2x$. The outputs of this system (which we are interested in hearing) are the changing values x and y, which we get by integrating (in this case with RK4) $\dot{x}$ and $\dot{y}$.

The amplitude and initial phase of the system are given by initial conditions. The amplitude is simply $A = \sqrt{x_0^2 + y_0^2}$ and the phase is $\phi = \tan^{-1}(y/x)$.

This implementation of simple harmonic motion is not new in digital audio. Mathews and Smith used oscillators in a similar form to design high-Q bandpass filters [22]; and, this oscillator is analyzed quite thoroughly in Chapter 3 of [21]. In that work it is presented as a basic building block for lumped networks.

By way of contrast, a more frequent implementation of a harmonic oscillator is the lookup oscillator (also called a wavetable oscillator). This consists of a phase signal that references values in a pre-computed table containing one period of a sinusoidal waveform. Oscillators of this variety are covered in most any introductory text to digital audio, for example Chapter 2 in [23]. The lookup oscillator can be generalized to lookup any arbitrary waveform, but we restrict ourselves to the sinusoidal case:

$$x(n) = \sin(\omega\phi(n)) \tag{15}$$

where $\phi(n)$ is a phase signal and (as usual) $\omega = 2\pi f$.

3.2. Reciprocal Sync

Oscillator 'sync' is a classic technique from modular analog systems in which a 'slave' oscillator has its phase reset to 0 if another 'master' oscillator's output crosses a certain threshold θ. This is a convenient way to create spectrally rich, periodic output.

If this routine is computed with lookup oscillators, the order of operations is important. Depending on whether or not the output of the master or slave oscillator is computed first, the output waveforms will be different. This scheduling issue, which results from delay in the feedback loop (depicted in Figure 2a), suggests that lookup oscillators are not ideal for representing two oscillators that sync each other in a tight feedback loop. However, if we model each oscillator as a dynamical system as in Equations (13) and (14), the discrete output of the whole system is computed at once. That is to say, there is no delay between the output of one oscillator and the input of another, as in Figure 2b.

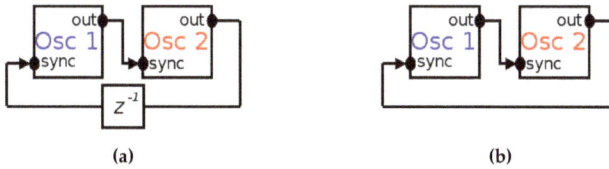

Figure 2. Block diagrams showing the signal flow for reciprocal sync. Using dyamical oscillators. (a) Using lookup oscillators; (b) Using dyamical oscillators.

Using a sync rule, we may couple two digital ODE + RK4 oscillators. The rule is that if the state of x_i for one of the oscillators crosses a threshold, the phase of the other will re-initialize. In the case of oscillator 1:

$$\dot{x}_1 = \begin{cases} \omega y_1 & \text{if } x_2 < \theta_1 \\ g_1(0 - x_1) & \text{if } x_2 \geq \theta_1 \end{cases} \tag{16}$$

$$\dot{y}_1 = \begin{cases} -\omega x_1 & \text{if } x_2 < \theta_1 \\ g_1(-1 - y_1) & \text{if } x_2 \geq \theta_1 \end{cases} \tag{17}$$

where there is another master/slave oscillator with a subscript of 2.

This is simply a nonlinear modification to the Equations (13) and (14). Here, g_i is a gain factor that speeds the return to the 'zero' phase ($x_i = 0$, $y_i = -1$). This rule is necessary, because if we are computing our state updates by adding approximate integrations of time derivatives (which is exactly what we do with RK4), then there is no simple way to describe instantaneous jumps from one state to another one that is far away. In the regime of reciprocal sync, each oscillator is both master and slave to the other. As can be seen in Figure 3, our dynamical models shows chaotic behavior which does not appear at all in the lookup oscillator implementation.

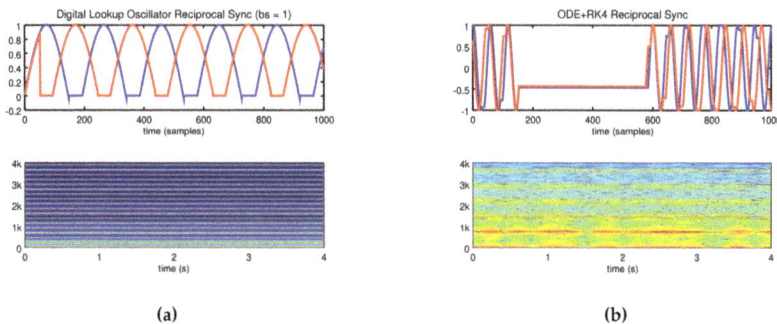

Figure 3. The output of (a) two digital lookup oscillators in a tight sync loop and (b) another pair given Equations (16) and (17) fed to the fourth order Runge-Kutta (RK4) method for solution. (a) 950 Hz and 900 Hz digital lookup oscillators in reciprocal sync. $\theta = 0.9$; (b) The same parameters but using the dynamical model in Equations (16) and (17).

3.3. Reciprocal Frequency Modulation

Frequency Modulation (FM) is a tried and true computer audio technique that has been used to simulate the human voice and many other real musical instruments [24]. In FM the scaled output of one oscillator (the modulator) affects the frequency of another (the carrier). In continuous-time, it can be expressed:

$$x(t) = \cos(\omega_c t + A \sin(\omega_m t)) \tag{18}$$

where A is the so-called modulation index, $\omega_c = 2\pi f_c$ is the angular frequency of the carrier, and $\omega_m = 2\pi f_m$ is the angular frequency of the modulator. A lookup oscillator digitization of Equation (18) would be:

$$x(n) = \cos(\omega\phi(n) + Ap(n)) \tag{19}$$

where $p(n)$ is simply the output of some other oscillator of the form shown in Equation (15). The frequencies in the spectrum of the carrier oscillator are predictable depending on ω_c and ω_m. If ω_c and ω_m are harmonically related, so are the resulting sidebands (the effects of foldover excepted). The amplitudes of these sidebands depend on A. For a thorough discussion, consult Chapter 5.5 in [23] or any other introductory text on digital audio effects. The implementation in Equation (19) is more precisely called 'phase modulation', because the phase signal is modulated [25].

We can implement a reciprocal FM network by again modifying the Equations (13) and (14) and solving with RK4.

$$\dot{x}_i = (\omega_i + \lambda_i)y_i \tag{20}$$

$$\dot{y}_i = -(\omega_i + \lambda_i)x_i \tag{21}$$

here, there are any number of oscillators that modulate one another, and λ_i is some linear combination of (the x part of) each other's outputs:

$$\lambda_i = \alpha_{i1}x_1 + \alpha_{i2}x_2 \ldots \alpha_{ij}x_j \tag{22}$$

The coefficients α_{ij} define the index of modulation acting on the frequency of oscillator i by oscillator j.

If we wish to modulate the modulating oscillator with the output of the carrier as in Figure 4, lookup oscillators will be insufficient for the same reasons as described above in the case of reciprocal sync. However we can incorporate a delay-free loop between oscillators by using RK4 to solve for any number of oscillators expressed by Equations (20)–(22).

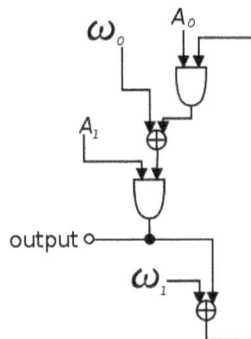

Figure 4. Signal flow diagram for a reciprocal frequency modulation (FM) scheme without delay.

Figures 5a and 5b compare reciprocal FM implemented as ODEs and as lookup oscillators. As the plots show, delay (in this case 0.0208 ms which is 1 sample at 48 kHz) between the lookup oscillators in a reciprocal FM network creates inharmonic distortion that is similar to broadband noise at higher modulation indices. On the contrary the ODE + RK4 implementation produces very different results. Broadband noise does not appear, and in the harmonic case, a low frequency periodicity emerges.

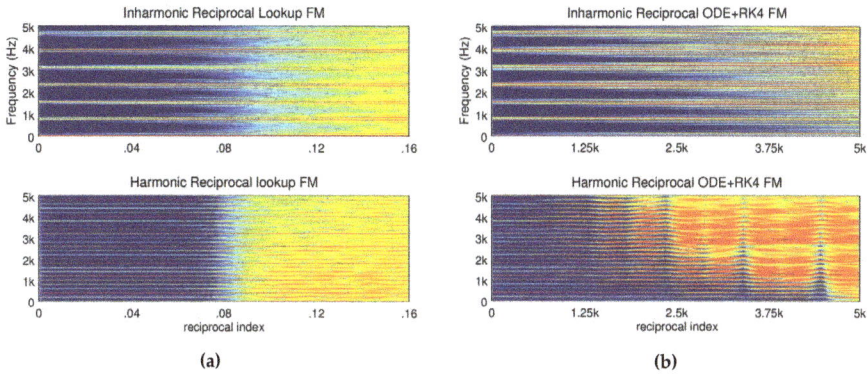

(a) (b)

Figure 5. Spectrograms of (a) reciprocal frequency modulation using lookup oscillators of frequencies 800 Hz and 777 Hz in the top and 800 Hz and 600 Hz in the bottom; In (b) the same network with the same frequencies, but using the ODE (ordinary differential equation) + RK4 implementation. In all four plots the modulation index of the 800 Hz oscillator is held fixed while the other modulation index is swept upwards.

The meaning of the modulation index is very different in the two implementations. The values given in Figure 5 were chosen by ear and by examining magnitude spectra so that they were roughly equivalent. There is a complex and nonlinear relationship between the equivalent indices in the two implementations.

As a final note, *feedback* FM (FBFM) has been known for some time [26]. FBFM differs from *reciprocal* FM in that FBFM needs only one lookup oscillator whose delayed output modulates its own phase input signal. As in the case with reciprocal FM, FBFM will cause a great deal of aliasing and foldover at higher indices and this comes across as broadband noise (Figure 6). In the ODE + RK4 implementation of FBFM we instead see that higher modulation indices bring about a flattening effect along with increased harmonic energy.

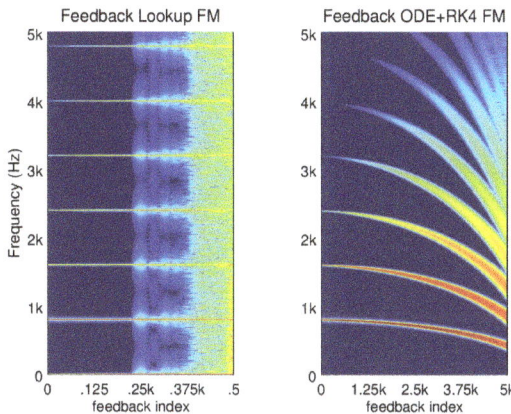

Figure 6. Sweeping the modulation index in feedback FM.

3.4. A Bowed Oscillator

In addition to arranging simple harmonic oscillators in delay-free feedback networks, we may use ODEs and RK4 as a means of introducing physically motivated nonlinearities. As an example, consider a harmonic oscillator driven by friction—such as a point on a bowed violin string.

The physics of bow-string interaction is well understood [27–29]. The effect of the bow on the point of bowing is usually modeled through a nonlinear function which is coupled to the point of bowing (here given as a single oscillator). The motion of this oscillator is

$$\ddot{u} = -\omega^2 u - F_b \Phi(v - v_b) \tag{23}$$

In this model u is displacement of the oscillator, $f = \omega/2\pi$ is the fundamental frequency, F_b is the force of the bow divided by mass (which is a quantity we neglect to include in our model in favor of using $\omega = \sqrt{k/m}$), v is the oscillator's velocity, v_b is the velocity of the bow, and Φ is a nonlinear friction model such as:

$$\Phi(v_{rel}) = \sqrt{2a} v_{rel} e^{-2a v_{rel}^2 + 1/2} \tag{24}$$

where a is a coefficient of friction and $v_{rel} = v - v_b$.

As is the case with the second order spring-mass system given in Equation (11), we can re-write Equation (23) as a pair of ODEs:

$$\dot{y} = -\omega^2 x - F_b \Phi(y - v_b)$$
$$\dot{x} = y \tag{25}$$

and solve with RK4.

A slightly different form of this model is discussed in Chapter 3 of [21]. There, the linear part of Equation (23) is solved with Backward Euler, and the nonlinearity Φ is solved with Newton-Raphson. This requires, of course, that Φ be differentiable and that the derivative be determined prior to computation. Furthermore, Newton-Raphson is an iterative solver so it may not be appropriate for real-time implementation due to convergence issues. These two implementations are compared in Figure 7. The code used to generate the Newton-Raphson version of the system was taken directly from Appendix A in [21].

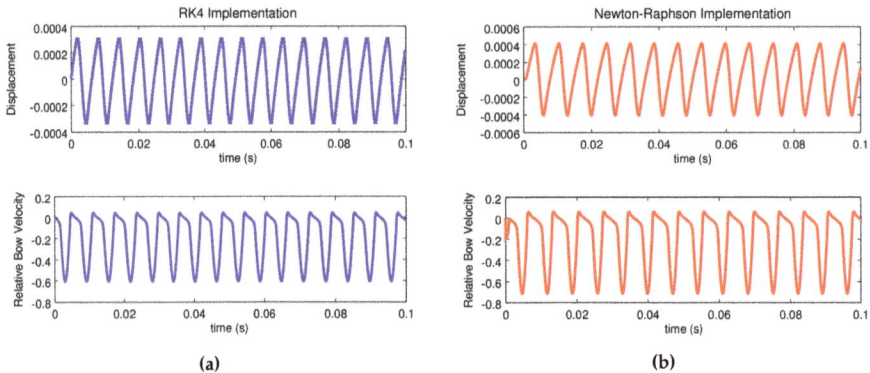

Figure 7. The bowed oscillator model given here (solved with RK4) and that given in [21]. In both implementations, oscillator frequency $f = 200$, $F_b = 500$, $v_b = .2$ and $a = 100$. (**a**) The system given in Equation (25) solved using RK4; (**b**) Using Newton-Raphson and Backward Euler to solve Equation (23).

Since RK4 puts us in a position to solve the system without having to differentiate Φ, we are free to experiment with friction models that are discontinuous. For example, we may study the discontinuous friction model given by

$$\Phi(v_{rel}) = \text{sign}(v_{rel})e^{-a|v_{rel}|} \tag{26}$$

A comparison of the bowed oscillator using this friction models and the one given in Equation (24) are compared for various values of a in Figure 8.

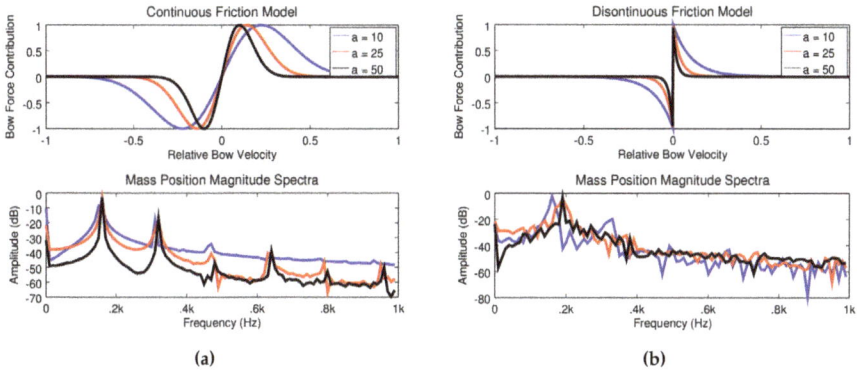

Figure 8. RK4 solving Equations (25) using (**a**) the continuous friction model in Equation (24) and (**b**) Equation (26). (**a**) Differentiable friction model; (**b**) Discontinuous friction model.

3.5. The Moog Ladder Filter

As an example of a musically interesting application of ODES and RK4 in audio synthesis that is not based on harmonic motion, we return to the 'ladder filter' invented by Robert Moog in 1965 [30]. This filter has been subject to numerous studies since [1]. Most digitizations use the bilinear transform [31] or some similar procedure [32] to discretize the linear portion (small signal response) of the analog transfer function. As mentioned above, since there is feedback in the circuit, digitizations of this filter often include unit sample feedback (z^{-1}). Again, this is problematic because the delay couples the gain parameter g (which controls the cutoff frequency of the filter) to the resonance parameter r (which controls the bandwidth of the filter). This causes additional phase shifting so that effect of the resonance parameter varies with frequency and vice versa. Many adjustments have been suggested to ameliorate this artifact of digitization [33] or to avoid the addition of delay to the feedback part of the filter without resorting to iteration. Examples of this include solution through the use of Volterra series [34,35] and linear compensation filters [36].

The Moog ladder filter is a four-pole network, wherein the voltage across each stage is defined:

$$\dot{y}_i = \omega(\tanh(y_{i-1}) - \tanh(y_i)) \tag{27}$$

where r is the gain that governs the cutoff frequency of the filter, $\dot{y}_i$ is the change in voltage across stage i, y_i is the present voltage across stage i, and y_{i-1} is the present voltage across the previous stage for $i = 1, 2, 3$. Ideally, the cutoff frequency has a ratio of $f_c = \omega/2\pi$ Hz for all frequencies.

Since each stage of the filter shifts the input signal phase by $\pi/4$ radians, the output of the filter is inverted and fed back into the input to create resonance. So, for $i = 0$ we have:

$$\dot{y}_0 = \omega(\tanh(V_{in} - rV_3) - \tanh(V_0)) \tag{28}$$

when $r = 4$, the filter saturates and will oscillate on its own. In the results presented here, we simply plug Equations (27) and (28) into RK4 and grind out the samples. As indicated by the results, no further analysis, linearization, or digitization than this is necessary to achieve good results.

Figure 9a shows a comparison of measured frequency and amplitude for impulses fed to the filter tuned to various values for the parameter f_c. This compare favorably with the results presented in [32] where compensation for the zero introduced by unit delay in the feedback path results in unbounded amplitude growth at higher values of f_c. The relationship between f_c and the measured frequency peaks in Figure 9b compares favorably with a similar analysis of another delay-free implementation given in [36]. Here there is noticeable flattening that results from lowering the value of r. This suggests that some coupling between this parameter and f_c persists.

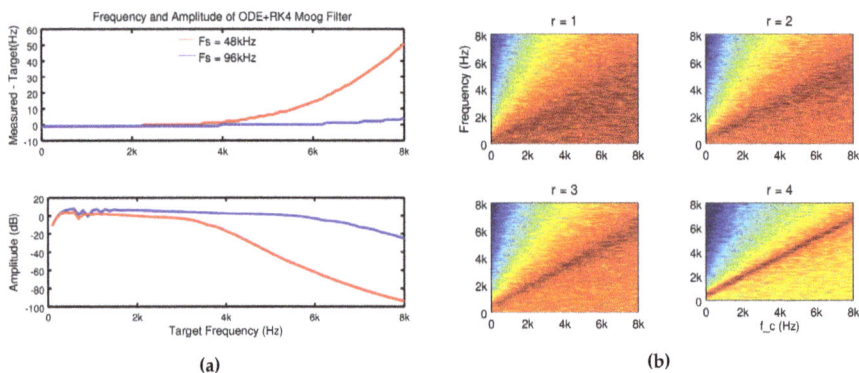

Figure 9. Plots illustrating the character of the ODE + RK4 implementation of the Moog filter. (**a**) Shown is the difference between target frequency and highest peak in the spectrum as well as amplitude for the ODE + RK4 Moog filter; (**b**) Sweeping f_c across a noise stimulated Moog filter for various values of r (sampling rate is 48 kHz).

4. Discussion

The above results show a small number of possible applications for the explicit solution of dynamical systems in audio synthesis. The benefit of the methods given here (defining a system as a network of ODEs then declaring that system to a software layer that can solve each equation in a time accurate manner) is simply that a user can easily create interesting and desirable sounds without having first to resort to deep analysis prior to synthesis. The drawback is that it cannot solve for implicit relations and that the numerical solver may fail when presented with stiff equations. There is no free lunch. However this method can be harnessed to synthesize implicit systems provided they can be expressed as coupled explicit ODEs. It is likely that by extending known systems that fit these requirements and by inventing new ones, many new interesting musical sounds and behaviors can be defined.

5. Conclusions

We present above a method for getting around (so to speak) the problem of delay-free loops in certain nonlinear digital synthesis routines. We do this by imposing the constraint that the systems we synthesize be ones that can be represented entirely through explicit first order ODEs. Explicit solvers can (more or less accurately) compute the output of each ODE even if there is instantaneous feedback in the network structure. There are many such systems that are of interest and the above shows only a few.

Conflicts of Interest: The author declares no conflict of interest.

References

1. Stilson, T.; Smith, J. Analyzing the Moog VCF with considerations for digital implementation. In Proceedings of the 1996 International Computer Music Conference, Hong Kong, China, 19–24 August 1996; pp. 398–401.
2. Nagel, L.W.; Pederson, D.O. *SPICE: Simulation Program with Integrated Circuit Emphasis*; No. UCB/ERL M382; Electronics Research Laboratory, College of Engineering, University of California: Berkeley, CA, USA, 12 April 1973.
3. Fettweis, A. Digital filters related to classical structures. *AEU Arch. Elektron. Ubertragungstechnik* **1971**, *25*, 78–89.
4. Pakarinen, J.; Tikander, M.; Karjalainen, M. Wave digital modeling of the output chain of a vacuum-tube amplifier. In Proceedings of the International Conference on Digital Audio Effects (DAFx), Como, Italy, 1–4 September 2009; pp. 1–4.
5. Pakarinen, J.; Karjalainen, M. Enhanced wave digital triode model for real-time tube amplifier emulation. *IEEE Trans. Audio Speech Lang. Process.* **2010**, *18*, 738–746.
6. Välimäki, V.; Bilbao, S.; Smith, J.O.; Abel, J.S.; Pakarinen, J.; Berners, D. Virtual Analog Effects. In *DAFX: Digital Audio Effects*, 2nd ed.; Wiley: Hoboken, NJ, USA, 2011.
7. Van Duyne, S.A.; Pierce, J.R.; Smith, J.O. Traveling wave implementation of a lossless mode-coupling filter and the wave digital hammer. In Proceedings of the International Computer Music Conference, Aarhus, Denmark, 12–17 September 1994; pp. 411–418.
8. Bensa, J.; Bilbao, S.; Martinet, K.R.; Smith, J. A power normalized non-linear lossy piano hammer. In Proceedings of the Stokholm Muisc Acoustics Conference, Stokholm, Sweden, 2–9 August 2003; pp. 365–368.
9. Van Walstijn, M.; Campbell, M. Discrete-time modeling of woodwind instrument bores using wave variables. *J. Acoust. Soc. Am.* **2003**, *113*, 575–585.
10. Borin, G.; de Poli, G.; Rocchesso, D. Elimination of delay-free loops in discrete-time models of nonlinear acoustic systems. *IEEE Trans. Speech Audio Process.* **2000**, *8*, 597–605.
11. Werner, K.; Nangia, V.; Smith, J., III; Abel, J. Resolving wave digital filters with multiple/multiport nonlinearities. In Proceedings of the Digital Audio Effects (DAFx), Trondheim, Norway, 30 November–3 December 2015.
12. Werner, K.; Smith, J., III; Abel, J. Wave digital filter adaptors for arbitrary topologies and multiport linear elements. In Proceedings of the Digital Audio Effects (DAFx), Trondheim, Norway, 30 November–3 December 2015.
13. Yeh, D.T.; Abel, J.S.; Smith, J.O., III. Automated physical modeling of nonlinear audio circuits for real-time audio effects—Part I: theoretical development. *IEEE Trans. Audio Speech Lang. Process.* **2010**, *18*, 728–737.
14. Yeh, D.T. Automated physical modeling of nonlinear audio circuits for real-time audio effects—Part II: BJT and vacuum tube examples. *IEEE Trans. Audio Speech Lang. Process.* **2012**, *20*, 1207–1216.
15. Dempwolf, K.; Holters, M.; Zölzer, U. Discretization of parametric analog circuits for real-time simulations. In Proceedings of the 13th International Conference on Digital Audio Effects, Graz, Austria, 4–6 September 2010.
16. Yeh, D.T.M. Digital Implementation of Musical Distortion Circuits by Analysis and Simulation. Ph.D. Thesis, Stanford University, Stanford, CA, USA, June 2009.
17. Puckette, M. Pure Data: Another integrated computer music environment. In Proceedings of the Second Intercollege Computer Music Concerts, Tachikawa, Japan, 7 May 1997; pp. 37–41.
18. Medine, D. Unsampled Digitial Synthesis: Computing the Output of Implicit and Non-Linear Systems. In Proceedings of the International Computer Music Conference, Denton, TX, USA, 25 September–1 October 2015; pp. 90-93.
19. Yeh, D.T.; Abel, J.; Smith, J.O. Simulation of the diode limiter in guitar distortion circuits by numerical solution of ordinary differential equations. In Proceedings of the 10th International Conference on Digital Audio Effects, Bordeaux, France, 10–15 September 2007; pp. 197–204.
20. Macak, J.; Schimmel, J. Nonlinear circuit simulation using time-variant filter. In Proceedings of the International Conference on Digital Audio Effects (DAFx), Como, Italy, 1–4 September 2009.
21. Bilbao, S. *Numerical Sound Synthesis: Finite Difference Schemes and Simulation in Musical Acoustics*; Wiley Online Library: Hoboken, NJ, USA, 2009.

22. Mathews, M.; Smith, J.O. Methods for synthesizing very high Q parametrically well behaved two pole filters. In Proceedings of the Stockholm Musical Acoustics Conference (SMAC), Stockholm, Sweden, 6–9 August 2003.
23. Puckette, M. *The Theory and Technique of Electronic Music*; World Scientific: Singapore, 2007.
24. Chowning, J.M. The synthesis of complex audio spectra by means of frequency modulation. *J. Audio Eng. Soc.* **1973**, *21*, 526–534.
25. Chowning, J.M. Method of synthesizing a musical sound. US Patent 4,018,121, April 1977.
26. Tomisawa, N. Tone production method for an electronic musical instrument. US Patent 4,249,447, February 1981.
27. McIntyre, M.; Woodhouse, J. On the fundamentals of bowed-string dynamics. *Acta Acust. United Acust.* **1979**, *43*, 93–108.
28. Cremer, L.; Allen, J.S. *The Physics of the Violin*; MIT Press: Cambridge, MA, USA, 1984.
29. Serafin, S. The Sound of Friction: Real-time Models, Playability and Musical Applications. Ph.D. Thesis, Stanford University, Stanford, CA, USA, June 2004.
30. Moog, R.A. A voltage-controlled low-pass high-pass filter for audio signal processing. In Proceedings of the 17th Audio Engineering Society Convention, New York, NY, USA, 11–15 October 1965.
31. D'Angelo, S.; Valimaki, V. An improved virtual analog model of the Moog ladder filter. In Proceedings of the 2013 IEEE International Conference on Acoustics, Speech and Signal Processing, Vancouver, BC, Canada, 26–31 May 2013; pp. 729–733.
32. Huovilainen, A. Nonlinear digital implementation of the Moog ladder filter. In Proceedings of the International Conference on Digital Audio Effects (DAFx), Naples, Italy, 5–8 October 2004; pp. 61–64.
33. Daly, P. A Comparison of Virtual Analogue Moog VCF Models. Master's Thesis, University of Edinburgh, Edinburgh, UK, August 2012.
34. Hélie, T. On the use of Volterra series for real-time simulations of weakly nonlinear analog audio devices: Application to the Moog ladder filter. In Proceedings of the International Conference on Digital Audio Effects (DAFx), Montreal, QC, Canada, 18–20 September 2006; pp. 7–12.
35. Hélie, T. Volterra series and state transformation for real-time simulations of audio circuits including saturations: Application to the Moog ladder filter. *IEEE Trans. Audio Speech Lang. Process.* **2010**, *18*, 747–759.
36. D'Angelo, S.; Valimaki, V. Generalized Moog ladder filter: Part II–Explicit nonlinear model through a novel delay-free loop implementation method. *IEEE/ACM Trans. Audio Speech. Lang. Process.* **2014**, *22*, 1873–1883.

Article

Chord Recognition Based on Temporal Correlation Support Vector Machine

Zhongyang Rao [1,2], Xin Guan [1,*] and Jianfu Teng [1]

[1] School of Electronic Information Engineering, Tianjin University, Tianjin 30072, China;
    yaozhongyang@sohu.com (Z.R.); jfteng@tju.edu.cn (J.T.)
[2] School of Information Science and Electronic Engineering, Shandong Jiaotong University,
    Ji'nan 250357, China
*   Correspondence: guanxin@tju.edu.cn; Tel.: +86-186-6890-9802

Academic Editor: Vesa Valimaki
Received: 11 February 2016; Accepted: 6 May 2016; Published: 19 May 2016

Abstract: In this paper, we propose a method called temporal correlation support vector machine (TCSVM) for automatic major-minor chord recognition in audio music. We first use robust principal component analysis to separate the singing voice from the music to reduce the influence of the singing voice and consider the temporal correlations of the chord features. Using robust principal component analysis, we expect the low-rank component of the spectrogram matrix to contain the musical accompaniment and the sparse component to contain the vocal signals. Then, we extract a new logarithmic pitch class profile (LPCP) feature called enhanced LPCP from the low-rank part. To exploit the temporal correlation among the LPCP features of chords, we propose an improved support vector machine algorithm called TCSVM. We perform this study using the MIREX'09 (Music Information Retrieval Evaluation eXchange) Audio Chord Estimation dataset. Furthermore, we conduct comprehensive experiments using different pitch class profile feature vectors to examine the performance of TCSVM. The results of our method are comparable to the state-of-the-art methods that entered the MIREX in 2013 and 2014 for the MIREX'09 Audio Chord Estimation task dataset.

Keywords: music information retrieval; hidden Markov models; robust principal component analysis; pitch class profile; chord estimation

---

1. Introduction

A musical chord can be defined as a set of notes played simultaneously. A succession of chords over time forms the harmony core in a piece of music. Hence, the compact representation of the overall harmonic content and structure of a song often requires labeling every chord in the song. Chord recognition has been applied in many applications such as the segmentation of pieces into characteristic segments, the selection of similar pieces, and the semantic analysis of music [1,2]. With its many applications, automatic chord recognition has been one of the main fields of interest in musical information retrieval in the last few years.

The basic chord recognition system has two main steps: feature extraction and chord classification. In the first step, the features used in chord recognition are typically variants of the pitch class profile (PCP) introduced by Fujishima (1999) [1]. Many publications have improved PCP features for chord recognition by addressing potentially negative influences such as percussion [2], mistuning [3,4], harmonics [5–7], or timbre dependency [5,8]. In particular, harmonic contents are abundant in both musical instrument sounds and the human singing voice. However, harmonic patterns in instrument sounds are more regular compared with the singing voice, which often includes ornamental features such as vibrato that lead to significant deviations of the frequency of partials from perfectly harmonic [9]. To attenuate the effect of the singing voice and consider the temporal correlations of music, we separate the singing voice from the accompaniment before obtaining the PCP.

Chord classification is computed once the feature has been extracted. Modeling techniques typically use template-fitting methods [1,3,10–13], the hidden Markov model (HMM) [14–20], and dynamic Bayesian networks [21,22] for this recognition process. The template-based method has some advantages, including the fact that it does not require annotated data and has a low computational time. However, its drawbacks include the problem of creating a model of templates of chroma vectors and the selection of a distance measure. The HMM method is a statistical model and its parameter estimation requires substantial training data. The recognition rate of the HMM method is typically relatively high, but it only considers the role of positive training samples without addressing the impact of negative training samples, thereby greatly limiting its discriminative ability. The support vector machine (SVM) method can achieve a high recognition rate, but encounters challenges in the presence of cross-aliasing that cannot be accurately judged. The main difference between HMM and SVM is in the principle of risk minimization [23]. HMM uses empirical risk minimization, which is the simplest induction principle. In contrast, SVM uses structural risk minimization as its induction principle. The difference in risk minimization leads to the better generalization performance of SVM compared with HMM [24]. We present a new method for chord estimation based on a hybrid model of HMM and SVM.

The remainder of this paper is organized as follows: Section 2 reviews related chord estimation work; Section 3 describes our PCP feature vector construction method; Section 4 explains our approach; Section 5 displays the results on the MIREX'09 (Music Information Retrieval Evaluation eXchange) dataset and a provides a comparison with other methods; and Section 6 concludes our work and suggests directions for future work.

2. Related Work

PCP is also called the chroma vector, which is often a 12-dimensional vector whereby each component represents the spectral energy or salience of a semi-tone on the chromatic scale regardless of the octave. The computation of the chroma representation of an audio recording is typically based either on the short-time Fourier transform (STFT) in combination with binning strategies [18,25–27] or on the constant Q transform [3,6,15,28,29]. The succession of these chroma vectors over time is often called the chromagram and this forms a suitable representation of the musical content of a piece.

Many features have been used for chord recognition including non-negative least squares [30], chroma DCT (Discrete Cosine Transform)-reduced log pitch (CRP) [31], loudness-based chromagram (LBC) [22], and Mel PCP (MPCP) [32]. In [1], Fujishima developed a real-time chord recognition system using a 12-dimensional pitch class profile derived from the discrete Fourier transform (DFT) of the audio signal, and performed pattern matching using binary chord-type templates. Lee [6] introduced a new input feature called the enhanced pitch class profile (EPCP) using the harmonic product spectrum. Gómez and Herrera [33] used harmonic pitch class profile (HPCP) as the feature vector, which is based on Fujishima's PCP, and correlated it with a chord or key model adapted from Krumhansl's cognitive study.

Variants of the pitch class profile (PCP) first introduced by Fujishima (1999) [1] address the potentially negative influences of percussion [2], mistuning [3,4], harmonics [5–7] or timbre dependency [5,8]. In addition to these factors, we explore ways to attenuate the influence of the singing voice. Weil introduced an additional pre-processing step for main melody attenuation [34]. To attenuate the negative influence of singing voices, we consider the amplitude similarity of neighborhood musical frames belonging to the same chord and obtain the enhanced PCP. Adding a pre-processing step consisting of robust principal component analysis (RPCA), we expect the low-rank matrix to contain the musical accompaniment and the sparse matrix to contain the vocal signals. Then, the low-rank matrix can be used to calculate the features. The pre-processing considers the temporal correlations of music.

The two most popular chord estimation methods are the template-based model and the hidden Markov model. For the audio chord estimation task of MIREX 2013 and 2014, one of the most popular methods is HMM [35–41]. A binary chord template with three harmonics was also presented [42].

Template-based chord recognition methods use the chord definition to extract chord labels from a musical piece. Neither training data nor extensive music theory knowledge is used for this purpose [43]. To smooth the resulting representation and exploit the temporal correlation, low-pass and median filters are used to filter the chromagram in the time domain [10]. The template-based method only outputs fragmented transcriptions without considering the temporal correlations of music.

The HMM can model sequences of events in a temporal grid considering hidden and visible variables. In the case of chord recognition, the hidden states correspond to the real chords that are being played, while the observations are the chroma vectors. In [35], Cho and Bello use a K-stream HMM which is then decoded using the standard Viterbi algorithm. Khadkevich and Omologo also use a multi-stream HMM, but the feature is a time-frequency reassignment spectrogram [36]. Steenbergen and Burgoyne present a chord estimation method based on the combination of a neural network and an HMM [40]. In [39], the probabilities of the HMM are not trained through expectation-maximization (EM) or any other machine learning technique, but are instead derived from a number of knowledge-based sub-models. Ni and McVicar proposed a harmony progression HMM topology that consists of three hidden and two observed variables [37]. The hidden variables correspond to the key K, the chord C, and the bass annotations B. These methods usually require a reference dataset for the learning period and entail more parameters to be trained.

In contrast, our method has only one parameter trained from the reference dataset, which is the state transition probability, and the other parameters are obtained from the SVM. The hybrid HMM and SVM model uses the respective advantages of these methods. Our novel method is called the temporal correlation support vector machine (TCSVM).

Our system is composed of two main steps: feature extraction and chord classification, as shown in Figure 1. We pre-process the audio to separate the singing voice. The system then tracks beat intervals in the music and extracts a set of vectors for the PCP. In the chord classification step, our method uses SVM classification and the Viterbi algorithm. Because we employ the temporal correlation of chords, the system can combine the SVM with the Viterbi algorithm, leading to a TCSVM. The Viterbi algorithm uses the transitions between chords to estimate the chords.

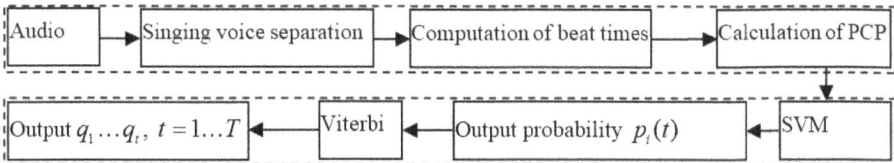

Figure 1. Chord estimation system.

3. Enhanced PCP Feature

3.1. Normalized Logarithmic PCP Feature

Our system begins by extracting suitable feature vectors from the raw audio. Like most chord recognition systems, we use a chromagram or a PCP vector as the feature vector. Müller and Ewert propose a 12-dimensional feature vector–quantized PCP [8,29] that determines the proper frequency resolution and is sufficient for separating musical notes by low-frequency components.

The calculation of PCP feature vectors can be divided into the following steps: (1) using the constant Q transform to calculate the 36-bin chromagram; (2) mapping the spectral chromagram to a particular semitone; (3) median filtering; (4) segmenting the audio signal with a beat-tracking algorithm; (5) reducing the 36-bin chromagram to a 12-bin chromagram based on beat-synchronous segmentation; (6) normalizing the 12-bin chromagram. The reader is referred to [15] for more detailed PCP calculation steps.

In the beat-synchronous (tactus) segmentation, we use the beat-tracking algorithm proposed by Ellis [44]. This method has proven successful for a wide variety of signals. Using beat-synchronous segments has the added advantage that the resulting representation is a function of the beat, or tactus, rather than time.

Unlike most of the traditional PCP methods, we determine the normalized value using the *p*-norm and logarithm. The formula is as follows:

$$QPCP_{\log}(p) = \log_{10}[C \cdot QPCP_{12}(p) + 1] \tag{1}$$

$$QPCP_{norm}(p) = QPCP_{\log}(p) / \| QPCP_{\log} \| \tag{2}$$

After applying the logarithm and normalization, the chromagram is called the LPCP.

The left image of Figure 2 shows a PCP of the **C** major triad and the right image shows its LPCP. The strongest peaks are found at C, E, and G because the **C** major triad comprises three notes at C (root), E (third), and G (fifth). Figure 2 demonstrates that LPCP more clearly approximates the underlying fundamental frequencies than PCP.

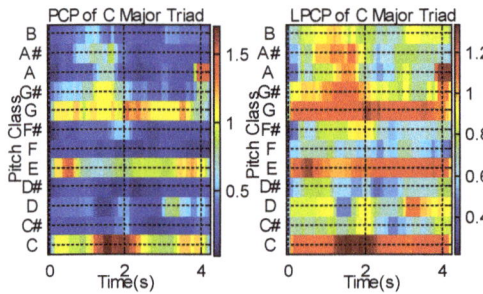

Figure 2. PCP (pitch class profile) (**left**) and LPCP (logarithmic pitch class profile) (**right**) of **C** major triad.

3.2. Enhanced PCP with Singing-Voice Separation

To attenuate the effect of the singing voice and consider the temporal correlations of the chord, we first separate the singing voice from the accompaniment before calculating the PCP or LPCP. This is denoted as enhanced PCP (EPCP) or enhanced logarithmic PCP (ELPCP). The framework of singing voice separation is shown in Figure 3.

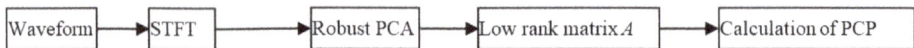

Figure 3. Calculation of enhanced PCP (EPCP).

In general, because of the underlying repeated musical structure, we assume that music is a low-rank signal. Singing voices offer more variation and have a higher rank but are relatively sparse in the frequency domain. We assume that the low-rank matrix *A* represents the music accompaniment and the sparse matrix *E* represents the vocal signals [45]. Then, we perform the separation in two steps. First, we compute the spectrogram of music signals in matrix *D*, which is calculated from the STFT. Second, we use the inexact augmented Lagrange multiplier (ALM) method [45], which is an efficient algorithm for solving the RPCA problem, to solve $A + E = |D|$, given the input magnitude of *D*. Then, using RPCA, we can separate matrices A and E. The low-rank matrix A can be exactly recovered from $D = A + E$ by solving the following convex optimization problem:

$$\text{Minimize} \ ||A||_* + \lambda ||E||_1 \ \text{Subject to} \ A + E = D \tag{3}$$

where λ is a positive weighting parameter. The inexact ALM method is as follows [45]:

Algorithm 1: Inexact ALM Algorithm

Input: matrix D, parameter λ

1: $Y_0{}^* = D/J(D); E_0 = 0; \mu_0 > 0; \rho > 1; k = 0.$
2: **while** not converged **do**
3:    // lines 4–5 solve $A_{k+1} = \underset{A}{\arg\min} L(A, E_k, Y_k, \mu_k).$
4: $(U, S, V) = svd(D - E_k + \mu_k^{-1} Y_k);$
5: $A_{k+1} = U S_{\mu_k^{-1}}[S] V^T.$
6:    // line 7 solves $E_{k+1} = \underset{E}{\arg\min} L(A_{k+1}, E, Y_k, \mu_k).$
7: $E_{k+1} = S_{\lambda\mu_k^{-1}}[D - A_{k+1} + \mu_k^{-1} Y_k].$
8: $Y_{k+1} = Y_k + \mu_k[D - A_{k+1} - E_{k+1}].$
9: $\mu_{k+1} = \rho\mu_k.$
10: $k = k + 1.$
11: **end while**
Output: $(A_k, E_k).$

Figure 4 shows the PCP and EPCP of an audio music piece (the music is "Baby It's You", from the Beatles' album *Please Please Me*). Figure 5 shows the LPCP and ELPCP of the same musical piece.

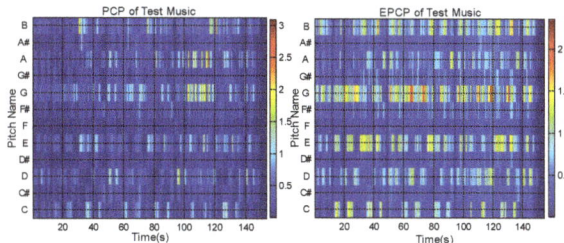

Figure 4. PCP (**left**) and EPCP (**right**) of test music.

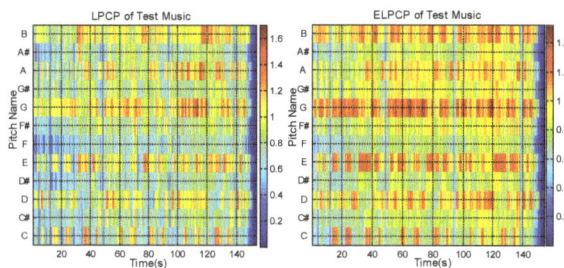

Figure 5. LPCP (**Left**) and ELPCP (**Right**) of test music.

Figure 4 shows that the EPCP has improved continuity compared with the PCP. Figure 5 shows that the ELPCP is further enhanced compared with the LPCP. In Figure 4, it is shown that the EPCP of audio music is more obvious than the PCP. Figure 5 shows that the ELPCP of audio music is clearer than the LPCP. Thus, PCP features for chord recognition are improved by singing voice separation before calculating the PCP or LPCP.

4. Automatic Chord Recognition

Our chord recognition system entails two parts: support vector machine classification and the Viterbi algorithm. For SVM classification, we use LIBSVM (Library for Support Vector Machines) to obtain the chord probability estimates [46]. Then the Viterbi algorithm uses the probability estimates and trained state transition probability to estimate the chord of the music. Because of the temporal correlation of chords, we combine the SVM classification with the Viterbi algorithm and call the system TCSVM (Temporal Correlation Support Vector Machine).

4.1. Support Vector Machine Classification

SVM is a popular machine learning method for classification, regression, and other learning tasks. LIBSVM is currently one of the most widely used SVM software packages. A classification task usually involves a training set where each instance contains the class labels and the features. The goal of SVM is to produce a model to predict the target labels of the test data given only the test data features.

Many methods are available for multi-class SVM classification [47,48]. LIBSVM uses the "one-against-one" approach for multiclass classification. The classification assumes the use of the radial basis function (RBF) kernel of the form $K(x_i, x_j) = e^{-\gamma||x_i - x_j||^2}$. The two parameters of an RBF kernel, C and γ, must be determined by a parameter search as the optimal values vary between tasks. We use the grid-search method to obtain the C and γ parameters.

Once the C and γ parameters are set, the class label and probability information can be predicted. This section discusses the LIBSVM implementation for extending the SVM to output probability estimates. Given K chord classes, for any x, the goal is to estimate $p_i = P(y = i|x)$, $i = 1, ..., K$.

4.2. Viterbi Algorithm in SVM

The SVM method recognizes the chord based on frame-level classification without considering the inter-frame temporal correlation of chord features. For multiple frames corresponding to the same chord, the recognition results of traditional SVM are independent and fluctuate. Accounting for the inter-frame temporal correlation in the recognition procedure can improve the overall chord recognition rate. Our system combines SVM with the Viterbi algorithm to introduce the temporal correlation prior to a chord. Suppose the system has hidden K states, and we denote each state as S_i, $i \in [1 : K]$ where the state refers to the chord type. The observed events are Q_t, $t \in [1 : T]$, which are PCP features. The current observed chord feature is $Q = \{Q_1, Q_2 ..., Q_T\}$, $t \in [1 : T]$. A_{ij} represents the transition probability from chord S_i to chord S_j. At an arbitrary time point t, for each of the states S_i, a partial probability $\delta_t(S_i)$ indicates the probability of the most probable path ending at the state S_i, given the current observed events $Q_1, Q_2 ..., Q_t$: $\delta_t(S_i) = \max_j(\delta_{t-1}(S_j) \cdot A(S_j, S_i) \cdot P(Q_t|S_i))$. Here, we assume that we already know the probability $\delta_{t-1}(S_j)$ for any of the previous states S_j at time $t - 1$. $P(Q_t|S_i)$ is $p_i(t)$, the current probability estimates of SVM. Once we have all of the objective probabilities for each state at each time point, the algorithm seeks from the end to the beginning to find the most probable path of states for the given sequence of observation events $\psi_t(i) = \arg[\max_{1 \leqslant j \leqslant N}(\delta_{t-1}(S_j) \cdot A(S_j, S_i))]$; $\psi_t(i)$ indicates the optimal state at time t based on the probability computed in the first stage.

The Viterbi algorithm is as follows:

Algorithm 2: Viterbi algorithm

---

1 Initialization:

$\delta_t(S_i) = \Pi_i P(Q_1|S_i), \; \psi_t(i) = 0, \; 1 \leqslant i \leqslant K;$

2 Recursion:

$\delta_t(S_i) = \max\limits_{1 \leqslant j \leqslant N} (\delta_{t-1}(S_j) \cdot A(S_j, S_i)) \cdot P(Q_t|S_i), 2 \leqslant t \leqslant T \; \psi_t(i) = \arg[\max\limits_{1 \leqslant j \leqslant N} (\delta_{t-1}(S_j) \cdot A(S_j, S_i))];$

3 Termination:

$q_T^* = \arg\max\limits_{1 \leqslant i \leqslant N} [\delta_t(S_i)], \; P^* = \max\limits_{i} [\delta_t(S_i)];$

4 Path backtracking: $q_t^* = \psi_{t+1}(q_{t+1}^*) \; t = T - 1, T - 2...1.$

In our method, we set the initialization observation probability Π_i to $1/24$. The observed events are PCP features y_t, where y_t is the PCP feature of the tth frame. Generally, SVM predicts only the class label without probability information. The LIBSVM implementation extends SVM to output the probability estimates. The current observation probability corresponds to the probability estimates of SVM and replaces the $P(Q_t|S_i)$ in the Viterbi algorithm. S_i represents the chord $i \in [1 : K]$, where K is the number of chords and is set to 24.

Figure 6 is the comparison of the ground truth chord and estimated chord for the Beatles song "Baby It's You". The top figure shows the result of using the SVM method to recognize the chord and the bottom figure uses TCSVM. The ground truth chord is represented in pink and the estimated chord labels are in blue. Figure 6 indicates that the estimation is more stable when using TCSVM.

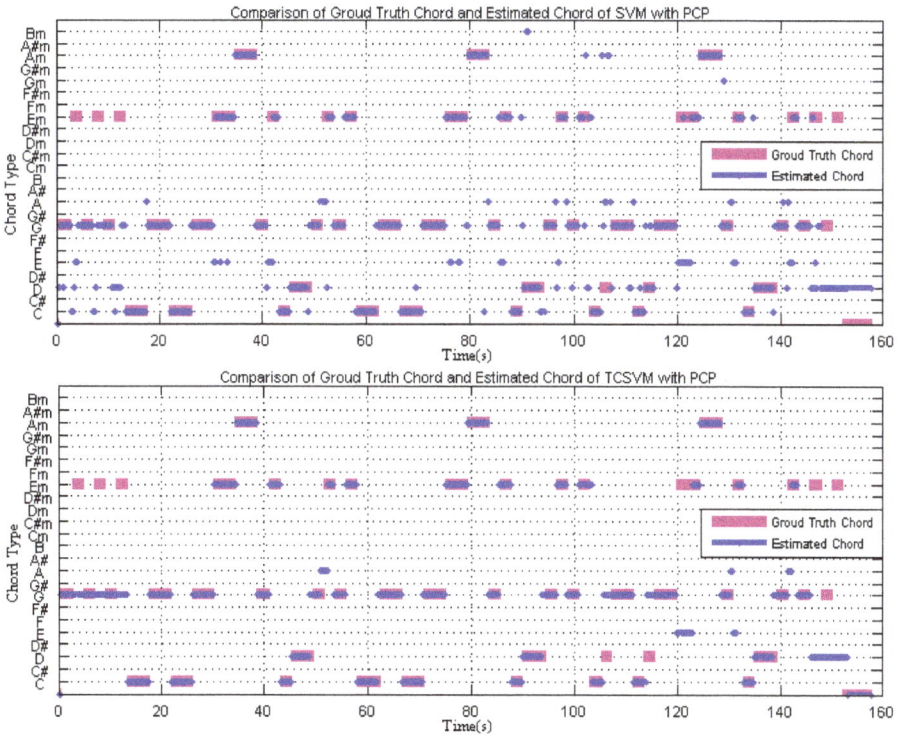

Figure 6. Comparison of ground truth and estimated chords using SVM (**top**) and TCSVM (**bottom**).

5. Experimental Results and Analysis

In this section, we compare the results of chord estimation using different features and methods. We compare our method with the methods that were submitted to MIREX 2013 and MIREX 2014 on the MIREX'09 dataset.

5.1. Corpus and Evaluation Results

For evaluation, we use the MIREX'09 Audio Chord Estimation task dataset which consists of 12 Beatles albums (180 songs, PCM 44 100Hz, 16 bits, mono). Besides the Beatles albums, in 2009, an extra dataset was donated by Matthias Mauch, which consists of 38 songs from Queen and Zweieck [21].

This database has been used extensively for the evaluation of many chord recognition systems, in particular those presented at MIREX 2013 and 2014 for the Audio Chord Estimation task. The evaluation is conducted based on the chord annotations of the Beatles albums provided by Harte and Sandler [49], and the chord annotations of Queen and Zweieck provided by Matthias Mauch [21].

According to [50], chord symbol recall (CSR) is a suitable metric to evaluate chord estimation performance. Since 2013, MIREX has used CSR to estimate how well the predicted chords match the ground truth:

$$CSR = \frac{t_E}{t_A} \qquad (4)$$

where t_E is the total duration of segments where annotation equals estimation, and t_A is the total duration of the annotated segments.

Because pieces of music vary substantially in length, we weight the CSR by the length of the song when computing the average for a given corpus. This final number is referred to as the weighted chord symbol recall. In this paper, the recognition rate and CSR are equivalent for a song and the recognition rate and weighted chord symbol recall are equivalent for a given corpus or dataset.

In the training stage, we randomly selected 25% of the songs from the Beatles albums to determine the parameters C and γ for the SVM kernel and the state transition probability matrix A. For SVM, the training dataset is composed of the PCP features of labeled musical fragments, which are selected from the training songs. The average estimation accuracies or recognition rates are reported.

First, we compare the recognition rates of SVM with PCP and EPCP features. The comparison between the ground truth and estimated chord is shown in Figure 7 for the example song (the Beatles song "Baby It's You"). The top and bottom figures show the results using the SVM method with PCP features and EPCP features, respectively. The ground truth chord is represented in pink and the estimated chord labels are in blue. Figure 7 indicates that EPCP improves the recognition rate. In Figure 8, the recognition rate using SVM with PCP features is 70.15%, while that of TCSVM with the same features is 75.07%. The top image of Figure 6 shows less reliable estimated chords at the times when the chords change. The bottom image considers the inter-frame temporal correlation of chord features and shows more stable estimated chords even when the chords change.

Second, we compare the recognition rates of SVM and TCSVM with different features. The recognition results of the TCSVM method with ELPCP are superior, as shown in Figure 8. Because EPCP and ELPCP consider the temporal correlation of music, the rates show few differences between the SVM and TCSVM.

5.2. Experimental Results Compared with State-of-the-Art Methods

We compare our method with the following methods from MIREX 2013 and MIREX 2014. MIREX 2013:

- CB4 and CB3: Taemin Cho and Juan P. Bello [35]
- KO1and KO2: Maksim Khadkevich and Maurizio Omologo [36]
- NMSD1 and NMSD2: Yizhao Ni, Matt Mcvicar, Raul Santos-Rodriguez [37]

- CF2 : Chris Cannam, Matthias Mauch, Matthew E. P. Davies [38]
- NG1 and NG2: Nikolay Glazyrin [42]
- PP3 and PP4: Johan Pauwels and Geoffroy Peeters [39]
- SB8: Nikolaas Steenbergen and John Ashley Burgoyne [40]

MIREX 2014:

- KO1: Maksim Khadkevich and Maurizio Omologo [51]
- CM3: Chris Cannam, Matthias Mauch [41]
- JR2: Jean-Baptiste Rolland [52]

More details about these methods can be found from the corresponding MIREX websites [53]. The results of this comparison are presented in Figure 9.

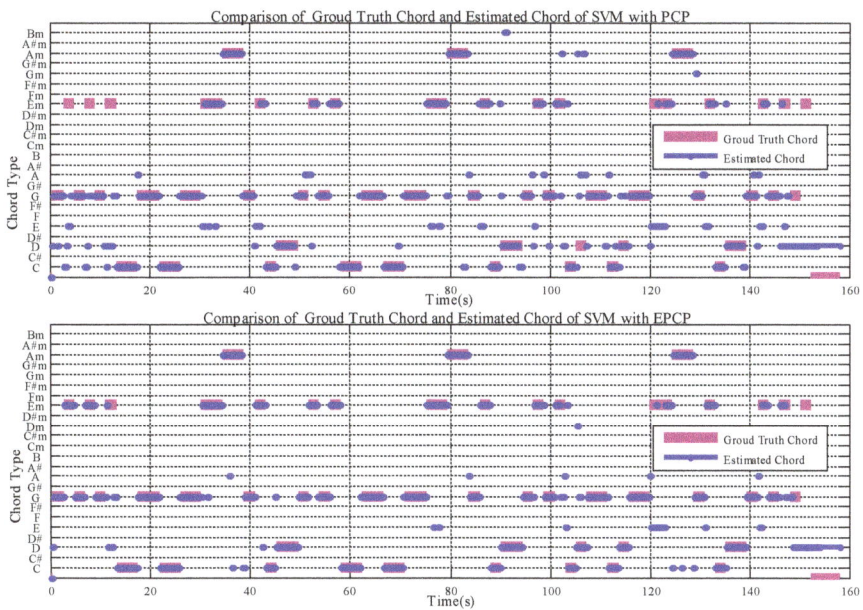

Figure 7. Comparison of ground truth and estimated chord with PCP (**top**) and EPCP (**bottom**).

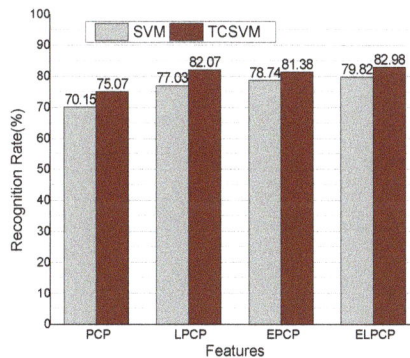

Figure 8. Recognition rates with different features.

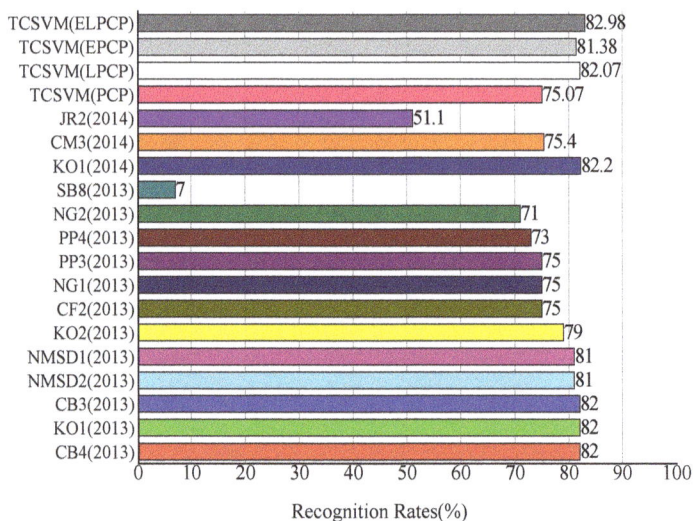

Figure 9. Comparison of recognition rates with state-of-the-art methods on Mirex'09 dataset.

The recognition rate of our TCSVM method with ELPCP is 82.98%. The recognition rate of our TCSVM (ELPCP) approach is similar to the best-scoring method (KO1) in MIREX 2014.

The results of the 2015 edition of MIREX automatic chord estimation tasks can be found on the corresponding websites [54]. Table 1 shows the comparison of the recognition rates of the 2015 edition on the Isophonics 2009 datasets. The chord classes are as follows: Major and minor(MajMin); Seventh chords(Sevenths); Major and minor with inversions(MajMinInv); Seventh chords with inversions(SeventhsInv). The recognition rate of our TCSVM method is higher than other methods except the rates of the MajMinInv chords.

Table 1. Comparison of recognition rates of the 2015 edition on Isophonics 2009 datasets.

Algorithm	MajMin	MajMinInv	Sevenths	SeventhsInv
CM3	54.65	47.73	19.29	16.17
DK4	67.66	64.61	59.56	56.92
DK5	73.51	68.87	63.74	59.72
DK6	75.53	63.56	64.70	54.01
DK7	75.89	70.38	58.37	53.53
DK8	75.89	64.77	66.89	56.94
DK9	76.85	74.47	68.11	66.08
KO1	82.19	79.61	76.04	73.43
TCSVM	**82.98**	79.22	**77.03**	**76.52**

Figure 10 shows the confusion between the chords using SVM and Figure 11 shows the confusion between the chords using TCSVM. The x-axis is the ground truth chord and the y-axis is the estimated chord. Comparing the two images suggests that TCSVM reduces the rate of erroneous identifications for a more reliable result.

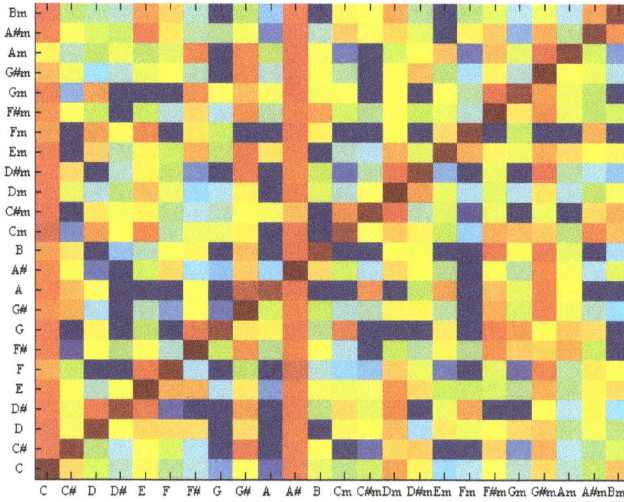

Figure 10. Confusion between chords with SVM.

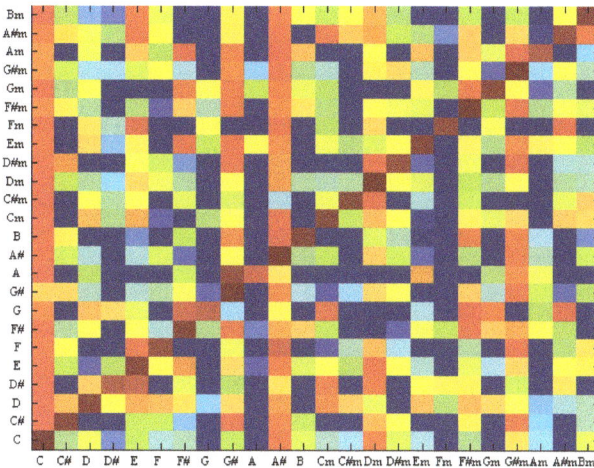

Figure 11. Confusion between chords with TCSVM.

6. Conclusions

We present a new feature called ELPCP and a machine learning model called TCSVM for chord estimation. We separate the singing voice from the accompaniment to improve the features and consider the temporal correlation of music. Temporal correlation SVM is used to estimate the chord. This system results in more accurate chord recognition and eliminates many spurious chord estimates appearing in the conventional recognition procedure.

Future work should address some limitations. First, this paper only involves common chord estimation as part of the audio chord estimation task. Future work will involve the recognition of more complex chords to increase the applicability of this work in the field of music information retrieval, including song identification, query by similarity, and structure analysis. Second, we consider the effect of the singing voice and the results of Figure 8 show that the recognition rate with singing voice separation is better than without it. There is more room for further improvement of the PCP features to make them more suitable for chord recognition. Finally, we will evaluate the TCSVM method for cases when the audio music contains noise or is corrupted by noise.

Acknowledgments: This work was supported by the national Natural Science Foundation of China (Grant No. 61101225).

Author Contributions: Zhongyang Rao and Xin Guan conceived and designed the experiments; Zhongyang Rao performed the experiments; Zhongyang Rao and Xin Guan analyzed the data; Jianfu Teng contributed materials and analysis tools; Zhongyang Rao and Xin Guan wrote the paper.

Conflicts of Interest: The authors declare that there is no conflict of interests regarding the publication of this paper.

References

1. Fujishima, T. Realtime Chord Recognition of Musical Sound: A System Using Common Lisp Music. In Proceedings of the International Computer Music Conference, Beijing, China, 22–27 October 1999; pp. 464–467.
2. Ueda, Y.; Uchiyama, Y.; Nishimoto, T.; Ono, N.; Sagayama, S. HMM-based approach for automatic chord detection using refined acoustic features. In Proceedings of the IEEE International Conference on Acoustics Speech and Signal Processing (ICASSP 2010), Dallas, TX, USA, 14–19 March 2010; pp. 5518–5521.
3. Harte, C.; Sandler, M. Automatic Chord Identifcation Using a Quantised Chromagram. In Proceedings of the Audio Engineering Society Convention 118, Barcelona, Spain, 28–31 May 2005.
4. Degani, A.; Dalai, M.; Leonardi, R.; Migliorati, P. Real-time Performance Comparison of Tuning Frequency Estimation Algorithms. In Proceedings of the 2013 8th International Symposium on Image and Signal Processing and Analysis (ISPA), Trieste, Italy, 4–6 September 2013; pp. 393–398.
5. Morman, J.; Rabiner, L. A system for the automatic segmentation and classification of chord sequences. In Proceedings of the 1st ACM Workshop on Audio and Music Computing Multimedia, Santa Barbara, CA, USA, 23–27 October 2006; pp. 1–10.
6. Lee, K. Automatic Chord Recognition from Audio Using Enhanced Pitch Class Profile. In Proceedings of the International Computer Music Conference, New Orleans, LA, USA, 6–11 November 2006.
7. Varewyck, M.; Pauwels, J.; Martens, J.-P. A novel chroma representation of polyphonic music based on multiple pitch tracking techniques. In Proceedings of the 16th ACM International Conference on Multimedia, Vancouver, BC, Canada, 26–31 October 2008; pp. 667–670.
8. Müller, M.; Ewert, S. Towards timbre-invariant audio features for harmony-based music. *IEEE Trans. Audio Speech Lang. Process.* **2010**, *18*, 649–662.
9. Nwe, T.L.; Shenoy, A.; Wang, Y. Singing voice detection in popular music. In Proceedings of the 12th Annual ACM International Conference on Multimedia, New York, NY, USA, 10–16 October 2004; pp. 324–327.
10. Oudre, L.; Grenier, Y.; Févotte, C. Template-based Chord Recognition: Influence of the Chord Types. In Proceedings of the International Society for Music Information Retrieval Conference, Kobe, Japan, 26–30 October 2009; pp. 153–158.
11. Rocher, T.; Robine, M.; Hanna, P.; Oudre, L.; Grenier, Y.; Févotte, C. Concurrent Estimation of Chords and Keys from Audio. In Proceedings of the International Society for Music Information Retrieval Conference, Utrecht, The Netherlands, 9–13 August 2010; pp. 141–146.
12. Cho, T.; Bello, J.P. A Feature Smoothing Method for Chord Recognition Using Rrecurrence Plots. In Proceedings of the Music Information Retrieval Evaluation eXchange (MIREX 2011), Miami, FL, USA, 24–28 October 2011.
13. Oudre, L.; Févotte, C.; Grenier, Y. Probabilistic template-based chord recognition. *IEEE Trans. Audio Speech Lang. Process.* **2011**, *19*, 2249–2259. [CrossRef]
14. Papadopoulos, H.; Peeters, G. Large-scale Study of Chord Estimation Algorithms Based on Chroma Representation and HMM. In Proceedings of the International Workshop on Content-Based Multimedia Indexing (CBMI'07), 25–27 June 2007; pp. 53–60.
15. Bello, J.P.; Pickens, J. A Robust Mid-Level Representation for Harmonic Content in Music Signals. In Proceedings of the International Society for Music Information Retrieval Conference, London, UK, 11–15 September 2005; pp. 304–311.

16. Lee, K.; Slaney, M. Acoustic chord transcription and key extraction from audio using key-dependent HMMs trained on synthesized audio. *IEEE Trans. Audio Speech Lang. Process.* **2008**, *16*, 291–301. [CrossRef]

17. Papadopoulos, H.; Peeters, G. Simultaneous Estimation of Chord Progression and Downbeats from an Audio File. In Proceedings of the IEEE International Conference on Acoustics, Speech and Signal Processing (ICASSP 2008), Bordeaux, France, 25–27 June 2008; pp. 121–124.

18. Sheh, A.; Ellis, D.P. Chord Segmentation and Recognition Using EM-Trained Hidden Markov Models. In Proceedings of the International Society for Music Information Retrieval Conference (ISMIR 2003), Maryland, MD, USA, 27-30 October 2003; pp. 185–191.

19. Scholz, R.; Vincent, E.; Bimbot, F. Robust Modeling of Musical Chord Sequences Using Probabilistic *N*-grams. In Proceedings of the IEEE International Conference on Acoustics, Speech and Signal Processing (ICASSP 2009), Taipei, Taiwan, 19–24 April 2009; pp. 53–56.

20. Yoshii, K.; Goto, M. A Vocabulary-Free Infinity-Gram Model for Nonparametric Bayesian Chord Progression Analysis. In Proceedings of the International Society for Music Information Retrieval Conference, Miami, FL, USA, 24–28 October 2011; pp. 645–650.

21. Mauch, M. Automatic Chord Transcription from Audio Using Computational Models of Musical Context. Ph.D. Thesis, University of London, London, UK, 23 March 2010.

22. Ni, Y.; McVicar, M.; Santos-Rodriguez, R.; de Bie, T. An end-to-end machine learning system for harmonic analysis of music. *IEEE Trans. Audio Speech Lang. Process.* **2012**, *20*, 1771–1783. [CrossRef]

23. Vapnik, V.N. An overview of statistical learning theory. *IEEE Trans. Neural Netw.* **1999**, *10*, 988–999. [CrossRef] [PubMed]

24. Miao, Q.; Huang, H.Z.; Fan, X. A comparison study of support vector machines and hidden Markov models in machinery condition monitoring. *J. Mech. Sci. Technol.* **2007**, *21*, 607–615. [CrossRef]

25. Bartsch, M.A.; Wakefield, G.H. Audio thumbnailing of popular music using chroma-based representations. *IEEE Trans. Multimed.* **2005**, *7*, 96–104. [CrossRef]

26. Gómez, E. Tonal description of polyphonic audio for music content processing. *Inf. J. Comput.* **2006**, *18*, 294–304. [CrossRef]

27. Khadkevich, M.; Omologo, M. Use of Hidden Markov Models and Factored Language Models for Automatic Chord Recognition. In Proceedings of the International Society for Music Information Retrieval Conference, Kobe, Japan, 26–30 October 2009; pp. 561–566.

28. Brown, J.C. Calculation of a Constant *Q* spectral Transform. *J. Acoust. Soc. Am.* **1991**, *89*, 425–434. [CrossRef]

29. Müller, M.; Ewert, S. Chroma Toolbox: MATLAB Implementations for Extracting Variants of Chroma-based Audio Features. In Proceedings of the 12th International Conference on Music Information Retrieval (ISMIR 2011), Miami, FL, USA, 24–28 October 2011.

30. Mauch, M.; Dixon, S. Approximate Note Transcription for the Improved Identification of Difficult Chords. In Proceedings of the International Society for Music Information Retrieval Conference (ISMIR 2010), Utrecht, The Netherlands, 9–13 August 2010; pp. 135–140.

31. Müller, M.; Ewert, S.; Kreuzer, S. Making chroma features more robust to timbre changes. In Proceedings of the International Conference on Acoustics, Speech and Signal Processing (ICASSP 2009), Taipei, Taiwan, 19–24 April 2009; pp. 1877–1880.

32. Wang, F.; Zhang, X. Research on CRFs in Music Chord Recognition Algorithm. *J. Comput.* **2013**, *8*, 1017. [CrossRef]

33. Gómez, E.; Herrera, P.; Ong, B. Automatic Tonal Analysis from Music Summaries for Version Identification. In Proceedings of the Audio Engineering Society Convention 121, San Francisco, CA, USA, 5–8 October 2006.

34. Weil, J.; Durrieu, J.-L. An HMM-based Audio Chord Detection System: Attenuating the Main Melody. In Proceedings of the Music Information Retrieval Evaluation eXchange (MIREX), Philadelphia, PA, USA, 14–18 September 2008.

35. Cho, T.; Bello, J.P. MIREX 2013: Large Vocabulary Chord Recognition System Using Multi-band Features and a Multi-stream HMM. In Proceedings of the Music Information Retrieval Evaluation eXchange (MIREX), Curitiba, Brazil, 4–8 November 2013.

36. Khadkevich, M.; Omologo, M. Time-frequency Reassigned Features for Automatic Chord Recognition. In Proceedings of the IEEE International Conference on Acoustics, Speech and Signal Processing (ICASSP 2011), Prague, Czech Republic, 22–27 may 2011; pp. 181–184.

37. Ni, Y.; McVicar, M.; Santos-Rodriguez, R.; de Bie, T. Harmony Progression Analyzer for MIREX 2013. In Proceedings of the Music Information Retrieval Evaluation eXchange (MIREX), Curitiba, Brazil, 4–8 November 2013.

38. Cannam, C.; Benetos, E.; Mauch, M.; Davies, M.E.P.; Dixon, S.; Landone, C.; Noland, K.; Stowell, D. MIREX 2015: Vamp Plugins from the Centre for Digital Music. In Proceedings of the Music Information Retrieval Evaluation eXchange (MIREX), Malaga, Spain, 26–30 October 2015.

39. Pauwels, J.; Peeters, G. The Ircamkeychord Submission for MIREX 2013. In Proceedings of the Music Information Retrieval Evaluation eXchange (MIREX), Curitiba, Brazil, 4–8 November 2013.

40. Steenbergen, N.; Burgoyne, J.A. MIREX 2013: Joint Optimization of an Hidden Markov Model-neural Network Hybrid Chord Estimation. In Proceedings of the Music Information Retrieval Evaluation eXchange (MIREX), Curitiba, Brazil, 4–8 November 2013.

41. Cannam, C.; Benetos, E.; Mauch, M.; Davies, M.E.; Dixon, S.; Landone, C.; Noland, K.; Stowell, D. MIREX 2014: Vamp Plugins from the Centre for Digital Music. In Proceedings of the Music Information Retrieval Evaluation eXchange (MIREX), Taipei, Taiwan, 27–31 October 2014.

42. Glazyrin, N. Audio Chord Estimation Using Chroma Reduced Spectrogram and Self-similarity. In Proceedings of the Music Information Retrieval Evaluation eXchange (MIREX), Curitiba, Brazil, 4–8 November 2013.

43. Oudre, L. Template-Based Chord Recognition from Audio Signals. Ph.D. Thesis, TELECOM ParisTech, Paris, France, 3 November 2010.

44. Ellis, D.P. Beat tracking by dynamic programming. *J. New Music Res.* **2007**, *36*, 51–60. [CrossRef]

45. The Augmented Lagrange Multiplier Method for Exact Recovery of Corrupted Low-Rank Matrices. Available online: http://arxiv.org/abs/1009.5055 (accessed on 18 October 2013).

46. Chang, C.C.; Lin, C.J. LIBSVM: A library for support vector machines. *ACM Trans. Intell. Syst. Technol.* **2011**, *2*, 389–396. [CrossRef]

47. Guo, H.; Wang, W. An active learning-based SVM multi-class classification model. *Pattern Recognit.* **2015**, *48*, 1577–1597. [CrossRef]

48. Tomar, D.; Agarwal, S. A comparison on multi-class classification methods based on least squares twin support vector machine. *Knowl.-Based Syst.* **2015**, *81*, 131–147. [CrossRef]

49. Harte, C.; Sandler, M.B.; Abdallah, S.A.; Gómez, E. Symbolic Representation of Musical Chords: A Proposed Syntax for Text Annotations. In Proceedings of the International Society for Music Information Retrieval Conference, London, UK, 11–15 September 2005; pp. 66–71.

50. Pauwels, J.; Peeters, G. Evaluating Automatically Estimated Chord Sequences. In Proceedings of the IEEE International Conference on Acoustics, Speech and Signal Processing (ICASSP 2013), Vancouver, BC, USA, 26–31 May 2013; pp. 749–753.

51. Khadkevich, M.; Omologo, M. Time-frequency Reassigned Features for Automatic Chord Recognition. In Proceedings of the Music Information Retrieval Evaluation eXchange (MIREX), Taipei, Taiwan, 27–31 October 2014.

52. Rolland, J.-B. Chord Detection Using Chromagram Optimized by Extracting Additional Features. In Proceedings of the Music Information Retrieval Evaluation eXchange (MIREX), Taipei, Taiwan, 27–31 October 2014.

53. MIREX HOME. Available online: http://www.music-ir.org/mirex/wiki/MIREX_HOME (accessed on 10 November 2015).

54. 2015:Audio Chord Estimation Results. Available online: http://www.music-ir.org/mirex/wiki/2015: Audio_Chord_Estimation_Results#Isophonics_2009 (accessed on 2 December 2015).

applied
sciences

MDPI

Article
Metrics for Polyphonic Sound Event Detection

Annamaria Mesaros *, Toni Heittola and Tuomas Virtanen

Department of Signal Processing, Tampere University of Technology, P.O. Box 553,
Tampere FI-33101, Finland; toni.heittola@tut.fi (T.H.); tuomas.virtanen@tut.fi (T.V.)
* Correspondence: annamaria.mesaros@tut.fi; Tel.: +358-50-300-5104

Academic Editor: Vesa Valimaki
Received: 26 February 2016; Accepted: 18 May 2016; Published: 25 May 2016

Abstract: This paper presents and discusses various metrics proposed for evaluation of polyphonic sound event detection systems used in realistic situations where there are typically multiple sound sources active simultaneously. The system output in this case contains overlapping events, marked as multiple sounds detected as being active at the same time. The polyphonic system output requires a suitable procedure for evaluation against a reference. Metrics from neighboring fields such as speech recognition and speaker diarization can be used, but they need to be partially redefined to deal with the overlapping events. We present a review of the most common metrics in the field and the way they are adapted and interpreted in the polyphonic case. We discuss segment-based and event-based definitions of each metric and explain the consequences of instance-based and class-based averaging using a case study. In parallel, we provide a toolbox containing implementations of presented metrics.

Keywords: pattern recognition; audio signal processing; audio content analysis; computational auditory scene analysis; sound events; everyday sounds; polyphonic sound event detection; evaluation of sound event detection

1. Introduction

Sound event detection is a rapidly developing research field that deals with the complex problem of analyzing and recognizing sounds in general everyday audio environments. It has many applications in surveillance for security, healthcare and wildlife monitoring [1–7], and audio and video content based indexing and retrieval [8–10].

The task of sound event detection involves locating and classifying sounds in audio—estimating onset and offset for distinct sound event instances and providing a textual descriptor for each. In general, sound event detection deals with the problem of detecting multiple sounds in an audio example, in contrast to the typical classification problem that assigns an audio example to one or more classes. Complexity of sound event detection tasks varies, with the simplest one being detection of a sequence of temporally separated sounds [11–14]. A more complex situation deals with detecting sound events in audio with multiple overlapping sounds, as is usually the case in our everyday environment. In this case, it is possible to perform detection of the most prominent sound event from the number of concurrent sounds at each time [15], or detection of multiple overlapping sound events [16–18]. We use the term *polyphonic sound event detection* for the latter, in contrast to *monophonic sound event detection* in which the system output is a sequence of non-overlapping sound events.

Quantitative evaluation of the detection accuracy of automatic sound event analysis systems is done by comparing the system output with a reference available for the test data. The reference can be obtained for example by manually annotating audio data [19] or by creating synthetic data and corresponding annotations using isolated sounds [20]. The most common format for the annotations is a list of sound event instances with associated onset and offset. The set of unique event labels form the event classes relevant to the sound event detection task, and evaluation of results takes into account both class name and temporal information. We use the term *polyphonic annotation* to mark annotations containing sounds that overlap in time.

There is no universally accepted metric for evaluating polyphonic sound event detection performance. For a monophonic annotation and monophonic output of sound event detection system, the system output at a given time is either correct or incorrect if the predicted event class coincides or not with the reference class. In polyphonic sound event detection, the reference at a given time is not a single class, and there can be multiple correctly detected and multiple erroneously detected events at the same time, which must be individually counted. A similar situation is encountered in polyphonic music transcription [21], where at a given time there can be overlapping notes played by different instruments.

In this paper, we review the use of metrics for measuring performance of polyphonic sound event detection, based on adapting metrics from neighboring fields to cope with presence of multiple classes at the same time. Special sessions at recent conferences and the AASP Challenge on Detection and Classification of Acoustic Scenes and Events 2013 [20] and 2016 [22] underline the acute need for defining suitable evaluation metrics. We introduce and analyze the most commonly used metrics and discuss the way they are adapted and interpreted in the case of polyphonic sound event detection. We also discuss the different possibilities for calculating each metric, and their advantages and disadvantages.

This paper is organized as follows: Section 2 presents a background for sound event detection and classification, and the components of a sound event detection system. Section 3 presents definitions of intermediate statistics that are used for computing all metrics, and Section 4 reviews the metrics. Section 5 discusses the choice of metrics when evaluating a sound event detection system and comparing performance of different systems, using a case study example. Finally, Section 6 presents conclusions and future work.

2. Background

2.1. Classification and Detection of Sound Events

Sound event classification in its simplest form requires assigning an event class label to each test audio, as illustrated in Figure 1A. Classification is performed on audio containing isolated sound events [10,13,14] or containing a target sound event and additional overlapping sounds [23]. In classification, the system output is a class label, and there is no information provided about the temporal boundaries of the sounds. Audio containing multiple, possibly overlapping sounds can be classified into multiple classes—performing audio *tagging*, illustrated in Figure 1B. Tagging of audio with sound event labels is used for example for improving the tags of Freesound audio samples [24], and has been proposed as an approach for audio surveillance of home environments [25]. Single-label classification is equivalent with tagging, when a single tag is assigned per test file.

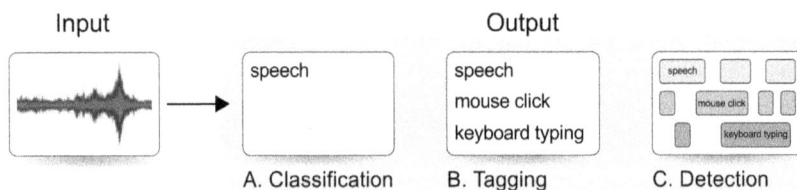

Figure 1. Illustration of sound event classification, tagging and detection.

Sound event detection requires detection of onsets and offsets in addition to the assignment of class labels, and it usually involves detection of multiple sound events in a test audio. This results in assigning sound event labels to selected segments of the audio as illustrated in Figure 1C. For overlapping sounds these segments overlap, creating a multi-level segmentation of the audio based on the number and temporal location of recognized events. Sound event detection of the most prominent event at each time provides a monophonic output, which is a simplified representation of the polyphonic output. These cases are presented in Figure 2, together with the polyphonic annotation of the audio. Polyphonic sound event detection can be seen as a frame by frame multi-class and multi-label classification of the test audio. In this respect, polyphonic sound event detection is similar to polyphonic music transcription, with sound events equivalent to musical notes, and the polyphonic annotation similar to the piano roll representation of music.

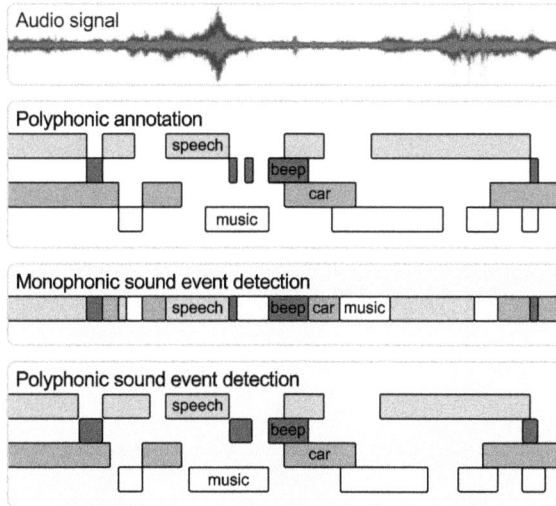

Figure 2. Illustration of the output of monophonic and polyphonic sound event detection systems, compared to the polyphonic annotation.

2.2. Building a Polyphonic Sound Event Detection System

In a multisource environment such as our everyday acoustic environment, multiple different sound sources can be active at the same time. For such data, the annotation contains overlapping event instances as illustrated in Figure 2, with each event instance having an associated onset, offset and label. The label is a textual description of the sound, such as "speech", "beep", "music". Sound event detection is treated as a supervised learning problem, with the event classes being defined in advance, and all the sound instances used in training belong to one of these event classes. The aim of sound event detection is to provide a description of the acoustic input that is as close as possible to the reference. In this case, the requirement is for the sound event detection system to output information consisting of detected sound event instances, having an associated onset, offset, and a textual label that belongs to one of the learned event classes.

The stages of a sound event detection system are illustrated in Figure 3. The training chain involves processing of audio and annotations, and the sound event detection system training. Acoustic features are extracted from audio, and the training stage finds the mapping between acoustic features and sound event activities given by the annotations. The testing chain involves processing of test audio in the same way as for training, the testing of the system, and, if needed, postprocessing of the system output for obtaining a representation similar to the annotations.

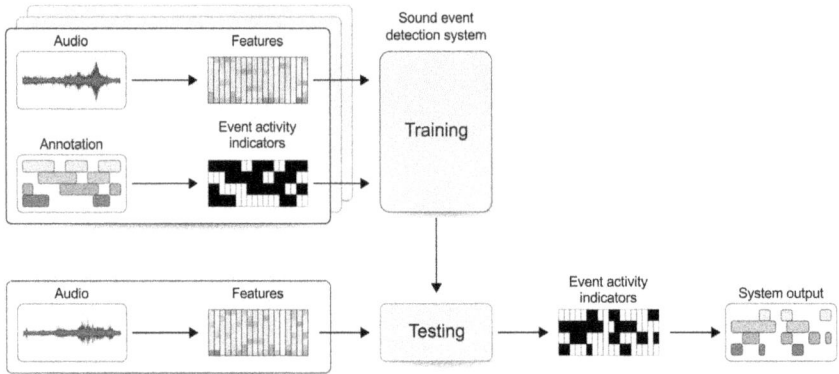

Figure 3. Sound event detection system overview.

The audio is processed in short frames of typically 20–200 ms to extract the audio features of choice. Often used in sound event detection are representations of the signal spectrum, such as mel-frequency cepstral coefficients [25,26], mel energies [17,27], or simply the amplitude or power spectrum [18,28]. The audio processing chain may include simple preprocessing of audio such as normalization and pre-emphasis before feature extraction, or more complex preprocessing such as sound source separation and acoustic stream selection for reducing the complexity of audio mixtures used in training [16]. The annotations are processed to obtain a representation suitable for the training method. In the illustrated example, the annotations are processed to obtain a binary activity representation that provides the class activity information for each frame in the system training.

Training uses the obtained audio features and the corresponding target output given by the reference for supervised learning. Possible learning approaches for this step include Gaussian mixture models [25], hidden Markov models [26], non-negative dictionaries [17], deep neural networks [18], etc. For testing, an audio recording goes through the same preprocessing and feature extraction process as applied in the training stage. Afterwards, the trained system is used to map the audio features to event class likelihoods or direct decisions, according to the employed method. A further step for postprocessing the system output may be needed for smoothing the output and obtaining a binary activity representation for the estimated event classes. Smoothing methods used include median filtering [27], use of a set length decision making window, majority voting, etc., while binarization is usually obtained by using a threshold [17,27].

2.3. Evaluation

Evaluation is done by comparing the system output with a reference available for the test data. Systems performing sound event classification are usually evaluated in terms of accuracy [2,4,5,11,13,14]. Studies involving both monophonic and polyphonic sound event detection report results using a variety of metrics, for example Precision, Recall and F-score [6] or only F-score [7,26], recognition rate and false positive rate [3], or false positive and false negative rates [1]. One of the first evaluation campaigns for sound event detection (CLEAR 2007) used the acoustic event error rate as the evaluation metric, expressed as time percentage [29]. In this case, however, the system output was expected to be monophonic, while the ground truth was polyphonic. Later, acoustic event error rate was redefined for frame-based calculation, and used for example in DCASE 2013 [20], while a similarly defined error rate was used as a secondary metric in MIREX Multiple Fundamental Frequency Estimation and Tracking task [30].

These metrics are well established in different research areas, but the temporal overlap in polyphonic sound event detection leads to changes in their calculation or interpretation. Simple metrics that count numbers of correct predictions, used in classification and information retrieval, must be defined to consider multiple classes at the same time. The evaluation of systems from neighboring fields of speech recognition and diarization dynamically align the system output with the ground truth, and evaluate the degree of misalignment between them. Polyphonic annotation and system output cannot be aligned in a unique way, therefore the error rate defined based on this misalignment must be adapted to the situation. In a similar way, evaluation of polyphonic music transcription uses metrics with modified definitions to account for overlapping notes played by different instruments.

Obtaining the reference necessary for training and evaluation of sound event detection systems is not a trivial task. One way to obtain annotated data is to create synthetic mixtures using isolated sound events—possibly allowing control of signal-to-noise ratio and amount of overlapping sounds [31]. This method has the advantage of being efficient and providing a detailed and exact reference, close to a true ground truth. However, synthetic mixtures cannot model the variability encountered in real life, where there is no control over the number and type of sound sources and their degree of overlapping.

Real-life audio data is easy to collect, but very time consuming to annotate. Currently, there are only few, rather small, public datasets consisting of real-world recordings with polyphonic annotations. DARES data [19] is one such example, but ill-suited for sound event detection due to the high amount of classes compared to amount of examples (around 3200 sound event instances belonging to over 700 classes). CLEAR evaluation data [29] is commercially available and contains audio recorded in controlled conditions in a meeting environment. TUT Sound Events [32] has recently been published for DCASE 2016 and contains sound events annotated using freely chosen labels. Nouns were used to characterize each sound source, and verbs to characterize the sound production mechanism, whenever this was possible, while the onset and offset locations were marked to match the perceived temporal location of the sounds. As a consequence, the obtained manual annotations are highly subjective.

For this type of data, no annotator agreement studies are available. One recent study on inter-annotator agreement is presented in [25], for tagging of audio recorded in a home environment. Their annotation approach associated multiple labels to each 4-s segment from the audio recordings, based on a set of 7 labels associated with sound sources present. With three annotators, they obtained three sets of multi-label annotations per segment. The work does not address subjectivity of temporally delimiting the labeled sounds. The authors observed strong inter-annotator agreement about labels "child speech", "male speech", "female speech" and "video game/TV", but relatively low agreement about "percussive sounds", "broadband noise", "other identifiable sounds" and "silence/background". The results suggest that annotators have difficulty assigning labels to ambiguous sound sources. Considering the more general task of sound event detection with a large number of classes, there is no sufficient data generated by multiple annotators to facilitate inter-annotator agreement assessment. For the purpose of this study, we consider the subjective manual annotation or automatically generated synthetic annotation as correct, and use it as a reference to evaluate the system performance.

Evaluation of the system output can be done at different stages illustrated in the example in Figure 3. It is possible to compare the event activity matrix obtained after preprocessing the annotation and the system output in the same form. If the system output is further transformed into separate event instances as in the annotations, the comparison can be performed at the level of individual events.

3. Intermediate Statistics and Averaging Options

The comparison between the system output and reference can be done in fixed length intervals or at event-instance level, as explained in the previous section. This results in two different ways of measuring performance: *segment-based metrics* and *event-based metrics*. For each, we need to define what constitutes correct detection and what type of errors the system produces. We refer to these as *intermediate statistics* that count separately the correct and erroneous outputs of the system, and define them based on the polyphonic nature of the problem.

3.1. Segment-Based Metrics

Segment-based metrics compare system output and reference in short time segments. Active/inactive state for each event class is determined in a fixed length interval that represents a segment. Based on the activity representation, the following intermediate statistics are defined:

- true positive: the reference and system output both indicate an event to be active in that segment;
- false positive: the reference indicates an event to be inactive in that segment, but the system output indicates it as active;
- false negative: the reference indicates an event to be active in that segment, but the system output indicates it as inactive.

Some metrics also count true negatives: when the reference and system output both indicate an event to be inactive. Total counts of true positives, false positives, false negatives and true negatives are denoted as *TP, FP, FN,* and *TN,* respectively.

The size of the time interval can be chosen based on the desired resolution needed for the application. For example, in [31] a segment of 100 ms was proposed, and the metric was referred to as frame-based. In [26] a length of one second was suggested, for calculating the metrics on longer time segments than the typical frame length used in analysis. By using a longer segment for evaluation, the activity indicator covers a longer chunk of audio, allowing a degree of misalignment between the reference and the system output. This alleviates issues related to annotator subjectivity in marking onset and offset of sound events.

3.2. Event-Based Metrics

Event-based metrics compare system output and corresponding reference event by event. The intermediate statistics are defined as follows:

- true positive: an event in the system output that has a temporal position overlapping with the temporal position of an event with the same label in the reference. A collar is usually allowed for the onset and offset, or a tolerance with respect to the reference event duration.
- false positive: an event in the system output that has no correspondence to an event with same label in the reference within the allowed tolerance;
- false negative: an event in the reference that has no correspondence to an event with same label in the system output within the allowed tolerance.

Event-based metrics have no meaningful true negatives, except in the case when measuring actual temporal errors in terms of length, in which case the total length of time segments where both system output and reference contain no active events is measured. The tolerance can be chosen depending on the desired resolution. For example in speaker diarization it was estimated that a ±250 ms collar is sufficient to account for inexact labeling of the data. In DCASE 2103 [31], the collar value was ±100 ms. The tolerance allowed for offset was the same ±100 ms collar or 50% of the reference event duration. The offset condition covers differences between very short and very long sound events, allowing long sound events to be considered correctly detected even if their offset is not within ±100 ms from the corresponding reference event.

3.3. Averaging Options in Calculating Metrics

Intermediate statistics can be aggregated in different ways: globally-irrespective of classes, or in two steps: first class-wise, then as an average over the results of the individual classes [33]. Highly unbalanced classes or individual class performance can result in very different overall performance when calculated with the two averaging methods.

Instance-based averaging or micro-averaging gives equal weight to each individual decision. In this case, each sound event instance (in event-based metrics) or active instance per segment (in segment-based metrics) has equal influence on the system performance. The number of true positives (TP), the number of false positives (FP) and the number of false negatives (FN) are aggregated over the entire test data, and the metrics are calculated based on the overall values. This results in performance values that are mostly affected by the performance on the larger classes in the considered problem.

Class-based averaging or macro-averaging gives equal weight to each class. Performance of the system on individual classes has equal influence on the overall system performance—in this case, the intermediate statistics are aggregated separately for each event class over the test data. Aggregated class-wise counts (TP, FP, FN) are used to calculate class-wise metrics. The overall performance is then calculated as the average of class-wise performance. This results in values that emphasize the system behavior on the smaller classes in the considered problem.

3.4. Cross-Validation

An important aspect of comparison and reproducibility of results is the experimental and cross-validation setup. Besides the measurement bias under high class imbalance, in multi-class problems it is not always possible to ensure that all classes are represented in every fold [34]. Missing classes will result in division by zero in calculations of some metrics. To avoid such situations, the recommendation is to count the total number of true positives and false positives over the folds, then compute the metrics [35]. This also avoids differences in the overall average values that arise from averaging over different subsets of the tested categories—such as differently splitting the data into train/test folds. In this respect, the cross-validation folds should be treated as a single experiment, with calculation of final metrics only after testing all folds. This ensures that the metrics are consistently calculated over the same data.

4. Metrics for Polyphonic Evaluation

4.1. Precision, Recall and F-Score

Precision (P) and Recall (R) were introduced in [36] for information retrieval purposes, together with the F-score derived as a measure of *effectiveness* of retrieval. They are defined as:

$$P = \frac{TP}{TP + FP}, \quad R = \frac{TP}{TP + FN} \tag{1}$$

Precision and Recall are the preferred metrics in information retrieval, but are used also in classification under the names *positive prediction value* and *sensitivity*, respectively. Adapting these metrics to polyphonic data is straightforward based on the definitions of the intermediate statistics. F-score is calculated based on P and R:

$$F = \frac{2 \cdot P \cdot R}{P + R} \tag{2}$$

or, alternatively, based on the intermediate measures:

$$F = \frac{2 \cdot TP}{2 \cdot TP + FP + FN} \tag{3}$$

Segment-based P, R and F are calculated based on the segment-based intermediate statistics, using instance-based averaging or class-based averaging. The calculation is illustrated in Figure 4, panel A. Event-based P, R and F are calculated the same way from the event-based intermediate statistics. The calculation is illustrated in Figure 5, panel A. Similarly defined event-based P, R, F are also used for evaluating polyphonic music transcription, with events being the musical notes [21].

The advantage of using F-score for evaluating sound event detection performance is that it is widely known and easy to understand. Its most important drawback is that the choice of averaging is especially important. Because the magnitude of F-score is mostly determined by the number of true positives, in instance-based averaging large classes dominate small classes. In class-based averaging it is necessary to ensure presence of all classes in the test data, to avoid cases when recall is undefined (when $TP + FN = 0$). Any dataset of real-world recordings will most likely have unbalanced event classes, so the choice of averaging will be always an important factor.

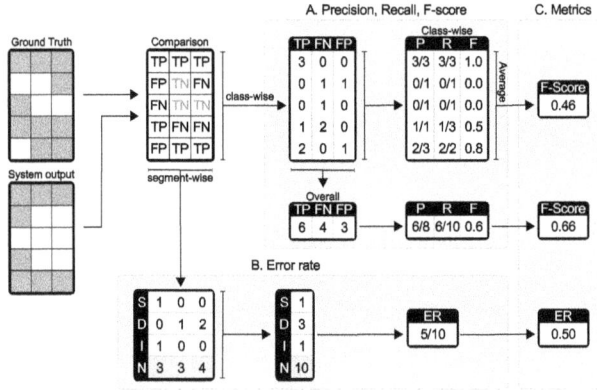

Figure 4. Calculation of segment-based metrics.

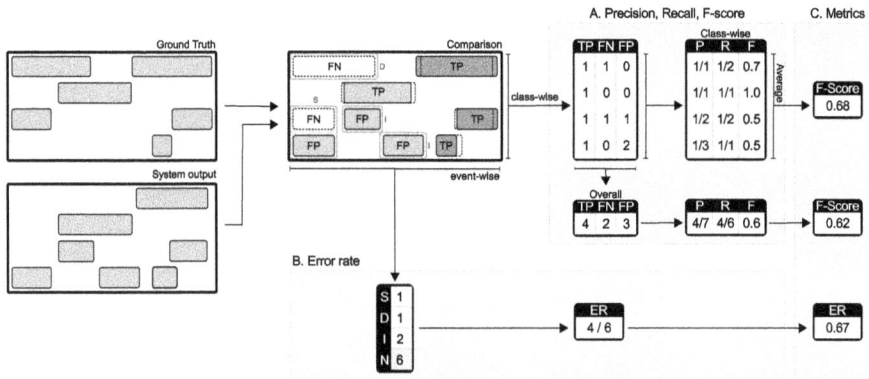

Figure 5. Calculation of event-based metrics.

4.2. Error Rate

Error rate measures the amount of errors in terms of *insertions* (I), *deletions* (D) and *substitutions* (S). To calculate segment-based error rate, errors are counted segment by segment. In a segment k, the number of substitution errors $S(k)$ is the number of reference events for which a correct event was not output, yet something else was. This is obtained by pairing false positives and false negatives, without designating which erroneous event substitutes which. The remaining events are insertions and deletions: $D(k)$—the number of reference events that were not correctly identified (false negatives after substitutions are accounted for) and $I(k)$—the number of events in system output that are not correct (false positives after substitutions are accounted for). This leads to the following formula:

$$S(k) = min(FN(k), FP(k))$$
$$D(k) = max(0, FN(k) - FP(k)) \tag{4}$$
$$I(k) = max(0, FP(k) - FN(k))$$

Total error rate is calculated by integrating segment-wise counts over the total number of segments K, with $N(k)$ being the number of sound events marked as active in the reference in segment k:

$$ER = \frac{\sum_{k=1}^{K} S(k) + \sum_{k=1}^{K} D(k) + \sum_{k=1}^{K} I(k)}{\sum_{k=1}^{K} N(k)} \tag{5}$$

The calculation is illustrated in Figure 4, panel B. A similar calculation of error rate for polyphonic detection was used in evaluating multi-pitch transcription in MIREX [30] and sound event detection in DCASE 2013 [20].

The event-based error rate is defined with respect to the sound events, by taking into account temporal position and label of each sound event in system output compared to the ones in the reference. Sound events with correct temporal position but incorrect class label are counted as substitutions, while insertions and deletions are the sound events unaccounted for as correct or substituted in system output or reference, respectively. Overall error rate is calculated according to Equation (5) based on the error counts. This calculation is illustrated in Figure 5, panel B. Class-wise measures of error rate (segment or event based) aggregate error counts individually for each sound event class, therefore they cannot account for substitutions, but can count only insertions and deletions.

In other specific sound recognition areas, error rate is defined differently to reflect the target of the evaluation. In speech recognition, *Word Error Rate* is defined with respect to word sequences, defining an insertion as a word present in the output but not in the reference, a deletion as a word present in the reference but not in the output, and a substitution as a word in the reference erroneously recognized (different word in the output). In practice, one substitution represents one insertion and one deletion (one word is missed, another inserted). Because N is the number of words in the reference, the word error rate can be larger than 1.0.

In speaker diarization, the error rate measures the fraction of time that is not attributed correctly to a speaker or to non-speech. Temporal errors are: S—the percentage of scored time that a speaker ID is assigned to the wrong speaker, I—the percentage of scored time that a non-speech segment from the reference is assigned to a hypothesized speaker, and D—the percentage of scored time that a speaker segment from the reference is assigned to non-speech. The total error is measured as the fraction of temporal error with respect to the total time of reference speech. This error rate was used for evaluating sound event detection in [29], after the multi-label reference was projected to single label for comparison with single label system output. The temporal error was calculated as in speaker diarization, with event classes instead of speaker ID.

The advantage of using total error rate to measure performance in polyphonic sound event detection is the parallel to established metrics in speech recognition and speaker diarization evaluation. Its disadvantage is that, being a score rather than a percentage, it can be over 1 in cases when the system makes more errors than correct predictions. This makes interpretation difficult, considering that it is trivial to obtain an error rate of 1 by outputting no active events.

4.3. Other Metrics

Sensitivity and specificity represent the true positive rate and true negative rate, respectively. They are used as a pair, to illustrate the trade-off between the two measured components.

$$Sensitivity = \frac{TP}{TP + FN}$$

$$Specificity = \frac{TN}{TN + FP} \tag{6}$$

A related measure is given by the Receiver Operating Characteristic (ROC) curve and Area Under the Curve (AUC). A ROC curve plots true positive rate *vs.* false positive rate, *i.e.*, sensitivity vs (1-specificity) at various thresholds. AUC allows summarizing the ROC curve into a single number, and is often used for comparing performance of binary classifiers. AUC was used to evaluate sound classification performance in [25] for a multi-label task by using binary classifiers for each class and calculating the average of the class-wise AUC values.

Accuracy measures how often the classifier takes the correct decision, as the ratio between the number of correct system outputs and the total number of outputs.

$$ACC = \frac{TP + TN}{TP + TN + FP + FN} \tag{7}$$

Accuracy does not characterize well the ability of the system to detect the true positives: when the events' activity is sparse, the count of true negatives will dominate the accuracy value. It is possible to assign weights to TP and TN, obtaining the *balanced accuracy*:

$$BACC = w \cdot \frac{TP}{TP + FN} + (1 - w) \cdot \frac{TN}{TN + FP} \tag{8}$$

With equal weights ($w = 0.5$), the balanced accuracy is the arithmetic mean of sensitivity and specificity. Specificity, accuracy and balanced accuracy are not defined in the event-based calculation, as there is no available count of true negatives.

Another definition for accuracy has been proposed in [37] for music transcription, and can be easily generalized for both segment-based and event-based sound event detection cases. It was defined as:

$$ACC_{MIR} = \frac{TP}{TP + FP + FN} \tag{9}$$

by discarding TN from Equation (7). A related metric is *Mean Overlap Ratio* [38] that was used to measure the average temporal overlap between correctly transcribed and reference notes. This is a temporal measure of accuracy as defined by Equation (9), as it measures the amount of overlap time—true positive duration—w.r.t. combined duration of the reference and output event—which for misaligned onsets and offsets consists of false negative, true positive and false positive durations. Mean overlap ratio for missed or inserted events is 0, leading to very low overall average for systems that output many errors, as is the case with polyphonic sound event detection systems at the moment.

Another accuracy measure is the *Relative Correct Overlap* used in chord transcription, measuring the proportion of the system output that matches the reference as relative frame count [39] or in seconds [40]. Because chords are combinations of individual notes, this metric measures frame-based accuracy of certain combinations of classes. In polyphonic sound event detection, this measure would represent the proportion of frames or time when the system detects correctly all events active in the reference. However, combinations of sound events have no specific meaning for the detection task, therefore such a measure is not conceptually meaningful for it.

Accuracy variants defined by Equations (7)–(9) have the advantage of being bounded between 0 and 1, with 1 in the case when the system makes no errors. The expressions in Equations (7) and (8) reach 0 when the system output contains only errors, while Equation (9) is 0 also when the system outputs nothing, as its value is not linked to the true negatives. Accuracy does not provide any information on the trade-off between the error types, which may be considered a disadvantage in certain situations. At the same time, it is capable of counting only insertions and deletions, while in some applications it is preferable to count substitutions.

4.4. Toolbox for Sound Event Detection Evaluation

Implementations of the presented metrics and variations for their calculation are provided as an open source software toolbox called `sed_eval` [41], implemented in Python. The toolbox aims to provide a transparent and uniform implementation of metrics. The toolbox contains segment-based calculation of precision, recall/sensitivity, F-score, specificity, error rate, accuracy, and balanced accuracy with chosen weight on the true positive rate and true negative rate, with instance-based and class-based averaging. Segment-based metrics can be calculated using the preferred segment length as a parameter. The toolbox also contains event-based calculation of precision, recall, F-score and error rate, with instance-based and class-based averaging. The event-based metrics can be calculated using the preferred collar size as a parameter, and using onset only comparison or onset and offset comparison.

5. Choosing a Metric

As a general issue of evaluating machine learning algorithms, the measure of performance would reflect the aim related to applicability of the proposed method. It is difficult to choose a measure without knowing the future usage of the system for example in terms of costs of missed events and false alarms, in which case using a single measure may provide an incomplete analysis of the results. One solution is to use a number of different metrics, in order to provide views of the multiple aspects. Often the different measures disagree, since each measure focuses on a different aspect of performance, in which case it may be impossible to draw a clear conclusion [42].

5.1. Measuring Performance of a System

We present an example illustrating how the different metrics and averaging options result in very different values, and discuss the meaning of each. The provided example is the output of the system presented in [28], using the development set of Office Synthetic task from DCASE 2013 in a leave-one out setup. The available data contains nine files at three different SNRs (−6 dB, 0 dB, 6 dB) and three different event densities (low, medium, high). The system uses coupled matrix factorization to learn dictionaries based on the spectral representation of the audio and the corresponding frame-based binary activity matrix. Based on the learned dictionaries, the system outputs an estimated activity of events classes. A complete description of the system can be found in [28].

The performance of the system is measured with evaluation parameters used in DCASE 2013 (10 ms segment for segment-based metrics, 100 ms collar for event-based metrics) and the ones proposed for DCASE 2016 (1 s segment for segment-based metrics, 250 ms collar for event-based metrics), to illustrate the differences in the choice of evaluation parameters.

5.1.1. Segment-Based Metrics

Segment-based metrics are presented in Table 1. Both instance-based and class-based averaging calculation methods are presented, to illustrate the effect of the averaging method on the metrics values. Detailed class-wise F-score is presented in Figure 6 for an evaluation segment of length 10 ms.

The averaging method influences the values of most metrics. The difference in F-score between instance-based and class-based averaging comes from the significantly different performance on different classes, as seen in Figure 6—there are few highly performing classes (F-score > 60%) and few low performing classes (F-score < 7%) which have equal weight on the class-based averaging value of F-score. An inspection of precision and recall values reveals that the system output contains mostly correctly detected events (high precision), but there is a large number of events undetected (low recall). This results in a small number of substitutions—hence the difference in error rate values when counting substitutions (instance-based averaging) or counting insertions and deletions separately (class-based averaging) is not very large. ACC is not influenced by the averaging type, due to the total number of evaluated segments being the same for all classes.

Table 1. Segment-based metrics calculated on the case study system.

Segment Length		F	ER	Sensitivity	Specificity	ACC	BACC
10 ms	class-based average	34.5	0.87	27.3	99.5	96.7	63.4
	instance-based average	**42.0**	**0.76**	**29.7**	99.5	96.7	**64.6**
1 s	class-based average	44.9	0.89	43.2	97.1	92.4	70.1
	instance-based average	**50.9**	**0.75**	**44.6**	97.0	92.4	**70.8**

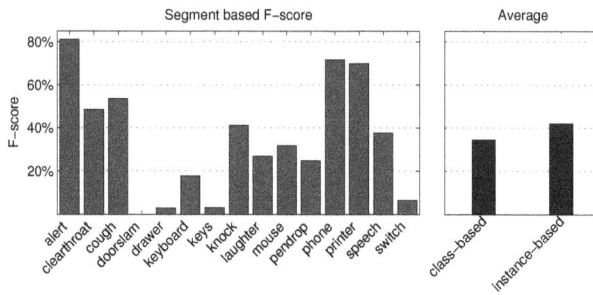

Figure 6. Segment-based F-score for the case study system, 10 ms evaluation segment.

The evaluation segment length also influences the values. A longer segment allows the system to produce a temporally less accurate output while still being considered correct. This is reflected in the measures that represent the amount of true positives—such as the F-score (TP being involved in calculation of both precision and recall) and sensitivity (which is the true positive rate). At the same time, the larger segment produces more false positives by marking an event active for the entire one second segment even if it was only active for a fraction of its length—resulting in lower accuracy and specificity values.

The choice between a short or a large segment in evaluation is dictated by the application—for applications that need precise detection of events' presence, a small evaluation segment should be used. However, this also depends on the quality of annotations, as sometimes annotations cannot be produced in the most objective way, and a 10 ms resolution is impossible to achieve. With a longer segment, the metric is more permissive on both the system output and subjectivity in onset/offset annotation.

5.1.2. Event-Based Metrics

Event-based F-score and error rate were calculated in two ways, first using the onset condition only, and second using both onset and offset conditions. The onset condition imposed the specific collar between the reference and system output events. The offset condition imposed either the same collar or a maximum difference between offsets of 50% of the reference event duration. The results are presented in Table 2.

First of all, it can be noticed that the performance of the system measured using the event-based metrics is much lower than when using segment-based metrics. The event-based metrics measure the ability of the system of detecting the correct event in the correct temporal position, acting as a measure of onset/offset detection capability.

Considering only the onset condition, the system is evaluated as having much better performance when the collar is larger. A larger collar behaves the same way as the longer segment in segment-based evaluation, allowing the system to produce a less exact output while still being considered correct. Figure 7 shows the class-wise F-score with 250 ms collar evaluated using onset only. It can be seen that for few event classes, the system is somewhat capable of detecting the onset within the allowed tolerance, while for other classes the system performs very poorly.

The offset condition allows a misalignment between the system output and ground truth event offsets. The permitted maximum misalignment is the collar length or 50% of the ground truth event length, whichever is higher. For very short events, this means that mostly the collar will be used, while long events benefit from the length condition. From Table 2 we can also notice that it is much easier to detect only onset of sound events compared to detecting both onset and offset. Similar behavior has been observed for algorithms used for note tracking in music transcription [30].

Table 2. Event-based F-score and ER calculated on the case study system.

Collar		Onset Only		Onset + Offset	
		F	ER	F	ER
100 ms	class-based average	7.5	2.10	4.7	2.10
	instance-based average	20.7	1.74	8.2	2.11
250 ms	class-based average	10.5	2.25	4.1	2.10
	instance-based average	30.8	1.49	10.1	2.06

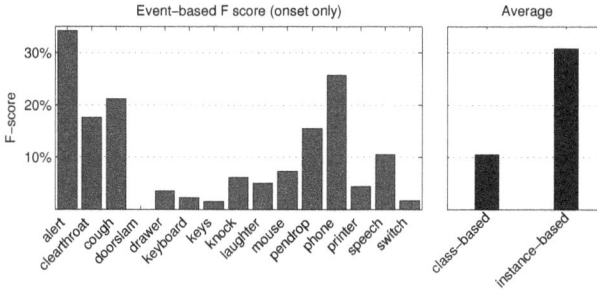

Figure 7. Event-based F-score for the case study system, 250 ms collar, onset only.

The choice between a small and a large collar, and the use of onset only condition or onset/offset is dictated by the application—in applications that need a precise detection of the sound event boundaries, a short collar and both onset and offset conditions should be used. For applications where it is more important to detect the presence of certain sound events rather than their exact temporal position and duration, evaluation with a large collar is suitable, and the offset condition can be omitted. A larger collar is also more permissive on the subjectivity in onset/offset annotation.

5.1.3. Discussion

The results from Tables 1 and 2 show the difficulty of characterizing the system performance through a single metric. It is important to know that there is always a trade off between measured aspects, and make a choice of metric based on what the requirements for the system are.

F-score is the first choice metric in many cases due to its widespread use in other research areas. From Table 1 we can conclude that when measured in one second segments, half of the activity of sound events has been correctly detected by the presented system. However, the system still makes many errors, indicated by ER. Error rate gives the measure of how much the system is wrong. In the provided example, the system makes a number of errors that is 75% of the number of events to be detected, which is relatively high.

Accuracy gives a measure of how well the system output matches the reference in both true positives and true negatives. The example system outputs a very small amount of false positives, which results in a very good rate of true negatives. Combining this with the fact that the used data has relatively sparse events' activity and in fact there are not too many positives to be detected, the system has an accuracy close to 100%. The ability to correctly output true negatives is important in applications where false positives are highly undesired. For example in an automatic alert system in an environment with sparse events, false positives will have immediate negative effect on user satisfaction. For such applications, specificity is a more appropriate measure, or true negative rate. On the other hand, there are monitoring applications in which it is more important to have very low rate of false negatives, such as gun shot detection or car crash detection. In this case the best choice of measure would be recall (sensitivity). However, if general accuracy is of interest for the application, balanced accuracy is a better choice for metric, because the true negatives have less weight in it.

Event-based metrics illustrate better the ability to correctly locate and label longer blocks of audio. For application where detection of correct onsets (and offsets) of sounds is important, the metric of choice should be event-based. The event-based error rate of the example system as seen in Table 2 is over 1, which means that the system makes much more errors than correct outputs. According to Equation (5), it is always possible to obtain an error rate of 1 by using a system that does not output anything. However, even with an error rate over 1, there are obviously many events correctly detected, according to F-score being 30.8%.

5.2. Comparing Performance of Different Systems

For comparing performance of different systems for sound event detection, the metric should be chosen depending on the application and the cost associated to different types of errors. We give an example in Table 3. System A is the one presented in the previous subsection. System B is the baseline system provided in DCASE 2016 [22]—it uses mel-frequency cepstral coefficients and Gaussian mixture models in a binary classification setup, and is trained using the polyphonic audio. System C is the submission to DCASE 2103 [43]—it uses mel-frequency cepstral coefficients and hidden Markov models with multiple Viterbi passes for detection, and is trained using isolated examples of sounds for all the classes. For a complete comparison, performance of the systems is also compared to a zero-output system (frame-wise activity matrix containing only zeros), all-active-output system (frame-wise activity matrix containing only ones) and random-output system. The random-output system generates in each frame for each event an activity level (0 or 1) based on the average sparseness of the dataset (priors 0.95 and 0.05 for 0 and 1, respectively). The random-output system was simulated 1000 times and the average calculated values over all runs are presented. The results in Table 3 are obtained using the same data as in Section 5, with one second evaluation segment and 250 ms collar.

Table 3. Comparing systems using different metrics calculated using instance-based averaging, 1 s segment, 250 ms collar.

Compared Systems	Segment-Based		Event-Based Onset Only		Event-Based Onset + Offset	
	F	ER	F	ER	F	ER
system A (NMF [28])	42.0	0.76	**30.8**	1.49	10.1	2.06
system B (GMM [22])	**68.5**	**0.49**	10.7	**1.42**	4.4	1.62
system C (HMM [43])	46.9	0.87	30.0	1.47	**20.8**	1.76
zero-output system	0.0	1.00	0.0	**1.00**	0.0	1.00
all-active-output system	15.4	10.93	0.0	1.44	0.0	1.44
random-output system	9.0	1.53	1.6	2.59	1.1	2.67

Faced with a choice between these systems, if we want a system that finds best the regions where certain event classes are active, we would choose system B for the 68.5% segment-based F-score. The system also makes less errors in the evaluated segments than both systems A and C, and is therefore a good choice for detecting most regions with active events. However, if we want a system that detects best the position in time of event instances, we would choose system A based on its 30.8% event-based F-score. This system makes approximately 1.5 times more mistakes than the number of events to be detected, while correctly detecting approximately one third of them. The event-based performance of system C is comparable with the performance of system A when using only the onset condition, but system C is much better at detecting the sound event offsets, with a 20.8% F-score. A choice of system based on ER would indicate system B as the best, making the smallest amount of errors in both segment-based and event-based calculation.

The zero output system has an F-score of 0 and ER of 1, as expected. The all-active output system achieves a nonzero segment-based F-score by finding all the true positives. It is however penalized by introducing false positives on all other instances in the event activity matrix. In event-based evaluation, this system outputs one long event active for the whole duration of the test audio, which counts as an insertion, with no correct output. The random-output system obtains a performance between the two extremes—but still very low in comparison with all the systems based on learning.

6. Conclusions

In this paper, we reviewed and illustrated calculation of the most common metrics as segment-based and event-based metrics, and the influence of the averaging method on the final result. The metrics were calculated on a case study example, and the results were analyzed from the perspective of choosing a metric to evaluate performance of sound event detection systems. We also provide a toolbox containing implementations of all presented metrics.

Evaluation of sound event detection systems performing polyphonic detection needs adaptation of the metrics used for other classification or detection problems. Polyphonic sound event detection brings and extra dimension to the problem through the presence of multiple outputs at the same time, and all existing metrics must be modified to account for this. Definitions of true/false positives and true/false negatives are modified to refer to the temporal position rather than the set of events output by the system, and subsequently the metrics measure the ability of the system to provide the target output in the target time frame. Segment-based metrics characterize the performance in finding most of the regions where a sound event is present, while event-based metrics characterize the performance in event onset and offset detection within a small tolerance. The two measuring methods represent different aspects and serve different needs, therefore different sound event detection approaches will be best suited for maximizing one or the other.

The importance of rigorous definitions for evaluation metrics cannot be overstated. Comparison of algorithms and reproducibility of results are dependent as much on benchmark databases as on uniform procedure for evaluation. This study and the associated toolbox are part of our effort to provide a reference point for better understanding and definition of metrics for the specific task of polyphonic sound event detection.

Acknowledgments: The research leading to these results has received funding from the European Research Council under the ERC Grant Agreement 637422 EVERYSOUND.

Author Contributions: Annamaria Mesaros drafted the main manuscript and planned the experiments. Toni Heittola assisted in the organization of the manuscript and implemented the associated metrics toolbox. Tuomas Virtanen helped in the preparation of the manuscript.

Conflicts of Interest: The authors declare no conflict of interest.

References

1. Clavel, C.; Ehrette, T.; Richard, G. Events Detection for an Audio-Based Surveillance System. In Proceedings of the IEEE International Conference on Multimedia and Expo, Amsterdam, The Netherlands, 6 July 2005; IEEE Computer Society: Los Alamitos, CA, USA, 2005; pp. 1306–1309.
2. Härmä, A.; McKinney, M.F.; Skowronek, J. Automatic surveillance of the acoustic activity in our living environment. In Proceedings of the IEEE International Conference on Multimedia and Expo (ICME), Amsterdam, The Netherlands, 6 July 2005; IEEE Computer Society: Los Alamitos, CA, USA, 2005; pp. 634–637.
3. Foggia, P.; Petkov, N.; Saggese, A.; Strisciuglio, N.; Vento, M. Reliable detection of audio events in highly noisy environments. *Pattern Recognit. Lett.* **2015**, *65*, 22–28.
4. Peng, Y.T.; Lin, C.Y.; Sun, M.T.; Tsai, K.C. Healthcare audio event classification using Hidden Markov Models and Hierarchical Hidden Markov Models. In Proceedings of the IEEE International Conference on Multimedia and Expo, New York, NY, USA, 28 June–3 July 2009; pp. 1218–1221.
5. Goetze, S.; Schröder, J.; Gerlach, S.; Hollosi, D.; Appell, J.; Wallhoff, F. Acoustic Monitoring and Localization for Social Care. *J. Comput. Sci. Eng.* **2012**, *6*, 40–50.
6. Guyot, P.; Pinquier, J.; Valero, X.; Alias, F. Two-step detection of water sound events for the diagnostic and monitoring of dementia. In Proceedings of the IEEE International Conference on Multimedia and Expo (ICME), San Jose, CA, USA, 15–19 July 2013; pp. 1–6.

7. Stowell, D.; Clayton, D. Acoustic Event Detection for Multiple Overlapping Similar Sources. In Proceedings of the IEEE Workshop on Applications of Signal Processing to Audio and Acoustics (WASPAA), New Paltz, NY, USA, 18–21 October 2015.

8. Cai, R.; Lu, L.; Hanjalic, A.; Zhang, H.J.; Cai, L.H. A flexible framework for key audio effects detection and auditory context inference. *IEEE Trans. Audio Speech Lang. Process.* **2006**, *14*, 1026–1039.

9. Xu, M.; Xu, C.; Duan, L.; Jin, J.S.; Luo, S. Audio Keywords Generation for Sports Video Analysis. *ACM Trans. Multimedia Comput. Commun. Appl.* **2008**, *4*, 1–23.

10. Bugalho, M.; Portelo, J.; Trancoso, I.; Pellegrini, T.; Abad, A. Detecting audio events for semantic video search. In Proceedings of the 10th Annual Conference of the International Speech Communication Association, Brighton, UK, 6–10 September 2009; pp. 1151–1154.

11. Chu, S.; Narayanan, S.; Kuo, C.C. Environmental Sound Recognition With Time-Frequency Audio Features. *IEEE Trans. Audio Speech Lang. Process.* **2009**, *17*, 1142–1158.

12. Tran, H.D.; Li, H. Sound Event Recognition With Probabilistic Distance SVMs. *IEEE Trans. Audio Speech Lang. Process.* **2011**, *19*, 1556–1568.

13. Dennis, J.; Tran, H.D.; Li, H. Combining robust spike coding with spiking neural networks for sound event classification. In Proceedings of the IEEE International Conference on Acoustics, Speech and Signal Processing (ICASSP), Brisbane, QLD, Australia, 19–24 April 2015; pp. 176–180.

14. Zhang, H.; McLoughlin, I.; Song, Y. Robust Sound Event Recognition Using Convolutional Neural Networks. In Proceedings of the IEEE International Conference on Acoustics, Speech and Signal Processing (ICASSP), Brisbane, QLD, Australia, 19–24 April 2015.

15. Zhuang, X.; Zhou, X.; Hasegawa-Johnson, M.A.; Huang, T.S. Real-world Acoustic Event Detection. *Pattern Recognit. Lett.* **2010**, *31*, 1543–1551.

16. Heittola, T.; Mesaros, A.; Virtanen, T.; Gabbouj, M. Supervised Model Training for Overlapping Sound Events based on Unsupervised Source Separation. In Proceedings of the 38th International Conference on Acoustics, Speech, and Signal Processing (ICASSP 2013), Vancouver, BC, Canada, 26–31 May 2013; pp. 8677–8681.

17. Mesaros, A.; Dikmen, O.; Heittola, T.; Virtanen, T. Sound event detection in real life recordings using coupled matrix factorization of spectral representations and class activity annotations. In Proceedings of the IEEE International Conference on Acoustics, Speech and Signal Processing (ICASSP); Brisbane, QLD, Australia, 19–24 April 2015; pp. 151–155.

18. Espi, M.; Fujimoto, M.; Kinoshita, K.; Nakatani, T. Exploiting spectro-temporal locality in deep learning based acoustic event detection. *EURASIP J. Audio Speech Music Process.* **2015**, *2015*, doi:10.1186/s13636-015-0069-2.

19. Grootel, M.; Andringa, T.; Krijnders, J. DARES-G1: Database of Annotated Real-world Everyday Sounds. In Proceedings of the NAG/DAGA Meeting, Rotterdam, The Netherlands, 23–26 March 2009.

20. Stowell, D.; Giannoulis, D.; Benetos, E.; Lagrange, M.; Plumbley, M. Detection and Classification of Acoustic Scenes and Events. *IEEE Trans. Multimedia* **2015**, *17*, 1733–1746.

21. Poliner, G.E.; Ellis, D.P. A Discriminative Model for Polyphonic Piano Transcription. *EURASIP J. Adv. Signal Process.* **2007**, *2007*, 048317.

22. Detection and Classification of Acoustic Scenes and Events 2016, IEEE AASP Challenge. Available online: http://www.cs.tut.fi/sgn/arg/dcase2016/ (accessed on 5 January 2016).

23. Mesaros, A.; Heittola, T.; Eronen, A.; Virtanen, T. Acoustic Event Detection in Real-life Recordings. In Proceedings of the 18th European Signal Processing Conference (EUSIPCO 2010), Aalborg, Denmark, 23–27 August 2010; pp. 1267–1271.

24. Martinez, E.; Celma, O.; Sordo, M.; Jong, B.D.; Serra, X. Extending the folksonomies of freesound.org using contentbased audio analysis. In Proceedings of the Sound and Music Computing Conference, Porto, Portugal, 23–25 July 2009.

25. Foster, P.; Sigtia, S.; Krstulovic, S.; Barker, J. CHiME-Home: A Dataset for Sound Source Recognition in a Domestic Environment. In Proceedings of the Worshop on Applications of Signal Processing to Audio and Acoustics (WASPAA), New Paltz, NY, USA, 18–21 October 2015.

26. Heittola, T.; Mesaros, A.; Eronen, A.; Virtanen, T. Context-Dependent Sound Event Detection. *EURASIP J. Audio Speech Music Process.* **2013**, *2013*, doi:10.1186/1687-4722-2013-1

27. Cakir, E.; Heittola, T.; Huttunen, H.; Virtanen, T. Polyphonic Sound Event Detection Using Multi Label Deep Neural Networks. In Proceedings of the International Joint Conference on Neural Networks 2015 (IJCNN 2015), Montreal, QC, Canada, 31 July–4 August 2015.

28. Dikmen, O.; Mesaros, A. Sound event detection using non-negative dictionaries learned from annotated overlapping events. In Proceedings of the IEEE Workshop on Applications of Signal Processing to Audio and Acoustics (WASPAA), New Paltz, NY, USA, 20–23 October 2013; pp. 1–4.

29. Temko, A.; Nadeu, C.; Macho, D.; Malkin, R.; Zieger, C.; Omologo, M. Acoustic Event Detection and Classification. In *Computers in the Human Interaction Loop*; Waibel, A.H., Stiefelhagen, R., Eds.; Springer: London, UK, 2009; pp. 61–73.

30. Music Information Retrieval Evaluation eXchange (MIREX 2016): Multiple Fundamental Frequency Estimation & Tracking. Available online: http://www.music-ir.org/mirex/wiki/2016:Multiple_Fundamental_Frequency_Estimation_&_Tracking (accessed on 18 April 2016).

31. Giannoulis, D.; Benetos, E.; Stowell, D.; Rossignol, M.; Lagrange, M.; Plumbley, M. Detection and classification of acoustic scenes and events: An IEEE AASP challenge. In Proceedings of the IEEE Workshop on Applications of Signal Processing to Audio and Acoustics (WASPAA), New Paltz, NY, USA, 20–23 October 2013; pp. 1–4.

32. Mesaros, A.; Heittola, T.; Virtanen, T. TUT Sound Events 2016. Available online: https://zenodo.org/record/45759 (accessed on 22 May 2016).

33. Sebastiani, F. Machine Learning in Automated Text Categorization. *ACM Comput. Surv.* **2002**, *34*, 1–47.

34. Sechidis, K.; Tsoumakas, G.; Vlahavas, I. On the Stratification of Multi-label Data. In *Machine Learning and Knowledge Discovery in Databases*; Lecture Notes in Computer Science; Springer: Berlin/Heidelberg, Geramny, 2011; Volume 6913, pp. 145–158.

35. Forman, G.; Scholz, M. Apples-to-apples in Cross-validation Studies: Pitfalls in Classifier Performance Measurement. *SIGKDD Explor. Newsl.* **2010**, *12*, 49–57.

36. Rijsbergen, C.J.V. *Information Retrieval*, 2nd ed.; Butterworth-Heinemann: Newton, MA, USA, 1979.

37. Dixon, S. On the computer recognition of solo piano music. In Proceedings of the Australasian Computer Music Conference, Brisbane, Australia, 10–12 July 2000; pp. 31–37.

38. Ryynanen, M.P.; Klapuri, A. Polyphonic music transcription using note event modeling. In Proceedings of the IEEE Workshop on Applications of Signal Processing to Audio and Acoustics, New Platz, NY, USA, 16–19 October 2005; pp. 319–322.

39. Sheh, A.; Ellis, D. Chord Segmentation and Recognition using EM-Trained Hidden Markov Models. In Proceedings of the 4th International Conference on Music Information Retrieval ISMIR, Baltimore, MD, USA, 27–30 October 2003.

40. Mauch, M.; Dixon, S. Simultaneous Estimation of Chords and Musical Context From Audio. *IEEE Trans. Audio Speech Lang. Process.* **2010**, *18*, 1280–1289.

41. Heittola, T.; Mesaros, A. sed_eval - Evaluation toolbox for Sound Event Detection. Available online: https://github.com/TUT-ARG/sed_eval (accessed on 22 May 2016).

42. Japkowicz, N.; Shah, M. *Evaluating Learning Algorithms*; Cambridge University Press: Cambridge, UK, 2011.

43. Diment, A.; Heittola, T.; Virtanen, T. *Sound Event Detection for Office Live and Office Synthetic AASP Challenge*; Technical Report; Tampere University of Technology: Tampere, Finland, 2013.

![applied sciences logo] *applied sciences*

MDPI

Article

Modal Processor Effects Inspired by Hammond Tonewheel Organs

Kurt James Werner * and Jonathan S. Abel

Center for Computer Research in Music and Acoustics (CCRMA), Department of Music, Stanford University, 660 Lomita Drive, Stanford, CA 94305-8180, USA; abel@ccrma.stanford.edu
* Correspondence: kwerner@ccrma.stanford.edu; Tel.: +1-650-723-4971

Academic Editor: Vesa Valimaki
Received: 16 March 2016; Accepted: 13 June 2016; Published: 28 June 2016

Abstract: In this design study, we introduce a novel class of digital audio effects that extend the recently introduced modal processor approach to artificial reverberation and effects processing. These pitch and distortion processing effects mimic the design and sonics of a classic additive-synthesis-based electromechanical musical instrument, the Hammond tonewheel organ. As a reverb effect, the modal processor simulates a room response as the sum of resonant filter responses. This architecture provides precise, interactive control over the frequency, damping, and complex amplitude of each mode. Into this framework, we introduce two types of processing effects: pitch effects inspired by the Hammond organ's equal tempered "tonewheels", "drawbar" tone controls, vibrato/chorus circuit, and distortion effects inspired by the pseudo-sinusoidal shape of its tonewheels and electromagnetic pickup distortion. The result is an effects processor that imprints the Hammond organ's sonics onto any audio input.

Keywords: audio signal processing; modal analysis; room acoustics; signal analysis; artificial reverberation; digital audio effects; virtual analog; musical instruments

1. Introduction

The Hammond tonewheel organ is a classic electromechanical musical instrument, patented by Laurens Hammond in 1934 [1]. Although it was intended as an affordable substitute for church organs [2], it has also become widely known as an essential part of jazz (where it was popularized by Jimmy Smith), R & B and rock music (where the Hammond playing of Keith Emerson of Emerson, Lake & and Palmer and Jon Lord of Deep Purple is exemplary). The most popular model is the Hammond B-3, although many other models exist [3]. The sound of the Hammond organ is rich and unusual. Its complexity comes from the Hammond organ's unique approach to timbre and certain quirks of its construction.

In this article, we describe a novel class of modal-processor-based audio effects which we call the "Hammondizer". The Hammondizer can imprint the sonics of the Hammond organ onto any sound; it mimics and draws inspiration from the architecture of the Hammond tonewheel organ. We begin by describing the architecture and sonics of the Hammond tonewheel organ alongside related work on Hammond organ modeling.

The Hammond organ is essentially an additive synthesizer. Additive synthesizers create complex musical tones by adding together sinusoidal signals of different frequencies, amplitudes, and phases [4]. In the Hammond organ, 91 sinusoidal signals are available. These sinusoids are created when "tonewheels"—ferromagnetic metal discs—spin and the pattern of ridges cut into their edges is transduced by electromagnetic pickups into electrical signals, a technique originated in Thaddeus Cahill's late-19th century instrument, the Telharmonium [5]. Hammond organ tonewheel pickups have not been studied much in particular, but modeling and simulation of electromagnetic pickups in general is an active research area [6–10]. Any nonlinearities in a pickup model will cause bandwidth expansion and add to the characteristic sound of the Hammond organ. In the case that this bandwidth expansion would go beyond the Nyquist limit, alias-suppression methods become relevant [11–14].

These 91 tonewheels are tuned approximately to the twelve-tone equal-tempered musical scale [15]—In scientific pitch notation, the lowest-frequency tonewheel on a Hammond organ is tuned to C1 ($\approx$32.7 Hz) and the highest-frequency tonewheel is tuned to F#7 ($\approx$5919.9 Hz) [16]. The lowest octave of tonewheels do not form sinusoids, but more complex tones that have strong 3rd and 5th harmonics, making them closer to square waves than sine waves [15]. Some aficionados have pointed to crosstalk between nearby tonewheel/pickup pairs as an important sonic feature of the Hammond organ [17,18].

The tone of the Hammond organ is set using nine "drawbars". Unlike traditional organs, where "stops" bring in entire complex organ sounds, the Hammond organ's drawbars set the relative amplitudes of individual sinusoids in a particular timbre. These nine sinusoids form a pseudo-harmonic series summarized in Table 1 [19]. This pseudo-harmonic series deviates from the standard harmonic series in three ways: (1) each overtone is tuned to the nearest available tonewheel; (2) certain overtones are omitted, especially the 6th harmonic, which would be between the 8th and 9th drawbar); and (3) new fictitious overtones are added (the 5th and sub-octave).

Table 1. Hammond Organ Drawbars—Pitch in organ stop lengths and musical intervals.

Pipe Pitch	16′	5⅓′	8′	.	4′	2⅔′	2′	1⅗′	1⅓′	1′
Scale Interval	sub-octave	5th	Unison		8th	12th	15th	17th	19th	22nd
Stop Name	Bourdon	Quint	Principal		Octave	Nazard	Block Flöte	Tierce	Larigot	Sifflöte
Semitone Offset	−12	+7	0		+12	+19	+24	+28	+31	+36
Error E (cents)	N/A	N/A	0		0	−1.955	0	+13.686	−1.955	0

The raw sound of the Hammond organ tonewheels is static. To enrich the sound, Hammond added a chorus/vibrato circuit [20]. Earlier models used a tremolo effect in place of the chorus/vibrato circuit [21]. The sound was further enriched by an electro-mechanical spring reverb device [22]. Although Hammond did not originally approve of the practice, it became customary to play Hammond organs through a Leslie speaker, an assembly with a spinning horn and baffle that creates acoustic chorus and tremolo effects. The Leslie speaker has been covered extensively in the modeling literature. Various approaches have involved interpolating delay lines [23,24] and amplitude modulation [25,26], perception-based models [28], and time-varying Finite Impulse Response (FIR) filters [29]. Recently, Pekonen et al. presented a novel Leslie model [18] using spectral delay filters [30]. Werner et al. used the Wave Digital Filter approach to model the Hammond vibrato/chorus circuit [31].

Although Hammond had stopped manufacturing their tonewheel organs by 1975, the Hammond sound remained influential. Many manufacturers developed clones of the Hammond tonewheel organ [17,32–35]. Commercial efforts have been accompanied by popular and academic work in virtual analog modeling [36]. Gordon Reid wrote a series of articles for *Sound on Sound* on generic synthesis approaches to modeling aspects of the Hammond organ [17,32–35]. Pekonen et al. studied efficient methods for digital tonewheel organ synthesis [18].

The Hammondizer audio effect is implemented as an extension to the recently-introduced "modal reverberator" approach to artificial reverberation [37–40]. Although there are many other approaches to modal sound synthesis in the literature (e.g., [41–44], the choice to extend the modal reverberator architecture to create the Hammondizer effect was a natural one for two reasons: (1) there are strong similarities between the system architecture of the Hammond organ and the system architecture of the modal reverberator; (2) the modal reverberator is already formulated as an audio effect which processes rather than synthesizes sound.

The rest of the article is structured as follows. Section 2 presents a simplified system architecture of the Hammond organ, Section 3 reviews relevant aspects of the modal processor approach,

Section 4 presents the novel Hammondizer digital audio effect, Section 5 demonstrates features of the Hammondizer through a series of examples, and Section 6 concludes.

2. Hammond Organ System Architecture

Here we extend the qualitative description above and present a mathematical formulation of the basic operation of the Hammond tonewheel organ. Referring to Figure 1, the player controls the organ by depressing keys on a standard musical keyboard shown on the left. Each of its 61 keys has a note on/off state $n_k(t) \in [0,1]$ that is indexed by a key number $k \in [1 \cdots 61]^\top$; these are collected into a column $n(t)$. Here and in the rest of the article, t is the discrete time sample index.

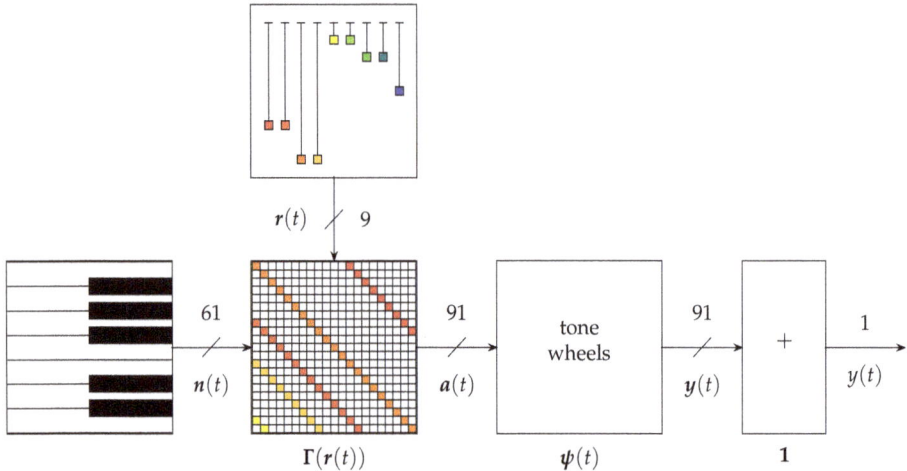

Figure 1. Hammond Tonewheel Organ block diagram.

The timbre is controlled by nine drawbars shown on the top. Each drawbar has a level $r_d(t) \in [0 \cdots 8]$ that is indexed by a drawbar number $d \in [1 \cdots 9]^\top$; these are collected into a column $r(t)$. The drawbars may be changed over time to alter the sounds of the Hammond organ. Each drawbar's level $r_d(t)$ is converted to an amplitude in -3 dB increments (Table 2) [45].

Table 2. Amplitude of each drawbar r_d, $d \in [1 \cdots 9]$.

r_d	0	1	2	3	4	5	6	7	8
amplitude (dB)	0	-3	-6	-9	-12	-15	-18	-21	$-\infty$

Furthermore, each drawbar has a tuning offset o_d, corresponding to the tuning offset in semitones of each pseudo-harmonic. The entire set of offsets is

$$o = [o_1 \cdots o_9]^\top = [-12, 7, 0, 12, 19, 24, 28, 31, 36]^\top \tag{1}$$

Each tuning offset (except the first two) approximates a harmonic overtone. This is discussed further at the end of the section.

Each tonewheel has a frequency f_w and amplitude $a_w(t)$ indexed by a tonewheel number $w \in [1 \cdots 91]^\top$; these are collected into columns f and $a(t)$. Each tonewheel is tuned to the twelve-tone equal-tempered scale

$$f_w = (440)2^{(w-45)/12} \text{ Hz} \tag{2}$$

In practice there are slight deviations according to the gearing ratios, producing deviations of up to 0.69 cents [15].

The outputs of all the tonewheels are summed by the 91×1 gain block $\mathbf{1} = [1 \cdots 1]^\top$ on the right to form the output signal $y(t)$:

$$y(t) = \mathbf{1}^\top \mathbf{y}(t) \tag{3}$$

The 91×61 routing matrix $\boldsymbol{\Gamma}(\mathbf{r}(t))$ forms the 91-tall column of tonewheel amplitudes $\mathbf{a}(t)$ from the 61-tall column of key on/off states $\mathbf{n}(t)$. This is accomplished by a matrix multiply

$$\mathbf{a}(t) = \boldsymbol{\Gamma}(\mathbf{r}(t))\mathbf{n}(t) \tag{4}$$

$\boldsymbol{\Gamma}(\mathbf{r}(t))$ is sparse (most entries are 0) and has a pseudo-convolutional form [46] in which the non-zero entries $r_1(t) \cdots r_9(t) \in [0 \cdots 8]$ are dictated by the 9-tall column of drawbar levels $\mathbf{r}(t)$. Denoting each entry in $\boldsymbol{\Gamma}(\mathbf{r}(t))$ as $\gamma_{w,k}(t)$, we have

$$\gamma_{w,k}(t) = \sum_{d=1}^{9} r_d(t)\, \delta\,(w - k - o_d) \tag{5}$$

where $\delta(x)$ is the Kronecker delta function

$$\delta(x) = \begin{cases} 1 & , x = 0 \\ 0 & , x \neq 0 \end{cases} \tag{6}$$

The tonewheel block is comprised of 91 tonewheel processors $\psi_w(t)$ in parallel. As shown in Figure 2, each individual tonewheel processor has a tonewheel producing a periodic signal $x_w(t)$ at a particular frequency f_w, an amplitude input $a_w(t)$ provided by the routing matrix $\boldsymbol{\Gamma}(\mathbf{r}(t))$, and an electromagnetic model $p_w\,()$. Each tonewheel processor forms an output $y_w(t)$ by

$$y_w(t) = a_w(t)\, p_w(x_w(t)) \tag{7}$$

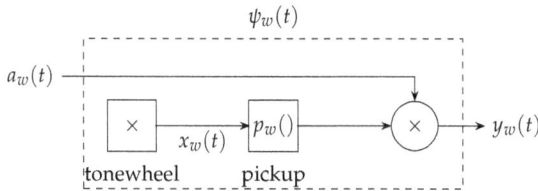

Figure 2. One tonewheel processor.

A block diagram of an individual tonewheel processor is shown in Figure 2. The matrix equation describing the entire bank of tonewheels is

$$\mathbf{y}(t) = \mathbf{p}(\mathbf{x}(t)) \odot \mathbf{a}(t) \tag{8}$$

where $\odot$ is the Hadamard (elementwise) product operator

$$(\mathbf{A} \odot \mathbf{B})_{i,j} = \mathbf{A}_{i,j}\mathbf{B}_{i,j} \tag{9}$$

where $\mathbf{A}_{i,j}$ denotes the ijth element of the matrix $\mathbf{A}$.

The lowest 12 tonewheels produce roughly square-wave signals and the rest produce essentially sinusoidal signals:

$$x_w(t) = \begin{cases} \frac{4}{\pi}\sin(2\pi\, f_w t) + \frac{4}{3\pi}\sin(2\pi\, 3f_w t) + \frac{4}{5\pi}\sin(2\pi\, 5f_w t) & , w \in [1 \cdots 12] \\ \sin(2\pi\, f_w t) & , w \in [13 \cdots 91] \end{cases} \qquad (10)$$

As a final note, we can discuss the pseudo overtone series of the Hammond organ in more detail. Equation (5) implies a certain relationship between any pressed key k and the set of frequencies that are produced. Here we state this relationship explicitly. Given Equations (1), (2) and (5), we can see that pressing any key k will, in general, drive a set of nine tonewheels with frequencies

$$f_{k,d} = (440)2^{(k+o_d-45)/12} \quad , \quad d \in [1 \cdots 9] \qquad (11)$$

Most wind and string instruments are characterized by a harmonic overtone series—i.e., one where overtone frequencies are integer multiples of a fundamental frequency. Most of the tonewheel frequencies given in Equation (11) approximate idealized harmonic overtones with frequencies given by

$$\tilde{f}_{k,d} = (440)2^{(k-45)/12} N_d \quad , \quad d \in [3 \cdots 9] \qquad (12)$$

The first two tonewheel frequencies $f_{k,1}$ and $f_{k,2}$ are the octave below the fundamental frequency and approximately a fourth below the fundamental frequency—they are not approximations of standard harmonic overtones.

In general, $\tilde{f}_{k,d} \neq f_{k,d}$. The error in "cents" (1/100 of a semitone) is given by

$$E_d = 1200 \log_2 \left(\tilde{f}_{k,d}/f_{k,d} \right) = 1200 \left[o_d/12 - \log_2 (N_d) \right] \quad , \quad d \in [3 \cdots 9] \qquad (13)$$

The tuning error of each tonewheel frequency is independent of k; it depends only on the drawbar index d—i.e., which overtone it is supposed to be approximating. These errors are given for each drawbar in Table 1. For the fundamental and octave overtones, the tonewheels are perfectly in tune. For the 12th and 19th, the tonewheels are ≈ -1.955 cents flat of the ideal overtones. The 19th is ≈ 13.686 cents sharp. This detuning is very unique to the Hammond organ.

3. Modal Processor Review

The Hammondizer effect involves decomposing an input signal into a parallel set of narrow-band signals, analogous to a bank of organ keys. Each of the "keys" is then pitch processed according to the drawbar settings, and distortion processed according to the tonewheel and pickup mechanics and electromagnetics. It turns out that this structure closely resembles that of the modal reverberator [37,38], which forms a room response as the parallel combination of room vibrational mode responses. In the following, we review the modal reverberator and adapt it to produce the needed pitch and distortion processing.

The impulse response $h(t)$ between a pair of points in an acoustic space may be expressed as the linear combination of normal mode responses [47,48],

$$h(t) = \sum_{m=1}^{M} h_m(t) \qquad (14)$$

where the system has M modes, with the mth mode response denoted by $h_m(t)$. The system output $y(t)$ in response to an input $x(t)$, the convolution $y(t) = h(t) * x(t)$, is therefore the sum of mode outputs

$$y(t) = \sum_{m=1}^{M} y_m(t), \quad y_m(t) = h_m(t) * x(t) \qquad (15)$$

where the mth mode output $y_m(t)$ is the mth mode response convolved with the input. The modal reverberator simply implements this parallel combination of mode responses (15), as shown in Figure 3. Denoting by $h(t)$ the M-tall column of complex mode responses, we have

$$y(t) = \mathbf{1}^\top (h(t) * x(t)) \tag{16}$$

with

$$h(t) = \psi(t) \odot (g(t) * \Gamma\varphi(t)) \tag{17}$$

and where convolution here obeys the rules of matrix multiplication, with each individual matrix operation replaced by a convolution.

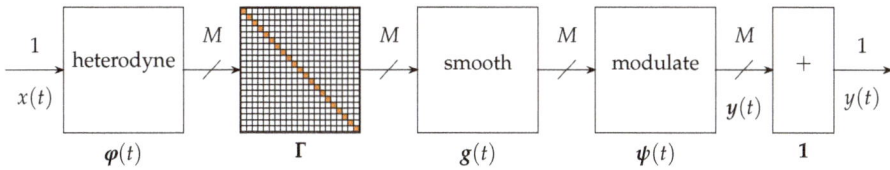

Figure 3. Basic modal reverberator architecture. The modal reverberator is the parallel combination of resonant filters matched to the modes of a linear system.

The mode responses $h_m(t)$ are complex exponentials, each characterized by a mode frequency $\omega_m = 2\pi f_w$, mode damping α_m, and mode complex amplitude γ_m,

$$h_m(t) = \gamma_m \exp\{(j\omega_m - \alpha_m)t\} \tag{18}$$

The mode frequencies and dampings are properties of the room or object; the mode amplitudes are determined by the sound source and listener positions (driver and pick-up positions for an electro-mechanical device), according to the mode spatial patterns.

Rearranging terms in the convolution $y_m(t) = h_m(t) * x(t)$, the mode filtering is seen to heterodyne the input signal to dc to form a baseband response, smooth this baseband response by convolution with an exponential, and modulate the result back to the original mode frequency,

$$y_m(t) = \sum_\tau e^{(j\omega_m - \alpha_m)(t-\tau)} x(\tau) = e^{j\omega_m t} \sum_\tau \gamma_m e^{-\alpha_m(t-\tau)} \left[e^{-j\omega_m \tau} x(\tau) \right] \tag{19}$$

All M γs are stacked into a diagonal gain matrix Γ. All the heterodyning sinusoids are stacked into a column $\varphi(t)$, and all of the modulating sinusoids into a column $\psi(t)$. The mode damping filters are stacked into a column $g(t)$. This process is shown in Figure 4. The heterodyning and modulation steps implement the mode frequency, and the smoothing filter generates the mode envelope, an exponential decay.

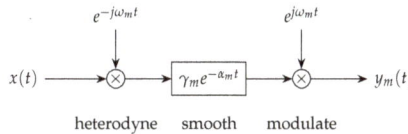

Figure 4. Mode response implementation. The mode response may be implemented as a cascade of heterodyning, smoothing, and modulation operations.

Using this architecture, rooms and objects may be simulated by tuning the filter resonant frequencies and dampings to the corresponding room or object mode frequencies and decay times.

The parallel structure allows the mode parameters to be separately adjusted, while Equation (19) provides interactive parameter control with no computational latency.

As described in [39], the modal reverberator architecture can be adapted to produce pitch shifting by using different sinusoid frequencies for the heterodyning and modulation steps in Equation (19), and adapted to produce distortion effects by inserting nonlinearities on the output of each mode or group of modes. The modal processor architecture has been used for other effects, including mode-wise gated reverb using Truncated Infinite Impulse Response (TIIR) filters [49], groupwise distortion, time stretching by resampling of the baseband signals, and manipulation of mode time envelopes by introducing repeated poles [39].

4. Hammondizer Modal Processor Implementation

The Hammondizer effect system architecture is shown in Figure 5. It turns out that this structure closely resembles that of the modal reverberator (Figure 3), which forms a room response as the parallel combination of room vibrational mode responses. Both have inputs designated by $x(t)$, a column of narrow-band outputs designated by $y(t)$, summed to form the system output $y(t)$.

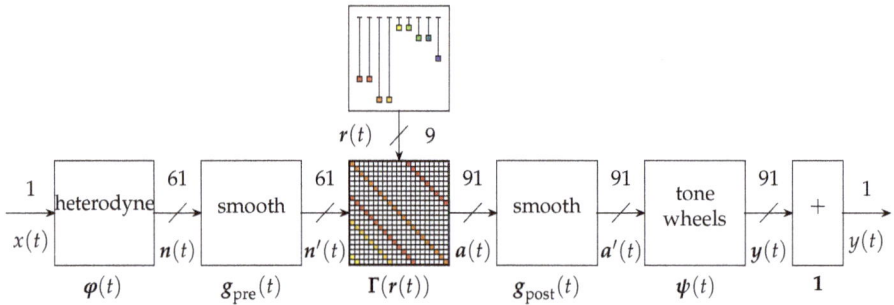

Figure 5. Block diagram of the Hammondizer effect.

In the Hammondizer, the input signal $x(t)$ is heterodyned to baseband by a column of modulating sinusoids $\boldsymbol{\varphi}(t)$:

$$n(t) = \boldsymbol{\varphi}(t)x(t) \tag{20}$$

These baseband signals are smoothed by a column of pre-smoothing filters $g_{\mathrm{pre}}(t)$

$$n'(t) = g_{\mathrm{pre}}(t) * n(t) \tag{21}$$

A column of tonewheel amplitudes $a(t)$ is formed by the drawbar routing matrix $\Gamma(r(t))$,

$$a(t) = \Gamma(r(t))n'(t) \tag{22}$$

and further smoothed by a column of post-smoothing filters $g_{\mathrm{post}}(t)$:

$$a'(t) = g_{\mathrm{post}}(t) * a(t) \tag{23}$$

A set of mode outputs $y(t)$ is formed by the tonewheel processing stages $\boldsymbol{\psi}(t)$, which include a column of pickup models $\boldsymbol{p}()$ and modulating signals $x(t)$

$$y(t) = p\left(x(t) \odot a'(t)\right) \tag{24}$$

An individual tonewheel processing stage is shown in Figure 6. Notice the slight change in architecture from the analogous Figure 2. In Figure 6, the pickup distortion has been moved to operate on the output rather than the raw tonewheel signal. The reason for this change is artistic—it disambiguates the effects of the memoryless pickup nonlinearities and the distortion of the tonewheel basis functions.

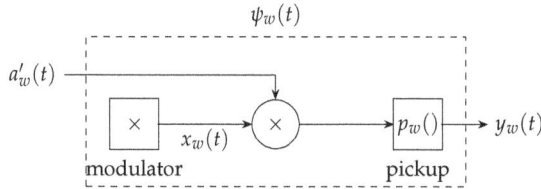

Figure 6. One tonewheel processor in the Hammondizer.

Finally, the output $y(t)$ is formed by summing all of the mode outputs:

$$y(t) = \mathbf{1}^\top \mathbf{y}(t) \tag{25}$$

In the rest of this section, we describe in detail how aspects of the modal processor are tuned and adapted to create the Hammondizer. Pitch processing adaptations include tuning the modes to the particular frequencies and frequency range of the Hammond organ Section 4.1), introducing drawbar-style controls to pitch processing (Section 4.2), adding vibrato to mode frequencies (Section 4.3), and adding crosstalk between nearby modes to simulate crosstalk between nearby tonewheels (Section 4.4). Distortion processing adaptations include adapting saturating nonlinearities for each mode to mimic the pickup distortion of each tonewheel (Section 4.5) and replacing modulation sinusoids with sums of sinusoids to mimic non-sinusoidal tonewheel shapes (Section 4.6).

4.1. Frequency Range

The first step of adapting the modal reverberator to create the Hammondizer effect is to pick the mode frequencies which specify the heterodyning and modulating sinusoids $\varphi(t)$ and $\psi(t)$. The unique sound of the Hammond organ is largely due to the tonewheels being tuned to the 12-tone equal tempered scale. Here we discuss how to preserve this feature in the context of the Hammondizer audio effect.

Since each mode of the modal reverberator is a narrow bandpass filter, a sufficient frequency density of modes is required to support typical wideband musical signals. In particular, unless each frequency component of the input is sufficiently close to a mode center, it may not contribute audibly to the output. For this reason, tuning the modal reverberator's frequencies to the 12-tone equal tempered scale used by the Hammond organ heavily attenuates the frequencies "in the cracks", producing an artificial sound (compare to composer Peter Ablinger's "Talking Piano" [50]).

To avoid this effect, we use many exponentially-spaced mode frequencies per semitone. Denoting the number of modes per semitone as S, the tuning of each mode is

$$f_w = f_1 \, 2^{w/(12\,S)} \text{ Hz} \tag{26}$$

(cf. Equation (2)). S is chosen to satisfy two subjective constraints. As S gets larger, the computational cost of the modal processor grows. As S becomes small, the modal density decreases and produces an artificial sound. We found by experimentation that $S = 14$ is a good setting that balances these two constraints.

Heterodyning and modulating sinusoids at constant frequencies are given by

$$\varphi_w(t) = \exp\{-j\omega_w t\} \tag{27}$$
$$\psi_w(t) = \exp\{+j\omega_w t\} \tag{28}$$

(cf. Equation (18)).

The next step of adapting the modal reverberator to create the Hammondizer effect is to choose the range of mode frequencies. The range of the Hammond organ is C1 ($\approx$32.7 Hz) to F#7 ($\approx$5919.9 Hz). For simplicity, we set $f_1 = 40$ Hz and let the modes range up seven octaves, up to $f_{1177} = 5120$ Hz; these modes are indexed by a tonewheel index $w \in [1 \cdots 1177]$. These round numbers correspond very closely to the range of the Hammond organ. Forty Hz corresponds to $k \approx 3.5$ and 5120 Hz to $k \approx 87.5$; therefore, this range technically cuts off $\approx$3 semitones from the top and bottom of the range of the Hammond organ tonewheel range. Nonetheless, it does not negatively affect the qualitative effect of the Hammondizer.

4.2. Tone Controls

The heart of the Hammondizer effect is the drawbar tone controls. As before, the drawbar settings give a column $r(t)$ of registrations, which drive the entries of the sparse matrix $\Gamma(r(t))$ according to

$$\gamma_{w,k}(t) = \sum_{d=1}^{9} r_d(t)\, \delta\left(w - k - o_d\, S\right) \tag{29}$$

The only difference from Equation (5) is the presence of S to account for the multiple modes per semitone.

In the Hammondizer context, the entries in $\Gamma(r(t))$ control a Hammond-style pitch shift. The structure of $\Gamma(r(t))$ means that energy in a smoothed baseband signal $n_w(t)$ (centered at some mode frequency f_w) contributes to *nine* different tonewheel amplitudes f_κ, $\kappa \in 1w + So$, according to $\gamma_{w,\kappa}(t)$.

4.3. Vibrato

A vibrato effect that can mimic Hammond organ vibrato is created when the frequencies of the modulating sinusoids $\psi(t)$ are varied. In this case, modulation sinusoids can be implemented with phase accumulators

$$x_w(t) = \exp\{-j\theta_w(t)\} \tag{30}$$

Each vibrato phase signal is given by

$$\theta_w(t) = \theta_w(t-1) + 2^{V_{\text{depth}}/1200 \sin(2\pi/f_s\, V_{\text{rate}}t)} 2\pi/f_s \tag{31}$$

where V_{depth} is the vibrato depth in cents and V_{rate} is the vibrato rate in Hz.

An early Hammond patent [20] praises "...a musical tone containing a vibrato, that is, a cyclical shift in frequency of approximately 1.5%, at a rate of about 6 per second..." To match that design criteria, we typically choose a vibrato depth of 26 cents $\approx$1.5% and a vibrato rate of 6 Hz. Of course, these can be parameterized as desired.

4.4. Crosstalk

Some aficionados point to crosstalk between tonewheels as an important part of Hammond organ sonics. We can consider that since mode filters are not "brick wall" filters, there is already a sort of crosstalk built into the Hammondizer effect.

Drawing inspiration from Pekonen et al. [18], we can explicitly simulate leakage between adjacent tonewheels by adding another matrix multiply between $g_{\text{post}}(t)$ and $\psi(t)$. This creates a new set of signals with crosstalk that includes modes one semitone away from the main modes with a crosstalk level C:

$$a_w''(t) = Ca_{w-S}'(t) + a_w'(t) + Ca_{w+S}'(t) \tag{32}$$

4.5. Memoryless Pickup Nonlinearities

As detailed in [39], distortion effects may be generated by passing a mode through a memoryless nonlinear function or by substituting a complex waveform for the modulation sinusoid waveform. Here we adapt both types of distortion to mimic aspects of the Hammond organ's sonics and design to the Hammondizer. Note that since both kinds of distortion are applied separately to each mode, the output will contain no intermodulation products.

Drawing inspiration from the Mustonen et al.'s model of a guitar pickup [10], we propose a memoryless nonlinearity of the form

$$y_w(t) = \left(1 - e^{-\alpha x_w(t) a'_w(t)}\right) / \alpha \qquad (33)$$

This memoryless nonlinearity is shown for values of $\alpha \in [0.1, 0.3, 0.9]$ in Figure 7. This has the property of maintaining unity gain around zero, but distorting signals with a large swing around zero by compressing positive signals and expanding negative signals. In this article, we will use a value of $\alpha = 0.3$.

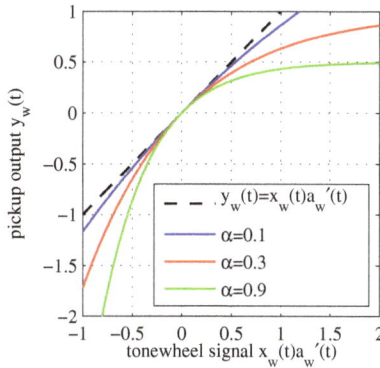

Figure 7. Memoryless tonewheel pickup nonlinearity.

Typically, memoryless nonlinearities like this will produce effects including "harmonic distortion" (new frequencies at multiples of existing frequencies) and "intermodulation products" (new frequencies at sums and differences of existing frequencies). Since this memoryless nonlinearity is applied to the output of a bandpass filter, mostly harmonic distortion will be created, since energy is concentrated at one frequency.

4.6. Tonewheel Basis Distortion

On the Hammond organ, tonewheels may not be perfectly sinusoidal. Also, the lowest octave of tonewheels are cut closer to a square wave shape than a sinusoid. This can be considered a distortion of the sinusoidal basis functions that the tonewheels represent. To approximate this distortion of the lower tonewheel basis functions, we can replace each modulating sinusoid $\psi_w(t)$ with a *sum* of sinusoids

$$\tilde{\psi}_w(t) = \frac{4}{\pi}\exp(jw_w t) + \frac{4}{3\pi}\exp(j3f_w t) + \frac{4}{5\pi}\exp(j5f_w t) \qquad (34)$$

(cf. Equation (10)).

Drawing inspiration from the Hammond organ, this should be done for the lowest octave of tonewheels. In practice, it can be useful to define the effect for a large range of modes.

Note that this distortion is very different in character from the saturating nonlinearities. Specifically, it has the unique feature of being amplitude-independent.

5. Results and Discussion

To demonstrate the features of the Hammondizer, we present a series of examples. Examples of the pitch processing and distortion processing Hammondizer components, operating on a pure tone input, are presented in Sections 5.1 and 5.2, respectively. Examples of the full Hammondizer, applied to program material, are described in Section 5.3. Aspects of the Hammondizer's sonics are visible in the spectrogram and explained in the text. To understand the full effect of the Hammondizer, it is necessary to listen to it. Audio recordings (.wav file format) of all these examples are available online [51].

For all of these examples, the Hammondizer is configured to have 1177 exponentially spaced modes, with 14 modes per semitone over the seven octave range from 40 Hz to 5120 Hz. The two columns of smoothing operations $g_{pre}(t)$ and $g_{post}(t)$ are set so that the gain of each mode during the smoothing operations is set to unity. $g_{pre}(t)$ is simply a column of ones. Except where noted, each mode is assigned a 200-ms decay time. We form $g_{post}(t)$ using smoothing filters which are applied twice, as suggested in [39]. This creates impulse responses with a linear ramp onset and a 200-ms decay (e.g., [52])—i.e., of the form $t \exp\{-\alpha t\}$.

Although we have not emphasized the variation of the mode dampings and complex amplitudes in this article—focusing rather on the novel aspects of the Hammondizer—the mode dampings and complex amplitudes can be set just as in the modal reverberator [37,38], creating hybrid Hammond/reverb effects. The different Hammond organ registrations shown in these results are given in Figure 8 and are taken from a Hammond owner's manual [53] and a Keyboard Magazine article [54].

5.1. Pitch Processing Examples

In this section, we demonstrate the Hammondizer's drawbar tone controls (Figure 9), its frequency range (Figure 10), and crosstalk and vibrato processing (Figure 11).

Figure 9 shows spectrograms of a pure tone input signal and versions processed with the Hammondizer. The input signal (Figure 9a) is a 1.75-second-long sine wave tuned to middle C (C3, ≈261.63 Hz). The output signal (Figure 9b) shows five different Hammondized versions of the input signal. Each of the five versions uses a different registration; the vibrato, crosstalk, and distortion were disabled. The different Hammond organ registrations shown in these results are given in Figure 8. Figure 9b uses the first five registrations of Figure 8 in order.

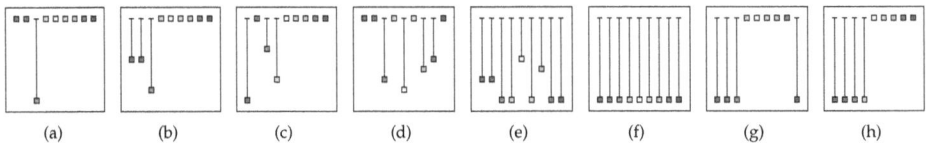

(a)   (b)   (c)   (d)   (e)   (f)   (g)   (h)

Figure 8. Various Hammond organ registrations and their names. (**a**) 008000000 fundamental; (**b**) 447000000 "bassoon"; (**c**) 803600000 "mellow-Dee"; (**d**) 006070540 "clarinet"; (**e**) 668848588 "shoutin' "; (**f**) 888888888 "all out"; (**g**) 888000008 "whistle stop"; (**h**) 888800000 "Jimmy Smith".

The C3 sine wave is tuned very close to the center frequency of mode $w = 455$. Knowing that the Hammondizer uses the matrix $\Gamma(r(t))$ to drive output modes that are offset from each analysis mode by the length-9 column o (recall Table 1 and Equation (1)), we expect that an input consisting of a single sinusoid will in general create output signals with nine sinusoidal components (recall Equation (11)) near modes $(287, 553, 455, 623, 721, 791, 847, 889, 959)$. However, since $\Gamma(r(t))$ is a function of the registration $r(t)$, the output behavior is heavily dependent on the registration. Notice that the 008000000 registration does not affect the signal much beyond a slight lengthening due to the decay time of the modes near C3. Since each $r(t)$ except $r_3(t)$ is zero, only one sinusoid comes out. The second setting, "bassoon" (447000000) produces three sinusoids in response to the input sinusoid, since it has three non-zero $r(t)$s. The amplitude of each sinusoid depends on its corresponding drawbar setting (recall Table 2). The "bassoon," "mellow-Dee," and "shoutin' " registrations have non-zero first drawbar settings—notice that they produce energy an octave below C3. The "shoutin' " and "all out" registrations have no non-zero drawbar settings—notice that the individual sine wave of the input has driven nine sine waves in the output, and that their relative amplitudes reflect the "shoutin' " and "all out" registrations (668848588 and 888888888, respectively).

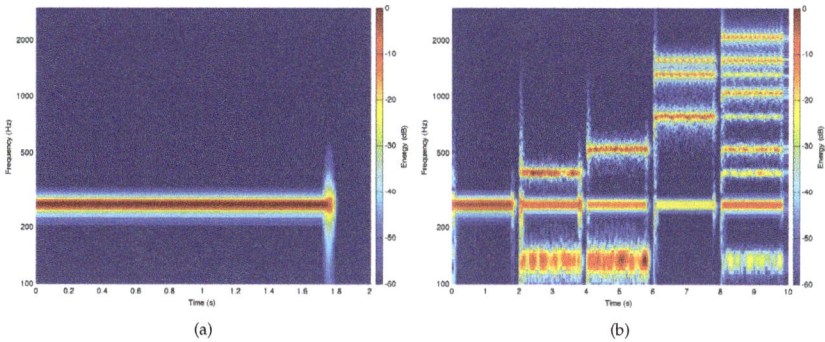

(a) (b)

Figure 9. (a) C3 sine wave input and (b) Hammondized version with five different registrations.

Figure 10 shows spectrograms of a sinusoidal input signal and its Hammondized response. The input signal (Figure 10a) is a series of nine 0.5-second-long sine waves, generated at octave intervals from C0 ($\approx$32.70 Hz) and to C8 ($\approx$8372.02 Hz). The Hammondized output (Figure 10a) used the 668848588 ("shoutin' ") registration, and the vibrato, crosstalk, and distortion were disabled. In a broad sense, the Hammondizer imprints the "shoutin' " partial structure onto the input sinusoids. Note, however, that since the Hammondizer does not have any modes outside the 40 Hz to 5120 Hz frequency range, the C0 and C8 inputs generate little output, though transients in the C0 sinusoid produce a ghostly "whoosh" sound.

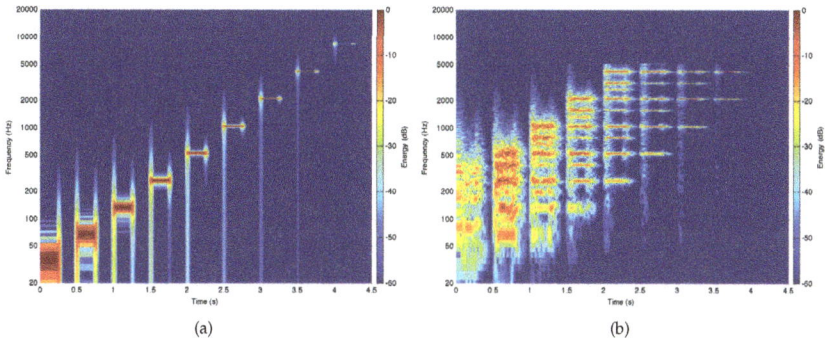

(a) (b)

Figure 10. Showing range of Hammond tonewheels. (a) C0–C8 input signal; (b) Hammondizer with "shoutin' " registration.

The Hammondizer crosstalk and vibrato components are now explored using the pure tone input of Figure 9a. In Figure 11a, the effect of crosstalk is illustrated using the "clarinet" registration with vibrato and distortion disabled. Crosstalk amplitudes of $-\infty$, -24, -18, -12, and -6 dB are simulated. Note the increased presence of energy in adjacent notes with increased crosstalk amplitude. In Figure 11b, the effect of vibrato is studied using a "whistle stop" (888000008) registration, with crosstalk and distortion disabled. Each output uses a 6 Hz vibrato, with (from left to right) vibrato depths of 0, 25, 50, 100, and 1200 cents, with a depth of 25 cents being typical for a Hammond tonewheel organ. As expected, there is a sinusoidal variation in the output frequency of each partial.

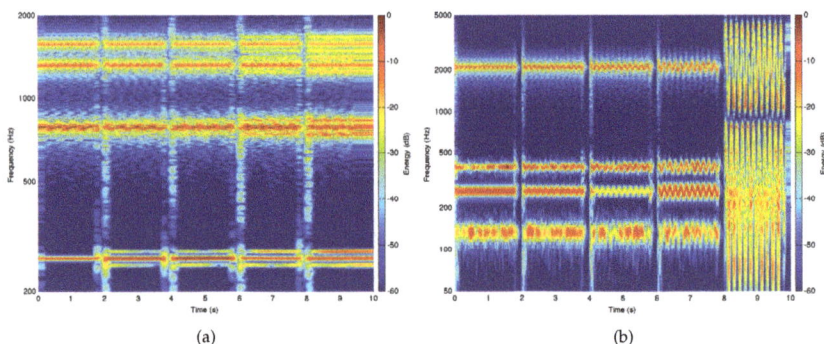

(a)   (b)

Figure 11. (**a**) Clarinet registration, various levels of crosstalk $\in [-\infty, -24, -18, -12, -6]$ dB and (**b**) Whistle stop registration, various levels of vibrato $\in [0, 26, 50, 100, 1200]$ cents on the right.

5.2. Distortion Processing Examples

Here, we demonstrate the Hammondizer's tonewheel shape distortion (Figure 12) and its mode-wise distortion (Figures 13 and 14).

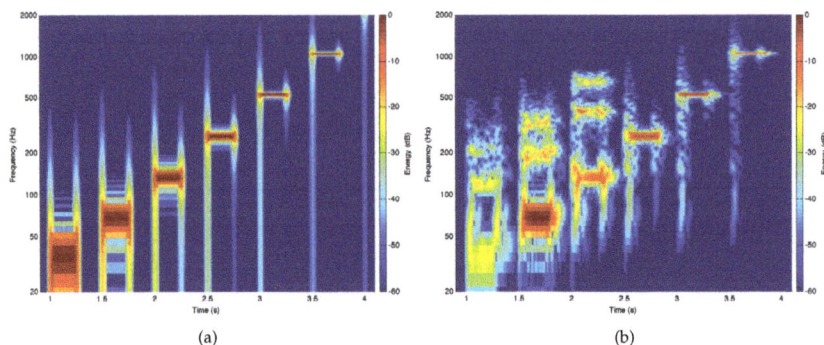

(a)   (b)

Figure 12. Keyboard split demonstration. (**a**) C0–C5 input signal; (**b**) Driving keyboard split.

Figure 12 shows an input signal spectrogram (Figure 12a) and a Hammondized version showing the tonewheel shape distortion (Figure 12b). The input signal is the collection of sinusoids C0 through C5. This is applied to the Hammondizer set to a fundamental-only registration (008000000), with vibrato and distortion disabled. As described above, the lowest two octaves of tonewheels are given 3rd and 5th harmonics. Notice how C0, C1, and C2 produce pronounced 3rd and 5th harmonics even though the registration is 008000000, but that C3–C5 don't generate harmonics.

Figure 13 shows spectrograms of an input signal and its Hammondized version. Figure 13a shows the input signal: five 1.75-second-long sinusoidal bursts, all tuned to C3. From left to right, the input sinusoid amplitudes are 0, −3, −6, −9, and −12 dB. Notice in the output (Figure 13b) that the degree of distortion decreases as the amplitude decreases, as is typical of saturating memoryless nonlinearities.

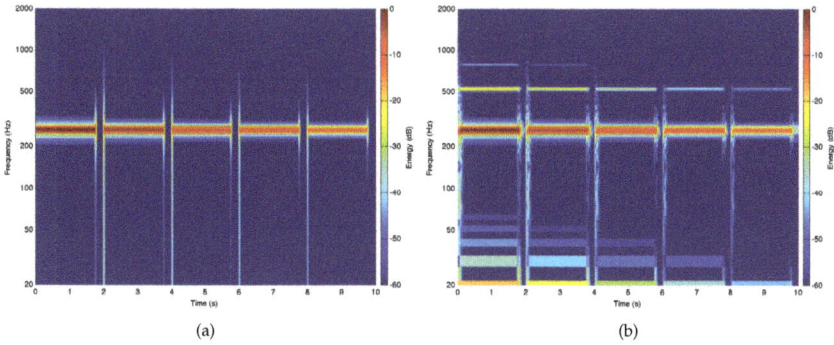

(a)   (b)

Figure 13. Decreasing amplitude interacting with distortion. (**a**) C3 input signal, various amplitudes; (**b**) Distorted.

Recall that the Hammond distortion is generated separately on each key, and accordingly there is no intermodulation distortion. To demonstrate this and to test the presence of intermodulation distortion in our Hammondizer process, we use a signal having C3 and E3 notes which appear both individually and overlapped—see Figure 14. Figure 14b shows the Hammondized result. Notice that there is little to no intermodulation distortion in the output; the response to the combination of C3 and E3 is very nearly equal to the sum of the response to C3 and the response to E3. Figure 14c shows the result of a modified algorithm $y(t) = p(\mathbf{1}^\top(x(t) \odot a'(t)))$ in which the hundreds of individual mode pickup distortions are replaced by a single pickup distortion that operates on the sum of all modes. (cf. Equations (24) and (25)). This more typical approach to implementing distortion produces heavy intermodulation distortion. This sort of intermodulation distortion can be considered unpleasant; its absence can be considered a unique feature of the Hammondizer.

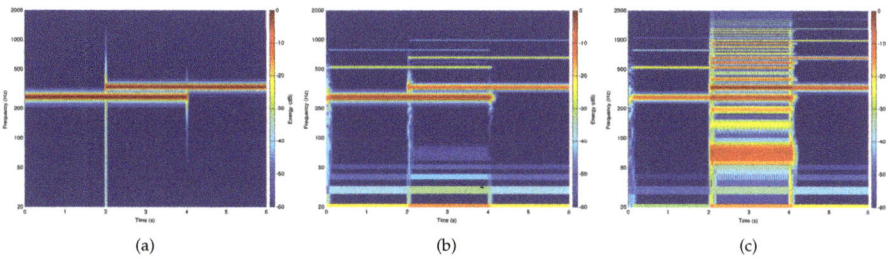

(a)   (b)   (c)

Figure 14. Showing how the Hammondizer mode-wise distortion does not cause intermodulation distortion. (**a**) Input signal; (**b**) Hammondizer with mode-wise distortion; and (**c**) Hammondizer with "global" distortion.

5.3. Full Examples

In this section, we present examples of the full Hammondizer processing program material, a guitar (Figure 15) and a violoncello (Figure 16).

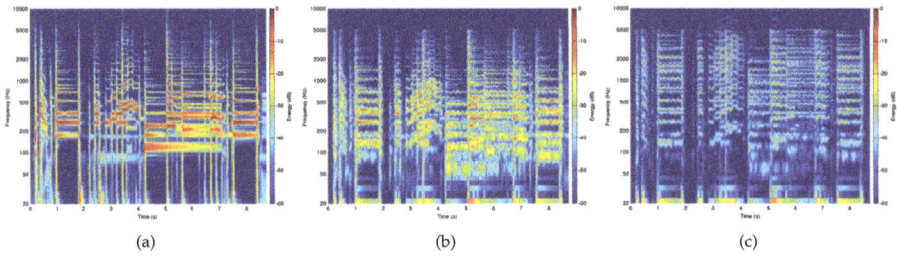

Figure 15. Blues guitar lick, original and two different Hammondized settings. (**a**) Input signal; (**b**) Hammondized, "Jimmy Smith" registration; (**c**) Hammondized, "all out" registration.

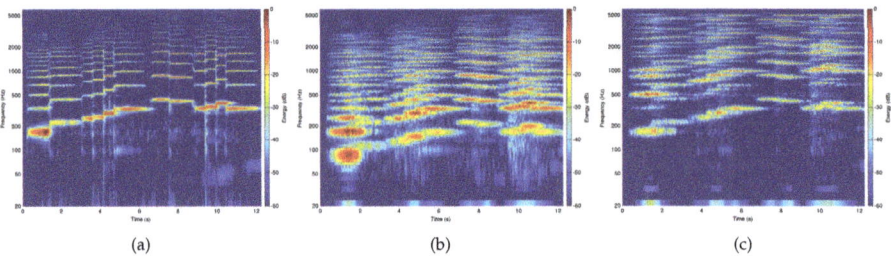

Figure 16. Beginning of "El Cant dels Ocells" [55], original and two different Hammondized settings. (**a**) Input signal; (**b**) Hammondized, "bassoon" registration; (**c**) Hammondized, "clarinet" registration.

Figure 15a shows a blues guitar lick, and two Hammondized versions, with a "Jimmy Smith" (888800000) registration in Figure 15b and an "all out" (888800000) registration in Figure 15c. Notice that the relatively full-range input of the guitar is mostly restricted to below 5120 Hz in the Hammondized examples. Especially from 1–2 s, the vibrato is visible. In the "all out" registration, some pickup distortion is visible above the 5120-Hz tonewheel limit.

Figure 16a shows a melody "El Cant dels Ocells" played on the violoncello, and two Hammondized versions, with a "bassoon" (447000000) registration in Figure 16b and a "clarinet" (006070540) registration in Figure 16c.

6. Conclusions

In this article, we've described a novel class of audio effects—the Hammondizer—that imprints the sonics of the Hammond tonewheel organ on any audio signal. The Hammondizer extends the recently-introduced modal processor approach to artificial reverberation and effects processing. We close with comments on two extensions to the Hammondizer audio effect.

We've discussed parameterizations of each aspect of the Hammondizer which are chosen to closely mimic the sonics of the Hammond organ. For example, the mode frequency range of the Hammondizer is chosen to match the range of tonewheel tunings on the Hammond organ, and the the vibrato rate and depth are chosen to mimic a standard Hammond organ vibrato tone. In closing, we wish to mention that these parameterizations can be extended to loosen the connection to the Hammond organ but widen the range of applicability of the Hammondizer. For instance, the mode frequencies can be tuned across the entire audio range ($\approx$20–20000 Hz) rather than being limited to 40–5120 Hz. In this context, some of the connection with the Hammond organ is relaxed, but the drawbar controls still give a powerful and unique interface for pitch shift in a reverberant context.

Although the Hammondizer is designed to process complex program material as a digital audio effect, it is possible to configure the Hammondizer so that it will act somewhat like a direct Hammond organ emulation. This can be done by driving the Hammondizer with only sinusoids (e.g., a keyboard set to a sinusoid tone) which act as control signals, effectively driving $n(t)$ directly. This is particularly effective using short mode dampings (as in this article). An example is given alongside the other audio online [51].

Acknowledgments: Thanks to Ross Dunkel for discussions on the Hammond Organ.

Author Contributions: Kurt James Werner drafted the main manuscript and helped write signal processing code. Jonathan S. Abel supervised the research, helped in the preparation of the manuscript, and wrote the signal processing code.

Conflicts of Interest: The authors declare no conflict of interest.

References

1.   Hammond, L. Electrical Musical Instrument. U.S. Patent 1,956,350, 24 April 1934.
2.   Ng, T.K. *The Heritage of the Future: Historical Keyboards, Technology, and Modernism.* Ph.D Thesis, University of California, Berkeley, CA, USA, 2015.
3.   Faragher, S. *The Hammond Organ: An Introduction to the Instrument and the Players Who Made It Famous*; Hal Leonard Books: Milwaukee, WI, USA, 2011.
4.   Smith, J.O., III. Spectral Audio Signal Processing. Additive Synthesis (Early Sinusoidal Modeling). 2011. Available online: https://ccrma.stanford.edu/~jos/sasp/Additive_Synthesis_Early_Sinusoidal.html (accessed on 21 March 2016).
5.   Bode, H. History of electronic sound modification. *J. Audio Eng. Soc. (JAES)* **1984**, *32*, 730–739.
6.   Ebeling, K.; Freudenstein, K.; Alrutz, H. Experimental investigation of statistical properties of diffuse sound fields in reverberation rooms. *Acta Acust. United Acust.* **1982**, *51*, 145–153.
7.   Jungmann, T. Theoretical and Practical Studies on the Behaviour of Electric Guitar Pickups. Diploma Thesis, Helsinki University of Technology, Helsinki, Finland, November 1994.
8.   Remaggi, L.; Gabrielli, L.; de Paiva, R.C.D.; Välimäki, V.; Squartini, S. A pickup model for the Clavinet. In Proceedings of the 15th International Conference on Digital Audio Effects (DAFx-12), York, UK, 17–21 September 2012.
9.   Gabrielli, L.; Välimäki, V.; Penttinen, H.; Squartini, S.; Bilbao, S. A digital waveguide-based approach for Clavinet modeling and synthesis. *EURASIP J. Adv. Signal Process.* **2013**, *2013*, 103, doi:10.1186/1687-6180-2013-103.
10. Mustonen, M.; Kartofelev, D.; Stulov, A.; Välimäki, V. Experimental verification of pickup nonlinearity. In Proceedings of the International Symposium on Musical Acoustics (ISMA), Le Mans, France, 7–12 July 2014; pp. 651–656.
11. Thornburg, H. Antialiasing for nonlinearities: Acoustic modeling and synthesis applications. In Proceedings of the International Computer Music Conference (ICMC), Beijing, China, 22–27 October 1999.
12. Horton, N.G.; Moore, T.R. Modeling the magnetic pickup of an electric guitar. *Am. J. Phys.* **2009**, *77*, 144–150.
13. Paiva, R.C.D.; Pakarinen, J.; Välimäki, V. Acoustics and modeling of pickups. *J. Audio Eng. Soc. (JAES)* **2012**, *60*, 768–782.
14. Esqueda, F.; Välimäki, V.; Bilbao, S. Aliasing reduction in soft-clipping algorithms. In Proceedings of the 23rd European Signal Processing Conference (EUSIPCO), Nice, France, 31 August–4 September 2015; pp. 2059–2063.
15. Wiltshire, T. Technical aspects of the Hammond Organ. Available online: http://electricdruid.net/technical-aspects-of-the-hammond-organ/ (accessed on 21 March 2016)
16. Hammond Tone Wheels. Available online: http://www.goodeveca.net/RotorOrgan/ToneWheelSpec.html/ (accessed on 21 March 2016)
17. Reid, G. *Synthesizing Hammond Organ Effects: Part 1*; Sound on Sound (SOS): Cambridge, UK, January 2004.
18. Pekonen, J.; Pihlajamäki, T.; Välimäki, V. Computationally efficient Hammond organ synthesis. In Proceedings of the 14th International Conference on Digital Audio Effects (DAFx-11), Paris, France, 19–23 September 2011.

19. An Introducion to Drawbars. Available online: http://www.hammond-organ.com/product_support/drawbars.htm (accessed on 21 March 2016).

20. Hanert, J.M. Electrical musical apparatus. U.S. Patent 2,382,413, 14 August 1945.

21. Leslie, D.J. Rotatable Tremulant Sound Producer. U.S. Patent 2,489,653, 29 November 1949.

22. Meinema, H.E.; Johnson, H.A.; Laube, W.C., Jr. A new reverberation device for high fidelity systems. *J. Audio Eng. Soc. (JAES)* **1961**, *9*, 284–326,

23. Smith, J.; Serafin, S.; Abel, J.; Berners, D. Doppler simulation and the Leslie. In Proceedings of the 5th International Conference on Digital Audio Effects (DAFx-02), Hamburg, Germany, 26–28 September 2002.

24. Smith, J.O, III. Physical Audio Signal Processing for Virtual Musical Instruments and Audio Effects. The Leslie. 2010. Available online: https://ccrma.stanford.edu/~jos/pasp/Leslie.html (accessed on 21 March 2016).

25. Disch, S.; Zölzer, U. Modulation and delay line based digital audio effects. In Proceedings of the 2nd COST G-6 Workshop on Digital Audio Effects (DAFx-99), Trondheim, Norway, 9–11 December 1999.

26. Dutilleux, P.; Holters, M.; Disch, S.; Zölzer, U. Modulators and Demodulators. In *DAFX: Digital Audio Effects*, 2nd ed.; John Wiley & Sons: West Sussex, UK, 2011; pp. 83–99.

27. Zölzer, U. *DAFX: Digital Audio Effects*, 2nd ed.; John Wiley & Sons: West Sussex, UK, 2011.

28. Kronland-Martinet, R.; Voinier, T. Real-time perceptual simulation of moving sources: Application to the Leslie cabinet and 3D sound immersion. *EURASIP J. Audio Speech Music Process.* **2008**, *2008*, 849696, doi:10.1155/2008/849696.

29. Herrera, J.; Hanson, C.; Abel, J.S. Discrete time emulation of the Leslie speaker. In Proceedings of the 127th Convention of the Audio Engineering Society (AES), New York, NY, USA, 9–12 October 2009.

30. Välimäki, V.; Abel, J.S.; Smith, J.O. Spectral delay filters. *J. Audio Eng. Soc. (JAES)* **2009**, *57*, 521–531.

31. Werner, K.J.; Dunkel, W.R.; Germain, F.G. A computational model of the Hammond organ vibrato/chorus using wave digital filters. In Proceedings of the 19th International Conference on Digital Audio Effects (DAFx-16), Brno, Czech, 5–9 September 2016.

32. Reid, G. *Synthesizing Tonewheel Organs*; Sound on Sound (SOS): Cambridge, UK, November 2003.

33. Reid, G. *Synthesizing Tonewheel Organs: Part 2*; Sound on Sound (SOS): Cambridge, UK, December 2003.

34. Reid, G. *Synthesizing the Rest of the Hammond Organ: Part 2*; Sound on Sound (SOS): Cambridge, UK, February 2004.

35. Reid, G. *Synthesizing the Rest of the Hammond Organ: Part 3*; Sound on Sound (SOS): Cambridge, UK, March 2004.

36. Pakarinen, J.; Välimäki, V.; Fontana, F.; Lazzarini, V.; Abel, J.S. Recent advances in real-time musical effects, synthesis, and virtual analog models. *EURASIP J. Adv. Signal Process.* 2011, *2011*, 940784, doi:10.1155/2011/940784.

37. Abel, J.S.; Coffin, S.; Spratt, K.S. A modal architecture for artificial reverberation. *J. Acoust. Soc. Am.* **2013**, *134*, 4220, doi:10.1121/1.4831495.

38. Abel, J.S.; Coffin, S.; Spratt, K.S. A modal architecture for artificial reverberation with application to room acoustics modeling. In Proceedings of the 137th Convention of the Audio Engineering Society (AES), Los Angeles, CA, USA, 9–12 October 2014.

39. Abel, J.S.; Werner, K.J. Distortion and pitch processing using a modal reverb architecture. In Proceeding of the 18th International Conference on Digital Audio Effects (DAFx-15), Trondheim, Norway, 30 November–3 December 2015.

40. Välimäki, V.; Parker, J.D.; Savioja, L.; Smith, J.O.; Abel, J.S. More than fifty years of artificial reverberation. In Proceedings of the 60th International Conference of the Audio Engineering Society (AES), Leuven, Belgium, 3–5 February 2016.

41. Bilbao, S. Sound synthesis and physical modeling. In *Numerical Sound Synthesis*; Wiley: Hoboken, NJ, USA, 2009.

42. Avanzini, F.; Marogna, R. A modular physically based approach to the sound synthesis of membrane percussion instruments. *IEEE Trans. Audio Speech Lang. Process.* **2010**, *18*, 891–902.

43. Morrison, J.D.; Adrien, J.-M. MOSAIC: A framework for modal synthesis. *Comput. Music J.* **1993**, *17*, 45–56.

44. Trautmann, L.; Rabenstein, R. Classical synthesis methods based on physical models. In *Digital Sound Synthesis by Physical Modeling Using the Functional Transform Method*, 1st ed.; Springer: New York, NY, USA, 2003.

45. Vorkoetter, S. The Science of Hammond Organ Drawbar Registration. Available online: http://www.stefanv.com/electronics/hammond_drawbar_science.html (accessed on 21 March 2016).

46. Smith, J.O., III. Mathematics of the Discrete Fourier Transform (DFT) with Audio Applications. Convolution. 2007. Available online: https://ccrma.stanford.edu/~jos/mdft/Convolution.html (accessed on 21 March 2016).

47. Morse, P.M.; Ingard, K.U. *Theoretical Acoustics*; Princeton University Press: Princeton, NJ, USA, 1987.

48. Fletcher, N.H.; Rossing, T.D. *Physics of Musical Instruments*, 2nd ed.; Springer: Berlin, Germany, 2010.

49. Wang, A.; Smith, J.O. On fast FIR filters implemented as tail-canceling IIR filters. *IEEE Trans. Signal Process.* **1997**, *45*, 1415–1427.

50. Barrett, G.D. Between noise and language: The sound installations and music of Peter Ablinger. *Mosaic J. Interdiscip. Study Lit.* **2009**, *42*, 147–164.

51. Werner, K.J.; Abel, J.S. "Modal Processor Effects Inspired by Hammond Tonewheel Organs"—Audio Examples. Available online: https://ccrma.stanford.edu/~kwerner/appliedsciences/hammondizer.html (accessed on 20 June 2016).

52. Abel, J.S.; Wilson, M.J. Luciverb: Iterated convolution for the impatient. In Proceedings of the 133rd Convention of the Audio Engineering Society (AES), San Francisco, CA, USA, 26–29 October 2012.

53. Hammond Suzuki, Ltd. Model: Sk1/ Sk2 stage keyboard owner's manual. Document ID: 00457-40173 V1.50-130218.

54. Finnigan, M. *5 Great B-3 Drawbar Settings*; Keyboard Magazine: San Bruno, CA, USA, 2012.

55. Xavier Serra. "El Cant dels Ocells." 27 January, 2013. Used under Creative Commons Attribution 3.0 Unported (https://creativecommons.org/licenses/by/3.0/), Available online: https://www.freesound.org/people/xserra/sounds/176098/ (accessed on 21 March 2016).

applied
sciences

MDPI

Article

Eluding the Physical Constraints in a Nonlinear Interaction Sound Synthesis Model for Gesture Guidance [†]

Etienne Thoret [‡], Mitsuko Aramaki, Charles Gondre, Sølvi Ystad * and Richard Kronland-Martinet

Laboratoire de Mécanique et d'Acoustique, Centre National de la Recherche Scientifique (CNRS), Unité Propre de Recherche (UPR) 7051—Aix Marseille Université, Centrale Marseille, 4 impasse Nikola Tesla, CS 40006, F-13453, Marseille cedex 13, France; etienne.thoret@mcgill.ca (E.T.); aramaki@lma.cnrs-mrs.fr (M.A.); gondre@lma.cnrs-mrs.fr (C.G.); kronland@lma.cnrs-mrs.fr (R.K.-M.)

* Correspondence: ystad@lma.cnrs-mrs.fr; Tel.: +33-4-91-16-42-59; Fax: +33-4-91-16-40-12

† This paper is an extended version of paper published in the 16th International Conference on Digital Audio Effects (DAFx), Maynooth, Ireland, 2–5 September 2013.

‡ Current address: Schulish School of Music, McGill University, Montreal, 555 Sherbrooke Street West, Montréal, QC H3A 1E3, Canada.

Academic Editor: Vesa Valimaki

Received: 15 March 2016; Accepted: 21 June 2016; Published: 30 June 2016

Abstract: In this paper, a flexible control strategy for a synthesis model dedicated to nonlinear friction phenomena is proposed. This model enables to synthesize different types of sound sources, such as creaky doors, singing glasses, squeaking wet plates or bowed strings. Based on the perceptual stance that a sound is perceived as the result of an action on an object we propose a genuine source/filter synthesis approach that enables to elude physical constraints induced by the coupling between the interacting objects. This approach makes it possible to independently control and freely combine the action and the object. Different implementations and applications related to computer animation, gesture learning for rehabilitation and expert gestures are presented at the end of this paper.

Keywords: friction sounds; intuitive control; gesture guidance

1. Introduction

Friction between two objects can produce a wide variety of sounds like squealing brakes, creaking doors, squeaking dishes or beautiful sounds from bowed strings in musical conditions. The understanding of such acoustical phenomena constitutes a historical challenge of physics that was subjected to early researches already from the 18th century [1]. Today friction phenomena are still debated and several mechanisms are under investigation, such as the conversion process from kinetic to thermal energy provoked by friction at the atomic scale [2]. In spite of the physical complexity of friction phenomena, physically inspired friction models have been developed within various domains [3,4]. Such models have for instance been used for musical purposes in the case of simulations of sounds from bowed string instruments intended for electronic cellos or violins [5] or unusual friction driven instruments based on both physical and signal models [6]. In the specific case of violin sounds, more elaborated models also considered the thermal effects that appear at the contact area between the bow and the string [7,8]. All these models strongly depend on physical considerations and require the control of specific physical parameters. For example in the model proposed by Avanzini et al. [3] seven physical parameters must be defined to synthesize a friction sound. Although these parameters well characterize the physical behavior of the friction phenomenon, they are not at all intuitive for naive users. In fact, to facilitate the control of such models, a better understanding of the perceptual impact of its parameters would be needed. Other approaches based on holistic observations have been proposed. Rather than simulating the physical behavior of the phenomena, these models generally focus on the perceptual relevance of the elements that contribute to the generated sound and whether they contribute to the evocation of a particular event. For example, Van den Doel et al. [9] considered that the noisy character of friction sounds was perceptually salient and proposed to synthesize noisy friction sounds by modeling the sound as band-pass filtered noise which central frequency varies with respect to the relative velocity of the two interacting objects. In addition to be computationally efficient, this model offers numerous control possibilities and provides a wide variety of controls of the friction sound that can be easily handled by naive users. In addition to the velocity control, the roughness of the surface can easily be modified by shaping the spectral content of the noise in filtering pre-processing step. Another advantage of using signal models that are unconstrained by mechanics is that physically impossible situations can be generated. Hence, unlimited sound generation possibilities are available, such as continuous morphing between sounding objects or perceptual cross-synthesis between sound properties.

In this article we therefore propose to synthesize nonlinear friction sounds from signal models using a paradigm based on perceptually relevant acoustic morphologies. This approach enables to elude the physical constraints and to separately control the nature of the interaction and the resonating object from a perceptual point of view. It should be noted that the underpinning terminology may therefore differ from the one used to describe the physics behind the friction phenomena. In particular, friction is generally modeled as an interaction between two objects while we here consider that a friction sound can be perceptually described as the consequence of an action of a moving, non-resonant exciter on a resonating object. Hence, the proposed approach makes it possible to freely combine actions and objects to simulate both real and physically non-plausible situations and to explore the ductility of synthesis in terms of control possibilities for various applications. For instance, from a perceptual point of view, squeaking or squealing sounds may convey the sense of "effort" or "annoyance" while tonal sounds may convey the sense of "success" or "wellness". Based on these evocative auditory percepts, new sonic devices aiming at guiding or learning specific gestures can then be developed with such flexible sound synthesis models. In particular, transitions between perceptually different acoustical behaviors can be defined according to arbitrary mapping specifications related to the used devices. Depending on the mapping, such models also enable to choose different levels of difficulty in a learning process. In the present article, we focus on a general model that simulates different acoustical behaviors from spectral considerations. Finally, several applications linked to gesture guidance and video game applications are presented.

2. Perceptually Informed Friction Sound Synthesis

To design a ductile synthesis tool simulating non-linear acoustical phenomena, we used a paradigm describing the auditory perception of sound events. This paradigm can be illustrated by the following example: when hearing the sound of a metal plate hitting the floor, the material of the plate can easily be recognized. Likewise, if the plate is dragged on the floor, the recognition of its metallic nature remains possible, but the continuous sound-producing dragging action can also be recognized. This example points out a general principle of auditory perception consisting in decomposing sounds into two main contributions: one that characterizes the resonating object properties and another linked to the action generating the vibrating energy. In other words, the produced sound can be defined as the consequence of an action on an object. It must be noted that this is not true from a physical point of view, since friction between two objects induces mutual energetic exchanges between interacting objects. Nevertheless from a perceptual point of view, the action/object approach is still relevant as long as the action part simulates the acoustic morphologies that characterize the nonlinear behavior provoked by the interaction.

Such descriptions of an auditory event have been highlighted in various psychological studies, pointing out that our auditory perception uses information from invariant structures included in the auditory signal to recognize the auditory events. Inspired from the ecological theory of visual perception and the taxonomy of invariant features of visual flow proposed by Gibson [10], McAdams [11] and Gaver [12,13] adapted this theory to auditory perception. It has since then been successfully used for sound synthesis purposes. Concerning the object, it has been shown that the damping behavior of each partial characterizes the perceived material of an impacted object [14]. In line with this result, a damping law defined by a global and a relative damping behavior was proposed to control the perceived material in an impact sound synthesizer [15]. Concerning the action, recent studies focused on the synthesis of continuous interaction sounds such as rubbing, scratching and rolling [16–18]. They revealed that the statistics of the series of impacts that model the continuous interaction between two objects characterize the nature of the perceived interaction. Additionally this enables to intuitively control the synthesis of such sounds. This ecological approach then provides a powerful framework to extract the acoustic information that makes sense when listening to a sound [19].

The framework of the action/object paradigm previously described supposes that actions and objects are perceived independently and can therefore be simulated separately. Practically, an implementation with a linear source-resonance filtering is well adapted to the present perceptual paradigm, see Figure 1. The output sound indeed results from a convolution between a source signal $e(t)$ and the impulse response $h(t)$ of a vibrating object. The resonator part $h(t)$ is implemented with a resonant filter bank, which central frequencies correspond to the eigen frequencies of the object. A control strategy of the perceptual attributes of the object by semantic descriptions for example the perceived material, size and shape is proposed in [20].

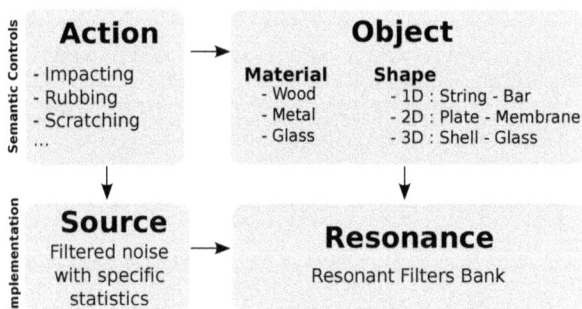

Figure 1. The Action/Object Paradigm.

2.1. Source Modeling

In this section, the modeling of the source signal related to the perception of actions such as squeaking or creaking is examined. For that purpose, basics physical considerations and empirical observations made on recorded signals are firstly presented. A signal model is finally proposed according to invariant morphologies revealed by these observations.

2.1.1. The Coulomb Friction Model

The Coulomb friction can be described in its first approximation by a simple phenomenological model sketched in Figure 2. When an object is rubbing another one, and when the contact force between the objects is large enough, the friction behavior can be described by the displacement of a mass held by a spring, and gradually moved from its equilibrium position by a conveyer belt moving at a velocity v. The object is then said to be in the sticking phase. When the friction force F_{fr} becomes smaller than the restoring force F_r, the mass slides in the opposite direction until F_{fr} becomes larger than F_r. The object is then said to be in the slipping phase. This model describes the so-called stick-slip motion, or the Helmoltz motion.

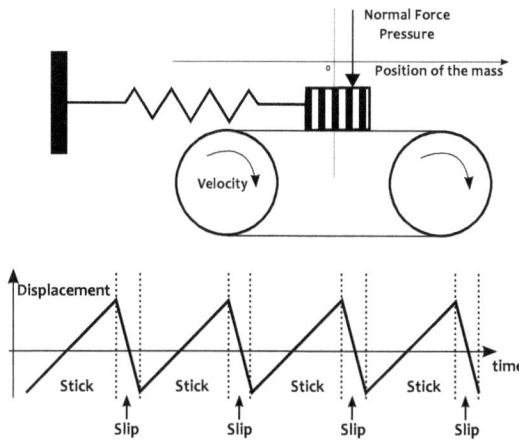

Figure 2. The conveyer-belt model and a typical stick-slip motion.

If we neglect the slipping phase, the resulting displacement $x(t)$ of the exciter on the conveyer belt surface corresponds to a sawtooth signal whose Fourier decomposition is:

$$x(t) = \frac{1}{2} - \frac{1}{\pi} \sum_{k=1}^{\infty} \frac{1}{k} sin(\pi f_0 kt) \tag{1}$$

Several models that take into account a non-zero sliding duration can be found, see Figure 2. The Fourier decomposition of such a signal is a harmonic spectrum with decreasing amplitude according to the harmonic order.

This model provides a first a priori about the behavior of the source signal to be simulated: the excitation should have a harmonic spectrum. This model allows us to determine a physically informed signal morphology associated to the nonlinear friction phenomenon, even if we are aware that it cannot be generalized to any friction situation.

2.1.2. Empirical Observations on Recorded Sounds

In this section, empirical observations of various nonlinear friction sounds are presented. Five situations are analyzed: a creaky door, a squeaking and singing wet wineglass, a squeaking wet plate, and a bowed cello string. These five situations were chosen as they are the consequence of a human movement and are therefore good candidates to be used in gesture guidance device. Except for the door creak and the cello, the recordings were made in an anechoic room using a cardioid Neumann-KM84i microphone positioned about 30 cm above the rubbed object. The recorded sounds are available online [21] from [22]. An inharmonicity analysis was performed for each sound. A partial tracking routine developed by Dan Ellis [23] was used to detect the set of peaks of the spectrum in each frame of the excerpt. A linear regression was applied to the inharmonicity value, $\gamma_n = \frac{f_n}{n f_0} - 1$, where f_n is the n-th peak detected, and f_0 the fundamental. If the spectrum is harmonic the inharmonicity γ_n is 0. The analysis of the three recordings revealed that they all had harmonic spectra.

- *Creaky Door.* The time-frequency representation of the sound is presented in Figure 3. The sound has a harmonic spectrum whose fundamental frequency varies over time. The fundamental frequency is related to the rotation speed and the pressure at the rotation axis of the door. The large range of variations of the fundamental frequency is also a noticeable characteristic of this signal morphology that varies from a very noisy like sound when f_0 is low, to a "singing" harmonic one for higher values.

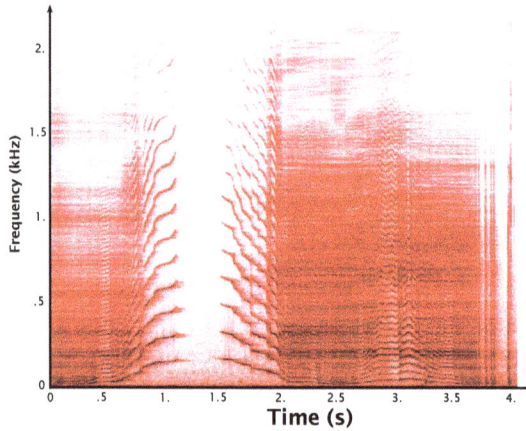

Figure 3. Short Term Fourier Transform of a creaky door sound.

- *Squeaking Wet Plate.* The spectrograms of the squeaking sound and the impulse response of the plate are shown in Figure 4. This sound is also harmonic but provides less variations of the fundamental frequency than the creaky door. In addition, the fundamental frequency f_0 slowly varies around a central value. The vibrating modes of the plate are clearly visible on the impulse response and are excited when f_0 gets close to the excitation frequency. This observation consolidates the stance to use the perceptual action/object paradigm based on a separate control of the exciter and the resonator.

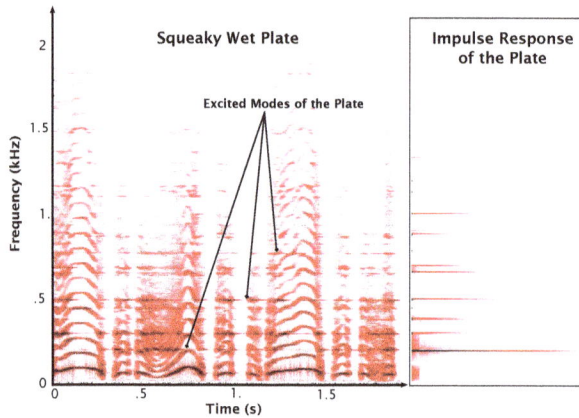

Figure 4. Short Term Fourier Transform of a sound produced by a squeaky wet plate and its impulse response.

- *Squeaking and Singing Glass.* When a glass is rubbed, the associated sound may reveal several behaviors qualified as squeaking and singing, see Figure 5. The squeaking phase provides a similar behavior as the squeaking wet plate: f_0 varies chaotically around a central value. The singing phase appears when f_0 falls on a wineglass mode. The transitions between the squeaking and singing situations are almost instantaneous and hardly predictable.

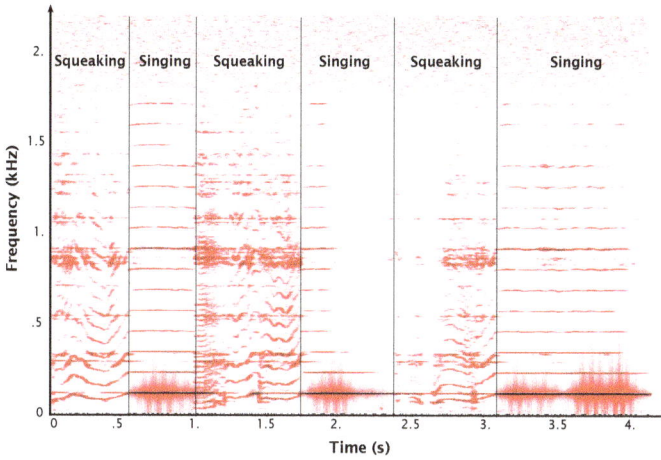

Figure 5. Short Term Fourier Transform of a sound produced by a squeaking and singing glass.

- *Bowed Cello String.* The sound of a bowed string shows behaviors similar to those of the singing wineglass, see Figure 6. The fundamental frequency is locked to a vibrating mode of the string according to the bow velocity and force. The main difference resides in the attack which is smoother than in the singing wineglass sound.

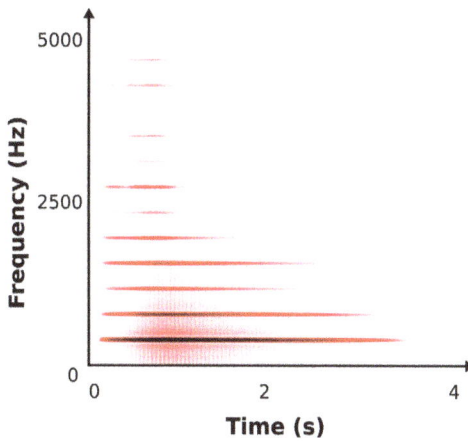

Figure 6. Short Term Fourier Transform of a bowed cello string—A3 (220 Hz).

These empirical observations lead us to hypothesize that the acoustical morphology that mainly characterizes such auditory events relies on the temporal evolution of the fundamental f_0 frequency of a harmonic spectrum. As it is well known for sustained musical instruments most of these acoustical behaviors can indeed be modeled by a harmonic spectrum whose fundamental frequency varies over time. The analyses of the previously recorded sounds confirmed that these assumptions also hold for other friction cases. Nevertheless, even if each spectrum locally is perfectly harmonic, each individual harmonic is not perfectly localized in frequency and should be considered as a band-pass filtered noise with a very thin bandwidth.

2.1.3. Signal Model

From the empirical signal observations presented in previous sections, the following general synthesis model of the source $e(t)$ is proposed:

$$e(t) = \sum_{k=1}^{N_{f0}} A_k(t) sin(\Omega_k(t)) \tag{2}$$

where $N_{f0} = \left\lfloor \frac{f_s}{2f_0} \right\rfloor$ and f_s the sample frequency. The instantaneous frequency is given by $\frac{d\Omega_k}{dt}(t) = 2\pi k f_0(t)$, and A_k the amplitude modulation law for the k-th harmonic. From a signal point of view, the sound morphologies for squeaking, creaking and singing objects differ by the temporal evolution of the fundamental frequency $f_0(t)$ and the amplitudes $A_k(t)$.

- *Fundamental frequency behavior.* The following general expression for f_0 is proposed:

$$f_0(t) = (1 + \epsilon(t))\Phi(\Gamma(t)) \tag{3}$$

  Φ represents the fundamental frequency defined from a given mapping $\Gamma(t)$. $\epsilon(t)$ is a low pass filtered noise with cutoff frequency defined below 20 Hz. This stochastic part models the chaotic variations of the fundamental frequency observed in the case of squeaking or creaking sounds. In the case of self-sustained oscillations, ϵ is null. For a more physically informed mapping, the fundamental frequency can be defined from the dynamical parameters of the two bodies in contact, in particular $\Gamma(t) = (p(t), v(t))$, by assuming that the friction sounds mainly depend on the dynamic pressure $p(t)$ and the relative velocity $v(t)$ between objects. The mapping can be defined with other descriptors according to the desired applications as mentioned in the introduction and developed below in the applications section.
- *Amplitude Modulation Behavior.* In the specific case of the singing wineglass low frequency beats are observed in the generated sounds. The frequency of the amplitude modulation is directly linked to the velocity $v(t)$ of the wet finger which is rubbing the glass rim and the diameter of the glass D. An explicit expression of $A_k(t)$ as been proposed [24,25]:

$$A_k(t) = \frac{1}{k} sin\left(2\pi \int_0^t \frac{v(t)}{\pi D} dt\right) \tag{4}$$

In other words, the smaller the diameter, the higher the velocity and the amplitude modulation, and vice versa. For other situation than the singing wineglass we proposed an amplitude modulation law defined by $A_k = \frac{1}{k}$.

The parameters of the model related to different friction phenomena are summarized in Table 1.

Table 1. Mapping between descriptors and signal parameters – $\Gamma(t)$ represents the mapping between the descriptors and the fundamental frequency, the velocity and the pressure – $\tilde{f}_{(n,0)}$ is the n-th mode of the vibrating object (wineglass or string). Note that in each case the resonant filter bank that models the material and the shape of the object has to be calibrated accordingly [15,26] .

Sound	Frequency Modulation $f_0(t)$	Amplitude Modulation $AM_k(t)$
Creaky Door Squeaky Vessel Squeaky Wineglass	$(1 + \epsilon(t))\Phi(\Gamma(t))$	$\frac{1}{k}$
Singing Wineglass	$\tilde{f}_{(n,0)}$	$\frac{1}{k}sin\left(2\pi \int_0^t \frac{v(t)}{\pi D} dt\right)$
Bowed String	$\tilde{f}_{(n,0)}$	$\frac{1}{k}$

2.2. Implementation

The signal model previously described proposes to reach the expected flexibility through a source-filter approach that enables to separate the perceptual control of the action from that of the object. The action properties are thus characterized by the source signal which corresponds to a harmonic spectrum whose temporal evolution of fundamental frequency and amplitude characterize the evoked nonlinear behavior. In this section, we propose two different ways to implement such a source.

2.2.1. Additive Synthesis

Additive synthesis is certainly the most natural way to simulate the harmonic spectrum. This approach consists in summing the outputs of sinusoidal oscillators whose frequencies f_n are driven by the different behaviors previously described.

The main flaw is that the synthesized sounds are sometimes perceived to be non natural. This is probably due to the fact that the harmonics are indeed not perfectly localized in frequency in the previously recorded sounds. They should therefore be considered as bandpass filtered noises with very thin bandwidths. Replacing the sinusoids by band-pass filtered noises could provide additional acoustic cues that are important for the perceived realism of the sound. Another possibility would be to expand the bandwidth around each harmonic by adding spectral components around each harmonic in an additive way. Nevertheless, this method drastically increases the complexity of the algorithm and makes it almost impossible to propose an intuitive control. In order to increase the realism of the synthesized sounds without loosing too much computational efficiency the following section proposes another way to implement the generation of a harmonic spectrum, and thus facilitating the shaping of the bandwidths around each harmonic.

2.2.2. Subtractive Synthesis

Subtractive synthesis emanated from the digital filtering theory [27] and has been extensively used for speech analysis and synthesis [28,29]. It consists in shaping a sound with a rich harmonic content such as a noise by a filtering process. It is thus possible to generate a harmonic spectrum by creating a resonant filter bank whose central frequencies are defined to construct the desired harmonic spectrum. When the source is a noise and the damping of each filter is set to zero, the spectrum is perfectly tonal and corresponds to the one obtained by additive synthesis. Contrary to the additive synthesis, the bandwidth around each harmonic can globally be expanded by increasing the damping values to make the sound more noisy, without increasing the computational cost. Practically, we implemented such a subtractive method by designing a filter bank composed of 2nd-order resonant filters [30] whose central frequencies are tuned to the frequencies of the harmonic spectrum. Their damping is then left as a control of the sound synthesis process. The source is finally generated by filtering a white noise with this resonant filter bank. The frequencies and amplitude modulation laws of the different friction behaviors described previously can then be applied in the same way as for the additive synthesis process.

3. Applications

The synthesis process and the control described above led to several applications. Three of them are presented here and constitute firstly the creation of a tool for motor disease rehabilitation and gesture learning, secondly the possibility to create flexible bowed string synthesizers enabling to choose their playability a priori, and finally the conception of an event-driven interactive sound synthesis tool in virtual audiovisual environments.

3.1. Auditory Guidance Tools for Motor Learning

The main goal behind the development of the flexible synthesis model described above was to create new devices based on sounds to guide gestures. In particular, this problem was motivated by the rehabilitation of a specific motor disease called dysgraphia which affects some children that aren't able to write fluidly. The idea was thus to provide an auditory feedback based on a metaphoric "singing wineglass" as a correct target. Such a feedback would continuously inform them on the correctness of their gesture and would influence them to intuitively modify their graphical movement to reach the required fluidity.

In order to evaluate the relevance of real-time auditory feedback to improve handwriting gestures, a study was performed on adults who were asked to write new characters with their non-dominant hand [31]. Thirty two French natives who were right-handed adults who had normal vision and hearing and presented no neurological or attentional deficits participated in this experiment. The task consisted in writing four new characters of Tamil script with their non-dominant hand on a sheet of paper that was affixed to a graphic tablet. A movie showing an example of the learning task is available online [32]. The writing was accompanied by synthetic sounds simulating a natural action on an object, which varied as a function of the handwriting quality. A rubbing sound evoking a chalk on a blackboard was associated to a correct handwriting. When the handwriting was too slow, the rubbing was transformed into a more unpleasant sound evoking a squeaking door, which urged the writers to increase the speed of their movements to obtain a more pleasant sound. The handwriting was considered too slow when the instantaneous tangential velocity was below $1.5\ \text{cm·s}^{-1}$. Consequently, the transition between the friction sound and the rubbing sound was made at this threshold.

In addition to the continuous guidance, impact sounds evoking cracking sounds were added whenever the handwriting was jerky. These additional discrete sounds, which were obtained from a synthesizer developed for intuitive control of impact sounds [15,26], evoked a lack of fluency of the handwriting action and could in a metaphorical way be associated with a chalk that breaks while writing or a cracking vinyl record. These sounds were generated when the difference between two velocity peaks was less than 40 ms and when the velocity difference between these two peaks was less than $1.0\ \text{cm·s}^{-1}$. These thresholds were determined empirically and validated from previous experiments on movement fluency [33].

The results were analyzed through two different types of variables: (1) kinematic variables on handwriting movement (2) spatial variables on the written trace. The kinematic variables were obtained from the mean tangential velocity on the tablet and the number of velocity peaks within a certain frequency range (5–10 Hz). The spatial variables corresponding to the trace length of the total trajectory was calculated in terms of a Euclidian distance between the subject's trace and a reference trace obtained from the recordings of a trained proficient adult using his dominant hand. The better the character matched with the reference, the higher the score.

The subjects were divided in two groups and were told to learn the four unknown characters with their non-dominant hand with or without real-time auditory feedback. Half of the participants first learned the two characters without the auditory feedback and the following two characters with the auditory feedback, while the second half first learned the two characters with the auditory feedback and the following without the auditory feedback.

Results revealed that sonifying handwriting during the learning process of unknown characters improved the fluency and speed of their production, despite a slight reduction of their short-term accuracy, hereby validating the use of sounds to inform about handwriting kinematics. By transforming kinematic variables into sounds, proprioceptive signals are translated into exteroceptive signals, thus revealing hidden characteristics of the handwriting movement to the writer. Although this study focused on adult participants, the encouraging results incited the authors to propose a similar approach for learning and rehabilitation of dysgraphic children which also appeared to be efficient [34]. In this study the difficulty of the task could be adjusted according to the degree of fluidity for each child. Results revealed that dysgraphic children increase their writing fluidity across training sessions by using this sonification tool. Current experiments are conducted to evaluate whether this sonification method is more efficient than existing ones based on visual and proprioceptive feedbacks.

3.2. Pedagogical Bowed String Instrument

The process of learning to play an instrument is generally long and cumbersome since specific and accurate gestures are involved. Instrument makers cannot easily simplify this learning process even if they can increase the playability to some extent by acting on the material of the instrument. However, they are always constrained by the physics and cannot ensure for example that a specific Shelleng diagram that is linked to the playability of a bowed string is associated to a given instrument. As the models proposed here enabled the generation of bowed string sounds that are not constrained by physics as traditional instruments, it is possible to imagine a new evolving pedagogical tool. We may indeed create a virtual instrument whose playability can be defined a priori in order to adapt the playability of the instrument to the progress of the performer. The violin string would be simulated by a metal string with controllable modal properties.

For instance, we defined the mapping function Γ between the dynamic descriptors $(v(t), p(t))$ and the fundamental frequency f_0 of the harmonic spectrum. The evolution of f_0 with respect to Γ is freely tunable and enables the simulation of sudden transitions occurring in bowed string instruments. An example of the proposed space is presented in Figure 7. This strategy fulfills the wanted requirements of flexibility. It enables to define behaviors which are theoretically not allowed by physics but that may be useful in the learning process. This mapping can further be adjusted with respect to the learning progress of a student.

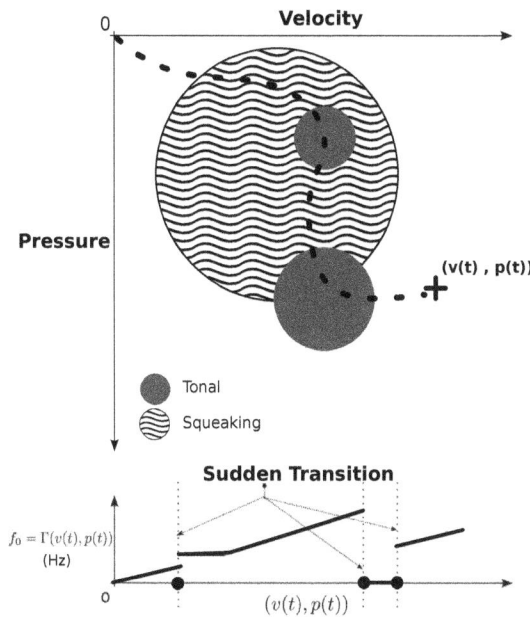

Figure 7. (**Top**) An example of control strategy that defines the vibrating behavior (squeaking and self-oscillating areas) according to Velocity/Pressure plane. The dotted line gives an example of possible trajectories in the Velocity/Pressure plane; (**Bottom**) Example of mapping function between the descriptors, e.g., velocity and pressure $v(t)$ and $p(t)$ and the fundamental frequency $f_0(t) = \Gamma(v(t), p(t))$.

3.3. Computer Animation

Most of the techniques used by current Foley artists are based on recorded samples which are triggered with the sound events occurring in the visual scene. Nevertheless, due to the use of pre-defined sound samples such techniques often lack of intrinsic timbre control and do not offer the possibility to precisely shape the sound continuously with respect to the user's actions. For video games, the sample storage moreover occupies a lot of storage space and the graphical data often takes up most available resources. Our model offers the possibility to synthesize such sounds in real-time, to freely modify their timbre independently of physical constraints, and to directly link the dynamic parameters of visual events to the sound generating parameters. Above all, sound synthesis doesn't use any data storage as sounds are directly generated on the fly. An example of such a control is available online [22,35].

In this context, a novel framework for interactive synthesis of solid sounds driven by a game engine has been developed in [36]. This work highlights the possibilities offered by the so-called "procedural audio" generation aiming at replacing the use of pre-recorded samples for sound effects. Such a tool is particularly adapted to handle evolutive and interactive situations like morphing or continuous transitions between objects and/or actions. Practically, a game engine in which the virtual scene is designed drives the sound synthesizer. The mapping between the visual and the sound parameters is defined so that the audio-visual coherency of the global scene is respected. Friction sounds like squeaking vessels or singing glasses constitute relevant cases to highlight interactive situations for which the user's actions have to be taken into account in a continuous way. An example of such an application to computer animation is available online [37].

The implemented sound model is able to smoothly transform rubbing sounds into squeaking or tonal sounds. The different regimes are reached with respect to the velocity value of the user's gesture. The squeaky effect is obtained by setting the filters so as to obtain a harmonic comb filter whose frequencies fluctuate with time. These random fluctuations happen around mean values which can be subject to sudden jumps depending on physical parameters of the interaction for example the relative speed and the normal force. Tonal source signals are obtained using a stationary and possibly inharmonic filter bank whose frequencies correspond to natural frequencies of the vibrating solid. An amplitude modulation is also added to the source signal. The transformation between rubbing and squeaking is achieved by a continuous increase of the filters' gain. A smooth transformation from squeaking to tonal is obtained by progressively moving the filters' frequencies from the frequencies of squeaking to natural frequencies of the solid. Simultaneously, the amplitude of the filter bank's frequency jitter is progressively brought to zero while the depth of the amplitude modulation is smoothly increased from zero to its maximal value.

4. Conclusions and Perspectives

In this article, a sound synthesis model of nonlinear friction sounds is proposed. Based on perceptual considerations, this model enables to elude physical considerations and to intuitively control and morph between the different friction behaviors. In particular, it enables to separately simulate actions and objects, in order to generate sounds that perceptually evoke different interactions, e.g., squeaking and singing combined with different objects. Various applications of these tools are proposed such as the auditory guidance of gestures for the rehabilitation of motor diseases, learning processes of expert musical gestures and sounds for video games.

The proposed model achieves the initial goal of control flexibility for the desired applications. Some improvements can nevertheless be envisaged. Concerning the synthesis models, it would be interesting to make further signal analyses on recorded sounds in order to define specific amplitude and frequency modulation laws for the different nonlinear behaviors. Moreover, in the case of bowing sounds, this model only takes into account the string vibrations and doesn't encompass the modeling of the body resonances of the instrument. These latter could be modeled by adding a filter bank to the output of the actual model and set up the filter parameters according to the modes and resonances of the instrument. It might also be of interest to provide a vibrato control. This could easily be achieved by defining a specific frequency modulation law of the fundamental frequency corresponding to such specific behaviors.

One other improvement of the synthesis model would be to give the possibility to simulate spectral behaviors such as multi-harmonic behaviors involved for instance in brake squeal sounds [38]. As implemented here, the synthesis model only enables to synthesize purely harmonic spectra. Nevertheless it is still possible to synthesize multi-harmonic spectral contents by changing the general definition of the additive model (cf. Equation (2)). For example, by considering two fundamental frequencies f_1 and f_2, a multi-harmonic set of partials can be defined by the following instantaneous frequencies: $\frac{d\Omega_{k,m}}{dt}(t) = \pm k f_1 \pm m f_2$, with m and k integers greater than or equal 0 [38]. More generally, the additive synthesis method offers the possibility to define any kind of spectral behaviors involving a sum of sinusoidal partials as those involved in many physical situations such as brake squeals [38,39]. The main difficulty resides in the high level of understanding of these phenomena, which may involve deterministic chaotic variations, in order to define intuitive controls of these behaviors for naive users.

The architecture based on a source-filter implementation is interesting as it is computationally efficient and also provides a modularity between the source and the resonance contributions. In the situations presented in this paper the source signal was a white noise but nothing prevents the user from employing other textures to create metaphoric situations such as a flow of water to which the same filtering process is applied. This enables to create physically impossible combinations between actions and objects such as squeaking on a liquid surface. This way of crossing the excitation with a resonating material could be seen as a kind of perceptual cross-synthesis [18,40]. Here, the crossing is not made between signal parameters of the two sounds like amplitudes and phases, but by crossing acoustical invariants that are separately related to different evocations. Such unheard-of sounds may be used to develop new worlds of sounds controlled by musical interfaces based on semantic descriptions.

Finally, in order to completely validate the perceptual relevance of the synthesized sounds, perceptual evaluations are planned. Such evaluations will necessitate several steps of investigations that are not only related to the sound quality, but also to the control issues and in particular the mapping strategies that determine how the gestural control of the synthesizer should be adapted to human gestures for a given application and how multimodal issues should be taken into account.

Acknowledgments: The first author is thankful to Simon Conan for the wide discussions and advices concerning the model implementation. The authors are also thankful to Jocelyn Rozé and Thierry Voinier for providing the cello recording. This work was funded by the French National Research Agency (ANR) under the MetaSon project (ANR-10-CORD-0003) in the CONTINT 2010 framework, the Physis project (ANR-12-CORD-0006) in the CONTINT 2012 framework, and the SoniMove project (ANR-14-CE24-0018).

Author Contributions: The authors contributed equally to this work.

Conflicts of Interest: The authors declare no conflict of interest.

References

1. Coulomb, C. The theory of simple machines. *Mem. Math. Phys. Acad.* **1785**, *10*, 4, ark:/12148/bpt6k1095299.
2. Akay, A. Acoustics of friction. *J. Acoust. Soc. Am.* **2002**, *111*, 1525–1548.
3. Avanzini, F.; Serafin, S.; Rocchesso, D. Interactive simulation of rigid body interaction with friction-induced sound generation. *IEEE Trans. Speech Audio Process.* **2005**, *13*, 1073–1081.
4. Rath, M.; Rocchesso, D. Informative sonic feedback for continuous human-machine interaction—Controlling a sound model of a rolling ball. *IEEE Multimed. Spec. Interac. Sonification* **2004**, *12*, 60–69.
5. Serafin, S. The Sound of Friction: Real-Time Models, Playability and Musical Applications. Ph.D. Thesis, Stanford University, Stanford, CA, USA, June 2004.
6. Serafin, S.; Huang, P.; Ystad, S.; Chafe, C.; Smith, J.O. Analysis and synthesis unusual friction-driven musical instruments. In Proceedings of the International Computer Music Conference, Gothenburg, Sweden, 16–21 September 2002.

7. Woodhouse, J.; Galluzzo, P.M. The bowed string as we know it today. *Acta Acust. United Acust.* **2004**, *90*, 579–589.
8. Maestre, E.; Spa, C.; Smith, J.O. A bowed string physical model including finite-width thermal friction and hair dynamics. In Proceedings of the Joint International Computer Music Conference and Sound and Music Computing, Athens, Greece, 14–20 September 2014.
9. Van Den Doel, K.; Kry, P.G.; Pai, D.K. FoleyAutomatic: Physically-based sound effects for interactive simulation and animation. In Proceedings of the 28th Annual Conference on Computer Graphics and Interactive Techniques, Los Angeles, CA, USA, 12–17 August 2008; pp. 537–544.
10. Gibson, J.J. *The Senses Considered as Perceptual Systems*; Houghton Mifflin: Oxford, UK, 1966.
11. McAdams, S.E.; Bigand, E.E. *Thinking in Sound: The Cognitive Psychology of Human Audition*; Clarendon Press/Oxford University Press: New York, NY, USA, 1993.
12. Gaver, W.W. How do we hear in the world? Explorations in ecological acoustics. *Ecol. Psychol.* **1993**, *5*, 285–313.
13. Gaver, W.W. What in the world do we hear?: An ecological approach to auditory event perception. *Ecol. Psychol.* **1993**, *5*, 1–29.
14. Wildes, R.P.; Richards, W.A. Recovering material properties from sound. In *Natural Computation*; Richards, W.A., Ed.; MIT Press: Cambridge, MA, USA, 1988.
15. Aramaki, M.; Besson, M.; Kronland-Martinet, R.; Ystad, S. Controlling the perceived material in an impact sound synthesizer. *IEEE Trans. Audio Speech Lang. Process.* **2011**, *19*, 301–314.
16. Conan, S.; Derrien, O.; Aramaki, M.; Ystad, S.; Kronland-Martinet, R. A synthesis model with intuitive control capabilities for rolling sounds. *IEEE/ACM Trans. Audio Speech Lang. Process.* **2014**, *22*, 1260–1273.
17. Conan, S.; Thoret, E.; Aramaki, M.; Derrien, O.; Gondre, C.; Ystad, S.; Kronland-Martinet, R. An Intuitive Synthesizer of Continuous-Interaction Sounds: Rubbing, Scratching, and Rolling. *Comput. Music J.* **2014**, *38*, 24–37.
18. Conan, S. Contrôle Intuitif de la Synthèse Sonore d'Interactions Solidiennes: Vers les Métaphores Sonores. Ph.D. Thesis, Ecole Centrale de Marseille, Marseille, France, December 2014. (In French)
19. Thoret, E.; Aramaki, M.; Kronland-Martinet, R.; Velay, J.L.; Ystad, S. From sound to shape: Auditory perception of drawing movements. *J. Exp. Psychol. Hum. Percept. Perform.* **2014**, *40*, doi:10.1037/a0035441.
20. Aramaki, M.; Kronland-Martinet, R. Analysis-synthesis of impact sounds by real-time dynamic filtering. *IEEE Trans. Audio Speech Lang. Process.* **2006**, *14*, 695–705.
21. Thoret, E.; Aramaki, M.; Gondre, C.; Kronland-Martinet, R.; Ystad, S. Recorded Sounds. Available online: http://www.lma.cnrs-mrs.fr/%7ekronland/thoretDAFx2013/ (accessed on 22 June 2016).
22. Thoret, E.; Aramaki, M.; Gondre, C.; Kronland-Martinet, R.; Ystad, S. Controlling a non linear friction model for evocative sound synthesis applications. In Proceedings of the 16th International Conference on Digital Audio Effects (DAFx), Maynooth, Ireland, 2–5 September 2013.
23. Ellis, D.P.W. Sinewave and Sinusoid + Noise Analysis/Synthesis in Matlab. Available online: http://www.ee.columbia.edu/%7edpwe/resources/matlab/sinemodel/ (accessed on 13 March 2015).
24. Rossing, T.D. Acoustics of the glass harmonica. *J. Acoust. Soc. Am.* **1994**, *95*, 1106–1111.
25. Inácio, O.; Henrique, L.L.; Antunes, J. The dynamics of tibetan singing bowls. *Acta Acust. United Acust.* **2006**, *92*, 637–653.
26. Aramaki, M.; Gondre, C.; Kronland-Martinet, R.; Voinier, T.; Ystad, S. Imagine the sounds: An intuitive control of an impact sound synthesizer. In *Auditory Display*; Ystad, S., Aramaki, M., Kronland-Martinet, R., Jensen, K., Eds.; Lecture Notes in Computer Science; Springer: Berlin/Heidelberg, Germany, 2010.
27. Rabiner, L.R.; Gold, B. *Theory and Application of Digital Signal Processing*; Prentice-Hall, Inc.: Englewood Cliffs, NJ, USA, 1975.
28. Atal, B.S.; Hanauer, S.L. Speech analysis and synthesis by linear prediction of the speech wave. *J. Acoust. Soc. Am.* **1971**, *50*, 637–655.
29. Flanagan, J.L.; Coker, C.; Rabiner, L.; Schafer, R.W.; Umeda, N. Synthetic voices for computers. *IEEE Spectr.* **1970**, *7*, 22–45.

30. Mathews, M.; Smith, J.O. Methods for synthesizing very high Q parametrically well behaved two pole filters. In Proceedings of the Stockholm Musical Acoustic Conference (SMAC), Stockholm, Sweden, 6–9 August 2003.

31. Danna, J.; Fontaine, M.; Paz-Villagrán, V.; Gondre, C.; Thoret, E.; Aramaki, M.; Kronland-Martinet, R.; Ystad, S.; Velay, J.L. The effect of real-time auditory feedback on learning new characters. *Hum. Mov. Sci.* **2015**, *43*, 216–228.

32. Danna, J.; Fontaine, M.; Paz-Villagrán, V.; Gondre, C.; Thoret, E.; Aramaki, M.; Kronland-Martinet, R.; Ystad, S.; Velay, J.L. Example of Learning Session. Available online: http://www.lma.cnrs-mrs.fr/%7ekronland/TheseEThoret/content/flv/chap5/newCharacters.mp4 (accessed on 22 June 2016).

33. Danna, J.; Paz-Villagrán, V.; Velay, J.L. Signal-to-Noise velocity peaks difference: A new method for evaluating the handwriting movement fluency in children with dysgraphia. *Res. Dev. Disabil.* **2013**, *34*, 4375–4384.

34. Danna, J.; Paz-Villagrán, V.; Capel, A.; Pétroz, C.; Gondre, C.; Pinto, S.; Thoret, E.; Aramaki, M.; Ystad, S.; Kronland-Martinet, R.; et al. Movement Sonification for the Diagnosis and the Rehabilitation of Graphomotor Disorders. In *Sound, Music, and Motion*; Aramaki, M., Derrien, O., Kronland-Martinet, R., Ystad, S., Eds.; Lecture Notes in Computer Science; Springer International Publishing: Cham, Switzerland, 2014.

35. Thoret, E.; Aramaki, M.; Gondre, C.; Kronland-Martinet, R.; Ystad, S. Example of Intuitive Control of Non Linear Friction Sounds. Available online: http://www.lma.cnrs-mrs.fr/%7ekronland/thoretDAFx2013/content/video/Thoretdafx2013.mp4 (accessed on 22 June 2016).

36. Pruvost, L.; Scherrer, B.; Aramaki, M.; Ystad, S.; Kronland-Martinet, R. Perception-based interactive sound synthesis of morphing solids' interactions. In Proceedings of the SIGGRAPH Asia 2015 Technical Briefs, Kobe, Japan, 2–5 November 2015; p. 17.

37. Pruvost, L.; Scherrer, B.; Aramaki, M.; Ystad, S.; Kronland-Martinet, R. Example of Application to Computer Animation. Available online: http://www.lma.cnrs-mrs.fr/%7ekronland/siggraph2015/index.html (accessed on 22 June 2016).

38. Sinou, J.J. Transient non-linear dynamic analysis of automotive disc brake squeal—On the need to consider both stability and non-linear analysis. *Mech. Res. Commun.* **2010**, *37*, 96–105.

39. Oberst, S.; Lai, J. Statistical analysis of brake squeal noise. *J. Sound Vib.* **2011**, *330*, 2978–2994.

40. Smith, J. Cross-Synthesis—CCRMA. Available online: https://ccrma.stanford.edu/%7ejos/sasp/Cross_Synthesis.html (accessed on 13 March 2015).

applied sciences

MDPI

Article

Adaptive Wavelet Threshold Denoising Method for Machinery Sound Based on Improved Fruit Fly Optimization Algorithm

Jing Xu [1], Zhongbin Wang [1,*], Chao Tan [1], Lei Si [1,2], Lin Zhang [1,3] and Xinhua Liu [1]

[1] School of Mechatronic Engineering, China University of Mining and Technology, No.1 Daxue Road, Xuzhou 221116, China; xujingcmee@cumt.edu.cn (J.X.); tccadcumt@126.com (C.T.); sileicool@163.com (L.S.); lin.zhang_2014@hotmail.com (L.Z.); l_xinhua_2006@126.com (X.L.)
[2] School of Information and Electrical Engineering, China University of Mining and Technology, No.1 Daxue Road, Xuzhou 221116, China
[3] Institute for Neural Computation, University of California, San Diego (UCSD), No.3950 Mahaila Ave, San Diego, CA 92093, USA
* Correspondence: wangzbpaper@126.com; Tel./Fax: +86-516-8388-4512

Academic Editor: Gino Iannace
Received: 5 May 2016; Accepted: 1 July 2016; Published: 6 July 2016

Abstract: As the sound signal of a machine contains abundant information and is easy to measure, acoustic-based monitoring or diagnosis systems exhibit obvious superiority, especially in some extreme conditions. However, the sound directly collected from industrial field is always polluted. In order to eliminate noise components from machinery sound, a wavelet threshold denoising method optimized by an improved fruit fly optimization algorithm (WTD-IFOA) is proposed in this paper. The sound is firstly decomposed by wavelet transform (WT) to obtain coefficients of each level. As the wavelet threshold functions proposed by Donoho were discontinuous, many modified functions with continuous first and second order derivative were presented to realize adaptively denoising. However, the function-based denoising process is time-consuming and it is difficult to find optimal thresholds. To overcome these problems, fruit fly optimization algorithm (FOA) was introduced to the process. Moreover, to avoid falling into local extremes, an improved fly distance range obeying normal distribution was proposed on the basis of original FOA. Then, sound signal of a motor was recorded in a soundproof laboratory, and Gauss white noise was added into the signal. The simulation results illustrated the effectiveness and superiority of the proposed approach by a comprehensive comparison among five typical methods. Finally, an industrial application on a shearer in coal mining working face was performed to demonstrate the practical effect.

Keywords: wavelet threshold denoising; sound signal; wavelet transform; improved fruit fly optimization algorithm; fly distance range

1. Introduction

Generally, the vibration and strain signals of a machine are mostly applied to provide dynamic information of the machine's working condition [1,2], even though they have some common disadvantages, such as contact measurement, limited detecting positions and difficult to maintain detectors in some severe situations. Therefore, vibration and strain measuring is inappropriate or sometimes even impossible in these cases. On the other hand, the sound signal of a machine can be a significant criterion for state recognition or fault diagnosis because it is convenient to collectand does not affect the machine [3]. Thus, acoustic-based diagnosis (ABS) has received much attention in recent years [4]. One of the most important preconditions for ABS is eliminating noise from the initial sound signal, and the performance of denoising directly influences the effect of subsequent processing [5,6].

Preliminary analysis of machinery sound shows that significant details distribute in both time and frequency domains, which implies that noise elimination methods considering both scales would perform better than those only focusing one. Throughout the development of signal processing, the most influential time–frequency joint analysis approaches are Fast Fourier Transform (FFT) and Wavelet Transform (WT) [7]. WT was firstly proposed by Mallat in 1989 [8]. As the window function of FFT is fixed, WT is obviously superior to FFT for non-stationary signal as the property of characterizing local features in both domains. Six years later, Donoho proposed hard-threshold and soft-threshold denoising solutions based on WT. The corresponding threshold value was selected by combining WT and Stein's unbiased risk estimate (SURE) [9]. However, since the derivative of standard threshold function is not continuous and lacks adaptability, many improved wavelet noise reduction methods have been proposed [10–12]. With the extensive application of artificial intelligence in recent years, adaptive threshold selecting approaches based on intelligent optimization algorithms, such as the particle swarm optimization (PSO), genetic algorithm (GA) and ant colony optimization (ACO) have been adopted gradually [13–15].

Fruit fly optimization algorithm (FOA) was proposed by Pan in 2012 [16–18]. As a meta-heuristic method, FOA simulates the intelligent foraging behavior of fruit fly group in food finding process [19]. The fruit fly is superior to other species with regard to its senses of osphresis and vision, it can even a smell food source from 40 km away and locate other flocks by its sensitive vision [20]. The FOA has many advantages compared with the above optimization algorithms, such as simple structure, immediately accessible for practical applications, ease of implementation and rapid convergence rate. Since the FOA was proposed, it has been widely applied in financial parameter optimization [21], forecasting [22], scheduling [23], etc. However, like other optimization algorithms, the basic FOA also has the possibility of falling into local extremes due to its fixed fly distance range [24].

Bearing the above observations in mind, an adaptive wavelet threshold denoising method for machinery sound based on an improved FOA (WTD-IFOA) is proposed. The rest of this paper is organized as follows. In Section 2, some related works are outlined based on the literature. In Section 3, the basic wavelet noise elimination method and optimization process of FOA are presented. In Section 4, an improved fly distance range, obeying normal distribution, is performed, and the denoising solution based on WTD-IFOA is elaborated. In Section 5, Gauss white noise is added into the motor sound signal to verify the effectiveness and superiority of the proposed method, and an industrial application is performed. Some conclusions and outlooks are summarized in Section 6.

2. Literature Review

Recent publications relevant to this paper are mainly concerned with two research streams: wavelet threshold denoising and fruit fly optimization algorithm. In this section, we try to summarize the relevant literature.

2.1. Wavelet Threshold Denoising

Traditional denoising methods, such as low-pass filter, Kalman filter and median filter, aim either at the time domain or the frequency domain. However, single-scale representations of signals are often inadequate when attempting to separate signals from noisy data. By combining the two scales, wavelet threshold denoising presents obvious superiority. According to the wavelet threshold denoising theory proposed by Donoho, the optimal threshold should diminish the noise but preserve the signal as much as possible [25]. The traditional hard-threshold function exhibits some discontinuities and may be unstable or more sensitive to small changes in the data, while in soft thresholding the wavelet coefficients are reduced by a quantity equal to the threshold value, which will induce the deviation when the filtered wavelet coefficients is reconstructed [26]. Moreover, the threshold is fixed once determined and adaptability is weak during the denoising process. In order to overcome the disadvantages of the original threshold functions proposed by Donoho, many adaptive denoising approaches have been elaborated by researchers. Improved solutions can be divided into two streams:

the first on the improvements of threshold function and the other focus on searching optimal threshold through intelligent algorithms. The threshold function-based methods aim at establishing appropriate function with continuous derivative and selecting thresholds based on gradient descent algorithm. In [10], a new adaptive denoising function with continuous first and second order derivative is presented based on SURE model. In [27], an adaptive logarithmic wavelet threshold denoising function was proposed to select optimal threshold for each decomposition level. Relative to hard and soft functions, the proposed approach increased the signal-to-noise ratio by 44.2% and 27.9%, and decreased processing time by 37.6% and 38.5%, respectively. With the rapid development of artificial intelligent optimization algorithm, intelligent searching-based wavelet denoising approaches have been widely applied in recent years. In [14], a PSO-based image denoising method was proposed for learning the parameters of the adaptive thresholding function required for optimum performance. Li et al. adopted an adaptive denoising solution for partial discharge signals based on threshold function and genetic algorithm (GA), and the result presents significantly smaller waveform distortion and magnitude errors than the Donoho's soft threshold estimation [15]. In order to eliminate noise components of satellite images, some stochastic global optimization techniques such as Cuckoo Search (CS) algorithm, artificial bee colony (ABC), and PSO as well as their different variants have been exploited for learning the parameters of adaptive thresholding function in [28].

2.2. Fruit Fly Optimization Algorithm

Although it is not long since FOA was put forward, it has aroused much attention and scored great academic achievements. In [21], FOA was adopted to optimize general regression neural network, and the simulation result showed the superiority compared with other intelligent optimization algorithms. In [29], an annual electric load forecasting method was proposed by the least squares support vector machine (LSSVM) model. The FOA was used to determine appropriate parameters of the model, and an experiment, with the mean absolute percentage error of 1.305%, proved the validity of the approach. Although the FOA has an extensive application in many fields, there still exists the possibility of getting into the local extreme [24]. The main reason lies in the fruit fly individuals move toward fixed fly distance range in the iteration of optimization. Once the fruit fly group fall into the local extreme and the fly distance range is not big enough, the optimization process is prone to fail [30]. On the other hand, excessive fly distance range may lead to slow convergence rate of the iteration process. In [23], an improved FOA was presented to solve the joint replenishment problems. In order to avoid local optimal solution, swarm collaboration and random perturbation were added into original FOA. Pan et al. presented a changeable fly distance range in FOA to eliminate the drawbacks lies with fixed values of search radius, and 29 benchmark functions were carried out to make a comparison with basic FOA [24]. Yuan et al. proposed a multi-swarm FOA, where several sub-swarms moving independently in the search space with the aim of simultaneously exploring global optimal and local behavior between sub-swarms is also considered [31]. In [32], an improved FOA, called linear generation mechanism of candidate solution fruit fly optimization algorithm (LGMS-FOA), was introduced for solving optimization problems. Four disadvantages of the original FOA were listed and some improvements were operated, and the simulation result showed local extreme could be avoided efficiently.

2.3. Discussion

Many valuable wavelet denoising methods have been proposed and applied by researchers in recent decades, which greatly pushes forward the development of this field. However, there are still some shortcomings including the following. Firstly, the disadvantage of weak adaptability seriously restricted the development of Donoho's wavelet threshold approaches. Secondly, adaptive noise elimination methods based on gradient descent algorithm were also limited because of the great amount of calculation. The above approaches were gradually replaced by intelligent optimization-based algorithms. Thirdly, the iterative process of the common optimization solution has the problems of slow convergence rate and high complexity of coding. In [15], the proposed method based on GA adaptive threshold cost more than 38.66 times the calculation time compared to soft threshold approach. Moreover, the FOA has great advantages in iteration rate and encoding efficiency, but still has the probability of falling into local extreme. Many improvements have been elaborated by past scholars, but few researchers could balance both local extreme and iterative rate.

Therefore, a novel wavelet threshold denoising method optimized by an improved FOA is proposed in this paper. The fly distance range obeying uniform distribution in the basic FOA is replaced by the following normal distribution. Both local extreme and iterative rate are taken into consideration. A series of simulations and an industrial application prove the effectiveness and superiority of the proposed method.

3. Basic Theory

3.1. Wavelet Threshold Denoising

Fundamental theory of wavelet denoising can be concluded as follows: wavelet decomposition is firstly conducted on the noisy signal, then wavelet coefficients that belong to useful signal are kept and others are eliminated, and finally inverse wavelet transform is operated to reconstruct the remainder coefficients.

Assume that the noisy signal series $x = \{x_1, x_2, x_3, ..., x_k\}$ can be expressed as follows:

$$x_i = s_i + n_i \tag{1}$$

where $i = 1, 2, 3, ..., k$, $s = \{s_1, s_2, s_3, ..., s_k\}$ is the useful initial signal and $n = \{n_1, n_2, s_3, ..., n_k\}$ is noise signal.

Then x is decomposed by J levels WT and the i-th wavelet coefficient in j-th can be presented as $d_{i,j}$, where $j = 1, 2, 3, ..., J$. Since WT is a kind of linear transform, wavelet coefficients of x are consisted of ones decomposed by s, denoted as $U_{i,j}$, and that of n, called $V_{i,j}$. The purpose of wavelet denoising is to eliminate $V_{i,j}$ and obtain the estimate signal $\hat{s}$ of the noisy signal. The ideal $\hat{s}$ has a minimum mean square error with s under the premise of eliminating noise component furthest. The mean square error (MSE) ξ can be calculated as follows:

$$\xi[\hat{s}, s] = \frac{1}{k} ||\hat{s} - s|| = \frac{1}{k} \sum_{i=1}^{k} [\hat{s}_i - s_i]^2 \tag{2}$$

The threshold during the denoising process is calculated according to SURE model, which can be obtained as follows:

$$\lambda_j = \mathrm{MAD}(|d_{i,j}|)/q \tag{3}$$

where λ_j denotes the threshold of j-th level, $\mathrm{MAD}(\cdot)$ is a median value function and the value range of q is $[0.4, 1]$ in general.

There are two typical threshold functions during the wavelet denoising process, the first called hard-threshold:

$$\hat{d}_{i,j} = \begin{cases} d_{i,j}, & \text{for } |d_{i,j}| \geq \lambda_j \\ 0, & \text{otherwise} \end{cases} \tag{4}$$

where $\hat{d}_{i,j}$ donates the wavelet coefficient of denoised signal. The other function is called soft-threshold:

$$\hat{d}_{i,j} = \begin{cases} \mathrm{sgn}(d_{i,j})(|d_{i,j}| - \lambda_j), & \text{for } |d_{i,j}| \geq \lambda_j \\ 0, & \text{otherwise} \end{cases} \tag{5}$$

where $\mathrm{sgn}(\cdot)$ is sign function, which returns 1 if the element is greater than 0; 0 if it equals 0; and -1 if it is less than 0.

The estimated signal is reconstructed through inverse wavelet transform on $\hat{d}_{i,j}$. It can be seen that the key point of the denoising process is selecting appropriate threshold to minimize Equation (2).

3.2. Fruit Fly Optimization Algorithm

FOA is a new interactive evolutionary computation method, which was proposed by Pan in 2012. By simulating the process of foraging behavior for fruit fly individuals and populations, global optimum can be obtained through appropriate iteration, as shown in Figure 1. Standard foraging process of FOA can be summarized as follows.

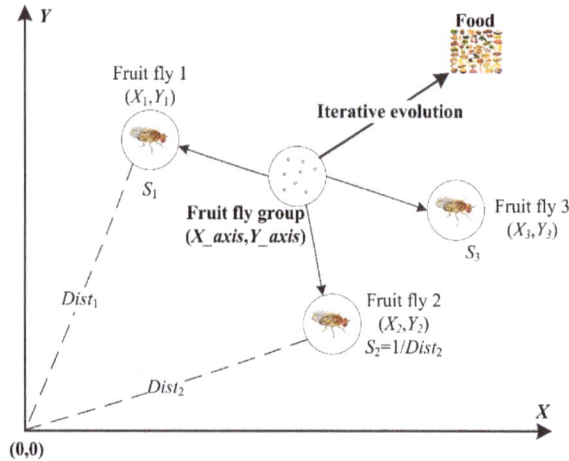

Figure 1. Process of foraging behavior for fruit fly group.

Step 1.1: The key initial parameters of the FOA are the population amount (*PA*), the fruit fly group location range (*LR*), the maximum iteration number (IN_{max}) and the random fly distance range (*FR*). The initial location of the fruit fly group can be presented as follows:

$$\begin{cases} X\_axis = \text{rand}\,(LR) \\ Y\_axis = \text{rand}\,(LR) \end{cases} \tag{6}$$

Step 1.2: The random direction and distance for the search of food using osphresis by an individual fruit fly is given as follows:

$$\begin{cases} X_i = X\_axis + \text{rand}\,(FR) \\ Y_i = Y\_axis + \text{rand}\,(FR) \end{cases} \tag{7}$$

Step 1.3: Since the food location cannot be known, the distance to the origin of coordinates (*Dist$_i$*) and the smell concentration judgment value (*S$_i$*) are calculated as follows:

$$Dist_i = \sqrt{X_i^2 + Y_i^2}, \quad S_i = 1/Dist_i \tag{8}$$

The fruit fly with maximal smell concentration among the fruit fly group can be searched according to the smell concentration judgment function (or called Fitness function), which can be presented as follows:

$$smell_i = \text{function}(S_i), \quad [bestsmell\ bestindex] = \max(smell) \tag{9}$$

where *bestsmell* donates the maximal smell concentration, *bestindex* is the corresponding fruit fly number and smell is the smell concentration set of the group.

Step 1.4: The smell concentration is compared with that of the former iteration. If it is inferior to the last generation, Steps 1.2 to 1.3 are repeated; else the best location and smell concentration can be presented as follows:

$$\begin{cases} smellbest = bestsmell \\ X\_axis = X(bestindex) \\ Y\_axis = Y(bestindex) \end{cases} \qquad (10)$$

Step 1.5: When the smell concentration reaches the preset precision value or the iteration number reaches the maximal *IN*, the circulation stops. Otherwise, Steps 1.2 to 1.4 are repeated.

4. The Proposed Method

In this section, the improved fruit fly optimization algorithm is proposed to enhance the capacity of global and local research. Then the flowchart of the improved method is designed and the process of the denoising approach based on WTD-IFOA is presented.

4.1. Improvement of FOA

The fly distance range (*FR*) of FOA is a random value in the range of $[-L, L]$, which can be presented as $FR \sim U(-L, L)$, where *L* named as step size. The value of *FR* is distributed evenly among the value range. If the step size is big enough, the global search capability will be improved remarkably, while the convergence speed will be decreased obviously. Otherwise, the FOA easily gets stuck in local optimal, while has a high convergence speed.

In order to balance the global search ability and convergence rate, the distribution function of FR is modified in this paper. The value of FR follows normal distribution, $FR \sim N(0, L^2)$. According to the characteristic of normal distribution, the probability of $FR \in [-L, L]$ is about 68.27%, the probability of $FR \in [-2L, 2L]$ is about 95.45% and the probability of $FR \in [-3L, 3L]$ is about 99.73%. The probability density distribution of original fly distance range and the proposed one are presented in Figure 2. Most individuals fly towards the present best location, while more than 30% fruit flies continuing searching at a larger scale. Moreover, individuals flying towards the present optimum tend to a more concentrated region according to the probability distribution condition. Thus, the capacity of global searching and partial location are both enhanced. The flowchart of the IFOA is shown in Figure 3.

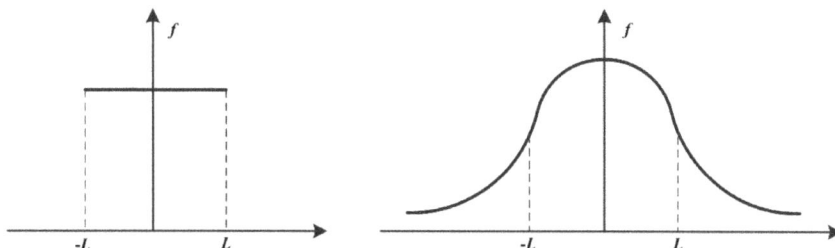

Figure 2. Probability density functions of basic fly distance range and the improved.

4.2. Flow of the Proposed Denosing Method

The adaptive threshold denoising method based on WTD-IFOA can be summarized as follows:

Step 2.1: For the sake of convenient calculation, the sound signal is first quantized into a certain range. Then, the initial signal is decomposed by a *J*-level wavelet transform. *i*-th coefficient in *j*-th level can be presented as $d_{i,j}$, where $i = 1, 2, 3, \dots, m, j = 1, 2, 3, \dots, J$, *m* is the length of the sound signal.

Step 2.2: The parameters of IFOA are initialized, such as *LR*, IN_{max} and *FR*, where $FR \sim N(0, L^2)$. For *J*-level WT, each level has an optimal threshold. So there are *J* groups of fruit flies, each group contains *PA* individuals. The initial location of the fruit fly group is obtained by Equation (6).

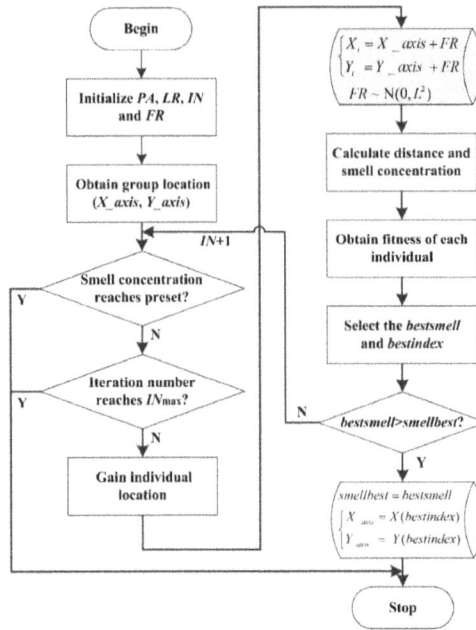

Figure 3. Flowchart of the improved fruit fly optimization algorithm.

Step 2.3: The location of each individual is gained through the fly group and *FR*. The distance and smell concentration of each fly are calculated according to Equation (8). Each smell concentration judgment is regarded as a potential threshold. Then, the useful signal and the noise component are separated according to the soft-threshold function. In order to judge the denoising performance of each fruit fly individual, the fitness function *f* is calculated as follows:

$$f = \frac{1}{1+g} \tag{11}$$

$$g = \frac{[r_{22} \times hr_{21}]^2}{r_{11}^2} \tag{12}$$

where r_{22} is an autocorrelation of noise. As noise is yielded randomly, it has neither high autocorrelation nor zero autocorrelation. However, an ascending value implies that more original signal adheres to noise, thus the restructured signal will not be a good recovery. hr_{21} is a high-order cross-correlation between useful signal and the noise. If these coefficients are descending, then it implies that both signals become more independent to each other. Thus, the original signal and noise are toward separation gradually. r_{11} is an autocorrelation of useful signal. An ascending value implies that its own component is more than component of noise. Hence, the restructured signal has a good recovery [33]. r_{22}, hr_{21} and r_{11} are defined as follows:

$$r_{ij} = \frac{C_{ij}}{\sqrt{C_{ii}C_{jj}}} = \frac{cov[s_i, s_j]}{\sqrt{cov[s_i]cov[s_j]}}, \quad i,j = 1,2 \tag{13}$$

$$hr_{ij} = \frac{HC_{ij}}{\sqrt{HC_{ii}C_{jj}}} = \frac{cov[\varphi(s_i), s_j]}{\sqrt{cov[\varphi(s_i)]cov[s_j]}}, \quad i,j = 1,2 \tag{14}$$

where s_1 is the useful signal, s_2 is the noise component, $cov[s_i, s_j] = E\{[s_i - E(s_i)][s_j - E(s_j)]\}$, $cov[s_i] = E\{[s_i - E(s_i)]^2\}$, $E(s_i)$ is mathematical expectation of s_i and $\varphi(s_i) = s_i^2 + s_i^3$. r_{22}, hr_{21} tend to minimum and r_{11} tends to the maximum when the location of each fruit fly group is placed in the best. Then g is the minimal and f is the maximal [34].

Step 2.4: Fruit fly with maximal fitness is selected as *bestsmell* and the corresponding fly number is named as *bestindex*. If the present *bestsmell* is bigger than that of the former, *smellbest*, the corresponding coordinates are updated. Otherwise, *smellbest*, $X_{\_axis}$ and $Y_{\_axis}$ are reserved.

Step 2.5: If the ending conditions are researched, *smellbest*, $X_{\_axis}$ and $Y_{\_axis}$ are treated as the optimum. Otherwise, Steps 2.3 and 2.4 are repeated.

Step 2.6: The wavelet coefficients are adjusted according to the soft-threshold function and inverse wavelet transform is conducted consequently to obtain denoised signal. The flowchart of the process is shown in Figure 4.

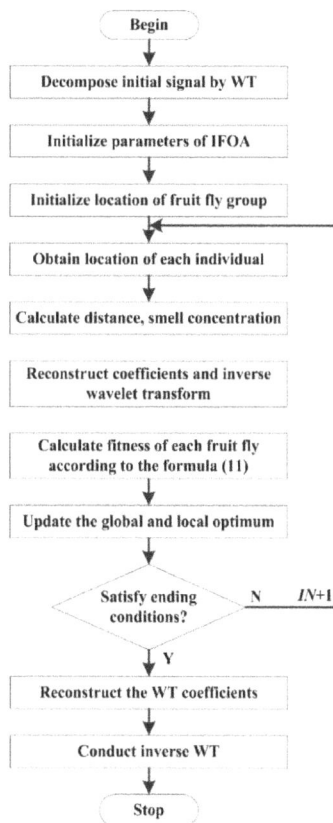

Figure 4. Process of the proposed denoising method.

5. Simulation and Industrial Application

In order to validate the effectiveness and superiority of the proposed method, a piece of pure sound signal of a motor was recorded. Then, Gauss white noise with different signal to noise ratio were added into the pure signal. The mean square error (MSE), peak value error (PVE) and the computation time (CT) were regarded as evaluation criteria of the noise elimination solutions. The denoising performance of standard soft threshold denoising method (SST), a threshold function-based noise elimination solution proposed in [10] (TFB), wavelet threshold denoising optimized by genetic algorithm (WTD-GA), wavelet threshold denoising optimized by fruit fly optimization algorithm (WTD-FOA) and the proposed WTD-IFOA were compared subsequently. Finally, an industrial application for the shearer of coal mining working face is exhibited. All calculations in this section were conducted on a workstation configured as shown in tab:applsci-06-00199-t001.

Table 1. Configuration of the workstation.

Operating System	Windows 7 (64 bits)
CPU	Intel Xeon E5-2690 (8 cores, 2.9 GHz)
Memory	16 GB (DDR3)
Hard disk space	SSD (512 GB)
Matlab version	8.0

5.1. Signal Acquisition

To test the performance of the denoising methods, a pure sound signal sequence was needed first. Because there was much background noise in industrial field, it was extremely difficult to collect the pure sound. Moreover, sound signal of a machine consisted of many frequency components, so it was also hard to synthesize a representative series with practical meaning artificially. In this paper, the sound of a motor working in a soundproof room was recorded as the original signal. Concretely, the sound signal was acquired from soundproof testing branch, Jiangsu Key Laboratory of Mining Mechanical and Electrical Equipment. The walls of the testing room were constructed of a special acoustic insulating material and echo cancellation was designed in the testing process. An AC servomotor with the rated out power of 1 kW and the corresponding electrical system were installed in the laboratory. The schematic of the testing room and the experiment site are shown in Figure 5.

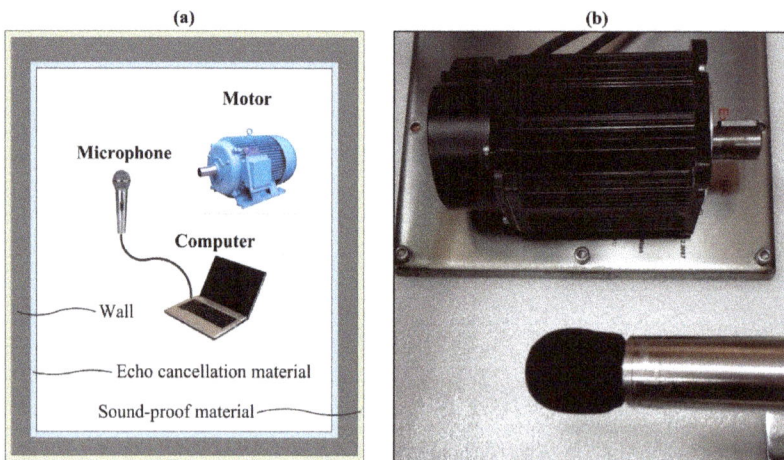

Figure 5. (a) Schematic of the soundproof testing branch; and (b) The sound recorded in site.

Echo cancellation material was installed on the inner wall and acoustic insulating equipment was placed on the external wall. The motor, microphone and computer were fixed in the room with the length of 6 m, width of 5 m and height of 4 m. Operators controlled the motor outside the room. The sound signal was recorded by the microphone and then transmitted to the computer. Sampling frequency of the sound signal was 44.1 kHz. The experiment was conducted as follows: start the motor remotely and keep it in no-load operation, then stop the motor after 10 min, pretreat and save the sound in the computer in wav format. Quantization was subsequently operated to convert the sound amplitude into the scope of [−1, 1]. Finally, a piece of relatively stable sound was extracted with the duration of 0.5 s, as shown in Figure 6.

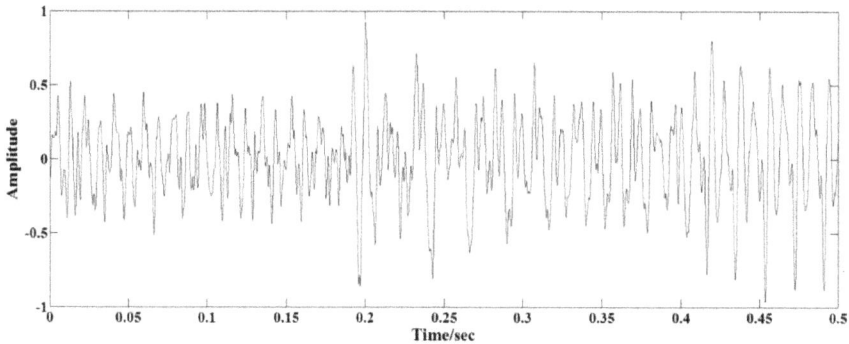

Figure 6. The extracted sound signal of the motor.

5.2. Signal Denoising

In this paper, the MSE ξ, the PVE η and the CT t were selected as evaluation indexes of the denoising methods. MSE and the PVE are calculated as follows:

$$\xi = \frac{1}{k}\sum_{i=1}^{k}[ds_i - s_i]^2 \tag{15}$$

$$\eta = \frac{|P_o - P_d|}{P_o} \times 100\% \tag{16}$$

where ds is denoised signal, s is original signal, k is the length of the signal, and P_o and P_d are, respectively, the peak value of the original signal and denoised signal. To test the performance of SST, TFB, WTD-GA, WTD-FOA and the proposed WTD-IFAO, Gauss white noise was added into the original sound. The signal to noise ratio (SNR) ζ (dB) was introduced to measure the degree of noise, and SNR was defined as follows:

$$\zeta = 10 \lg \frac{\sum_{i=1}^{k} s_i^2}{\sum_{i=1}^{k} n_i^2} \tag{17}$$

where s_i was the amplitude of the original signal and n_i was that of the added noise.

The denoising process was conducted as follows:

(1) Add Gaussian white noise into the original sound and conduct wavelet decomposition. A noisy signal with ζ = 5 dB was firstly analyzed and the synthesis was finished in Matlab 8.0 (MathWorks Inc., Natick, MA, USA, 2012). Then the synthetic signal was decomposed by wavelet decomposition with db2 wavelet at 5 levels [6]. The decomposition result is shown in Figure 7.

(2) Denoise the noisy signal by SST. The value range of the wavelet threshold was firstly calculated according to Equation (3). The recommended threshold λ_{rec} was adopted according to Donoho, where q = 0.6745. The wavelet coefficients of each level were shrunk according to Equation (5). Then the signal was reconstructed by inverse WT.

(3) Denoise the noisy signal by TFB. An improved threshold function with continuous first and second order derivative was introduced in [10], and is presented as follows:

$$
\hat{d}_{i,j} = \left\{
\begin{array}{ll}
d_{i,j} + \lambda - \frac{\lambda}{2k+1}, & d_{i,j} < -\lambda \\
\frac{1}{(2k+1)\lambda^{2k}} \cdot d_{i,j}^{2k+1}, & |d_{i,j}| < \lambda \\
d_{i,j} - \lambda + \frac{\lambda}{2k+1}, & d_{i,j} > \lambda
\end{array}
\right\}
\tag{18}
$$

where λ is determined according to Equation (3) and $k = 3$.

(4) Denoise the noisy signal by WTD-GA. The maximum λ_{max} appeared at $q = 1$ and λ_{min} obtained when $q = 0.4$, so $\lambda \in [\lambda_{min}, \lambda_{max}]$. The population size was 100, each chromosome was a five-dimensional vector, the crossover probability was 0.7, the mutation probability was 0.01 and the most iteration generation was 100, as recommended by ref. [15].

(5) Denoise the noisy signal by WTD-FOA. The parameters were set as follows: $\lambda \in [\lambda_{min}, \lambda_{max}]$. λ_{min} and λ_{max} were calculated as previously. The fly distance obeys uniform distribution, $FR \sim U(-0.2, 0.2)$. The population number was 5, each group contained 20 individuals and the iteration number was 100. The fitness of each fruit fly was calculated according to Equation (11).

(6) Denoise the noisy signal by WTD-IFOA. The parameters were set as follows: $\lambda \in [\lambda_{min}, \lambda_{max}]$, the fly distance obeys normal distribution, $FR \sim N(0, 0.2^2)$. The population number was 5, each group contained 20 individuals and the iteration number was 100. The fitness of each fruit fly was calculated according to Equation (11).

Subsequently, a comprehensive comparison was made for the five methods. The ξ, η and t at $\zeta = 5$ dB of the average value of the five simulation results are presented in tab:applsci-06-00199-t002. And the denoised signals of the 5 solutions were shown in Figure 8. It can be seen in the table that adaptive denoising methods based on intelligent optimization had a better comprehensive performance than SST and TFB. MSE of the WTD-FOA was smaller than that of WTD-GA while the PVE was contrary, which indicated the two methods had no optimal solution, both in global and local. The WTD-IFOA overcame the disadvantage and contributed a superior scheme. The MSE of WTD-IFOA was decreased about 35.36% compared with SST, and the PVE decreased about 9.40%. As optimal threshold was obtained through the iterative process, the last four methods were much more time-consuming. Among these approaches, TFB cost the most time due to its complex calculation. The optimization process of WTD-GA was much more complex than the other intelligent solutions from the table. Moreover, the WTD-IFOA, respectively, saved 26.96% and 12.47% time compared with WTD-GA and WTD-FOA because of its stronger addressing ability.

Table 2. Comparison of the five methods (SNR = 5 dB).

Index \ Method	SST	TFB	WTD-GA	WTD-FOA	WTD-IFOA
MSE ($\times 10^{-4}$)	12.19	9.61	11.78	8.53	7.88
PVE (%)	10.56	7.39	2.18	2.28	1.16
CT (s)	0.87	24.61	11.24	9.38	8.21

In order to research the denoising performance at different SNR, a further comparison was made and the average value of five simulation results are shown in Figure 9. Four noisy signal with SNR = 5 dB, 10 dB, 15 dB and 20 dB were synthetized and handled. Then the MSE, PVE and CT of the denoising processes were presented in the figure. It can be roughly obtained that the three evaluation parameters decreased with the SNR. As the threshold value of each level and wavelet coefficients were determined directly by Equations (3) and (4) in SST, the denoising process could be finished in a short time, while the TFB based on gradient descent algorithm was time-consuming in different SNR as its complex calculation process. Denoising methods using optimization algorithm had a comprehensive denoising performance. The computation time was decreased sharply compared with TFB while it still cost more time than SST. In detail, the WTD-GA did not obviously vary from WTD-FOA in PVE, while it had a distinct weakness in MSE and CT compared with FOA-based methods. It could be seen from the simulation results that the TFB, WTD-GA and WTD-FOA fell into local extreme during the parameters optimization. Moreover, the WTD-IFOA exhibited obvious superiority in the denoising effect, which revealed its terrific global and local ability compared to the other methods in different SNR.

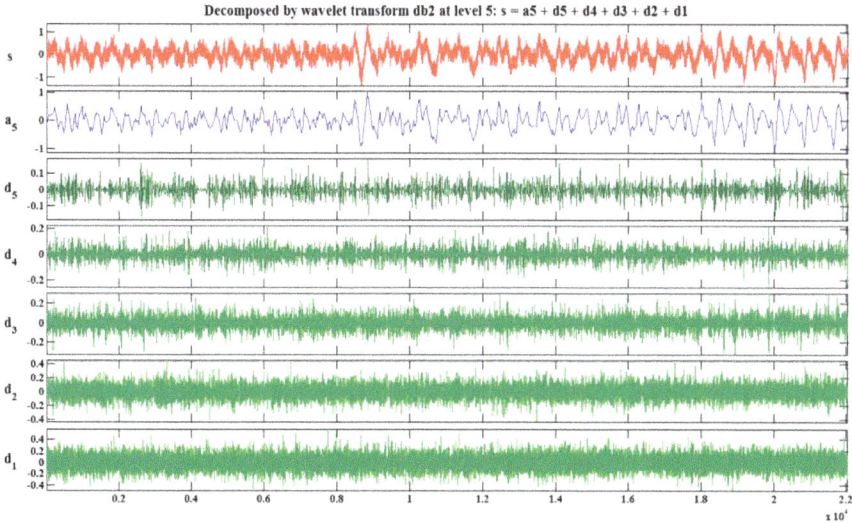

Figure 7. Gauss white noise-added signal and its wavelet coefficients (SNR = 5 dB).

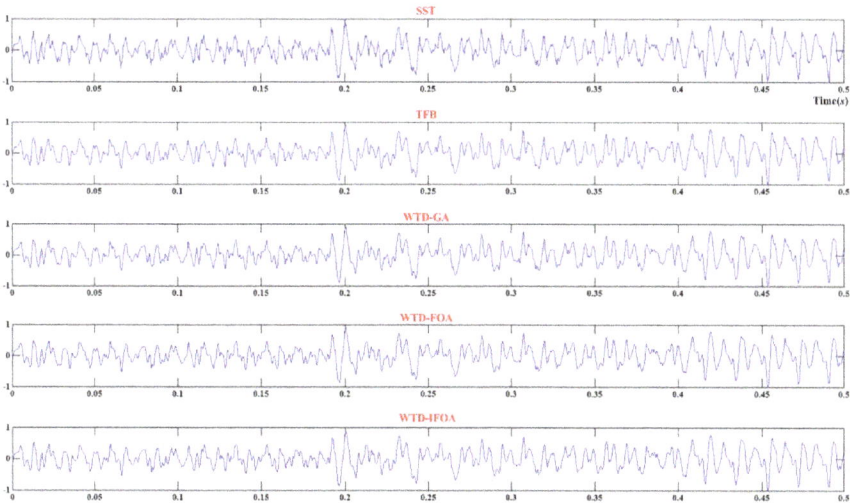

Figure 8. Denoise result of the five methods (SNR = 5 dB).

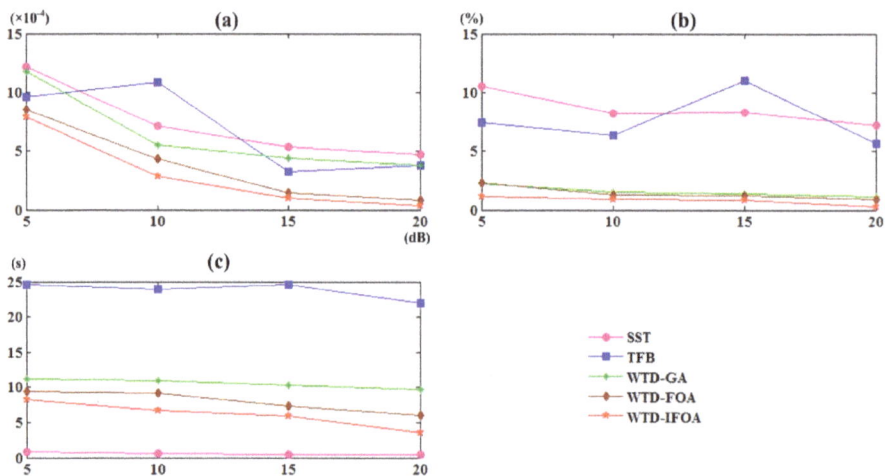

Figure 9. Comprehensive comparison of the five methods: (**a**) Mean square error of the five methods; (**b**) Peak value error of the five methods; and (**c**) Computation time of the five methods.

5.3. Application

In order to test practical effect of the proposed adaptive denoising method for the machinery sound signal based on WTD-IFOA, an industrial application was operated in a fully-mechanized coal mining working face. The shearer is an important machine in automatic coal mining, and working condition monitoring for the shearer is of great necessity. Traditional monitoring methods are mainly based on the vibration signal [35], even though the working life of the vibration sensors are very short due to the bad working condition and contact-measurement. Coal output is seriously restricted by the frequent maintenance. In the August 2015, an online monitoring system through the shearer cutting sound signal was built in the 71,507 coal mining face in the NO.2 Mine of Yangquan Coal Industry Group Corporation. However, there existed a large number of noise signal among the initial signal because of the harsh working environment. To eliminate the background noise from the sound, an industrial microphone was installed and WTD-IFOA was applied. The three-dimensional model of the coal mining shearer and field sound collection is presented in Figure 10.

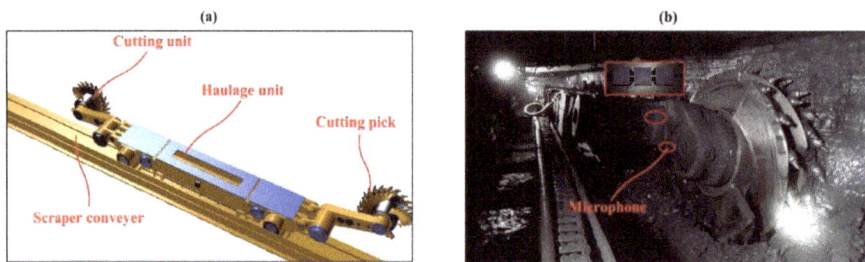

Figure 10. (**a**) Three-dimensional model of the shearer; and (**b**) collection of field cutting sound.

To illustrate the effectiveness of the proposed method, a piece of field sound signal with the length of 0.5 s was extracted and denoised. The original sound and the denoised one are shown in Figure 11. Then, 4096 points FFT was conducted to analyze the frequency components of the two signals, as presented in Figure 12. It can be seen in Figure 12 that the processed signal was sharp decreased in amplitude compared with the original field signal. The reason lies in that the scope saltation caused by the noise component was removed. Moreover, frequency components of original signal shown in Figure 12 had a disordered distribution, and it was difficult to identify the working state of the shearer. On the contrary, it was regular for the denoised signal. Some wave peaks appeared in the spectrogram, and other areas were stable. Different spectra of the collected signal reflected different working conditions of the shearer. Thus, the working state could be identified according to the wave peaks.

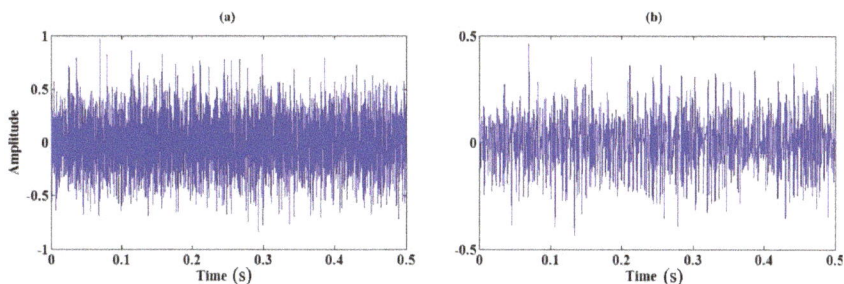

Figure 11. (**a**) The orginal sound signal; and (**b**) the denoised sound signal.

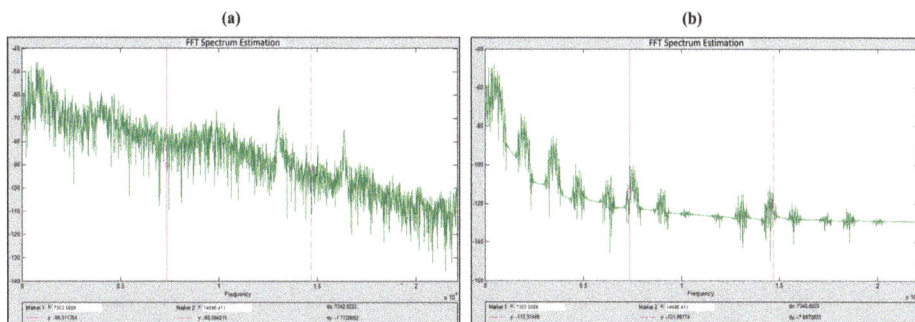

Figure 12. (**a**) Frequency components of the original signal; and (**b**) frequency components of the denoised signal.

6. Conclusions and Future Work

In order to eliminate noise components from the sound signal of a working machine, this paper proposes a novel approach based on wavelet threshold denoising and an improved FOA. Improved strategy on the basis of fly distance range obeying normal distribution was applied in the denoising process. To verify the feasibility and superiority of the proposed WTD-IFOA, a simulation example was provided and some comparisons were conducted. The simulation example and comparison results showed that the adaptive denoising method could effectively eliminate noise components and the proposed approach outperformed the others. Finally, an industrial application was performed on the shearer in a fully-mechanized coal mining face to test practical effect.

However, there are also some deficiencies and shortcomings in this method including the following. On the one hand, although the proposed WTD-IFOA is much more timesaving than other adaptive denoising approaches, the calculation duration is still a serious problem that cannot be neglected. On the other hand, parameters during the optimization process are determined according to past scholars and a large number of simulation experiments, while strict mathematical deduction are lacking. In future studies, the authors plan to investigate some improvements to the proposed approach. These may include an improved algorithm code with higher execution efficiency and appropriate scheme for determining optimization parameters.

Acknowledgments: The support of Joint Funds of the National Natural Science Foundation of China (No. U1510117), National Key Basic Research Program of China (No. 2014CB046301) and the Priority Academic Program Development (PAPD) of Jiangsu Higher Education Institutions in carrying out this research are gratefully acknowledged.

Author Contributions: Zhongbin Wang and Jing Xu contributed the new processing method; Chao Tan and Lin Zhang designed the simulations and experiments; Lei Si and Xinhua Liu performed the experiments; and Jing Xu wrote the paper.

Conflicts of Interest: The authors declare no conflict of interest.

References

1. Lin, J.; Qu, L.S. Feature extraction based on Morlet wavelet and its application for mechanical fault diagnosis. *J. Sound Vib.* **2000**, *234*, 135–148. [CrossRef]
2. Li, X.; Bassluny, A.M. Transient dynamical analysis of strain signals in sheet metal stamping processes. *Int. J. Mach. Tool. Man.* **2008**, *48*, 576–588. [CrossRef]
3. Xue, W.F.; Chen, J.; Li, J.Q.; Liu, X.F. Acoustical feature extraction of rotating machinery with combined wave superposition and blind source separation. *Proc. Inst. Mech. Eng. C J. Mech. Eng. Sci.* **2006**, *220*, 1423–1431. [CrossRef]
4. Benko, U.; Petrovcic, J.; Juricic, D.; Tavcarb, J.; Rejec, J. An approach to fault diagnosis of vacuum cleaner motors based on sound analysis. *Mech. Syst. Signal Process.* **2005**, *19*, 427–445. [CrossRef]
5. Ning, D.Y.; Gong, Y.J. Shocking fault component of abnormal sound signal in the fault engine extract method based on linear superposition method and cross-correlation analysis. *Adv. Mech. Eng.* **2015**, *7*. [CrossRef]
6. Abbasiona, S.; Rafsanjania, A.; Farshidianfarb, A.; Iranic, N. Rolling element bearings multi-fault classification based on the wavelet denoising and support vector machine. *Mech. Syst. Signal Process.* **2007**, *21*, 2933–2945. [CrossRef]
7. White, D.J.; William, P.E.; Hoffman, M.W.; Balkir, S. Low-Power Analog Processing for Sensing Applications: Low-Frequency Harmonic Signal Classification. *Sensors* **2013**, *13*, 9604–9623. [CrossRef] [PubMed]
8. Mallat, F.G. A Theory for Multiresolution Signal Decomposition: The Wavelet Representation. *IEEE Trans. Pattern Anal.* **1989**, *11*, 674–693. [CrossRef]
9. Donoho, D.L. De-Noising by Soft-Thresholding. *IEEE Trans. Inform. Theory* **1995**, *41*, 613–627. [CrossRef]
10. Zhang, X.P.; Desai, M.D. Adaptive Denoising Based on SURE Risk. *IEEE Signal Proc. Lett.* **1998**, *5*, 265–267. [CrossRef]
11. Nasri, M.; Nezamabadi-pour, H. Image denoising in the wavelet domain using a new adaptive thresholding function. *Neurocomputing* **2009**, *72*, 1012–1025. [CrossRef]
12. Zhang, B.; Sun, L.X.; Yu, H.B.; Xin, Y.; Cong, Z.B. A method for improving wavelet threshold denoising in laser-induced breakdown spectroscopy. *Spectrochim. Acta B* **2015**, *107*, 32–44. [CrossRef]
13. Tian, J.; Yu, W.Y.; Ma, L.H. AntShrink: Ant colony optimization for image shrinkage. *Pattern Recognit. Lett.* **2010**, *31*, 1751–1758. [CrossRef]
14. Bhutada, G.G.; Anand, R.S.; Saxena, S.C. PSO-based learning of sub-band adaptive thresholding function for image denoising. *Signal Image Video Proc.* **2012**, *6*, 1–7. [CrossRef]
15. Li, J.; Cheng, C.; Jiang, T.; Grzybowski, S. Wavelet De-noising of Partial Discharge Signals Based on Genetic Adaptive Threshold Estimation. *IEEE Trans. Dielectr. Electr. Insul.* **2012**, *19*, 543–549.
16. Pan, W.T. A new Fruit Fly Optimization Algorithm: Taking the financial distress model as an example. *Knowl. Based Syst.* **2012**, *26*, 69–74. [CrossRef]
17. Pan, W.-T. Using modified fruit fly optimisation algorithm to perform the function test and case studies. *Connect. Sci.* **2013**, *25*, 151–160. [CrossRef]
18. Pan, W.-T. Mixed modified fruit fly optimization algorithm with general regression neural network to build oil and gold prices forecasting model. *Kybernetes* **2014**, *43*, 1053–1063. [CrossRef]
19. Xu, W.J.; Deng, X.; Li, J. A New Fuzzy Portfolio Model Based on Background Risk Using MCFOA. *Int. J. Fuzzy Syst.* **2015**, *17*, 246–255. [CrossRef]

20. Xing, Y.F. Design and optimization of key control characteristics based on improved fruit fly optimization algorithm. *Kybernetes* **2013**, *42*, 466–481. [CrossRef]

21. Lin, S.M. Analysis of service satisfaction in web auction logistics service using a combination of Fruit fly optimization algorithm and general regression neural network. *Neural Comput. Appl.* **2013**, *22*, 783–791. [CrossRef]

22. Wang, W.C.; Liu, X.G. Melt index prediction by least squares support vector machines with an adaptive mutation fruitfly optimization algorithm. *Chemometr. Intell. Lab.* **2015**, *141*, 79–87. [CrossRef]

23. Wang, L.; Shi, Y.L.; Liu, S. An improved fruit fly optimization algorithm and its application to joint replenishment problems. *Expert Syst. Appl.* **2015**, *42*, 4310–4323. [CrossRef]

24. Pan, Q.K.; Sang, H.Y.; Duan, J.H.; Gao, L. An improved fruit fly optimization algorithm for continuous function optimization problems. *Knowl. Based Syst.* **2014**, *62*, 69–83. [CrossRef]

25. Donoho, D.L.; Johnstone, I.M. Adapting to unknown smoothness via wavelet shrinkage. *J. Am. Stat. Assoc.* **1995**, *90*, 1200–1224. [CrossRef]

26. Yi, T.H.; Li, H.N.; Zhao, X.Y. Noise Smoothing for Structural Vibration Test Signals Using an Improved Wavelet Thresholding Technique. *Sensors* **2012**, *12*, 11205–11220. [CrossRef] [PubMed]

27. Meng, B.; Li, Z.P.; Wang, H.H.; Li, Q.S. An improved wavelet adaptive logarithmic threshold denoising method for analysing pressure signals in a transonic compressor. *Proc. Inst. Mech. Eng. C J. Mech. Eng. Sci.* **2015**, *229*, 2023–2030. [CrossRef]

28. Soni, V.; Bhandari, A.K.; Kumar, A.; Singh, G.K. Improved sub-band adaptive thresholding function for denoising of satellite image based on evolutionary algorithms. *IET Signal Process.* **2013**, *7*, 720–730. [CrossRef]

29. Li, H.Z.; Guo, S.; Zhao, H.R.; Su, C.B.; Wang, B. Annual Electric Load Forecasting by a Least Squares Support Vector Machine with a Fruit Fly Optimization Algorithm. *Energies* **2012**, *5*, 4430–4445. [CrossRef]

30. Ramachandran, B.; Bellarmine, G.T. Improving observability using optimal placement of phasor measurement units. *Int. J. Elec. Power* **2014**, *56*, 55–63. [CrossRef]

31. Yuan, X.F.; Dai, X.S.; Zhao, J.Y.; He, Q. On a novel multi-swarm fruit fly optimization algorithm and its application. *Appl. Math. Comput.* **2014**, *233*, 260–271. [CrossRef]

32. Shan, D.; Cao, G.H.; Dong, H.J. LGMS-FOA: An Improved Fruit Fly Optimization Algorithm for Solving Optimization Problems. *Math. Probl. Eng.* **2013**. [CrossRef]

33. Liu, C.C.; Sun, T.Y.; Tsai, S.J.; Yu, Y.H.; Hsieh, S.T. Heuristic wavelet shrinkage for denoising. *Appl. Soft Comput.* **2011**, *2011*, 256–264. [CrossRef]

34. Lou, S.T.; Zhang, X.D. Fuzzy-based learning rate determination for blind source separation. *IEEE Trans. Fuzzy Syst.* **2003**, *11*, 375–383.

35. Zhao, L.J.; Tian, Z. Vibration characteristics of thin coal seam shearer. *Chin. J. Vib. Shock* **2015**, *34*, 195–199.

applied
sciences

MDPI

Article

Passive Guaranteed Simulation of Analog Audio Circuits: A Port-Hamiltonian Approach

Antoine Falaize *,† and Thomas Hélie

Project-team S3 (Sound Signals and Systems) and Analysis/Synthesis team, Laboratory of Sciences and Technologies of Music and Sound (UMR 9912), IRCAM-CNRS-UPMC, 1 Place Igor Stravinsky, Paris 75004, France; thomas.helie@ircam.fr
* Correspondence: antoine.falaize@ircam.fr; Tel.: +33-1-44-78-13-14
† The PhD thesis of A. Falaize is funded by UPMC, ED 130 (ÉDITE), Paris.

Academic Editor: Vesa Valimaki
Received: 25 April 2016 ; Accepted: 13 September 2016; Published: 24 September 2016

Abstract: We present a method that generates passive-guaranteed stable simulations of analog audio circuits from electronic schematics for real-time issues. On one hand, this method is based on a continuous-time power-balanced state-space representation structured into its energy-storing parts, dissipative parts, and external sources. On the other hand, a numerical scheme is especially designed to preserve this structure and the power balance. These state-space structures define the class of port-Hamiltonian systems. The derivation of this structured system associated with the electronic circuit is achieved by an automated analysis of the interconnection network combined with a dictionary of models for each elementary component. The numerical scheme is based on the combination of finite differences applied on the state (with respect to the time variable) and on the total energy (with respect to the state). This combination provides a discrete-time version of the power balance. This set of algorithms is valid for both the linear and nonlinear case. Finally, three applications of increasing complexities are given: a diode clipper, a common-emitter bipolar-junction transistor amplifier, and a wah pedal. The results are compared to offline simulations obtained from a popular circuit simulator.

Keywords: simulation; analog circuits; network modeling; passive system

1. Introduction

The characteristic input-to-output behavior of analog audio circuits (timbre, transitory) rests on the possibly highly nonlinear components appearing in such systems. These components make the stability of the simulations difficult to guarantee. The motivation of this work stems from the following observations:

1. Analog circuits combine energy-storing components, dissipative components, and sources.
2. Storage components do not produce energy, and dissipative components decrease it.

In this sense, analog circuits can be considered as passive systems with external power supply. We shall exploit this passivity property by transposing it to the digital domain, ensuring the stability of the simulations (see [1–3]).

The available approaches for the automated derivation of physical modeling and numerical simulation of audio circuits can be divided in two classes [4]: *wave scattering methods* (WS) and *Kirchhoff's variables methods* (KV). Mixed WS/KV methods have also been proposed in references [5,6]. The well-established *wave-digital filter* (WDF) formalism [7] belongs to the class of WS methods. For linear circuits, it provides a computationally realizable system of equations: First, by defining parametric wave variables for each elementary component and multiports (serial and parallel);

Second, by discretizing the corresponding constitutive laws with the bilinear transform; and Third, by choosing the wave's parameters so as to reduce the computational complexity and to avoid instantaneous feedback loops. An extension to nonlinear circuits has been considered in [8,9] and applied to, for example, the real-time simulation of vacuum-tube guitar amplifiers in [10,11]. WDF ensures the passivity of the resulting digital system [7,12], including systems with scalar nonlinearity [13]. However, the passivity property of WDF structures is not ensured for circuits with more than one nonlinear element (e.g., [14] §3.2 and [15] §6).

The class of KV methods for audio circuits encompasses nonlinear state-space representations [16]. Several modeling techniques are available to derive the discrete-time state-space model, either from the global time-continuous model (e.g., [17]) or from the interconnection of discretized elementary components (e.g., [18]). The resulting set of nonlinear implicit equations solved at each sample can be structured so as to obtain a computationally realizable system by applying the *K-method* introduced in [19] with developments in [18]. However, this structure does not encode the passivity of the original circuit naturally, and it must be investigated on a case-by-case basis [20].

In this paper, we consider the *port-Hamiltonian systems* (PHS) approach, introduced in the 1990's [21–23]. PHS are extensions of classical Hamiltonian systems [24], specifically defined to address open dynamical systems made of energy storage components, dissipative components, and some connection ports through which energy can transit. This approach leads to a state-space representation of physical systems structured according to energy flow, thus encoding the passivity property, even for nonlinear cases. This class of physical systems encompasses not only electrical circuits, but also multi-domain systems, such as loudspeakers, which involves electrical, magnetic, mechanical, acoustical, and thermodynamical phenomena.

The port-Hamiltonian structure is derived by applying the Kirchhoff's laws to a given schematic, similarly to other existing approaches (e.g., WDF and K-method). Here, the advantage of the PHS formulation is the direct encoding of the underlying passive structure. This passivity property is transposed to the discrete-time domain by appropriate numerical methods, so as to ensure the numerical stability. For linear storage components (inductors and capacitors), the combination of the PHS structure with any of the trapezoidal rule or the mid-point rule yields the same numerical scheme that preserves the passivity in discrete-time. For nonlinear storage components, we propose the use of the *discrete gradient method* [25] combined with the PHS structure to achieve this goal. This result is compared to the aforementioned methods. As a second result, we provide an automated method that derives the PHS structure from a given analog circuit, based on an especially designed graph analysis.

This paper is organized as follows. Section 2 presents the class of the port-Hamiltonian systems; Section 3 is devoted to the automated derivation of algebraic-differential equations in the continuous time domain from the electronic schematics; Section 4 presents the numerical scheme which provides a discrete-time version of the power balance; Then, applications are presented in Section 5 and results are compared to LT-Spice simulations, before conclusions and perspectives.

2. Port-Hamiltonian Systems

First and foremost, we provide an introduction to the *port-Hamiltonian systems* (PHS) formalism. It is shown how this structure guarantees the passivity of the model in continuous time. Second, for the sake of intuition, we give an introductory example.

2.1. Formalism and Property

Denote $E(t) \geq 0$ the energy stored in an open physical system (an electronic circuit). If the system is conservative, its time variation $\frac{dE}{dt}(t)$ reduces to the power $S(t)$ received from the sources through the external ports. If the system includes dissipative phenomena, the power $D(t) \geq 0$ is dissipated, and the evolution of energy is governed by the following *power balance*:

$$\frac{dE}{dt}(t) = -D(t) + S(t). \tag{1}$$

The port-Hamiltonian approach is used to decompose such open physical systems in (i) a set of *components* that are combined according to (ii) a *conservative interconnection network*. These two ingredients are detailed below in the case of electronic circuits.

2.1.1. Components

Electronic circuit components are sorted as (or can be a combination of):

n_S internal components that store energy $E \geq 0$ (capacitors or inductors),

n_D internal components that dissipate power $D \geq 0$ (resistors, diodes, transistors, etc.),

n_P external ports that convey power S ($\in \mathbb{R}$) from sources (voltage or current generators) or any external system (active, dissipative, or mixed).

The behavior of each component is described by a relation between two *power variables*: the current i and the voltage v, defined in *receiver convention* (the received power is $P = v \cdot i$).

The energy E_s stored in storage component $s \in [1 \cdots n_S]$ is expressed as a *storage function* h_s of an appropriate *state* x_s: $E_s(t) = h_s(x_s(t)) \geq 0$. Typically, for a linear capacitor with capacitance C, the state can be the charge $x = q$ and the positive definite function is $h(q) = q^2/(2C)$. Storage power variables (v_s, i_s) are related to the variation of the state $\frac{dx_s}{dt}$ and the gradient of the storage function $h'_s(x_s)$, the product of which is precisely the *received power*: $v_s \cdot i_s = \frac{dE_s}{dt} = h'_s \cdot \frac{dx_s}{dt}$. For the capacitance, these *constitutive laws* are $i = \frac{dq}{dt} = \frac{dx}{dt}$ and $v = q/C = h'$. Note that these definitions apply equally for non-quadratic storage functions $h(x) \geq 0$ for which $h''(x)$ is not constant.

The power D_d instantaneously dissipated by the dissipative component $d \in [1 \cdots n_D]$ is expressed with respect to an appropriate *dissipation variable* w_d: $D_d(t) \equiv D_d(w_d(t)) \geq 0$. Typically, for a linear resistance R, w can be a current $w = i$ and $D(i) = R \cdot i^2$. As for storage components, a mapping of the dissipative power variables (v_d, i_d) is provided, based on the factorization $D_d(w_d) = w_d \cdot z_d(w_d)$, introducing a *dissipation function* z_d. For the resistance, $i = w$ and $v = R \cdot i = z(w)$.

The power instantaneously provided to the system through external port $p \in [1 \cdots n_P]$ is $S_p(t)$, and we arrange the source variables (v_p, i_p) in two vectors: one is considered as an *input* u_p, and the other as the associated *output* y_p, so that the power received from sources on port p is $S_p = y_p \cdot u_p = -v_p \cdot i_p$ (receiver convention, with $v_p \cdot i_p$ the power received by the sources).

2.1.2. Conservative Interconnection

The interconnection of the components is achieved by relating all the voltages and currents through the application of the Kirchhoff's laws to the interconnection network (schematic). This defines a conservative interconnection, according to Tellegen's theorem recalled below (see also [26] and [27] §9.4).

Theorem 1 (Tellegen). *Consider an electronic circuit made of N edges defined in same convention (here receiver), with individual voltages $v = (v_1, \cdots, v_N)^\mathsf{T}$ and currents $i_n = (i_1, \cdots, i_N)^\mathsf{T}$ which comply with the Kirchhoff's laws. Then*

$$\mathbf{v}^\mathsf{T} \cdot \mathbf{i} = 0. \tag{2}$$

A direct consequence of (2) is that no power is created nor lost in the structure: $\mathbf{v}^\mathsf{T} \cdot \mathbf{i} = \sum_{i=1}^{N} P_n = 0$, with $P_n = v_n \cdot i_n$ the power received by edge n, thus defining a *conservative interconnection* (Tellegen's theorem is a special case of a more general interconnection structure, namely, the *Dirac structure* (see [23] §2.1.2 for details)). Now, denote $(\mathbf{v_s}, \mathbf{i_s})$, $(\mathbf{v_d}, \mathbf{i_d})$, and $(\mathbf{v_p}, \mathbf{i_p})$ the sets of all the power variables associated with storage components, dissipative components, and sources (respectively), and $\mathbf{v} = (\mathbf{v}_s^\mathsf{T}, \mathbf{v}_d^\mathsf{T}, \mathbf{v}_p^\mathsf{T})^\mathsf{T}$, $\mathbf{i} = (\mathbf{i}_s^\mathsf{T}, \mathbf{i}_d^\mathsf{T}, \mathbf{i}_p^\mathsf{T})^\mathsf{T}$ the vectors of all the power variables. Then, Tellegen's theorem restores the power balance (1) with

$$
\begin{aligned}
\mathbf{v}^\mathsf{T} \cdot \mathbf{i} &= \mathbf{v}_s^\mathsf{T} \cdot \mathbf{i}_s + \mathbf{v}_d^\mathsf{T} \cdot \mathbf{i}_d + \mathbf{v}_p^\mathsf{T} \cdot \mathbf{i}_p \\
&= \underbrace{\nabla \mathcal{H}^\mathsf{T}(\mathbf{x}) \cdot \frac{d\mathbf{x}}{dt}}_{\frac{dE}{dt}} + \underbrace{\mathbf{z}(\mathbf{w})^\mathsf{T} \cdot \mathbf{w}}_{D} - \underbrace{\mathbf{u}^\mathsf{T} \cdot \mathbf{y}}_{S},
\end{aligned}
\tag{3}
$$

where $\nabla \mathcal{H} : \mathbb{R}^{n_S} \to \mathbb{R}^{n_S}$ denotes the gradient of the total energy $E = \mathcal{H}(\mathbf{x}) = \sum_{s=1}^{n_S} h_s(x_s)$ with respect to (w.r.t.) the vector of the states $[\mathbf{x}]_s = x_s$, and function $\mathbf{z} : \mathbb{R}^{n_D} \to \mathbb{R}^{n_D}$ denotes the collection of functions z_d w.r.t. the vector $\mathbf{w} \in \mathbb{R}^{n_D}$ of $[\mathbf{w}]_d = w_d$ so that $\mathbf{z}(\mathbf{w})^\mathsf{T} \cdot \mathbf{w} = \sum_{d=1}^{n_D} D_d(w_d)$ is the total dissipated power.

The above description of storage components, dissipative components, and source, along with the conservative interconnection stated by the Kirchhoff's laws, constitute the minimal definition of a port-Hamiltonian system (PHS) (see [23] §2.2). In this work, we focus on circuits that admit an *explicit realization* of PHS, for which the quantities $\mathbf{b} = (b_1, \cdots, b_N)^\mathsf{T} = (\frac{d\mathbf{x}}{dt}, \mathbf{w}, -\mathbf{y})^\mathsf{T}$ (with $b_n = v_n$ or $b_n = i_n$) can be expressed as linear combinations of the remaining N powers variables organized in the dual vector $\mathbf{a} = (a_1, \cdots, a_N)^\mathsf{T} = (\nabla \mathcal{H}(\mathbf{x}), \mathbf{z}(\mathbf{w}), \mathbf{u})^\mathsf{T}$ (with $a_n = i_n$ if $b_n = v_n$ or $a_n = v_n$ if $b_n = i_n$):

$$
\mathbf{b} = \mathbf{J} \cdot \mathbf{a}.
\tag{4}
$$

Then, $\mathbf{a}^\mathsf{T} \cdot \mathbf{b} = \mathbf{a}^\mathsf{T} \cdot \mathbf{J} \cdot \mathbf{a} = 0$ from Tellegen's theorem, so that the matrix $\mathbf{J}$ is necessarily skew-symmetric ($\mathbf{J}^\mathsf{T} = -\mathbf{J}$). More precisely, we consider the following algebraic-differential system of equations

$$
\underbrace{\begin{pmatrix} \frac{d\mathbf{x}}{dt} \\ \hline \mathbf{w} \\ \hline -\mathbf{y} \end{pmatrix}}_{\mathbf{b}} = \underbrace{\begin{pmatrix} \mathbf{J_x} & -\mathbf{K} & -\mathbf{G_x} \\ \hline \mathbf{K}^\mathsf{T} & \mathbf{J_w} & -\mathbf{G_w} \\ \hline \mathbf{G_x}^\mathsf{T} & \mathbf{G_w}^\mathsf{T} & \mathbf{J_y} \end{pmatrix}}_{\mathbf{J}} \cdot \underbrace{\begin{pmatrix} \nabla \mathcal{H}(\mathbf{x}) \\ \hline \mathbf{z}(\mathbf{w}) \\ \hline \mathbf{u} \end{pmatrix}}_{\mathbf{a}},
\tag{5}
$$

where matrices $\mathbf{J_x}, \mathbf{J_w}, \mathbf{J_y}$ are skew-symmetric. The significance of the structure matrices is the following:

$\mathbf{J_x} \in \mathbb{R}^{n_S \times n_S}$ expresses the conservative power exchanges between storage components (this corresponds to the so-called $\mathbf{J}$ matrix in classical Hamiltonian systems);

$\mathbf{J_w} \in \mathbb{R}^{n_D \times n_D}$ expresses the conservative power exchanges between dissipative components;

$\mathbf{J_y} \in \mathbb{R}^{n_P \times n_P}$ expresses the conservative power exchanges between ports (direct connections of inputs to outputs);

$\mathbf{K} \in \mathbb{R}^{n_S \times n_D}$ expresses the conservative power exchanges between the storage components and the dissipative components;

$\mathbf{G_x} \in \mathbb{R}^{n_S \times n_P}$ expresses the conservative power exchanges between ports and storage components (input gain matrix);

$\mathbf{G_w} \in \mathbb{R}^{n_D \times n_P}$ expresses the conservative power exchanges between ports and dissipative components (input gain matrix).

The PHS (5) fulfills the definition of passivity (e.g., [16]) according to the following property.

Property 1 (Power Balance). *The variation of the total energy* $E = \mathcal{H}(\mathbf{x})$ *of a system governed by (5) is given by (1), with* $D = \mathbf{z}(\mathbf{w})^\mathsf{T} \cdot \mathbf{w} \geq 0$ *the total dissipated power, and* $S = \mathbf{u}^\mathsf{T} \cdot \mathbf{y}$ *the total power incoming on external ports.*

Proof. We have $\mathbf{a}^\mathsf{T} \cdot \mathbf{b} = \frac{dE}{dt} + D - S$. Now $\mathbf{a}^\mathsf{T} \cdot \mathbf{b} = \mathbf{a}^\mathsf{T} \cdot \mathbf{J} \cdot \mathbf{a} = 0$ since $\mathbf{J}$ is skew-symmetric. $\square$

Remark 1 (Power variables). *This work is devoted to the treatment of electronic circuits for which power variables are chosen as current and voltage. However, all the aforementioned definitions apply equally to multiphysical systems, provided an adapted set of power variables, generically denoted by flux (currents, velocities, magnetic flux variations) and efforts (voltages, forces, magnetomotive force), the product of which is a power (see [23] Table 1.1). This follows the bond-graph modeling approach [28,29], on which the PHS formalism is built (see [23] §1.6 and 2.1). The treatment of multiphysical audio systems in the PHS formalism can be found in [30] (electromechanical piano that includes mechanical, electrical, and magnetic phenomena) and [31] (§III.B) (modulated air flow for musical acoustics applications that includes mechanical and acoustical phenomena).*

2.2. Example

Consider the resistor-inductor-capacitor (RLC) circuit in Figure 1, with $n_S = 2$, $n_D = 1$, and $n_P = 2$, described as follows. For the linear inductance L, the state and the positive definite function can be the magnetic flux $x_1 = \phi$ and $h_1(\phi) = \phi^2/(2L)$, so that $v_L = dh_1/dx_1$ and $i_L = \frac{dx_1}{dt}$. For the capacitance and the resitance, quantities are defined with $x_2 = q$ and $\mathbf{w} = [i_R]$. Port variables are arranged as input $\mathbf{u} = [v_1, v_2]^\mathsf{T}$ and output $\mathbf{y} = [-i_1, -i_2]^\mathsf{T}$ (edges receiver convention).

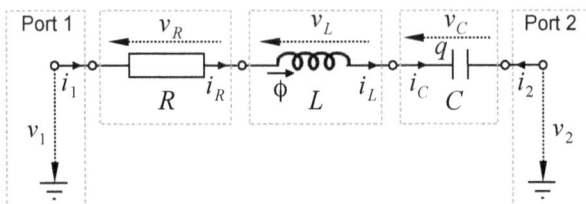

Figure 1. Resistor-inductor-capacitor (RLC) circuit (notations and orientations).

Applying Kirchhoff's laws to this simple serial circuit yields

$$
\begin{pmatrix} v_L \\ i_C \\ i_R \\ i_1 \\ i_2 \end{pmatrix} =
\begin{pmatrix}
0 & -1 & -1 & -1 & +1 \\
+1 & 0 & 0 & 0 & 0 \\
+1 & 0 & 0 & 0 & 0 \\
+1 & 0 & 0 & 0 & 0 \\
-1 & 0 & 0 & 0 & 0
\end{pmatrix}
\cdot
\begin{pmatrix} i_L \\ v_C \\ v_R \\ v_1 \\ v_2 \end{pmatrix}.
$$

From the constitutive laws of components, this equation restores the form (5) exactly, block by block. It provides the algebraic-differential equations that govern the system with input $\mathbf{u}$ and output $\mathbf{y}$.

This work aims at simulating such passive systems by firstly generating Equation (5) associated to a given circuit, and secondly by deriving its numerical version so that a discrete power balance is satisfied.

Remark 2 (Reduction). *The system (5) can be reduced by decomposing function $\mathbf{z}$ into its linear and nonlinear parts. See Appendix A for details.*

3. Generation of Equations

This section provides a method to translate the description of a circuit (components and interconnections) from a netlist in a Spice-style [32] to the Formulation (5). Compared to standard methods that express all the currents as a function of all the voltages (see [18,27,32]), Formulation (5) expresses vector $\mathbf{b}$ of selected power variables (voltage or current) as a function of the vector $\mathbf{a}$

of complementary power variables (if $[\mathbf{b}]_n$ is a voltage of a branch, $[\mathbf{a}]_n$ is the associated current with receiver convention). To derive the matrix $\mathbf{J}$ that relates the voltages and the currents arranged in vectors $\mathbf{a}$ and $\mathbf{b}$ according to Kirchhoff's laws (as in example §2.2), we propose a two-step method:

Step 1 : from a netlist ($\mathcal{L}$) to a graph ($\mathcal{G}$) that represents the Kirchhoff's laws for a chosen orientation (convention);

Step 2 : from ($\mathcal{G}$) to the skew-symmetric matrix $\mathbf{J}$ in (5).

Step 1 is standard. The presentation focuses on convention choices and details our procedure. In step 2, we propose an algorithm that analyzes if Formulation (2) is available (that is, the circuit is realizable into the PHS formalism) and delivers the matrix $\mathbf{J}$ in this case. Otherwise, the circuit corresponds to an implicit formulation that is not addressed in this paper. In practice, such cases appear for serial(/parallel) connection of voltage(/current)-controlled components. In this case, port-Hamiltonian Formulation (5) requires extension (see [22,23]).

3.1. Graph Encoding

3.1.1. Netlists

Each line of a netlist describes an element of the corresponding schematic, with: identification label, list of connection nodes, type of element, and list of parameters. We divide netlists into two blocks: internal components (dissipative and storage) and external ports (supplies and ground). In the first block (components), each line includes a reference to the appropriate entry in the dictionary and a list of the parameters for the corresponding model. Each line of the second block (external ports) provides the label of the externalized node, the type of supply (voltage or current), and the symbol~if the supply is modulated (typically, the input signal), or a value if constant (typically, a battery).

As an example, the netlist corresponding to the circuit in Figure 2 is given in Table 1. Here, the components are given lines ℓ_1 to ℓ_3 (gray). The first two lines describe dipoles: a linear capacitor between N_1 and N_2 with label $C1$ and capacitance value $20e^{-9}$ F; and a resistor between N_3 and N_4 with label $R1$ and resistance value $1.5e^3$ Ω. The third line describes a npn bipolar-junction transistor. From the dictionary (Appendix B), the base terminal appears to be connected to the circuit's node N_2, the emitter terminal to N_3, and the collector terminal to N_5. For this component, the list of parameters is: forward and reverse common emitter current gain, reverse saturation current, and thermal voltage. External ports are given lines ℓ_4 to ℓ_7. Line ℓ_4 describes a constant 9 V voltage supply (labeled Vcc) on the circuit's node N_4. ℓ_5 describes a modulated voltage supply (here considered as the input signal) on N_4. ℓ_6 describes a constant 0 A current supply on node N_3; this permits the recovery of the voltage on that node N_3 as an output to the circuit. ℓ_7 describes the connection of the circuit's node N_5 to the ground.

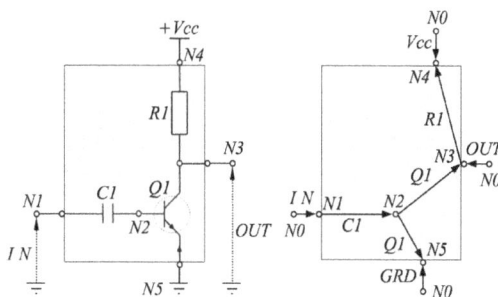

Figure 2. Schematic and corresponding graph of a simple bipolar-junction transistor (BJT) amplifier with feedback. The grey part corresponds to the components, and the outer elements correspond to the external ports, or sources (as in Table 1).

Table 1. Example of a netlist corresponding to the circuit in Figure 2. The grey part corresponds to the components, and the other elements correspond to the external ports, or sources (as in figure 2).

Line	Label	Node List	Type	Parameters
ℓ_1	C1	N_1, N_2	CapaLin	$20e^{-9}$
ℓ_2	R1	N_3, N_4	Resistor	$1.5e^3$
ℓ_3	Q1	N_2, N_3, N_5	NPN_ Type1	*List of parameters*
ℓ_4	Vcc	N_4	Voltage	9
ℓ_5	IN	N_1	Voltage	$\sim$
ℓ_6	OUT	N_3	Current	0
ℓ_7	GRD	N_5	Voltage	0

3.1.2. Graph

A graph $G = \{N, B\}$ is defined by two lists of *nodes* N (also called *vertices*) and *branches* B (also called *edges*), with $B \subset N^2$ (each element of B is an object defined on two elements of N, see [33] for details). The dictionary (Appendix B) encodes the graph of each elementary component. The branches of such an elementary graph contain the constitutive laws of the corresponding component:

- *Dipoles* are made of two nodes and a single branch, defining a single couple of state x and storage function $h(x)$ (storage component), or dissipative variable w and scalar relation $z(w)$ (dissipative components).
- More generally, *n-ports multipole* are made of n nodes and at least $n - 1$ branches, defining $n - 1$ couples of variables and functions. Typically, the graph for the bipolar junction is made of two branches (base-emitter and base-collector).

The graph corresponding to a given circuit is derived from its netlist description in two steps:

1. build the *internal* graph by connecting the elementary graph of the components from the first block of the netlist,
2. introduce a reference node N_0 (or datum, see [27] §10) to define the *external* branches from the second block.

Typically, N_0 corresponds to the ground or any local electrostatic potential which does not impact the currents nor voltages. Then, N is built from the list of nodes appearing at least once in the netlist, plus the reference node $N \triangleq [N_0, N_1, \cdots, N_{n_N}]$. According to Section 2, the set of branches is organized as $B = \{B_S, B_D, B_P\}$, with B_S the n_S energy storage branches, B_D the n_D dissipative branches, and B_P the n_P sources.

As an example, the construction of the graph in Figure 2 from its netlist 1 is as follows. Firstly, the internal graph is built. It is made of $n_N = 5$ nodes $\{N_1, \cdots, N_5\}$ and four branches $B_S = \{C_1\}$ and $B_D = \{R_1, Q_{1,bc}, Q_{1,be}\}$. Secondly, we introduce the (virtual) reference node N_0 to define the four branches corresponding to the external ports $B_P = \{IN, Vcc, OUT, GRD\}$.

3.1.3. Kirchhoff's Laws on Graphs

We assign to each branch b both a voltage v_b and a current i_b in *receiver convention*, the direction of the branch indicating the direction of the current. Note that the power supplied to the system on port p is the power *emitted* by the port branch $S_p = u_p \cdot y_p = -v_p \cdot i_p$. For a circuit made of $n_N + 1$ nodes and $n_B = n_S + n_D + n_P$ branches, we define: the set of electrostatic potentials on the nodes $\mathbf{e} = (e_1 \cdots e_{n_N})^\mathsf{T}$, the set of voltages $\mathbf{v} = (v_1 \cdots v_{n_B})^\mathsf{T}$, and the set of currents $\mathbf{i} = (i_1 \cdots i_{n_B})^\mathsf{T}$. The orientation of an entire graph is encoded in its *incidence matrix* $\Gamma \in \mathbb{R}^{(n_N+1) \times n_B}$, defined below [27] (§9).

$$[\Gamma]_{n,b} = \begin{cases} -1 & \text{if branch } b \text{ is outgoing node } n, \\ 1 & \text{if branch } b \text{ is ingoing node } n, \\ 0 & \text{otherwise.} \end{cases} \tag{6}$$

As an example, the incidence matrix for the circuit described in Table 1 is given equation below (0 are replaced by dots). Notice the grey columns correspond to the components, and the other columns correspond to the external ports, or sources (as in Table 1 and Figure 2).

$$\Gamma = \begin{array}{c} \\ \\ \\ \\ \\ \\ \end{array} \begin{pmatrix} \begin{array}{cccccccc} \text{C1} & \text{R1} & \text{Q1,bc} & \text{Q1,be} & \text{Vcc} & \text{IN} & \text{OUT} & \text{GRD} \\ \vdots & \vdots & \vdots & \vdots & 1 & 1 & 1 & 1 \\ 1 & \vdots & \vdots & \vdots & \vdots & -1 & \vdots & \vdots \\ -1 & \vdots & 1 & 1 & \vdots & \vdots & \vdots & \vdots \\ \vdots & 1 & -1 & \vdots & \vdots & \vdots & -1 & \vdots \\ \vdots & -1 & \vdots & \vdots & -1 & \vdots & \vdots & \vdots \\ \vdots & \vdots & \vdots & -1 & \vdots & \vdots & \vdots & -1 \end{array} \end{pmatrix} \begin{array}{c} N_0 \\ N_1 \\ N_2 \\ N_3 \\ N_4 \\ N_5 \end{array} \, . \tag{7}$$

Since the reference potential e_0 does not influence the voltages nor the currents, it is not taken into account in the Kirchhoff's laws, and we define the reduced incidence matrix $\widehat{\Gamma} \in \mathbb{R}^{n_N \times n_B}$ obtained by deleting the row corresponding to the datum N_0 in Γ. This leads to the following matrix formulation of Kirchhoff's Voltage Law (KVL) and Kirchhoff's Current Law (KCL) [27] (§10), from which the structure (5) is derived.

$$\begin{cases} \widehat{\Gamma}^{\mathsf{T}} \cdot \mathbf{e} = \mathbf{v}, & \text{(KVL)} \\ \widehat{\Gamma} \cdot \mathbf{i} = 0. & \text{(KCL)} \end{cases} \tag{8}$$

3.2. Realizability Analysis

The PHS structure (5) relies on (i) an arrangement of currents $\mathbf{i}$ and voltages $\mathbf{v}$ in two vectors $\mathbf{a}$ and $\mathbf{b}$ and (ii) a set of linear relations encoded in the skew-symmetric matrix $\mathbf{J}$ that corresponds to the conservation laws (8) applied on $(\mathbf{i}, \mathbf{v})$. For storage and sources components, step (i) is straightforward with the constraints given in Table 2. For dissipative components, this step is achieved by selecting each component as voltage-controlled or current-controlled in order to satisfy a criterion on the matrix description of the interconnection scheme. This *realizability criterion* is given in Section 3.2.1, assuming the control type of every edge is known. A method of choosing the control type of dissipative edges so as to satisfy the realizability criterion is addressed in Section 3.2.2. This leads to Algorithm 1, which solves (i) and (ii).

Table 2. Sorting components according to their realizability.

Component type	Current-Controlled $[a]_b = i_b$ $[b]_b = v_b$	Voltage-Controlled $[a]_b = v_b$ $[b]_b = i_b$
storages	capacitor	inductor
resistors	resistance	conductance
nonlinear		diodes, transistors
sources	voltage source	current source

3.2.1. A Criterion for Realizability

In Section 3.1, the set of edges B has been partitioned based on the differentiation between *internal edges* (or component edges $\{B_S, B_D\}$, grey) and *external edges* (or ports edges B_P, white), in order to build the complete graph from the netlist. In this section, we are interested in the PHS Formulation (5) associated to a given complete graph. To that end, we leave out from here the differentiation between internal and external edges in order to focus on the differentiation between voltage-controlled and current-controlled edges.

Suppose the control type of every edge is known, and the set of edges is split according to $B = \{B_1, B_2\}$, with B_1 the set of n_1 *voltage-controlled* edges and B_2 the set of n_2 *current-controlled* edges (see Table 2). Correspondingly, the sets of power variables are split as $\mathbf{v} = (\mathbf{v}_1, \mathbf{v}_2)^\mathsf{T}$ and $\mathbf{i} = (\mathbf{i}_1, \mathbf{i}_2)^\mathsf{T}$, and we define $\tilde{\mathbf{a}} = (\mathbf{v}_1, \mathbf{i}_2)^\mathsf{T}$ and $\tilde{\mathbf{b}} = (\mathbf{i}_1, \mathbf{v}_2)^\mathsf{T}$. Since the reference potential e_0 defined on node N_0 does not influence the voltages nor the currents, it is not considered in the sequel, and the incidence matrix splits as follows:

$$\Gamma = \left(\begin{array}{c|c} \gamma_0 & \\ \hline \gamma_1 & \gamma_2 \end{array} \right), \quad \text{with } \gamma_0 \in \mathbb{R}^{1 \times n_B}, \ \gamma_1 \in \mathbb{R}^{n_N \times n_1}, \ \gamma_2 \in \mathbb{R}^{n_N \times n_2}.$$

This leads to a rewrite of the Kirchhoff laws (8) as:

$$(\gamma_1, \gamma_2)^\mathsf{T} \mathbf{e} = \mathbf{v}, \tag{9}$$

$$(\gamma_1, \gamma_2) \mathbf{i} = 0. \tag{10}$$

Proposition 1 (Realizability). *If γ_2 is invertible, then the port-Hamiltonian structure (5) provides a realization of the graph $G = \{N, (B_1, B_2)\}$.*

Proof. From the relation on the voltages in (9), we get $\mathbf{v}_1 = \gamma_1^\mathsf{T} \mathbf{e}$ and $\mathbf{v}_2 = \gamma_2^\mathsf{T} \mathbf{e}$. From the relation on the currents (10), we get $\gamma_2 \mathbf{i}_2 = -\gamma_1 \mathbf{i}_1$. Now, if γ_2 is invertible, we denote $\gamma = \gamma_2^{-1} \gamma_1$ and

$$\underbrace{\left(\begin{array}{c} \mathbf{v}_1 \\ \mathbf{i}_2 \end{array} \right)}_{\tilde{\mathbf{a}}} = \underbrace{\left(\begin{array}{cc} 0 & \gamma^\mathsf{T} \\ -\gamma & 0 \end{array} \right)}_{\tilde{\mathbf{J}}} \underbrace{\left(\begin{array}{c} \mathbf{i}_1 \\ \mathbf{v}_2 \end{array} \right)}_{\tilde{\mathbf{b}}}. \tag{11}$$

□

The PHS (5) is obtained by rearranging the edges according to their role with respect to the power balance, according to the permutation of vector elements $\Pi(\tilde{\mathbf{a}}) = \left(\frac{d\mathbf{x}}{dt}, \mathbf{w}, \mathbf{y} \right)^\mathsf{T} = \mathbf{a}$ (and correspondingly $\Pi(\tilde{\mathbf{b}}) = (\nabla\mathcal{H}, \mathbf{z}, \mathbf{u})^\mathsf{T} = \mathbf{b}$), which is also applied on rows and columns of $\tilde{\mathbf{J}}$ to yield $\mathbf{a} = \mathbf{J} \mathbf{b}$.

From the invertibility condition on γ_2 in Proposition 1, we state the following remark, which is used in the sequel to derive the *realizability analysis algorithm*.

Remark 3 (Necessary condition for realizability). *A necessary condition for the graph G to be realizable as a PHS (5) is that it includes as many current-controlled edges as nodes n_N, with $\gamma_2 \in \mathbb{R}^{n_N \times n_N}$.*

3.2.2. Algorithm

This section introduces an algorithm that selects the appropriate control type for each dissipative edge so that the partition $B = \{B_1, B_2\}$ satisfies Proposition 1. From Remark 3, the total number n_2 of current-controlled edges should be exactly equal to the number of nodes n_N. From the special structure of the incidence matrix Γ, this in turn ensures that the potential on each node is uniquely defined by a linear combination of the voltages $\mathbf{v}_2$ (elements in $\mathbf{a}$ associated to current-controlled edges); i.e., γ_2 is invertible.

Consider the current-controlled edge b from node i to node j in Figure 3. If the potential on node j is known, the remaining potential is obtained from $e_i = v_b \pm e_j$, where the sign depends on the orientation. In this case, we say edge b *imposes* the potential on node i. Now, the objective is to perform this analysis globally so that $\mathbf{e}$ is a linear combination of $\mathbf{v}_2$ with $\mathbf{e} = (\gamma_2^{\mathsf{T}})^{-1} \cdot \mathbf{v}_2$. To that end, we introduce the *realizability matrix* Λ defined element-wise as follows

$$[\Lambda]_{n,b} = \begin{cases} 1 & \text{if branch } b \text{ imposes potential on node } n, \\ 0 & \text{else,} \end{cases} \tag{12}$$

$$\Lambda = \left(\begin{array}{c} \lambda_0 \\ \hline \lambda_1 \;\big|\; \lambda_2 \end{array} \right), \quad \text{with } \lambda_0 \in \mathbb{R}^{1 \times n_B}, \; \lambda_1 \in \mathbb{R}^{n_N \times n_1} \; \lambda_2 \in \mathbb{R}^{n_N \times n_2}. \tag{13}$$

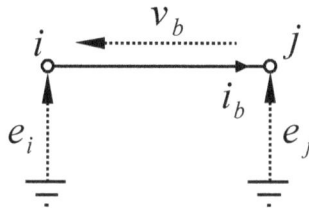

Figure 3. Definitions and orientations for a single current-controlled edge b from node i to node j, with nodes potentials e_i and e_j, respectively. The knowledge of the potential e_j is transferred to node i with $e_i = v_b + e_j$.

Then, a given graph is realizable if the type of each resistor can be selected so that the following set of constraints is fulfilled.

(C1) The potential on each node $n \in [1, \cdots, n_N]$ is uniquely defined so that $\sum_{b=1}^{n_B} [\Lambda]_{n,b} = 1$.
(C2) Each Current-controlled edge $b \in [n_1 + 1, \cdots, n_N]$ propagates the knowledge of the potential on one node to the other, so that $\sum_{n=1}^{n_N} [\Lambda]_{n,b} = 1$.
(C3) No edge imposes the reference potential e_0 so that $\lambda_0 = \mathbb{0}_{1 \times n_B}$.
(C4) No voltage-controlled edge $b \in [1, \cdots, n_1]$ imposes any potential so that $\lambda_1 = \mathbb{0}_{n_N \times n_2}$.

Constraints (C1–C2) ensure that γ_2 is invertible, so that $\mathbf{e} = (\gamma_2^{\mathsf{T}})^{-1} \cdot \mathbf{v}_2$. Constraint (C3) ensures that the reference potential on datum does not contribute to the system's dynamics. Constraint (C4) ensures that inputs of voltages-controlled edges are explicitly given by a linear combination of the nodes potentials so that $\mathbf{v}_1 = \gamma_1^{\mathsf{T}} \mathbf{e} = (\gamma_2^{-1} \cdot \gamma_1)^{\mathsf{T}} \cdot \mathbf{v}_2$. To build and analyze the matrix Λ, we start from the adjacency matrix A of the graph, defined as follows:

$$[A]_{b,n} = \begin{cases} 1 & \text{if branch } b \text{ is connected to node } n, \\ 0 & \text{else.} \end{cases}$$

Then the non-zero elements in A are analyzed to cope with the realizability constraints (C1–C4). This yields Algorithm 1. The PHS (5) is finally recovered as discussed in the proof of Property 1.

Algorithm 1: Analysis of realizability. If successfully complete, the resulting PHS structure is given by the procedure in the proof of Proposition 1

Input: A graph $G = \{N, (B_x, B_w, B_y)\}$ corresponding to the interconnection of storage, dissipative, and source edges.

Output: Sets of voltage-controlled edges B_1 and current-controlled edges B_2.

$\Lambda \leftarrow$ adjacency matrix of G

$B_1, B_2 \leftarrow \varnothing, \varnothing$

foreach $b \in (B_x, B_y)$ // See Table 2

do
> **if** *b is voltage-controlled* **then**
> > $|\quad B_1 \leftarrow B_1 \cup b$
>
> **end**
> **if** *b is current-controlled* **then**
> > $|\quad B_2 \leftarrow B_2 \cup b$
>
> **end**

end

$B_i \leftarrow B_w$

$\Lambda(N_0, :) \leftarrow 0$

foreach $b \in B_1$ // See constraint (C4)

do
> $|\quad \Lambda(:, b) \leftarrow 0$

end

repeat
> $\Lambda^* \leftarrow \Lambda$
> **foreach** $b \in B_2$ // See constraint (C2)
> **do**
> > **if** $\sum \Lambda(:, b) = 0$ **then**
> > > $|\quad$ **break**: G is not realizable
> >
> > **end**
> > **else if** $\sum \Lambda(:, b) = 1$ **then**
> > > $n \leftarrow \{n \text{ s.t. } \Lambda(n, b) = 1\}$
> > > $\Lambda(n, : \backslash b) \leftarrow 0$
> >
> > **end**
>
> **end**
> **foreach** $b \in B_i$ **do**
> > **if** $\sum \Lambda(:, b) = 0$ // See constraint (C2)
> > **then**
> > > $B_1 \leftarrow B_1 \cup b$
> > > $B_i \leftarrow B_i \backslash b$
> >
> > **end**
> > **else if** $\sum \Lambda(:, b) = 1$ // See constraint (C4)
> > **then**
> > > $n \leftarrow \{n \text{ s.t. } \Lambda(n, b) = 1\}$
> > > $\Lambda(n, : \backslash b) \leftarrow 0$
> > > $B_2 \leftarrow B_2 \cup b$
> > > $B_i \leftarrow B_i \backslash b$
> >
> > **end**
>
> **end**
> **foreach** $n \in N \backslash N_0$ // See constraint (C1)
> **do**
> > **if** $\sum \Lambda(n, :) = 0$ **then**
> > > $|\quad$ **break**: G is not realizable
> >
> > **end**
>
> **end**

until $\Lambda = \Lambda^*$

if $B_i \neq \varnothing$ **then**
> $b \leftarrow$ first edge in B_i
> $B_1 \leftarrow B_1 \cup b$
> $B_i \leftarrow B_i \backslash b$
> **go to 10**

end

return B_1 *and* B_2

3.2.3. Example

As an example, the realizability analysis for the system in Figure 2 with the choice of inputs/outputs in Table 1 is as follows. In step 1, the realizability matrix Λ is initialized with the adjacency matrix A, which is built by taking the absolute value of incidence matrix $[A]_{n,b} = \mathrm{abs}\left([\Gamma]_{n,b}\right)$:

$$
\Lambda = A =
\begin{array}{c}
\begin{array}{cccccccc}
C1 & R1 & Q1,bc & Q1,be & Vcc & IN & OUT & GRD
\end{array} \\
\left(\begin{array}{cccccccc}
\vdots & \vdots & \vdots & \vdots & 1 & 1 & 1 & 1 \\
1 & \vdots & \vdots & \vdots & \vdots & 1 & \vdots & \vdots \\
1 & \vdots & 1 & 1 & \vdots & \vdots & \vdots & \vdots \\
\vdots & 1 & 1 & \vdots & \vdots & \vdots & 1 & \vdots \\
\vdots & 1 & \vdots & \vdots & 1 & \vdots & \vdots & \vdots \\
\vdots & \vdots & \vdots & 1 & \vdots & \vdots & \vdots & 1
\end{array}\right)
\begin{array}{c}
N_0 \\ N_1 \\ N_2 \\ N_3 \\ N_4 \\ N_5
\end{array}
\end{array}.
\tag{14}
$$

In steps 3–7, the set of edges $B = \{C_1, R_1, Q_{1,bc}, Q_{1,be}, IN, Vcc, OUT, GRD\}$ is split as $B = \{B_1, B_i, B_2\}$ according to the definition of components with voltage-controlled edges $B_1 = \{Q_{1,bc}, Q_{1,be}, OUT\}$, current-controlled edges $B_2 = \{C_1, Vcc, IN, GRD\}$, and indeterminate edge $B_i = \{R_1\}$:

$$
\Lambda =
\begin{array}{c}
\begin{array}{cccccccc}
Q1,bc & Q1,be & OUT & R1 & C1 & Vcc & IN & GRD
\end{array} \\
\left(\begin{array}{ccc|ccccc}
\vdots & \vdots & 1 & \vdots & \vdots & 1 & 1 & 1 \\
\vdots & \vdots & \vdots & \vdots & 1 & \vdots & 1 & \vdots \\
1 & 1 & \vdots & \vdots & 1 & \vdots & \vdots & \vdots \\
1 & \vdots & 1 & 1 & \vdots & \vdots & \vdots & \vdots \\
\vdots & \vdots & \vdots & 1 & \vdots & 1 & \vdots & \vdots \\
\vdots & 1 & \vdots & \vdots & \vdots & \vdots & \vdots & 1
\end{array}\right)
\begin{array}{c}
N_0 \\ N_1 \\ N_2 \\ N_3 \\ N_4 \\ N_5
\end{array}
\end{array}.
\tag{15}
$$

The realizability matrix after step 11 in Algorithm 1 is

$$
\Lambda =
\begin{array}{c}
\begin{array}{cccccccc}
Q1,bc & Q1,be & OUT & R1 & C1 & Vcc & IN & GRD
\end{array} \\
\left(\begin{array}{ccc|ccccc}
\vdots & \vdots & \vdots & \vdots & \vdots & \vdots & \vdots & \vdots \\
\vdots & \vdots & \vdots & \vdots & 1 & \vdots & 1 & \vdots \\
\vdots & \vdots & \vdots & \vdots & 1 & \vdots & \vdots & \vdots \\
\vdots & \vdots & \vdots & 1 & \vdots & \vdots & \vdots & \vdots \\
\vdots & \vdots & \vdots & 1 & \vdots & 1 & \vdots & \vdots \\
\vdots & \vdots & \vdots & \vdots & \vdots & \vdots & \vdots & 1
\end{array}\right)
\begin{array}{c}
N_0 \\ N_1 \\ N_2 \\ N_3 \\ N_4 \\ N_5
\end{array}
\end{array}.
\tag{16}
$$

After step 19, the algorithm concludes that the potential on node N_1 is imposed by edge B_{IN} so that the potential on node N_2 is imposed by the capacitor B_{C1}. After step 28, the algorithm concludes that the potential on node N_4 is imposed by the edge B_{Vcc} so that the resistor is current-controlled (so as to impose the potential on node N_3).

$$\Lambda = \left(\begin{array}{ccc|ccccc} & \text{Q1,bc} & \text{Q1,be} & \text{OUT} & \text{R1} & \text{C1} & \text{Vcc} & \text{IN} & \text{GRD} \\ \vdots & \vdots & \vdots & \vdots & \vdots & \vdots & \vdots & \vdots & N_0 \\ \vdots & \vdots & \vdots & \vdots & \vdots & \vdots & 1 & \vdots & N_1 \\ \vdots & \vdots & \vdots & \vdots & 1 & \vdots & \vdots & \vdots & N_2 \\ \vdots & \vdots & \vdots & 1 & \vdots & \vdots & \vdots & \vdots & N_3 \\ \vdots & \vdots & \vdots & \vdots & \vdots & 1 & \vdots & \vdots & N_4 \\ \vdots & \vdots & \vdots & \vdots & \vdots & \vdots & \vdots & 1 & N_5 \end{array}\right). \tag{17}$$

This concludes the realizability analysis. To recover the associated port-Hamiltonian structure, we return to the incidence matrix Γ. With the new edges ordering $B = \{B_1, B_2\}$ prescribed by the above analysis, it is rewritten as

$$\Gamma = \left(\begin{array}{ccc|ccccc} \text{Q1,bc} & \text{Q1,be} & \text{OUT} & \text{R1} & \text{C1} & \text{Vcc} & \text{IN} & \text{GRD} & \\ \vdots & \vdots & 1 & \vdots & \vdots & 1 & 1 & 1 & N_0 \\ \vdots & \vdots & \vdots & \vdots & 1 & \vdots & -1 & \vdots & N_1 \\ 1 & 1 & \vdots & \vdots & -1 & \vdots & \vdots & \vdots & N_2 \\ -1 & \vdots & -1 & 1 & \vdots & \vdots & \vdots & \vdots & N_3 \\ \vdots & \vdots & \vdots & -1 & \vdots & -1 & \vdots & \vdots & N_4 \\ \vdots & -1 & \vdots & \vdots & \vdots & \vdots & \vdots & -1 & N_5 \end{array}\right). \tag{18}$$

Finally, the structure (5) is recovered by computing the matrix $\gamma = \gamma_2^{-1} \cdot \gamma_1$ in (11) with

$$\gamma_1 = \begin{pmatrix} 0 & 0 & 0 \\ 1 & 1 & 0 \\ -1 & 0 & -1 \\ 0 & 0 & 0 \\ 0 & -1 & 0 \end{pmatrix}; \quad \gamma_2 = \begin{pmatrix} 0 & 1 & 0 & -1 & 0 \\ 0 & -1 & 0 & 0 & 0 \\ 1 & 0 & 0 & 0 & 0 \\ -1 & 0 & -1 & 0 & 0 \\ 0 & 0 & 0 & 0 & -1 \end{pmatrix}. \tag{19}$$

4. Guaranteed-Passive Simulation

This section is devoted to the discrete-time simulation of the algebraic-differential system (5); that is, the computation of $x(k) \equiv x(k \cdot T)$ from $u(k) \equiv u(k \cdot T)$, with $k \in \mathbb{N}$, for the constant sampling frequency $f_s = 1/T$.

First, we present the design of a numerical scheme that properly transposes the power balance (1) to the discrete time domain: this choice makes the passivity property preserved, from which stability issues stem. Second, a numerical method is used to solve the implicit equations due to the numerical scheme (on x) and the algebraic equations (on w).

4.1. Numerical Scheme

To ensure the stable simulation of stable dynamical system $\frac{dx}{dt} = f(x)$, many numerical schemes focus on the approximation quality of the time derivative (or integration), combined with operation of the vector field f. Here, we adopt an alternate point of view, by transposing the power balance (1) into the discrete time-domain to preserve passivity. This is achieved by numerical schemes that provide a discrete version of the chain rule for computing the derivative of the composite function $E = \mathcal{H}(x)$.

This is the case of the forward difference scheme, for which first order approximation of the differential applications $dx(t, dt) = \frac{dx}{dt}(t) \cdot dt$ and $d\mathcal{H}(\mathbf{x}, d\mathbf{x}) = \nabla\mathcal{H}(\mathbf{x})^\mathsf{T} \cdot d\mathbf{x}$ on the sample grid $t \equiv kT$, $k \in \mathbb{Z}$ are given by

$$\delta\mathbf{x}(k, T) = \mathbf{x}(k+1) - \mathbf{x}(k), \tag{20}$$

$$\delta\mathcal{H}(\mathbf{x}(k), \delta\mathbf{x}(k, T)) = \mathcal{H}(\mathbf{x}(k) + \delta\mathbf{x}(k, T)) - \mathcal{H}(\mathbf{x}(k)) \tag{21}$$

$$= \nabla_d\mathcal{H}(\mathbf{x}(k), \mathbf{x}(k) + \delta\mathbf{x}(k, T))^\mathsf{T} \cdot \delta\mathbf{x}(k, T).$$

where, for mono-variate energy storing components $(\mathcal{H}(\mathbf{x}) = \sum_{n=1}^{n_S} h_n(x_n))$, the n-th coordinate is given by

$$[\nabla_d\mathcal{H}(\mathbf{x}, \mathbf{x} + \delta\mathbf{x})]_n = \begin{cases} \frac{h_n(x_n + \delta x_n) - h_n(x_n)}{\delta x_n} & \text{if } \delta x_n \neq 0, \\ h_n'(x_n) & \text{otherwise.} \end{cases} \tag{22}$$

A discrete chain rule is indeed recovered

$$\frac{\delta E(k, T)}{T} = \nabla_d\mathcal{H}(\mathbf{x}(k), \mathbf{x}(k+1))^\mathsf{T} \cdot \frac{\delta\mathbf{x}(k, T)}{T} \tag{23}$$

so that the following substitution in (5)

$$\begin{aligned} \frac{dx}{dt}(t) &\rightarrow \frac{\delta\mathbf{x}(k, T)}{T} \\ \nabla\mathcal{H}(\mathbf{x}) &\rightarrow \nabla_d\mathcal{H}(\mathbf{x}(k), \mathbf{x}(k+1)) \end{aligned} \tag{24}$$

leads to

$$\begin{aligned} 0 &= \mathbf{b}(k)^\mathsf{T} \cdot \mathbf{J} \cdot \mathbf{b}(k) = \mathbf{b}(k)^\mathsf{T} \cdot \mathbf{a}(k) \\ &= \underbrace{\left[\nabla_d\mathcal{H}^\mathsf{T} \cdot \frac{\delta\mathbf{x}}{\delta t} \right](k)}_{\frac{\delta E(k, T)}{T}} + \underbrace{\mathbf{z}(\mathbf{w}(k))^\mathsf{T} \cdot \mathbf{w}(k)}_{D(k)} - \underbrace{\mathbf{u}(k)^\mathsf{T} \cdot \mathbf{y}(k)}_{S(k)}. \end{aligned} \tag{25}$$

Remark 4 (Multi-variate components). *The case of mono-variate energy storing components covers most of the applications in electronics. Additionally, a generalization of the discrete gradient for multi-variate Hamiltonians such that Equations (20) and (21) are satisfied is given in Appendix C.*

In this paper, we consider the class of the PHS composed of a collection of linear energy storing components, with quadratic Hamiltonian $h_n(x_n) = \frac{x_n^2}{2C_n}$ (C_n is a capacitance or an inductance and we define $\mathbf{Q} = \text{diag}(C_1 \cdots C_{n_S})^{-1}$). Then the discrete gradient (22) reads

$$\nabla_d\mathcal{H}(\mathbf{x}, \mathbf{x} + \delta\mathbf{x}) = \mathbf{Q}\left(\mathbf{x}(k) + \frac{\delta\mathbf{x}(k)}{2}\right), \tag{26}$$

which restores the midpoint rule that coincides in this case with the trapezoidal rule. For nonlinear cases, (22) leads to another numerical scheme depending on the nonlinearity, still preserving passivity (see (25) and §4.3).

4.2. Solving the Implicit Equations

Injecting the numerical scheme (26) in (5) and solving for the quantity $\delta\mathbf{x}(k) = \mathbf{x}(k+1) - \mathbf{x}(k)$ leads to the following energy-preserving numerical system:

$$\begin{pmatrix} \delta\mathbf{x}(k) \\ \mathbf{w}(k) \\ \mathbf{y}(k) \end{pmatrix} = \begin{pmatrix} \mathbf{A_x} & \mathbf{B_x} & \mathbf{C_x} \\ \mathbf{A_w} & \mathbf{B_w} & \mathbf{C_w} \\ \mathbf{A_y} & \mathbf{B_y} & \mathbf{C_y} \end{pmatrix} \cdot \begin{pmatrix} \mathbf{x}(k) \\ \mathbf{z}(\mathbf{w}(k)) \\ \mathbf{u}(k) \end{pmatrix}, \tag{27}$$

where matrices are related to **J** in (5) as follows.

$$
\begin{aligned}
\mathbf{D} &= \left(\tfrac{\mathbf{I_d}}{T} - \tfrac{\mathbf{J_x \cdot Q}}{2}\right)^{-1}, \quad & \mathbf{A_w} &= \tfrac{1}{2}\mathbf{K^T \cdot Q} \cdot (2\mathbf{I_d} + \mathbf{A_x}), \\
\mathbf{A_x} &= \mathbf{D \cdot J_x \cdot Q}, & \mathbf{B_w} &= \mathbf{J_w} + \tfrac{1}{2}\mathbf{K^T \cdot Q \cdot B_x}, \\
\mathbf{B_x} &= -\mathbf{D \cdot K}, & \mathbf{C_w} &= -\mathbf{G_w} + \tfrac{1}{2}\mathbf{K^T \cdot Q \cdot C_x}, \\
\mathbf{C_x} &= -\mathbf{D \cdot G_x}, & \mathbf{A_y} &= \tfrac{1}{2}\mathbf{G_x^T \cdot Q} \cdot (2\mathbf{I_d} + \mathbf{A_x}), \\
& & \mathbf{B_y} &= \mathbf{G_w^T} + \tfrac{1}{2}\mathbf{G_x^T \cdot Q \cdot B_x}, \\
& & \mathbf{C_y} &= \mathbf{J_y} + \tfrac{1}{2}\mathbf{G_x^T \cdot Q \cdot C_x}.
\end{aligned}
$$

Given $\mathbf{u}(k)$, the solution of (27) is obtained from the solution of the static nonlinear implicit function $\mathbf{f}(\mathbf{w}(k)) = \mathbf{p}(\mathbf{x}(k), \mathbf{u}(k))$, with

$$
\begin{aligned}
\mathbf{f}(\mathbf{w}(k)) &= \mathbf{w}(k) - \mathbf{B_w} \cdot \mathbf{z}(\mathbf{w}(k)), \\
\mathbf{p}(\mathbf{x}(k), \mathbf{u}(k)) &= \mathbf{A_w} \cdot \mathbf{x}(k) + \mathbf{C_w} \cdot \mathbf{u}(k).
\end{aligned}
\tag{28}
$$

Remark 5 (Explicit mapping). *From the global inverse function theorem (see [34]), there exists an explicit mapping* $\mathbf{w}(k) = \mathbf{f}^{-1}(\mathbf{p}(\mathbf{x}(k), \mathbf{u}(k)))$ *provided the Jacobian matrix* $\mathcal{J}_f(\mathbf{w}(k)) = \mathbf{I_d} - (\mathbf{J_w} - \tfrac{1}{2}\mathbf{K^T \cdot Q \cdot D} \cdot \mathbf{K}) \cdot \mathcal{J}_z(\mathbf{w}(k))$ *is invertible for all* $\mathbf{w}(k)$, *connecting the proposed method to the K method [18,19]. This is true since* $\mathbf{Q}$, $\mathbf{D}$ *and the Jacobian of* $\mathbf{z}$ *(for the components of the dictionary in Table B1) prove positive definite, and* $\mathbf{J_w}$ *is skew-symmetric.*

In this paper, we use the Newton–Raphson algorithm, which iteratively approximates the nearest root of function $\mathbf{r}: \mathbf{w}(k) \in \mathbb{R}^{n_D} \to \mathbf{r}(\mathbf{w}(k)) \in \mathbb{R}^{n_D}$ with the following update rule: $\mathbf{w}_{n+1}(k) = \mathbf{w}_n(k) - \mathcal{J}_r(\mathbf{w}_n(k))^{-1} \cdot \mathbf{r}(\mathbf{w}_n(k))$, where $\mathcal{J}_r(\mathbf{w})$ is the Jacobian matrix of $\mathbf{r}(\mathbf{w}(k)) = \mathbf{f}(\mathbf{w}(k)) - \mathbf{p}(k)$. Once a solution $\mathbf{w}(k)$ to the implicit equation is available, the output and state updates are given by:

$$
\begin{aligned}
\mathbf{y}(k) &= \mathbf{A_y} \cdot \mathbf{x}(k) + \mathbf{B_y} \cdot \mathbf{z}(\mathbf{w}(k)) + \mathbf{C_y} \cdot \mathbf{u}(k), \\
\delta \mathbf{x}(k) &= \mathbf{A_x} \cdot \mathbf{x}(k) + \mathbf{B_x} \cdot \mathbf{z}(\mathbf{w}(k)) + \mathbf{C_x} \cdot \mathbf{u}(k), \\
\mathbf{x}(k+1) &= \mathbf{x}(k) + \delta \mathbf{x}(k).
\end{aligned}
$$

Finally, denoting by n_t the number of time-steps and n_{NR} the number of Newton–Raphson iterations per time-step, the simulation is performed according to Algorithm 2.

Algorithm 2: Simulation, with n_t the number of time-steps and n_{NR} the (fixed) number of Newton–Raphson iterations.

$\mathbf{x}_1 \leftarrow 0$
$\mathbf{w}_0 \leftarrow 0$
for $k = 1$ *to* n_t **do**
    $\mathbf{w}_{k,0} \leftarrow \mathbf{w}_{k-1}$
    $\mathbf{p}_k \leftarrow \mathbf{A_w} \cdot \mathbf{x}_k + \mathbf{C_w} \cdot \mathbf{u}_k$
    for $n = 0$ *to* $n_{NR} - 1$ **do**
        $\mathbf{r}_n \leftarrow \mathbf{w}_{k,n} - \mathbf{B_w} \cdot \mathbf{z}(\mathbf{w}_{k,n}) - \mathbf{p}_k$
        $\mathcal{J}_n \leftarrow \mathbf{I_d} - \mathbf{B_w} \cdot \mathcal{J}_z(\mathbf{w}_{k,n})$
        $\mathbf{w}_{k,n+1} \leftarrow \mathbf{w}_{k,n} - \mathcal{J}_n^{-1} \cdot \mathbf{r}_n$
    end
    $\mathbf{w}_k \leftarrow \mathbf{w}_{k,n_{NR}}$
    $\mathbf{y}_k \leftarrow \mathbf{A_y} \cdot \mathbf{x}_k + \mathbf{B_y} \cdot \mathbf{z}(\mathbf{w}_k) + \mathbf{C_y} \cdot \mathbf{u}_k$
    $\delta \mathbf{x}_k \leftarrow \mathbf{A_x} \cdot \mathbf{x}_k + \mathbf{B_x} \cdot \mathbf{z}(\mathbf{w}_k) + \mathbf{C_x} \cdot \mathbf{u}_k$
    $\mathbf{x}_{k+1} \leftarrow \mathbf{x}_k + \delta \mathbf{x}_k$
end

4.3. Comparison with Standard Methods

In this section, the proposed approach (PHS structure combined with the discrete gradient method) is compared with two standard methods: the *trapezoidal rule* (average of the vector field at $\mathbf{x}(k)$ and $\mathbf{x}(k+1)$, used in the WDF approach [7]) and the *midpoint rule* (evaluation of the vector field at $\frac{\mathbf{x}(k)+\mathbf{x}(k+1)}{2}$, suitable for any differential-algebraic system of equations). Both are known to preserve the passivity of linear undamped systems (see [35] for a detailed analysis). The updates associated with these three methods are given in Table 3. These methods are applied on the same conservative system $\frac{d\mathbf{x}}{dt} = \mathbf{J_x} \cdot \nabla\mathcal{H}(\mathbf{x})$, the power balance of which is given by $\frac{d\mathbb{E}}{dt} = 0$ (with $\mathbb{D} = \mathbb{S} = 0$). The comparison measure is then the relative error on energy $\varepsilon(k) = \frac{\left|\mathcal{H}\left(\mathbf{x}(k+1)\right) - \mathcal{H}\left(\mathbf{x}(k)\right)\right|}{\mathcal{H}\left(\mathbf{x}(0)\right)}$ for $k \geq 0$.

Table 3. Updates for the three methods considered in §4.3. PHS stands for port-Hamiltonian system.

Method	Update
Trapezoidal rule	$\mathbf{x}(k+1) = \mathbf{x}(k) + T \cdot \mathbf{J_x} \cdot \frac{\nabla\mathcal{H}\left(\mathbf{x}(k)\right) + \nabla\mathcal{H}\left(\mathbf{x}(k+1)\right)}{2}$
Midpoint rule	$\mathbf{x}(k+1) = \mathbf{x}(k) + T \cdot \mathbf{J_x} \cdot \nabla\mathcal{H}\left(\frac{\mathbf{x}(k)+\mathbf{x}(k+1)}{2}\right)$
PHS with discrete gradient	$\mathbf{x}(k+1) = \mathbf{x}(k) + T \cdot \mathbf{J_x} \cdot \nabla_d\mathcal{H}\left(\mathbf{x}(k), \mathbf{x}(k+1)\right)$

First, notice that for quadratic Hamiltonian $\mathcal{H}(\mathbf{x}) = \frac{\mathbf{x}^\mathsf{T} \cdot \mathbf{Q} \cdot \mathbf{x}}{2}$ with linear gradient $\nabla\mathcal{H}(\mathbf{x}) = \mathbf{Q} \cdot \mathbf{x}$, the three methods yield the same update:

$$\mathbf{x}(k+1) = \mathbf{x}(k) + T \cdot \mathbf{J_x} \cdot \mathbf{Q}\left(\mathbf{x}(k) + \frac{\delta\mathbf{x}(k)}{2}\right). \tag{29}$$

As a consequence, these three methods induce the same frequency warping (see [36] for the analysis of the bilinear transform derived from the trapezoidal rule).

We focus on the nonlinear case. For comparison, we choose a simple nonlinear conservative system with state $\mathbf{x} = (x_1, x_2)^\mathsf{T}$, non-quadratic Hamiltonian

$$\mathcal{H}(\mathbf{x}) = 10 \log\left(\cosh(x_1)\right) + \left(\cosh(x_2) - 1\right), \tag{30}$$

and canonical skew-symmetric matrix $\mathbf{J_x} = \begin{pmatrix} 0 & -1 \\ 1 & 0 \end{pmatrix}$.

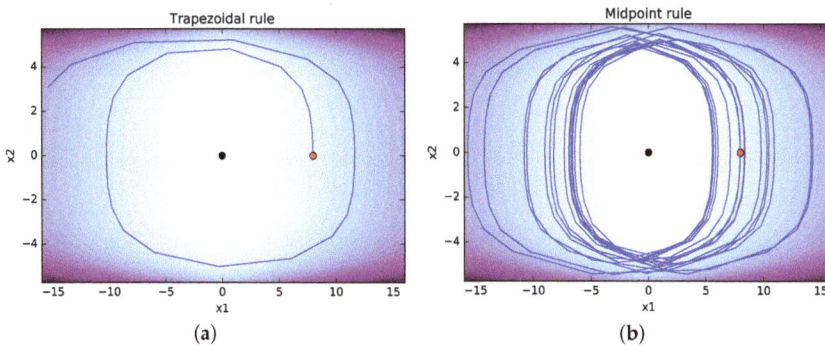

(a)　　　　(b)

Figure 4. *Cont.*

401

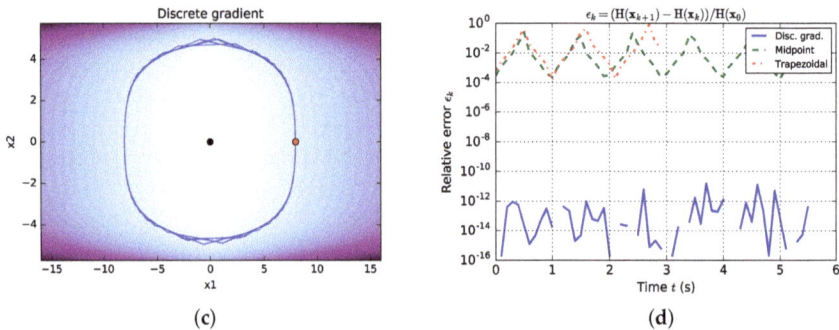

(c)

(d)

Figure 4. Simulation results and comparison of the methods in Table 3, for a nonlinear conservative system $\frac{dx}{dt} = \mathbf{J_x} \cdot \nabla \mathcal{H}(\mathbf{x})$ with $\mathcal{H}(\mathbf{x})$ given in (30): (**a**) Trapezoidal rule; (**b**) Midpoint rule; (**c**) PHS combined with discrete gradient; (**d**) Relative error on energy balance. The comparison measure is the relative error on the power balance defined by $\epsilon(k) = \frac{\left| \mathcal{H}(x(k+1)) - \mathcal{H}(x(k)) \right|}{\mathcal{H}(x(0))}$. We see from Figure 4d that the error associated with the proposed method (PHS approach combined with the discrete gradient method) is low compared to the two other methods (with machine precision $\simeq 10^{-16}$). The accumulation of these errors is responsible for the apparently unstable behavior of the trapezoidal rule.

In each case, the resulting implicit equations are solved by Python iterative solver (see [37]). The Python code is available at the url given in [38]. In order to exhibit the behavior of each method in the worst case, simulations are performed with an especially low sample rate of $f_s = 10$ Hz. The results for each method are given in Figure 4, with comparison in Figure 4d. We see that the error of the proposed method is low (close to machine precision $\simeq 10^{-16}$) compared to standard methods.

5. Applications

This section is devoted to the simulation of three analog audio circuits by the application of Algorithms 1 and 2. Those circuits are a diode clipper, a common-emitter BJT audio amplifier, and a wah-pedal as a full device. Results obtained with (i) the method in Section 4 and (ii) with the offline circuit simulator LT-Spice [32] are compared.

5.1. Diode Clipper

Diode clipper circuits can be found in several audio-distortion devices. They are made of one resistor and two diodes ($n_S = 0$, $n_D = 3$) connected to the ground in reversed bias (see Figure 5a). The external ports are the input/output and the ground ($n_P = 3$). The resistor is current-controlled and the ground is removed. The vectors $(\mathbf{a}, \mathbf{b})$ and the structure $\mathbf{J}$ returned by Algorithm 1 are:

$$
\underbrace{\begin{pmatrix} i_R \\ v_{D1} \\ v_{D2} \\ \hline i_{IN} \\ v_{OUT} \end{pmatrix}}_{\mathbf{a}} = \underbrace{\left(\begin{array}{ccc|cc} 0 & -1 & 1 & 0 & -1 \\ 1 & 0 & 0 & 1 & 0 \\ -1 & 0 & 0 & -1 & 0 \\ \hline 0 & -1 & 1 & 0 & -1 \\ 1 & 0 & 0 & 1 & 0 \end{array} \right)}_{\mathbf{J}} \cdot \underbrace{\begin{pmatrix} v_R \\ i_{D1} \\ i_{D2} \\ \hline v_{IN} \\ i_{OUT} \end{pmatrix}}_{\mathbf{b}}.
$$

The simulation is performed according to Algorithm 2 at the sample rate $f_s = 96$ kHz, with three Newton–Raphson iterations (shown to be enough to converge in practice). We apply a linearly increasing 1 kHz sinusoidal excitation $u_{IN} = -v_{IN}$ during 10 ms with maximum amplitude 2 V ($i_{OUT} = 0$ A, $v_{GRD} = 0$ V). The output $y_{OUT} = -v_{OUT}$ is given in Figure 5b. We see the signal is clamped between ± 0.6 V, in accordance with LT-Spice results.

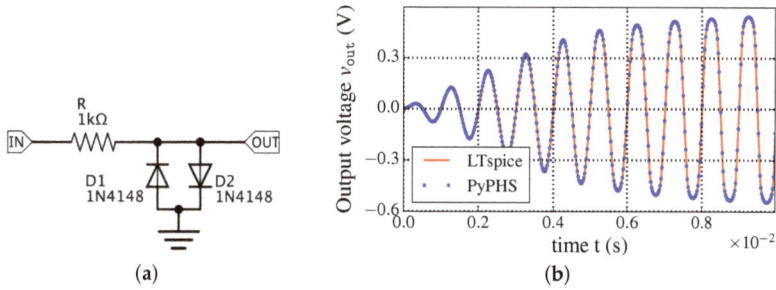

Figure 5. Simulation (Figure 5b) of a dissipative diode clipper (Figure 5a) at the sample rate $f_s = 96$ kHz, with three Newton–Raphson iterations, for a 10 ms sinusoidal excitation at 1 kHz with linearly increasing amplitude between 0 V and 2 V. (**a**) Diode clipper schematic; (**b**) Simulation of the diode clipper of Figure 5a.

5.2. Common-Emitter BJT Audio Amplifier

Common-emitter bipolar-junction transistor (BJT) amplifiers are widely used as amplification stages in analog audio processing. They are made of two capacitors ($n_S = 2$), two resistors, and one NPN transistor which is made of two nonlinear dissipative branches ($n_D = 4$, see Figure 6a and the dictionary in Table B1). The external ports are the input/output signals, the 9 V supply, and the ground ($n_P = 4$). Note that the ground is removed. The resistor Rc is current-controlled, and the resistor Rf is voltage-controlled. The vectors $(\mathbf{a}, \mathbf{b})$ and the structure $\mathbf{J}$ returned by Algorithm 1 are given in Equations (31) and (32).

$$
\begin{aligned}
\mathbf{a} &= (i_{Ci}, i_{Co} | v_{Rf}, i_{Rc}, v_{Bc}, v_{Be} | i_{IN}, v_{OUT}, i_{VCC}) \\
\mathbf{b} &= (v_{Ci}, v_{Co} | i_{Rf}, v_{Rc}, i_{Bc}, i_{Be} | v_{IN}, i_{OUT}, v_{VCC})
\end{aligned}
\tag{31}
$$

$$
\mathbf{J} =
\left(
\begin{array}{cc|cccc|ccc}
0 & 0 & 1 & 0 & -1 & 1 & 0 & 0 & 0 \\
0 & 0 & 0 & 0 & 0 & 0 & 0 & -1 & 0 \\
\hline
-1 & 0 & 0 & -1 & 0 & 0 & -1 & 0 & 1 \\
0 & 0 & 1 & 0 & -1 & 0 & 0 & 1 & 0 \\
1 & 0 & 0 & 1 & 0 & 0 & 1 & 0 & -1 \\
-1 & 0 & 0 & 0 & 0 & 0 & -1 & 0 & 0 \\
\hline
0 & 0 & 1 & 0 & -1 & 1 & 0 & 0 & 0 \\
0 & 1 & 0 & -1 & 0 & 0 & 0 & 0 & 1 \\
0 & 0 & -1 & 0 & 1 & 0 & 0 & -1 & 0
\end{array}
\right)
\tag{32}
$$

The system is reduced according to Appendix A, and the simulation is performed according to Algorithm 2 at the sample rate $f_s = 384$ kHz, with 10 Newton–Raphson iterations. The reason for increasing the sample-rate and the number of Newton–Raphson iterations is twofold. Firstly, it attenuates the effect of aliasing for input signals limited to the audio range (see [39] for details). Secondly, it ensures that the iterative solver converges, which is difficult due to the *numerical stiffness* of the problem; that is, the Lipschitz constant associated to the inital Cauchy problem is very high, see [40]. At first, we turn the supply $v_{VCC} = -9$ V on, and we wait 0.3 s for the system to reach its steady state. Then, we apply a 10 ms sinusoidal excitation $u_{IN} = -v_{IN}$ at 1 kHz with linearly increasing amplitude between 0 V and 0.2 V ($i_{OUT} = 0$ A). The resulting output $y_{OUT} = -v_{OUT}$ is given in Figure 6b. We see that the signal is amplified between 0 V and 9 V, with a strong asymmetrical saturation, in accordance with LT-Spice results. Additionally, spectrograms obtained for an exponential chirp on the audio range are given in Figure 7 (see Figure 16 in [39] for comparison).

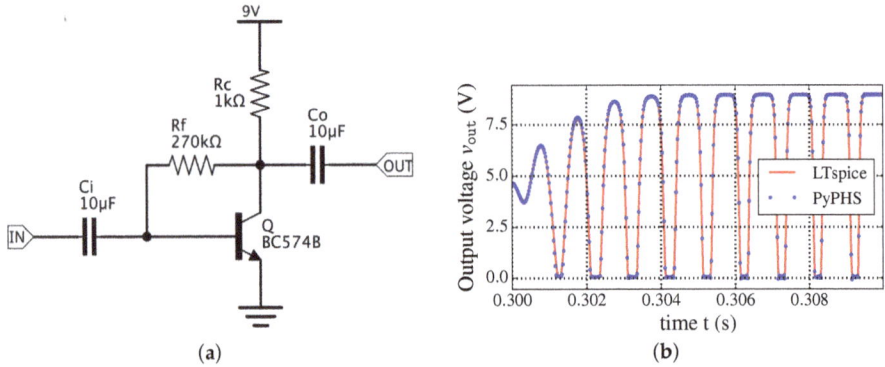

(a) (b)

Figure 6. Simulation (Figure 6b) of the common-emitter bipolar-junction transistor (BJT) amplifier with feedback (Figure 6a) at the sample rate $f_s = 384$ kHz, with 10 Newton–Raphson iterations, for a 10 ms sinusoidal excitation at 1 kHz with linearly increasing amplitude between 0 V and 0.2 V. (a) Schematic of a common-emitter BJT amplifier with feedback; (b) Simulation of the BJT amplifier in Figure 6a.

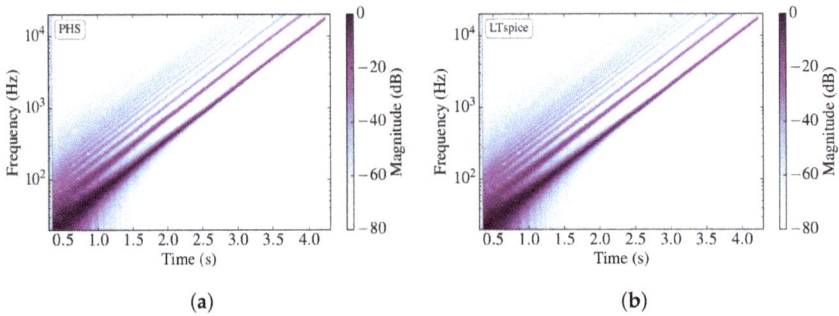

(a) (b)

Figure 7. Results for the common-emitter BJT amplifier with feedback (Figure 6a). The input voltage signal is a 4 s exponential chirp on the audio range (20 Hz–20 kHz) with amplitude 0.05 V (logarithmic frequency scale). Simulation starts at 0.3 s (after the switching transient). (a) Spectrogram of output v_{OUT} obtained with the proposed method; (b) Spectrogram of output v_{OUT} obtained with LT-Spice.

5.3. Wah Pedal

This section addresses the simulation of a full device (namely the Dunlop Cry-Baby wah pedal) to be used in real time. The circuit is given in Figure 9. It provides a continuously varying characteristic *wah* filtering of the input signal. This circuit has been treated with the nodal discrete K-method in [41] and with the PHS framework in [42]. It is composed of $n_S = 7$ storage branches (6 capacitors and 1 inductor), $n_D = 18$ dissipative branches (11 resistors, 1 PN diode, 2 NPN transistors and a potentiometer), and $n_P = 3$ ports (input/output signals and battery, discarding the 5 grounds). The *wah* parameter is the potentiometer's coefficient α. Notice this circuit includes several edges that do not contribute to the device input-to-output behavior, as analyzed in [41]. In this work, we consider the complete original schematic. From Algorithm 1, the resistors R_1, $R_6 \cdots R_9$ and R_{11} are considered as conductances, and the others as resistances. The structure **J** is not shown here. The sets of PHS variables are:

$$\dot{\mathbf{x}} = [i_{C1}, \cdots, i_{C6}, v_{L_1}]^{\mathsf{T}},$$
$$\nabla \mathcal{H}(\mathbf{x}) = [v_{C1}, \cdots, v_{C6}, i_{L_1}]^{\mathsf{T}},$$

$$\mathbf{w} = \left[\mathbf{w}_R | v_d | v_{BC_1}, v_{BE_1} | v_{BC_2}, v_{BE_2} | v_{p1}, i_{p2}\right]^{\mathsf{T}},$$
$$\mathbf{z}(\mathbf{w}) = \left[\mathbf{z}_R | i_d | i_{BC_1}, i_{BE_1} | i_{BC_2}, i_{BE_2} | i_{p1}, v_{p2}\right]^{\mathsf{T}},$$

where $\mathbf{w}_R$ is the set of dissipative states and $\mathbf{z}_R$ the set of characteristics according to each resistor's type, and

$$\textit{Inputs} \quad \mathbf{u} = \left[v_{in}, i_{out}, v_{cc}\right]^{\mathsf{T}},$$
$$\textit{Outputs} \quad \mathbf{y} = \left[i_{in}, v_{out}, i_{cc}\right]^{\mathsf{T}}.$$

The system is reduced according to Appendix A (with potentiometer's time varying resitors kept in $\mathbf{w}, \mathbf{z}$). Firstly, we realize an offline simulation (in Python) with Algorithm 2 for the sampling rate $f_s = 96$ kHz, and three Newton–Raphson iterations. We apply a white noise normalised to 1 V on the input $u_{IN} = -v_{IN}$ ($i_{OUT} = 0$ A). The magnitudes of transfer functions obtained from fast Fourier transform are given in Figure 8 for the two extreme positions of the pedal. These results are in accordance with LT-Spice.

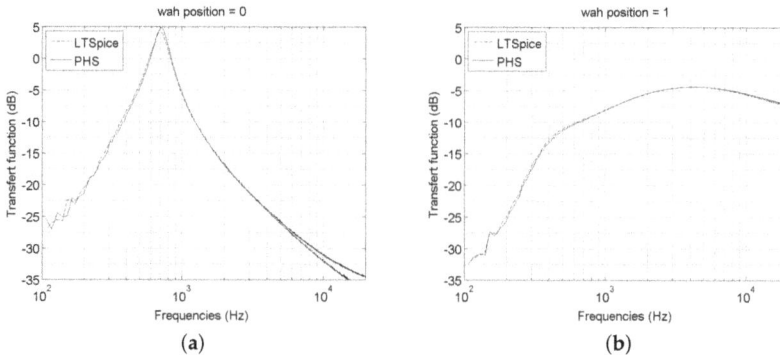

Figure 8. Simulations of the Cry-Baby's circuit of Figure 9, for the potentiometer parameter $\alpha = 0$ (**a**) and $\alpha = 1$ (**b**) in the frequency domain, compared with LT-Spice simulations on the audio range 20 Hz–20 kHz.

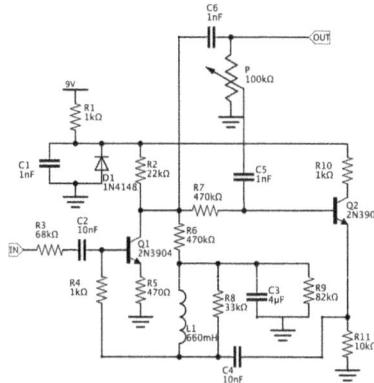

Figure 9. Schematic of the Cry-Baby wah pedal. Note the IN/OUT terminals and the 9 V supply. The potentiometer P controls the effect.

Secondly, a VST plugin [43] to simulate the Cry-Baby in real-time is made from Algorithm 2. First, a C++ code is automatically generated; Second, this code is encapsulated in a Juce template to compile the audio plugin (see [44]). The sample rate f_s is imposed by the host digital audio workstation (here Ableton Live!), and we force five Newton–Raphson iterations. The simulation performed well (audio examples are available at the url [45]). The CPU load on a laptop (Macbook 2.9 GHz Intel Core i7 with 8Go RAM) is 37% for $f_s = 96$ kHz, and 20% for $f_s = 48$ kHz.

Remark 6 (Time-varying stability). *The use of the Newton–Raphson method can hamper the stability of the numerical solution for time-varying systems, especially in the case of fast variations (here, of the potentiometer) for which the Jacobian matrix of the implicit Function (28) can be ill-conditioned. For linear storage components, a solution is to use the K-method instead (see Remark 5).*

6. Conclusions

We have established a method to automatically recast an analog audio circuit into PHS formalism, which guarantees passivity of the continuous time model. The generation of the PHS from a given schematic lies on two points:

1. the graph theory to describe the interconnection network of a given circuit's schematic,
2. a dictionary of elementary components which are conformable with PHS formalism.

Then, we transposed this physical principle to the digital domain by properly defining the discrete gradient of the Hamiltonian, such that a discrete time version of the power balance is satisfied. The resulting stable numerical scheme is of second order (restoring the midpoint rule for linear systems). It has been shown that the K-method is always applicable to PHS (providing efficient implementations of the implicit relation due to the proposed numerical scheme).

Offline simulations are consistent with LT-Spice results. The whole method allows the automatic generation of C++ simulation code to be used in the core of a real-time VST audio plug-insimulating the Dunlop Cry-Baby wah pedal.

A first perspective on this work is to consider higher-order numerical schemes (namely, the class of Runge–Kutta schemes). Moreover, it would be possible to symmetrize the roles of the voltages and the currents at the interconnection by applying the *Cayley transform* to the PHS structure, thus adopting *wave variables*, with possible connection with the WDF formalism. Additionally, an automated analysis of the original schematic could be developed, so as to identify the unimportant or degenerate states and to reduce the dimensionality of the system. Finally, it could be possible to exploit the compatibility of the proposed method with the K-method to alleviate the numerical cost due to Newton–Raphson iterations.

Acknowledgments: The authors acknowledge the members of the french National Research Agency project HaMecMoPSys for support in port-Hamiltonian theory. Also, the authors acknowledge Robert Piechaud and Mattia Bergomi for careful proofreading.

Author Contributions: The scientific work has been achieved by both authors at IRCAM Laboratory, Paris, within the context of the French National Research Agency sponsored project HaMecMoPSys. Further information is available at http://www.hamecmopsys.ens2m.fr. The numerical experiments have been performed by Antoine Falaize.

Conflicts of Interest: The authors declare no conflict of interest.

Appendix A. Reduction

The dimension of system (5) can be reduced, considering the following decomposition of variable **w** and function **z**:

$$\mathbf{w} = \begin{pmatrix} \mathbf{w_L} \\ \mathbf{w_N} \end{pmatrix}, \quad \mathbf{z}(\mathbf{w}) = \begin{pmatrix} \mathbf{z_L} \cdot \mathbf{w_L} \\ \mathbf{z_N}(\mathbf{w_N}) \end{pmatrix},$$

with z_L a diagonal matrix whose elements are the resistance or conductance of linear dissipative components, and z_N a collection of the nonlinear dissipative relations. Correspondingly, the structure matrices are decomposed as

$$\mathbf{K} = \begin{pmatrix} \mathbf{K_L} \\ \mathbf{K_N} \end{pmatrix}, \; \mathbf{G_w} = \begin{pmatrix} \mathbf{G_L} \\ \mathbf{G_N} \end{pmatrix}, \; \mathbf{J_w} = \begin{pmatrix} \mathbf{J_{LL}} & -\mathbf{K_{LN}} \\ \mathbf{K_{LN}}^\mathsf{T} & \mathbf{J_{NN}} \end{pmatrix}.$$

Defining $\mathbf{L} = (\mathbf{K_L}^\mathsf{T}, -\mathbf{K_{LN}}, -\mathbf{G_L})$, and $\mathbf{M} = (\mathbf{z_L}^{-1} - \mathbf{J_{LL}})^{-1}$ a positive definite matrix, the system (5) is reduced to

$$\underbrace{\begin{pmatrix} \frac{d\mathbf{x}}{dt} \\ \hline \mathbf{w_N} \\ \hline -\mathbf{y} \end{pmatrix}}_{\mathbf{b_N}} = (\tilde{\mathbf{J}} - \mathbf{R}) \cdot \underbrace{\begin{pmatrix} \nabla \mathcal{H}(\mathbf{x}) \\ \hline \mathbf{z_N}(\mathbf{w_N}) \\ \hline \mathbf{u} \end{pmatrix}}_{\mathbf{a_N}} \tag{A1}$$

with $\tilde{\mathbf{J}}$ a skew-symmetric matrix given by

$$\tilde{\mathbf{J}} = \left(\begin{array}{c|c|c} \mathbf{J_x} & -\mathbf{K_N} & -\mathbf{G_x} \\ \hline \mathbf{K_N}^\mathsf{T} & \mathbf{J_{NN}} & -\mathbf{G_N} \\ \hline \mathbf{G_x}^\mathsf{T} & \mathbf{G_N}^\mathsf{T} & \mathbf{J_y} \end{array} \right) - \tfrac{1}{2}\mathbf{L}^\mathsf{T} \cdot (\mathbf{M} - \mathbf{M}^\mathsf{T}) \cdot \mathbf{L}$$

and $\mathbf{R}$ is a symmetric positive definite matrix given by

$$\mathbf{R} = \left(\begin{array}{c|c|c} \mathbf{R_x} & \mathbf{R_{xn}} & \mathbf{R_{xy}} \\ \hline \mathbf{R_{xn}}^\mathsf{T} & \mathbf{R_n} & \mathbf{R_{ny}} \\ \hline \mathbf{R_{xy}}^\mathsf{T} & \mathbf{R_{ny}}^\mathsf{T} & \mathbf{R_y} \end{array} \right) = \tfrac{1}{2}\mathbf{L}^\mathsf{T} \cdot (\mathbf{M} + \mathbf{M}^\mathsf{T}) \cdot \mathbf{L}.$$

Indeed, matrix $\tilde{\mathbf{J}}$ (respectively $\mathbf{R}$) corresponds to the conservative (respectively resistive) interconnection of dynamical storage components, nonlinear dissipative components, and sources.

The system (A1) is simulated by Algorithm 2, with

$$
\begin{aligned}
\tilde{\mathbf{D}} &= \left(\tfrac{\mathbf{I_d}}{T} - \tfrac{(\tilde{\mathbf{J}}_x - \mathbf{R}_x)\cdot\mathbf{Q}}{2} \right)^{-1} \\
\tilde{\mathbf{A}}_x &= \tilde{\mathbf{D}} \cdot (\tilde{\mathbf{J}}_x - \mathbf{R}_x), \\
\tilde{\mathbf{B}}_x &= -\tilde{\mathbf{D}} \cdot (\tilde{\mathbf{K}} + \mathbf{R}_{xn}), \\
\tilde{\mathbf{C}}_x &= -\tilde{\mathbf{D}} \cdot (\tilde{\mathbf{G}}_x + \mathbf{R}_{xy}), \\
\tilde{\mathbf{A}}_w &= \tfrac{1}{2}(\tilde{\mathbf{K}} - \mathbf{R}_{xn})^\mathsf{T} \cdot \mathbf{Q} \cdot (2\mathbf{I_d} + \tilde{\mathbf{A}}_x), \\
\tilde{\mathbf{B}}_w &= \tilde{\mathbf{J}}_w - \mathbf{R}_n + \tfrac{1}{2}(\tilde{\mathbf{K}} - \mathbf{R}_{xn})^\mathsf{T} \cdot \mathbf{Q} \cdot \tilde{\mathbf{B}}_x, \\
\tilde{\mathbf{C}}_w &= -\tilde{\mathbf{G}}_w - \mathbf{R}_{ny} + \tfrac{1}{2}(\tilde{\mathbf{K}} - \mathbf{R}_{xn})^\mathsf{T} \cdot \mathbf{Q} \cdot \tilde{\mathbf{C}}_x, \\
\tilde{\mathbf{A}}_y &= \tfrac{1}{2}(\tilde{\mathbf{G}}_x - \mathbf{R}_{xy})^\mathsf{T} \cdot \mathbf{Q} \cdot (2\mathbf{I_d} + \tilde{\mathbf{A}}_x), \\
\tilde{\mathbf{B}}_y &= (\tilde{\mathbf{G}}_w - \mathbf{R}_{ny})^\mathsf{T} + \tfrac{1}{2}(\tilde{\mathbf{G}}_x - \mathbf{R}_{xy})^\mathsf{T} \cdot \mathbf{Q} \cdot \tilde{\mathbf{B}}_x, \\
\tilde{\mathbf{C}}_y &= \tilde{\mathbf{J}}_y - \mathbf{R}_y + \tfrac{1}{2}(\tilde{\mathbf{G}}_x - \mathbf{R}_{xy})^\mathsf{T} \cdot \mathbf{Q} \cdot \tilde{\mathbf{C}}_x,
\end{aligned}
\tag{A2}
$$

Remark A1 (Reduced explicit mapping). *From the global inverse function theorem (see [34]), there exists an explicit mapping $\mathbf{w_N}(k) = \mathbf{f}^{-1}(\mathbf{p}(\mathbf{x}(k), \mathbf{u}(k)))$ provided*

$$\det \left((\tilde{\mathbf{K}} - \mathbf{R}_{xn})^\mathsf{T} \cdot \mathbf{Q} \cdot \tilde{\mathbf{D}} \cdot (\tilde{\mathbf{K}} + \mathbf{R}_{xn}) \right) > 0.$$

Appendix B. Dictionary of Elementary Components

The dictionary is given in Table B1. We choose variables **x** and **w** so that matrices of the PHS (5) are canonical, that is, they do not involve any physical constants.

Table B1. Dictionary of elementary components.

2-Ports					
Storage	**Diagram**	**x**	**Stored Energy** $E = h(x)$	**Voltage** v	**Current** i
Inductance		ϕ	$\frac{\phi^2}{2L}$	$\frac{d\phi}{dt}$	$\frac{dh}{d\phi}$
Capacitance		q	$\frac{q^2}{2C}$	$\frac{dh}{dq}$	$\frac{dq}{dt}$
Dissipative	**Diagram**	**w**	**Dissipated Power** $D(w) = w.z(w)$	**Voltage** v	**Current** i
Resistance		i	$R.i^2$	$z(w)$	w
Conductance		v	v^2/R	w	$z(w)$
PN Diode		v	$v.I_S\left(\exp\left(\frac{v}{\mu v_0}\right) - 1\right) + v^2.G_{min}$	w	$z(w)$

3-Ports			
Dissipative	**Diagram**	**w**	$\mathbf{z(w)}$
NPN Transistor		$\begin{pmatrix} v_{BC} \\ v_{BE} \end{pmatrix}$	$\begin{pmatrix} i_{BC} \\ i_{BE} \end{pmatrix} = \begin{pmatrix} \alpha_R & -1 \\ -1 & \alpha_F \end{pmatrix} . \begin{pmatrix} I_S\left(e^{v_{BC}/v_t} - 1\right) + v_{BC}.G_{min} \\ I_S\left(e^{v_{BE}/v_t} - 1\right) + v_{BE}.G_{min} \end{pmatrix}$
Potentiometer		$\begin{pmatrix} v_{p1} \\ i_{p2} \end{pmatrix}$	$\begin{pmatrix} i_{p1} \\ v_{p2} \end{pmatrix} = \begin{pmatrix} v_{p1}/(1 + \alpha.R_p) \\ i_{p2}.(1 + (1 - \alpha).R_p) \end{pmatrix}$

Appendix B.1. Storage Components

Such components are defined by their storage function h associated with the constitutive laws of Table B1. In this paper, all storage components are linear dipoles. However, nonlinear components can also be considered if appropriate state and corresponding energy function can be found.

Appendix B.2. Linear Dissipative Components

The characteristics of dissipative components are algebraic relations on **w**. Potentiometers are modeled as two time-varying resistors, the sum of which is R_p. To avoid 0 value of the resistors, 1 Ω have been added to those characteristics. The modulation parameter is $\alpha \in [0,1]$. As an example, in Table B1, we choose a conductance between N_1 and N_2, and a resistance between N_2 and N_3, so that $\mathbf{w}_P = [v_{p1}, i_{p2}]^\mathsf{T}$.

Appendix B.3. Nonlinear Dissipative Components

PN junctions are modeled as voltage-controlled components by the Shockley equation: $z_D(w_D) = i_D = I_S\left(e^{\frac{v_D}{\mu v_0}} - 1\right)$, where I_S is the saturation current, μ is an ideality factor, and v_0 the reference voltage, specified for each diode type. Note that the passivity property is fulfilled

$(\mathbf{z}_D(\mathbf{w}_D)^\top.\mathbf{w}_D \geq 0)$. As in LT-Spice simulators, a minimal conductance G_{min} is added in Table B1. This helps convergence in the simulation process.

NPN junctions are passive 3-ports, with dissipated power $D_Q = v_B \cdot i_B + v_C \cdot i_C + v_E \cdot i_E \geq 0$. Here we use the Ebers–Moll model, which preserves this passivity property. I_S is the saturation current, β_R and β_F are respectively the reverse and forward common emitter current gains, and v_t is the thermal voltage. The corresponding voltage-controlled dissipative characteristic $\mathbf{z}_Q(\mathbf{w}_Q) = [i_{BC}, i_{BE}]^\top$ is given in Table B1, denoting $\alpha_R = \frac{\beta_R+1}{\beta_R}$, $\alpha_F = \frac{\beta_F+1}{\beta_F}$, and including minimal conductances. Note that $D_Q = \mathbf{z}_Q(\mathbf{w}_Q)^\top.\mathbf{w}_Q \geq 0$. In our final simulations, some resistors are added to model the resistance of contacts in the nonlinear components, choosing the same values as in LT-Spice models.

Appendix B.4. Incidence Matrices Γ

Incidence matrices for 2-ports, potentiometer P, and transistor Q with conventions of Table B1.

$$
\Gamma_{\text{2-port}} = \begin{pmatrix} 1 \\ -1 \end{pmatrix} \begin{matrix} N_1 \\ N_2 \end{matrix} \ , \Gamma_P = \begin{matrix} B_{p1} & B_{p2} \\ \begin{pmatrix} 1 & 0 \\ -1 & 1 \\ 0 & -1 \end{pmatrix} \end{matrix} \begin{matrix} N_1 \\ N_2 \\ N_3 \end{matrix} \ , \Gamma_Q = \begin{matrix} B_{BC} & B_{BE} \\ \begin{pmatrix} 1 & 1 \\ -1 & 0 \\ 0 & -1 \end{pmatrix} \end{matrix} \begin{matrix} N_B \\ N_C \\ N_E \end{matrix} \ . \tag{B1}
$$

Appendix C. Discrete Gradient for Multi-Variate Hamiltonian

A generalization of the discrete gradient for multi-variate Hamiltonians such that Equations (20) and (21) are satisfied is given by replacing definition (22) by (see [46]):

$$
[\nabla_d \mathcal{H}(\mathbf{x}, \mathbf{x} + \delta \mathbf{x})]_n = \frac{\Delta_n \mathcal{H}(\mathbf{x}, \mathbf{x} + \delta \mathbf{x})}{\delta x_n}. \tag{C1}
$$

with

$$
\begin{aligned}
\Delta_n \mathcal{H}(\mathbf{x}, \tilde{\mathbf{x}}) = \ & \mathcal{H}(\tilde{x}_1, \ldots, \tilde{x}_{n-1}, \tilde{x}_n, x_{n+1}, \ldots, x_{n_S}) \\
& - \mathcal{H}(\tilde{x}_1, \ldots, \tilde{x}_{n-1}, x_n, x_{n+1}, \ldots, x_{n_S}).
\end{aligned} \tag{C2}
$$

For mono-variate components, (22) and (C1) coincide and yield (discrete) constitutive laws that are insensible to the ordering of the state variables. For multi-variate components, this last property is lost, but can be restored by replacing (C2) by the averaged operator: $\Delta_n \mathcal{H}(\mathbf{x}, \tilde{\mathbf{x}}) = \frac{1}{n_S!} \sum_{\pi \in \mathcal{P}(n_S)} \Delta_n \mathcal{H}^\pi(\mathbf{x}_\pi, \tilde{\mathbf{x}}_\pi)$, where for all permutation $\pi \in \mathcal{P}(n_S)$, $\mathbf{x}_\pi = (x_{\pi(1)}, \ldots, x_{\pi(n_S)})^\top$, and $\mathcal{H}^\pi(\mathbf{x}_\pi) = \mathcal{H}(\mathbf{x})$.

References

1. Bilbao, S. Sound Synthesis and Physical Modeling. In *Numerical Sound Synthesis: Finite Difference Schemes and Simulation in Musical Acoustics*; John Wiley & Sons Ltd.: Chichester, UK, 2009.
2. Bilbao, S. Conservative numerical methods for nonlinear strings. *J. Acoust. Soc. Am.* **2005**, *118*, 3316–3327.
3. Chabassier, J.; Joly, P. Energy preserving schemes for nonlinear Hamiltonian systems of wave equations: Application to the vibrating piano string. *Comput. Methods Appl. Mech. Eng.* **2010**, *199*, 2779–2795.
4. Välimäki, V.; Pakarinen, J.; Erkut, C.; Karjalainen, M. Discrete-time modelling of musical instruments. *Rep. Prog. Phys.* **2006**, *69*, 1–78.
5. Petrausch, S.; Rabenstein, R. Interconnection of state space structures and wave digital filters. *IEEE Trans. Circuits Syst. II Express Br.* **2005**, *52*, 90–93.
6. Yeh, D.T.; Smith, J.O. Simulating guitar distortion circuits using wave digital and nonlinear state-space formulations. In Proceedings of the 1st International Conference on Digital Audio Effects (DAFx'08), Espoo, Finland, 1–4 September 2008; pp. 19–26.
7. Fettweis, A. Wave digital filters: Theory and practice. *Proc. IEEE* **1986**, *74*, 270–327.
8. Sarti, A.; De Poli, G. Toward nonlinear wave digital filters. *IEEE Trans. Signal Process.* **1999**, *47*, 1654–1668.
9. Pedersini, F.; Sarti, A.; Tubaro, S. Block-wise physical model synthesis for musical acoustics. *Electron. Lett.* **1999**, *35*, 1418–1419.

10. Pakarinen, J.; Tikander, M.; Karjalainen, M. Wave digital modeling of the output chain of a vacuum-tube amplifier. In Proceedings of the 12th International Conference on Digital Audio Effects (DAFx'09), Como, Italy, 1–4 September 2009; pp. 1–4.

11. De Paiva, R.C.D.; Pakarinen, J.; Välimäki, V.; Tikander, M. Real-time audio transformer emulation for virtual tube amplifiers. *EURASIP J. Adv. Signal Process.* **2011**, *2011*, 1–15.

12. Fettweis, A. Pseudo-passivity, sensitivity, and stability of wave digital filters. *IEEE Trans. Circuit Theory* **1972**, *19*, 668–673.

13. Bilbao, S.; Bensa, J.; Kronland-Martinet, R. The wave digital reed: A passive formulation. In Proceedings of the 6th International Conference on Digital Audio Effects (DAFx-03), London, UK, 8–11 September 2003; pp. 225–230.

14. Schwerdtfeger, T.; Kummert, A. A multidimensional approach to wave digital filters with multiple nonlinearities. In Proceedings of the 22nd European Signal Processing Conference (EUSIPCO), Lisbon, Portugal, 1–5 September 2014; pp. 2405–2409.

15. Werner, K.J.; Nangia, V.; Bernardini, A.; Smith, J.O., III; Sarti, A. An Improved and Generalized Diode Clipper Model for Wave Digital Filters. In Proceedings of the 139th Convention of the Audio Engineering Society (AES), New York, NY, USA, 29 October–1 November 2015.

16. Khalil, H.K. *Nonlinear Systems*; Prentice Hall: Upper Saddle River, NJ, USA, 2002; Volume 3.

17. Cohen, I.; Helie, T. Real-time simulation of a guitar power amplifier. In Proceedings of the 13th International Conference on Digital Audio Effects (DAFx-10), Graz, Austria, 6–10 September 2010.

18. Yeh, D.T.; Abel, J.S.; Smith, J.O. Automated physical modeling of nonlinear audio circuits for real-time audio effects—Part I: Theoretical development. *IEEE Trans. Audio Speech Lang. Process.* **2010**, *18*, 728–737.

19. Borin, G.; De Poli, G.; Rocchesso, D. Elimination of delay-free loops in discrete-time models of nonlinear acoustic systems. *IEEE Trans. Audio Speech Lang. Process.* **2000**, *8*, 597–605.

20. Hélie, T. Lyapunov stability analysis of the Moog ladder filter and dissipativity aspects in numerical solutions. In Proceedings of the 14th International Conference on Digital Audio Effects DAFx-11, Paris, France, 19–23 September 2011; pp. 45–52.

21. Maschke, B.M.; Van der Schaft, A.J.; Breedveld, P.C. An intrinsic Hamiltonian formulation of network dynamics: Non-standard Poisson structures and gyrators. *J. Frankl. Inst.* **1992**, *329*, 923–966.

22. Van der Schaft, A.J. Port-Hamiltonian systems: An introductory survey. In Proceedings of the International Congress of Mathematicians, Madrid, Spain, 22–30 August 2006; pp. 1339–1365.

23. Stramigioli, S.; Duindam, V.; Macchelli, A. *Modeling and Control of Complex Physical Systems: The Port-Hamiltonian Approach*; Springer: Berlin, Germany, 2009.

24. Marsden, J.E.; Ratiu, T.S. *Introduction to Mechanics and Symmetry: A Basic Exposition of Classical Mechanical Systems*; Springer: New Yor, NY, USA, 1999; Volume 17.

25. Itoh, T.; Abe, K. Hamiltonian-conserving discrete canonical equations based on variational difference quotients. *J. Comput. Phys.* **1988**, *76*, 85–102.

26. Tellegen, B.D.H. A general network theorem, with applications. *Philips Res. Rep.* **1952**, *7*, 259–269.

27. Desoer, C.A.; Kuh, E.S. *Basic Circuit Theory*; Tata McGraw-Hill Education: Noida, India, 2009.

28. Karnopp, D. Power-conserving transformations: Physical interpretations and applications using bond graphs. *J. Frankl. Inst.* **1969**, *288*, 175–201.

29. Breedveld, P.C. Multibond graph elements in physical systems theory. *J. Frankl. Inst.* **1985**, *319*, 1–36.

30. Falaize, A.; Hélie, T. Guaranteed-passive simulation of an electro-mechanical piano: A port-Hamiltonian approach. In Proceedings of the 18th International Conference on Digital Audio Effects (DAFx), Trondheim, Norway, 30 November–3 December 2015.

31. Falaize, A.; Lopes, N.; Hélie, T.; Matignon, D.; Maschke, B. Energy-balanced models for acoustic and audio systems: A port-Hamiltonian approach. In Proceedings of the Unfold Mechanics for Sounds and Music, Paris, France, 11–12 September 2014.

32. Vladimirescu, A. *The SPICE Book*; John Wiley & Sons, Inc.: New Yor, NY, USA, 1994.

33. Diestel, R. *Graph Theory*, 4th ed.; Springer: New Yor, NY, USA, 2010.

34. Do Carmo, M.P. *Differential Geometry of Curves and Surfaces*; Prentice-Hall: Englewood Cliffs, NJ, USA, 1976; Volume 2, p. 131.

35. Hairer, E.; Lubich, C.; Wanner, G. *Geometric Numerical Integration: Structure-Preserving Algorithms for Ordinary Differential Equations*; Springer Science & Business Media: New York, NY, USA, 2006; Volume 31.

36. Oppenheim, A.V.; Schafer, R.W. *Discrete-Time Signal Processing*, 3rd ed.; Pearson Higher Education: San Francisco, CA, USA, 2010.
37. Jones, E.; Oliphant, E.; Peterson, P. SciPy: Open Source Scientific Tools for Python, function scipy.optimize.root. Available online: http://docs.scipy.org/doc/scipy/reference/generated/scipy.optimize.root.html#scipy.optimize.root (accessed on 22 September 2016).
38. Falaize, A. A comparison of numerical methods. Available online: http://recherche.ircam.fr/anasyn/falaize/applis/comparisonnumschemes/ (accessed on 22 September 2016).
39. Yeh, D.T. Automated physical modeling of nonlinear audio circuits for real-time audio effects—Part II: BJT and vacuum tube examples. *IEEE Trans. Audio Speech Lang. Process.* **2012**, *20*, 1207–1216.
40. Butcher, J.C. *Numerical Methods for Ordinary Differential Equations*; John Wiley & Sons, Ltd.: London, UK, 2008; p. 26.
41. Holters, M.; Zölzer, U. Physical Modelling of a Wah–Wah Effect Pedal as a case study for Application of the nodal DK Method to circuits with variable parts. In Proceedings of the 14th International Conference on Digital Audio Effects (DAFx-11), Paris, France, 19–23 September 2011.
42. Falaize-Skrzek, A.; Hélie, T. Simulation of an analog circuit of a wah pedal: A port-Hamiltonian approach. In Proceedings of the 135th Convention of the Audio Engineering Society, New York, NY, USA, 17–20 October 2013.
43. Steinberg Media Technologies GmbH. Virtual Studio Technology. Available online: http://www.steinberg.net/en/company/technologies/vst3.html (accessed on 22 September 2016).
44. ROLI Ltd. The JUCE framework. Available online: http://www.juce.com (accessed on 22 September 2016).
45. Falaize, A. Companion web-site to the present article entitled "Passive Guaranteed Simulation of Analog Audio Circuits: A port-Hamiltonian Approach". Available online: http://recherche.ircam.fr/anasyn/falaize/applis/analogcircuits/ (accessed on 22 September 2016).
46. Aoues, S. Schémas d'intégration dédiés à l'étude, l'analyse et la synthèse dans le formalisme Hamiltonien à ports. Ph.D. Thesis, INSA, Lyon, France, December 2014; pp. 32–35.

applied
sciences

MDPI

Article

Sinusoidal Parameter Estimation Using Quadratic Interpolation around Power-Scaled Magnitude Spectrum Peaks [†]

Kurt James Werner * and François Georges Germain

Center for Computer Research in Music and Acoustics (CCRMA), Department of Music, Stanford University,
660 Lomita Drive, Stanford, CA 94305-8180, USA; fgermain@stanford.edu
* Correspondence: kwerner@ccrma.stanford.edu; Tel.: +1-650-723-4971
† This paper is an extended version of our paper published in the 41st International Computer Music Conference
(ICMC), Denton, TX, USA, 25 September–1 October 2015, entitled "The XQIFFT: Increasing the accuracy of
quadratic interpolation of spectral peaks via exponential magnitude spectrum weighting." .

Academic Editor: Vesa Valimaki
Received: 16 March 2016; Accepted: 11 October 2016; Published: 21 October 2016

Abstract: The magnitude of the Discrete Fourier Transform (DFT) of a discrete-time signal has a limited frequency definition. Quadratic interpolation over the three DFT samples surrounding magnitude peaks improves the estimation of parameters (frequency and amplitude) of resolved sinusoids beyond that limit. Interpolating on a rescaled magnitude spectrum using a logarithmic scale has been shown to improve those estimates. In this article, we show how to heuristically tune a power scaling parameter to outperform linear and logarithmic scaling at an equivalent computational cost. Although this power scaling factor is computed heuristically rather than analytically, it is shown to depend in a structured way on window parameters. Invariance properties of this family of estimators are studied and the existence of a bias due to noise is shown. Comparing to two state-of-the-art estimators, we show that an optimized power scaling has a lower systematic bias and lower mean-squared-error in noisy conditions for ten out of twelve common windowing functions.

Keywords: acoustics; discrete Fourier transforms; frequency estimation; interpolation; signal analysis; sinusoidal modeling

1. Introduction

Sinusoidal parameter estimation is a central task in engineering applications including sonar, power systems, measurement and instrumentation [1], array signal processing and radar signal processing [2], wireless communication, and speech analysis [3]. Sinusoidal parameter estimation is also central to musical signal processing, specifically spectral audio signal processing [4]. Tonal aspects of standard musical signals including the human singing voice and musical instrument sounds can be effectively modeled as the sum of multiple time-varying sinusoids [5]. In the musical signal processing context, the number of sinusoids is generally not known a priori, making it difficult to apply techniques like MUSIC [6] and ESPRIT [7] which jointly estimate the parameters of a number of sinusoids; as well these techniques can be computationally expensive.

The problem of detecting multiple sinusoids in a musical signal is often framed as a successive search for a single sinusoid in the Short-Time Fourier Transform (STFT) domain [8]. Framing the problem in this way essentially ignores the effect of interferers—nearby sinusoids whose sidelobes may tilt magnitude spectrum peaks slightly so that they no longer correspond exactly to sinusoidal components. Despite this, the approach of a successive search for sinusoids allows a complex problem to be broken down into many manageable simple problems without, e.g., any assumption of harmonic structure in the signal. Therefore, in this article, we formalize the problem as the search for a single sinusoid—in practical situations the search will be repeated in each frame until all sinusoids of interest are detected and their parameters estimated [8].

Sinusoidal parameter estimation in the STFT domain is exploited heavily in multi-sinusoid analysis/synthesis methods [5,9–12]. For these systems, as well as fundamental frequency detection [13], audio coding, and music information retrieval (MIR), improving the estimation of sinusoidal parameters is very important.

Early work presents interpolation methods only for signals windowed by the rectangular window or Rife–Vincent Class I windows. For the case of arbitrary windows, only a few methods have been presented. The state of the art for arbitrary windows is represented by the estimators of Duda [14] and Candan [15].

In [14], Duda builds on previous magnitude spectrum interpolators [16,17] derived for the Rife—incent Class I windows in order to build a first coarse estimator, whose error is compensated by an optimized polynomial remapping of the coarse estimated value to obtain a finer estimate, with the accuracy of the method controlled through the polynomial order of the remapping. In [15], Candan proposes a method derived from his previous work on the rectangular window case [18], itself building on a previous complex spectrum interpolator [19], where a corrected interpolator is derived from the Taylor expansion of the signal's Discrete-Time Fourier Transform (DTFT) around the sinusoid frequency location. This work also proposes an iterative mechanism to update the correction factor using the previous estimate, further improving its accuracy.

Another approach to this problem which is applicable to arbitrary windows involves parabolic interpolation over magnitude spectrum peaks [4]. In [12] it was shown that logarithmic scaling improves parabolic interpolation significantly. Inspired by this finding, we extend the basic parabolic interpolation approach by fitting a parabola to a *power-scaled* magnitude spectrum—an approach we call the *XQIFFT*. Intuitively, this approach nonlinearly scales the shape of a window transform's main lobe so that it very closely resembles a parabola and parabolic fitting will be very accurate. This approach was introduced in [20], where a coarse brute-force approach was used to optimize a power scaling coefficient and the XQIFFT was shown to greatly outperform parabolic fitting on a linear or logarithmic scale [20] also contained a rudimentary noise analysis.

We extend the XQIFFT approach by replacing the brute force search with a cheaper and more accurate search algorithm that leverages Fibonacci search and Simpson quadrature to achieve optimal search speed and controlled numerical accuracy. Invariance properties of the XQIFFT are studied and it is shown that linear and logarithmic scalings can be considered special cases of the XQIFFT. We extend the noise analysis to a proper study relating estimator mean-squared error to signal-to-noise ratio. We compare the proposed estimator to two state-of-the-art estimators and the Cramér—Rao bound for unbiased estimators, and analyze the systematic bias due to noise. A case study on Hann-windowed signals shows that the optimal power-scaling coefficient is related to window length in a structured way.

In another study, we compare the XQIFFT to the state-of-the-art estimators of Duda [14] and Candan [15] as well as the linear- and logarithmically-scaled versions of the QIFFT for twelve common windowing functions. Although Duda [14] and Candan [15] do not directly compare their methods to the QIFFT family, our study shows that indeed their methods usually outperform the linear- and logarithmically-scaled versions of the QIFFT. Our main finding is that the XQIFFT outperforms the state-of-the-art estimators of Duda [14] and Candan [15] at high and low SNRs for ten out of twelve common windowing functions.

The structure of the remainder of this article is as follows. Section 2 defines the problem of single sinusoidal parameter estimation. Section 3 reviews previous work on extracting sinusoidal parameters from the three spectral bins surrounding a magnitude peak using parabolic interpolation. Section 4 introduces our proposed method: the XQIFFT. Section 5 contains an error definition, a study on the error properties, and a heuristic method for optimizing the power scaling of the XQIFFT. In Section 6, we run experiments to study different aspects of the XQIFFT and compare it to two state-of-the-art estimators.

2. Problem Statement

A single complex discrete-time (sample index n) sinusoid with a sampling period of T is given by

$$x_s[n] = \mathcal{A}\, e^{j(2\pi \mathcal{F} nT + \phi)} = \mathcal{A}\, e^{j\phi}\, e^{j2\pi \mathcal{F} nT}. \tag{1}$$

In the most general case, the goal of sinusoid parameter estimation methods is to estimate the frequency $\mathcal{F}$, amplitude $\mathcal{A}$, and phase offset ϕ of a sinusoid. In this work, we restrict ourselves to the case of estimating $\mathcal{F}$ and $\mathcal{A}$ only. It is common practice to reduce spectral leakage by applying a length-N smoothing window w to x_s to form a windowed signal x with Discrete Fourier Transform (DFT) X given by

$$x[n] = x_s[n]\, w[n], \; n = 0, \ldots, N-1, \tag{2}$$
$$X[k] = \mathcal{A}\, e^{j\phi}\, W(\mathcal{F} - k/NT), \; k = 0, \ldots, N-1, \tag{3}$$

where k is the discrete frequency bin index and W is the Discrete-Time Fourier Transform (DTFT) of the window w defined by

$$W(f) = \sum_{n=0}^{N-1} w[n]\, e^{j(2\pi f nT)}. \tag{4}$$

The DTFT is defined for continuous rather than discrete frequencies, and in fact the DFT of a signal can be seen as a sampling of the DTFT at the frequencies indexed by k.

The window magnitude DTFT $|W|$ for typical smoothing windows is characterized by a main lobe centered around dc. After shifting and scaling, the main lobe is centered at the fractional bin number $\mathcal{K}$ with a maximum amplitude $\mathcal{X}$ defined as:

$$\mathcal{K} = \mathcal{F} NT \quad \text{and} \quad \mathcal{X} = \mathcal{A}\, W(0) \tag{5}$$

In spectral analysis, it is natural to assume that the center of each main lobe in the magnitude spectrum is associated with a complex sinusoid. Other available techniques build an estimate using all the magnitude spectrum samples to increase peak-estimation accuracy [21–26], but these tend to be computationally costly.

Sinusoidal parameter estimation methods which operate on the DFT exploit the assumption that the peak in the magnitude spectrum is associated with a complex sinusoid—in this context finding the fractional bin index $\mathcal{K}$ and peak magnitude $\mathcal{X}$ of the window transform is considered equivalent to recovering the frequency $\mathcal{F}$ and amplitude $\mathcal{A}$ of the sinusoid. This allows the problem of parameter estimation to be abstracted into the problem of estimating underlying DTFT peaks. We consider only the parametrization $(\mathcal{K}, \mathcal{X})$ in the rest of the article to preserve the clarity and conciseness of the discussion independently of changes in the chosen window, sampling interval or DFT length, knowing that in the idealized case of a single sinusoid $(\mathcal{F}, \mathcal{A})$ are readily recovered from $(\mathcal{K}, \mathcal{X})$ using (5).

3. Estimating Sinusoidal Parameters from 3 Bins of Spectrum

The methods in this article estimate $(\mathcal{K}, \mathcal{X})$ using some form of parabolic fit on the magnitude spectrum maximum at peak bin k_m and its lower and upper neighbors at bins $k_m - 1$ and $k_m + 1$. For compactness, we introduce the magnitude spectrum value substitutions [12]

$$\alpha = |X[k_m - 1]| \tag{6}$$
$$\beta = |X[k_m]| \tag{7}$$
$$\gamma = |X[k_m + 1]|. \tag{8}$$

A coarse way of estimating sinusoid parameters is to find peaks in the DFT magnitude spectrum $|X[k]|$ and use their frequency and amplitude as estimates. Here the sinusoid's fractional bin index and magnitude $(\mathcal{K}, \mathcal{X})$ are estimated as $(\hat{\mathcal{K}}, \hat{\mathcal{X}})$. The values associated with a peak bin k_m in the magnitude spectrum are

$$\hat{\mathcal{K}} = k_m \tag{9}$$

$$\hat{\mathcal{X}} = |X[k_m]| . \tag{10}$$

We call this method the *nearest bin method*, or "nearest" for short. Here the fractional bin estimate $\mathcal{K}$ will always be an integer; hence the bin estimate of the nearest bin method can have up to ± 0.5 bins of error. The nearest bin method is inherently limited by the finite frequency definition of the DFT. This lack of accuracy can be problematic, especially if typical DFT frequency definition falls below a satisfactory level (e.g., human perception in the lower frequency domain).

A typical engineering solution is to refine this coarse estimate by a fine search over ± 0.5 bins around this peak [2,14,15,18,27,28]. Approaches to refining the coarse estimate can be divided into *iterative* approaches and *direct* approaches [27]. Iterative approaches (e.g., [29]) involve multiple rounds of function evaluations or a series of DTFT evaluations that narrow in on the solution. Direct methods form the estimate $(\hat{\mathcal{K}}, \hat{\mathcal{X}})$ using a single function evaluation on $X[k_m - 1]$, $X[k_m]$, and $X[k_m + 1]$. For many spectral audio applications, efficiency is important. This is especially true for real-time applications and applications based on processing huge data sets, e.g. MIR. For this reason, we focus on direct methods in particular. It is worth noting that as with most direct methods, the direct method presented in this article can be further refined into an interactive method using correction functions [20,30].

A popular class of techniques, sometimes referred to as the *Quadratically-Interpolated FFT* (QIFFT) in the literature [30–33], is the subject of this article. Assuming that over a range of ± 1 bins around a peak bin k_m (i.e., between $k_m - 1$ and $k_m + 1$), the underlying DTFT of a window transform is reasonably smooth, QIFFT methods fit a parabola to the magnitude spectrum and use the vertex of this parabola as a refined estimate of the true spectral peak. (Some other direct methods fit a predetermined parabola shape to the spectrum peak bin and its left and right neighbors, e.g., using least-squares regression [34].) Parabolic fitting in the QIFFT is done by writing the Lagrange interpolating polynomial [4,35] that fits the three points $(k_m - 1, \alpha)$, (k_m, β), and $(k_m + 1, \gamma)$, rearranging it into the vertex form of a parabola $\chi(\kappa) = a(\kappa - \hat{\mathcal{K}})^2 + \hat{\mathcal{X}}$ with dummy bin index κ, and then considering the vertex of the parabola $(\hat{\mathcal{K}}, \hat{\mathcal{X}})$ to be an estimate of the sinusoidal parameters. In terms of k_m, α, β, and γ, $(\hat{\mathcal{K}}, \hat{\mathcal{X}})$ are:

$$\hat{\mathcal{K}} = k_m + \frac{1}{2} \frac{\alpha - \gamma}{\alpha - 2\beta + \gamma} \tag{11}$$

$$\hat{\mathcal{X}} = \beta - \frac{1}{8} \frac{(\alpha - \gamma)^2}{\alpha - 2\beta + \gamma} . \tag{12}$$

Since it is performed directly on the magnitude spectrum, we denote this technique the "magnitude-spectrum QIFFT" or MQIFFT. The parabolic fitting is shown in Figure 1. The thick dashed blue line is the DTFT of a length-4096 Hann-windowed sinusoid and the thick solid green line is a parabola fit to the top three bins (k_{m-1}, k_m, and k_{m+1}) of the magnitude DFT, shown as stemmed circle. The vertical dot-dashed lines mark the true fractional bin index $\mathcal{K}$ and the estimate $\hat{\mathcal{K}}$; the horizontal dot-dashed lines mark the true amplitude $\mathcal{X}$ and the estimate $\hat{\mathcal{X}}$. The QIFFT can be considered a perfect match to a truncated (order-2) Taylor series expansion of the window transform.

QIFFT-derived methods are very attractive, as they all have the following desirable properties: (1) they can greatly reduce estimation error; (2) they are inexpensive (requiring only a few multiplies to find the refined peak frequency and amplitude estimates); (3) they can be used with *any* window type with the notable exception of the non-zero-padded rectangular window as its main lobe width less that three bins [30]; and (4) they can be combined seamlessly with zero padding to further increase their accuracy.

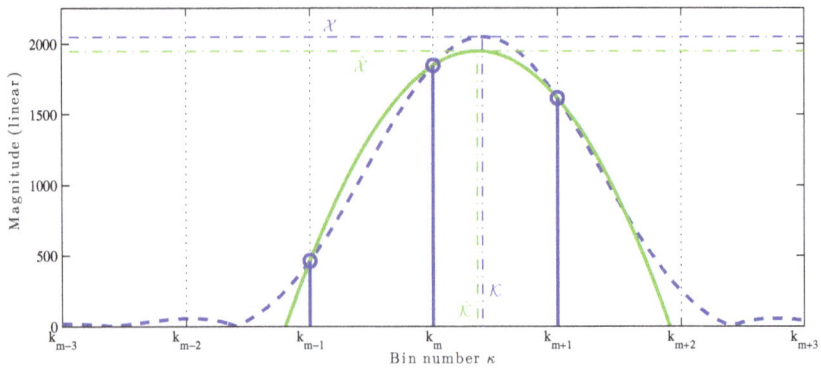

Figure 1. Demonstration of the MQIFFT.

The property of being applicable to any window type is one of the main advantages of QIFFT-derived methods. Other popular direct parameter estimation techniques are derived from properties of the rectangular window (no windowing) [2,8,18,19,36] or a particular window (e.g., [1], Cosine window) or class of windows (e.g., [16], Rife–Vincent class I windows). A few state-of-the-art methods are fully general and can be applied to any window, including methods by Duda [14] and Candan [15]. General methods have the distinct advantage of being portable to any window type, even those without analytic descriptions such as the Slepian (DPSS) window [4]. For instance, Duda gives examples of applying his method to Dolph–Chebyshev and Kaiser–Bessel windowed signals [14], which can only be approximated with Rife–Vincent class I windows and hence can't take advantage of the approaches of, e.g., [16].

4. Proposed Method: XQIFFT

The QIFFT fits a parabola to the magnitude spectrum directly, however, it is possible to implement variations of the QIFFT where we fit parabolas to a magnitude spectrum scaled by a nonlinear function $f(\Theta) = \Phi$. To qualify as a scaling, the function f needs to be invertible (and thus monotonic) with its inverse denoted as $f^{-1}(\Phi) = \Theta$. This property is desirable for the fact that we may need to retrieve the amplitude value estimate on the original linear scale, and for the fact that this preserves the concavity property of the main lobe. These variations are based on nonlinear scalings of the magnitude spectrum estimate $(\hat{\mathcal{K}}, \hat{\mathcal{X}})$ by:

$$\hat{\mathcal{K}} = k_m + \frac{1}{2}\frac{f(\alpha) - f(\gamma)}{f(\alpha) - 2f(\beta) + f(\gamma)} \tag{13}$$

$$\hat{\mathcal{X}} = f^{-1}\left(f(\beta) - \frac{1}{8}\frac{(f(\alpha) - f(\gamma))^2}{f(\alpha) - 2f(\beta) + f(\gamma)}\right). \tag{14}$$

Rather than operating on α, β, and γ directly, these methods operate on nonlinear scalings $f(\alpha)$, $f(\beta)$, and $f(\gamma)$. Since $\hat{\mathcal{X}}$ is a linear scale estimate, we must also un-scale the vertex y-coordinate by $f^{-1}(\Phi)$.

4.1. Logarithmically-Scaled QIFFT

A variant of the QIFFT approach is available in the literature which is calculated through a parabolic fit on a logarithmically-scaled magnitude spectrum, which we denote the "logarithmically-scaled QIFFT" or *LQIFFT*.

The LQIFFT uses (13) and (14) with the following nonlinear scaling and unscaling functions:

$$f(\Theta) = \log(\Theta) = \Phi \tag{15}$$

$$f^{-1}(\Phi) = \exp(\Phi) = \Theta. \tag{16}$$

The intuition leading to the use of logarithmic scaling in the QIFFT is that since the Fourier Transform (FT) of a continuous-time Gaussian window is also a Gaussian, the FT of a sinusoid will have a Gaussian-shaped main lobe. Since the logarithm of a Gaussian is exactly a parabola, using a parabolic fit on a logarithmically-scaled magnitude response of a Gaussian-windowed signal would give a perfect estimate of $(\mathcal{K}, \mathcal{X})$. Considering that many popular smoothing window have a shape that can be seen as more closely approximating the shape of a Gaussian than a parabola, the logarithmic scaling was introduced in [12,31], and was shown to outperform the QIFFT on a linear scale for a variety of typical windows (Rectangular, Hann, Hamming, Blackman–Harris, Kaiser–Bessel). Note that changing the base of the logarithm has no impact on the resulting estimation—we arbitrarily choose the natural logarithm.

4.2. Power-Scaled QIFFT (XQIFFT)

While applying the QIFFT on a logarithmically-scaled magnitude spectrum was found to outperform the QIFFT on a linearly-scaled magnitude spectrum, some systematic error remains [30]. To further reduce the estimation error of the QIFFT method, we review our proposed variation on the QIFFT method that performs a parabolic fit to a magnitude spectrum rescaled by a power function [20].

The power-scaled QIFFT—the *XQIFFT*—uses (13) and (14) and has a tuning parameter p that controls the following nonlinear scaling and un-scaling functions:

$$f(\Theta) = \Theta^p \tag{17}$$

$$f^{-1}(\Phi) = \Phi^{1/p}. \tag{18}$$

The intuition leading to the XQIFFT is simply to try to scale the window transform shape to resemble a parabola. The performance of the XQIFFT depends heavily on choosing the tuning parameter p properly to scale the shape of the main lobe to be maximally parabolic. The shape of the main lobe is highly dependent on window type, as well as zero-padding factor and window length. Hence the proper value of p is also window-dependent. A simple procedure for choosing p is presented later, in Section 5.2.

4.3. Invariance Properties of QIFFT Methods

In this section, we show that for the XQIFFT along with the MQIFFT and LQIFFT, the estimate $\hat{\mathcal{K}}$ is insensitive to amplitude and that the amplitude estimate $\hat{\mathcal{X}}$ scales linearly as a function of the true amplitude $\mathcal{A}$. This property shows that the XQIFFT and its cousins are suitable for estimating sinusoids of any amplitude (other nonlinear scaling functions may not have these properties).

All the scalings discussed so far verify that the estimate in fractional bin index is insensitive to the true amplitude $\mathcal{A}$, while the estimate $\hat{\mathcal{X}}$ scales linearly as a function of the true amplitude $\mathcal{A}$. Indeed, if we have another sinusoid of fractional bin number $\mathcal{K}$ and amplitude $\mathcal{A}' = \lambda \mathcal{A}$, the QIFFT estimates (11) and (12) yield:

$$\hat{\mathcal{K}}' = k'_m + \frac{1}{2}\frac{\alpha' - \gamma'}{\alpha' - 2\beta' + \gamma'} = k_m + \frac{1}{2}\frac{\lambda\alpha - \lambda\gamma}{\lambda\alpha - 2\lambda\beta + \lambda\gamma} \qquad \Rightarrow \hat{\mathcal{K}}' = \hat{\mathcal{K}} \tag{19}$$

$$\hat{\mathcal{X}}' = \beta' - \frac{1}{8}\frac{(\alpha' - \gamma')^2}{\alpha' - 2\beta' + \gamma'} = \lambda\beta - \frac{1}{8}\frac{(\lambda\alpha - \lambda\gamma)^2}{\lambda\alpha - 2\lambda\beta + \lambda\gamma} \qquad \Rightarrow \hat{\mathcal{X}}' = \lambda\hat{\mathcal{X}} \tag{20}$$

For the log-scaled QIFFT (13)–(16), we have the property that $f(\lambda\cdot) = f(\cdot) + f(\lambda)$, and $f^{-1}(f(\lambda) + \cdot) = \lambda f^{-1}(\cdot)$, so that:

$$\hat{\mathcal{K}}' = k_m' + \frac{1}{2}\frac{f(\alpha') - f(\gamma')}{f(\alpha') - 2f(\beta') + f(\gamma')} = k_m + \frac{1}{2}\frac{f(\alpha) - f(\gamma)}{f(\alpha) - 2f(\beta) + f(\gamma)} \quad \Rightarrow \quad \hat{\mathcal{K}}' = \hat{\mathcal{K}} \tag{21}$$

$$\hat{\mathcal{X}}' = f^{-1}\left(f(\beta') - \frac{1}{8}\frac{(f(\alpha') - f(\gamma'))^2}{f(\alpha') - 2f(\beta') + f(\gamma')} \right) = f^{-1}\left(f(\lambda\beta) - \frac{1}{8}\frac{(f(\alpha) - f(\gamma))^2}{f(\alpha) - 2f(\beta) + f(\gamma)} \right)$$

$$= f^{-1}\left(f(\lambda) + f(\beta) - \frac{1}{8}\frac{(f(\alpha) - f(\gamma))^2}{f(\alpha) - 2f(\beta) + f(\gamma)} \right) \quad \Rightarrow \quad \hat{\mathcal{X}}' = \lambda\hat{\mathcal{X}}. \tag{22}$$

For the power-scaled QIFFT (13), (14), (17) and (18), we have the property that $f(\lambda\cdot) = \lambda^p f(\cdot)$, and $f^{-1}(\lambda^p f(\lambda)) = \lambda f^{-1}(\cdot)$, so that:

$$\hat{\mathcal{K}}' = k_m' + \frac{1}{2}\frac{f(\alpha') - f(\gamma')}{f(\alpha') - 2f(\beta') + f(\gamma')} = k_m + \frac{1}{2}\frac{\lambda^p f(\alpha) - \lambda^p f(\gamma)}{\lambda^p f(\alpha) - 2\lambda^p f(\beta) + \lambda^p f(\gamma)} \quad \Rightarrow \quad \hat{\mathcal{K}}' = \hat{\mathcal{K}} \tag{23}$$

$$\hat{\mathcal{X}}' = f^{-1}\left(f(\beta') - \frac{1}{8}\frac{(f(\alpha') - f(\gamma'))^2}{f(\alpha') - 2f(\beta') + f(\gamma')} \right) = f^{-1}\left(\lambda^p f(\beta) - \frac{\lambda^{2p}}{8\lambda^p}\frac{(f(\alpha) - f(\gamma))^2}{f(\alpha) - 2f(\beta) + f(\gamma)} \right)$$

$$= f^{-1}\left(\lambda^p \left[f(\beta) - \frac{1}{8}\frac{(f(\alpha) - f(\gamma))^2}{f(\alpha) - 2f(\beta) + f(\gamma)} \right] \right) \quad \Rightarrow \quad \hat{\mathcal{X}}' = \lambda\hat{\mathcal{X}}. \tag{24}$$

5. Estimation Error, Properties, and Reduction

In this section we provide error definitions (Section 5.1) and study interpolation bias (Section 5.2). Using these definitions, we propose a heuristic to find the optimal p value for a given windowing situation (Section 5.3). Finally, we mention correction functions which can unbias remaining systematic error (Section 5.4) and noise sensitivity properties (Section 5.5).

5.1. Error Definition

Following the error definition in [31], we compute the error in a particular fractional bin index $e_{\mathcal{K}}(\mathcal{K})$ and a particular magnitude estimate $e_{\mathcal{X}}(\mathcal{K})$ as follows. For each true fractional bin $\mathcal{K}$, the error in a particular fractional bin estimate is defined as

$$e_{\mathcal{K}}(\mathcal{K}) = \hat{\mathcal{K}} - \mathcal{K} \tag{25}$$

and the error in a magnitude estimate is defined proportionally by

$$e_{\mathcal{X}}(\mathcal{K}) = \frac{\hat{\mathcal{X}} - \mathcal{X}}{\mathcal{X}}. \tag{26}$$

Because of the invariance of $\hat{\mathcal{K}}$ to amplitude, and the linear dependency of $\hat{\mathcal{X}}$ to amplitude, these errors are thus independent of the true $\mathcal{X}$, so that we only need to index these by $\mathcal{K}$.

We also define the *worst-case* bin estimates of the fractional bin as $e_{\mathcal{K}}^{\text{w.c.}}$ and the magnitude as $e_{\mathcal{X}}^{\text{w.c.}}$, meaning the maximum errors obtained across all the possible fractional bin indices:

$$e_{\mathcal{K}}^{\text{w.c.}} = \max_{\Delta_{\mathcal{K}} \in [0,\frac{1}{2}]} |e_{\mathcal{K}}(\Delta_{\mathcal{K}})| \tag{27}$$

$$e_{\mathcal{X}}^{\text{w.c.}} = \max_{\Delta_{\mathcal{K}} \in [0,\frac{1}{2}]} |e_{\mathcal{X}}(\Delta_{\mathcal{K}})|. \tag{28}$$

These errors have *means* $E[|e_{\mathcal{K}}(\mathcal{K})|]$ and $E[|e_{\chi}(\mathcal{K})|]$:

$$E[|e_{\mathcal{K}}(\kappa)|] = 2 \int_0^{1/2} |e_{\mathcal{K}}(\Delta_{\kappa})| d\Delta_{\kappa} \qquad (29)$$

$$E[|e_{\chi}(\kappa)|] = 2 \int_0^{1/2} |e_{\chi}(\Delta_{\kappa})| d\Delta_{\kappa} \qquad (30)$$

and *variances* $\mathrm{Var}\,(e_{\mathcal{K}}(\mathcal{K}))$ and $\mathrm{Var}\,(e_{\chi}(\mathcal{K}))$.

5.2. Interpolation Bias

We know that the MQIFFT and the LQIFFT present systematic fractional bin index and amplitude errors that are strongly dependent [30] on the distance of the sinusoid's true fractional bin index $\mathcal{K}$ from a bin center $\lfloor\mathcal{K}\rceil$. $\lfloor\cdot\rceil$ indicates the *rounding* function, which returns its argument rounded to the nearest integer. This distance, denoted by $\Delta_{\mathcal{K}}$, is defined as

$$\Delta_{\mathcal{K}} = \mathcal{K} - \lfloor\mathcal{K}\rceil. \qquad (31)$$

By construction, bin estimates have odd symmetry around $\Delta_{\mathcal{K}} = 0$ and magnitude estimates have even symmetry around $\Delta_{\mathcal{K}} = 0$ (symmetries exploited in the previous error mean and worst-case definitions). The errors in estimates associated with the XIQFFT have all these properties as well. These can be seen in the *bias curves* in Figure 2, which show $e_{\mathcal{K}}$ and e_{χ} as a function of bin offset $\Delta_{\mathcal{K}}$ for a length-4096 Hann window.

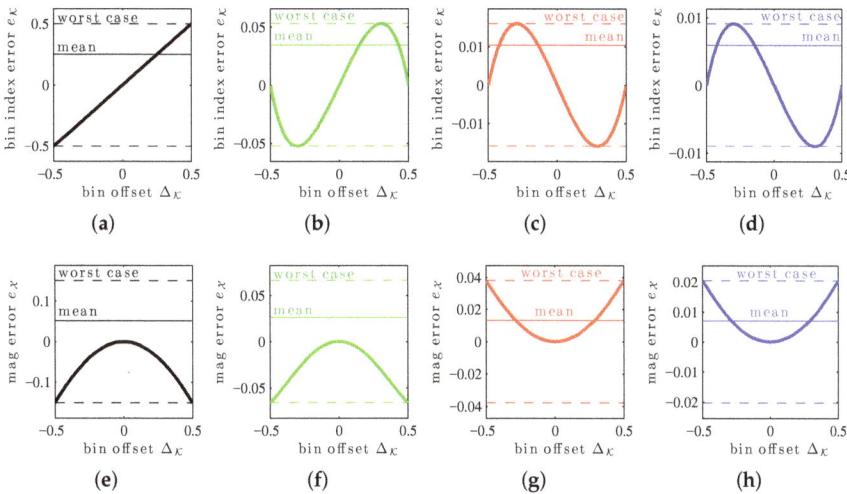

Figure 2. Mean and worst-case of absolute value of errors for each method as a function of bin offset. (a) nearest bin method; (b) MQIFFT; (c) LQIFFT; (d) XQIFFT, $p = 0.1$; (e) nearest bin method; (f) MQIFFT; (g) LQIFFT; (h) XQIFFT, $p = 0.1$.

Note that while it may sometimes (for certain windows) be possible to obtain closed form solutions for those two statistics, these solutions are generally very complex, even for simple smoothing windows. In the context of this article, we propose instead the following heuristic to evaluate those statistics numerically with arbitrary precision

Finding the worst-case errors $e_{\mathcal{K}}^{\text{w.c.}}$ and $e_{\mathcal{X}}^{\text{w.c.}}$ requires finding the global maximum on the appropriate bias curve in Figure 2. Doing so numerically requires first to identify local maxima as the bias curves do not necessarily exhibit a single local maximum. We identify the regions containing a local maximum by computing the empirical error at a coarse sampling of the bin index space and finding the different local peaks that we associate with a single neighboring local maximum. We then run a Fibonacci search [37–39] (pp. 275–278, [40]) in the neighborhood of each found peak to find a local maximum. Using the Fibonacci search algorithm allows us to compute the location of that maximum up to a preset precision. We finally pick the global maximum as the largest local maximum.

Finding the mean errors $E[|e_{\mathcal{K}}(\mathcal{K})|]$ and $E[|e_{\mathcal{X}}(\mathcal{K})|]$ requires to evaluate the integral of the error function across the range of all possible bin indices (i.e., between -0.5 and $+0.5$). To do so numerically, we run an adaptive integration algorithm, the Simpson's quadrature algorithm [41], on the empirical error function. This algorithm allows us to control the desired numerical precision of the final result.

For both methods, the procedure described above is aimed at obtaining estimates with arbitrary numerical precision (allowed by the Fibonacci search and the adaptive integration algorithms), while minimizing the number of empirical error function evaluations required during the extremum search (allowed by the Fibonacci search algorithm) and not requiring the derivation of a closed form error function.

5.3. Choice of the Power-Scaling Factor p

To demonstrate the XQIFFT and the process of choosing p, we study one particular representative window in detail: a length-4096 ($N = 4096$) Hann window (p. 98, [42]) [43], defined as

$$ w_{\text{Hann}}(n) = \left[\frac{1}{2} + \frac{1}{2}\cos\left(\frac{2\pi n}{N}\right) \right]. \tag{32} $$

When computing the worst-case and mean errors for various values of power-scaling factor p (Figure 3), we can observe that the XQIFFT strongly outperforms the other QIFFT-derived methods if p is picked appropriately. In particular, we see that there seems to exist a single value p for which each statistic is minimized. This optimal value of p is typically different for each statistic, as well as for the chosen smoothing window, DFT length, and zero-padding factor.

As the function exhibits a single minimum (i.e., it is unimodal), we propose here to find an optimal p value numerically finding that minimum. Finding the minimum of a unimodal function can be done numerically by running a Fibonacci search on the desired error statistic. Such a heuristic yields a choice of p minimizing the desired error statistic up to an arbitrary precision order, while minimizing the number of power-scaling factors p for which we need to evaluate the statistic (using the proposed approach in Section 5.2), and still without requiring the complex derivation of a closed form solution for the different statistics as a function of p. Other estimators which allow for arbitrary windows also favor a computational rather than analytic approach [14,15].

Table 1 shows the different error statistics from Figure 3 for the four different optimal cases. From the results, we can see that (1) for each error statistic, its associated optimal p yields an error much smaller than the MQIFFT and LQIFFT; and (2) a p value optimized for a given error statistic still yields small errors for the other ones, in particular, p values that yield small fractional bin index estimation error also yield small amplitude estimation error.

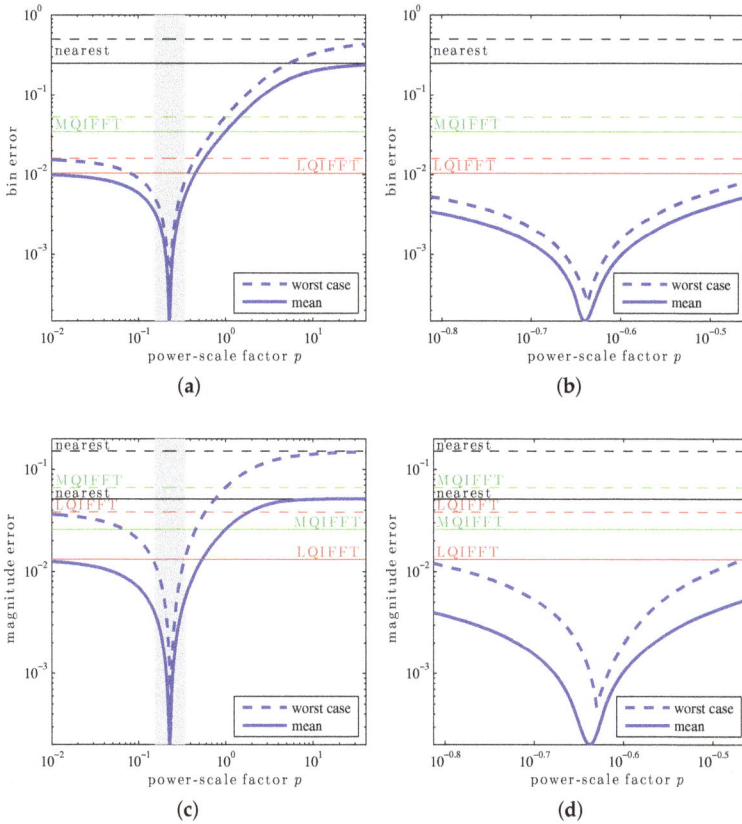

Figure 3. Mean and worst-case for absolute value of errors for the QIFFT method on a length-4096 Hann window, including bin error and detail on bin error, magnitude error and detail on magnitude error. Horizontal lines indicate the worst case and mean errors for the nearest bin, MQIFFT, and LQIFFT methods for comparison. Shaded regions indicate the range of the detail in corresponding figures to the right. The thick lines indicate the XQIFFT. (**a**) Bin error; (**b**) Bin error (zoom); (**c**) Magnitude error (zoom); (**d**) Magnitude error (zoom).

Table 1. Comparing $e_{\mathcal{K}}(\mathcal{K})$ and $e_{\chi}(\mathcal{K})$ for various estimation approaches on a length-4096 Hann window. XQIFFT cases which heuristically minimize worst case and mean error bin and magnitude errors are shown. Numbers are given to five significant figures and minimized values are shaded.

Case	$e_{\mathcal{K}}^{\text{w.c.}}$	$e_{\chi}^{\text{w.c.}}$	$E[|e_{\mathcal{K}}(\mathcal{K})|]$	$E[|e_{\chi}(\mathcal{K})|]$	Minimizes
nearest	5.0000×10^{-1}	1.5110×10^{-1}	2.5000×10^{-1}	5.1688×10^{-2}	n/a
MQIFFT	5.2764×10^{-2}	6.6237×10^{-2}	3.4221×10^{-2}	2.5601×10^{-2}	n/a
LQIFFT	1.5997×10^{-2}	3.7932×10^{-1}	1.0392×10^{-2}	1.3121×10^{-2}	n/a
XQIFFT, $p = 0.23086$	2.4484×10^{-4}	9.5196×10^{-4}	1.5693×10^{-4}	2.0239×10^{-4}	$e_{\mathcal{K}}^{\text{w.c.}}$
XQIFFT, $p = 0.23437$	4.4380×10^{-4}	4.7735×10^{-4}	2.3462×10^{-4}	2.5251×10^{-4}	$e_{\chi}^{\text{w.c.}}$
XQIFFT, $p = 0.22917$	3.1861×10^{-4}	1.1803×10^{-3}	1.4645×10^{-4}	2.0637×10^{-4}	$E[|e_{\mathcal{K}}(\mathcal{K})|]$
XQIFFT, $p = 0.23039$	2.6445×10^{-4}	1.0149×10^{-3}	1.5203×10^{-4}	2.0170×10^{-4}	$E[|e_{\chi}(\mathcal{K})|]$

Notice that in Figure 3, the mean and worst-case bin and magnitude errors of the XQIFFT approach the performance of the nearest neighbor method for large values of p. This can be explained by the asymptotic behavior of the XQIFFT estimator as p approaches ∞. Because we have $0 \leq \alpha/\beta < 1$ and $0 \leq \gamma/\beta < 1$, $\lim_{p \to \infty} (\alpha/\beta)^p = 0$ and $\lim_{p \to \infty} (\gamma/\beta)^p = 0$ and the XQIFFT estimate (13), (14), (17) and (18) reduces to the nearest neighor method:

$$\hat{\mathcal{K}} = \lim_{p \to \infty} \left[k_m + \frac{1}{2} \frac{\alpha^p - \gamma^p}{\alpha^p - 2\beta^p + \gamma^p} \right] = \lim_{p \to \infty} \left[k_m + \frac{1}{2} \frac{(\alpha/\beta)^p - (\gamma/\beta)^p}{(\alpha/\beta)^p - 2 + (\gamma/\beta)^p} \right] = k_m \tag{33}$$

$$\hat{\mathcal{X}} = \lim_{p \to \infty} \left[\left(\beta^p - \frac{1}{8} \frac{(\alpha^p - \gamma^p)^2}{\alpha^p - 2\beta^p + \gamma^p} \right)^{1/p} \right] = \lim_{p \to \infty} \left[\beta \left(1 - \frac{1}{8} \frac{[(\alpha/\beta)^p - (\gamma/\beta)^p]^2}{(\alpha/\beta)^p - 2 + (\gamma/\beta)^p} \right)^{1/p} \right] = \beta. \tag{34}$$

Additionally, we can show that, as suggested by Figure 3, the XQIFFT converges towards the LQIFFT as p approaches 0. Indeed, from the Taylor expansion of the bin estimate (13) using the XQIFFT scaling (17) and (18), we have that $\alpha^p = \exp(p \log \alpha) \sim 1 + p \log \alpha + \mathcal{O}(p^2)$, $\beta^p \sim 1 + p \log \beta + \mathcal{O}(p^2)$ and $\gamma^p \sim 1 + p \log \gamma + \mathcal{O}(p^2)$. Hence, we get:

$$\hat{\mathcal{K}} = \lim_{p \to 0} \left[k_m + \frac{1}{2} \frac{\alpha^p - \gamma^p}{\alpha^p - 2\beta^p + \gamma^p} \right] \sim \lim_{p \to 0} \left[k_m + \frac{1}{2} \frac{p \log \alpha - p \log \gamma + \mathcal{O}(p^2)}{2 p \log \alpha - 2p \log \beta + p \log \gamma + \mathcal{O}(p^2)} \right]$$

$$= k_m + \frac{1}{2} \frac{\log \alpha - \log \gamma}{2 \log \alpha - 2 \log \beta + \log \gamma} \tag{35}$$

This matches (13) for a logarithmic scaling, verifying that the XQIFFT's bin estimate (17) approaches the LQIFFT as $p \to 0$.

For the amplitude estimate (14) using the XQIFFT scaling (17) and (18) we have

$$\beta^p - \frac{1}{8} \frac{(\alpha^p - \gamma^p)^2}{\alpha^p - 2\beta^p + \gamma^p} \sim 1 + p \left(\log \beta + \mathcal{O}(p) - \frac{1}{8} \frac{(\log \alpha - \log \gamma)^2 + \mathcal{O}(p)}{\log \alpha - 2 \log \beta + \log \gamma + \mathcal{O}(p)} \right) \tag{36}$$

Similarly, $\log(1 + x) \sim x + \mathcal{O}(x^2)$, such that:

$$\log \left[\beta^p - \frac{1}{8} \frac{(\alpha^p - \gamma^p)^2}{\alpha^p - 2\beta^p + \gamma^p} \right] \sim p \left(\log \beta + \mathcal{O}(p) - \frac{1}{8} \frac{(\log \alpha - \log \gamma)^2 + \mathcal{O}(p)}{\log \alpha - 2 \log \beta + \log \gamma + \mathcal{O}(p)} \right) + \mathcal{O}(p^2) \tag{37}$$

Furthermore, since $(\cdot)^{1/p} = \exp \left[\frac{1}{p} \log(\cdot) \right]$, we have:

$$\left[\beta^p - \frac{1}{8} \frac{(\alpha^p - \gamma^p)^2}{\alpha^p - 2\beta^p + \gamma^p} \right]^{1/p} = \exp \left[\frac{1}{p} \log \left[\beta^p - \frac{1}{8} \frac{(\alpha^p - \gamma^p)^2}{\alpha^p - 2\beta^p + \gamma^p} \right] \right]$$

$$\sim \exp \left[\log \beta - \frac{1}{8} \frac{(\log \alpha - \log \gamma)^2 + \mathcal{O}(p)}{\log \alpha - 2 \log \beta + \log \gamma + \mathcal{O}(p)} + \mathcal{O}(p) \right] \tag{38}$$

As $p \to 0$, we have $\mathcal{O}(p) \to 0$ and we then get the following limit:

$$\hat{\mathcal{X}} = \lim_{p \to 0} \left[\left[\beta^p - \frac{1}{8} \frac{(\alpha^p - \gamma^p)^2}{\alpha^p - 2\beta^p + \gamma^p} \right]^{1/p} \right] = \exp \left[\log \beta - \frac{1}{8} \frac{(\log \alpha - \log \gamma)^2}{\log \alpha - 2 \log \beta + \log \gamma} \right] \tag{39}$$

This matches (14) for a logarithmic scaling, verifying that the XQIFFT's amplitude estimate (18) approaches the LQIFFT as $p \to 0$.

5.4. Correction Functions

Correction functions are known to unbias some of the systematic error of interpolation on linear and logarithmic scales. Figure 4 shows systematic error curves for the MQIFFT, LQIFFT, and XQIFFT as a function of the bin estimate offset $\Delta_{\hat{\mathcal{K}}}$. Similar to the bin offset (31), the bin estimate offset is defined by

$$\Delta_{\hat{\mathcal{K}}} = \hat{\mathcal{K}} - \lfloor \hat{\mathcal{K}} \rceil. \tag{40}$$

The reason to index the correction curves by $\Delta_{\hat{\mathcal{K}}}$ instead of $\Delta_{\mathcal{K}}$ is that when implementing correction functions, the true value $\mathcal{K}$ (and hence $\Delta_{\mathcal{K}}$ which depends on it according to (31)) is not known. Only the estimate $\hat{\mathcal{K}}$ (and hence $\Delta_{\hat{\mathcal{K}}}$) is known.

Notice that all of the error curves shown in Figure 4 are highly structured. This means that it is possible to design parametric correction functions to match these curves and unbias the estimates somewhat, leading to further refinements. It should also be possible to create a tabulation of these curves when it is not convenient to develop a parametric equation. The limit of the efficacy of correction functions is controlled by the level of fitness of the fit curve or the density of a tabulation. Abe and Smith report correction functions for the LQIFFT [30] and Werner proposes magnitude and bin correction functions for the XQIFFT [20]. As we are focused solely on direct methods, a discussion of the particular form or performance of XQIFFT correction functions is outside the scope of this article since correction functions are considered to be iterative methods [27].

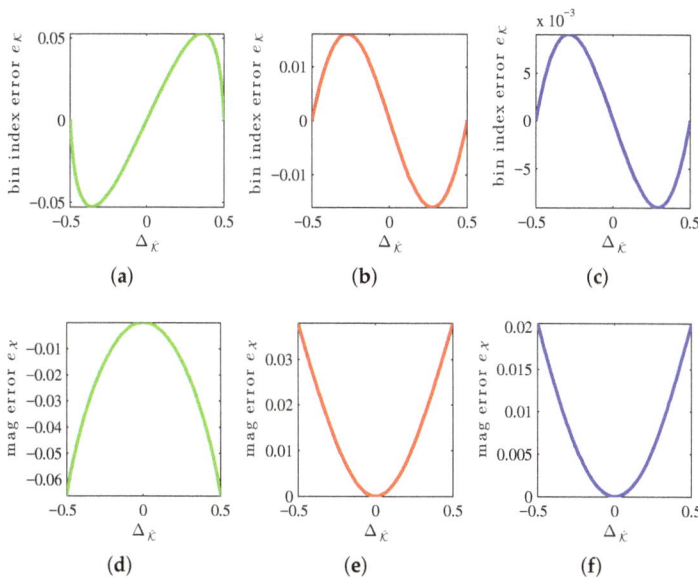

Figure 4. Correction curves. (**a**) MQIFFT; (**b**) LQIFFT; (**c**) XQIFFT, $p = 0.1$; (**d**) MQIFFT; (**e**) LQIFFT; (**f**) XQIFFT, $p = 0.1$.

5.5. Sensitivity to Noise

We analyze how the parameter estimation is affected by the presence of noise in the signal. To do so we corrupt the signal $A\, e^{j(2\pi \mathcal{F} nT + \phi)}$ with different realizations of an additive white Gaussian noise random process (with zero-mean and standard deviation σ) at different levels of signal-to-noise ratio (SNR). The SNR is computed as the ratio of the amplitude of the sinusoid by the standard deviation of the noise:

$$\text{SNR} = A/\sigma \tag{41}$$

In this analysis, we quantify the effect of added noise by looking at the distribution of the noisy estimates $\hat{\mathcal{K}}_G$ around the noiseless estimates $\hat{\mathcal{K}}$. As such, we decorrelate the influence of the systematic bias associated with the estimator from the influence of the noise by reporting the bias and variance of $\hat{\mathcal{K}}_G - \hat{\mathcal{K}}$ for various bin index offsets .

In the following experiments, we consider a length-64 Hann window. We sample uniformly the bin index $\Delta_{\mathcal{K}}$ in $[-0.5, 0]$. For each bin index, we generated 10^9 noise trials to get reliable estimates of the bin index error bias and variance.

Figure 5 shows the bias of the bin index estimate (25) due to the noise corruption alone at various signal-to-noise ratios (SNR). As expected, for each estimator the bias decreases as the SNR increases. Across all $\Delta_{\mathcal{K}}$ and tested SNRs, the XQIFFT ($p = 0.2776$) has a noise bias between that of the MQIFFT and the LQIFFT. Considering that the XQIFFT tends towards the LQIFFT as p approaches zero (as shown in Section 5.3) and is identical to the MQIFFT when $p = 1$, the XQIFFT with a value of $p = 0.2776$ can be considered an intermediate between the MQIFFT and LQIFFT. So, it is unsurprising that its noise bias also falls between the two. Interestingly, the bias is nonzero for most offsets $\Delta_{\mathcal{K}}$. This means that the presence of noise actually shifts the mean error of our estimator away from the systematic bias already present in the noiseless estimation.

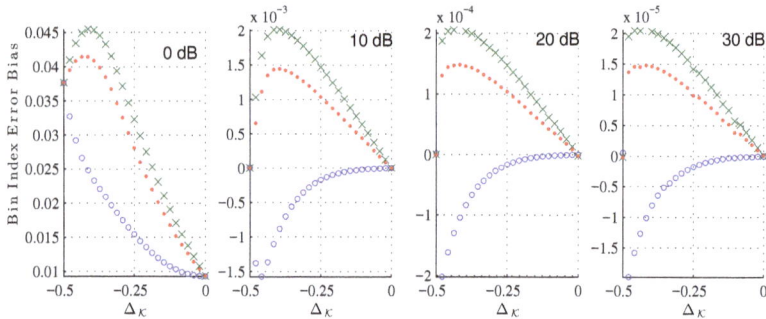

Figure 5. Bin index error bias for the MQIFFT ($\bigcirc$), the LQIFFT ($\times$), and the XQIFFT ($\cdot$, $p = 0.2276$) at various SNR: 0, 10, 20, and 30 dB.

Figure 6 shows the bin index error variance around the error mean. Again, the variance of the XQIFFT ($p = 0.2776$) is between the MQIFFT and the LQIFFT, and we can notice that its variance has less variation across the different bin index offsets $\Delta_{\mathcal{K}}$.

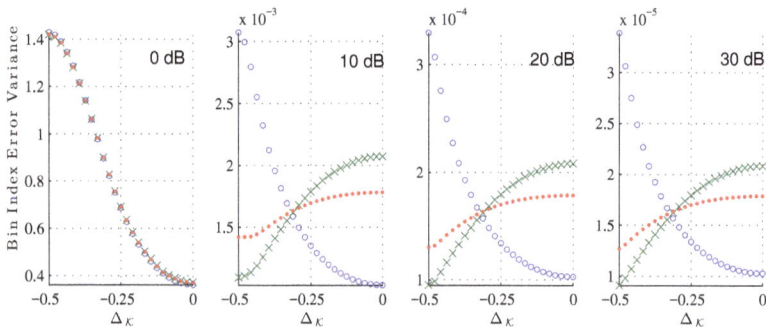

Figure 6. Bin index error variance for the MQIFFT ($\bigcirc$), the LQIFFT ($\times$), and the XQIFFT ($\cdot$, $p = 0.2276$) at various SNR: 0, 10, 20, and 30 dB.

Figure 7 shows the bin index error variance of each estimator plotted against SNR, compared against the Cramér–Rao bound for unbiased estimators. Normalizing to bin width, the Cramér–Rao bound for a discrete-time signal corrupted by additive white Gaussian noise, with unknown phase and unknown amplitude is [44]

$$\text{Var}(\hat{\mathcal{K}}) = \frac{12\sigma^2 N}{4\pi^2 \mathcal{A}^2 (N^2 - 1)}, \tag{42}$$

where $N = 64$ is the signal length, σ is the noise standard deviation, and $\mathcal{A}$ the sinusoid amplitude. To compare against the Cramér–Rao bound for unbiased estimation, we recompute the variance of the bin index estimation error under the assumption of a zero bias. Though we showed in Figure 5 that the estimation is not actually unbiased, the bias is commonly neglected for the noise robustness analysis of sinusoidal parameter analysis. Notice that all the QIFFT-based methods have a similar robustness against noise independently of the chosen magnitude spectrum scaling function. This result highly contrasts against the systematic estimation bias, which is highly sensitive to the choice of scaling function. We can also observe the expected thresholding effect at 0 dB [44], as the peak finding initial step becomes unreliable. The thresholding effect can also be observed in Figures 5 and 6 where the behavior of the error bias and variance are very different for the 0 dB case.

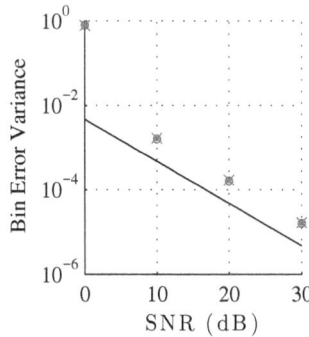

Figure 7. Comparison of the estimation variance under the unbiased hypothesis for the MQIFFT ($\bigcirc$), the LQIFFT ($\times$), and the XQIFFT ($\cdot$, $p = 0.2276$) against the Cramér–Rao bound at various SNRs.

From this noise sensitivity study, we can see that as long as the systematic error dominates, the best estimator will be the estimator with the lowest systematic bias—the XQIFFT with a properly-chosen value of p. At low SNR, the noise variance dominates and the methods all have a very similar compounded bias. A study on the compounded mean-squared error (MSE) comparing the XQIFFT to state-of-the-art estimators is given in the following Section.

6. Results

In this results section, the XQIFFT is compared to other QIFFT variants and two state-of-the-art estimators which, like the XQIFFT, are direct methods which can be used on any window. These two estimators are Duda's [14] and the first step of Candan's [15]. Candan also has a second refinement step that we do not compare to, since it corresponds to an iterative method and we are limiting our comparison to the class of direct methods.

First, we study the optimal p as a function of window length for the Hann window; Second, we study the MSE of each bin estimator for the case of a length-4096 Han window; Third, we expand that study to twelve common windows, showing that the XQIFFT minimizes MSE across all SNRs for two of the twelve tested windows.

6.1. Optimal p as a Funtion of Window Length

In this first study, the search for values of p which optimize $e_{\mathcal{K}}^{\text{w.c.}}$, $e_{\mathcal{X}}^{\text{w.c.}}$, $\text{E}[|e_{\mathcal{K}}\left(\mathcal{K}\right)|]$, or $\text{E}[|e_{\mathcal{X}}\left(\mathcal{K}\right)|]$ is repeated for Hann windows of various lengths, powers of two from 16 to 8192. The result is a family of curves, given in Figure 8, showing the optimal value of p for each window length and optimized metric.

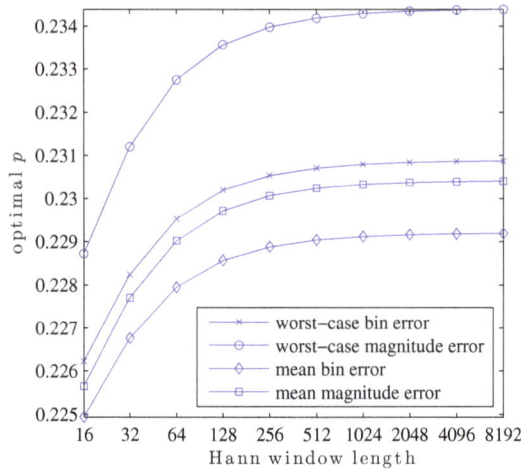

Figure 8. Value of p to minimize four different metrics ($e_{\mathcal{K}}^{\text{w.c.}}$, $e_{\mathcal{X}}^{\text{w.c.}}$, $\text{E}[|e_{\mathcal{K}}\left(\mathcal{K}\right)|]$, $\text{E}[|e_{\mathcal{X}}\left(\mathcal{K}\right)|]$) for different length Hann windows.

This plot exemplifies a few interesting aspects of the search for optimal p. For any particular metric, the optimal p is a function of the shape of the main lobe within the region ±1.5 bins of its center. The reason for this is that the true DTFT peak may not be further than ±0.5 bins from a DFT bin. Since only the three closest bins are used to calculate any QIFFT-derived estimate, parts of the main lobe shape further than ±1.5 DFT bins of the DTFT peak will never be used directly in a QIFFT-derived method.

Main lobe shape is, of course, defined by the window type. A subtler but important point is that main lobe shape changes appreciably as a function of window length. This is due to the aliasing of sidelobe components of the window transform. For the Hann window, aliased sidelobe components have a decreasing effect on the main lobe shape as the window gets long due to the sidelobe rolloff of 18 dB/octave—in Figure 8 this is visible as a "flattening out" of the curves as window length increases.

Another interesting facet of this family of curves is its structure. The optimal p depends in a structured way on the window length. The same way that, e.g., Candan fits a polynomial to tabulated bias correction factors to avoid tabulating and storing every single possibility [15], it would be possible to fit a function to this family of curves to predict the optimal p for a particular metric and window shape without performing the optimization each time.

6.2. Bin Estimate MSE as a Function of SNR (Hann)

In this second study, the performance of the XQIFFT and other QIFFT-derived methods is compared to the state-of-the-art direct methods of Duda [14] and Candan [15] using again a length-4096 Hann window. Here, we choose $p = 0.22917$ to minimize $\text{E}[|e_{\mathcal{K}}\left(\mathcal{K}\right)|]$ as we found earlier. As before, the complex sinusoids are corrupted by additive white Gaussian noise. For each SNR from -40 to 120, in intervals of 5 dB, 20,000 sinusoids are tested with bin offsets of −0.5 to +0.5. The MSE of bin estimate is plotted against SNR for each method and the Cramér–Rao bound (CRB) in Figure 9a.

This plot shows a "thresholding" effect at low SNR (below −15 db). All of the tested methods use as their coarse estimate the maximum of the magnitude spectrum. The thresholding effect arises when the DFT samples are so noisy that the coarse estimate no longer reliably picks the DFT bin closest to the true fractional bin index $\mathcal{K}$. At low SNRs, the MSE of each method is dominated by noise—as SNR increases, the variance decreases. At high SNRs, the variance of each method is dominated by its systematic bias—noise no longer contributes much to the MSE so the traces "flatten out."

For this particular window, the length-4096 Hann window, Figure 9a shows that Candan has the best performance at high SNRs (above 30 dB), followed by Duda, followed by the XQIFFT. However, at low SNRs (between −15 and 30 dB), the XQIFFT has better performance than both Duda and Candan and gets closer to the Cramér–Rao bound—a zoomed detail on this region is shown in Figure 9b.

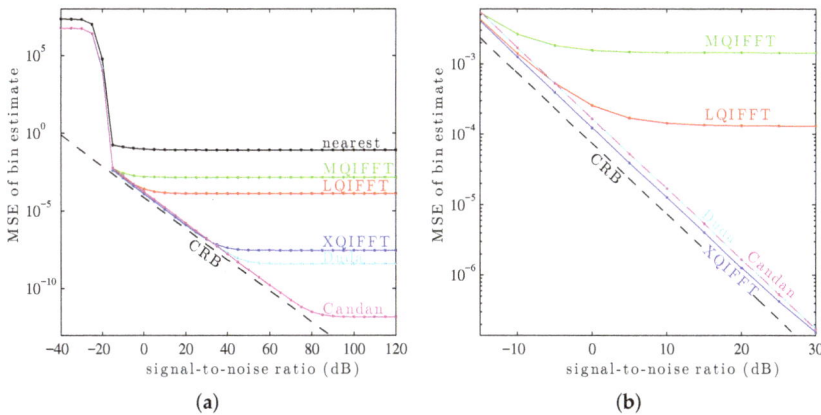

Figure 9. MSE of bin error as a function of singal-to-noise ratio. (**a**) Full SNR range; (**b**) Zoom.

6.3. Bin Estimate MSE as a Function of SNR (Twelve Windows)

In this third study, the performance of the XQIFFT and other QIFFT-derived methods is compared to the state-of-the-art direct methods of Duda [14] and Candan [15] for *twelve* common length-4096 windows. These windows include the Hann window, the Bartlett–Hann window [45], the Bartlett window [4], the Hamming window [4], the Blackman window [4], the Blackman–Harris window [43], the Gaussian window [43], the Digital Prolate Spheroidal Sequence (DPSS/Slepian) window (with time–halfbandwidth product $NW = 3$) [4], the Kaiser–Bessel window ($\beta = 0.5$) [46], the Nuttall window [47], the Dolph–Chebyshev window (sidelobes at −100 dB) [43], and the Tukey window (with a 50% taper) [48]. For each of the twelve windows, p is chosen to minimize $\mathrm{E}[|e_{\mathcal{K}}(\mathcal{K})|]$ using the procedure explained earlier. The values of p used for each window are given in Table 2.

Again, the complex sinusoids are corrupted by additive white Gaussian noise. For each SNR from −40 to 120, in intervals of 5 dB, 20,000 sinusoids are tested with bin offsets of −0.5 to +0.5. The MSE of the bin estimate is plotted against SNR for each method and the Cramér–Rao bound (CRB) in Figure 10.

Table 2. The values of p that minimize $E[|e_{\mathcal{K}}(\mathcal{K})|]$ for the twelve length-4096 windows (highlighted column) used for tests in Figure 10, as well as length-512, -1024, and -2048 versions of the same windows. p is rounded to five decimal points.

Window	Length-512	Length-1024	Length-2048	Length-4096
Hann	0.22903	0.22911	0.22915	0.22917
Bartlett–Hann	0.21635	0.21642	0.21645	0.21647
Bartlett	0.22530	0.22535	0.22538	0.22539
Hamming	0.18505	0.18575	0.18611	0.18628
Blackman	0.13056	0.13057	0.13058	0.13058
Blackman–Harris	0.08552	0.08553	0.08553	0.08554
Gaussian	0.12024	0.12074	0.12099	0.12112
DPSS/Slepian ($NW = 3$)	0.11144	0.11144	0.11144	0.11144
Kaiser–Bessel ($\beta = 0.5$)	0.28214	0.28270	0.28298	0.28312
Nuttall	0.08153	0.08155	0.08157	0.08157
Dolph–Chebyshev (sidelobes at -100dB)	0.08403	0.08403	0.08404	0.08404
Tukey window (50% taper)	0.50592	0.50609	0.50618	0.50622

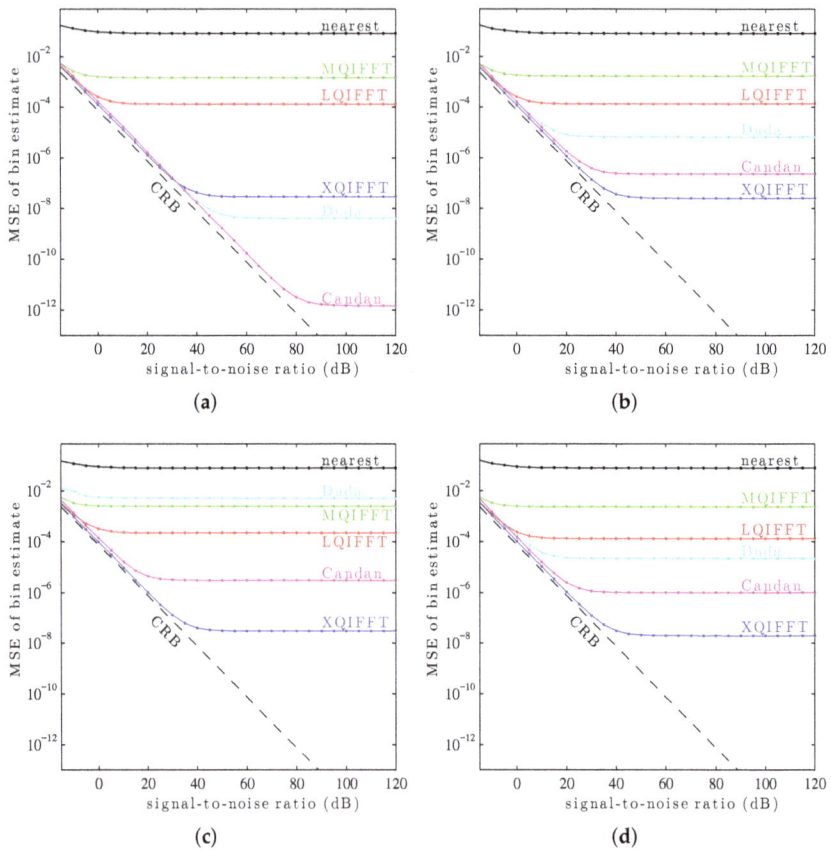

(a)

(b)

(c)

(d)

Figure 10. *Cont.*

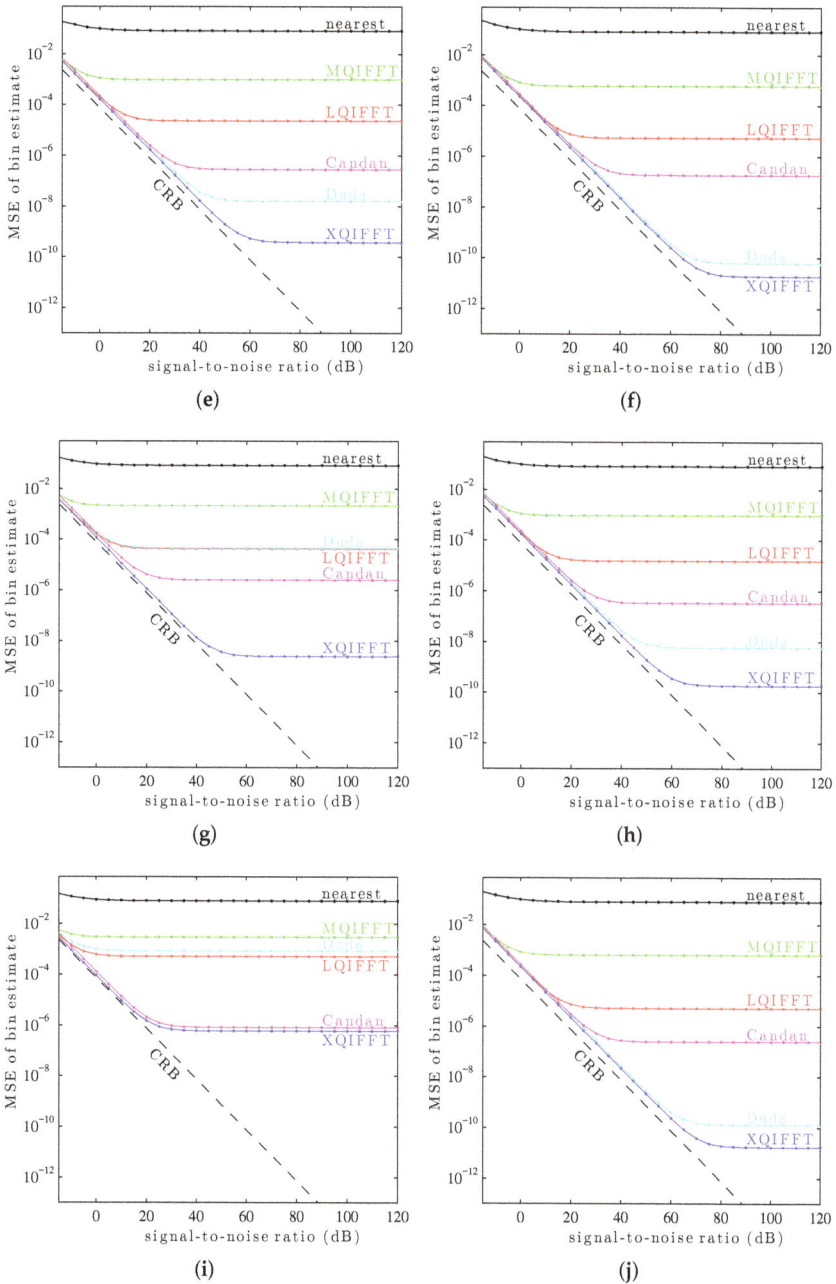

Figure 10. *Cont.*

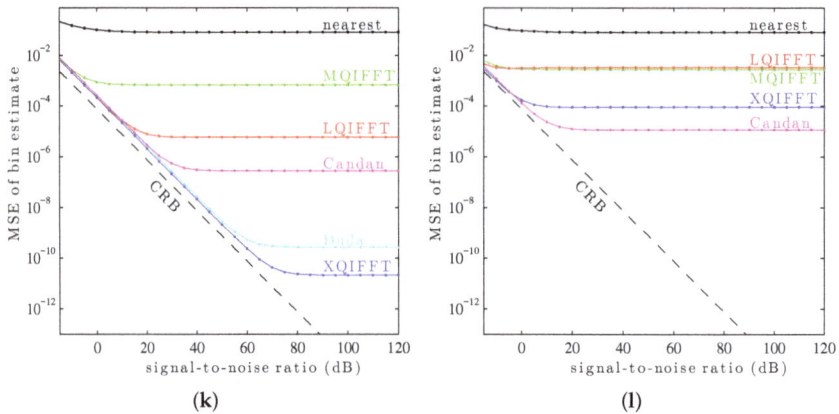

Figure 10. Testing different window types. (**a**) Hann window; (**b**) Bartlett–Hann window; (**c**) Bartlett window; (**d**) Hamming window; (**e**) Blackman window; (**f**) Blackman–Harris window; (**g**) Gaussian window; (**h**) DPSS window; (**i**) Kaiser–Bessel window; (**j**) Nuttall window; (**k**) Chebyshev window; (**l**) Tukey window (50%).

It can be seen in Figure 10 that the XQIFFT (with a properly-chosen p according to Table 2) has the lowest MSE error across the entire range of SNRs for every tested window except the Hann window and the Tukey window (50% taper). In some cases the XQIFFT outperforms its closest competitor by several orders of magnitude. Figure 10 demonstrates that the XQIFFT is applicable to a wide range of common windows and that it even outperforms the state of the art for most tested windows. In a broader sense, this demonstrates that the best choice of method is highly dependent on the window type which is used—the performance ordering of different methods is highly variable, although for most windows tested the XQIFFT has the best performance. It is also interesting to note that of all the windows tested, the Nuttall window using the XQIFFT has the lowest MSE of bin estimate at high SNR. This implies that Nuttall using XQIFFT achieves the lowest systematic bias of the windows and estimators tested.

7. Conclusions

In this article, we detailed the concept of sinusoidal parameter estimation using quadratic interpolation over three bins of a power-scaled DFT surrounding a magnitude spectrum peak. This method, the XQIFFT, improves on existing quadratic interpolation methods based on linear and logarithmic scaling. We presented a heuristic for optimizing the XQIFFT power-scaling exponent p for a particular window type/window length/zero-padding factor. We presented an analysis of noise sensitivity of the proposed estimator, the dependency of the optimal exponent on window length, and a comparison with competing methods in noisy conditions under twelve common window types. We observed that across the entire tested SNR range, the proposed XQIFFT had a lower bin estimate MSE than the state-of-the-art direct estimates of Duda and Candan for ten of the twelve tested windows. From examining the high SNR regions we can also say that for ten of the twelve tested windows the XQIFFT has a lower systematic bias than the state-of-the-art methods as well. For the remaining two windows tested (Hann and Tukey), the XQIFFT has a lower MSE only at low SNR.

When implementing the XQIFFT, the reader may choose the correct value of p in one of two ways depending on their application. For common windows whose lengths are between 512 and 4096 samples, linear interpolation between the values of p reported in Table 2 may be used to approximate the optimal value of p to five decimal places. For even more precision or for window types or lengths not reported in Table 2, the reader may use the generalized procedure reported in Section 5.3 to produce an optimal value of p.

Future work should consider applying the XQIFFT-style magnitude spectrum scaling to types of estimators that rely on techniques other than parabolic interpolation. We expect that for other approaches that rely in some way on main lobe shape, the main-lobe-shaping philosophy of the XQIFFT may yield improvements.

Acknowledgments: Thank you to the anonymous reviewers who contributed valuable suggestions for refining this work.

Author Contributions: Kurt James Werner contributed the central concept of the article. François Georges Germain was the main contributor to the "Problem Statement," "Invariance Properties of QIFFFT Methods," and "Sensitivity to Noise" sections and the material on the asymptotic behavior of the XQIFFT. Kurt James Werner was the main contributor to the "Results" section and created Figures 1–4. Otherwise the authors contributed equally to the manuscript.

Conflicts of Interest: The authors declare no conflict of interest.

References

1. Belega, D.; Petri, D. Frequency estimation by two- or three-point interpolated Fourier algorithms based on cosine windows. *Signal Process.* **2015**, *117*, 115–125.
2. Candan, Ç. A method for fine resolution frequency estimation from three DFT samples. *IEEE Signal Process. Lett.* **2011**, *18*, 351–354.
3. Liao, Y. Phase and Frequency Estimation: High-Accuracy and Low-Complexity Techniques. Master's Thesis, Worcester Polytechnic Institute, Worcester, MA, USA, 2011.
4. Smith, J.O., III. *Spectral Audio Signal Processing*; W3K Publishing: Standford, UK, 2011.
5. Serra, X. Musical sound modeling with sinusoids plus noise. In *Musical Signal Processing*; Roads, C., Pope, S., Picialli, A., Poli, G.D., Eds.; Routledge: London, UK, 1997; pp. 91–122.
6. Schmidt, R.O. Multiple emitter location and signal parameter estimation. *IEEE Trans. Antennas Propag.* **1986**, *34*, 276–280.
7. Roy, R.; Kailath, T. ESPRIT—Estimation of signal parameter via rotational invariance techniques. *IEEE Trans. Acoust. Speech Signal Process* **1989**, *37*, 984–995.
8. Macleod, M.D. Fast nearly ML estimation of the parameter of real of complex single tones or resolved multiple tones. *IEEE Trans. Signal Process* **1998**, *46*, 141–148.
9. McAulay, R.; Quatieri, T.F. Speech analysis/synthesis based on a sinusoidal representation. *IEEE Trans. Acoust. Speech Signal Process* **1986**, *34*, 744–754.
10. Serra, X. A System for Sound Analysis/Transformation/Synthesis Based on a Deterministic Plus Stochastic Decomposition. Ph.D. Thesis, Stanford University, Stanford, CA, USA, October 1989.
11. Serra, X.; Smith, J.O., III. Spectral modeling synthesis: A sound analysis/synthesis system based on a deterministic plus stochastic decomposition. *Comput. Music J.* **1990**, *14*, 12–24.
12. Smith, J.O., III; Serra, X. PARSHL: A program for the analysis/synthesis of inharmonic sounds based on a sinusoidal representation. In Proceedings of the International Computer Music Conference (ICMC), Champaign–Urbana, IL, USA, 23–26 August 1987.
13. Maher, R.C.; Beauchamp, J.W. Fundamental frequency estimation of musical signals using a two-way mismatch procedure. *J. Acoust. Soc. Am.* **1994**, *95*, 2254–2263.
14. Duda, K. DFT interpolation algorithm for Kaiser–Bessel and Dolpha–Chebyshev windows. *IEEE Trans. Instrum. Meas.* **2011**, *60*, 784–790.
15. Candan, Ç. Fine resolution frequency estimation from three DFT samples: Case of windowed data. *Signal Process.* **2015**, *114*, 245–250.
16. Agrež, D. Weighted multipoint interpolated DFT to improve amplitude estimation of multifrequency signal. *IEEE Trans. Instrum. Meas.* **2002**, *51*, 287–292.
17. Offelli, C.; Petri, D. Interpolation techniques for real-time multifrequency waveform analysis. *IEEE Trans. Instrum. Meas.* **1990**, *39*, 106–111.
18. Candan, Ç. Analysis and further improvement of fine resolution frequency estimation method from three DFT samples. *IEEE Signal Process. Lett.* **2013**, *20*, 913–916.

19. Jacobsen, E.; Kootsookos, P. Fast, accurate frequency estimators. *IEEE Signal Process. Mag.* **2007**, *24*, 123–125.
20. Werner, K.J. The XQIFFT: Increasing the accuracy of quadratic interpolation of spectral peaks via exponential magnitude spectrum weighting. In Proceedings of the 41st International Computer Music Confernece (ICMC), Denton, TX, USA, 25 September–1 October 2015.
21. Auger, F.; Flandrin, P. Improving the readability of time-frequency and time-scale representations by the reassignment method. *IEEE Trans. Signal Process.* **1995**, *43*, 1068–1089.
22. Degani, A.; Dalai, M.; Leonardi, R.; Miglorati, P. Time-frequency analysis of musical signals using the phase coherence. In Proceedings of the International Conference on Digital Audio Effects (DAFx-13), Maynooth, Ireland, 2–5 September 2013.
23. Fitz, K.; Haken, L. On the use of time-frequency reassignment in additive sound modeling. *J. Audio Eng. Soc.* **2002**, *50*, 879–893.
24. Fitz, K.R. The Reassigned Bandwidth-Enhanced Method of Additive Synthesis. Ph.D. Thesis, University of Illinois at Urbana–Champaign, Urbana, IL, USA, 1999.
25. Keiler, F.; Marchand, S. Survey on extraction of sinusoids in stationary sounds. In Proceedings of the International Conference on Digital Audio Effects (DAFx-02), Hamburg, Germany, 26–28 September 2002; pp. 51–58.
26. Nam, J.; Mysore, G.J.; Ganseman, J.; Lee, K.; Abel, J.S. A super-resolution spectrogram using coupled PLCA. In Proceedings of the Conference of the International Speech Communication Association (Interspeech), Makuhari, Japan, 26–30 September 2010; Volume 11, pp. 1696–1699.
27. Liao, J.-R.; Chen, C.-M. Phase correction of discrete Fourier transform coefficents to reduce frequency estimation bias of single tone complex sinusoid. *Signal Process.* **2014**, *94*, 108–117.
28. McLeod, P. Fast, Accurate Pitch Detection Tools for Music Analysis. Ph.D Thesis, University of Otago, Dunedin, New Zealand, 30 May 2008.
29. Aboutanios, E.; Mulgrew, B. Iterative frequency estimation by interpolation on Fourier coefficients. *IEEE Trans. Signal Process.* **2005**, *53*, 1237–1242.
30. Abe, M.; Smith, J.O., III. *CQIFFT: Correcting Bias in a Sinusoidal Parameter Estimator Based on Quadratic Interpolation of FFT Magnitude Peaks*; STAN-M 117; CCRMA, Department of Music, Stanford University: Stanford, CA, USA, 2004.
31. Abe, M.; Smith, J.O., III. *Design Criteria for the Quadratically Interpolated FFT Method (I): Bias Due to Interpolation*; STAN-M 114; CCRMA, Department of Music, Stanford University: Stanford, CA, USA, 2004.
32. Goto, Y. Highly accurate frequency interpolation of apodized FFT magnitude-mode spectra. *Appl. Spectrosc.* **1998**, *52*, 134–138.
33. Smith, J.O., III; Serra, X. *PARSHL: An Analysis/Synthesis Program for Non-Harmonic Sounds Based on a Sinusoidal Representation*; STAN-M 43; CCRMA, Department of Music, Stanford University: Stanford, CA, USA, 1985.
34. McIntyre, M.C.; Dermott, D.A. A new fine-frequency estimation algorithm based on parabolic regression. In Proceedings of the International Conference on Acoustics, Speech, and Signal Processing (ICASSP), San Francisco, CA, USA, 23–26 March 1992; Volume 2, pp. 541–544.
35. Archer, B.; Weisstein, E.W. Lagrange interpolating polynomial. MathWorld—A Wolfram Web Resource. Available online: http://mathworld.wolfram.com/LagrangeInterpolatingPolynomial.html (accessed on 18 October 2016).
36. Quinn, B.G. Estimating frequency by interpolation using Fourier coefficients. *IEEE Trans. Signal Process.* **1994**, *42*, 1264–1268.
37. Berman, G. Minimization by successive approximation. *SIAM J. Numer. Anal.* **1966**, *3*, 123–133.
38. Kiefer, J. Sequential minimax search for a maximum. *Proc. Am. Math. Soc.* **1953**, *4*, 503–506.
39. Mathews, J.H.; Fink, K.D. *Numerical Methods Using Matlab*, 4th ed.; Pearson: Harlow, UK, 2004.
40. Ortega, J.M.; Rheinboldt, W.C. *Iterative Solution of Nonlinear Equations in Several Variables*; Academic Press: New York, NY, USA, 1970.
41. Moin, P. *Fundamentals of Engineering Numerical Analysis*; Cambridge University Press: Cambridge, UK, 2010.
42. Blackman, R.B.; Tukey, J.W. *The Measurement of Power Spectra from the Point of View of Communications Engineering*, 2nd ed.; Dover Publications, Inc.: New York, NY, USA, 1959.
43. Harris, F.J. On the use of windows for harmonic analysis with the discrete Fourier transform. *Proc. IEEE* **1978**, *66*, 51–83.

44. Rife, D.C.; Boorstyn, R.R. Single tone parameter estimation from discrete-time observations. *IEEE Trans. Inf. Theory* **1974**, *20*, 591–598.

45. Ha, Y.H.; Pearce, J.A. A new window and comparison to standard windows. *IEEE Trans Acoust. Speech Signal Process.* **1989**, *37*, 298–301.

46. Kaiser, J.F.; Schafer, R.W. On the use of the i_0-sinh window for spectrum analysis. *IEEE Trans. Acoust. Speech Signal Process.* **1980**, *28*, 105–107.

47. Nuttall, A.H. Some windows with very good sidelobe behavior. *IEEE Trans. Acoust. Speech Signal Process.* **1981**, *29*, 84–91.

48. Bloomfield, P. *Fourier Analysis of Time Series: An Introduction*, 2nd ed.; John Wiley & Sons: New York, NY, USA, 2000.